KT-567-552

ENVIRONMENTAL ECOLOGY

SECOND EDITION

The Ecological Effects of Pollution, Disturbance, and Other Stresses

Bill Freedman
Department of Biology and
School for Resource and Environmental Studies
Dalhousie University
Halifax, Nova Scotia, Canada

ACADEMIC PRESS
San Diego   New York   Boston   London   Sydney   Tokyo   Toronto

Front cover — View of the 381-m-tall "superstack" at the Copper Cliff smelter near Sudbury, Ontario. Most of the ecological devastation apparent in the vicinity was caused by emissions of toxic sulfur dioxide and metals from ground-level roast beds until the late 1920s, and from the now-unused smokestacks (present in the photo) that subsequently replaced the roast beds. Local air quality has greatly improved since the superstack was commissioned in 1972, because the smelter emissions are injected relatively high into the atmosphere, making ground-level fumigations infrequent. However, the continued emissions contribute to regional problems associated with long-range transported air pollutants, such as acidic precipitation. (See Chapters 2, 3, and 4.) Photo: B. Freedman. *Inset*, A scene of a small village located on a logging road in West Kalimantan, Indonesian Borneo. This road was constructed to allow commercial extraction of selectively harvested logs of tropical hardwoods from the rainforest. These logs were used to make plywood for export to Japan. However, unregulated secondary logging of residual trees to manufacture lumber for more local use took place soon after the completion of commercial logging. Afterward, poor, landless farmers converted the remaining forest to subsistence and local-market agriculture, using a slash-and-burn method. In this case, the economic resource and much of the integrity of the tropical forest may have been sustainable under regulated commercial logging. However, secondary logging and subsequent agricultural conversion resulted in deforestation. This represents a longer-term, probably permanent degradation of the tropical forest ecosystem in terms of biodiversity, productivity, and other values. (See Chapters 9, 10, and 12.) Photo: B. Freedman. Back cover — A view of water draining a pile of waste rock and slag from mining and smelting. The water has been extremely acidified through the oxidation of mineral sulfides, a phenomenon known as acid mine drainage. The reddish color is associated with a loose precipitate of iron oxide that develops as acidity is neutralized by inorganic chemical reactions. At increasing distance from the source of acidification, the acidity of the flowing water is progressively decreased. This establishes a gradient of environmental stress that will be differentially exploited by species, largely depending on their tolerance of the toxicity of acidity and associated metals. (See Chapters 2 and 4.) Photo: B. Freedman.

This book is printed on acid-free paper. ∞

Copyright © 1995, 1989 by ACADEMIC PRESS, INC.
All Rights Reserved.
No part of this publication may be reproduced or transmitted in any form or by any means, electronic or mechanical, including photocopy, recording, or any information storage and retrieval system, without permission in writing from the publisher.

Academic Press, Inc.
A Division of Harcourt Brace & Company
525 B Street, Suite 1900, San Diego, California 92101-4495

United Kingdom Edition published by
Academic Press Limited
24-28 Oval Road, London NW1 7DX

Library of Congress Cataloging-in-Publication Data

Freedman, Bill.
     Environmental ecology : the ecological effects of pollution,
  disturbance, and other stresses / Bill Freedman. – 2nd. ed.
         p.   cm.
     Includes bibliographical references (p.   ) and indexes.
     ISBN 0-12-266542-2
       1. Pollution–Environmental aspects.   2. Man–Influence on nature.
  3. Ecology.   I. Title.
  QH545.A1F74   1993
  574.5'222--dc20   94-7665
                                                    CIP

PRINTED IN THE UNITED STATES OF AMERICA
94 95 96 97 98 99 EB 9 8 7 6 5 4 3 2 1

KING ALFRED'S COLLEGE
WINCHESTER

333.7
FRE   0169 6823

"Now it seems as though our mother planet is telling us, 'My children, my dear children, behave in a more harmonious way. My children, please take care of me.'"

The Dalai Lama

CONTENTS

PREFACE

As considered in this book, environmental ecology includes the study of the ecological effects of pollution, disturbance, and other stressors. Particular attention is paid to stressors associated with the activities of humans, although important knowledge is also gained from the study of stressors that are natural in origin.

For obvious reasons, the subject matter of environmental ecology is pivotal to understanding the causes and consequences of ecological damages that humans are causing in so many places. This knowledge can also contribute to the design of management practices that avoid or alleviate those harms. These subjects are important to society because their understanding is essential if ecologically sustainable pathways of economic development are to be identified, rationalized, and pursued.

Many universities and colleges offer classes and programs in ecology and/or environmental studies. However, remarkably few institutions offer classes that focus on the subject matter of environmental ecology. One reason for the scarcity of classes in this subject may be the paucity of suitable textbooks to complement lectures, seminars, and laboratory sessions.

Since 1979, I have been teaching senior undergraduate and graduate classes in environmental ecology at Dalhousie University. Since I was unable to find a textbook that, in my opinion, comprehensively dealt with the subject matter, I relied on reprints and other primary materials to supplement my lectures and seminars—hence, my perception of a need for a book such as the one offered here. This book, now in its second edition, was written primarily as a basic, broad-reaching text for classes and seminars in environmental ecology. I also hope that the availability of this book will facilitate the offering of more classes dealing with this subject.

This book will be useful to teachers of classes in the subject matter, and also to working ecologists who would like to gain an overview of environmental ecology, or of particular issues within this field. The book can also supply supplemental information and case materials for more general, university-level classes in ecology and environmental studies. Other professionals with interests in the multi- and interdisciplinary realms of environmental science, including environmental impact assessment, will also find useful background information in this book.

This second edition of *Environmental Ecology* is substantially different from the first edition. All previously existing chapters are extensively updated with recent literature, and many new case studies have been included. A new chapter has been written, dealing with the use and abuse of biological resources and the emerging field of ecological economics. In addition, the last chapter of the book on applications of environmental ecology has been greatly expanded. New sections deal with environmental impact assessment; ecological monitoring; and the responsibilities of ecologists in environmental issues, environmental education, and the design of sustainable economic systems.

Readers who have not experienced the equivalent of introductory classes in ecology may consult the book's glossary to understand the specialized terminology that is sometimes used.

Bill Freedman

ACKNOWLEDGMENTS

A number of acknowledgments are in order. First, I gratefully thank my first academic mentor, Tom Hutchinson of Trent University, formerly of the University of Toronto. Tom initially helped me appreciate the importance of the anthropogenic ravaging of ecosystems that has been and is now occurring. He also gave me an opportunity to build a career focusing on the ecological effects of human activities. Of course, my thinking about ecology has also been strongly influenced and molded by other friends and colleagues. I am especially grateful to Paul Catling of the Biosystematics Research Institute in Ottawa and to Josef Svoboda of the University of Toronto for their contributions to my professional development.

Many colleagues provided an invaluable service by reviewing one or more chapters of this book and by making important reference materials known to me. Inevitably, I was not always able to incorporate all of their criticisms and suggestions because they did not always jibe with my own ideas or world view. However, the great majority of suggestions made by these people resulted in changes in my preliminary manuscripts, and they improved the quality, and sometimes the accuracy, of my writing. These helpful colleagues are: J. Bamwoya, H. Beach, S. Beauchamp, P. Brodie, B. Brunsdon, P. Burton, D. Busby, E. Caldwell, A.R.O. Chapman, T. Clair, R.P. Cote, R. Cox, C. Cronan, L. Cwynar, J. Dale, R.W. Doyle, D. Eidt, W. Ernst, T. Fleming, S. Fudge, M. Gardner, J. Gilbert, E. Greene, M. Havas, A. Hebda, T. Herman, E. Hickey, S. Holmes, R. Hughes, T.C. Hutchinson, J. Kerekes, D. Lachance, D. MacKinnon, D. MacLean, I.A. McLaren, P. Mandell, I. Methven, P. Mineau, R. Myers, D.N. Nettleship, G. Parker, D. Patriquin, M. Paterson, T. Pollock, R. Scheibling, N. Shackell, C. Staicer, J. Svoboda, J. Vandermeulen, R. Wassersug, P. Wells, A. Westing, H. Whitehead, S. Wood, N. Yan, and A. Zakaria. Eight anonymous reviewers of chapter manuscripts also provided helpful criticisms. Todd Keith and Annette Lutterman assisted with preparation of the indices. If any readers have comments or criticisms of the material or approach used in this book, I'd be most grateful if you would communicate your ideas and suggestions to me.

A number of the professional staff at Academic Press have also been of great assistance in the preparation of this book. The editorial suggestions of Charles Arthur have made the book more readable, and the design work of Michael Remener has given the book a pleasing and interesting appearance. Production was overseen by editors Mike Rutherford and Holly Garrett.

Finally, I thank my two children, Jonathan and Rachael. They sometimes help me with my fieldwork, but they were of no direct help with this book. Still, they inspire and motivate me in mysterious ways, and they kindly permitted me to work on the manuscript when they were in school or asleep. I am also, of course, grateful to my spouse George-Anne Merrill, for her uncomplaining support, for her assistance with library-related activities, for permitting me to go to far-away places to do fieldwork, and for patiently allowing me to work when Rachael and Jonathan were not in school or asleep.

1

THE ECOLOGICAL EFFECTS OF POLLUTION, DISTURBANCE, AND OTHER STRESSES

1.1
THE SCOPE OF THIS BOOK

Like ecology, environmental science is multi- and interdisciplinary. The three major subdisciplines of environmental science can be appreciated from the title of an early, influential textbook on the subject, "Ecoscience: Population, Resources, Environment" (Ehrlich *et al.*, 1977).

1. *Population*. The size of the human population has increased tremendously in recent centuries, and it continues to do so. This population growth could arguably be considered the central cause of the environmental crisis. The cumulative, anthropogenic impact on the biosphere is a function of two factors: (a) the size of the human population and (b) the environmental impact per person, which varies greatly among and within countries, largely depending on the nature and degree of industrialization.

2. *Resources*. The natural resources that are available to sustain human endeavors include (a) nonrenewable resources, which are present in a finite supply that is diminished by mining and use (although the availability of some resources, especially metals, can be extended by recycling materials that have been discarded), and (b) renewable natural resources, which if managed suitably can provide flows of materials or energy that are sustainable over the longer term. However, the inappropriate management or overexploitation of any potentially renewable resource can degrade it irretrievably. In such a situation, the resource is effectively mined—that is, it is managed as if it were a nonrenewable resource.

3. *Environment*. This subdiscipline of environmental science deals with the ways in which the integrity of the biosphere, and of purely human environments, are affected by various anthropogenic and natural influences. The effects of pollution, disturbance, and other stresses on natural and managed ecosystems are the primary subject

material of this book. Although the purview of "environment" also includes the direct effects of pollution and other stressors on the health of humans, and on their constructed environments, in some respects these are not ecological issues because they do not deal with natural ecosystems and wild species. Nevertheless, several key aspects of these anthropocentric topics will be addressed, notably with respect to air pollution, toxic elements, pesticides, and warfare.

Of the above three major subdisciplines within environmental science, this book is most concerned with the third—the ecological effects of stressors, with particular reference to those associated with the activities of humans.

1.2
GROWTH OF THE HUMAN POPULATION

To give a broader perspective to a book that focuses on only one aspect of environmental science, it is important to consider briefly the remarkable increases that have occurred in the size of the human population during recent centuries, as well as the anticipated changes. This population growth is critically important, because it essentially drives the aggregate environmental impact of human activities. However, the consideration of this topic in this book is fairly superficial. For more detailed information on past and projected changes in the human population, see Ehrlich *et al.* (1977), Barney (1980); Peterson (1984), Shuman (1984), Ehrlich (1985), World Resources Institute (WRI) (1990, 1992), or El-Badry (1992).

At present, *Homo sapiens* is by far the most abundant large animal on Earth, with a total population in 1993 of more than 5.6 billion individuals, growing at 1.7%/year or 93 million/year (Anonymous, 1993a).

Some of the domesticated, large-animal symbionts of humans have also become enormously abundant, together comprising a population of more than 4 billion individuals. Among these companion species are about 1.72 billion sheep and goats (*Ovis aries* and *Capra hircus*), 1.27 billion cows (*Bos taurus* and *B. indica*), 0.85 billion pigs (*Sus scrofa*), 0.12 billion equines (mostly horses, *Equus caballus*), and 0.16 billion camels and water buffaloe (*Camelus dromedarius* and *Bubalus bubalis*) (Durning and Brough, 1992; WRI, 1992). Some smaller, animal symbionts of humans are even more abundant, including an estimated 10–11-billion fowl (mostly chickens, *Gallus gallus*) (Durning and Brough, 1992; WRI, 1992).

It is notable that, within historical times, no other large animals have achieved such a great abundance. Furthermore, it is quite possible that for such large animals, the tremendous population of humans and some of our large-animal symbionts is unprecedented on Earth. It has been estimated that prior to having been overhunted, American bison (*Bison bison*) may have numbered as many as 60 million individuals (McClung, 1969; Wood, 1972). At present, the most populous large animals in the wild are probably the crabeater seal (*Lobodon carcinophagus*) of the Antarctic, with an estimated 15–30 million individuals (Bengston and Laws, 1985), and white-tailed deer (*Odocoileus virginianus*) of North America, with an estimated 40–60 million individuals (McCabe and McCabe, 1984; Smith, 1991). The abundance of both of these species is about 1% or less of the present abundance of humans.

Some other large, wild animals whose populations number in the millions include: (1) gray and red kangaroos (*Macropus gigantea* and *M. rufus*) of Australia, with an aggregate abundance of about 20 million individuals; (2) ringed seal (*Phoca hispida*) of the Arctic, with 6–7 million; (3) caribou or reindeer (*Rangifer tarandus*) of the Arctic and sub-Arctic, with about 3 million; (4) harp seal (*P. groenlandica*) of the North Atlantic, with 3–4 million; (5) striped and spotted dolphins (*Stenella* spp. and *Lagenorhynchus* spp.) of the east Pacific, with >2 million each; (6) northern fur seal (*Callorhinus ursinus*) of the North Pacific, with 2 million; (7) sperm whale (*Physeter macrocephalus*) of all oceans, with 1.9 million; and (8) wildebeest (*Connochaetes taurinus*) of Africa, with 1.4 million [Walker *et al.*, 1968; Roff and Bowen, 1983; Northridge, 1984; Organization for Economic Cooperation and Development (OECD), 1985; Dobson and Hudson, 1986; WRI, 1987; Bergerud, 1988; Grigg, 1989; National Marine Fisheries Service (NMFS), 1991a]. Some species of large fish may also have populations that number in the millions, but good data are not available. Clearly, for a large animal, humans are atypically populous on Earth.

Prior to the beginnings of agriculture around 10 thousand years ago, the global abundance of humans was about 5 million individuals (after Ehrlich *et al.*, 1977). The progressive development of agricultural technologies allowed slow but steady growths of population, because farming could support more people than could the hunting and gathering of wild animals and plants. The rates of population increase were further influenced by additional technological innovations that enhanced the abilities of humans to control and exploit their environment. For example, the discovery of the properties of metals and their alloys allowed the development of superior tools; the discovery of the wheel made it possible to more easily move large quantities of materials; and the progressive domestication and genetic modification of certain plants and animals increased agricultural yields.

Each of these innovations increased the effective carrying capacity of the environment for humans and permitted additional population growth. As a result, there were about 300 million people in 0 A.D., and 500 million in 1650. At about this time the rate of population growth quickened markedly, and this trend has continued to the present. The most important reasons for the recent, unparalled growth of the human population have been (1) the development of increasingly more-effective medical treatments and sanitation, which substantially decrease death rates; (2) technological advances in resource extraction, manufacturing, and agriculture, which further increase the carrying capacity of the environment; and (3) the "discovery" by Europeans of conquerable lands rich in natural resources, which were quickly colonized and used to sustain and absorb additional population growth. As a result of these developments, the global population of humans increased to 1 billion by 1850, then to 2 billion by 1930, to 4 billion in 1975, and 5 billion in 1987 (Ehrlich *et al.*, 1977; WRI, 1986, 1987). In 1993, the human population was more than 5.6 billion individuals (Anonymous, 1993a).

Regional and more local increases of some human populations have recently been even more intense than the global averages. For example, between 1917 and 1977 the human population in central Sudan increased by a factor of 6.4 times, to 18.4 million (Olsson and Rapp, 1991). The populations of agricultural large mammals increased at even-greater rates in this area—cattle by 20 times to 16 million, camels by 16 times to 3.7 million, sheep by 12.5 times to 16 million, and goats by 8.5 times to 10.4 million. The enormously increased abundances of humans and their large-mammal symbionts have caused substantial degradations of the carrying capacity of drylands in that region of Africa (Olsson and Rapp, 1991). Similarly, between 1970 and 1988 the human population of the province of Rondonia in the Brazilian Amazon increased by about 12 times, largely by immigration, while the population of cattle increased by 30 times (Hecht, 1989).

The rate of growth of the human population has decreased from its maximum global rate in the late 1960s of 2.1% per year, capable of doubling the total population in only 33 years, to 1.7% in the early 1990s (doubling in 41 years) (El-Badry, 1992). Still, the human population continues to grow rapidly, with about 93 million people added to the population each year in the early 1990s (Anonymous, 1993a).

Predictions of future growth of human populations must take into account the likely rates of change in fecundity, death, and other uncertain, demographic variables. Hence, it is impossible to foretell accurately the future abundance of humans. Nevertheless, the most sophisticated population models have projected that the global population of humans could reach about 6.2 billion by the year 2000, 8.5 billion by 2025, 9.5 billion by 2050, and 10.2 billion by 2100, and that it could then stabilize at about that level (Peterson, 1984; WRI, 1986, 1987; Anonymous, 1993a). Therefore, barring some intervening catastrophe such as a collapse of the carrying capacity of the environment, an unprecedented pandemic, or a nuclear holocaust, it appears that the global population of humans will approximately double from its late-1980s level before it stabilizes.

Clearly, the human population has been growing exponentially in recent centuries, although this is unlikely to continue at present rates into the foreseeable future. This population increase has caused environmental degradations through deforestation, overharvesting of wild animals and plants, pollution, rapid mining of nonbiological resources, and other detrimental effects. It is clearly apparent that the cumulative, anthropogenic impact on the environment must increase with an increase in the abundance of our species.

It must also be stressed, however, that there has been a great intensification of the per-capita environmental impact of humans during the course of evolution of our socioeconomic systems. This trend can be illustrated by changes in energy use, which has often been used as an indicator of the per-capita environmental effects of humans. It has been estimated that the average per-capita energy consumption of a hunting society is about 20 MJ/day, compared with 48 MJ/day for primitive agriculture, 104 MJ/day for advanced agriculture, 308 MJ/day for an industrial society, and 1025 MJ/day for an advanced industrial society (Goldemberg, 1992). Changes in per-capita energy usage have been espe-

cially large during the most recent century or so of vigorous technological development and innovation. Trends in the intensity of energy use clearly show that, during the present century, global per-capita economic output and fuel consumption have both increased more rapidly than has the human population (Table 1.1). Of course, this trend has been most substantial in relatively industrialized countries. The per-capita energy consumption in industrialized countries averaged 199 GJ/year in 1980, compared with 17 GJ/year in less-developed countries (Goldemberg, 1992). At that time, industrialized countries accounted for 25% of the human population, but 80% of global energy use (Goldemberg, 1992).

Growth of the human population and the resulting more-than-proportional environmental deterioration have, in combination, changed the biosphere on a scale and intensity that is comparable to the effects of such massive geological events as glaciations.

The enormous growth of the human population should be kept in mind whenever environmental problems are considered. The environmental effects of people and their societies can be modified by technological strategies such as pollution control and the conservation of natural resources. However, the size of the human population remains a root cause of the ecological degradations caused by our species.

Table 1.1   Global population, economic output, and consumption of fossil fuels in 1900, 1950, and 1986[a]

Year	Population ($\times 10^9$)	Economic activity			Fossil fuel consumption (10^9 tonnes coal equiv.)	
		Gross world product (GWP) (1980 dollars $\times 10^{12}$)	GWP per capita (dollars $\times 10^3$)		Total	Per capita
1900	1.6	0.6	0.4		1	0.6
1950	2.5	2.9	1.2		3	1.2
1986	5.0	13.1	2.6		12	2.4

[a]Modified from Brown and Postel (1987).

1.3

THE ECOLOGICAL EFFECTS OF STRESS

Stress is simplistically defined in this book as "environmental influences that cause measurable ecological changes, or that limit ecological development." The stressing agent (i.e., stressor) can be associated with environmental factors that are physical, chemical, or biological in nature. Stressors can originate externally to the ecosystem, as is often typical for toxic chemicals, such as pesticides and gases. Stress can also result from an intensification of some preexisting, internal factor beyond a threshold of ecological tolerance, as can occur when there are substantial increases in nutrient loading, windspeed, or temperature. Often, implicit within the notion of stress is an anthropocentric judgement about the quality of the ecological change, that is, whether the resulting effect is "good" or (more usually) "bad."

Several classes of environmental stress can be defined according to the causative factor. The most prominent of these include the following.

1. *Physical stress* is an acute, episodic stressor (or a disturbance), associated with an intense, often brief, loading of kinetic energy to an ecosystem. Examples of physical stressors are the blast from a volcanic eruption or other explosions, hurricanes and tornadoes, oceanic tidal waves or tsunamis, earthquakes, and (over geological times) glacial advance.

2. *Wildfire* is another episodic disturbance, caused by the rapid combustion of much of the biomass of an ecosystem. Once ignited by lightning or by humans, the biomass oxidizes in an uncontrolled fashion, until the fire either runs out of fuel or is quenched.

3. *Pollution* occurs when a chemical agent is present in the environment at a concentration sufficient to elicit physiological responses by organisms and thereby cause ecological changes. Chemicals that are commonly involved in toxic pollution include the gases sulfur dioxide and ozone, elements such as mercury and arsenic, and

pesticides. Pollution by nutrients can distort ecological processes such as productivity, as in eutrophication and its consequences. Note that environmental contamination by a potentially toxic agent does not necessarily imply pollution. Because of the sophistication of modern analytical chemistry, contamination by potentially toxic chemicals can often be measured in concentrations that are smaller than the thresholds of exposure, or dose, beyond which physiological or ecological changes can be demonstrated. (See Chapter 8 for further discussion of this topic.)

4. *Thermal stress* occurs when the release of heat energy into an ecosystem causes ecological change. Thermal pollution is usually associated with the dissipation of waste heat from power plants and other industrial facilities, but it is also naturally present in the vicinity of hot springs and hot submarine vents.

5. *Radiation stress* occurs when there is an excessive load of ionizing energy. Examples include the effects of radiation flux from nuclear waste or explosions and the experimental exposure of ecosystems to ionizing gamma radiation.

6. *Climatic stress* is caused by an insufficient or excessive regime of temperature, moisture, solar radiation, or combinations of these.

7. *Biological stress* is associated with interactions among organisms, exerted through competition, herbivory, predation, parasitism, and disease. For example, individuals of the same or different species may compete for access to essential resources such as nutrients, water, insolation, or space. Competition occurs when the capability of the environment to supply resources is smaller than the potential biological demand, so that organisms interfere with each other. Competition-related stresses are relatively symmetric if they involve organisms of similar abilities to use these resources. Competition is asymmetric if there are substantial differences in these abilities, as in the case of stresses exerted by forest-canopy trees on plants of the understory.

Herbivory, predation, parasitism, and disease are all trophic interactions, in which individuals of one species exploit individuals of other species. Exploitative stresses can broadly refer to the selective removal of particular species or size classes of organisms from a community. Exploitation can be by humans, as in the harvesting of wild animals or trees, or exploitation can be natural, as in widespread mortality caused by defoliating insects or disease-causing pathogens. The primary effect of exploitative stress is the removal of individuals, populations, stands, or communities. There may also, however, be secondary consequences of harvesting. For example, the removal of nutrients contained in agricultural biomass may be sufficiently large to cause a reduction in site fertility. Large, indirect effects also occur if the harvested organisms are disproportionately important in the structure and function of their ecosystem.

When an ecosystem is disrupted by a substantial event of physical stress, wildfire, or harvesting, it can quickly suffer a great deal of mortality of its component species, along with structural disruptions and other ecological damages. Primary or secondary succession then begins and may restore an ecosystem similar to the that present prior to the disturbance. A process of ecological succession also follows after alleviation of the intensity of longer-term, chronic stresses.

Stress by pollution, thermal energy, and ionizing radiation results in ecological changes by causing physiological effects on organisms. If the intensity of the stressor exceeds thresholds of tolerance, organisms may suffer chronic or acute toxicity, and individuals may be less successful and may die prematurely. If there is genetic variation in the vulnerability of individuals to the stressor, then evolution will result in an increased tolerance at the population level. At the community level, relatively vulnerable species may be reduced in abundance or eliminated from stressed sites. Their ecological role may then be assumed by other more tolerant members of the community or by invading species that can exploit a stressful but weakly competitive habitat.

An important aspect of toxic chemicals is that, if a sufficiently large dose is achieved, any chemical can poison any organism. This suggests that all chemicals are potentially toxic, even water, carbon dioxide, table sugar, sodium chloride, and other substances that are routinely encountered (see Table 8.2). In contrast, very small doses of even extremely toxic chemicals may not cause a measurable, poisonous effect. The exposure must exceed physiological thresholds of tolerance before an effect can be demonstrated, as was noted previously when distinguishing between contamination and pollution. In view of this interpretation, it is best to refer to "potentially toxic chemicals" in contexts in which the actual environmental exposures are not known. (Note, however, that some scientists believe that exposures to even single molecules of some chemicals may be of toxicological importance, and that dose–response curves can therefore be linearly extrapolated to zero dosages.) These topics are discussed in Chapter 8, Section 3.1.

If an ecosystem experiences a prolonged exposure to intense pollution, thermal energy, or ionizing radiation, there will be longer-term, ecological adjustments. Consider a case in which chronic, severe, atmospheric or soil pollution is caused by the operation of a smelter in a remote, forested landscape. If the ensuing toxic stresses are sufficiently severe and/or cumulative, they will cause a progressive replacement of the original mature forest by an open woodland, followed later by shrub-sized vegetation and then a herbaceous community. Ultimately, a devegetated landscape can result. The most important features of this type of ecological collapse are changes in species composition, in the spatial distribution of biomass, and in ecosystem functions, such as productivity, litter decomposition, and nutrient cycling (Smith, 1981, 1984; Bormann, 1982; Rapport *et al.*, 1985; Schindler, 1987). Because pollution-related stresses decrease exponentially with increasing distance from point sources, these ecosystem changes eventually become manifest in continuous gradients of community change along transects that originate at the source of pollution.

Several ecologists have described the generic attributes of ecosystems that have been stressed for shorter or longer periods of time. An intensification of stress often results in increased rates of mortality, reductions of species richness, increased exports of nutrients and biomass, rates of community respiration that exceed net production; replacements of sensitive species with tolerant ones, and other responses typical of degrading ecosystems (Auerbach, 1981; Odum, 1981, 1985; Bormann, 1982; Smith, 1981, 1984; Rapport *et al.*, 1985; Schindler, 1988; Kelly and Harwell, 1989; see also Table 13.1 in Chapter 13). In comparison, chronically stressed ecosystems are relatively stable, simple in terms of their species richness and structural complexity, and composed of longer-lived species with a small standing crop of biomass and have small rates of productivity, decomposition, and nutrient cycling (Grime, 1979; Freedman *et al.*, 1993e).

Because many cases of natural pollution and other stresses are ancient, their ecological effects can give insights into the potential, longer-term effects of modern, anthropogenic stressors. For this reason, natural cases are dealt with frequently in this book, and several are described as detailed case studies. Some notable examples of natural environmental stresses include the following:

1. Erupting volcanos can emit large quantities of sulfur dioxide, particulates, and other pollutants into the atmosphere. The global volcanic emission of SO_2–S averages 2–5 million tonnes/year, and individual eruptions can emit more than 1×10^6 tonnes (Cullis and Hirschler, 1980; Moller, 1984). In 1883, the cataclysmic eruption of the volcano Krakatau in the Sunda Strait of Indonesia injected an estimated 18–21 km³ of particulates as high as 50–80 km into the atmosphere, causing spectacular sunsets and other visual phenomena throughout the world for several years (Thornton, 1984). The even larger eruption of Tambora in Indonesia in 1815 injected an estimated 300–1000 km³ of ash into the atmosphere. Some of the finer particulates made their way into the stratosphere, where they formed a thin, reflective veil that increased the Earth's albedo and caused global cooling. The year 1816 was known as the "year without a summer" in Europe and North America because of its unusually cool

and wet weather, including frost and snowfall events during the summer (Stothers, 1984). A volcanic event can also initiate a devastating oceanic wave or tsunami—the 30-m-high wave associated with the eruption of Krakatau travelled as quickly as 25 m/sec and killed an estimated 36 thousand people (Thornton, 1984). In addition, a large eruption can physically damage great expanses of forest and other ecosystems. For example, the 1980 explosion of Mount St. Helens in Washington blew down 21.1 thousand ha of conifer forest, killed another 9.7 thousand ha of forest by heat injury, and otherwise damaged another 30.3 thousand ha. There was also devastation caused by great mudslides, and a large area was covered by particulate tephra that settled from the atmosphere to a depth of up to 50 cm (Rosenfeld, 1980; Del Moral, 1983; Means and Winjum, 1983).

2. A large wildfire can kill mature trees over great areas, after which a secondary succession ensues. Fire is especially frequent in boreal forests and in other seasonally droughty ecosystems, such as prairie and savannah. For example, an average of about 3 million ha of forest burns each year in Canada, mostly as a result of natural ignition by lightning (Wein and MacLean, 1983; Honer and Bickerstaff, 1985). The well-publicized Yellowstone fires of 1988 burned 570 thousand ha, including 45% of Yellowstone National Park (Christensen *et al.*, 1989). Even moist tropical rainforest will sometimes burn, as occurred over 3.5 million ha of Borneo during relatively dry conditions in 1982–1983 (Malingreau *et al.*, 1985). In addition to the direct effects on ecosystems, forest fires emit large quantities of gases and particulates into the atmosphere (Smith, 1984). In Canada, for example, forest fires emit an estimated 730×10^6 kg/year of particulates to the atmosphere (Anonymous, 1981). Wildfires also oxidize most of the organic nitrogen of the combusted biomass, and this is emitted to the atmosphere as gaseous oxides of nitrogen (Chapter 2).

3. Other natural agents that cause severe physical disturbances of ecosystems include hurricanes and tornadoes, flooding, and, over geologically long periods of time, glaciation.

4. Outbreaks of pests can devastate vulnerable ecosystems. For example, in 1975 spruce budworm (*Choristoneura fumiferana*) severely defoliated more than 55 million ha of conifer forest in eastern North America, causing much tree mortality and other ecological changes (Kettela, 1983; Chapter 8). Outbreaks of herbivorous invertebrates can damage mature marine ecosystems in ways that parallel the effects of forest defoliators. For example, off the rocky coast of Nova Scotia, an outbreak of a sea urchin (*Strongylocentrotus droebachiensis*) overgrazed a mature kelp "forest" dominated by *Laminaria* spp. and *Agarum* spp., and converted the ecosystem to a "barren ground" with a much smaller standing crop and productivity (Breen and Mann, 1976; Mann, 1977; Chapman, 1981). After a collapse of the urchin population was caused by a disease induced by unusually warm water, the kelp forest rapidly reestablished (Scheibling, 1984; Scheibling and Stephenson, 1984).

5. Outbreaks of pathogens can also devastate vulnerable ecosystems. An example of the effects of pathogens is the introduced chestnut blight fungus (*Endothia parasitica*) that eliminated the American chestnut (*Castanea dentata*) as a canopy species in the deciduous forest of eastern North America, resulting in its replacement by other shade-tolerant species (Hepting, 1971; Spurr and Barnes, 1980). A similar pandemic of the introduced Dutch elm disease fungus (*Ceratocystis ulmi*) is removing native elms (*Ulmus* spp., especially *Ulmus americana*) from deciduous forests of North America and elsewhere (Fowells, 1965; Hepting, 1971).

6. Sometimes, certain species can cause ecological damage by the synthesis and release of toxic chemicals. For example, toxin-releasing marine phytoplankton occasionally bloom and cause ecological damage by the synthesis of biochemicals that are toxic to a broad spectrum of animals, which are exposed through the food web. For example, a bloom of the flagellated phytoplankton *Chrysochromulina polylepis* in the Baltic Sea in 1988 caused a mass mortality of various species

of macroalgae, invertebrates, and fish (Rosenberg *et al.*, 1988).

7. Other natural sources of toxic stress include surface occurrences of metal-rich mineralizations (Chapter 3); the bioaccumulation of toxic elements such as mercury by large fish and selenium by certain plants (Chapter 3); the spontaneous combustion of bituminous material with resulting releases of toxic SO_2 gas (Chapter 2); and the natural acidification of land and water caused by biological processes or by the oxidation of sulfide minerals in soil (Chapter 4).

While discussion of natural stressors is useful and informative in environmental ecology, this book emphasizes ecological damages caused by human influences. This treatment is justified by the increasing prominence of anthropogenic stressors and the need to manage their effects.

As described below, the various chapters are aggregated by the type of stressor, rather than by separate consideration of topics by ecosystem or landscape type (e.g., aquatic versus terrestrial pollution). The book is organized such that the individual chapters stand alone. However, where necessary there is integration across chapter topics. For example, the topic of acidification (Chapter 4) cannot be considered without some knowledge of the biogeochemistry and ecological effects of gaseous air pollutants (Chapter 2) and toxic elements (Chapter 3). Furthermore, all three of these topics are relevant to understanding the declines of forests in many parts of the industrialized world (Chapter 5).

The first class of stressors to be considered is gaseous air pollutants (Chapter 2). For each of the most important pollutant gases (i.e., sulfur dioxide, oxides of nitrogen, ozone, and some others), the sources of emission, chemical transformations, and toxicity are described. This is followed by a brief consideration of air pollution effects on human health. Three detailed case studies of the ecological effects of toxic gases are then described. The first concerns a situation in the Arctic where a natural emission of SO_2 is damaging the tundra; the second case describes ecological damage in the vicinity of

SO_2-emitting smelters near Sudbury, Ontario; and the third examines regional forest damages caused by ozone in southern California. The last topic in this chapter deals with the phenomenon of increasing concentrations of carbon dioxide and some other gases in the atmosphere and the ecological effects that may be taking place through global climatic warming and CO_2 fertilization of plants.

Toxic elements are initially described in Chapter 3 in terms of their biogeochemistry and toxicity, especially to plants. Several examples of the ecological effects of natural exposures to toxic elements are followed by descriptions of some of the most important types of anthropogenic pollution, with particular emphasis on the effects of agriculture and of industries that mine and process metals.

Consideration of acidification (Chapter 4) begins with a description of the chemistry of precipitation and the dry deposition of acidifying substances from the atmosphere. Changes in water chemistry are then described, as precipitation percolates through the forest canopy, the soil, and then to surface waters. This is followed by an examination of the effects of acidification on terrestrial and aquatic biota and of the use of management practices such as liming or fertilization to counter acidification.

The phenomenon of forest decline is examined in Chapter 5. This syndrome has recently occurred in forests in many parts of the industrialized world. The causal agents may include gaseous air pollutants, toxic elements, and acidic deposition. The precise etiology of forest declines are usually uncertain, however, and sometimes the declines appear to be natural in origin.

Oil pollution is dealt with in Chapter 6, by consideration of the nature of petroleum, the sources and weathering of oil in the environment, and the toxicity of crude oil and its common refined products. The ecological effects of oil spills are then described as case studies, with particular reference to marine spills from wrecked supertankers and drilling platforms and the effects of oil spilled in arctic environments.

The excessive enrichment of ecosystems with nutrients, especially the eutrophication of fresh wa-

ters, is the topic of Chapter 7. The causes of eutrophication are considered by the examination of whole-lake fertilization experiments, the anthropogenic enrichment of Lake Erie and some other lakes, and the recovery of eutrophied waterbodies after the diversion of waste nutrients.

The ecological effects of pesticides are described in Chapter 8. Pesticides are first classified and described by their chemical characteristics and uses. Their ecological effects, both direct and indirect, are then examined using case studies of the insecticide DDT and its chlorinated-hydrocarbon relatives, some agricultural pesticide uses, the spraying of spruce budworm-infested forests with insecticides, and the use of herbicides in forestry.

The ecological effects of harvesting forests are discussed in Chapter 9. The major topics considered here are the effects of nutrient removals on site fertility; the effects of disturbance on leaching of nutrients, erosion, and hydrology; and the effects of harvesting on species and their communities.

Changes in global species richness caused by natural and anthropogenic extinctions are examined in Chapter 10. Case studies are used to illustrate species that have been made extinct or endangered by overharvesting or the destruction of their habitat. The chapter closes with examples of endangered species that have been brought back from the brink of extinction by vigorous conservation efforts, and a broad discussion of ecological reserves.

The environmental effects of warfare are examined in Chapter 11. This topic is divided into three sections by separate consideration of the effects of conventional, chemical, and nuclear warfare.

Effects of the excessive harvesting of biological, natural resources are discussed in Chapter 12. Case studies of particular biological resources that were depleted by overharvesting are presented, with a focus on whales, seals, and marine fish. More appropriate management, based on the principles of resource ecology, is then discussed. The chapter closes with a comparison of conventional and ecological economics and a discussion of the notion of ecologically sustainable systems.

The final chapter of the book discusses some

applications of environmental ecology. The first part demonstrates common patterns among the ecological effects of various types of stressors. Major topics include the nature of the damages caused by stressors on both the shorter and the longer term, the spatial patterns of ecological damages that are observed around point sources of stress, and ecological changes resulting from the intensification of stress and from its alleviation. The relationships of environmental ecology to environmental impact assessment, and to ecological and environmental monitoring, are then described. Finally, there is a discussion of the roles that ecologists must play in helping society to deal with the most important predicament facing humanity today and into the future—the environmental crisis, and its many ecological manifestations.

2

AIR POLLUTION

2.1

INTRODUCTION

There are many natural sources of gaseous air pollutants. These include volcanoes, forest fires, and outgassings from anaerobic wetlands. In some cases the magnitudes of natural emissions of air pollution can rival or exceed those associated with human activities. In other cases anthropogenic emissions are more important and are increasing in quantity because of growth of the human population and because of technological developments that result in the emission of pollutants. Of particular note in the latter respect are emissions of waste gases from fossil-fueled power plants and automobiles; neither of these sources existed prior to the present century.

Even in ancient times, however, the activities of humans caused air pollution. Smoky wood fires used for cooking and space heating were an early source of air pollution that must have impaired air quality inside caves, sod houses, and other poorly ventilated dwellings.

Later on, when industrialization became an increasingly dominant feature of human endeavors, air pollution became much more extensive. In particular, the burning of coal caused severe air pollution by sulfur dioxide and soot in the cities of Europe. This problem became intense in the

nineteenth century, when it was first noted as a cause of damage to human health, buildings, and ecosystems. After these effects were recognized, mitigative actions were taken progressively, including: (1) the construction of tall smokestacks that spread emissions over a wider area, so that ground-level fumigations were less frequent and less intense (this is the "dilution solution to pollution"); (2) the switching to "cleaner" hydrocarbon fuels such as methane and oil; (3) a reduction of total emissions by the centralization of energy production (e.g., the construction of power plants, to supplant most of the relatively dirty burning of coal in fireplaces to heat homes); and (4) the removal of pollutants from waste gases before they are vented to the atmosphere.

A contamination of the atmosphere by sulfur dioxide and soot was characteristic of the initial phases of urban air pollution during and following the industrial revolution of the 1800s. This type of air pollution is often called reducing, or London-type, smog (the word "smog" is a composite of "smoke" plus "fog"). Smog occurs during episodes of stability of the lower atmosphere, which prevents the mixing of polluted ground-level air masses with cleaner air from higher altitudes. As recently as the 1950s, reducing smog caused hundreds of human deaths and a high frequency of acute respiratory distress during pollution episodes in London, Glasgow, and other industrial centers of Europe and around Pittsburgh in the United States.

To some degree, reducing smog has now been supplanted in importance by oxidizing, or Los Angeles-type, smog. Oxidizing smog occurs in sunny locations where there are large emissions of hydrocarbons and nitric oxide from automobiles and industrial sources and where atmospheric temperature inversions are frequent. Oxidizing smog is formed when the primary, emitted pollutants are transformed by complex photochemical reactions into secondary pollutants, most notably ozone and peroxyacetyl nitrate. It is these secondary gases that are most harmful to people and vegetation exposed to oxidizing smog.

Secondary pollutants are also formed from emitted sulfur dioxide and oxides of nitrogen. Ultimately, these gases are transformed to the ions sulfate and nitrate, respectively, and in these forms they can be delivered to terrestrial and aquatic ecosystems as acidic deposition (described in Chapter 4).

In this chapter, the chemical characteristics, transformations, toxicity, sources, and sinks of gaseous air pollutants are examined. This is followed by descriptions of case studies of ecological damage caused by environmental pollution with toxic gases. The chapter ends with a consideration of the increasing atmospheric concentrations of carbon dioxide and other radiatively active gases. Because these gases can interfere with the processes by which the Earth dissipates absorbed solar radiation, they may be capable of causing global climatic changes, with important implications for natural and anthropogenic ecosystems.

2.2
EMISSION, TRANSFORMATION, AND TOXICITY OF AIR POLLUTANTS

The most important gaseous pollutants are sulfur dioxide (SO_2), hydrogen sulfide (H_2S), oxides of nitrogen (NO_x), ammonia (NH_3), carbon monoxide (CO), carbon dioxide (CO_2), methane (CH_4), ozone

(O_3), and peroxyacetyl nitrate (PAN). In addition, there are pollutant vapors of hydrocarbons and elemental mercury and small-diameter (<1 μm) particulates that behave aerodynamically like gases and remain suspended in the atmosphere for a long time. These tiny particulates include inert siliceous or other minerals; dusts containing toxic elements such as arsenic, lead, copper, nickel, etc.; organic aerosols emitted as smoke from combustions; and high-molecular-weight condensed hydrocarbons such as polycyclic aromatics.

Many of these air pollutants have both natural and anthropogenic sources of emission. O_3 and PAN, however, are not emitted directly into the atmosphere, but are produced secondarily in the atmosphere by complex photochemical reactions.

Sulfur Gases

Characteristics, Emissions, and Transformations

Gaseous sulfur is largely emitted as SO_2 and H_2S. SO_2 is a colorless but pungent gas that can be tasted at 0.3–1 ppm. H_2S is a gas with the foul smell of rotten eggs, which humans can typically detect by smell at <1 ppb (Urone, 1976). In the atmosphere H_2S has a residence time of <1 day, as it is rapidly oxidized to SO_2 (Table 2.1).

Atmospheric SO_2 is transformed ultimately to the anion sulfate (SO_4^{-2}). The rate of oxidation of SO_2 ranges from <1 to 5%/hr during the day, and the process is influenced by the intensity of sunlight, humidity, and the presence of nitrogen oxides, hydrocarbons, strong oxidants, and catalytic metal-containing particulates (Meszaros, 1981; Newman, 1981; Wilson, 1981; Anlauf et al., 1982; Liebsch and de Pena, 1982; Fox, 1986). Because of its moderately long residence time (about 4 days), most SO_2 is transported a long distance from its point of emission before it is oxidized or deposited to the surface of landscapes.

The atmospheric reactions by which SO_2 is oxidized to SO_4^{-2} are as follows (Hicks, 1990; Venkatram, 1990):

$$SO_2 + OH \rightarrow HO \cdot SO_2 \qquad (1)$$

Table 2.1 Global emission and other characteristics of important air pollutants[a]

Pollutant	Anthropogenic emissions (10^6 tonnes/year)	Natural emissions (10^6 tonnes/year)	Background concentration (ppb)	Atmospheric residence time	Typical concentration (ppm) Clear air	Polluted air
SO_2	63–72	16–60	0.2	4 days	0.0002	0.2
H_2S	3	100	0.2	<1 day	0.0002	—
CO	304	33	100	<3 years	0.1	40–70
NO/NO_2	53 (as NO_2)	NO: 430 NO$_2$: 658	0.2–2 0.5–4	5 days	<0.002 <0.004	0.2 (as NO_2)
NH_3	4	1160	6–20	7 days	0.01	0.02
N_2O	6	18	300	4 years	0.3	—
Hydrocarbons	88	200	<1	?	<0.001	—
CH$_4$		1600	1.5×10^3	4 years	1.5	2.5
CO_2	14,000	1,000,000	340×10^3	2–4 years	340	400
Particulates	3900	3700	—	—	—	—
O_3	—	—	—	—	0.03	0.5

[a]Modified from Kellogg *et al.* (1972), Robinson and Robbins (1972), Urone (1976), Whelpdale and Munn (1976), Cullis and Hirschler (1980), Moller (1984), Matson and Vitousek (1990), and Anonymous (1993b).

$$HO \cdot SO_2 + O_2 \rightarrow HO_2 + SO_3 \qquad (2)$$

$$SO_3 + H_2O \rightarrow H_2SO_4 \qquad (3)$$

$$H_2SO_4 \rightarrow 2H^+ + SO_4^{-2} \quad \text{(in aqueous solution)}. \qquad (4)$$

Note that reaction (4) only occurs in aqueous solutions, such as a raindrop. The atmospheric SO_4^{-2} produced by the oxidation of SO_2 is balanced electrochemically by various cations. In eastern North America, most particulate sulfate occurs as ammonium sulfate [$(NH_4)_2SO_4$], a major component of the pollution haze that impairs visibility in cities and remote areas subject to air pollutants transported over long distances (Ferman *et al.*, 1981; Hosker and Lindberg, 1982; Anonymous, 1993b). If there are not sufficient cation equivalents other than H^+ to balance the SO_4^{-2}, the latter will be present as a strongly acidic and hygroscopic sulfuric acid aerosol (H_2SO_4), which contributes to the acidity of precipitation (see Chapter 4).

The largest natural sources of SO_2 emissions are volcanos and forest fires. Emissions of SO_2 by the oxidation of organic sulfur during forest fires have not been well quantified, but the volcanic emissions are estimated up to 12 million tonnes of sulfur per year (i.e., 12×10^6 tonnes/year), and in particular, large eruptions are estimated to emit more than 1 million tonnes (Cullis and Hirschler, 1980; Moller, 1984; Berresheim and Jaeschke, 1983). About 90% of the global volcanic emission of sulfur occurs as SO_2 and 10% as H_2S. For example, the 1980 eruption of Mount St. Helens, Washington, resulted in the emission of about 0.15×10^6 tonnes of SO_2-S and 0.02×10^6 tonnes of H_2S-S. The more substantial eruption of El Chichon in Mexico in 1982 resulted in the emission of 1.6×10^6 tonnes of SO_2-S (Placet, 1990), while the even larger eruption of Mount Pinatubo in the Philippines in 1991 may have emitted $7–10 \times 10^6$ tonnes of SO_2-S (Anonymous, 1991, 1993b).

The anthropogenic emission of SO_2 to the atmosphere is much larger than the natural emission and has been estimated as $63–72 \times 10^6$ tonnes/year (Table 2.1; Anonymous, 1993b). The largest source of SO_2 is the burning of fossil fuels, accounting for 54% of the total anthropogenic emission (Moller, 1984). Fossil fuels contain sulfur in both mineral and organic forms, and during combustion more than 90% of the sulfur is oxidized to gaseous sulfur dioxide. Typical sulfur concentrations of fossil fuels are: hard coals from eastern North America, 1–12%;

softer coals from western North America, <0.3–
1.5%; lignite, 0.7–0.9%; crude oil, 0.8–1.0%; residual fuel oils, 0.3–0.4%; kerosene, 0.4%; and
motor fuels, 0.04–0.05%. Manufacturing processes
(23% of anthropogenic emissions) and the smelting
of sulfide ores (7%) are other large sources of SO_2
emission (Dvorak and Lewis, 1978; Moller, 1984).

The global, anthropogenic emission of SO_2 has
increased greatly in the past century (Table 2.2),
from approximately 5 million tonnes in 1860 to as
much as 130 million tonnes/year in the mid-1980s
(see Figure 2.1 for U.S. trends in SO_2 emission). In
the near future, increasing demands for electric
power will be at least partly met by the construction
of additional fossil-fueled power plants. This will
result in even larger emissions of SO_2, unless there
are increased efforts toward emission reductions by
the removal of SO_2 from flue gases, fuel desulfurization, fuel switching, and energy conservation
(Moller, 1984).

Changes in the nature and scale of industrialization and urbanization during the last century have
also increased the area of terrain that is affected by
SO_2 pollution. This is due partly to the use of increasingly taller smokestacks as a means of SO_2
dispersal, a practice that causes a more regionalized
pollution as a consequence of the long-distance
transport of emissions.

Table 2.2 Changes in the global, anthropogenic
emission of SO_2 over time[a]

Year	Coal	Oil	Others	Total
	Emissions of SO_2-S (10^6 tonnes/year)			
1860	2.4	0.0	0.1	2.5
1880	5.6	0.0	0.5	6.1
1900	12.6	0.2	1.3	14.1
1920	21.2	0.7	3.4	25.0
1940	24.2	2.3	6.2	32.7
1960	30.4	8.3	10.7	48.6
1970	32.4	17.6	12.0	62.0
1977	37.2	24.0	13.7	74.9
1985	48	25	17	90
2000	55	23	22	100

[a]Modified from Moller (1984).

The emissions of SO_2 and other pollutants vary
greatly between and within countries because of
differences in population density, degree and type of
industrialization, quantity and type of fuels used,
etc. These differences are particularly large in any
comparison between relatively developed and less-
developed nations. However, neighboring industrialized countries can also differ markedly. For
example, in 1985 the total U.S. emission of SO_x
(almost all of which is SO_2) was 5.7 times larger
than that of Canada (Table 2.3). However, because
the U.S. population is about 10 times greater (248.7
million versus 26.8 million, in 1990), the Canadian
per capita emission was about 2 times larger. Of the
total U.S. emission, 70% was from stationary
sources such as fossil-fueled power plants and 24%
from large industrial sources. In Canada, only 19%
of the total SO_2 emission was from power plants,
while 76% was from industrial sources, especially
sulfide metal smelters. The most important reasons
for these differences in emissions sectors are: (1) a
relatively large proportion of Canadian electricity
generation is by nuclear and hydroelectric technology, which do not emit SO_2, and (2) the metal smelting and refining industry is relatively important in
Canada.

The most important SO_2-emitting states in 1985
were Ohio (11.1% of the national U.S. emission of
21×10^6 tonnes of SO_2), Indiana (8.4%), Texas
(6.4%), Pennsylvania (6.2%), and Illinois (6.0%), a
pattern that reflects the regional prominence of coal
burning for electricity and other industrial purposes
(Placet, 1990).

Rapidly industrializing countries have a requirement for increasingly large quantities of inexpensive energy, and this is often satisfied by the combustion of high-sulfur fuels such as coal. In China,
for example, coal is the major source of industrial
energy, comprising 72% of the total energy demand
in 1980 and 76% in 1986 (Hongfa, 1989). Chinese
emissions of sulfur dioxide increased correspondingly during that period, from 10 million tonnes in
1980, to 14 million tonnes in 1985, leading to an
intensification of reducing smogs, acidic precipitation, and agricultural damages (Hongfa, 1989).

In contrast to the pattern for SO_2, the global

Table 2.3 Comparison of United States and Canadian emissions of air pollutants by source (1985 data, 10^6 tonnes/year)[a]

Source	SO$_x$ (as SO$_2$)	%	NO$_x$ (as NO$_2$)	%	Hydrocarbons	%	CO[b]	%	Particulates[b]	%
United States										
Transportation	—	—	8.0	42.8	8.0	40.0	73.5	77.7	1.3	6.7
Stationary fuel combustion	14.6	69.5	6.0	32.1	0.8	4.0	0.9	1.0	5.9	30.2
Industrial processes	5.1	24.3	3.8	20.3	1.5	7.5	12.7	13.4	11.0	56.4
Solid waste disposal	<0.1	<0.1	0.1	0.5	—	—	2.4	2.5	0.5	2.6
Miscellaneous	1.3[c]	6.2	0.8	4.3	9.7[d]	48.5	5.1	5.4	0.8	4.1
Total	21.0		18.7		20.0		94.6		19.5	
Canada										
Transportation	—	—	1.2	63.2	1.0	45.5	10.8	70.1	0.1	4.2
Stationary fuel combustion	0.7	18.9	0.3	15.8	0.1	4.5	0.2	1.3	0.3	12.5
Industrial processes	2.8	75.7	0.3	15.8	0.2	9.0	1.3	8.5	1.5	62.5
Solid waste disposal	<0.1	<0.1	<0.1	<0.1	<0.1	<0.1	0.4	2.6	<0.1	<0.1
Miscellaneous	0.2[c]	5.4	0.1	5.2	0.9[d]	40.9	2.7	17.5	0.5	20.8
Total	3.7		1.9		2.2		15.4		2.4	

[a] Modified from Council on Environmental Quality (CEQ) (1975), Anonymous (1976a), and Placet (1990); see Benkovitz (1982) for a more detailed comparison. %, percentage total by country.
[b] 1974 data.
[c] Includes transportation, plus commercial and residential fuel use.
[d] Includes oil and gas production and storage.

emission of gaseous sulfides is predominantly from natural sources, especially emissions of H_2S from anaerobic sediments of shallow coastal and inland waters and the emission of dimethyl sulfide by marine phytoplankton (Moller, 1984; Charlson et al., 1987). The total natural emission of H_2S-S has been estimated as 100×10^6 tonnes/year (Table 2.1) and dimethyl sulfide about 15×10^6 tonnes/year (expressed as sulfur; Bates et al., 1987). Anthropogenic emissions of H_2S are from certain chemical industries, sewage treatment facilities, and animal manure. These total about 3×10^6 tonnes/year (Table 2.1).

By combining information for SO_2, H_2S, and other sulfur gases, the global emissions of sulfurous gases to the atmosphere can be estimated. Of the total sulfur emission of 251×10^6 tonnes/year in 1976, 59% was from natural sources (Table 2.4).

The natural emissions were approximately equally divided between the northern and southern hemispheres, and the biogenic emission was evenly split between oceanic and terrestrial sources. Ninety-four percent of the anthropogenic emission was from the northern hemisphere, reflecting the patterns of the global distribution and industrialization of the human population. Interestingly, about 80% of the average sulfur content of the atmosphere is estimated to be in the form of a relatively unreactive gas of natural marine origin, carbonyl sulfide, even though this gas comprises less than 3% of the global emission of sulfur gases (Moller, 1984).

The typical concentrations of SO_2 and H_2S in clean, unpolluted air are each less than about 0.2 ppb (Table 2.2). Concentrations of SO_2 in polluted air are extremely variable, but they average about 0.2 ppm in urban air (Table 2.1) and can range to

Table 2.4 Global emission of sulfur in 1976[a]

| Source | Emission of S (10^6 tonnes/year) | | |
	Northern Hemisphere	Southern Hemisphere	Total
Natural			
Volcanoes	3	2	5
Sea spray	19	25	44
Biogenic (land)	32	16	48
Biogenic (oceans)	22	28	50
Total	76	71	147
Anthropogenic			
Coal	59	2	61
Petroleum	24	1	25
Nonferrous ores	8	3	11
Others	7	<1	7
Total	98	6	104
Total emission	174	77	251

[a]Modified from Cullis and Hirschler (1980).

more than 3 ppm near major sources of emission (Shriner, 1990).

Toxicity

H_2S is seldom present in a concentration sufficiently large to damage plants. The phytotoxicity of SO_2 is well known, and there are many field and laboratory observations of injuries and reduced yields of cultivated and wild plants. The toxic effects of SO_2 include (Jacobson and Hill, 1970; Heck and Brandt, 1978; Roberts, 1984; Shriner, 1990): (1) *acute injury*, in which there are tissue damages such as necrosis, usually in response to shorter-term exposures to large SO_2 concentrations; (2) *chronic injury*, in which losses of yield are associated with less severe injuries such as chlorosis or premature abscission of foliage, usually in response to longer-term exposures to smaller SO_2 concentrations; and (3) *hidden injuries*, in which a loss of yield occurs in the absence of visible symptoms of injury.

Roberts (1984) reviewed the effects of SO_2 on plants. He concluded that it is difficult to generalize about phytotoxic thresholds of SO_2 in air, because of large differences in susceptibility (1) among species, (2) among varieties within species, and (3) under varying conditions of stress caused by drought, nutrient supply, and other pollutants. Studies of the effects on plant yield of short-term exposures to SO_2 are most relevant to conditions near point sources, where fumigations are relatively sporadic but usually intense. For example, a 1-hr exposure to 0.7 ppm SO_2 is sufficient to cause acute injury to most plant species, as is an 8-hr exposure to 0.18 ppm (Shriner, 1990). Roberts (1984) concluded that where fumigations occurred during about 10% of the growth period, the average concentration of SO_2 must be >185 ppb to affect most crop species and >170 ppb to affect most tree species (for particularly sensitive species or varieties the thresholds are lower than these).

Roberts also reviewed studies involving the continuous exposure of plants to ambient concentrations of SO_2. These studies are most relevant to field conditions where the concentrations of SO_2 are regionally elevated, for example, in a large city. Studies of the pasture grass *Lolium perenne* exposed to unfiltered or charcoal-filtered (which removes SO_2) air in open-top chambers at various localities in Britain demonstrated that longer-term average concentrations of >38 ppb SO_2 were required to measurably reduce yield. This loss of productivity occurred as a hidden injury and was not accompanied by obvious symptoms of tissue damage.

Roberts also reviewed laboratory and field studies of the effects of controlled SO_2 fumigation on the yields of a wide range of species and concluded that exposure to (1) 76–150 ppb SO_2 for 1–3 months produced significant decreases in yield in most species; (2) 38–76 ppb for several months produced small decreases in yield for some but not all species; and (3) <38 ppb variously produced beneficial, no, or minor effects on yield. Notably, the concentrations described in (1) and (2) are considerably smaller than the U.S. air quality standard for SO_2 in the atmosphere, which is 500 ppb for 3 hr. This standard is based on threshold concentrations that cause acute foliar injury to a variety of plants (Roberts, 1984).

Therefore, SO_2 can be phytotoxic to some plants, even in moderately polluted atmospheres. In very polluted environments SO_2 has caused substantial ecological damages, as is described later in several case studies. Other consequences of SO_2 exposure are (1) selection for SO_2-tolerant populations of plants in chronically polluted environments (Roose *et al.*, 1982); and (2) sulfur fertilization in certain agricultural environments with high nitrogen-phosphorus-potassium (NPK) fertilization. In such cases, sulfur deficiencies may limit crop productivity, and sulfur uptake from the atmosphere may partially compensate for this as long as the sulfur does not occur at phytotoxic concentrations (Coleman, 1966; Beaton *et al.*, 1971, 1974; Terman, 1978). In North America, situations of sulfur deficiency are most frequent in western grain agriculture and on calcareous soils. Sulfur deficiency in agriculture is rare in the eastern United States and southeastern Canada, where sulfur dioxide pollution and sulfate deposition are relatively important environmental problems (see also Chapter 4).

Humans are generally more tolerant of exposure to SO_2 than are most plants. The American Conference of Governmental Industrial Hygienists (ACGIH) has recommended a concentration no larger than 2 ppm SO_2 as an average occupational exposure and 5 ppm for a short-term exposure (ACGIH, 1982; Henderson-Sellers, 1984). However, exposure to SO_2 concentrations of <1 ppm can cause respiratory distress in sensitive individuals, especially people with a history of asthma or other respiratory diseases (Goldsmith, 1986). It should be noted that there is controversy over the effects of chronic exposure of large human populations to the relatively small concentrations of SO_2 and sulfate-particulate aerosols that commonly occur in many urban environments. Some epidemiological and other medical studies have demonstrated statistically significant effects on human health at ambient urban concentrations of sulfurous and other urban-air pollutants, while other studies have not (Amdur, 1980; Goldsmith, 1986; Chestnut and Rowe, 1989; Graham, 1990). More research is required to sort out this potential environmental problem.

Nitrogen Gases

Characteristics, Emissions, and Transformations

From a pollution perspective, the most important nitrogen gases are ammonia (NH_3), nitric oxide (NO), nitrogen dioxide (NO_2), and nitrous oxide (N_2O). Together, NO and NO_2 are often abbreviated as NO_x.

Ammonia is a colorless gas. Its major source is natural emissions from wetlands, where NH_3 is produced during the decomposition of biological materials. Total natural emissions of NH_3 are estimated as $>10^9$ tonnes/year (Table 2.1). Anthropogenic emissions of NH_3 are much smaller, and sources include coal combustion (3×10^6 tonnes/year), oil and gas combustion (1×10^6 tonnes/year), and cattle feedlots (0.2×10^6 tonnes/year) (Whelpdale and Munn, 1976). Ammonia is oxidized to NO_x in the atmosphere, where it has an average residence time of 7 days (Table 2.1).

N_2O is a colorless and nontoxic gas. N_2O is used in medicine as a mild anesthetic, and it is sometimes called laughing gas because it produces a mild euphoria. The background concentration of N_2O in the atmosphere is 0.3 ppm, and because it is relatively unreactive it has a long residence time of 4 years (Table 2.1). Industrial emissions of N_2O are mostly associated with fuel combustions and amount to about 6×10^6 tonnes/year, but there are larger biological emissions of about 18×10^6 tonnes/year, largely resulting from anaerobic, microbial denitrification of nitrate in soil and water [Box, 1982; Aneja *et al.*, 1984; World Meteorological Organization (WMO), 1986; Matson and Vitousek, 1990]. Fertilized agricultural soils can have particularly large rates of N_2O emission. It has been estimated that modern agriculture has increased the global emission of N_2O by 50% and that the concentrations of atmospheric N_2O have been increasing by 0.2%-0.3% per year during the most recent 20–30 years (Aneja *et al.*, 1984; Matson and Vitousek, 1990).

NO is a colorless, odorless, and tasteless gas, while NO_2 is reddish-brown, pungent, and irritating to respiratory membranes. Background atmospheric concentrations of NO are 0.2–2 ppb, while that of NO_2 is 0.5–4 ppb. In polluted atmospheres, these gases are present at ca. 0.2 ppm (expressed as NO_2). In the atmosphere, NO is oxidized relatively rapidly to NO_2 by reactions that are described later, when photochemical air pollution is discussed. NO_2 is eventually oxidized photochemically and catalytically to nitrate, an anion that contributes substantially to the acidity of precipitation (Chapter 4).

NO and NO_2 are eventually oxidized in the atmosphere to NO_3^-, by the following atmospheric reactions (Hicks, 1990; Venkatram, 1990):

$$NO + HO_2 \rightarrow NO_2 + OH \qquad (5)$$

$$NO_2 + OH \rightarrow HNO_3 \qquad (6)$$

$$HNO_3 \rightarrow H^+ + NO_3^- \quad \text{(in aqueous solution)}. \qquad (7)$$

Both NO and NO_2 have large natural sources, which in aggregate exceed the total anthropogenic emissions. However, there is considerable uncertainty over the magnitude of the natural flux of NO_x to the atmosphere, largely because of difficulties in the estimation of emissions resulting from denitrification in soil. The global natural emissions of NO have been estimated as 430×10^6 tonnes/year and natural emissions of NO_2 as 658×10^6 tonnes/year (Table 2.1; see also Guicherit and van den Hout, 1982; Logan, 1983; Hov, 1984). The largest natural sources of NO_x are the following:

1. Emissions from soil and water as a result of bacterial denitrification of nitrate. The global emission from soil has been estimated to be about 40×10^6 tonnes/year (as NO_2; range of estimates $11–60 \times 10^6$ tonnes/year; Placet, 1990).
2. Dinitrogen fixation by lightning, i.e., the oxidation of N_2 to NO_x at high temperature and pressure. The global fixation by this process has been estimated to be about 9×10^6 tonnes/year (as NO_2; Borucki and Chameides, 1984).
3. Combustions of biomass, i.e., the oxidation of organic N to NO at high temperature, plus some oxidation of N_2 during the combustion.

Recent estimates of anthropogenic emissions of NO_x range from 36 to 60×10^6 tonnes/year (ex-

pressed as NO_2; Urone, 1976; Whelpdale and Munn, 1976; Logan, 1983; Hov, 1984; Elsom, 1987). The largest anthropogenic sources are associated with combustions of fossil fuels, which produce NO_x by the oxidation of organic N during combustion, and by oxidation of atmospheric N_2 to NO_x. The latter process is especially important if combustions occur at a high temperature (i.e., $>1000°C$) and pressure, as occurs in the internal combustion engine of automobiles. NO is the primary NO_x gas that is produced by these oxidations; NO_2 is produced secondarily in the atmosphere by the oxidation of emitted NO.

The most important NO_x-emitting states in 1985 were Texas (11.7% of the national U.S. emission of 19.5×10^6 tonnes of NO_x), California (5.8%), Ohio (4.7%), Illinois (4.5%), Pennsylvania (4.5%), and Indiana (4.1%), reflecting the regional prominence of large-scale combustions for electricity and other industrial purposes, plus the population of automobiles (Placet, 1990). Temporal changes in the rates and sourcing of NO_x emissions in the U.S. are described in Fig. 2.1.

Toxicity

NH_3 and the NO_x gases are capable of causing injuries to plants. However, the concentrations required to cause this effect are considerably larger than the usual exposures in the field, except in rare, exceptionally polluted environments near industrial point sources (Taylor et al., 1975; Amundson and MacLean, 1982). Thresholds for acute injuries to plants by NO_2 are 20 ppm for a 1-hr exposure and 2 ppm for a 48-hr exposure (Shriner, 1990). As discussed later, the major importance of the NO_x gases with respect to plant damage is associated with their roles in the secondary, photochemical synthesis of ozone.

Similarly, the ambient concentrations of NH_3 and NO_x gases are rarely large enough to affect humans. The guidelines for longer-term occupational exposures are 25 ppm for NO and 5 ppm for NO_2, while those for short-term exposures are 35 ppm and 5 ppm, respectively (Henderson-Sellers, 1984).

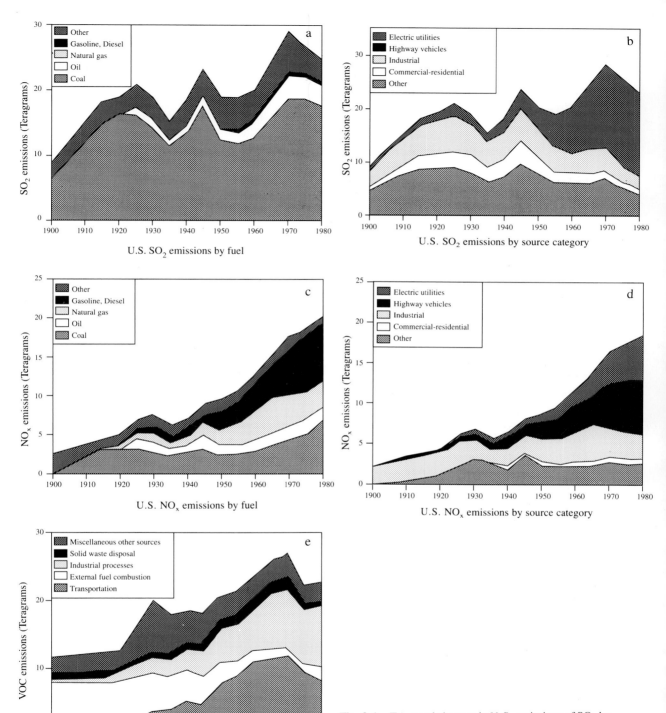

Fig. 2.1 Temporal changes in U.S. emissions of SO_2 by (a) fuel and (b) source, NO_x by (c) fuel and (d) source, and (e) nonmethane hydrocarbons (VOC) by source. Modified from Gschwandtner *et al.* (1985, 1988) and Gschwandtner and Wagner (1988).

Hydrocarbons

Characteristics, Emissions, and Transformations

Hydrocarbons are a chemically diverse group of air pollutants, ranging from gaseous methane (CH_4), through various vapor-phase hydrocarbons, to complex, large-molecular-weight compounds such as polycyclic aromatic hydrocarbons (see Chapter 6).

A typical background concentration of CH_4 in the atmosphere is 1.5 ppm, while all other hydrocarbons together comprise <1 ppb (Table 2.1). Emissions of CH_4 are largely natural. The most important sources of CH_4 emission are microbial fermentations in anaerobic wetlands, along with smaller outgassings from natural gas and coal deposits and methane produced by incomplete oxidations of organic matter during forest fires (Urone, 1976, 1986; Whelpdale and Munn, 1976; Altshuller, 1983). Estimates of the global emissions of CH_4 range from about 300×10^6 tonnes/year (Whelpdale and Munn, 1976; Altshuller, 1983) to 1600×10^6 tonnes/year (Table 2.1, from Urone, 1976). The wide variation reflects typical uncertainties involved in estimation of the global rates of cycling of air pollutants.

Estimates of the natural emissions of nonmethane hydrocarbons range from 200×10^6 tonnes/year (Table 2.1, after Urone, 1976) to 830×10^6 tonnes/year (Zimmerman, 1979), while anthropogenic emissions have been estimated as 65×10^6 tonnes/year (Duce, 1978). The largest natural emissions of nonmethane hydrocarbons are from living vegetation, especially forests, along with outgassings of light hydrocarbons such as ethane, propane, butane, and pentane from fossil-fuel deposits (Dimitriades, 1981; Altshuller, 1983; Salop *et al.*, 1983; Yokouchi *et al.*, 1983).

In view of the substantial, natural emission from forests of hydrocarbons, which are chemicals considered to be important pollutants if present in large quantities, former U.S. president Ronald Reagan stated in 1987: "Eighty percent of pollution is caused by plants and trees" (Luinenberg and Osborne, 1990).

The most important hydrocarbons emitted by forest vegetation are isoprene and various terpenoids. The rates of emission in the temperate zones are much larger during the growing season than in cooler months, especially on hot, sunny days (Altshuller, 1983; Salop *et al.*, 1983; Isidorov *et al.*, 1985). An estimated 68% of the annual emission of nonmethane hydrocarbons in the United States occurs during the warmest months of May through August (Placet, 1990).

Emissions of nonmethane hydrocarbons vary greatly among plant species and forest types. Salop *et al.* (1983) studied the emissions of hydrocarbons from forested terrain in Virginia and found much larger rates for angiosperm stands or species, compared with pines (Table 2.5). The specific hydrocarbons emitted differ greatly among plant species (Altshuller, 1983; Isidorov *et al.*, 1985; Placet, 1990).

The most important anthropogenic emissions of nonmethane hydrocarbons are from automobiles, aircraft, and petroleum mining and refining industries (Table 2.3), along with the use of solvents and oil-based paints.

The rates and types of hydrocarbon emission from vehicles depend on the type of engine (e.g., diesel versus spark-ignition engine), vehicle weight, and specific engine conditions such as spark timing, exhaust-gas recirculation, air : fuel ratio, and engine temperature. These all affect the effi-

Table 2.5 Emission factors for nonmethane hydrocarbons from forest types and tree species of a 6.3×10^5-ha area of Virginia[a]

Forest/Tree	Emission rate	
Deciduous forest	58	g/ha·hr
Mixed-wood forest	59	g/ha·hr
Conifer forest	17	g/ha·hr
Sweetgum (*Liquidambar styraciflua*)	61	μg/g·hr
Oaks (Quercus lauriflora, Q. michauxii, Q. nigra, Q. alba)	25	μg/g·hr
Black gum (*Nyssa sylvatica*)	10	μg/g·hr
Red maple (*Acer rubrum*)	6.5	μg/g·hr
Loblolly pine (*Pinus taeda*)	5.5	μg/g·hr

[a]Modified from Salop et al. (1983).

ciency of combustion and therefore the rate of hydrocarbon emission. Nelson and Quigley (1984) found that the most prominent hydrocarbons emitted by automobiles were ethylene, 11% of total emissions; toluene, 10%; acetylene, 9%; *m,p*-xylenes, 7%; benzene, propylene, and *i*-pentane, each 5%; and a wide variety of other species comprising the remainder.

Another anthropogenic source of hydrocarbon emissions are power plants that burn fossil fuels. Although they are a relatively small source [<2% of the total anthropogenic emission; National Research Council (NRC), 1976], power plants emit some toxic hydrocarbons. These include polycyclic aromatic hydrocarbons such as benzo[a]pyrene, which are mainly vented to the atmosphere as small particulates <3 μm in diameter (Freedman, 1981b).

The most important hydrocarbon-emitting states in 1985 were Texas (10.5% of the national U.S. emission of 20.0×10^6 tonnes of nonmethane hydrocarbons), California (9.9%), Ohio (4.5%), New York (4.5%), Illinois (4.3%), and Pennsylvania (4.1.%). This pattern reflects the regional distributions of humans, automobiles, and petroleum-refining capacity (Placet, 1990). Temporal changes in the rates and sourcing of emissions of nonmethane hydrocarbons to the atmosphere in the U.S. are described in Fig. 2.1.

Toxicity

The natural and anthropogenic emissions of hydrocarbons are large. However, except in the extremely polluted vicinities of large point sources, there is no evidence of acute damage caused to plants or animals by exposure to these chemicals in the atmosphere. In terms of ecological effects, atmospheric hydrocarbons are most important for the roles they play in secondary, photochemical reactions involving NO_x and O_3 (described below).

Photochemical Air Pollutants
Characteristics and Transformations

Ozone (O_3) is by far the most damaging of the photochemical air pollutants. Peroxy acetyl nitrate, hydrogen peroxide (H_2O_2), and aldehydes and other oxidants play relatively minor roles. All of these are secondary pollutants; that is, they are not emitted, but are synthesized in the atmosphere by photochemical reactions involving emitted gases, especially NO_x and hydrocarbons.

Ozone is a bluish gas, 1.6 times as heavy as air, and very reactive as an oxidant. Ozone is naturally present in relatively large concentrations in the stratosphere, an upper-atmospheric layer higher than 8–17 km altitude. Strictly speaking, stratospheric ozone should not be considered a photochemical air pollutant, a phrase whose use should be restricted to ozone issues associated with the lower atmosphere, or troposphere. However, issues related to stratospheric ozone are environmentally important and are considered here briefly.

Stratospheric O_3 concentrations typically average ca. 0.2–0.3 ppm, compared with <0.02–0.03 ppm in background situations closer to ground level (Urone, 1976, 1986; Singh *et al.*, 1980; Grennfelt and Schjoldager, 1984; Skarby and Sellden, 1984). Stratospheric O_3 is formed naturally by ultraviolet photochemical reactions (after Haagen-Smit and Wayne, 1976), as follows:

$$O_2 + h\nu \ (200 \ nm) \rightarrow O + O \qquad (8)$$

$$O + O + M \rightarrow O_2 + M \qquad (9)$$

$$O + O_2 + M \rightarrow O_3 + M \qquad (10)$$

$$O_3 + h\nu \ (290–200 \ nm) \rightarrow O_2 + O, \qquad (11)$$

where M is an energy-accepting third body. To summarize, O_2 interacts with ultraviolet radiation to form O atoms [reaction (8)], which can either recombine to O_2 [reaction (9)] or combine with O_2 to form O_3 [reaction (10)]. Ozone can be consumed by a variety of reactions, including an ultraviolet photodissociation [reaction (11)], or reaction with trace gases such as NO_x, N_2O, and ions or simple molecules of chlorine, bromine, and fluorine.

Because there have been large increases in the anthropogenic emissions of some of these O_3-consuming substances or their precursors, concerns over potential upsets of the dynamic equilibria among stratospheric-O_3 reactions, with resulting decreases in O_3 concentration, have been raised. One atmospheric-process model estimated a 16.5% reduction in stratospheric concentrations of O_3 in response to late-1970s rates of chlorofluorocarbons (CFCs) emission [National Academy of Sciences (NAS), 1976, 1979a].

Beginning around the mid-1980s, late-winter–early-springtime decreases in the concentrations of stratospheric O_3 (called ozone "holes") were observed at high latitudes. This phenomenon is most notable over Antarctica, where the holes develop under intensely cold conditions between September and November and where the average decreases in springtime stratospheric ozone between the late 1970s and the late 1980s were 30–40%. In October 1987, the average decrease of stratospheric ozone over Antarctica was 50%, and it was 95% in the hardest-hit altitudinal zone of the lower stratosphere, at 15–20 km (Kerr, 1988b). The cause of the ozone depletions is believed to involve O_3-consuming chlorine atoms or simple chlorine compounds such as ClO, which have an indirect, anthropogenic origin through the emissions of CFCs (Farman et al., 1985; Crutzen and Arnold, 1986; Farmer et al., 1987; Hoffman et al., 1987; Solomon, 1987, 1990; Solomon et al., 1987; de Zafra et al., 1987; Bowman, 1988; Rowland, 1988).

Although the seasonal ozone holes are restricted to high-latitude regions, stratospheric-ozone concentrations at lower latitudes can also be affected. This occurs by dilutions of the normal ozone concentrations, by low-ozone air that is dispersed widely when the ozone holes break up in the late springtime. During the 1980s, the seasonal concentrations of stratospheric ozone over 50° S latitude may have decreased by 3–8% (Solomon, 1990).

Because stratospheric O_3 absorbs much of the incoming solar ultraviolet (UV) radiation [ozone is an especially effective absorber at 200–290 nm and partially effective at 290–320 nm (Rowland, 1988; Longstreth, 1991)], it serves as a UV shield. As such, stratospheric O_3 helps to protect organisms on the Earth's surface from some of the deleterious effects of this high-energy electromagnetic radiation. If not intercepted, the ultraviolet radiation could disrupt genetic material, since DNA is an effective absorber of UV, especially wavelengths less than 320 nm. Damage to DNA could cause increased incidence of skin cancers, including deadly melanoma. Other health effects could include increases in the incidence of cataracts and other corneal damage such as snowblindness, and suppression of the immune system (NAS, 1979a; Cicerone, 1987; Longstreth, 1991). Ecological damage could include inhibitions of primary productivity in ultraviolet-stressed regions, possibly related to UV-caused pigment degradation (Smith et al., 1992).

During thunderstorms and other events of great turbulence in the upper atmosphere, stratospheric O_3 can enter the troposphere. Usually this affects the upper troposphere, although there have been observations of incursions reaching ground level for short (<2 hr) periods of time (Singh et al., 1980; Chung and Dann, 1985).

However, most tropospheric O_3 is formed and consumed by endogenous photochemical reactions, and these are the source of oxidants in Los Angeles-type smogs. The most important of these reactions are summarized below (following Grennfelt and Schjoldager, 1984). As described previously, the O_3-forming reaction is

$$O + O_2 + M \rightarrow O_3 + M. \qquad (12)$$

Atomic O, required for the formation of ozone, is formed by the photodissociation of NO_2:

$$NO_2 + h\nu \ (<440 \text{ nm}) \rightarrow NO + O. \qquad (13)$$

NO reacts with O_3, regenerating NO_2 and consuming ozone:

$$NO + O_3 \rightarrow NO_2 + O_2. \qquad (14)$$

If other reactions convert NO to NO_2 [as in reaction (15), below] the O_3 can accumulate, since this operates in competition with reaction 14 for NO:

$$NO + RO_2 \rightarrow NO_2 + RO. \qquad (15)$$

The reaction species RO_2 includes various peroxy radicals. These are formed by the degradation of organic molecules (RH) by reaction with hydroxyl radicals (OH) [reaction (16a)], followed by the addition of molecular O_2 [reaction (16b)]:

$$RH + OH \rightarrow R + H_2O \qquad (16a)$$

$$R + O_2 \rightarrow RO_2. \qquad (16b)$$

The concentration of OH is maintained by photodissociation of O_3 to produce atomic O [reaction (17a)], followed by reaction with H_2O to form OH [reaction (17b)]:

$$O_3 + h\nu \ (<320 \ nm) \rightarrow O + O_2 \qquad (17a)$$

$$O + H_2O \rightarrow 2OH. \qquad (17b)$$

Other photochemical reactions, including those by which PAN and aldehydes are formed, have been described, but these are not described here (see Haagen-Smit and Wayne, 1976; Guicherit and van den Hout, 1982; Grennfelt and Schjoldager, 1984; Fox, 1986).

Therefore, the formation of oxidizing smogs involves a complex group of photochemical reactions between anthropogenically emitted pollutants (NO and hydrocarbons) and secondarily produced chemicals (O_3, NO_2, PAN, aldehydes). The concentra-

tions of these chemicals exhibit a pronounced diurnal pattern, depending on the rates of emission, the intensity of solar radiation, and atmospheric stability at different times of the day.

This pattern is illustrated in Figure 2.2 for important air pollutant gases in Los Angeles. NO is the emitted NO_x, and it has a morning peak of concentration at 06:00–07:00, largely due to emissions from vehicles during the morning rush of traffic. Hydrocarbons are emitted from vehicles and refineries, and they have a temporal pattern similar to that of NO except that their concentration peaks slightly later. In bright sunlight the NO is photochemically oxidized to NO_2 [reactions (14) and (15)], resulting in a decrease in NO concentration and a peak of NO_2 at 07:00–09:00. Photochemical reactions involving NO_2 produce O atoms [reaction (13)], which react with O_2 to form O_3 [reaction (12)]. These result in a net decrease in NO_2 concentration and an increase in O_3 concentration, peaking broadly at 12:00–15:00. Aldehydes are also formed photochemically, but they peak earlier than O_3. As

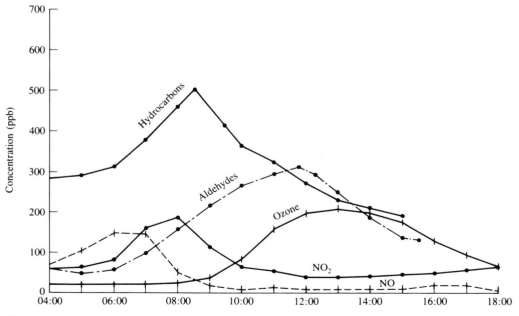

Fig. 2.2 The average concentration of various air pollutants in the atmosphere of Los Angeles during days of eye irritation. Modified from Haagen-Smit and Wayne (1976).

the day proceeds, the various gases decrease in concentration as they are diluted by fresh air masses blowing in from the Pacific Ocean or are consumed by photochemical reactions [e.g., reaction (17) and others], including the formation of nitrated organics, peroxides, aerosols, and other terminal products. This cycle is repeated daily, and it is typical of an area that experiences photochemical smog (Urone, 1976, 1986).

Hov (1984) calculated a global tropospheric O_3 budget (Table 2.6). The considerable range of the estimates reflects uncertainties in the calculation of the O_3 fluxes. On average, stratospheric incursions account for about 18% of the total O_3 flux to the troposphere, while endogenous photochemical production accounts for the remaining 82%. About 31% of the tropospheric O_3 is consumed by oxidative reactions at vegetative and inorganic surfaces at ground level, while the other 69% is consumed by photochemical reactions in the atmosphere.

Many countries have set standards for allowable O_3 concentrations, below which little acute damage to vegetation is expected (Table 2.7). Some standards have been set rather low in comparison to the ambient concentrations in certain areas, and this means that the limits may be frequently exceeded. Until 1979 the U.S. standard for the maximum 1-hr average O_3 concentration was 80 ppb. However, the federal standard was increased in that year to 120 ppb, because 80 ppb was so frequently exceeded in many jurisdictions (Skarby and Sellden, 1984). Because the original ozone standard was not enforceable in a practical sense, it was raised. Even now,

many regions of the United States cannot meet the maximum 1-hr average O_3 concentration of 120 ppb.

In the Los Angeles basin, the 1-hr peak concentration of O_3 can reach 580 ppb, and it typically exceeds 100 ppb for more than 15 days during the growing season (Roberts, 1984). Elsewhere O_3 concentrations are usually smaller. In other North American cities the maximum annual 1-hr peak concentration is typically 150–250 ppb, and in London, England, it is 90–180 ppb (Roberts, 1984). During the dry and sunny summer of 1976, however, O_3 concentrations >200 ppb occurred in southern England (Skarby and Sellden, 1984). In West Germany, an hourly mean concentration of 50–125 ppb O_3 is frequent over a large area during the growing season, but during the dry and sunny summer of 1976 maximum 1-hr concentrations of >150 ppb occurred (Skarby and Sellden, 1984).

Toxicity

Ozone is the most toxic constituent of photochemical smog, and it has caused considerable damage to agricultural and native plants in many locations. Ozone causes a distinctive, acute injury to plants (Jacobson and Hill, 1970), resulting in a loss of photosynthetic area. It is often assumed that the resulting yield decrement is proportional to the percentage loss of leaf area to acute injury. However, in some cases, this relationship is moderated by compensatory increases of photosynthetic rates in undamaged foliar tissue (Jacobson, 1982a; Roberts, 1984).

In general, most plants are acutely injured by an exposure to 200–300 ppb ozone for 2–4 hr (Shriner, 1990). A general threshold for yield decrease caused by a longer-term exposure of crops to O_3 is about 100 ppb (Roberts, 1984; Shriner, 1990). However, in sensitive species, both acute and hidden injuries can be caused at smaller concentrations (Jacobson, 1982a; Roberts, 1984; Skarby and Sellden, 1984). For example, in a laboratory fumigation experiment, tobacco (*Nicotiana tobacum*) was acutely injured by 50–60 ppb during 2- to 3-hr exposures, and spinach (*Spinacea oleracea*) was in-

Table 2.6   A tropospheric ozone budget[a]

	Northern Hemisphere (kg/ha·year)	Southern Hemisphere (kg/ha·year)
Transport from stratosphere	13–20	8–10
Photochemical production	48–78	28–73
Destruction at ground	18–35	10–20
Photochemical destruction	48–55	28–30

[a]Modified from Hov (1984).

Table 2.7 Ozone standards that have been set or recommended by different countries to prevent significant damage to vegetation[a]

Country	Maximum 1-hr average concentration (ppb)	Frequency of exceeding	Status
United States	120	Once a year	Legal standard
Japan	60	Not to be exceeded	Long-term goal
Canada	50	Max. desirable	Guideline
	80	Max. acceptable	Guideline
	150	Max. tolerable	Guideline
Sweden	60	Once a month	Proposed guideline
Norway	50–100		Recommended
World Health Organization	60		Recommended

[a]Modified from Grennfelt and Schjoldager (1984).

jured by 60–80 ppb for 1–2 hr (Skarby and Sellden, 1984). Sensitive conifers can be injured by ozone exposures of only 80 ppb over 12 hr (Shriner, 1990).

A number of field studies have examined the growth of agricultural plants in chambers receiving either ambient air or charcoal-filtered air without O_3. These studies have frequently observed substantial decreases in yield caused by exposures to O_3 in unfiltered, ambient air. Thompson and Taylor (1969) found that the yield of fruit of lemon and orange trees (*Citrus limonia* and *C. sinensis*) was ca. 50% smaller in unfiltered air in southern California. Studies in New York reported a 30% decrease in the yield of tomato (*Lycopersicum esculentum*) and snap beans (*Vicia faba*) (MacLean and Schneider, 1976). Studies of soybean (*Glycine max*) in Maryland found a 20% reduction in yield (Howell *et al.*, 1979). Other studies in Maryland between 1972 and 1979 found average reductions in yield of 12% (range 4–20%) among potato (*Solanum tuberosum*), soybean, corn (*Zea mays*), snap bean, and tomato (Heggestad, 1980).

In a detailed series of field experiments at five sites in the United States, important crop species were exposed in open-top chambers to either ambient air or 25 ppb O_3, a typical background concentration (Heck *et al.*, 1982; Lefohn, 1990). The average annual concentrations of O_3 in ambient air at the study sites were: northeast site, 40 ppb; southeast-

A, 56 ppb; southeast-B, 31 ppb; central, 42 ppb; and southwest, 106 ppb. Symptoms of acute O_3 injury were observed in crop species at all of these sites. Across all sites, the yield decreases due to ozone were 10% in soybean, 14–17% in peanut (*Arachnis hypogaea*), 7% in turnip (*Brassica napa*), 53–56% in lettuce (*Lactuca sativa*), and 2% in kidney bean (*Phaseolus vulgaris*). It was estimated that at the current U.S. standard for O_3 in air (i.e., 120 ppb for a 1-hr exposure), the crop losses would be equivalent to about 2–4% of the total potential yield, with a direct economic value in 1982 dollars of as much as $5 billion per year (Heck, 1989; Heck *et al.*, 1982, 1986).

Because the threshold concentrations for yield reductions caused by O_3 toxicity are so close to or even less than the ambient concentrations in many parts of North America, ozone is by far the most economically damaging air pollutant in agriculture (Heck *et al.*, 1982, 1983, 1986; Jacobson, 1982b; Howitt *et al.*, 1984; Skarby and Sellden, 1984; Reich and Amundson, 1985; Rowe and Chestnut, 1985; Mackenzie and El-Ashry, 1989). Effects of ozone on forests are described later as a case study.

Humans are also sensitive to ozone. This gas causes irritation and damage to membranes of the respiratory system and eyes. The guideline for a longer-term, occupational exposure to ozone is 0.1 ppm, while for a short-term exposure it is 0.3 ppm

(ACGIH, 1982). However, relatively sensitive people can suffer respiratory distress at concentrations smaller than these, including exposures that commonly occur during events of oxidizing smog.

Fluorides

Characteristics and Emissions

Atmospheric fluorides exist in a variety of gaseous and particulate forms. The gases that are emitted anthropogenically in largest quantities are hydrogen fluoride (HF) and silicon tetrafluoride (SiF_4). Fluorides are also vented naturally from volcanos and fumaroles; identified gases include ammonium fluoride (NH_4F), silicon tetrafluoride, ammonium fluorosilicate [$(NH_4)_2SiF_6$], sodium fluorosilicate (Na_2SiF_6), potassium fluoroborate (KBF_4), and potassium fluorosilicate (K_2SiF_6) (NAS, 1971; Urone, 1976, 1986). A secondary aerosol is the strong acid, fluorosilicic acid (H_2SiF_6), which is formed in the atmosphere by the hydrolysis of SiF_4. Important particulates include the minerals fluorospar (CaF_2), cryolite (Na_3AlF_6), and fluorapatite [$Ca_{10}F_2(PO_4)_6$], which can be released as dusts from industrial sources (NAS, 1971).

Volcanoes are the only quantified natural source of fluoride, emitting about 7.3×10^6 tonnes/year (Urone, 1976). The anthropogenic emissions of fluoride are much smaller, amounting to about 0.4×10^6 tonnes/year. The total anthropogenic emission of fluoride in the United States was less than 0.1×10^6 tonnes/year in 1985, with the largest sources being coal combustion (74% of the total emission), phosphate fertilizer processing (12%), aluminum reduction (8%), and brick and tile manufacturing (<6%) (Placet, 1990).

The atmospheric concentrations of fluoride in rural areas are generally small (<0.3 ppb; Urone, 1976). Urban air has a somewhat larger concentration, generally <3 ppb (NAS, 1971; Urone, 1976). Close to an industrial point source, the concentrations in air can be much larger; up to 250 ppb was measured in an area in Florida that was contaminated with dust from the mining and processing of phosphate rock that contained 3–4% F (NAS, 1971).

Toxicity

The threshold concentrations that cause injuries and reductions of vegetation yield depend on the susceptibility of plant species or varieties, and on environmental conditions. For a 1-day exposure, the threshold concentration of HF in air required to cause acute injury to sensitive plants is 4–5 ppb, and >13 ppb for more tolerant species, while for a 1-month exposure the concentrations are 0.6 ppb and 1–4 ppb, respectively (NAS, 1971). Sensitive plants exhibit symptoms of acute injury at foliar concentrations of 20–150 ppm F, while resistant accumulator species can tolerate >4000 ppm (NAS, 1971; Shriner, 1990). Gladiolus (*Gladiolus communis*) is sensitive to F, and it is injured by only >20 ppm in foliage, whereas tea (*Camellia sinensis*) is fairly tolerant, accumulating as much as 200 ppm without suffering acute injury. Cotton (*Gossypium hirsutum*) is tolerant, and it can accumulate as much as 4000 ppm in its foliage without suffering acute injury (NAS, 1971).

2.3
AIR POLLUTION AND HUMAN HEALTH

On rare occasions, natural emissions of toxic gases have caused human deaths. A remarkable incident occurred in 1986, when a large volume of CO_2 was unexpectedly released from Lake Nyos in Cameroon (Freeth and Kay, 1987; Kling *et al.*, 1987). This release was apparently triggered by a large slump of sediment on a steep bank of the lake, which caused an overturn of the stratified, 200-m-deep volcanic lake. The hypolimnetic waters of Lake Nyos are supercharged with CO_2 as a result of inputs from soda springs, and when the deeper water was brought to the surface, it quickly degassed. The resulting dense, atmospheric mass was laden with water vapor and CO_2, and it flowed into a surrounding lowland and killed 1.7 thousand people, 3 thousand cattle, and uncounted wildlife by CO_2 asphyxiation. Most of the humans were killed as they slept. No plant damage was caused by this event of CO_2 pollution.

There have also been incidents in which anthropogenic air pollution has caused marked increases in human mortality, particularly within high-risk groups of people with chronic respiratory or heart disease. These toxic pollution events occurred during periods of prolonged atmospheric stability, which prevented the dispersion of emissions. This resulted in accumulations of large concentrations of SO_2 and particulates, often accompanied by fog. The term smog was originally coined to label these coincident events of air pollution by SO_2 and particulates.

Coal smoke has long been recognized as a pollution problem in England, from at least year 1500 (Wise, 1968; Chambers, 1976). In London, dirty, pollutant-laden fog occurred quite frequently and was known as "peasoupers." The first conclusive linkage of an air pollution event with a large increase in human mortality was in Glasgow in 1909, when 1063 deaths were attributed to a noxious accumulation of air pollutants during a period of atmospheric stagnation (Chambers, 1976). Another episode occurred in the Meuse Valley of Belgium in 1930 and resulted in 60 deaths. In that toxic episode, a stable atmospheric inversion developed. The continued emissions of toxic gases and particulates from steel works, a sulfuric acid plant, glass factories, and zinc works caused the accumulation of large concentrations of pollutants, especially SO_2 (Perkins, 1974).

Another notable episode occurred in October 1948 in Donora, a town located in a valley some 45 km from Pittsburgh, Pennsylvania. In that case, a fog was coincident with a temperature inversion that persisted for 4 days. Continued emissions from a steel mill and zinc and sulfuric acid plants caused severe air pollution, with SO_2 estimated at 0.5–2 ppm, and large but unmeasured concentrations of particulates. A high rate of mortality was associated with the smog (20 deaths out of a population of 14,100). There was also great morbidity, with about 43% of the population made ill, 10% of them severely so. Typical symptoms were irritation of the eyes, nose, throat, and lower respiratory tract, along with coughing, headache, and vomiting (Perkins, 1974; Davidson, 1979).

In terms of causing toxicity to humans, the best-known air pollution event is the so-called "killer smog" that affected London, England, in December of 1952. In this case, an extensive temperature inversion was accompanied by a "white fog" throughout southern England. In the vicinity of London, this was transformed into a noxious "black fog" with virtually zero visibility, as the concentrations of SO_2 and particulates built up as a result of emissions from coal combustion for the generation of electricity and other industrial purposes and to heat homes because of the cold temperatures at the time. Visibility was so poor that many people lost their way while driving cars, others walked off wharves and fell into rivers, and theaters were closed because the projection screens were not visible to most of the audience. Of course, there were also serious disruptions of activities in workplaces and in commerce. During the 4-day episode, the maximum daily concentration of SO_2 was 1.3 ppm, and total suspended particulates were 4.5 mg/m^3. In total, there were 18 days of greater-than-usual human mortality. Overall, 3900 "excess" deaths were attributed to the episode, mainly among the elderly, the young, and persons with preexisting respiratory or coronary disease (Wise, 1968; Perkins, 1974; Amdur, 1980).

Pollution episodes such as this were frequent in the industrialized cities of Britain, Europe, and the northeastern United States, and they were in large part due to the burning of coal. Apart from effects on human health, there were many instances of deaths of livestock and of damage to vegetation. It was, for example, widely known that only plants that were tolerant of pollution could be grown in most large cities. Such species included privet (*Ligustrum vulgare*), plane tree (*Platanus acerifolia*), Norway maple (*Acer platanoides*), and tree-of-heaven (*Ailanthus altissima*). In Britain, the passing (and enforcement) of the Clean Air Act in 1956 resulted in great reductions in the local emissions of air pollutants. This caused an improvement of air quality in cities such as London, so that severe reducing smogs may now be historical occurrences, at least in those places.

It should be stressed that the above cases represent severe episodes of air pollution. The effects on human health of more typical, nonepisodic, urban air quality have also been investigated. However, the conclusions with respect to enhanced morbidity or mortality are much more tentative and controversial (Coffin and Stokinger, 1978; Goldsmith and Friberg, 1977; Imai *et al.*, 1985; Goldsmith, 1986; Chestnut and Rowe, 1989; Graham, 1990).

2.4
CASE STUDIES OF THE ECOLOGICAL EFFECTS OF TOXIC GASES

In the following, selected case studies of ecological damage caused by gaseous air pollution are examined. The first case describes the effects of naturally occurring SO_2 pollution at a remote location in the Arctic. Damage caused by anthropogenic emissions of SO_2 from point sources is then examined, focusing on the large smelters at Sudbury, Canada. The next case describes extensive damages to coniferous forest in California, caused by regional ozone pollution. Finally, consideration is made of

the potential effects of climatic and other changes that may be caused by increasing concentrations of CO_2 and some other radiatively active gases in the atmosphere.

Effects of Natural SO_2 Pollution at the Smoking Hills

The Setting

The Smoking Hills are at 70°N on the mainland seacoast of the Canadian Arctic. The area is remote, pristine, and virtually uninfluenced by humans. At a number of locations along 30 km of seacoast, strata of bituminous shales in a 100-m-high seacliff have spontaneously ignited. As a result, the adjacent tundra is being fumigated with SO_2 and other pollutants (Hutchinson *et al.*, 1978; Havas and Hutchinson, 1983a; Freedman *et al.*, 1990).

The area has a bedrock of shale, covered with 10 m or less of calcareous glacial drift and fluvial deposits. Interbedded in the shale are dark-colored,

View of a plume contaminated with sulfur dioxide at the Smoking Hills in the Canadian Arctic. Bituminous shales have ignited in a discrete slump of the 100-m-high coastal seacliff. An older burned-out slump is located to the left of the burn. The plume is about 100 m wide where it reaches the top of the cliff. Note that the plume fumigates the ground surface as it moves inland. (Photo: B. Freedman.)

A sparsely vegetated site at the Smoking Hills. All of the plant clumps are the herb *Artemisia tilesii*, which is very tolerant of sulfur dioxide, acidity, and toxic metals such as aluminum. (Photo: B. Freedman.)

bituminous strata. When this material is freshly exposed to the atmosphere by slumping of the seacliff, pyritic sulfur undergoes an exothermic oxidation to sulfate. In some cases the heat can accumulate sufficiently to ignite the bituminous shale, which proceeds to burn into the cliff until the oxygen supply becomes insufficient to support further combustion. This mechanism is similar to that proposed for the spontaneous ignition of bituminous materials in coal-waste piles and mines, where pyrites are also exposed to oxygen (Sussman and Mulhern, 1964; Mathews and Bustin, 1984).

The earliest, documented sighting of the Smoking Hills was by British explorers in 1826, but the burns are much more ancient than this. The area was not covered by ice during the most recent glaciation more than about 10 thousand years ago, and the tremendous piles of ash and roasted shale at the bottom of the seacliff below the combustions suggest great antiquity.

Atmospheric Pollution

During the growing season, the prevailing wind direction at the Smoking Hills is onshore. As a result, the sulfurous plumes are carried inland where they affect the tundra ecosystem. Gizyn (1980) found that the largest concentrations of SO_2 occurred within 20 m of the edge of the seacliff, just where the plume begins to roll inland (10-min average values were as large as 1.7 ppm SO_2). The longer-term average concentrations of SO_2 decreased rapidly along transects running inland from the sea (Table 2.8).

Atmospheric sulfur has also been sampled by exposing *Sphagnum* moss in nylon bags to the atmosphere. With this technique, gaseous and particulate pollutants are adsorbed, filtered, and otherwise accumulated. Moss bags placed for 14 days near an active burn increased in sulfur concentration from an initial 0.05% to 8–21%. The sulfur was so

Table 2.8   Concentration of SO_2 along a transect running inland from the edge of the seacliff at one of the burn sites at the Smoking Hills[a]

Distance from edge of cliff (m)	SO_2 concentration (ppm)	
	1975	1977
20	0.27	0.61
40	0.20	0.53
80	0.17	0.43
160	0.10	0.25
320	0.06	0.16
640	0.02	0.11
1280	0.02	0.06
1920	0.02	0.05
2560	0.01	0.04

[a]SO_2 concentration was calculated from the rate of sulfation, a simple field measurement in which SO_2 reacts with PbO_2 to form $PbSO_4$ at a rate proportional to the concentration of SO_2 [Huey, 1968; American National Standards Institute (ANSI), 1976]. The data are mean values for an 8-day sampling period in 1975, and 14 days in 1977. Modified from Gizyn (1980).

acidic that the nylon bag and *Sphagnum* were partially digested! Along a transect from the cliff edge in a major plume, sulfur concentrations in the moss decreased from 2.5% at the cliff edge to 1.4% at 80 m, 1.1% at 160 m, 0.39% at 640 m, and 0.08% at 4.8 km (Hutchinson *et al.*, 1978). Sulfate deposition was measured with open-bucket bulk collectors and was 0.39 mg m^{-2} per 30 days near the cliff edge, compared with 0.041 mg/m^2 per 30 days at a remote site (Gizyn, 1980). The bulk-collected rainwater was very acidic near the cliff edge (pH 2.6–2.7). The acidity was due to sulfuric acid, since the concentration of equivalents of SO_4^{-2} was virtually equal to that of H^+ (Gizyn, 1980).

Suspended particulates in the atmosphere were also sampled. Sites within a major plume had 120–160 ng S/m^3, compared with <0.3 ng S/m^3 at a nonfumigated site. Other elements with elevated concentrations were selenium (130–270 ng/m^3 versus <30), arsenic (13–36 ng/m^3 versus <1), manganese (53–109 ng/m^3 versus 2), and bromine (55–73 ng/m^3 versus 5) (Hutchinson *et al.*, 1978).

A grassland community at the Smoking Hills, in a zone of moderate air and soil pollution. The dominant plant is the pollution-tolerant grass *Arctagrostis latifolia*. (Photo: B. Freedman.)

An acidic pond at the base of the seacliffs at the Smoking Hills. This pond is frequently fumigated by plumes having a large concentration of sulfur dioxide, and it also receives acidic drainage from its small watershed. Consequently, the pond has a very acidic pH of 1.8 and very large concentrations of sulfate and of soluble, toxic forms of aluminum, manganese, zinc, and other metals. The observer is standing on a slope of roasted shales. At the far upper left are more whitish, roasted shales and an active burn. Beyond the observer is sea ice. (Photo: B. Freedman.)

Soil and Water Pollution

The most important chemical effects of air pollution at the Smoking Hills have been the acidification of soil and fresh water and the subsequent solubilization of toxic metals. Surface (0–2 cm) soils in fumigated areas have pHs as acidic as 2.7–3.2, compared with pH 7.2 at reference, nonfumigated sites (Gizyn, 1980). Sulfur concentrations are also large in soil at fumigated sites (1.0–1.5%, compared with 0.44% for reference soil). The acidic condition has caused the leaching of basic cations from surface soils. This effect is especially obvious for calcium, which decreased from 1.8% at a reference site to 0.5% in fumigated surface soils (Gizyn, 1980).

Some ponds at the Smoking Hills are very acidic, with pH values as low as 1.8 occurring in ponds located at the base of the seacliff (Table 2.9).

These ponds receive acidic water that percolates through deposits of shale and ash, where pyrite is actively being oxidized (this is analogous to acid mine drainage, described in Chapter 4). The only reports of similarly acidic pH values in natural waters are for volcanic crater lakes in Japan, where pH values as low as 1.0 to 1.4 have been measured (Yoshimura, 1933; Ueno, 1958; Takano, 1987) and surface waters with pH <2 caused by drainage from coal mines and coal waste disposal areas (Riley, 1960; King *et al.*, 1974).

Tundra ponds at the top of the seacliff at the Smoking Hills do not receive the acidic drainages that have percolated through pyrite-containing shale and ash. Instead, these water bodies have been acidified by: (1) direct depositions of sulfuric acid mist, rain, and snow from the atmosphere; (2) dissolution of atmospheric SO_2 into water, followed by its oxidation to sulfuric acid (Terraglio and Man-

Table 2.9   Chemistry of shallow tundra ponds in the vicinity of the Smoking Hills[a]

pH range	n	Al	Fe	Mn	Zn	Ni	Cd	As	Ca	SO$_4$
1.8–2.5	4	270	500	61	14	6.3	0.52	0.13	301	8200
2.5–3.5	14	5.5	18	15	0.45	0.21	0.022	0.005	157	890
3.5–4.5	9	1.1	1.2	3.6	0.12	0.04	0.011	0.005	44	156
4.5–5.5	1	<0.6	0.5	2.3	0.03	0.04	0.001	0.004	182	813
5.5–6.5	1	<0.2	0.2	1.8	0.05	0.06	0.012	0.006	249	713
6.5–7.5	4	<0.8	<0.04	0.7	0.08	0.02	0.001	0.004	90	360
7.5–8.5	8	<0.7	0.1	<0.5	0.04	0.004	0.003	0.005	49	106
8.5–9.5	3	<0.2	<0.04	<0.2	<0.03	0.007	<0.001	0.006	31	34
9.5–9.7	2	<0.2	0.2	1.2	5.3	0.01	<0.001	0.005	16	29

[a]The sample size refers to the number of ponds within each pH class. The data are geometric means (mg/liter). Modifed from Hutchinson *et al.* (1978) and Havas and Hutchinson (1983a).

ganelli, 1967; Beilke and Gravenhorst, 1978; Meszaros, 1981); and (3) acidic surface runoff from the small watersheds of the ponds. The most acidic pondwater pH at the top of the seacliff is 2.4. The waters of the tundra ponds are either acidic with pH <4.5 or nonacidic with pH >6.5, with few ponds in the weakly buffered pH range of 4.5–6.5. A bimodality of pH distribution in surface waters has also been observed in areas affected by acid mine drainage (King *et al.*, 1974) and by acidic deposition from the atmosphere (Wright and Gjessing, 1976).

Other chemical constituents are present in large concentrations in the acidic pondwaters (Table 2.9). The large concentrations of sulfate result from atmospheric deposition and in some cases from leaching from the terrestrial part of the watershed. The large concentrations of other elements such as aluminum result from the acidic dissolution of minerals in pond sediments and watershed soils—this process causes large concentrations of soluble metals in all acidic environments (Chapter 4).

Ecological Effects

Intensely fumigated sites at the Smoking Hills are toxic to most species. The most important factors that affect the terrestrial vegetation are SO$_2$, acidic soils, and large concentrations of soluble aluminum, manganese, and other metals. Within 40 m of the seacliff in the major fumigation zone there is no

vegetation whatsoever—ecological degradation is complete (Hutchinson *et al.*, 1978; Gizyn, 1980; Freedman *et al.*, 1990). From about 80 to 320 m there is a depauperate plant community, with less than 3% cover and few species present. At 80 m the only plant species is the perennial, dicot herb *Artemisia tilesii*; at 320 m this species is accompanied by the grass *Arctagrostis latifolia* and the lichen *Cladonia bellidiflora*, with the moss *Pohlia nutans* present at low frequency. All of these plants are widespread in this region of the Arctic, but they are minor components of the typical, reference vegetation. Therefore, species that are tolerant of pollution stress have replaced plants that are characteristic of unpolluted habitat, such as the mountain avens (*Dryas integrifolia*), arctic willow (*Salix arctica*), and at least 70 other plant species.

A pollution-tolerant biota has also developed in the acidic ponds. Even the most acidic pond (pH 1.8) has six species of algae. The most prominent alga is *Euglena mutabilis*, which can be present as a green benthic edging just below the water surface (Sheath *et al.*, 1982; Havas and Hutchinson, 1983a). Laboratory experiments with *E. mutabilis* from the Smoking Hills showed that it could survive at pH 1.0, although the optimum pH for its growth was 4.0 (Hutchinson *et al.*, 1981). Other algae in the pH 1.8 pond are *Chlamydomonas acidophila*, *Eunotia glacialis*, *Nitzschia communis*, *N. palea*, and *Cryptomonas* sp. (Sheath *et al.*, 1982). Some of these are part of a widely distributed, acid-tolerant algal flora. *Euglena mutabilis* has been reported at pH 1.8 in coal-mine drainage in England and the United States (Van Dach, 1943; Hargreaves *et al.*, 1975; Hargreaves and Whitten, 1976). Similarly, *Chlamydomonas acidophila* occurs in pH-1.7 volcanic lakes in Japan (Ueno, 1958) and in pH-1.8 mine drainage in Britain (Hargreaves and Whitten, 1976). In contrast to the acidic ponds at the Smoking Hills, the reference ponds are alkaline (pH 8.1–8.2) and have a rich algal flora comprising at least 90 taxa (Sheath *et al.*, 1982).

Both the acidic and the alkaline pond floras are well adapted to their respective environments. In a field experiment involving the *in situ* manipulation of the pH of pond water within plexiglass enclosures, the live phytoplankton volume collapsed from 1.49 mm^3/liter at the natural pH of 2.8, to 0.0 mm^3/liter at pH 8 (Sheath *et al.*, 1982). Similarly, acidification of an alkaline pond water caused cell volume to decrease from 17.4 mm^3/liter at pH 8.1, to 0.0 mm^3/liter at pH 3.

A few invertebrates are found in acidic ponds above pH 2.8 (Havas and Hutchinson, 1982, 1983b). The most prominent zooplankton species is the rotifer *Brachionus urceolaris*, also reported from a pH 3.0 volcanic pond in Japan (Ueno, 1958). The benthos is dominated by the chironomid *Chironomus riparius*. Neither of these species are present in alkaline ponds, which have a much more diverse fauna dominated by acid-sensitive crustaceans such as the water fleas *Daphnia middendorffiana* and *Diaptomus arcticus*, the tadpole shrimp *Lepidurus arcticus*, and the fairy shrimp *Branchinecta paludosa*.

An important lesson to be learned from the Smoking Hills is that "natural" gaseous pollution can cause an intensity of ecological damage that is as severe as anything caused by anthropogenic pollution. (This does not, of course, in any way justify anthropogenic pollution and its ecological effects.) Therefore, SO$_2$ can have a damaging effect on ecosystems irrespective of its source. Because pollution at the Smoking Hills has been present for a long time, the rate of ecosystem change and adaptation by organisms has probably stabilized. Under a

longer-term condition of toxic pollution stress, the tundra ecosystem is characterized by biological simplification and by disrupted productivity and nutrient cycling. In addition, the pollution-tolerant biota is composed of species that are not normally present or are rare in the reference, unpolluted habitat.

Effects of Emissions from the Sudbury Smelters and Other Large Point Sources of SO$_2$

The Setting

In 1883, a worker employed in the construction of Canada's first transcontinental railroad made the first discovery of a commercial ore body in the vicinity of Sudbury, Ontario, in a surface bedrock cut. The principal metals that are mined and processed at Sudbury are nickel and copper, although iron, cobalt, gold, and silver are also produced, as are the nonmetals sulfur and selenium. The Sudbury area now contains one of the world's largest mining and metal-processing complexes, with many shaft mines, an open pit, two ore-crushing mills with associated tailings disposal areas, two smelters, several metal refineries, an iron-ore recovery plant, sulfuric-acid plants, and a host of secondary and supporting industries. These activities are the primary economic base for a population of more than 100,000 people.

Atmospheric Emissions and Environmental Pollution

From the beginning of the mining and smelting developments to the present, SO$_2$ and heavy metals have been emitted to the atmosphere in large quantities (Costescu, 1974; Freedman and Hutchinson, 1980c). Prior to 1928, smelting was conducted in open pits using a primitive roasting technique. Large heaps of sulfide ore were placed over locally cut timber. The wood was ignited to initiate an oxidation of the metal sulfides. These exothermic reactions soon generated enough heat to cause a self-sustaining oxidation of the ore. The roast beds

A view of the O'Donnell roast bed in the vicinity of Sudbury, Ontario, circa 1925. The roast bed was prepared with a bottom layer of locally cut cordwood (foreground), much of which was salvage harvested from nearby fume-killed forests. Above this was placed a layer of sulfide nickel–copper ore (background), using the track-mounted gantry with its continuous-feed apparatus. (Photo courtesy of INCO Archives.)

typically burned for several months, after which the nickel and copper concentrates were collected and taken elsewhere for refining into pure metals.

This roasting process generated choking, ground-level plumes that contained SO_2, acidic mists, and metal particulates. The fumigations devastated the surrounding terrestrial vegetation. The devegetation caused severe erosion of soil from slopes and the exposure of granitic-gneissic bedrock, which was then blackened and pitted by the sulfurous fumes. During this early phase of smelting by the use of roast beds, an estimated 2.7×10^5 tonnes/year of SO_2 plus large but undocumented quantities of metal particulates were emitted at ground level, from as many as 30 roast beds (Holloway, 1917).

The open roast beds were eventually replaced by more efficient smelters, which vented waste gases to the atmosphere through tall smokestacks and thereby alleviated the ground-level pollution (Freedman and Hutchinson, 1980a,c). By 1928, the use of roast beds was forbidden by legislation, and all subsequent roasting was carried out at three smelters located near Sudbury at Copper Cliff, Coniston, and Falconbridge. Operation of the largest smelter at Copper Cliff began in 1929, when most of its pollutants were vented through a 155-m smokestack. Two additional smokestacks were added in 1936, and in 1956 an iron-ore recovery plant with a 191-m stack was built. In 1972 the three smokestacks at the Copper Cliff smelter were replaced by a single 381-m "superstack," the world's tallest. At various times during the development of the Copper Cliff smelting complex, some degree of pollution abatement was achieved by the installation of flue-dust recovery units, wet scrubbers, electrostatic precipitators, and sulfuric acid plants. These various units remove some of the SO_2 and particulate

An aerial view of the O'Donnell roast bed while ore is being oxidized, circa 1925. Once initiated, the roasting would proceed for several months. When the metal-concentrated substrate had cooled sufficiently, it was collected and sent on to a refinery. The ground-level, phytotoxic emissions that resulted from this primitive roasting procedure devastated the surrounding forests and acidified freshwaters. (Photo courtesy of INCO Archives.)

pollution before the waste gases are vented to the atmosphere.

Two other smelters were constructed in the Sudbury area. The Coniston smelter operated from 1913 to 1972 and had smokestacks of 114 and 122 m. The third smelter has operated since 1928 at Falconbridge and has smokestacks of 93 and 140 m.

The emissions of SO_2 in the Sudbury area peaked around 1965–1970, when the total emissions from all three smelters were about 2.7 million tonnes per year. This was equivalent to more than 4% of the global anthropogenic emission of SO_2 (cf. Table 2.2) and ranked Sudbury as the world's largest point source of SO_2 emission. Emissions have decreased substantially since then, to still-large quantities of about 1.4×10^6 tonnes/year over 1973–1977 and 0.4–0.9×10^6 tonnes/year over 1980–1984 (Freedman and Hutchinson, 1980a; Chan and Lusis, 1985; D. Yap, Ontario Ministry of

the Environment, personal communication). The large decreases in emissions were caused by a downturn in nickel markets and hence decreased production, by interruptions of operations by strikes, by a shift to lower-sulfur ores in order to reduce SO_2 emissions per unit of metal produced, and by other pollution-control measures such as the building of flue-gas desulfurization facilities. However, the emissions of SO_2 and other pollutants are still large (Table 2.10).

Several researchers have computed deposition budgets for the emitted SO_2 and metals. Chan *et al.* (1984a) estimated deposition of sulfur compounds within 40 km of the superstack between 1978 and 1980. Their study indicates that the tall stack disperses pollutants effectively; locally, the "dilution solution to pollution" works quite well. This is especially true of gaseous sulfur. Only 1.3% of the total sulfur emission was deposited within 40 km of

A view of the "supertrack" at the Copper Cliff smelter near Sudbury. This 381-m-tall smokestack vents pollutants very high into the atmosphere, and it has greatly alleviated the local air pollution since it was commissioned in 1972. The ecological damage that is visible in the foreground is due to presuperstack fumigations. (Photo: B. Freedman.)

the superstack, and only 0.2% if background deposition is accounted for to isolate the smelter influence (Table 2.11). Similar observations were made by Muller and Kramer (1977) and Freedman and Hutchinson (1980a). Therefore, more than 99% of the sulfur emitted from this tall stack is exported beyond 40 km; these long-range transported air pollutants contribute to regional acidic precipitation (Chapter 4).

There are several reasons why so little of the emitted sulfur is deposited close to the superstack:

1. The tall stack emits pollutants sufficiently high into the atmosphere that local surface impactions are infrequent. Ground-level fumigations only tend to occur on warm, sunny days during May to September when large-scale convective turbulence can cause a certain type of plume behavior, termed "looping." Individual plume loops typically impact the surface at 3–10 km from the stack, and they cause half-hour average SO_2 concentrations of 100–300 ppb. At any particular site within this distance, these ground-level events occur several times per year (Chan and Lusis, 1985). During ground-level SO_2 episodes, dry deposition is a relatively important mechanism of sulfur deposition (Tang et al., 1987);

2. In the absence of a ground-level fumigation, SO_2 is not efficiently deposited to the surface by precipitation. The solubility of SO_2 in rainwater is rather low, particularly if the solution

Table 2.10 Average annual emissions (tonnes/year) of pollutants in the Sudbury area between 1973 and 1981[a]

Source of emission	SO_2	H_2SO_4	Total particulates	Fe	Cu	Ni	Pb	As
INCO								
381-m "superstack"	886,000	7270	11,400	990	245	228	184	114
194-m stack of iron ore recovery plant	55,000	1664	2400	643	171	226	6	4
Low-level sources	12,000	88	600	70	242	31	1	—
Two 45-m stacks of pelletizer plant	—	—	4100	2354	—	—	—	—
Falconbridge, 93-m stack	173,000	438	900	98	11	10	13	6
Total	1,126,000	9460	19,400	4155	669	495	204	124

[a]Modified from Chan and Lusis (1985). In addition to the pollutants summarized above, the total emissions of NO_x from these sources were $\sim 3 \times 10^3$ tonnes/year, and the total emissions of HCl were $\sim 0.50 \times 10^3$ tonnes/year.

A closer view of a devastated hillside located about 2 km from the Copper Cliff smelter. The worst of this damage was caused prior to 1929, during the period of roast bed use, and during the early smelter era before tall smokestacks were built (prior to about 1935). However, revegetation was also prevented by frequent fumigations from the smelters, even after relatively tall stacks had been constructed and by severe soil toxicity caused by acidity and large plant-available concentrations of aluminum, nickel, copper, and other metals. Following the demise of the forest and the salvage clearcutting of dead trees to fuel local roast beds, there was severe erosion of soil from slopes. This exposed the pinkish gneissic and granitic bedrock, which was then blackened and pitted by reaction with the acidic, sulfurous fumes. (Photo: B. Freedman.)

Table 2.11 Emission and wet and dry deposition of pollutants within a 40-km radius of the 381-m "superstack" at Copper Cliff, Ontario[a]

| | Emission (1) | Total deposition | | | Background deposition | | | Total deposition as percentage of emissions (4/1) | Plume deposition as percentage of emissions (4 − 7)/1 |
		Wet (2)	Dry (3)	W + D (4)	Wet (5)	Dry (6)	W + D (7)		
H⁺	449	2762	—	2762	2741	—	2741	615	4.7
SO_2	2,334,000	—	16,100	16,100	—	12,700	12,700	0.69	0.15
SO_4^{-2}	49,100	21,300	2250	23,550	19,800	2200	22,000	48	3.2
S	1,183,000	7100	8794	15,894	6600	7080	13,680	1.3	0.19
Fe	1626	415	1344	1759	344	1132	1476	108	17
Ni	798	29.2	30.1	59.3	10.8	12.2	23.0	7.4	4.5
Cu	591	61.2	22.8	84.0	21.5	9.1	30.6	14	9.0
Al	406	313	533	846	292	502	794	208	13
Pb	480	86.7	37.2	123.9	76.5	35.0	111.5	26	2.6

[a]The background deposition does not include any smelter-derived pollutants. Emission and deposition data are in kg/day. Modified from Chan *et al.* (1984a).

An open community dominated by shrub-sized individuals of red maple (*Acer rubrum*) and white birch (*Betula papyrifera*), located about 6 km from the Copper Cliff smelter. The sparse ground vegetation is dominated by the metal-tolerant moss *Pohlia nutans*. Note that the photo was taken in October, after the autumn leaf-fall. (Photo: B. Freedman.)

A forest dominated by *Betula papyrifera*, located about 15 km from the Copper Cliff smelter. Conifers are relatively sensitive to sulfur dioxide, and they are notably absent from this stand. Note that the photo was taken after the autumn leaf-fall. (Photo: B. Freedman.)

has a pH less than ca. 5.5. As a result, precipitation that has passed through the superstack plume is only slightly elevated in SO_2 concentration, and therefore wash-out is not an effective mechanism of deposition for this gas (Millan et al., 1982; Lusis et al., 1983);

3. The deposition of emitted sulfate is more efficient than that of SO_2, averaging 3% overall (Table 2.11), and about 48% during a precipitation event (Chan et al., 1984b). The plume sulfate that is washed out close to the superstack is mainly emitted sulfate. Sulfate produced secondarily in the plume by the oxidation of emitted SO_2 is of minor importance close to the source, since the SO_2 oxidation rate only averages 1%/hr or less (Millan et al., 1982; Chan et al., 1983).

These represent the modern, post-superstack mechanisms of sulfur deposition. Ground-level fumigations by SO_2 and acidic mists were much more frequent and intense during the era of roast bed use, and to a lesser extent prior to 1972 when smelter emissions were from shorter stacks. The great ecological damage in the Sudbury area was caused by these earlier emissions of pollutants, along with other disturbances such as the harvesting of forests for lumber and to fuel the roast beds and fires started by prospectors and by railroad-engine sparks (Winterhalder, 1978). Although modern emissions have been as large or greater than earlier emissions, they are dispersed into the atmosphere rather effectively, so that acute injuries to vegetation are now infrequent.

Unfortunately, there are no data from the roast bed era for SO_2 concentrations in air. Without question, the ground-level roast bed plumes were extremely phytotoxic, and they devastated the surrounding vegetation. More recently, air-quality data collected prior to the commissioning of the superstack in 1972 showed that the local SO_2 pollution was considerably worse than at present. Dreisinger (1967) summarized 3×10^5 hr of air quality monitoring between 1954 and 1963. He reported no detectable SO_2 for 87% of the time, >0.25 ppm SO_2 for 2% of the time (6000 hr), and >1 ppm for <0.1% of the time (257 hr). Before the superstack was commissioned, the half-hour SO_2 concentration during fumigation episodes reached 3.0 ppm at a station 7.7 km from Copper Cliff and 3.6 ppm at a site 26 km away (McGovern and Balsillie, 1972). During 1970, a 5074-km^2 area surrounding Sudbury had a mean-annual SO_2 concentration of <5 ppb, 3100 km^2 had 5–10 ppb, 1157 km^2 had 10–20 ppb, and 266 km^2 had >30 ppb. These concentrations were sufficient to frequently cause acute damage to vegetation close to the SO_2 sources (Dreisinger and McGovern, 1971).

Ecological Damage

Several studies have demonstrated that the severity of the ecological damage decreases geometrically with increasing distance from the Sudbury point sources. The most important cause of the damage has been SO_2, but toxic effects of nickel, copper, aluminum, and acidification are also important (Hutchinson and Whitby, 1974; Whitby and Hutchinson, 1974; see also Chapter 3).

No studies of vegetation damage were made during the period of roast-bed use. However, a palynological study of annually laminated sediment in a meromictic lake demonstrated that beginning in 1880–1885 there was a 51% decrease in the influx of *Pinus* pollen, a 14% decrease in *Picea* pollen, and a decrease in total pollen of 67% (Huhn, 1974). At the same time there was a 400% increase in smoke microspherules, indicating an increased frequency of wildfire. These observations suggest that a general decrease in forest cover was coincident with the onset of industrial activity in the Sudbury area. The forests were affected by several factors, including toxic fumes, a lumber industry, clearcutting to fuel the roastbeds, and fires in fume-killed vegetation and logging debris (Winterhalder, 1978; Freedman and Hutchinson, 1980c). Huhn (1974) also demonstrated an increased influx of tree pollen (especially *Pinus*, *Picea*, and *Betula papyrifera*) beginning in 1930–1935. This coincides with a time of abatement of ground-level pollution, when the use of roast beds was stopped and most SO_2 was vented to the atmosphere through smokestacks.

Other studies of vegetation damage in the Sudbury area documented the cumulative effects of the roast-bed era, plus the effects of the smelters that replaced them. Watson and Richardson (1972) recorded 103 km² of "severely barren" land around the Sudbury smelters and 363 km² of terrain with "impoverished" vegetation (notably lacking conifers). Linzon (1971) related atmospheric SO_2 concentrations to damages caused to white pine (*Pinus strobus*), a sensitive conifer. He observed reduced growth rates, high mortality, and severe foliar injuries within an area of 1840 km² and foliar injuries alone over 4100 km²; almost all of the white pine observations were made beyond the impoverished zone of Watson and Richardson (1972). Therefore, in the pre-superstack period, vegetation was injured over at least 6400 km² of terrain around the Sudbury smelters.

The first detailed study of plant communities was that of Gorham and Gordon (1960a), who examined a transect running northeast from the Falconbridge smelter. They observed a sharp decrease in species richness within 6.4 km of the smelter, with the most tolerant taxa being *Sambucus pubens*, *Polygonum cilinode*, *Acer rubrum*, and *Quercus rubra*. *Pinus strobus* and *Vaccinium myrtilloides* were only present at sites further than 18 km, indicating their particular sensitivity to pollutants. Needles of *P. strobus* were injured as far as 30 km from the smelter. At about the same time, a qualitatively similar pattern of damage was observed among epiphytic, arboreal lichens in the Sudbury area (LeBlanc and Rao, 1966; LeBlanc *et al.*, 1972).

Freedman and Hutchinson (1980c) documented longer-term effects on vegetation by examining stands along a transect running south from the Copper Cliff smelter (Figs 2.3 and 2.4). There were no forest remnants within 3 km of the smelter. Within this zone hilltops and slopes were denuded of vegetation and severely eroded, and the bedrock was pitted and blackened by reaction with the fumigations. The ground vegetation within this inner zone was depauperate, with <1% plant cover. The most prominent vascular species were the grasses *Deschampsia caespitosa*, *Agrostis hyemalis*, and *A. gigantea*, and the dicots *Polygonum cilinode* and *Acer rubrum*.

From 3 to 8 km, forest remnants occurred in sites where soil moisture was available and the topography gave some protection from fumigations. These sites were either beside water bodies and dominated by *Populus tremuloides*, or they were on relatively protected, mesic, lower slopes of hills where woody angiosperms that are relatively tolerant of pollution were dominant (e.g., *Acer rubrum*, *Quercus rubra*, *Populus tremuloides*, *P. grandidentata*, *Betula papyrifera*). The individuals of these "tree" species had a stunted growth form, usually with dead upper branches and much stump sprouting (see also James and Courtin, 1985). Denuded, blackened hilltops were common within this patchily vegetated zone.

Beyond 8 km the forest cover was almost continuous, but the tree-species composition and basal area were relatively depauperate, as was the understory vegetation. Stands beyond about 20 km were little affected by pollution, and the forest was a mixed conifer–hardwood association, typical of the region. In this reference forest, conifers contributed >50–75% of the tree basal area—mostly of *Pinus strobus* and *Picea glauca* (see also Amiro and Courtin, 1981).

The pattern of vegetation damage around the Copper Cliff smelter clearly parallels the patterns of pollution intensity (Freedman and Hutchinson, 1980a). Pollutants that had accumulated in soil (e.g., nickel, copper, and to a lesser degree sulfur) exhibited well-defined decreases in concentration with increasing distance from the smelter, as did the contemporary patterns of deposition of nickel, copper, iron, and sulfate in precipitation and dustfall and SO_2 in air (see also Chapter 3).

Various studies have shown that lakes close to the Sudbury smelters have been acidified and that they have large concentrations of sulfate, nickel, copper, and other elements (Gorham and Gordon, 1960b; Conroy *et al.*, 1975; Whitby *et al.*, 1976; Dillon and Smith, 1984). In parallel with the damage to terrestrial vegetation, the biota of the stressed water bodies has responded by a general reduction of standing crop and productivity, a small species richness and diversity, and dominance by pollution-tolerant species. These general effects are particularly marked in the planktonic algae, zooplankton,

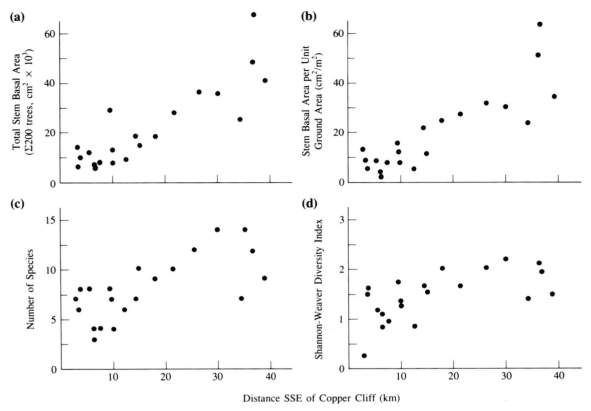

Fig. 2.3 Overstory vegetation characteristics as they vary with distance along a transect south of the Copper Cliff smelter. (a) Total stem basal area; (b) stem basal area per unit ground area; (c) species richness (total number species occurring among 200 randomly chosen trees; (d) Shannon–Weaver diversity ($-\Sigma\, p_i \ln p_i$, where p_i is approximated by relative frequency). [Modified from Freedman and Hutchinson (1980c); © 1980 National Research Council of Canada, with permission.]

and macrophytes; fish are not present in the most polluted lakes (Gorham and Gordon, 1963; Whitby *et al.*, 1976; Yan, 1979; Yan and Strus, 1980; Yan and Miller, 1984).

Ecological Recovery After the Abatement of Pollution

As mentioned earlier, there have been great decreases in ground-level atmospheric pollution in the Sudbury area since 1972. This has resulted in dramatic recoveries of terrestrial and aquatic ecosystems. In particular, there has been a vigorous spread of a natural grassland in mesic and wet sites where

soils from eroded hilltops have accumulated and along the borders of water bodies. The most important grasses are *Deschampsia caespitosa* and *Agrostis gigantea*, both of which have been shown to be ecotypically tolerant of the residual soil toxicity associated with nickel and copper (see Chapter 3). In addition, lakes in the vicinity of Coniston have recovered dramatically since the closing of that smelter. Over the period 1972–1984, the pH of a monitored, acidic lake increased from 4.1 to 5.8, while the concentrations of sulfate, copper, nickel, cobalt, manganese, and zinc decreased by 60–90% (Hutchinson and Havas, 1985). Studies of the biota of the recovering lakes have not yet been made.

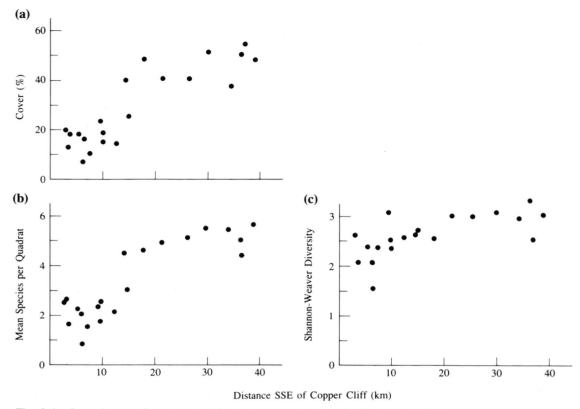

Fig. 2.4 Ground vegetation characteristics in the same stands as in Fig. 2.3. (a) Total cover per quadrat; (b) species richness (mean number species per 0.5×0.5-m quadrat); (c) Shannon–Weaver diversity ($-\Sigma p_i \ln p_i$). [Modified from Freedman and Hutchinson (1980c); © 1980 National Research Council of Canada, with permission.]

Other Large Point Sources of Atmospheric Emissions

The Sudbury example is one of the world's best-documented cases of extensive ecological damage caused primarily by emissions of toxic gases from smelters. Other metal smelters have also been studied. Although the ecological damage around them is smaller in extent than near Sudbury, the effects are comparable in a qualitative sense. Examples of SO_2 damage include a copper smelter at Ducktown, Tennessee (Smith, 1981), copper smelters near Superior, Arizona, and Anaconda, Montana (Wood and Nash, 1976; Miller, 1989), a lead-zinc smelter at

Trail, British Columbia (Katz, 1939; Miller, 1989), and an iron-sintering plant at Wawa, Ontario (Gordon and Gorham, 1963; Scale, 1982). Less severe damage to terrestrial ecosystems has been described around sources of gaseous fluoride, particularly aluminum-reduction smelters and phosphorus plants (Bunce, 1979, 1984; Thompson *et al.*, 1979; Murray, 1981; Taylor and Basake, 1984). In eastern Europe, intensive forest damage has also been attributed to air pollution from local, industrial sources. In an industrial part of the Czech Republic, for example, a forest dieback caused by air pollution has affected an extensive, mountainous area in the Erzgebirge region, where atmospheric-sulfur deposition can exceed 100 kg/ha·yr (Molden and Schnoor, 1992).

In all of the studies of point sources, there is a generic observation of ecological damage becoming progressively less severe as distance from the source increases. This pattern closely tracks the gradients of pollution, which are characterized by a roughly exponential decrease in intensity as distance from the source increases.

If a forested ecosystem is being affected, then the tree stratum is generally impacted first and is "stripped" away. As trees decline, shrubs, and then the ground vegetation, are affected. This syndrome of sequential death of horizontal strata of the terrestrial vegetation was described as a "peeling" or "layered vegetation effect" by Gordon and Gorham (1963; see also Woodwell, 1970). This pattern of damage is probably caused by physical, exposure effects related to target surfaces; i.e., the canopy trees are affected first, while the understory vegetation is initially protected by a boundary-layer effect. Of lesser importance are differences in physiological susceptibility to pollution by the component species of the vegetation strata.

Interestingly, the few studies that have been made in the vicinity of coal-fired power plants in North America and western Europe have only shown relatively minor ecological damage, even though these facilities are large, point-source emitters of SO_2. For example, Jacobson and Showman (1984) surveyed vegetation in an area of the Ohio River valley with four coal-fired power plants. During a 12-year study period only one episode of acute injury to vegetation was observed, affecting only a few susceptible species in a 14×6-km oval around one of the power plants and causing no longer-term damages. In addition, a survey of lichen distributions showed no effects that could be attributed to the power plants. McClenahan (1978) studied seven forest stands along another 50-km stretch of the Ohio River, with gradients of pollution by SO_2, Cl, and F from a coal-fired power plant, three chemical factories, and an aluminum reduction facility. There was only a minor decrease in the species diversity of trees, shrubs, and ground vegetation in the vicinity of the sources. The tree *Acer saccharum* was less

dominant in polluted sites, possibly indicating sensitivity to the gases, while tolerant woody plants included the tree *Aesculus octandra* and the shrub *Lindera benzoin*.

Rosenberg *et al.* (1979) studied a potentially more damaging situation in Pennsylvania, in which a coal-burning power plant is located in a river valley and has a 100-m stack that is barely higher than the surrounding ridgetops. Eight stands were examined along an upwind–downwind transect, and there were decreases in tree-species richness, diversity, and basal area close to the source. However, these effects were relatively minor in extent because they only occurred in stands within 2 km of the source. The maples *Acer saccharum* and *A. rubrum* were relatively prominent close to the power plant, while the conifers *Pinus strobus* and *Tsuga canadensis* were most prominent further away.

Because many lichen taxa are sensitive to toxic gases, these "species" (actually, a fungal–algal symbiosis) have been surveyed in many studies of regional and point-source air pollution. Only minor damage has been reported in studies done around coal-fired power plants. Showman (1975) surveyed 128 sites in the vicinity of a coal-fired power plant in Ohio. He found that the occurrence of terricolous and saxicolous species (growing on soil and rock substrates, respectively) was unaffected by proximity to the power plant. In contrast, distribution maps showed that two corticolous epiphytes (growing on tree bark) were sensitive—*Parmelia caperata* and *P. rudecta* had voids of occurrence of ca. 90 km² around the power plant. This area corresponded to mean-annual SO_2 concentrations of >20 ppb. Other corticolous species were more tolerant. Showman (1981) resurveyed the study area following an abatement of local pollution by the construction of taller stacks. He found a recolonization by *P. caperata* into its previous void and a return to a near-normal distribution 8 years after the decrease in pollution. Interestingly, in the studies of Jacobson and Showman (1984) of coal-fired power plants along the Ohio River, no effects were observed on the distributions of *Parmelia caperata* or *P. rudecta*, in contrast to their sensitivity in the Showman (1975) study.

In another study, Will-Wolf (1980a,b) examined lichen distributions and injuries around a coal-fired power plant in Wisconsin. When she divided her 29 stands into high and low SO_2 exposure classes, she found minor effects at the community level. Two species appeared to be relatively tolerant of SO_2: on average, *Bacidia chloroantha* increased in frequency by 200% in the high-SO_2 sites and decreased by 31% in the low-SO_2 sites, while *B. chlorococca* differed by +180% and +24%, respectively. The most sensitive lichens were *Lepraria membranacea* (−82% at high-SO_2 sites but −7% at low-SO_2 sites) and *Parmelia caperata* (−17% and +39%, respectively).

Overall, these observations of lichens represent measurable but ecologically minor effects on vegetation in the vicinity of coal-fired power plants. In contrast, relatively severe damage to lichens has been widely reported in urban areas with chronic air pollution (LeBlanc and De Sloover, 1970; Hawksworth and Rose, 1976; Johnsen and Sochting, 1980).

Oxidant Air Pollution and Forest Damage in California

The Setting

Ozone is a secondary pollutant that is of regional importance, rather than being concentrated around point sources of emission. Large areas of natural and agricultural vegetation can be injured by ozone in sunny regions with large concentrations of NO_x and hydrocarbons, coupled with topographic and/or meteorologic factors that restrict air circulation.

A well-documented case of ozone-caused damage has been demonstrated in conifer forests in southern California, especially along the western slopes of the Sierra Nevada and San Bernardino Mountains. The ozone is produced photochemically from NO_x and hydrocarbons emitted in Los Angeles and other population centers to the west, and the polluted air masses are transported eastward to the mountains where vegetation is dam-

aged (P.R. Miller *et al.*, 1972; Williams *et al.*, 1977; Williams, 1980, 1983; Barnard, 1990).

Air Pollution

During the growing season, the diurnal concentration of ozone is largest between ca. 12:00 and 22:00, with a much smaller concentration at night and in the early morning (Williams *et al.*, 1977). There can be especially large ozone concentrations at the top of the atmospheric inversion layer that frequently develops on sunny days. As a result, sampling stations at this altitude measure large concentrations of oxidants, and forest damage is most severe at about 600–2600 m (P.R. Miller *et al.*, 1972; Williams *et al.*, 1977). Between 1973 and 1982, a sampling station at San Bernardino had an annual average O_3 concentration of 36 ppb, an average daily maximum of 158 ppb between May and October, 650 hr/year greater than 120 ppb, and 163 hr/year greater than 200 ppb (Walker, 1985; note that the Environmental Protection Agency (EPA) standard for the maximum 1-hr average concentration of ozone is 120 ppb). In comparison, the state of California overall has an annual average of 25 ppb O_3, an average daily maximum between May and October of 72 ppb, 85 hr/year >120 ppb, and 17 hr/year >200 ppb (Walker, 1985).

There has been an intensive effort to control the emissions of oxidant precursors in California. For example, between 1973 and 1982 the emission standards for new vehicles were reduced by 88% for hydrocarbons and by 50% for NO_x. These effects resulted in a 33% decrease in atmospheric CO concentrations between 1975 and 1981, a 29% reduction in the emission of transportation-derived hydrocarbons between 1975 and 1982, and a 7% reduction in ambient NO_x between 1976 and 1982. However, these changes did not result in significant decreases in average O_3 concentration during that period. For example, between 1977 and 1981, the maximum hourly ozone concentration in the San Francisco air-management basin ranged from 170 to 230 ppb, while between 1982 and 1985 the range was 150–200 ppb (Miller, 1989). Moreover, only 10 of 65 air-quality monitoring sites in California were able to meet the 120-ppb EPA standard between 1972 and 1982 (Walker, 1985).

Damages to Vegetation

Several conifer species are especially susceptible to ozone damage in the impacted forests. The most sensitive species are ponderosa pine (*Pinus ponderosa*), the dominant tree in most affected stands, and Jeffrey pine (*P. jeffreyi*). Within *P. ponderosa*, hypersensitive genotypes comprise 7–10% of the population (Williams, 1983). Less susceptible conifers in these forests include *Abies concolor*, *Pinus lambertiana*, and *Calocedrus decurrens* (Miller, 1973, 1989; Williams *et al.*, 1977; McBride *et al.*, 1985; Barnard, 1990).

The smog damage was first noticed in the 1950s, and it was first recognized as a disease attributable to ozone in 1963, when the damage to *Pinus ponderosa*-dominated forest was about 10 thousand hectares, increasing to 40–65 thousand hectares by 1969 (MacKenzie and El-Ashry, 1989). The ozone damages are characterized by an initial chlorotic mottling of foliage, followed by a necrosis that spreads backward from the leaf tip, premature abscission of older foliage, and ultimately, death of the tree. Diseased trees are relatively susceptible to secondary damage caused by bark beetle infestation and fungal root pathogens, and these are often the ultimate cause of death of weakened trees (Williams *et al.*, 1977; Williams, 1980; Miller, 1989).

The degree of forest damage varies greatly among sites, depending on their relative exposure to ozone. In a survey of eight stands in the Sequoia and Los Padres National Forests, Williams (1980) found that 8–100% of the dominant trees were injured. At severely affected sites, 15–25% of the current needles had acute injury symptoms, while 27–75% of 1- and 2-year-old needles were damaged and exhibited premature senescence and leaf drop. In relatively polluted sites, there were no leaves older than 4 years, whereas in less contaminated sites, foliage could be 6 years old and older. Williams (1980) estimated that ozone caused more than 3% excess mortality per year in these forests of

southern California and a reduction of up to 83% in the volume of wood production.

In spite of vigorous efforts to control the emissions of air pollutants from automobiles and other sources in southern California, the ozone damage to *Pinus ponderosa* forests increased between 1975 and 1983. Williams and Williams (1986) resurveyed permanent forest plots in the Sierra Nevada mountains. They found that the incidence of ozone damages to the current foliage of *P. ponderosa* had increased from 15% of trees in 1975 to 24% in 1983, while on 1-year-old needles, the incidence had increased from 44 to 61%. Only 21% of the trees had 2 years of needle retention in 1983, and there was a 14% decrease in annual ring width between the two samplings. These observations indicate that a substantial decrease of vigor is taking place in *P. ponderosa* populations in the half million hectares of forest that are affected by oxidants in the Sierra Nevada.

Since tree species differ markedly in their susceptibility to O_3 damage, it is not surprising that there have been large changes in forest-community composition. McBride *et al.* (1985) found that the dominance of *Pinus ponderosa* was decreasing at sites in the San Bernardino Mountains, where the O_3 concentrations are large, and that this pine was being replaced by relatively O_3-tolerant conifers such as *Abies concolor* and *Calocedrus decurrens*, to form an oxidant-caused disclimax.

Large changes have also occurred in the community of epiphytic lichens in these conifer forests. Sigal and Nash (1983) found that about 50% of the species recorded in a survey done in the early 1900s were now locally extirpated and that species richness was especially depauperate in relatively polluted sites. *Hypogymnia enteromorpha* was markedly injured at polluted sites. Identical symptoms developed when healthy specimens of this species were transplanted into high-oxidant sites.

Ozone Elsewhere

DeBauer *et al.* (1985) described a syndrome of damage and decline in high-elevation (3000–3500 m) pine forests in the mountains around Mexico City.

The ozone injury symptoms and ecological effects were similar to those observed with *Pinus ponderosa* in southern California. *Pinus hartwegii* was the most sensitive species near Mexico City, while *Pinus montezumae* var. *lindleyi* was less susceptible.

Well-documented ozone injury to trees in the northeastern and midwestern United States has also been described. *Pinus strobus* is a relatively sensitive species, but individuals within populations display a varying degree of damage depending on genotypic sensitivity (Berry, 1973; Houston and Stairs, 1973; Houston, 1974). Sensitive trees have a relatively small growth rate (Benoit *et al.*, 1982), but these effects have not been accompanied by a widespread mortality of white pine. During an episode of pollution, the O_3 concentration in this region of the United States is frequently 50–70 ppb, with episodal peaks ranging up to 100 to 200 ppb (Lioy and Samson, 1979; Benoit *et al.*, 1982; Lefohn, 1990). Similarly, ozone events of up to 150 ppb occur in western Europe (Colbeck and Harrison, 1985). These oxidant events are as severe as, but less frequent than, those that occur in southern California.

2.5
CARBON DIOXIDE, CLIMATE CHANGE, AND ECOLOGICAL RESPONSES

The concentration of CO_2 in the atmosphere has been increasing steadily for at least the last century. The phenomenon has been caused by the release of gaseous CO_2 as a result of many human activities, but especially: (1) the burning of fossil fuels, during which virtually all of the hydrocarbon-carbon is oxidized to CO_2, and (2) the conversion of mature forests characterized by large standing crops of organic-carbon, to ecosystems with much smaller biomass, with the difference in carbon content being balanced by CO_2 emitted to the atmosphere through detritivore activity or fire.

The changes in concentration of atmospheric CO_2 are well documented, but there is much uncertainty about eventual sinks for the emitted CO_2,

feedback mechanisms, and ecological implications of the atmospheric changes. As discussed below, the most likely effects of increasing CO_2 in the atmosphere are: (1) global warming and other climatic changes caused by enhancement of the "greenhouse effect," and (2) direct, possibly beneficial influences on the physiology of some plants, especially a CO_2-fertilization effect, and decreases in transpiration. These topics are discussed in the following sections.

Anthropogenic Influences on Global Carbon Geochemistry

Emissions of CO_2 by the Burning of Fossil Fuels

The rate of fossil fuel utilization has increased remarkably during the last century, mostly to supply expanding demands for the production of energy for space heating, transportation, and industrial purposes (Fig. 2.5). During the decade of 1860–1869, the global emission of CO_2 from the combustion of fossil fuels averaged 118×10^6 tonnes of CO_2-C/year, almost entirely resulting from the burning of coal and lignite. By 1973–1982, the global emission had increased by a factor of 41, to 5×10^9 tonnes/year, and to 5–6×10^9 tonnes/year in the late 1980s (Rotty and Masters, 1985; Mooney *et al.*, 1991; WRI, 1992). Of the total modern emission of CO_2, 40% results from the burning of liquid hydrocarbons, 40% from solids, 17% from gases, and 3% from the manufacture of cement (WRI, 1992). There is much uncertainty in the prediction of future emissions of CO_2 by the combustion of fossil fuels, but the best-guess scenario is for a rate of about 8–15×10^9 tonnes/year, or as much as three times larger than the current rate (Edmonds and Reilly, 1985).

Of course, countries differ greatly in their per-capita emissions of carbon dioxide by fuel combustions. The global average emission of CO_2-C is about 1.15 tonnes/person-year (WRI, 1992). The rates of emission are largest in relatively industrialized and fuel-intensive countries, e.g., United States, 5.37 tonnes/person-year; Canada, 4.72;

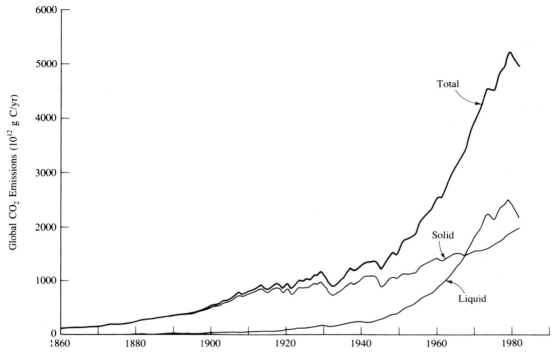

Fig. 2.5   The global emission of CO_2 as a result of the combustion of fossil fuels. Data are from Rotty and Masters (1985).

Czechoslovakia, 3.95; Russia, 3.63; Poland, 3.14; United Kingdom, 2.70 (WRI, 1992). The citizens of Qatar are the world's greatest per-capita emitters of CO_2-C at 10.3 tonnes/person-year, because of a large influence of gas flaring in that petroleum producing country. Less-developed countries emit substantially smaller quantities of fuel-derived CO_2-C to the atmosphere, e.g., Uganda, 0.01 tonnes/person-year; Chad 0.01; Cambodia, 0.02; Bangladesh, 0.03; Afghanistan, 0.11. Clearly, citizens of industrialized countries have much more substantial, per-capita CO_2 emissions from fuel combustions than do citizens of poorer countries.

Emissions of CO_2 by the Clearing of Mature Forests

A mature, forested ecosystem has a relatively large content of organic carbon, present in its living and dead biomass, and in the organic matter of the forest floor and mineral soil. This quantity of "fixed" carbon is considerably larger than the amounts in either (1) relatively young, successional forests or (2) any other type of ecosystem, including anthropogenic ones. Therefore, when a unit of land covered by mature forest is cleared for any purpose (but usually to provide fuel or fiber, or to create new agricultural land), it is replaced at least temporarily by an ecosystem that contains a much smaller quantity of fixed carbon. The difference in the carbon contents of the ecosystems is balanced by a net flux of CO_2 to the atmosphere, caused by the progressive oxidation of much of the mature-forest carbon by fire and detritivore activity (Woodwell *et al.*, 1978; Cooper, 1983; Houghton *et al.*, 1983, 1985; Detwiler and Hall, 1988).

It is well established that human activities have caused a large reduction in the global coverage of mature forest and that this effect has resulted in a large flux of CO_2 to the atmosphere. Prior to any substantial forest clearing on Earth, terrestrial vegetation may have stored 900×10^9 tonnes of carbon, of which 90% occurred as forest carbon and 50% in tropical forests. At the present time, about 560×10^9 tonnes of carbon are stored in terrestrial vegetation, and that quantity is diminishing further as more mature ecosystems are converted to lower-carbon systems (Fig. 2.6; Olson *et al.*, 1985; Dale *et al.*, 1991).

Between 1850 and 1980, the global reduction in area of mature forest, coupled with other anthropogenic changes in land use, caused a net CO_2-carbon emission to the atmosphere of about $108-120 \times 10^9$ tonnes. This is similar in quantity to the $170-190 \times 10^9$ tonnes CO_2-C that were emitted by the burning of fossil fuels over the same period of time (Houghton *et al.*, 1983, 1991; Dale *et al.*, 1991).

A roughly similar pattern of CO_2 emission has been observed in more recent years. Between 1958 and 1980, the combustion of fossil fuels resulted in an estimated CO_2-C flux to the atmosphere of 85.5×10^9 tonnes, while the net release from terrestrial ecosystems was 57.3×10^9 tonnes (Houghton *et al.*, 1983). During the late 1980s, the combustion of fossil fuels was annually injecting $5.0-5.5 \times 10^9$ tonnes of carbon into the atmosphere, while deforestation was accounting for another $1.5-3.0 \times 10^9$ tonnes/year (Fig. 2.7; Dale *et al.*, 1991; Houghton *et al.*, 1991; WRI, 1992).

The global net flux of CO_2-C caused by land-use practices between 1958 and 1980 is summarized in Table 2.12. The most important net source of carbon during that time period was the clearing of forests followed by development of the land for cultivation or grazing, because this practice causes a longer-term landscape conversion and large decreases in the amounts of carbon present on sites.

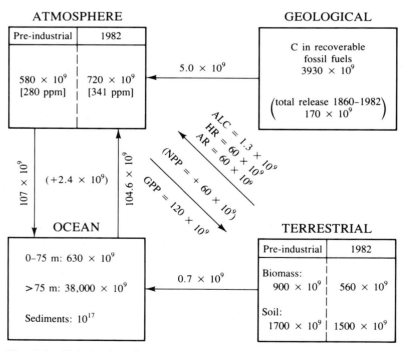

Fig. 2.6 Global values for some key compartments and fluxes of the global carbon cycle. The magnitude of compartments is in units of tonnes of carbon; fluxes are in units of tonnes C/year. Data are from Blasing (1985) and Solomon *et al.* (1985). See text for explanation of symbols.

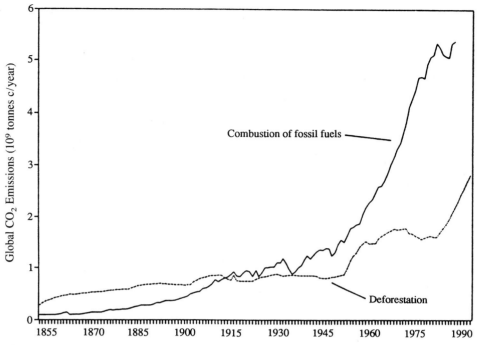

Fig. 2.7 The global emission of CO_2 from deforestation and the burning of fossil fuels. Modified from Houghton (1991).

Table 2.12 Estimated net flux of CO_2 carbon between 1958 and 1980 as a result of (1) changes in carbon storage in the soil and vegetation of terrestrial ecosystems caused by anthropogenic effects on ecosystem cover type and (2) the combustion of fossil fuels[a]

Category	1958–1980 Net flux (10^9 tonnes C)	Total reserves (10^9 tonnes C)
Terrestrial biota and soil		2500
Forest clearing and cultivation	−29.1	
Decay of wood from forests	−11.1	
Forest clearing and grazing	−5.9	
Harvest of forests	−58.3	
Regrowth of harvested forest	+45.8	
Net abandonment and afforestation	+1.3	
Terrestrial net flux	−57.3	
Burning of fossil fuels	−85.5	10,000

[a]Modified from Houghton *et al.* (1983).

The net effect of forest harvesting followed by regeneration back to forest is smaller, since the successional regrowth of a cut stand restores much of the carbon removed during the harvest (this is discussed later in more detail; see also Fig. 2.11).

It is important to note that the global data of Table 2.12 conceal large regional differences in the net CO_2 fluxes from terrestrial ecosystems (Table 2.13). The net fluxes of carbon from terrestrial ecosystems of North America and Europe were large in 1920 and before, but they are currently small or negative. The present, relatively small net changes occur because (1) the carbon content of forests within those regions is regenerating about as fast as it is being harvested and (2) a large area of poor-quality agricultural land has been abandoned and is regenerating to forest. In contrast, the net fluxes of carbon in low-latitude, tropical regions of Latin America, Africa, and Asia have increased greatly because of substantial deforestation since 1920 (Detwiler and Hall, 1988; Houghton *et al.*, 1991).

Table 2.13   Regional net flux of carbon in 1920 and 1980 resulting from forest harvest and other changes in land use[a]

	Net flux (10^6 tonnes/year)	
Region	1920	1980
North America	290	10
Europe	100	−50
Latin America	140	800
Tropical Africa	100	700

[a] Modified from Houghton *et al.* (1983).

Effects on Global Carbon Geochemistry

The burning of fossil fuels and anthropogenic changes of vegetation cover have caused large and approximately equal fluxes of CO_2 to the atmosphere. These anthropogenic effects are summarized in Fig. 2.6, as are estimates of the magnitudes of other global CO_2 fluxes and the sizes of the largest compartments of carbon. The most important conclusions to be drawn from Fig. 2.6 are the following:

1. Because the atmospheric CO_2 compartment is relatively small in magnitude, the anthropogenic emissions have caused a large change in its size. The atmospheric content of CO_2-C has increased from about 580×10^9 tonnes prior to the industrial revolution to 720×10^9 tonnes in 1982, a 25% increase. This change is reflected by an increase in the atmospheric concentration of CO_2 over that time period, from about 280 ppm to 341 ppm (see below for further details).

2. The global net flux of carbon to the atmosphere is believed to have been close to zero prior to the era of intense anthropogenic influences on the nature of the vegetation on the global landscape. Therefore, global gross primary production (GPP) approximately equalled global ecosystem respiration (ER), and there was no net accumulation of biologically fixed carbon. [Note that ER is

the sum of respirations of all autotrophs (AR) and heterotrophs (HR). HR plus wildfire ultimately converts almost all global net primary production (NPP) to CO_2, except for a small percentage that may be geologically stored, for example, as peat, coal, petroleum, or limestone.] This dynamic balance has been upset by anthropogenic conversions of forested landscapes (ALC in Fig. 2.6) to ecosystems with smaller carbon contents. At present, the global flux of carbon from terrestrial ecosystems (ALC + ER) is about 2% larger than global NPP. Because of this effect, the carbon content of the global terrestrial landscape is smaller today than it was prior to the industrial era, by an estimated factor of 38% in the biota, and 12% in soil.

3. Ultimately, the most important sink for anthropogenically emitted CO_2 is the ocean, which currently has a positive net flux of carbon of 2.4 $\times 10^9$ tonnes/year. This flux to the oceanic sink is large, but it is smaller than the total rate of anthropogenic emission, and hence the atmospheric CO_2 pool is increasing in magnitude. The ocean has a tremendous capacity to absorb atmospheric CO_2 by the formation of carbonic acid (after Baes *et al.*, 1985; Moore and Bolin, 1987):

$$CO_2(g) + H_2O \leftrightarrows H_2CO_3 \text{ (aq)}. \quad (18)$$

The carbonic acid dissociates to bicarbonate and carbonate, which respectively comprise about 90 and 10% of the total oceanic concentration of inorganic carbon:

$$H_2CO_3(aq) \leftrightarrows H^+ + HCO_3^- \quad (19)$$

$$HCO_3^- \leftrightarrows H^+ + CO_3^{-2}. \quad (20)$$

The ionic carbon species can be taken up by certain biota and used for the construction of shells and other structures containing calcium carbonate (or calcite), most of which eventually deposit to a sediment sink. The most frequent calcite-forming reaction is

$$Ca^{2+} + 2HCO_3^- \leftrightarrows CaCO_3 + CO_2 + H_2O. \quad (21)$$

The progressive increase in the concentration of atmospheric CO_2 is one of the best-documented, longer-term trends in environmental science. The phenomenon is illustrated in Fig. 2.8, for an observatory at high altitude on the island of Hawaii. The annual periodicity of CO_2 concentration is caused by the large net uptake of CO_2 by vegetation during the growing season of temperate and higher latitudes of the northern hemisphere. This effect is illustrated in more detail in Fig. 2.9, which shows that the amplitude of the variation is largest at high latitude in the northern hemisphere (Barrow, Alaska) and smallest in the southern hemisphere (Samoa, near the equator, and the South Pole). In spite of the large differences in annual variation of atmospheric CO_2 concentration among these four stations, between 1976 and 1982 they all exhibited an average increase in CO_2 of 9.0 ppm (Gammon *et al.*, 1985).

The Greenhouse Effect and Global Climate Change

A potentially important consequence of the increasing concentration of CO_2 in Earth's atmosphere is that of global warming, caused by an intensification of an existing physical process known as the "greenhouse effect."

As a physical phenomenon, Earth's greenhouse effect is well understood and is believed responsible for maintenance of the planet's surface at an average temperature of about 25°C, or about 33° warmer than would be possible if certain gases and vapors, especially carbon dioxide and water, did not interfere with the rate of dissipation of absorbed solar radiation. The energy budget of Earth, and the influence of the greenhouse effect, are briefly described below (after Gates, 1962, 1985; Schneider and Mesirow, 1976; Odum, 1983; Luther and Ellingson, 1985; Solomon *et al.*, 1985; Schneider, 1989; Houghton *et al.*, 1991).

Almost all of the extraterrestrial radiation incident to the Earth has been radiated by its closest star, the sun. The rate of influx is called the solar constant, and it has a magnitude of 2 cal/cm²/min[1]. About half of the incoming energy occurs as relatively short "visible" radiation within a wavelength band of about 0.4–0.7 μm, and half is composed of

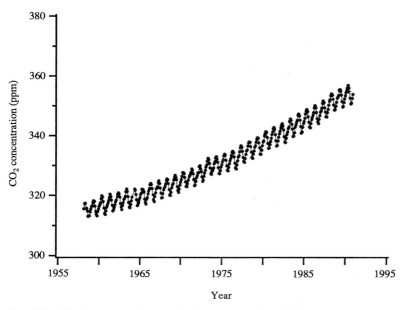

Fig. 2.8   The time series of atmospheric concentration of CO_2 measured at the Mauna Loa Observatory, Hawaii. The dots represent monthly averages. From Keeling and Whorf (1991).

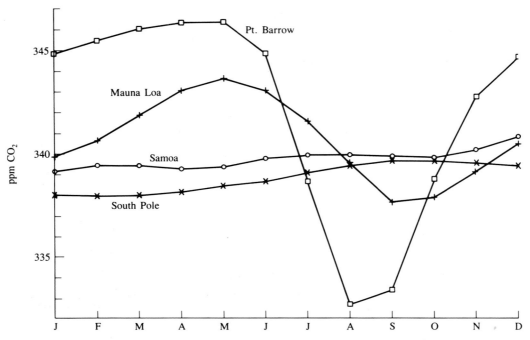

Fig. 2.9   Annual variation in the mean monthly concentration of CO_2 at four remote sites. Data are means over 1980–1983 and were taken from Gammon *et al.* (1985).

"near-infrared" wavelengths between about 0.7 and 2.0 μm. There is a virtually perfect energetic balance between the quantity of electromagnetic energy that is incident to the Earth and the amount that is ultimately dissipated back to outer space. Important components of Earth's energy budget are the following (Fig. 2.10):

1. An average of about one-third of the incident solar energy is reflected back to outer space by the atmosphere or by Earth's surface. This process is related to Earth's albedo, which is strongly influenced by cloud cover, by the quantity of small particulates suspended in the atmosphere, and by the character of Earth's surface, especially the types and amount of plant and water (including ice) cover.

2. About one-third of the incident energy is absorbed by atmospheric gases, converted to thermal kinetic energy (i.e., energy of molecular vibration), and then reradiated to space or to Earth's

surface as longer-wavelength (7–14 μm) electromagnetic radiation (see also 3 below).

The remaining one-third of the incident electromagnetic energy from the sun is transformed or dissipated by the following processes:

3. Much of the incident radiation is absorbed at Earth's surface by various inorganic and living materials, mostly resulting in a conversion to thermal energy, which increases the temperature of the absorbing surfaces. Over the medium term (i.e., days) and longer term (i.e., years), there is little net storage of heat. This occurs because virtually all of the thermal energy is reradiated by the absorbing surface, as electromagnetic radiation of a longer wavelength than that of the incident radiation. The wavelength spectrum of typical, reradiated electromagnetic energy from Earth's surface peaks at about 10 μm.

4. Some of the electromagnetic energy that

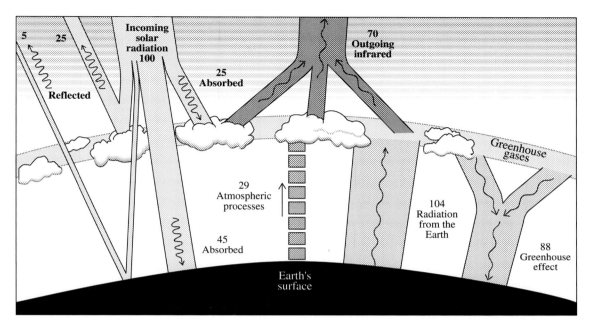

Fig. 2.10 Diagrammatic representation of important components of Earth's physical energy budget. About 30% of the incoming solar radiation is reflected by atmospheric clouds and particulates or by Earth's surface. The remaining 70% is absorbed and must be dissipated in various ways. Much of the absorbed energy serves to heat up the atmosphere and surfaces and is then mostly reradiated as long-wave infrared radiation. Atmospheric moisture and radiatively active gases interfere with this dissipative process. This so-called "greenhouse effect" keeps Earth's surface about 33°C warmer than it would otherwise be. Numbers refer to percentage of incoming solar radiation. See text for further discussion. Modified from Schneider. Copyright © 1989 by Scientific American, Inc.

penetrates to Earth's surface is absorbed by surfaces and transformed to heat. Much of this energy subsequently causes water to evaporate from plant and inorganic surfaces (as evapotranspiration), or it causes ice and snow to melt.

5. A small amount (<1%) of the absorbed radiation drives mass-transport processes that redistribute some of the unevenly distributed thermal energy of Earth's surface. The most important of these processes are wind, water currents, and waves on the surface of water bodies.

6. A small but ecologically critical quantity of incoming energy, averaging about 1% or less of the total, is absorbed by plant pigments (especially chlorophyll) and used to drive photosynthesis. As a result of this autotrophic fixation, some of the absorbed solar energy is "tempo-

rarily" stored in the interatomic bonds of biochemical molecules.

If Earth's atmosphere was transparent to the reradiated, long-wave infrared energy from dissipation mechanisms 2 and 3, then that energy would travel unobstructed to outer space. However, certain so-called radiatively active gases in the atmosphere absorb effectively within this infrared-wavelength band, and these gases thereby slow the rate of radiative cooling of Earth. The most important of these atmospheric constituents are carbon dioxide and water, but the trace gases methane, nitrous oxide, ozone, and chlorofluorocarbons are also important (Marland and Rotty, 1985; Ramanathan, 1988).

Some of these trace gases absorb infrared wavelengths more effectively than carbon dioxide, and they are therefore more radiatively active per molecule. For example, each molecule of methane is about 11–25 times as effective as carbon dioxide at absorbing long-wave infrared radiation, while nitrous oxide is 200–270 times as effective and tropospheric ozone 2000 times. The synthetic chlorofluorocarbons (CFCs) are especially intense absorbers of infrared wavelengths, with CFC-11 being 3–12 thousand times as effective as carbon dioxide as a greenhouse gas and CFC-12 7–15 thousand times (Barnaby, 1988; Blake and Rowland, 1988; Rodhe, 1990; Hengeveld, 1992).

Similarly to the pattern for CO_2, the atmospheric concentrations of various of these other radiatively active gases have increased significantly since pre-industrial times. Prior to 1850, the concentration of CO_2 in the atmosphere was about 280 ppm, while in 1985 it was 345 ppm. Over the same time period, CH_4 increased from 0.7 to 1.7 ppm; N_2O from 0.285 to 0.304 ppm; $CFCl_3$ from essentially zero to 0.22 ppb; CF_2Cl_2 from zero to 0.38 ppb; and CCl_4 from zero to 0.12 ppb (Ramanathan, 1988). For most of these gases, the rates of increase of their concentration in the atmosphere have been especially great since about 1950 (Blake and Rowland, 1988; Ramanathan, 1988; Rowland, 1988; Wigley and Raper, 1992).

The contribution of each of the greenhouse gases to the total, anthropogenic enhancement of Earth's greenhouse effect is a function of both their effectiveness at absorbing long-wave infrared radiation and their actual atmospheric concentration. Overall, carbon dioxide accounts for an estimated 60% of the anthropogenic greenhouse effect, methane 15%, nitrous oxide 5%, tropospheric ozone 8%, CFC-11 4%, and CFC-12 8% (Rodhe, 1990).

When these various substances absorb long-wave infrared radiation, they develop a larger content of thermal energy, which is then dissipated by another reradiation (again, of a longer wavelength than that of the electromagnetic energy that was absorbed). Because some of the secondarily reradiated energy is directed back to Earth's surface, the net effect of the radiatively active gases is to slow the rate of cooling of the planet. This process is called the greenhouse effect because its physical mechanism is similar to the one by which a glass-enclosed space is heated by solar radiation; that is, the encasing glass and humid atmosphere of a greenhouse are transparent to incoming solar radiation, but they absorb much of the outgoing, long-wave infrared radiation and slow down the rate of radiative cooling of the interior.

The physical mechanism of the greenhouse effect is conceptually simple and well understood, and as noted previously, this process is believed to have helped to make Earth habitable, by maintaining the average surface temperature at about 25°C, or some 33° warmer than would have been possible with a nongreenhouse atmosphere. It is also well known that the concentrations of CO_2 and some other radiatively active gases in Earth's atmosphere are increasing markedly. However, it has been difficult to conclusively demonstrate a warming trend of Earth's surface or lower atmosphere, due to a hypothesized intensification of the greenhouse effect, caused by the increases of CO_2 and other radiatively active gases.

It appears that since the initiation of detailed instrumental recordings of surface air temperatures around 1880, the four warmest years (up to 1988) all occurred in the 1980s (namely, 1980, 1981, 1983, and 1987), with 1987 averaging about 0.8°C warmer than the average for the decade of the 1880s (Hansen and Lebedev, 1987, 1988). Overall, it appears that there has been a net increase in Earth's surface air temperature of about 0.5°C since about 1850 (Blasing, 1985; Jones *et al.*, 1986).

However, the empirical temperature records on which these conclusions are based suffer from several problems: (1) many historical data are less accurate than modern records; (2) weather stations tend to be located in or near urban areas, so their data can be influenced by "heat island" effects; (3) air temperature is extremely variable on both temporal and spatial scales (that is, there is a small ratio of signal : noise), and this can make it difficult to interpret longer-term trends of temperature data; and (4) on the medium term (i.e., several years) global climate can change for many reasons other than a "greenhouse" response to increased CO_2 concentration, including the albedo-related effects of volcanic irruptions that inject large quantities of sulfur dioxide, sulfate, and fine particulates into the upper atmosphere (Harrington, 1987; Kerr, 1988a).

There is some rather compelling evidence for paleoclimatic linkages between variations of atmospheric carbon dioxide and Earth's surface temperature. Some of the most interesting evidence comes from data based on a core of glacial ice from Antarctica, representing a 160-thousand-year period of time (Fig. 2.11). The concentrations of carbon dioxide were determined by microanalysis of air bubbles extracted from age-dependent depth strata of the ice core, while air-temperature deviations were inferred from variations of isotopic ratios. The strong correlations of these two variables, carbon

dioxide and air-temperature deviation, obviously suggest a possible causal mechanism. However, it is not clear from such a relationship whether increased CO_2 caused the warming via an intensified greenhouse effect, or vice versa (i.e., warming could increase CO_2 release from certain ecosystems by enhancing rates of decomposition of fixed carbon, as is discussed below as a potential effect of climatic warming at high latitudes).

Because of the "problems" associated with the measurement and interpretation of ongoing climatic change using empirical data from the real world, sophisticated computer models have been used to predict the potential climatic effects of changes in atmospheric CO_2. The most complex simulations are the so-called "three-dimensional general circulation models" (GCMs), which must be run on high-speed supercomputers. The GCM models attempt

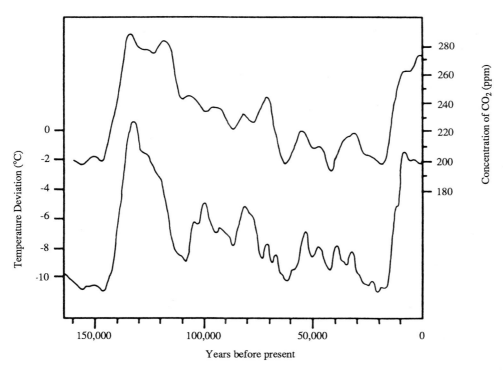

Fig. 2.11 Variations of atmospheric concentrations of carbon dioxide and of surface-temperature deviations, based on data from a 160,000-year glacial record from Vostoc, Antarctica. Modified from Barnola *et al.* (1987) and Lorius *et al.* (1988).

to simulate many of the complex, mass-transport fluxes and other processes that are involved in atmospheric circulation and the interaction of these with other variables that contribute to climate (NRC, 1982, 1983; Blasing, 1985; MacCracken and Luther, 1985; Bolin *et al.*, 1986). In order to perform simulation "experiments" with a GCM model, particular components are parameterized to take into account the likely physical effects of particular scenarios of increased CO_2 concentration in the atmosphere.

Many simulation experiments have been run, using a variety of GCM models, and the results have varied according to the specifics of the experiment. However, an emerging, central tendency of experiments involving a widely accepted (but not necessarily correct) CO_2 scenario (i.e., an eventual doubling of CO_2 from the present concentration of about 355 ppm) is for an increase in average global-surface temperature of about 1.3–3.9°C. The climatic warming is predicted to be relatively large in polar regions, where the temperature increase could be about two to three times larger than in the tropics (NRC, 1983; Blasing, 1985; Wigley and Raper, 1992).

One particular, frequently cited model was designed and used to run various scenario "experiments" by the International Panel on Climate Change (IPCC) (Wigley and Raper, 1992). This model made various assumptions concerning population and economic growth, resource availability, and abatement and mitigation options, and used these to develop scenarios for emissions of: (1) carbon dioxide from fossil-fuel combustion and from land-use changes, (2) other radiatively active gases (methane, nitrous oxide, and halocarbons), and (3) sulfate aerosols, which can cool the atmosphere by contributing to albedo and by affecting cloud formation and longevity. For a simple doubling of the concentration of atmospheric carbon dioxide, the best-guess IPCC estimate was for a 2.5°C increase in average surface temperature (range 1.5–4.5°C). The best guesses for the most advanced IPCC scenarios were similar and involved a 1.5–3.0°C increase in average surface temperature by the year 2100, compared with 1990 (Wigley and Raper, 1992).

Possible Ecological Consequences of Anthropogenic Increases in Atmospheric CO_2

Effects of Climatic Change

Many of the direct effects of climate changes caused by an increased concentration of CO_2 would likely be restricted to plants. Animals would also, of course, be affected secondarily by any large changes in their habitat.

The anticipated increases in air temperature might be of little direct consequence to plants. However, if the temperature changes were to cause large changes in the amounts, spatial distribution, or seasonality of precipitation, then there would be a large ecological effect. There is much uncertainty about the specifics of the potential changes in global rainfall, and therefore the effects on soil moisture and vegetation are uncertain. However, it can reasonably be expected that any large changes in precipitation would cause a fundamental restructuring of the vegetation mosaic on the landscape (Blasing, 1985; Solomon and West, 1985; Bolin *et al.*, 1986; Harrington, 1987; Melillo *et al.*, 1990; Wetherald, 1991).

Paleoecological studies of vegetation changes during warming climates of the Holocene, which followed the most recent, Pleistocene glaciation, indicate that plant species appear to have responded in individualistic ways, because of their differing tolerances of the climatic changes and different abilities to migrate into newly suitable habitats (Davis, 1983; Clark, 1991; Pielou, 1991). Consequently, the species composition and organization of plant communities were different from what is seen today. This situation remains dynamic, of course, because many species have not completed their post glacial reoccupation of suitable habitats.

In any area where the climate becomes drier, it could be expected that there would be a decrease in the extent of closed-canopy forest, and a concomitant expansion of the area of savannah or prairie. This sort of landscape change is believed to have occurred in the neotropics during the Pleistocene glaciations (Haffer, 1969, 1982; Prance, 1982; see also Chapter 10). At that time, the relatively dry climate may have caused a contraction of the presently continuous tropical rainforest into a number of relatively small and isolated refugia. It is believed that these forest remnants were interbedded within a landscape dominated by savannah and grassland. Obviously, such a great restructuring of the vegetational character of the tropical landscape would have a tremendous effect on the habitats of the multitude of rare and endangered species that occur in moist tropical forests (Peters and Darling, 1985; Peters, 1991; see also Chapter 10).

There would also be profound changes in the ability of the landscape to support certain crop plants. This is especially true of the great expanses of land cultivated in regions that are marginal from a rainfall perspective and that are vulnerable to drought and to longer-term degradation by desertification (Sheridan, 1981; Mabbutt, 1984; Dekker *et al.*, 1985; WRI, 1986; Table 2.14). For example, important agricultural crops, particularly wheat (*Triticum aestivum*), are grown in areas of the interior of western North America that were formerly shortgrass prairie or semidesert. An estimated 40% of this 400×10^6-ha semiarid region has already been desertified as a result of anthropogenic activ-

Table 2.14 Extent of desertification in various geographic areas[a]

Region	Total area of productive dry land (10^6 ha)	Percentage desertified
Sudano-Sahelian Africa	473	88
Southern Africa	304	80
Mediterranean Africa	101	83
Western Asia	142	82
Southern Asia	359	70
U.S.S.R. in Asia	298	55
China and Mongolia	315	69
Australia	491	23
Mediterranean Europe	76	39
South America and Mexico	293	71
North America	405	40

[a]Modified from World Resources Institute (WRI) (1986).

ities (Table 2.14), and crop-threatening droughts occur sporadically over this entire area. Within limits, this critical climatic limitation can be overcome by irrigation of the land, but there is a shortage of water for this purpose, and irrigation can cause secondary environmental problems such as salinization (Sheridan, 1981).

Wildfire regimes would also be affected by changes in the amounts and distribution of precipitation. For example, based on the predictions of several climate models and CO_2 scenarios, it has been suggested that the area of forest burned annually in Canada, typically about 1–2 million hectares per year, might increase by a factor of about one-half (Flannigan and Van Wagner, 1991).

Some marine biota might be adversely affected by increases in seawater temperature. Corals may be especially vulnerable to prolonged increases in water temperature, because under various conditions of environmental stress they can lose their pigmented, symbiotic algae (zooxanthellae), sometimes leading to death of the coral. This syndrome can be induced by abnormally high or low water temperatures, by changes in salinity, and by excessive insolation or shading (Smith and Buddemeier, 1992). Widespread coral bleachings were observed in response to warming associated with an El Nino event in 1982–1983 (Smith and Buddemeier, 1992).

Another likely effect of climatic warming would be a change in sea level, caused by a combination of (1) a volumetric expansion of warmed seawater and (2) possible melting of polar glaciers. The IPCC climate-change models referred to above predicted that sea level would be 27–53 cm higher in 2100 than in 1990 (Wigley and Raper, 1992). A change of sea level of this magnitude might cause important problems for low-lying coastal cities and agricultural areas. However, the intensity of the ecological effects at the sea–land interface would be determined substantially by the rate of change.

As noted previously, it has been predicted that the intensity of climatic warming will be greatest at high latitudes. Some ecologists have suggested that the warming of northern regions could result in a self-accelerating feedback of climate change, occurring through the conversion of large areas of the subarctic forest and low-arctic tundra from net CO_2 sinks to CO_2 sources (Billings *et al.*, 1982; Peterson *et al.*, 1984). The mechanism of this effect would involve a CO_2-forced climate warming, resulting in an increased depth of annual thawing of surface soils (i.e., the active layer), exposing large quantities of previously frozen, organic-rich substrates to microbial oxidation, thereby substantially increasing the flux of CO_2 and possibly CH_4 to the atmosphere.

Direct Effects of CO_2

In addition to the indirect, climatic effects discussed above, it is likely that increased concentrations of atmospheric CO_2 would have direct effects on certain plants (Alcock and Allen, 1985; Strain and Cure, 1985). The most important of these effects would probably be a stimulation of photosynthesis in some species, caused by CO_2 fertilization.

Plants with a C3 photosynthetic system would benefit especially, because of an enhanced activity of the carboxylating enzyme *rubisco* (ribulose 1,5-biphosphate carboxylase/oxygenase) (Mooney *et al.*, 1991; Grodzinski, 1992). *Rubisco* is probably the world's most abundant enzyme, as it comprises about 50% of the soluble chloroplast protein of many plants. This large concentration is required because the low affinity of *rubisco* for CO_2, a primary substrate in photosynthesis, means that it is a relatively inefficient enzyme (Grodzinski, 1992). The activity of *rubisco* is, however, significantly facilitated by an increased CO_2-diffusion gradient between the chloroplasts and the substomatal cavity. This diffusion gradient would be increased by larger concentrations of CO_2 in air.

The CO_2 fertilization of C3 plants is pronounced under the optimized condition of laboratory studies, in which there is an adequate supply of other important nutrients that potentially limit plant productivity (e.g., nitrogen, phosphorus, potassium) and where there is a minimal climatic limitation, particularly with respect to the availability of water. However, under relatively limiting conditions in the field, the CO_2-fertilization response is usually considerably smaller or absent.

Experimental studies in the laboratory and to a lesser extent in the field have shown a marked CO_2-fertilization effect on the productivity of many agricultural crop species and of some native plants as well (Alcock and Allen, 1985; Bazzaz *et al.*, 1985; Cure, 1985; Oechel and Strain, 1985; Mooney *et al.*, 1991). In fact, it is a frequent operational practice to enrich the atmosphere of commercial greenhouses with CO_2 to concentrations of 600–2000 ppm to increase the productivity of such C3-crops as cucumbers (*Cucumis sativus*), tomatoes (*Lycopersicum esculentum*), and some ornamentals (Enoch and Kimball, 1986; Porter and Grodzinski, 1985; Grodzinski, 1992).

It should be noted, however, that often there is a longer-term acclimation of the CO_2-fertilization response of many C3 plants, probably due to reduced activity of *rubisco* (Bazzaz, 1990; Mooney *et al.*, 1991; Grodzinski, 1992).

Plants with a C4 photosynthetic system do not benefit from CO_2 fertilization, because their initial CO_2-fixing enzyme, phosphoenolpyruvate carboxylase, is efficient even at small concentrations of CO_2 (Bazzaz, 1990; Mooney *et al.*, 1991; Grodzinski, 1992). For example, Curtis *et al.* (1989) exposed stands of temperate salt marsh to average CO_2 concentrations of 350 ppm or 686 ppm. The rush *Scirpus olneyi*, a C3 species, responded by increases in productivity and shoot density and by delayed senescence, while the grass *Spartina patens*, a C4 species, was unaffected by the CO_2 concentration.

Another important influence of increased CO_2 on plants is that of decreased transpiration (Alcock and Allen, 1985; Strain and Cure, 1985; Mooney *et al.*, 1991). This effect occurs because stomatal aperture is partly under the control of the atmospheric (substomatal) concentration of CO_2, so that at a relatively large CO_2 concentration, the stomata tend to close somewhat or entirely. A decreased loss of water from vegetation, coupled with increased productivity due to CO_2 fertilization, would lead to an increase in the so-called "water-use efficiency" of the ecosystem, that is, the amount of water evapotranspired per unit of biomass produced. In general, this effect would be looked upon favorably in agriculture, and it might also be beneficial in many natural ecosystems.

The CO_2 fertilization and decreased-transpiration effects are illustrated in Table 2.15 for soybean (*Glycine max*) grown under laboratory conditions. The data show that the experimental enrichment of the atmosphere with CO_2 stimulated productivity, decreased transpiration, and increased water-use efficiency. However, this effect was somewhat smaller if the plants had previously been acclimated to 800 ppm of CO_2 and if they grew in a relatively competitive canopy with a large leaf-area index.

It is, of course, impossible to accurately predict the ecological consequences of the well-documented increases of CO_2 and other radiatively active gases in Earth's atmosphere. However, there is a general consensus among scientists that there would be a major readjustment of ecosystem boundaries and community composition in response to any substantial changes in rainfall and other climatic effects, and there could be an increase in the amount of desertification of presently semiarid regions. These ecological changes would probably be of much greater importance than any beneficial effects on plant physiology of CO_2 fertilization and decreased transpiration.

Reducing Atmospheric Carbon Dioxide

Because of the potential climatic and ecological consequences of a potential enhancement of the greenhouse effect, many scientists and citizens are advocating reductions in the anthropogenic emissions of CO_2 and other radiatively active gases. A substantial abatement of emissions will be an important component of the strategy that society eventually develops for coping with changes in the greenhouse effect.

Any such strategy of reduced emissions will, however, require substantial adjustments by societies and economies, because there will undoubtedly be a requirement for: (1) large reductions in energy use and (2) the use of different, possibly new technologies for the generation of energy. These technologies would, of course, have to emit much smaller quantities of radiatively active gases to the atmosphere.

The development and implementation of such a mitigative strategy will be difficult for society, especially in industrialized countries, because of the

Table 2.15 Effect of CO_2 acclimation, leaf area index, and experimental CO_2 enrichment on CO_2 exchange, transpiration, and water-use efficiency of soybeans[a]

CO_2 acclimation treatment[b] (ppm)	Experimental CO_2 exposure (ppm)	Total daily CO_2 exchange (mol CO_2/m^2)	Total daily transpiration (mol H_2O/m^2)	Water-use efficiency (10^{-3} mol CO_2/mol H_2O)
800a	800	1.64	346	4.73
	330	1.07	494	2.17
800b	800	1.79	380	4.71
	330	0.95	495	1.92
330a	800	1.84	294	6.26
	330	0.95	364	2.61
330b	800	1.72	282	6.10
	330	0.79	339	2.62

[a]Modified from Jones *et al.* (1985).
[b]Note that 800a,b had an average leaf area index (LAI) of 6.0, while 330a,b had an LAI of 3.3.

great changes that will be required in economic systems, resource use, industrial capitalization, and expectations of living standards. Societal changes of this magnitude and quality are revolutionary and will be a tremendous challenge to design. The implementation of such changes will require enlightened and forceful leadership.

A complementary, ecological strategy to emissions mitigation would be to actively remove some CO_2 from the atmosphere by increasing the net fixation of this gas by growing plants, especially through afforestation (Marland, 1988; WRI, 1988; Sedjo, 1989a,b; Houghton, 1990, 1991; Vitousek, 1991; Freedman et al., 1992). (There are also, of course, smaller CO_2 emissions to the atmosphere if natural forests are not converted into low-carbon agricultural systems.)

Consider a case of a medium-sized, 200-MW, coal-fired generating station in a region of temperate forest in New Brunswick, Canada (Freedman et al., 1992). The power plant would consume about 14×10^{15} J of energy per year which, because coal would be burned, would result in a CO_2-C emission of 0.34×10^6 tonnes/year. To offset these emissions through the storage of carbon in the biomass of aggrading forest would require a carbon reserve with an area of: (1) about 5×10^5 ha of typical, unmanaged forest or (2) as little as 0.7×10^5 ha of intensively managed forest established on moderately fertile, abandoned farmland.

Of course, to truly offset the emissions of CO_2 from a power plant, the forest must be kept as a reserve. It should not be harvested, since such a disturbance would result in the oxidation of much of the organic C of the forest (Fig. 2.12). Some factors that influence the storage of organic C and atmospheric fluxes of CO_2 after disturbance of a forest are discussed below.

1. After disturbance the site can be allowed (and encouraged, through silviculture) to regenerate back to forest. The system would eventually reattain a standing crop similar to that present before harvesting, assuming the secondary succession proceeds for a long enough time. For temperate forests of eastern North America, the net accumulation of organic C in the biomass of plants, litter, and soil can proceed for at least 300 years of uninterrupted succession, while old-growth conifer forests of coastal western North America can accumulate organic C for perhaps 1000 years. Large-dimension, dead, woody debris is important in the longer-term net accumulation of carbon in temperate old-growth forests (Lugo and Brown, 1986; Maser et al., 1988; Harmon et al., 1990).

2. During any combustion, there is a rapid oxidation of much of the organic C on the site, particularly of above-ground biomass. This factor is relevant to (a) natural wildfire, (b) prescribed burns for silvicultural purposes, (c) burns used to prepare the site for conversion to agriculture, and (d) the combustion of forest biomass to produce heat for industrial steam or residential space-heating.

3. After a forest is harvested, the wood or paper products that are manufactured and the debris that is left on site are variously oxidized by decomposition processes, ultimately resulting in an emission of CO_2 to the atmosphere. For tissues such as foliage, this results in a relatively rapid emission of CO_2, with most of the organic C oxidized in no longer than several years. For paper products, the oxidation process can require a few months, or many decades if the paper is discarded into an anaerobic landfill. For products such as lumber, it may take many decades or centuries before the organic C of the harvested wood is returned to the atmosphere, but such oxidation eventually occurs. The typical life span of a wooden building in North America has been estimated as 80–100 years (Harmon et al., 1990; Row and Phelps, 1990). Ultimately, about two-thirds of the original organic C is eventually released to the atmosphere as CO_2 when boreal or temperate forests are harvested by humans (Woodwell et al., 1978; Harmon et al., 1990).

4. After a forest is harvested, the site may be converted to some agricultural land use. Because the agricultural ecosystem has a much smaller standing crop of organic C than the previous forest, the net effect is a large flux of CO_2-C to the

(a)   **(b)**

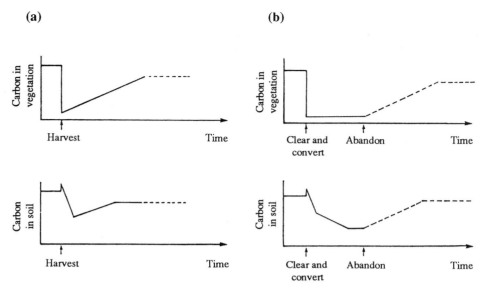

Fig. 2.12 Qualitative models of changes in carbon density (tonnes/ha) after forest harvest. (a) The site is allowed to regenerate back to forest, so there is only a moderate decrease in soil organic matter, and carbon stored in vegetation reaccumulates rapidly during secondary succession. (b) The harvested site is converted to an agricultural use for some time, resulting in a very small carbon pool in the site vegetation and a relatively large decrease in organic matter stored in the soil. After abandonment of the agricultural land, followed by an old field succession to forest, there is a rapid reaccumulation of organic matter stored in vegetation and soil. Modified from Houghton *et al.* (1983) and Freedman *et al.* (1992).

atmosphere. The difference in organic C is especially great for plant biomass and less so for organic matter in the soil. In general, the carbon density (i.e., carbon per square meter) of the plant biomass of agricultural land is considerably less than 10% that of forest, while the carbon density of agricultural soil is typically 60–80% that of forest soil (Schlesinger, 1977; Freedman *et al.*, 1992). Because deforested sites are starting from a relatively small base of fixed C, reforested agricultural lands have a relatively great potential utility as sites for the compensatory fixation and storage of CO_2-C emitted by fossil-fueled generating stations. Relatively speaking, already existing, mature, natural or managed forests can store less "additional" carbon per unit ground area, than can reforested agricultural land (Houghton, 1990; Birdsey, 1990; Vitousek, 1991; Freedman *et al.*, 1992).

Considering the above, it is clear that the harvesting of forested carbon reserves would mostly result in a net deferral of CO_2 emissions from the power plant and not a longer-term offset of the emissions. Similarly, the reserve forest would have to be protected against catastrophic disturbance by wildfire, insect defoliation, etc., since those destructive events also result in large transformations of organic C to atmospheric CO_2, thereby detracting from the C-storage function.

It is also clear that the establishment of forest-carbon reserves to offset industrial emissions would require large areas of land and would preclude many types of economically productive land use and ecosystem management. Such a strategy would therefore require a substantial commitment by society, as would any decision to substantially decrease the emissions of greenhouse gases. There are no easy solutions to problems of this magnitude.

3

TOXIC
ELEMENTS

3.1
INTRODUCTION

All chemicals are potentially toxic if organisms are exposed to a sufficiently large dose. Among the naturally occurring elements, the ones that have most frequently been associated with toxicity from environmental exposures include the heavy metals silver (Ag), cadmium (Cd), chromium (Cr), cobalt (Co), copper (Cu), iron (Fe), mercury (Hg), molybdenum (Mo), nickel (Ni), lead (Pb), tin (Sn), and zinc (Zn), as well as lighter elements such as aluminum (Al), arsenic (As), and selenium (Se).

All of the naturally occurring elements are ubiquitous in the environment in at least trace concentrations. This means that there is a universal contamination of soil, water, air, and biota with all of the natural elements. This contamination will always be detectable, if the analytical–chemical methodology has detection limits that are sufficiently small.

Some of these elements are required by plants and/or animals as essential micronutrients, for example, copper, iron, molybdenum, zinc, and possibly aluminum, nickel, and selenium. Under certain environmental conditions, however, these same elements may bioaccumulate to toxic concentrations and cause ecological damages.

Some instances of elemental "pollution" are natural in origin. Such occurrences often involve surface exposures of minerals containing large concentrations of particular toxic elements, resulting in the pollution of soil, biota, and water. In some cases, these natural situations can have an intensity of pollution by toxic elements that exceeds that caused by anthropogenic emissions.

However, natural pollution is usually relatively local in extent. It was not until the modern era of industrialization that there was a widespread environmental contamination by toxic elements (and by other types of pollutants). This modern phenomenon can be illustrated by the increasing lead contamination of forests in large areas of the northeastern United States and elsewhere. As discussed later in this chapter, the most likely cause of the regional lead contamination has been emissions from automobiles that use leaded gasoline.

In some cases, anthropogenic emissions of toxic elements from large point sources can cause severe but relatively localized environmental pollution. For example, the contamination by metals in the vicinity of the large smelters at Sudbury has contributed to the damage caused to terrestrial and aquatic ecosystems. The effects of toxic metals near Sudbury are described in a case study later in this chapter.

There have also been instances where human health has been affected by toxic elements. Historians have suggested that the decline of the Roman empire may have been partly caused by a decrease in the mental abilities of the population, caused by chronic lead intoxication. The Romans used water piping manufactured of lead, and they stored wine in pottery that was glazed with lead, some of which was leached by the acidic beverage and subsequently ingested. Another, nineteenth century example involves people working in the felt-hat industry in Britain. These workers frequently suffered neurological damage caused by their exposure to mercury compounds used as a finish to top hats— hence Lewis Carroll's phrase, "mad as a hatter."

More recent examples of mercury poisoning are the deaths in the 1960s of hundreds of people in Iraq, Iran, India, Pakistan, and elsewhere, caused by eating seed grains that had been treated with mercury-based fungicides (discussed further under "Inorganic Pesticides," this chapter). The poisonous grain was intended for planting, not eating. The toxicity of the treated seed had been clearly labeled on the bags, but many of the victims were illiterate, or they did not understand the implications of the message. Mercury poisoning has also caused deaths of humans in other situations. At Minamata, Japan, hundreds died and many more were acutely poisoned as a result of eating mercury-contaminated fish. In that situation, an acetaldehyde factory had discharged elemental mercury, a relatively nontoxic form of mercury, into Minamata Bay. However, microbes in anaerobic sediment of the bay converted the elemental mercury into methylmercury. This toxic and bioavailable compound of mercury entered the aquatic food web and caused a wide-scale poisoning of fish-eating birds, domestic cats, and humans. Other toxic elements that have poisoned modern humans after being assimilated from food or from some other environmental exposure (most commonly occupational) include arsenic, cadmium, lead, nickel, and vanadium.

In this chapter several cases of natural and anthropogenic contaminations of the environment with toxic elements and the resulting ecological consequences are described.

3.2
BACKGROUND CONCENTRATIONS IN THE ENVIRONMENT

As used here, "background concentration" refers to concentrations of elements that occur in the environment in situations that have not been significantly influenced by anthropogenic emissions or by unusual natural exposures. Background concentrations of elements in selected components of the environment are listed in Table 3.1. In all cases, concentrations in soil and rock are much larger than in water and are generally larger than in organisms. However, it is important to note that since the chemical form of elements dissolved in water is usually relatively available for uptake by biota, even a seemingly small aqueous concentration may exert a powerful toxic effect. In contrast, the relatively large concentrations in soil and rock are largely insoluble. Hence, in solid substrates such as soil and rock, the availability of toxic elements to plants and animals is much less than might be suggested by the "total" concentration, and there may not be a toxic effect. The topic of availability is further discussed below.

Most elements are consistently present in small concentrations in the environment, for example, cadmium, chromium, cobalt, copper, lead, mercury, molybdenum, nickel, selenium, silver, tin, and uranium (Table 3.1). However, all of these elements are potentially toxic, and they can affect many species of plants and animals at water-soluble concentrations smaller than about 1 ppm.

In contrast, other elements can occur in large concentrations in some compartments of the biosphere. Most notable in this respect are aluminum and iron, both of which are important constituents of rock and soil. Aluminum averages 8% of the mass of Earth's crust, in which it is the third-most abundant element after oxygen (47%) and silicon (28%), while iron averages 3–4% (Hausenbuiller, 1978; Bowen, 1979). As noted above, however, almost all of the Al, Fe, and other potentially toxic elements in soil and minerals is present in insoluble forms that are not readily assimilated by biota and are considered to be generally "unavailable."

Table 3.1 Typical background concentrations (ppm = mg/kg, d.w.) of elements in selected environmental compartments[a]

	Rocks					Soil
	Granite	Basalt	Shale	Limestone	Sandstone	
Ag	0.04	0.1	0.07	0.12	0.25	0.05
Al	77,000	87,600	88,000	9000	43,000	71,000
As	1.5	1.5	13	1	1	6
Cd	0.09	0.13	0.22	0.028	0.05	0.35
Co	1	35	19	0.1	0.3	8
Cr	4	90	90	11	35	70
Cu	13	90	39	5.5	30	30
F	1400	510	800	220	180	200
Fe	27,000	56,000	48,000	17,000	29,000	40,000
Hg	0.08	0.012	0.012	0.18	0.29	0.06
Mn	400	1500	850	620	460	1000
Mo	2	1	2.6	0.16	0.2	1.2
Ni	0.5	150	68	7	9	50
Pb	24	3	23	5.7	10	35
Se	0.05	0.05	0.5	0.03	0.01	0.4
Sn	3.5	1	6	0.5	0.5	4
U	4.4	0.43	3.7	2.2	0.45	2
V	72	250	130	45	20	90
Zn	52	100	120	20	30	90

Aluminum, for example, is present in the soil in many chemical forms, most notably: (1) tetra- and octahedral crystals of primary aluminosilicate minerals; (2) amorphous and crystalline clays and sesquioxides; (3) aluminum phosphates; and (4) ions bound to exchange sites on organic matter and clay surfaces or occurring freely in the soil solution.

In general, the amount of aluminum that is freely dissolved in the soil water is strongly influenced by acidity, with much larger soluble concentrations being present in highly acidic solutions (aluminum solubility also increases markedly in alkaline solutions). This characteristic of enhanced solubility in acidic solutions holds generally for most of the toxic elements, especially the metals. Moreover, the particular ionic species of aluminum are strongly dependent on solution pH (Fig. 3.1). Al^{3+} is the dominant ionic species in acidic solutions with pH less than about 5.0, while $AlOH^{2+}$ and $Al(OH)_2^+$ are relatively important in less acidic solutions with pH 4.5–5.5, $Al(OH)_3$ dominates at pH 5.2–9, and $Al(OH)_4^-$ in strongly alkaline conditions with pH greater than about 8.5 (Hausenbuiller, 1978; Spry and Wiener, 1991). It is principally the ionic forms of aluminum that are toxic to biota, largely because the other forms are essentially insoluble in water.

Therefore, consideration of the "total" quantity of aluminum in soil or in some other environmental compartment (usually determined after solubilization of the sample using a vigorous, heat-assisted, strong-acid digest) gives little direct information about biological availability or about potential toxicity [availability is usually measured in some sort of aqueous extract; see Black *et al.* (1965), Allen *et al.* (1974), or Allen (1989) for a discussion of this topic]. In general, the available quantity of a toxic element in soil is a small part of the total quantity—usually <10% and frequently <1%.

3.3

TOXICITY

In toxicology, any toxic effect is related to two factors: (1) the exposure or dose and (2) the susceptibility of the organism to the specific poison that is being considered. The dose received by a target

Table 3.1 (*Continued*)

Seawater	Freshwater	Terrestrial plants	Mammals Muscle	Mammals Bone	Marine fish
0.00004	0.0003	0.01–0.8	0.009–0.28	0.01–0.4	0.04–0.1
0.002	0.3	90–530	0.7–28	4–27	20
0.0037	0.0005	0.2–7	0.007–0.09	0.08–1.6	0.2–10
0.0001	0.0001	0.1–2.4	0.1–3.2	1.8	0.1–3
0.00002	0.0002	0.005–1	0.005–1	0.01–0.04	0.006–0.05
0.0003	0.001	0.03–10	<0.002–0.84	0.1–33	0.03–2
0.0003	0.003	5–15	10	1–26	0.7–15
1.3	0.1	0.02–24	0.05	2000–12,000	1400
0.002	0.5	70–700	180	3–380	9–98
0.0003	0.0001	0.005–0.02	0.02–0.7	0.45	0.4
0.0002	0.008	20–700	0.2–2.3	0.2–14	0.3–4.6
0.01	0.0005	0.06–3	0.02–0.07	<0.7	1
0.00058	0.0005	1–5	1.2	<0.7	0.1–4
0.00003	0.003	1–13	0.2–3.3	3.6–30	0.001–15
0.0002	0.0002	0.03	0.4–1.9	1–9	0.2
0.000004	0.000009	0.2–2	0.01–2	1.4	—
0.0032	0.0004	0.005–0.04	0.001–0.003	0.0002–0.7	0.04–0.08
0.0025	0.0005	0.001–0.5	0.002–0.02	0.003–0.03	0.3
0.005	0.015	20–400	240	75–170	9–80

[a]Modified from Bowen (1979).

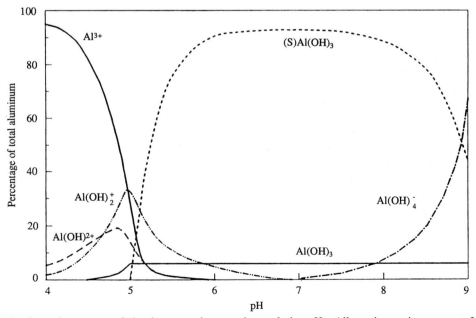

Fig. 3.1 Occurrence of aluminum species at various solution pHs. All species are ions, except for $(S)Al(OH)_3$, which is a precipitate. Modified from Martell and Motekaitis (1989) and Spry and Wiener (1991).

organism is influenced by the available concentration of the poison in the environment and by the period of exposure. Therefore, in certain situations a long-term exposure to a small available concentration of some elements may cause a toxic effect. Often this occurs because of a progressive bioaccumulation of the element, until some toxic dose or threshold is exceeded.

Organisms can vary widely in their tolerance of exposures to toxic elements and to other poisons. In any environmental exposure to a large dose of some potentially toxic chemical, certain species will suffer damage, while other species will be more tolerant and may even benefit from the demise of more sensitive species. There can also be genetically based variance for metal tolerance within species, as is described later for certain plant populations, known as ecotypes, which have extreme tolerances to particular metals.

The mechanism of toxicity of metals is frequently damage to an enzyme system. This occurs when metal ions bind to the enzyme and cause a change in its three-dimensional configuration, with a resulting change or loss of its specific catalytic function. Another common mechanism of toxicity is damage to DNA via metal binding, resulting in genetic damages by a disruption of transcription, by an inability to produce specific proteins (especially enzymes), or by some other toxic effect. Symptoms of acute toxicity to biota can include abnormal patterns or amounts of growth and development, impaired reproduction, disease, and death. Chronic toxicity can also occur, but the symptoms are more difficult to detect and may only consist of small decreases in productivity.

Particular elements differ greatly in their ability to cause toxic effects. Toxicity is also influenced by a number of other factors, such as those listed below:

1. *The chemical form of the toxic element.* This is important because of its influence on aqueous solubility and hence on biological availability and because of the inherently different toxicities of various ionic and molecular species. For example, in the ionic series previously described for aluminum, the most toxic species are Al^{3+} and $AlOH^{2+}$ (Havas, 1986).

2. *The chemical environment.* This factor can reduce or exacerbate toxicity. For example: (a) the co-occurrence of calcium in a large concentration may decrease the toxicity of many metals; (b) soil organic matter and clay can bind and thus partially immobilize ionic metals; (c) acidity frequently increases solubility and therefore enhances the exposure to metals; (d) the co-occurrence of two or more toxic elements may cause a more-than-additive effect, that is, they are synergistic; (e) depending on the element, oxygen concentrations and redox potentials may have large influences on total solubility and speciation of ions; (f) the osmolarity of the soil solution, which is a function of the total concentration of dissolved substances, may be a factor in the solubility of potentially toxic ions (solutions with strong osmolarity are more effective at extracting ions from exchange surfaces in soil); and (g) solubility may increase at higher temperatures.

3. *Differences in susceptibility among individuals, populations, and species.* These differences may be related to: (a) different exposure pathways, which are a reflection of environmental influences on toxicity; (b) varying susceptibility among life-history stages or sexes; or (c) genetically based differences in physiological susceptibility, as is the case of metal-tolerant plant ecotypes, discussed later in this chapter.

These various influences on elemental toxicity are not described in detail in this book. For in-depth treatments see Goodman *et al.* (1973), Foy *et al.* (1978), Bowen (1979), Lepp (1981), or Freedman and Hutchinson (1986).

3.4
NATURALLY OCCURRING POLLUTION AND ITS ECOLOGICAL EFFECTS

Sources and Intensity of Natural Pollution

Natural emissions of toxic elements to the atmosphere can occur by volcanic outputs and by vapor-phase outgassings of relatively volatile elements such as As, Hg, and Se. The entrainment of contaminated soil dust into the atmosphere may also be important. Estimates of the natural rates of emission of various elements are summarized in Table 3.2, together with estimates of their anthropogenic emissions. These two source categories are similar in magnitude (i.e., within a factor of 4) for Cr, Co, Mn, Ni, and V, while anthropogenic sources exceed natural ones by a factor of more than 10 for the other elements.

Surface and near-surface mineralizations containing toxic elements can give rise to localized areas of natural, terrestrial pollution. The occurrence of metal anomalies in soil, vegetation, and surface water and sediment overlying an ore body, sometimes associated with the presence of indicator-plant species, has given rise to an exploration technique known as biogeochemical prospecting (Cannon, 1960; Allan, 1971; Wolfe, 1971; Levinson, 1974; Fortescue, 1980; Brooks, 1983).

These natural elemental pollutions can be comparable in concentration to the worst cases of anthropogenic pollution. For example, Boyle (1971) and Stone and Timmer (1975) reported copper con-

Table 3.2 Global emission of selected elements[a]

Element	Emissions (10^8 g/year)	
	Natural	**Anthropogenic**
Sb	9.8	380
As	28	780
Cd	2.9	55
Cr	580	940
Co	70	44
Cu	190	2600
Pb	59	20,000
Mn	6100	3200
Hg	0.40	110
Mo	11	510
Ni	280	980
Se	4.1	140
Ag	0.6	50
Sn	52	430
V	650	2100
Zn	360	8400

[a]Modified from Galloway et al. (1982b).

centrations as large as 10% in surface peat that was filtering Cu-rich spring water emerging into a marsh in New Brunswick, Canada. Forgeron (1971) described surface soil with up to 3% Pb + Zn at a site on Baffin Island, Canada. Warren et al. (1966) reported mercury concentrations of 1–10 ppm in soil overlying a cinnabar (HgS) deposit in British Columbia, compared with <0.1 ppm in reference soil.

Predictably, the severe pollution of soil at some metalliferous sites usually results in large metal concentrations in plants. This phenomenon is especially pronounced in genetically adapted, hyperaccumulator species, which are often endemic to metalliferous sites. For example, nickel concentrations as large as 10% have been found in *Alyssum bertolanii* and *A. murale* in Russia (Mishra and Kar, 1974, citing Malyuga, 1964). A nickel concentration as large as 25% occurs in the blue-colored latex of *Sebertia acuminata* from the Pacific island of New Caledonia (Jaffre et al., 1976). In a chemical survey of 181 herbarium specimens of 70 *Rhinorea* species, specimens of *R. bengalensis* were found with as much as 1.8% nickel and *R. javanica* with up to 0.22% (Brooks et al., 1977). In comparison, background concentrations of nickel in plant tissues are 1–5 ppm (Table 3.1).

Similarly, copper indicators and copper endemics have been described. The labiate *Becium homblei* was important in the discovery of copper deposits in Zambia and Zimbabwe, where its presence is confined to soil with more than 1000 ppm Cu (Cannon, 1960). *Becium homblei* is a cuprophile, tolerant of >7% copper in soil (Reilly, 1967; Reilly and Reilly, 1973). Copper mosses were originally described from Scandinavia, and later from Alaska, Russia, and elsewhere (Persson, 1948; Shacklette, 1965). These bryophytes are specific to mineral substrates that have large concentrations of copper, and they have been used by prospectors as botanical indicators of surface mineralizations of this metal.

In another case, arsenic enrichment of aquatic macrophytes has been described for surface waters in New Zealand that are contaminated by geothermal springs (Reay, 1972). An arsenic concentration as large as 970 ppm was found in *Ceratophyllum demersum*, compared with 1.4 ppm in reference plants.

Serpentine Soil and Vegetation

Soil derived in part from serpentine minerals is a well-known example of natural metal contamination. This soil type is derived from ultramafic minerals, particularly olivine, pyroxenes, hornblende, and their secondary products such as serpentine minerals, fibrous amphiboles, and talc (Aumento, 1970; Douglas, 1970; Proctor and Woodell, 1975; Brooks, 1987). The serpentine minerals are a complex of layered silicates with the general formula $Mg_6Si_4O_{10}(OH)_8$. However, there is much substitution for magnesium by other elements, especially by nickel, cobalt, chromium, and iron. Substantial amounts of nickeliferous pyrite and chromite may also be present in some serpentine-derived soils. In many cases, serpentine-derived soils are toxic and inhibitory to the growth and survival of nonadapted plants.

The toxicity of serpentine-derived soils is largely due to an imbalance of the nutrients calcium and magnesium (the Ca:Mg ratio is small), and the large concentrations of available nickel, chromium, and cobalt. Small concentrations of the nutrients po-

A chaparral community on serpentine-influenced soil in northern California. The codominant shrubs are *Quercus durata* and *Ceanothus jepsonii*, both of which are endemics tolerant of the stresses associated with serpentine minerals, whose distributions are restricted to such sites. On more normal soils without serpentine minerals, the local landscape is typically dominated by a conifer forest of pines (*Pinus ponderosa* and *P. lambertiana*) and fir (*Abies concolor*) (photo courtesy of A. R. Kruckeberg).

tassium, phosphorus, nitrogen, and molybdenum may also be important (Whittaker, 1954; Proctor and Woodell, 1975; Roberts and Proctor, 1992).

Typical serpentine soils contain about 2000 ppm nickel, but as much as 25 thousand ppm (i.e., 2.5%) can be present at some sites (Proctor and Woodell, 1975; Brooks, 1987). Peterson (1977) described soil from a location in Zimbabwe with as much as 12.5% chromium and 0.55% nickel. As might be expected, a fraction of the toxic elements in serpentine soils is available for plant uptake, and a range of nickel concentration of 600–10,000 ppm (0.06–1%) in foliage ash has been observed, compared with 20–70 ppm in reference samples (Proctor and Woodell, 1975). Concentrations as large as 16.4% dry weight (d.w.) nickel have been reported in tissues of the hyperaccumulator *Streptanthus polygaloides*, a cruciferous serpentine endemic from California (Reeves *et al.*, 1981). As much as 11–25% nickel has been found in the blue-colored latex of the hyperaccumulator *Sebertia acuminata* of the Pacific island of New Caledonia (Jaffre *et al.*, 1976). Poor crop growths on moderately serpentinized soil in Scotland, Rhodesia, and elsewhere have been ascribed to nickel and chromium toxicity (Hunter and Vergnano, 1952; Vergnano and Hunter, 1952; Soane and Saunder, 1959).

The natural vegetation that develops on serpentine soils is often strikingly low growing and stunted (Brooks, 1987). On serpentine sites in northern California the typical vegetation is an open chaparral, characterized by endemic sclerophyllous shrubs (however, the range of vegetation types on Californian serpentine sites is from sparse barrens to open forest) (Kruckeberg, 1984). Serpentine tablelands in eastern Quebec and Newfoundland can be tundra-like islands in an otherwise-forested landscape (Scoggan, 1950).

Sites with soil having a strong serpentine influence may support a unique and biogeographically interesting plant community (Proctor and Woodell, 1975; Kruckeberg, 1984; Brooks, 1987). These sites often have a large percentage of endemics and ecotypes in their flora. On more productive and less toxic sites, the serpentine-adapted endemics and ecotypes are quickly eliminated through competition with species or populations that are better adapted to more moderate habitat conditions.

In eastern Canada, serpentine barrens are refugia for arctic species, but the vegetation also has disjunct temperate taxa and a few serpentine endemics (Scoggan, 1950). A notable example of the latter is the rare, serpentine-indicating lip-fern, *Cheilanthes siliquosa*.

The vegetation of the eastern Canadian serpentine barrens is relatively young, since these sites were only deglaciated about 10 thousand years ago. There is a much larger frequency of endemism in the much older serpentine vegetation of northern California. Kruckeberg (1984) estimated that 215 taxa of vascular plants are endemic to Californian serpentine formations, representing 14% of the total serpentine flora of 1544 taxa. Of the 1329 nonendemics, Kruckeberg classified 17% as local or regional indicators of serpentine sites, while the other 83% are indifferent.

The most diverse genus of serpentine endemics in California is the crucifer *Streptanthus*, with 16 endemic taxa. Three of these have a particularly narrow distribution: *Streptanthus niger*, *S. batrachopus*, and *S. brachiatus* are all restricted to a few sites. *Streptanthus hesperidis* and *S. polygaloides* have a wider distribution, but they are nevertheless restricted to serpentine sites. *Streptanthus glandulosus* is particularly widespread and occurs on virtually all Californian serpentine sites, as does the endemic oak *Quercus durata*. In addition, the conifers *Pinus jeffreyi* and *Calocedrus decurrens* have altitudinally disjunct populations in some Californian serpentine sites, probably due to a relatively low intensity of competition in this type of chemically stressed habitat. In the Sierra Nevada mountains the usual range of *P. jeffreyi* is about 1830–2745 m, whereas on serpentine it can occur as low as 915 m (Kruckeberg, 1984).

Not surprisingly, nickel-tolerant ecotypes of widespread species have been identified from serpentine sites in North America and Europe (Kruckeberg, 1954; Proctor, 1971a,b; Proctor and Woodell, 1975). Nickel tolerance is also, of course, a key feature of species that are endemic to serpentine sites.

A very sparsely vegetated serpentine barrens in the inner south Coast Range of central California. The widely spaced trees are digger pine (*Pinus sabiniana*). In the absence of soil influenced by serpentine minerals, the vegetation at this location would be an oak-dominated woodland (photo courtesy of A. R. Kruckeberg).

Seleniferous Soil and Vegetation

In many semiarid areas, soils with large concentrations of selenium support indicator plants that can hyperaccumulate this element. These plants are poisonous to livestock that graze upon them, causing a toxic syndrome known as alkali disease or "blind staggers."

In North America, the most important selenium-accumulating plants are legumes in the genus *Astragalus*. Of the 500 species of *Astragalus* in North America, 25 are known to be selenium accumulators. These can contain thousands of parts per million of Se in foliage, to a maximum concentration of about 15 thousand ppm (Trelease and Trelease,

1938, 1939; Rosenfeld and Beath, 1964; Davis, 1972; Stadtman, 1974). Often, both accumulator and nonaccumulator species of *Astragalus* grow together on the same seleniferous site. In a location in Nebraska with 5 ppm of selenium in soil, *A. bisulcatus* had 5560 ppm Se in its foliage, while *A. missouriensis* had only 25 ppm (Shrift, 1969).

The selenium-accumulating species of *Astragalus* also co-occur with unrelated indicator species, including *Mentzelia decapetala*, *Oonopsis condensata*, *Stanleya* spp., and *Xylorhiza* spp., which can also contain selenium in concentrations of thousands of parts per million (Trelease and Martin, 1936). Selenium accumulators also occur outside of North America: the endemic Australian le-

The perennial herb *Castilleja neglecta*, an example of a serpentine endemic. This species has a highly restricted distribution in the Bay Area of California, and it only grows on sites where the soil is strongly influenced by serpentine minerals (photo courtesy of A. R. Kruckeberg).

gume *Neptunia amplexicaulis* can contain as much as 5000 ppm Se and is known to poison cattle (Peterson and Butler, 1962, 1967). In addition, various cultivated varieties of cabbage (*Brassica rapa*) can accumulate selenium to concentrations greater than 1000 ppm (Trelease and Martin, 1936).

Plants that hyperaccumulate a toxic element from the environment frequently turn out to have a physiological requirement for that element. This pattern has been demonstrated for *Astragalus racemosus* and *A. pattersonii*, which grow better in the presence of selenium in experimental growth media (Trelease and Trelease, 1938; Frost and Lish, 1975). Studies of some of the selenium-accumulating species of *Astragalus* have shown that they sequester much of their tissue Se in specialized amino-acid analogues, especially se-

lenomethionine (Trelease *et al.*, 1960; Virupaksha and Shrift, 1965). In addition, the accumulator species of *Astragalus* are capable of emitting selenium-containing biochemicals to the atmosphere, including dimethyl selenide and dimethyl diselenide (Evans *et al.*, 1968). These specialized volatiles are responsible for the distinctive and unpleasant odor of these plants.

Mercury in the Aquatic Environment

In remote oceanic situations, mercury can accumulate in marine fish and in piscivorous birds and mammals. In offshore waters of North America, fish species that are known to accumulate large concentrations of mercury include Atlantic swordfish

(*Xiphias gladius*), Pacific blue marlin (*Makaira ampla*), bluefin tuna (*Thunnus thynnus*), yellowfin tuna (*Thunnus albacares*), skipjack tuna (*Euthynnus pelamis*), Atlantic and Pacific halibut (*Hippoglossus hippoglossus* and *H. stenolepis*), and Atlantic and Pacific dogfish (*Squalus* spp.) and other sharks. All of these species accumulate mercury from trace concentrations in seawater [i.e., <0.1 ppb; Armstrong (1979)] to concentrations in edible flesh (Table 3.3) that exceed the maximum acceptable concentration in fish for human consumption of 0.5 ppm fresh weight (f.w.) (Rivers *et al.*, 1972; Armstrong, 1979).

The contamination of oceanic fish by mercury is apparently natural and is not a modern phenomenon. For example, no difference was found between the mercury contaminations of contemporary tuna and seven museum specimens collected between 1878 and 1909 (G.E. Miller *et al.*, 1972), or in the mercury concentrations of feathers of pre-1930 and post-1980 seabirds collected from islands in the northeast Atlantic (Thompson *et al.*, 1992).

Within species of marine fish, there is a strong tendency for larger, older fish to have relatively large concentrations of mercury. For example, in a sample of 224 Atlantic swordfish, the average mercury concentration of animals weighing less than 23 kg was 0.55 ppm, for those between 23 and 45 kg

Table 3.3 Average mercury concentration in the muscle tissue of some commercially important marine fish species[a]

Species	Mercury concentration (ppm f.w.)
Swordfish (>45 kg)	1.08
Bluefin tuna (>14 kg)	0.89
Yellowfin tuna (>32 kg)	0.62
Skipjack tuna (>4 kg)	0.21
Atlantic dogfish	0.41
Pacific dogfish	0.70
Pacific halibut (>45 kg)	0.42
Atlantic halibut (>45 kg)	0.80

[a] After Armstrong (1979).

the average was 0.86 ppm, and for those heavier than 45 kg it was 1.1 ppm (Armstrong, 1979).

In some fish species, only a small proportion of the total mercury is present as the particularly toxic compound, methylmercury. This probably indicates that demethylation reactions are an important adaptation that decreases the toxicity of mercury to those fish, since methylmercury is the form in which most mercury is actually absorbed from the aqueous environment (Rivers *et al.*, 1972; Bryan, 1976). In addition, concentrations of selenium tend to vary in direct proportion to mercury in these fish species. It has been suggested that selenium may somehow function to ameliorate the toxicity of mercury to fish, and also to fish-eating predators (Ganther *et al.*, 1972; Koeman *et al.*, 1973; Ganther and Sundi, 1974).

Large concentrations of mercury have also been found in piscivorous marine mammals at the top of the marine food web (Gaskin *et al.*, 1972; Buhler *et al.*, 1975; Jones *et al.*, 1975; Smith and Armstrong, 1978; Armstrong, 1979). The concentration of mercury in adult harp seals (*Phoca groenlandica*) in the Canadian Atlantic averaged 0.34 ppm in muscle but 5.1 ppm in the liver, while pups had 0.29 and 0.73 ppm, respectively (Armstrong, 1979). Since less than 10% of the total mercury in seal liver was present as highly toxic methylmercury, it is likely that demethylation is a detoxification mechanism (Armstrong, 1979).

Analysis of feathers collected from fish-eating birds off coastal Peru revealed average concentrations of 2.0 ppm in Inca tern (*Larosterna inca*), 1.0 ppm in red-legged cormorant (*Phalacrocorax bougainvillii*), 0.81 ppm in Peruvian booby (*Sula variegata*), and 0.72 ppm in sooty shearwater (*Puffinus griseus*) (Gochfeld, 1980). These concentrations are smaller than the 5–10 ppm that was measured in three species of tern (*Sterna* spp.) on Long Island, New York, where anthropogenic pollution is an important contributor to environmental mercury (Gochfeld, 1980). Large concentrations of mercury have also been measured in feathers of fish-eating seabirds in the north Atlantic, with an average of 1–2 ppm among fulmar (*Fulmarus glacialis*), kittiwake (*Rissa tridactyla*), razorbill (*Alca torda*), and common murre (*Uria aalge*), 5 ppm in puffin (*Fratercula arctica*), and 7 ppm in northern skua (*Catharacta skua*) (Thompson *et al.*, 1991)

In some remote freshwater ecosystems, a similar phenomenon of mercury contamination of fish has been observed, and in relatively remote lakes this may be a natural occurrence. For example, mercury concentrations exceeding 0.5 ppm f.w. are regularly measured in fish collected from remote lakes in many parts of Canada (McKay, 1985). About three-quarters of 1500 lakes monitored in Ontario had at least some fish that exceeded 0.5 ppm f.w. in flesh (Gilmour and Henry, 1991). In a particular, remote lake in northern Manitoba, the average concentration of mercury in muscle was 2 ppm f.w. among a sample of 53 northern pike (*Esox lucius*), and one individual had a concentration of 5 ppm f.w. (McKay, 1985). In general, species of freshwater fish that are top predators have the largest concentrations of mercury, and within species, older and larger individuals tend to be the most contaminated (MacCrimmon *et al.*, 1983; McKay, 1985; Grieb *et al.*, 1990; Spry and Wiener, 1991). Virtually all (95–99%) of the mercury in these fish is methylmercury (Grieb *et al.*, 1990; Spry and Wiener, 1991).

Because of the frequent occurrence of large mercury concentrations in fish in certain areas, some governments regularly monitor this contamination and have developed fish-consumption advisements and/or restrictions for problematic waterbodies. For example, in Sweden about 250 lakes have been "blacklisted" in terms of consumption of their fish, and another 9400 lakes are candidates for that status, while in Ontario about 1200 lakes have fish-consumption restrictions (Spry and Wiener, 1991).

Some fish-eating wildlife may also be affected by mercury in their food (Scheuhammer, 1987, 1991). Breeding success of common loons (*Gavia immer*) in remote lakes in northwestern Ontario appears to be negatively influenced by mercury concentrations of 0.3–0.4 ppm f.w. in their food (Barr, 1986).

The bioaccumulation of mercury in fish of remote lakes is enhanced if the lake is acidic, since this condition favors the production of bioavailable methylmercury in anaerobic sediments (Winfrey and Rudd, 1990; Wiener *et al.*, 1990; Scheuhammer, 1991). In less acidic lakes, the formation of dimethylmercury is favored, but this chemical species is much less bioaccumulative because it tends to volatilize from the waterbody into the atmosphere (Winfrey and Rudd, 1990), as does aqueous elemental mercury (Gilmour and Henry, 1991).

In less remote lakes closer to sources of mercury emissions to the atmosphere, anthropogenic sources are more important than natural inputs of mercury from watershed soils and sediment. For example, Harp Lake is a small (71 ha), oligotrophic waterbody in a region of south-central Ontario that is relatively close to sources of mercury emissions, such as smelters, power plants, and other industrial facilities, and atmospheric deposition accounts for about 57% of mercury inputs to that lake (Mierle, 1990). Mercury bioaccumulation is also enhanced when reservoirs are created. In this case, mercury is mobilized from flooded soils, methylated in anaerobic sediments, and then bioaccumulated by fish and other biota (Bodaly *et al.*, 1984; Strange *et al.*, 1991).

3.5
ANTHROPOGENIC SOURCES OF TOXIC ELEMENTS

In this section, pollution caused by human activities is described. The effects of certain agricultural practices, the metal mining and processing industries, the use of lead shot by hunters, and the emission of lead from automobiles are highlighted.

Agricultural Practices

The contamination of agricultural lands by toxic elements has been caused by the longer-term use of inorganic pesticides, by the use of contaminated sewage sludges as a soil conditioner, and by irrigation systems that cause toxic elements to accumulate in large concentrations in certain wetlands. These are discussed below.

Inorganic Pesticides

The use of inorganic pesticides has been particularly important in fruit orchards, where chemicals such as lead arsenate, calcium arsenate, and copper sulfate were used to control fungal pathogens and arthropods for more than a century (their use has now been largely supplanted by synthetic organic pesticides). In orchards in Ontario, the annual spray rates were as great as 8.7 kg/ha/year of lead, 2.7 of arsenic, 7.5 of zinc, and 3.0 of copper, depending on the crop, the pest, and the pesticide formulation (Frank *et al.*, 1976a,b). Depending on the pesticide-use patterns, all of these elements could be deposited in the same orchards.

Because these elements are bound by organic matter in soil and by other ion-exchange surfaces and under such conditions are rather insoluble in water, they tend to accumulate in treated soil. For example, concentrations as large as 890 ppm lead and 126 ppm arsenic were found in surface soils of apple (*Malus pumila*) orchards in Ontario, compared with background levels of these elements of <25 and <10 ppm, respectively (Frank *et al.*, 1976b). These accumulations in soil were caused by

as many as 70 years of use of lead arsenate as a pesticide, particularly against the codling moth (*Laspeyresia pomonella*), which causes "wormy" apples. The progressive accumulation of lead and arsenic residues in soil is illustrated in Fig. 3.2. In a similar study, an apple orchard near Amherst, Massachusetts, had a surface-soil contamination of up to 1400 ppm lead and 330 ppm arsenic and a strong positive correlation between the concentrations of these elements (Veneman *et al.*, 1983).

Agricultural soils can also be contaminated by mercury. The most important sources have been the use of organic mercurials as a seedcoat dressing to prevent fungal diseases of seedlings just after germination, of mercury sulfate as a root dip for cruciferous crops, and of phenyl mercuric acetate for the treatment of apple scab (Frank *et al.*, 1976a,b). Mercurial compounds have also been used for the control of fungal turfgrass diseases and to reduce infestations of crabgrass (*Digitaria* spp.). Mercury concentrations ranging from 24 to 120 ppm were found in surface soils of golf course putting greens, where intense efforts are made to maintain uniformly weed- and disease-free lawns (A. J. MacLean *et al.*, 1973).

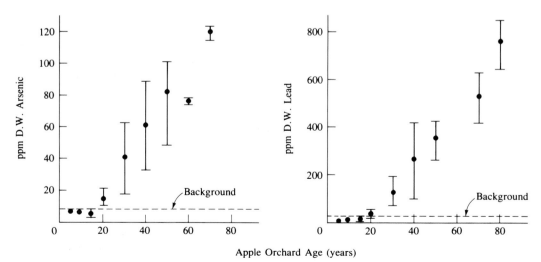

Fig. 3.2   The accumulation of arsenic and lead in the surface soil (0–15 cm) of 31 apple orchards in Ontario, Canada. Lead arsenate was applied at various rates to the orchards. Mean values and ranges are indicated. After Frank *et al.* (1976b).

Mercury compounds have been widely used as a seedcoat dressing. The intent of this practice is to prevent fungal diseases of newly germinated seedlings, especially "damping-off," a fungal infection that initiates at the soil–air interface and that causes the weakened seedling to fall over and die. The planting of mercury-coated seed has had an effect beyond that of soil contamination. There have been observations of mercury accumulation and poisoning of wild birds and mammals that consume the planted seed and of their predators (Tejning, 1967; Fimreite, 1970; Fimreite et al., 1970; Johnels et al., 1979). The use of alkyl mercury compounds such as methylmercury was particularly damaging, since Hg in this form is toxic and readily assimilated by animals from their food. Fimreite et al. (1970) examined the mercury concentrations of tissues of rodents and birds and found a large difference between alkyl mercury-sprayed and unsprayed areas of western Canada (Table 3.4).

The use of alkyl mercury compounds as a seed dressing was prohibited in most industrialized countries in the late-1960s, after the associated ecological problems of these chemicals became recognized, particularly the poisoning of animal wildlife. In Sweden, their use was prohibited in 1966, and much less toxic alkoxyl-alkyl mercury compounds were approved as replacements. This led to almost immediate decreases in the mercury contamination of previously affected wildlife, such as raptorial birds (Figs. 3.3a and 3.3b), and has contributed (along with the banning of DDT, dieldrin, and other chlorinated hydrocarbons; see Chapter 8) to increased populations of some species (Wallin, 1984).

Humans have also been poisoned by the ingestion of mercury-treated seed that was intended for planting. One of the worst cases of a human population experiencing direct toxicity from exposure to a toxic metal occurred in Iraq in the 1960s, when about 6500 people were poisoned (about 500 people died) by the consumption of mercury-treated grain that had been supplied as crop seed by foreign aid. This happened in spite of toxic-hazard warnings on the bags in several languages, including Arabic. In some cases the warnings had actually been read, but were ignored by people who did not comprehend

Table 3.4 Mercury concentration in tissue of seed-eating rodents and birds and in their avian predators, in alkylmercury-treated and untreated agricultural areas of western Canada[a]

Organism	Mercury concentration (ppm d.w., mean ± SD)	
	Treated area	Untreated Area
Rodents[b]	1.25 ± 0.68 (n = 6)	0.18 ± 0.15 (n = 5)
Songbirds[b]	1.63 ± 1.00 (n = 10)	0.03 ± 0.01 (n = 3)
Upland-game birds[b]	1.88 ± 0.44 (n = 19)	0.35 ± 0.22 (n = 12)
All seed eaters[b]	1.70 ± 0.38 (n = 35)	0.26 ± 0.14 (n = 20)
Predatory birds[c]	0.24 ± 0.04 (n = 89)	0.10 ± 0.02 (n = 34)
Seed-eating prey[d]	1.16 ± 0.23 (n = 61)	0.37 ± 0.15 (n = 32)

[a] After Finreite et al. (1970).
[b] Mercury in liver; treated vs. untreated in Alberta.
[c] Mercury in egg; treated areas in Alberta vs untreated areas in Saskatchewan.
[d] Mercury in liver; treated areas in Alberta vs untreated areas in Saskatchewan.

the concept that grain treated with a pesticide would be toxic to humans. Similar poisonings by the ingestion of mercury-treated grain took place in Iran, Pakistan, Guatemala, and elsewhere (Goldwater and Clarkson, 1972; Ehrlich et al., 1977; Murphy, 1980; Ziff, 1985).

Sewage Sludge

The application of metal-containing sewage sludges can cause a contamination of agricultural soil and crops. Sewage sludge is a by-product of the secondary treatment of municipal sewage. Sewage sludge has a favorable soil-conditioning property due to its large concentration of humified organic matter, and it also contains substantial concentrations of the macronutrients nitrogen and phosphorus. As a result, sewage sludge is frequently disposed of by application to agricultural land. For example, of the

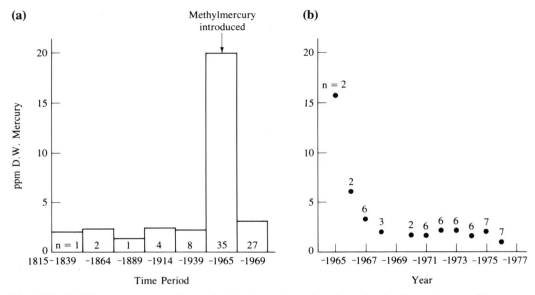

Fig. 3.3 (a) Mercury concentration in feathers taken from female goshawks (*Accipiter gentilis*) collected at nests in Sweden during April to June. (b) Mercury in tail feathers of young marsh harriers (*Circus aeruginosus*) from central Sweden. After Johnels *et al.* (1979).

5.6×10^6 tonnes/year of sludge produced in the United States, about 42% is applied to farmland, as is about 40% of the 5.9×10^6 tonnes/year produced in western Europe (Brown and Jacobson, 1987). Similarly high rates of usage of sewage sludge are typical of most industrialized countries where wastewater treatment facilities are frequent.

Unfortunately much of this sewage sludge, especially that having significant industrial inputs, contains considerable quantities of toxic elements. Table 3.5 summarizes typical concentrations of toxic elements in sludge from several industrialized countries. Clearly, the metal concentrations in sludges vary tremendously. This variation largely reflects differences in the nature of industrial inputs to particular wastewater systems. Of the various elements in Table 3.5, cadmium, copper, nickel, and zinc are most likely to cause phytotoxicity when the sludges are applied to agricultural land (Page, 1974; Keeney, 1975).

Only a portion of the "total" quantity of toxic elements present in sewage sludges is available for plant uptake. The uptake depends on such variables

as soil cation exchange capacity and pH, the amount of sludge applied and its elemental composition, and plant species or variety. Gaynor and Halstead (1976) showed large increases in "available" metals in soil treated with sewage sludge (Table 3.6), and other researchers have reported similar results (Page, 1974; Lagerwerff *et al.*, 1976; MacLean and Dekker, 1978). Crop plants grown on sludge-amended soils can have increased metal concentrations in their tissue, and metal toxicity has occasionally been reported (Table 3.7; Cunningham *et al.*, 1975a,b; Webber and Beauchamp, 1977; Sidle *et al.*, 1976; Cohen *et al.*, 1978; MacLean and Dekker, 1978). The assimilated metals can enter the human food chain through the direct ingestion of produce, or indirectly if the crop is first fed to livestock.

Clearly, it is important to monitor the amounts of toxic elements applied to agricultural land in sewage sludge, and the resulting contaminations of crop plants, so that tracts of otherwise fertile cropland are not significantly contaminated or rendered toxic by this agricultural practice. The U.S. Environmen-

Table 3.5   Average and/or range of elemental concentration (ppm d.w.) in sewage sludge from a variety of locations

Element	Sweden[a]	Michigan[b]	Britain[c]	North America, Europe[d]	Ontario[e]
Ag			32 (5–150)	5–150	25 (4–60)
As		7.8 (1.6–18)		1–18	
B			70 (15–1000)		
Ba			1700 (150–4000)		
Cd	13 (2.3–171)	74 (2–1100)	<200 (<60–1500)	1–1500	29 (2–147)
Co	15 (2.0–113)		24 (2–260)	2–260	
Cr	872 (20–40,615)	2030 (22–30,000)	980 (40–8800)	20–40,000	4200 (16–16,000)
Cu	791 (52–3300)	1020 (84–10,400)	970 (200–8000)	52–11,700	1100 (162–3000)
Hg	6.0 (<0.1–55)	5.5 (0.1–56)		0.1–56	9 (1–24)
Mn	517 (73–3860)		500 (150–2500)	60–3860	310 (60–500)
Mo			7 (2–30)	2–1000	
Ni	121 (16–2120)	371 (12–2800)	510 (20–5300)	10–53,000	390 (7–1500)
Pb	281 (52–2910)	1380 (80–26,000)	820 (120–3000)	15–26,000	1200 (85–4000)
Sn			160 (40–700)		
V			75 (20–400)	40–700	
Zn	2060 (705–14,700)	3320 (72–16,400)	4100 (700–49,000)	74–49,000	4500 (610–19,000)

[a] 93 Sewage works; after Bergren and Oden (1972).
[b] 57 Sewage works; after Blakeslee (1973).
[c] 42 Sewage works; after Berrow and Webber (1972).
[d] 300 Sewage works; after Page (1974).
[e] 10 Sewage works; after Van Loon (1974).

tal Protection Agency has developed standards for the maximum concentrations of metals in sewage sludge intended for application to agricultural land, as well as maximum annual loading rates and maximum cumulative loading rates (Table 3.8).

Selenium in Agricultural Wetlands

Many irrigation systems require the use of subsurface water, which may have substantial concentrations of various chemicals, including selenium. In a few cases the drainage of excess irrigation water from fields into nearby wetlands has caused the accumulation of large concentrations of selenium, which has poisoned wildlife, especially waterfowl.

The best-known example of this toxic syndrome is the Kesterton Reservoir, which has been used as a hydrologic sink for drainage from a large area of the extensively agricultural, western San Joachin Valley of central California (Ohlendorf, 1986, 1989; Hoffman *et al.*, 1988; Paveglio *et al.*, 1992). The

Kesterton Reservoir consists of 12 shallow (1.0–1.5 m deep) ponds, totalling about 500 ha, that were constructed in 1968 and 1975 to serve as evaporation and holding basins for agricultural drainage, while benefitting wildlife through the provision of productive, shallow-water habitat. Regrettably, it was not foreseen that selenium would be leached in large amounts from the fields and that this element would progressively accumulate in the reservoir and then bioaccumulate to concentrations that would poison waterfowl and other wildlife.

Inflow waters to Kesterton Reservoir had an average concentration of selenium of 0.3 mg/liter during 1983–1985, while downstream holding ponds had 0.05–0.2 mg/liter (Ohlendorf, 1986, 1989). Concentrations of selenium were much larger in plants, as follows.

1. *Submerged plants.* Geometric mean of 73 mg/kg (range 18–340 mg/kg) in foliage of widgeongrass (*Ruppia maritima*) and

Table 3.6 Influence of sludge addition on available metal concentration in three soil types[a]

Soil type	Treatment[b]	Available metals (ppm d.w.)[c]			
		Zn	Cu	Pb	Cd
Sandy loam (a)	Control	7.7	4.4	3.4	0.37
	+sludge	52.8	11.5	6.2	0.70
Sandy loam (b)	Control	1.0	0.7	1.6	0.07
	+sludge	31.1	4.9	3.5	0.25
Clay	Control	5.7	2.2	1.8	0.14
	+sludge	74.6	12.8	5.5	0.72

[a] After Gaynor and Halstead (1976).
[b] 3.3 kg d.w. sludge was added to 82 kg of soil, followed by an 8-week incubation. The metal concentration in the sludge was Cd, 30 ppm; Pb, 200 ppm; Ni, 360 ppm; Cu, 539 ppm; and Zn, 3200 ppm.
[c] The concentrations of available metals were determined after extraction with 0.005 M diethylenetriamine pentaacetic acid (DTPA).

horned pondweed (*Zannichellia palustris*), compared with 0.2 mg/kg in plants from Volta Wildlife Area, a reference wetland. Note that these are important waterfowl-food species.

2. *Filamentous algae*. Average 69 mg/kg (range 12–330 mg/kg), compared with 0.9 mg/kg in the reference wetland.
3. *Emergent plants*. Average 37 mg/kg (range

17–160 mg/kg) in leaves and 154 mg/kg (89–320 mg/kg) in roots of cattail (*Typha domingensis*) and bulrush (*Scirpus maritimus*), compared with <0.1 mg/kg (maximum 2 mg/kg) in the reference wetland.

Planktonic invertebrates at Kesterton Reservoir had an average selenium concentration of 85 mg/kg (maximum 300 mg/kg), compared with 2.0 mg/kg in the reference wetland. The mosquitofish *Gambusia affinis* averaged 170 mg/kg (maximum 332 mg/kg), compared with a reference value of 1.3 mg/kg. Bullfrogs (*Rana catesbeiana*) had 45 mg/kg in their liver, compared with 6 mg/kg in the reference wetland.

Selenium toxicity to aquatic birds was studied during the early 1980s (Ohlendorf, 1986, 1989; Hoffman *et al.*, 1988). American coot (*Fulica americana*) at Kesterton Reservoir had an average selenium concentration in liver of 43 mg/kg, while dabbling ducks averaged 20 mg/kg (mallard, *Anas platyrhynchos*; cinnamon teal, *A. cyanoptera*; gadwall, *A. strepera*; and pintail, *A. acuta*), black-necked stilt (*Himantopus mexicanus*) had 63 mg/kg, and grebes had 127 mg/kg (pied-billed grebe, *Podilymbus podiceps*; eared grebe, *Podiceps nigrocollis*), compared with values for these birds in the reference wetland of 4–9 mg/kg. At least 39% of 578 monitored nests had at least one dead or

Table 3.7 Increase in the metal concentration of plants grown in soil containing sewage sludge at various rates of application: Average of four different sewage sludges and two experiments using corn (*Zea mays*) and one using rye (*Secale cereale*)[a]

Application rate (tonnes/ha)	Metal concentration in plant tissue (ppm d.w.)					
	Cd	Cr	Cu	Mn	Ni	Zn
Control (0)	0.4	<3.0	7.4	33.3	<4.5	38
63	1.6	<3.0	14.4	44.1	<4.5	149
125	3.7	<3.0	15.8	110	<4.5	206
251	4.1	3.2	19.1	376	6.5	251
502	4.8	5.5	23.3	346	16.3	289
Average metal concentration in sludge (ppm d.w.)	194	9200	1200	600	860	5100

[a] After Cunningham *et al.* (1975a).

Table 3.8 Standards for toxic elements in sewage sludges intended for application to agricultural lands in the United States[a]

Metal	Maximum concentration (mg/kg d.w.)	Maximum cumulative loading rate (kg/ha)	Maximum annual loading rate (kg/ha·year)
As	75	41	2.0
Cd	85	39	1.9
Cr	3000	3000	150
Cu	4300	1500	75
Pb	840	300	15
Hg	57	17	0.85
Mo	75	18	0.90
Ni	420	420	21
Se	100	100	5.0
Zn	7500	2800	140

[a]These criteria were developed from risk assessments that were mostly based on metal uptake by agricultural plants, and the subsequent toxic hazards to people eating the produce (after Goldstein 1993).

deformed embryo or chick, often with brain damage or developmental abnormalities such as missing or abnormal eyes, beak, wings, legs, or feet. Mortality of fledgling and adult birds at Kesterton Reservoir was also attributed to selenium toxicity.

Clearly, in spite of the superficial attractiveness of the highly productive wetlands at Kesterton Reservoir to waterfowl and other wildlife, selenium originating with agricultural drainage has rendered it a toxic place in which to live and attempt to raise avian families.

Pollution from Metal Mining and Processing

The Industrial Metal Cycle

Pollution by toxic elements is associated with various aspects of the industries by which metals are mined and processed (Fig. 3.4). Mining produces ore as a product, along with a considerable amount of metal-contaminated wastes. The ore is taken to a mill, which crushes the ore and separates mineral fractions, producing a metal-rich concentrate plus a large quantity of waste tailings. The tailings are

usually discharged as a slurry into a contained dump, which is usually located in a natural basin or lake. Concentrate from the mill is processed in a primary smelter. Smelter wastes include a molten waste called slag that is disposed on land, along with atmospheric emissions of SO_2, other gases, and metal-containing particulates. Concentrate from the primary smelter is taken to a refinery, where pure grades of metal are produced, along with further emissions of gases and metal-containing particulates to the atmosphere. The pure, refined metal is used in diverse manufacturing processes, which may also emit pollution to the atmosphere as dusts and gases. After their useful lifetime, manufactured products can be recycled through a secondary smelter and refinery, they can be discarded into a managed, solid-waste disposal site, or regrettably, they may be summarily dumped in some convenient location. All of these processes produce environmental contamination, as is described below.

Mining Residues

Pollution around metal mine sites is caused by the dumping of contaminated overburden, excavation wastes, etc. (see Table 3.9 for examples). Because of toxicity from the large metal concentrations, the development of vegetation on the wastes can be limited to an early-successional grassland–herb community. In some cases, the toxicity can be so severe that even after a long period of time, only a few pioneer species of plants manage to establish, as has been the case with mine wastes from old Roman lead workings in Britain (Bradshaw and Chadwick, 1981).

Metal-Tolerant Plant Ecotypes

Sites that are polluted by toxic elements from mine wastes or other sources, including natural ones, are often dominated by genetically adapted populations of plants, or metal-tolerant ecotypes as they are often called. These tolerant individuals can survive and grow in metal-contaminated environments, whereas nontolerant individuals are quickly elimi-

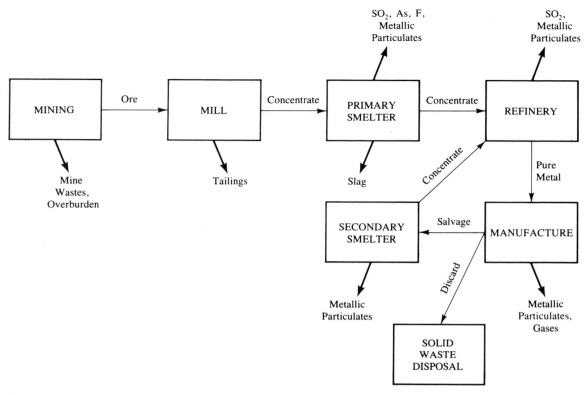

Fig. 3.4 Sources of pollution in the metal mining and processing industries.

Table 3.9 Metal concentrations in surface soil in studies of contaminated mine waste disposal sites

Study site	Site	Metal concentration in soil (ppm d.w.)			Reference[a]
		Cu	Zn	Pb	
Near operating mine,	Site 1	90	1500	500	(1)
British Columbia, Canada	Site 2	110	390	180	
Near abandoned mines,	Mine 1	680	120	2	(1)
British Columbia, Canada	Mine 2	120	130	40	
	Mine 3	160	390	300	
Near abandoned mine,		240	1750	960	(1)
Devonshire, U.K.					
Near abandoned mine,		—	1270	21,300	(2)
Wales, U.K.					
Near abandoned mines,	Mine 1	—	—	1100–1750	(3)
Wales, U.K.	Mine 2	—	4500–5000	—	
	Mine 3	680–2700	—	—	
Near abandoned mines,		90–2300	75–40,000	80–3600	(4)
Wales, U.K.					

[a]Referenes: (1) Warren *et al*. (1966); (2) Williams *et al*. (1977); (3) Jain and Bradshaw (1966); (4) Gregory and Bradshaw (1965).

nated by the toxic stress. Conversely, metal-tolerant ecotypes are usually poor competitors, and they are rare components of the vegetation of sites that are not stressed by metal toxicity.

Several studies have shown that individual plants with a genetically based tolerance of toxic metals can be present at a small frequency ($<1\%$) in populations growing in nonpolluted sites (Jowett, 1964; Bradshaw et al., 1965; Wu and Bradshaw, 1972; Cox and Hutchinson, 1980). Because of this preexisting tolerance within the population, the frequency of tolerant genotypes can increase rapidly when toxic stress is caused by anthropogenic metal pollution (Bradshaw et al., 1965; Antonovics et al., 1971; Wu et al., 1975; Bradshaw, 1977). These population differences can be maintained over short distances (less than 100 m) by steep gradients of toxic stress associated with metal pollution, because tolerant individuals are highly favored on polluted soil but are poor competitors on nonpolluted soil (Jain and Bradshaw, 1966; Antonovics and Bradshaw, 1970; Antonovics et al., 1971).

In the Sudbury region of Ontario, there is an extensive and severe pollution of soils by metals. Especially important pollutants are nickel and copper from mining and smelting activities. In addition, large concentrations of available soil aluminum have resulted from acidification of the soil. The vegetation on these sites largely consists of metal-tolerant grasses, especially *Agrostis gigantea* and *Deschampsia caespitosa* (Hogan et al., 1977a,b; Cox and Hutchinson, 1979, 1980). Grasslands dominated by these species developed rapidly on barren, mesic-wet sites after a reduction in SO_2 pollution in 1972, even though the soil was still acidic and contaminated by nickel, copper, and other metals (Cox and Hutchinson, 1980).

The likely source of seed of *Deschampsia caespitosa* was a site about 75 km from Sudbury where this grass grows in calcareous meadows beside Lake Huron. Huge piles of coal presently occur in the midst of one of these meadows at the port of Little Current, where coal was stored prior to its shipment to the Sudbury smelters. Mature seed-

Clones of a metal-tolerant ecotype of the grass *Deschampsia caespitosa*, growing in nickel- and copper-contaminated soil near the Coniston smelter at Sudbury, Ontario (photo: B. Freedman).

producing plants of *Deschampsia caespitosa* are present on this acidic coal substrate, and seed of these plants must have been transported with the coal to Sudbury. Metal tolerance occurs at a low level in the natural, calcareous population. Cox and Hutchinson (1980) grew seedlings from two populations of *Deschampsia caespitosa* on metal-polluted soils collected near a smelter (122 ppm Ni, 89 ppm Cu, pH 3.5) and near an old roast bed (2900 ppm Ni, 867 ppm Cu, pH 3.0). Seedlings from the nonpolluted Little Current site survived at a rate of 46% on the smelter soil and 2.6% on the roast bed soil, compared with rates for a smelter population of 94 and 14%, respectively. The above information suggests the following scenario regarding the history of *Deschampsia caespitosa* at Sudbury: Seeds from a reference population having some individuals tolerant of acidity and metal stress were transported to the vicinity of the Sudbury smelters with the coal. The metal-tolerant individuals were positively selected at the metal-stressed sites near Sudbury, and tolerance became much more frequent in the population.

Cox and Hutchinson (1979, 1980) compared the metal tolerance of mature plants growing near Sudbury with those from a nonpolluted reference site (Table 3.10). The Sudbury population was markedly more tolerant of nickel and copper. These were the two most important metals in their soil environment, occurring at average concentrations of 423 ppm Ni and 359 ppm Cu, compared with 22 ppm Ni and 19 ppm Cu at the reference site. The smelter population was also more tolerant of aluminum. This is explained by the acidic nature of soils near the smelter (pH 3.5–3.9, compared with pH 6.8–7.2 at the reference site), coupled with the fact that aluminum solubility and toxicity are much greater in acidic soil (Freedman and Hutchinson, 1986). A moderate but statistically significant increase in tolerance to lead and zinc by the smelter population was unexpected, since the soil in which they grew was not contaminated by these metals. This observation of cotolerance implies that there may be similar physiological mechanisms of tolerance for several metals in *Deschampsia caespitosa*.

Table 3.10   Tolerance indices[a] to several metals, for the grass *Deschampsia caespitosa* collected from a metal-contaminated site near Sudbury, and from an uncontaminated reference site

Metal	Contaminated Sudbury site	Uncontaminated reference site
Ni	0.96	1.97[b]
Cu	1.06	2.01[b]
Al	1.06	1.21[c]
Cd	1.24	1.36
Pb	1.37	1.90[b]
Zn	1.55	1.98[b]

[a]Note that the smaller the index, the greater the metal tolerance, since the tolerance index is $1 - \log(R_c/R_m)$, where R_c is the average length of the longest root of 20 tillers in a control nutrient solution and R_m is the average length of the longest root of 20 tillers in a metal-spiked solution. A tolerance index of 1.0 means that the growth of roots in the metal solution was the same as that in the reference solution without the metal. Modified from Cox and Hutchinson (1980).
[b]Statistically significant difference at $p < 0.001$.
[c]Indicates $p < 0.05$.

Metals in the Terrestrial Food Web

In certain situations, some toxic elements have a substantial capability for both bioaccumulation and food-web accumulation. Several examples of these phenomena were described earlier in this chapter for mercury and selenium in aquatic ecosystems and for terrestrial plants that are hyperaccumulators of certain elements. Compared with these situations, the biota of more typical terrestrial systems have a smaller propensity to accumulate metals and other toxic elements.

One relatively detailed study examined the concentrations of copper and cadmium in various components of grassland ecosystems, at sites variously influenced by emissions from a copper refinery near Liverpool, England (Hunter *et al.*, 1987a,b,c, 1989). Soils close to the refinery had copper concentrations as large as 52,200 ppm (average 11,000 ppm) and cadmium up to 59 ppm (average 15.4 ppm), compared with reference values of 15 ppm Cu and 0.8 ppm Cd (Table 3.11). Metal concentra-

A reclaimed tailings area near the Copper Cliff mill at Sudbury. Most of the herbaceous vegetation was established using a mixture of pasture species, but the woody shrubs are seeding in from the nearby hardwood forest. The tailings pond, visible in the background, is used by ducks for breeding and during migration (photo: B. Freedman).

tions in plants were also larger close to the refinery, but were much smaller than the concentrations in soil. Within the arthropod fauna, metal concentrations were also larger closer to the refinery, generally larger than in plants, smaller than in soil, and substantially larger in soil-dwelling detritivores than in predators or herbivores (although data for the detritivores, in particular, may have been influenced by the presence of contaminated soil in their gut). Herbivorous and carnivorous small mammals had much smaller metal concentrations than arthropods, and only cadmium was present in consistently larger concentrations in small mammals at the most polluted, refinery site.

Compared with many aquatic situations, the biota of this grassland had a smaller tendency to bioaccumulate metals, and both bioaccumulation and food-web accumulations were much less pronounced than are observed with persistent, lipid-soluble, chlorinated hydrocarbons (see Chapter 8).

Metalliferous Tailings and Their Reclamation

Tailings are a waste by-product of milling. In this process, raw ore is ground to a fine powder, which is then separated either magnetically or by physicochemical flotation into (1) a metal-rich fraction that is roasted and then smelted or (2) waste tailings. Although the tailings are a waste product, they can still contain large concentrations of metals, so that toxicity can make it difficult to establish plants for the eventual revegetation and stabilization of old tailings-disposal sites. Additional toxicity can arise if the metals are present as sulfides, because acidity is generated when sulfides are oxidized by microbes. Table 3.12 lists metal analyses of tailings from

Table 3.11   Average concentrations of copper and cadmium in various components of grasslands at different distances from a copper refinery near Liverpool, England[a]

Component	Site:	Copper (ppm)			Cadmium (ppm)		
		Reference	1 km	Refinery	Reference	1 km	Refinery
Soil		15	543	11,000	0.8	6.9	15
Plants							
Agrostis stolonifera (grass)		10	25	122	0.6	1.3	3.3
Equisetum arvense (horsetail)		14	29	140	0.6	1.0	2.6
Tussilago farfara (dicot)		19	26	260	0.3	1.1	4.7
Detritivorous arthropods							
Collembola (springtail)		50	175	2,370	2.1	12	52
Isopoda (woodlouse)		78	836	2,390	15	130	231
Oligochaeta (annelid worm)		24	155	1,170	4.0	34	107
Herbivorous arthropods							
Orthoptera (cricket)		38	66	333	0.2	0.3	20
Formicidae (ant)		33	131	731	1.2	5.4	38
Lepidoptera (moth)		14	69	160	0.6	7.1	22
Curculionidae (beetle)		30	66	421	0.6	3.6	15
Predatory arthropods							
Staphylinidae (beetle)		29	58	522	0.6	4.9	14
Carabidae (beetle)		21	46	460	0.7	5.6	15
Lycosidae (spider)		58	160	887	2.6	35	102
Linphiidae (spider)		89	200	1,020	2.4	19	89
Herbivorous small mammals							
Apodemus sylvaticus (mouse)		11	11	11	0.9	1.2	2.1
Microtus agrestis (vole)		11	12	13	1.1	1.4	3.1
Carnivorous small mammal							
Sorex araneus (shrew)		13	18	29	3.9	19	46

[a]All data are averages of large samples. Plant samples were live, green foliage. All animal analyses are for whole-body samples, except for cadmium in small mammals, which are for liver plus kidney. Modified from Hunter *et al.* (1987a,b,c, 1989).

a number of disposal sites in North America. All of these tailings contain large concentrations of various toxic elements, depending on the nature of the ore that had been fed to the mill. Toxicity of the acidic tailings is especially severe, since most metals have an increased solubility and bioavailability at acidic pHs.

Much research has been done on the establishment of vegetation on metal-contaminated tailings after they have been abandoned. Apart from the esthetic and economic problems associated with large areas of derelict land, tailings dumps can be important sources of water pollution and of windborne, metal-containing dusts. These problems can

be substantially mitigated if the tailings are rehabilitated with a stable cover of vegetation. The establishment of vegetation on tailings usually involves some combination of liming to raise pH and reduce metal availability, fertilizing to alleviate nutrient deficiency, incorporation of organic matter to improve soil structure and water-holding capacity, and planting with seeds of various plant species. Occasionally, a relatively novel approach is required, such as the use of acid- or metal-tolerant plant ecotypes, inoculation of the seed mix with metal-tolerant mycorrhizal fungi, or the overlaying of the entire tailings area with some sort of nontoxic, locally available overburden, which is then vegetated (LeRoy

Table 3.12 Chemical analyses of a variety of metal-contaminated tailings

Site	pH	Concentration (ppm d.w.)						S (%)	Reference[a]
		As	Cd	Cu	Ni	Pb	Zn		
Yukon, Canada, gold mine tailings	1.9	5200	18	140	13	952	2400	1.1	(1)
Yukon, Canada, gold tailings	3.2	400	15	613	14	130	9600	44.4	(1)
Yukon, Canada, gold tailings	6.0	1350	102	172	15	3600	6200	5.7	(1)
Yukon, Canada, gold tailings	6.9	50,000	114	330	20	2300	18,000	13.5	(1)
Yukon, Canada, tungsten tailings	7.0	—	0.3	1420	22	12	288	6.5	(1)
Yukon, Canada, gold tailings	7.1	7	2.6	33	21	1130	1060	4.0	(1)
Yukon, Canada, copper tailings	9.0	15	0.7	1710	21	9	178	0.1	(1)
Ontario, Canada, nickel tailings	3.2	—	0.4	392	290	18	291	0.9	(1)
New Mexico, U.S.A., copper tailings[b]	2.2	—	—	>600	—	<5	21	—	(2)
Utah, U.S.A., uranium tailings[b]	2.9	—	—	475	—	16	103	—	(2)
Montana, U.S.A., copper tailings[b]	5.2	—	—	195	—	<5	24	—	(2)

[a]References: (1) after Kuja (1980); (2) after Peterson and Nielson (1973).
[b]In ppm in saturation water extract.

and Keller, 1972; Moore and Zimmermann, 1972; Dean *et al.*, 1973; Peterson and Nielson, 1973; Harris and Jurgensen, 1977; Bradshaw and Chadwick, 1981; Peters, 1984).

As a case study, consider the rehabilitation of tailings from a large mill at Sudbury (Peters, 1984). The mill can produce 54,000 tonnes/day of tailings, which are piped as a slurry for disposal in low-lying, contained depressions in the landscape. In 1983, there was about 1120 ha of active disposal area, 485 ha undergoing vegetation establishment, and 600 ha with an established grassland.

The mineral constituents of the Sudbury tailings are feldspar (>50%), chlorite (20%), quartz (10%), pyroxenes (7%), biotite (7%), and pyrites (7%). These are not toxic minerals. However, the acidity that is produced by oxidation of the pyrites can cause a pH lower than 3.7. In addition, there are relatively large concentrations of some metals, especially copper (1–81 ppm in an "available" extract of the tailings measured with an acetic-acid leachate), nickel (1–87 ppm), and iron (59–441 ppm).

Over the years, various vegetation-establishment schemes were used to try to stabilize the Sudbury tailings, largely because they caused severe dust problems during dry weather. The techniques that are currently used result in a well-established grass-

land, which is being secondarily colonized by seedlings of native trees and other plants. The methods used to achieve this include: (1) the application of limestone ($CaCO_3$) at ca. 900 kg/ha to raise the tailings pH to 4.5–5.5; (2) the application of nitrogen and other macronutrients several times during establishment of the grassland; and (3) sowing with a mixture of pasture grasses (25% *Agrostis gigantea*, 25% *Poa compressa*, 15% *Phleum pratense*, 15% *Poa pratensis*, 10% *Festuca arundinacea*, and 10% *Festuca rubra*). Legumes have also been used to a limited extent; the most successful species have been *Medicago sativa* and *Lotus corniculatus*. An important innovation was the sowing of a short-lived nurse crop of annual rye (*Secale cereale*) to provide initial shading and reduced wind stress for the tender seedlings of pasture grasses and legumes as they were becoming established. In addition, there has been a natural invasion by native herbaceous plants, especially of Asteraceae, and also by seedlings of tree species such as birch (*Betula papyrifera*), trembling aspen (*Populus tremuloides*), and willow (*Salix* spp.). The establishment of trees has been supplemented by the planting of various conifer species.

As the ecosystem developed on the reclaimed tailings at Sudbury, animals began to invade. At least 90 bird species, most of which are migrants, have been observed on the vegetated tailings area and its central pond. Birds that regularly breed in the tailings habitat include mallard and black duck (*Anas platyrhynchos* and *A. rubripes*), American kestrel (*Falco sparverius*), killdeer (*Charadrius vociferus*), and savannah sparrow (*Passerculus sandwichensis*).

Primary Smelters

Several detailed studies have been made of the environmental consequences of pollution in the vicinity of metal smelters. Among the better-documented cases are the Sudbury nickel–copper smelters (Hutchinson and Whitby, 1974; Whitby and Hutchinson, 1974; Freedman and Hutchinson, 1980c; Amiro and Courtin, 1981; see also Chapter 2), a zinc smelter at Palmerton, Pennsylvania (Buchauer, 1973; Jordan, 1975; Jordan and Lechevalier, 1975; Strojan, 1978a,b; Beyer *et al.*, 1985), a brassworks at Gusum, Sweden (Tyler, 1984), and a lead–zinc smelter at Avenmouth, England (Little and Martin, 1972; Hutton, 1984).

Common features of these studies are (Freedman and Hutchinson, 1981): (1) the occurrence of severe environmental pollution with metals close to the point sources, especially in surface soils; (2) exponential decreases in the intensity of pollution with increasing distance from the point sources; (3) the presence of damaged plant communities in the most polluted sites, characterized by a small biomass, little productivity, short stature of the community dominants, altered species composition, and depauperate species richness and diversity; and (4) disruptions of nutrient cycling by toxic metals, including an impoverished abundance and species richness of soil-dwelling organisms, decreased litter decomposition, and slower mineralization and other transformations of nitrogen and phosphorus in the forest floor.

Sudbury. The total atmospheric emission of metal-containing particulates from the smelters near Sudbury averaged 1.9×10^4 tonnes/year between 1973 and 1981, including 4.2×10^3 tonnes/year of iron, 6.7×10^2 tonnes/year of copper, 5.0×10^2 tonnes/year of nickel, 2.0×10^2 tonnes/year of lead, and 1.2×10^2 tonnes/year of arsenic (Chan and Lusis, 1985). The emitted particulates can be effectively removed from the atmosphere during wet precipitation events (60–80% removal efficiency), but less so by dry deposition (<15%) (Chan and Lusis, 1985). Overall, about 50% of the particulates emitted from the smelters are deposited nearby (Muller and Kramer, 1977; Freedman and Hutchinson, 1980a; Chan and Lusis, 1985). The deposition rates are especially large close to the point sources, and they decline exponentially with increasing distance (Fig. 3.5).

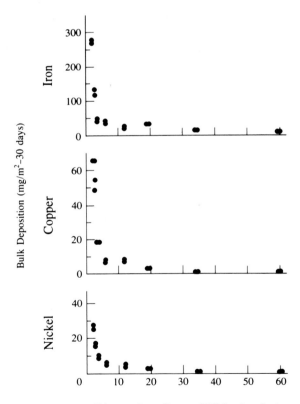

Fig. 3.5 Atmospheric deposition of metals along a transect from the Copper Cliff smelter at Sudbury. The sampling period was July 18 to August 16, 1977. Modified from Freedman and Hutchinson (1980a).

This pattern of atmospheric deposition has caused a parallel pattern of environmental pollution, as illustrated by the concentrations of nickel and copper in the forest floor of sites along a transect from the Copper Cliff smelter (Fig. 3.6). Concentrations of nickel and copper as large as 4900 ppm each were present in the forest floor of sites close to the smelter, while up to 370 ppm Ni and 260 ppm Cu occurred in plant foliage (Freedman and Hutchinson, 1980a). The large metal-binding capacity of the organic forest floor makes it a virtual sink for deposited metals. The concentrations of metals in mineral soil are much smaller because of its limited cation-exchange capacity (Freedman and Hutchinson, 1980a).

Because sulfur dioxide has been such an important pollutant in the Sudbury area, it is impossible to attribute ecological damage directly or solely to toxic metals. However, the importance of metals has been demonstrated through experiments in which bioassay plants were grown in polluted soils, but in the absence of sulfur dioxide. In a greenhouse experiment, Whitby and Hutchinson (1974) grew a variety of plant species in metal-polluted soils collected in the vicinity of a Sudbury smelter. They observed large reductions in the productivity and

root elongation of all bioassay species, compared with their growth in reference soil. This toxic effect was due to large concentrations of soluble nickel, copper, cobalt, and aluminum, and it persisted to a substantial degree when soil acidity was neutralized by liming. Similar toxic symptoms were observed when the bioassay plants were grown in nutrient solutions containing pure salts of these metals. In a field experiment, the concentrations of toxic metals in limed soil precluded the establishment of test species by seed at sites within 12 km of the smelter.

Nickel- and copper-tolerant populations of the grasses *Deschampsia caespitosa* and *Agrostis gigantea* were previously described. The presence of these ecotypes indicates that metal stress is important in sites close to the smelter. However, the rapid spread of a grassland dominated by these species plus *Agrostis scabra* only occurred after the local emissions of SO_2 were greatly reduced by building the tall superstack in 1972 (Cox and Hutchinson, 1981; Chapter 2). This observation underscores the important role of that toxic gas in restricting the development of terrestrial plant communities near the Sudbury smelters.

In addition to their effects on plants, toxic metals in soils near the Sudbury smelters have been shown

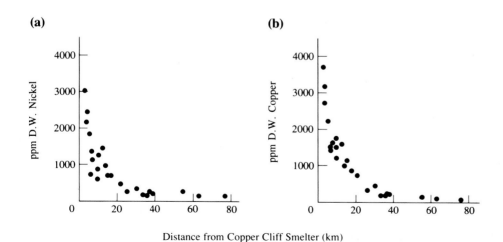

Distance from Copper Cliff Smelter (km)

Fig. 3.6 (a) Total nickel and (b) total copper in the forest floor of stands in a transect from the Copper Cliff smelter. Each point is the average of 10 replicate determinations. Modified from Freedman and Hutchinson (1980a).

to inhibit litter decomposition and other nutrient-cycling processes (Freedman and Hutchinson, 1980b). In part, the effects on nutrient cycling explain the relatively small concentrations of some plant macronutrients in soil and foliage at polluted sites near Sudbury (Lozano and Morrison, 1981).

Gusum.   Gusum, Sweden, is another case where metals emitted from a point source have damaged terrestrial vegetation. In this situation, the emission of SO_2 was unimportant, so that all ecological damage can be ascribed to toxic metals, primarily zinc and copper. The brassworks at Gusum has operated continuously since 1661, and there are now large concentrations of metals in soil near the point source (Figure 3.7) (see also Tyler, 1974, 1975, 1976). Zinc is the most abundant metal, reaching

16,000–20,000 ppm (1.6–2.0%) in the surface organic matter of sites within 0.3 km of the source, compared with <200 ppm at reference sites 7–9 km away. The pattern of pollution with copper is similar to that of zinc (copper was 11,000–17,000 ppm at <0.3 km, compared with <20 ppm at 7–9 km). The high soil pH at sites close to the source is due to the saturation of most of the cation exchange capacity by the large metal concentrations.

The metal pollution has caused local ecological damage. Effects on vegetation (Folkeson, 1984; Tyler, 1984) include the death or decline of most pine (*Pinus sylvestris*) and birch (*Betula* spp.) trees close to the source and substantial reductions in the covers of lichens and the feather mosses *Hylocomium splendens*, *Pleurozium schreberi*, and *Dicranum* spp. (<1% cover, compared with 25–90% in reference stands). Decreases in the species

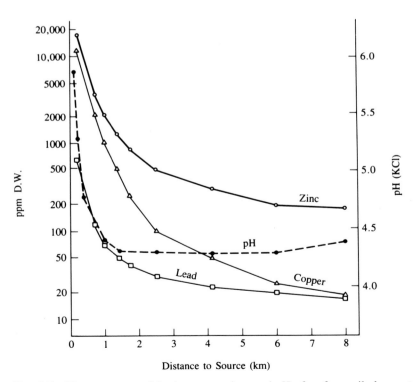

Fig. 3.7 Zinc, copper, and lead concentrations and pH of surface soil along a transect from the brassworks at Gusum. After Tyler (1984).

richness of many groups of organisms are clearly related to the metal concentrations of surface soils (Fig. 3.8).

Not unexpectedly, there are large differences in the tolerances of plant species to the metal pollution. For example, *Hylocomium splendens* is normally an abundant species of feather moss on the forest floor, but it is virtually eliminated from sites within 1.5 km of the brassworks, whereas *Pohlia nutans* is relatively abundant in polluted sites (*P. nutans* is also abundant in Ni- and Cu-polluted sites near Sudbury). The grass *Deschampsia flexuosa* is more abundant close to the brassworks than in reference sites. This species may have experienced a competitive release after the death

or dieback of the overstory and most of the ground vegetation. Some fungi are relatively abundant close to the source, notably *Laccaria laccata* and *Paecilomyces farinosus*. However, some 25 species of macrofungi are less abundant in polluted sites, while 11 species are indifferent (Ruhling *et al.*, 1984; Fig. 3.8). Soil-dwelling invertebrates are also affected; populations of oligochaete worms are absent 175 m from the brassworks, compared with a density of about $25/m^2$ at 8 km and $50/m^2$ at 20 km (Bengston *et al.*, 1983; Fig. 3.8).

Studies were also made of the effects of metal pollution on litter decomposition and nutrient cycling in the forest floor of sites near Gusum (Fig. 3.9; see also Ebregt and Boldewijn, 1977). Rates of

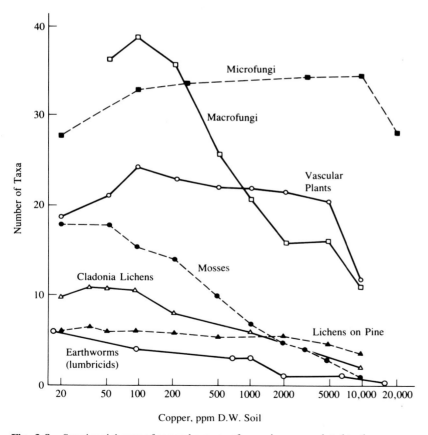

Fig. 3.8 Species richness of several groups of organisms, as related to the copper concentration of the forest floor in the Gusum area. After Tyler (1984).

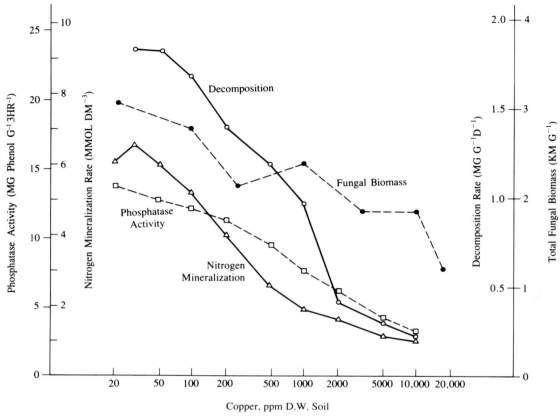

Fig. 3.9 Total fungal biomass, phosphatase activity, and decomposition and mineralization rates as related to the copper concentration of the forest floor in the Gusum area. Modified from Tyler (1984).

decomposition, activity of phosphatase enzymes, and nitrogen mineralization by ammonification were especially reduced by toxic metals.

Birds and Lead Shot

An important source of lead toxicity to birds is spent shotgun pellets, which are ingested during feeding and retained in the gizzard, where they are progressively abraded and then dissolved by acidic gastric fluids and absorbed into the bloodstream. Waterfowl are especially widely affected by lead shot, with an estimated 2–3% of the North American

autumn and winter population, or about 2–3 million individuals dying each year from lead toxicity (Bellrose, 1959, 1976). Only one or two pellets retained in the gizzard can be sufficient to kill a duck, by causing a wasting disease characterized by the eventual loss of 30–50% of the body mass, neurological toxicity, and ultimately death (Sanderson and Bellrose, 1986). A compilation of U.S. data for 1973–1984 indicated that 8.9% of about 172 thousand manually examined duck gizzards contained at least one lead-shot pellet (Sanderson and Bellrose, 1986). It should be noted, however, that manual examination of gizzards underestimates the presence of lead shot by about 20–25%, compared with more reliable X-ray examinations (Sanderson and Bellrose, 1986).

Waterfowl are not the only birds that are killed by the toxicity of lead shot. Upland-game birds, such as mourning doves (*Zenaida macroura*), have also been poisoned by ingested shotgun pellets (Kendall, 1982; Kendall and Scanlon, 1982). In a related syndrome, birds that scavenge dead carcasses can be poisoned by ingested lead shot and bullets. For example, three of the five known deaths in the wild of the critically endangered California condor (*Gymnogyps californianus*) between 1980 and 1986 were attributed to lead toxicity from ingested bullet fragments (Wiemeyer *et al.*, 1988).

Lead fishing weights may also be retained in the gizzard of larger birds, such as swans. Lead sinkers or shot were attributed as the cause of 20% of the known mortality of trumpeter swans (*Cygnus buccinator*) in Indiana, Montana, and Wyoming, and 50% of their deaths in eastern Washington state (Blus *et al.*, 1989). Lead sinkers have also poisoned mute swans (*C. olor*) in England and elsewhere in Europe (Simpson *et al.*, 1979; Birkhead, 1982) and wintering tundra swans (*C. columbianus*) in the eastern United States (Bellrose, 1959).

To reduce the toxic hazards to waterfowl and other birds, steel shot is rapidly replacing lead shot.

There has been controversy about the possibility that steel shot might cause more crippling deaths than lead shot, but field tests have shown that this effect is marginal and that the inferior ballistics of steel shot can be compensated for by shooting at closer distances and by using a larger size of shot (Sanderson and Bellrose, 1986).

Automobile Emissions and Lead

Many studies have documented the distributions of heavy metals in soil, vegetation, and the atmosphere along transects perpendicular to busy highways. All have found well-defined gradients of lead pollution, with the intensity of the effect being related to traffic volume. Some studies have also demonstrated less-well-defined contaminations by other metals, including cadmium, chromium, copper, nickel, vanadium, and zinc (Lagerwerff and Specht, 1970; Page *et al.*, 1971; Hutchinson, 1972; Ward *et al.*, 1977; Peterson, 1978; Dale and Freedman, 1982). This pattern is illustrated in Table 3.13 for soil near four urban roads of differing traffic density. A secondary contamination occurs in other biota living beside roads, for example, small mammals (Clark, 1979).

Lead emitted by automobiles has also been a major contributor to the general lead contamination

Table 3.13   Lead and zinc in surface (0–2 cm) soil collected at various distances from roads of different traffic density (ADT, average daily traffic) in Halifax, Canada[a]

Distance from road edge (m)	Lead concentration (ppm d.w.) and ADT				Zinc concentration (ppm d.w.) and ADT			
	Road A 50,000	Road B 18,500	Road C 16,000	Road D 3,000	Road A 50,000	Road B 18,500	Road C 16,000	Road D 3,000
0	3045	858	1075	465	880	422	272	106
1	2813	402	457	118	700	198	167	69
5	342	177	136	32	144	122	79	75
15	223	75	163	26	150	55	92	59
30	—	45	63	26	—	63	—	65
50	223	45	95	38	95	69	58	64
100	—	—	60	21	—	—	76	78
Background	14				60			

[a]Modified from Dale and Freedman (1982).

that occurs in cities and secondarily causes contaminations of urban wildlife with lead. For example, substantial lead burdens have been measured in pigeons (*Columba livia*) living in London, England, and the most contaminated birds exhibited symptoms of acute lead poisoning (Hutton, 1980; Hutton and Goodman, 1980).

The cause of lead emission from automobiles is the use of leaded gasoline. Since 1923 and especially since 1946, tetraethyl lead [$Pb(CH_2CH_3)_4$] was added to gasoline as an antiknock compound at a rate of about 0.8 g lead/liter. This was done to increase engine efficiency and gasoline economy, while decreasing engine wear caused by "knock" (Atkins, 1969; Lavallee and Fedoruk, 1989). Most of the lead in gasoline is emitted through the automobile tailpipe, at a rate of about 0.07 g/km and accounting for up to 40% of automobile particulate emissions (Atkins, 1969; Cantwell *et al.*, 1972). In 1975, about 95% of gasoline sold in North America was leaded to a maximum concentration of about 770 mg/liter, compared with 35% at a maximum of 290 mg/liter in 1987, and the almost exclusive use of unleaded gasoline after 1990 when its general use was banned [except for the use of low-lead fuels containing up to 26 mg Pb/liter, for some farm vehicles, marine engines, and large trucks; Lavallee and Fedoruk (1989)].

It was estimated that 98% of the total 1970 U.S. lead emission of 1.6×10^5 tonnes/year originated with automobile exhaust (NAS, 1972). Some 61% of the emitted lead particulates have a diameter >3 μm, and most of this emission settles gravitationally within about 50 m of the roadway. The remainder is well dispersed and contributes to regional lead pollution (Cantwell *et al.*, 1972).

As noted previously, the degree of urban lead contamination is being alleviated in many countries by forbidding the use of leaded gasoline. This trend is related in part to the increasing use of catalytic converters to reduce the emissions of other automobile pollutants, especially carbon monoxide and hydrocarbons. Automobiles equipped with these devices require the use of unleaded gasoline, because the converter catalysts, usually based on platinum, are inactivated by lead. The contamination of

urban environments has decreased greatly in direct response to decreased use of leaded gasoline (Hoggan *et al.*, 1978; Eisenreich *et al.*, 1986; R. A. Smith *et al.*, 1987).

Sources of other metals that are roadside contaminants are less well defined than for lead. Some sources are presumably the wear of metallic automobile parts containing these metals, the erosion of tires, and the use of metals such as nickel as a gasoline or oil additive (Lagerwerff and Specht, 1970; Hutchinson, 1972). For example, it was estimated that in 1972 about 800 tonnes of zinc was released into the environment in Canada from tire erosion (zinc is used in the vulcanization of rubber) (Anonymous, 1973).

As noted earlier, a substantial fraction of the lead emitted from automobiles occurs as small particulates that are dispersed far from roadways. This dispersed lead (plus lead emitted from other sources, such as power plants) is largely removed from the atmosphere by precipitation. The rate of deposition can be especially great at high-elevation forested sites, because of relatively large rates of precipitation due to an orographic effect, and because of direct cloud-water deposition.

In the late 1970s and early 1980s, the lead accumulation rate (i.e., atmospheric input minus streamwater export) in forests of the northeastern United States ranged from <0.1 to 0.7 kg/ha/year, with the largest recorded rate occurring at a 1000-m-elevation site on Camel's Hump, Vermont (Siccama and Smith, 1978; Andresen *et al.*, 1980; Smith and Siccama, 1981; Friedland and Johnson, 1985). This accumulation caused a substantial lead contamination of certain components of the forest, especially the organic forest floor (Andresen *et al.*, 1980; Smith and Siccama, 1981; Johnson *et al.*, 1982; Friedland *et al.*, 1984, 1985). The forest floor acts as a chemical filter for lead and other metals, through the formation of relatively strong and immobile organometallic complexes with humic substances (Schnitzer, 1978; Friedland *et al.*, 1984).

The distribution and movement of lead in a high elevation forest in Vermont are shown in Table 3.14 (Friedland and Johnson, 1985). Within the vegetation, the largest lead concentrations (23–33 ppm)

Table 3.14 Distribution and movement of lead in a high-elevation forest of *Abies balsamea, Picea rubens,* and *Betula papyrifera* on Camel's Hump, Vermont[a]

Lead distribution	Amount (kg/ha)[b]	Concentration (ppm)
Vegetation		
Foliage	0.045	3
Twigs	0.475	28
Bark	0.160	23
Wood	0.200	3
Herbs	0.005	16
Root wood	n.d.	10
Root bark	n.d.	33
Forest floor	20	219
Soil		
0–10 cm	14	26
10–20 cm	24	31
>20 cm	25	35

Lead flux	Amount (kg/ha-year)	Concentration (ppm in water)
Rain water	0.300	0.017
Cloud water	0.400	0.051
Forest floor at 3 cm	0.085	0.004
at 12 cm	0.039	0.002
Mineral soil at 25 cm	0.023	0.001
at 40 cm	0.020	0.001
Streams	<0.012	<0.001

[a]Modified from Friedland and Johnson (1985).
[b]n.d., no data.

were present in tree bark and twigs (the latter have a large proportion of bark). However, the quantity of lead also depends on the standing crop of the biomass compartment. For example, wood at this site only had a lead concentration of 3 ppm, but because of the great mass of wood in the forest, this component contained 0.20 kg/ha of lead, next only to twigs at 0.48 kg/ha. By far the largest lead concentration was in the forest floor at 219 ppm, and this component contained 20 kg/ha of lead. However, because of the large mass of the mineral soil, its quantity of lead (63 kg/ha) was larger. The influx of lead from the atmosphere was 0.70 kg/ha/year, while streamwater export was <0.012 kg/ha/year. This yields a net accumulation rate of about 0.7 kg/ha/year. The average residence time of lead in the forest floor was estimated to be about 500 years, and lead was accumulating at a rate of 3.3% per year, for a doubling time of 21 years.

Subsequent to the banning of the use of leaded gasoline in North America in the late 1980s and early 1990s, the lead contamination of high-elevation sites in the northeastern United States has decreased (Friedland *et al.*, 1992). The forest floor of sites remeasured in 1990 had 17% smaller concentrations of lead than in 1980, suggesting a net loss of lead during that decade.

The ecological consequences, if any, of the contamination of remote forests with lead (and other metals) are not understood. This is because of a lack of information on toxic thresholds of lead in foliage and in the soil solution and because most of the soil lead is bound as relatively immobile and unavailable organometallic complexes. However, there is an important message in the observation of extensive lead contaminations of forests—even in fairly remote situations, the trend is toward increasing contamination, with uncertain ecological effects.

4

ACIDIFICATION

4.1
INTRODUCTION

The environmental effects of the deposition of acidifying substances from the atmosphere, popularly known as "acid rain," are high-profile issues in both scientific and public forums. Only about 20 years ago, acid-rain research was essentially nonexistent, and few people were aware of or concerned about the potential effects of this cause of acidification of natural ecosystems. This state of affairs changed quickly in the early 1970s, beginning with a catalytic conference held in 1970 in Stockholm: the United Nations Conference on the Human Environment. At this international scientific/environmental gathering, the Swedish government presented as a case study the results of an innovative research program, summarized in a report titled: "Air Pollution across National Boundaries: The Impact on the Environment of Sulfur in Air and Precipitation" [Royal Ministry of Foreign Affairs, Royal Ministry of Agriculture (RMFA/RMA), 1971].

Partly because of the attention that this report focused on the anthropogenic acidification of ecosystems, this field of investigation blossomed. Scientific research became vigorous and relatively well funded, there was intense lobbying by nongovernment environmental organizations and by industry (of course, these groups lobbied in different directions), and various levels of government even enacted some pollution-control legislation. Because of these activities, acid-rain effects are now one of the better understood ways by which humans degrade their environment. Equally important, acid rain is now one of the most identifiable environmental problems to the "person-on-the-street," at least in north-temperate industrialized countries.

An important reason why acid rain has received relatively focused attention from governments is the international, transboundary nature of the phenomenon. Because acidifying substances and their emitted, gaseous precursors are transported over long distances in the atmosphere, often far from the sources of emission, these chemicals do not respect political boundaries (see also Chapter 2). Therefore, emissions in one country or region can degrade ecosystems and economically valuable resources in other jurisdictions.

In Europe, for example, the Scandinavians have justifiably blamed England and Germany for most of the emissions of SO_2 and NO_x, the acid precursors that have acidified parts of their northern landscape. In North America, one atmospheric-transport model estimated that emissions in the United States were responsible for 90% of the wet deposition and 43% of the dry deposition of nitrogen to eastern Canada, along with 63% of the wet and 24% of the dry depositions of sulfur (Shannon and Lecht, 1986). In return, Canada contributed less than 5% of the wet and dry depositions of sulfur and nitrogen in the eastern United States. These North American patterns reflect both the intensity of emissions in the two countries and the prevailing air-flow patterns.

It is now known that the most important acidifying substances that are deposited from the atmosphere are weak solutions of sulfuric and nitric acids, which arrive in the form of acidic precipitation. In addition, the dry depositions of gaseous sulfur dioxide and oxides of nitrogen, and of certain particulates such as ammonium sulfate and ammonium nitrate, can contribute substantial acidifying potential to ecosystems. There is evidence that the areas of North America and Europe that are experiencing acidic deposition have increased in the most recent half century or so.

Certain ecosystems are more vulnerable to acidification by these atmospheric inputs than others. These ecosystems usually have thin, often coarse-grained, noncalcareous soils overlying a mantle of hard, slowly weathering bedrock of such minerals as granite, gneiss, and quartzite. Although the evidence is less certain that terrestrial ecosystems have been degraded by acidifying depositions, there is incontrovertible evidence of damage to freshwater ecosystems. There is evidence that many lakes, streams, and rivers have acidified fairly recently and that fish populations have declined or become extirpated in many of these surface waters.

In this chapter, the chemical characteristics of acidic precipitation and dry deposition are examined, as are the effects of acidifying deposition on the chemistry and biota of terrestrial and aquatic ecosystems, and possible methods for use in the reclamation of acidic water bodies.

4.2
DEPOSITION OF ACIDIFYING SUBSTANCES FROM THE ATMOSPHERE

There are several pathways by which acidifying substances can be deposited from the atmosphere to aquatic and terrestrial ecosystems. These are: (1) the wet deposition of materials entrained in rain, snow, and fog, that is, "acidic precipitation"; (2) the uptake of certain gases by vegetation, soil, and water surfaces; and (3) the dry deposition of particulates. These subjects are discussed below.

Chemistry of Precipitation

Acidic precipitation is usually defined functionally as having a pH less than 5.65. This is the degree of acidity that is produced by carbonic acid (H_2CO_3) at its equilibrium concentration that occurs when atmospheric CO_2 at ca. 350 ppm is in contact with pure water:

$$CO_2 + H_2O \leftrightarrows H_2CO_3 \leftrightarrows$$
$$H^+ + HCO_3^- \leftrightarrows H^+ + CO_3^{2-}.$$

Therefore, the slightly acidic pH 5.65 is considered to be the acid-rain threshold, and not pH 7.0, which is the pH corresponding to zero acidity (Cogbill and Likens, 1974; Reuss, 1975).

Atmospheric moisture is not, however, merely distilled water in a pH equilibrium with CO_2. Because of the neutralizing influence of other atmospheric gases, and especially of soil-derived cations such as Ca^{2+} and Mg^{2+} that naturally occur in trace concentrations in precipitation, the pH of non-acidified rainwater can be higher than pH 5.65 in some areas, especially in agricultural and prairie landscapes (Kramer and Tessier, 1982; Liljestrand, 1985).

However, in some remote areas, precipitation is not strongly influenced by either pollutants or calcareous dusts. For example, over much of northern North America over Precambrian bedrock, the pH of precipitation is close to 5.0 because of the presence of natural, mineral acids (Schindler, 1988). Consider also the average rainwater pHs of some remote places in various parts of the world: Amsterdam Island in the southern Indian Ocean, pH 4.9; Poker Flats in central Alaska, pH 5.0; Katherine in northern Australia, pH 4.8; San Carlos in Venezuelan Amazonia, pH 4.8; and St. George's in Bermuda, pH 4.8 (Galloway *et al.*, 1982a).

The most abundant cations in precipitation are

usually H^+, NH_4^+, Ca^{2+}, Mg^{2+}, and Na^+, while the most abundant anions are SO_4^{2+}, Cl^-, and NO_3^-. Other ions are also present, but in relatively small concentrations. The acidity of precipitation is due to the presence of H^+ ions, and it is measured in units of pH. Since the pH scale is logarithmic to base 10, a one-unit difference in solution pH implies a 10-fold difference in the concentration of H^+.

The quantity of H^+ in solution is directly related to the difference in concentration (in units of equivalents) of the sum of all anions and the sum of all cations other than H^+. If there are more equivalents of anions than of cations other than hydrogen ion, then H^+ goes into solution to balance the cation "deficit," that is (in microequivalents),

$$H^+ = SO_4^{2-} + NO_3^- + Cl^- - Na^+ - NH_4^+ - Ca^{2+} - Mg^{2+}.$$

This follows from the *principle of conservation of electrochemical neutrality of aqueous solutions*: the total number of cation equivalents must equal the total number of anion equivalents, so that the aqueous solution does not have a net electrical charge (Reuss, 1975). The above equation has been used to calculate the pH of precipitation samples that were collected and analyzed prior to about 1955, for which the reported pH values were suspected to be inaccurate, but there were reliable determinations of the concentrations of all other important ions (Cogbill, 1976).

Table 4.1 summarizes longer-term data for precipitation chemistry at Hubbard Brook, New Hampshire, located in a region of the northeastern United States that is currently experiencing severe acidic deposition, with a mean-annual pH of 4.1–4.2. This site also has the longest continuous record of precipitation chemistry in North America, with weekly collections having been made from 1963 to the present. The Hubbard Brook data set therefore represents an important record of precipitation chemistry.

The average pH of precipitation at Hubbard Brook between 1963 and 1982 was 4.16, and hydrogen ion comprised 71% of the total cation equivalents (Likens *et al.*, 1984). The most important anions were sulfate and nitrate, which were present in a 2:1 equivalent ratio and together accounted for

Table 4.1 Precipitation chemistry at Hubbard Brook, New Hampshire: Average $\pm$ SE of the annual volume-weighted concentration of weekly bulk-collected precipitation over a 19-year study period (1963–1982)[a]

Constituent	µeq/l	%[b]
H	69.3 ± 2.1	71.1
NH₄	10.6 ± 0.6	10.9
Ca	6.5 ± 0.8	6.7
Na	4.8 ± 0.5	4.9
Mg	3.0 ± 0.5	3.1
Al	1.8 ± 0.2	1.8
K	1.5 ± 0.3	1.5
SO₄	54.0 ± 2.1	60.6
NO₃	23.5 ± 1.0	26.4
Cl	11.2 ± 1.2	12.6
PO₄	0.3 ± 0.1	0.3
HCO₃	0.1	0.1
Sum of cations	97.5 ± 2.5	
Sum of anions	89.1 ± 2.6	
pH	4.16	

[a]Modified from Likens *et al.* (1984).
[b]Percentage of total cation or total anion equivalents.

87% of the total anion equivalents. Therefore, most of the acidity in precipitation at Hubbard Brook is in the form of sulfuric and nitric acids. During the 19-year period between 1963 and 1982, there were no statistically significant trends in the mean-annual, volume-weighted concentrations of either hydrogen ion or nitrate in precipitation, but there were significant decreases of 34% for sulfate, 34% for ammonium, 63% for chloride, 79% for magnesium, and 86% for calcium. The specific reasons for these changes in precipitation chemistry are not known.

Not surprisingly, the chemistry of precipitation events at Hubbard Brook is influenced by the direction of movement of the precipitation air mass. Munn *et al.* (1984) presented a trajectory analysis of 69 precipitation events from the period 1975 to 1978. The largest concentrations of hydrogen ion, sulfate, and nitrate were associated with storms tracking from the SSE–SSW sector. These air masses had passed over the large metropolitan conurbations of Boston and New York. Approximately two-thirds of the wet deposition of H^+ was associ-

(a)

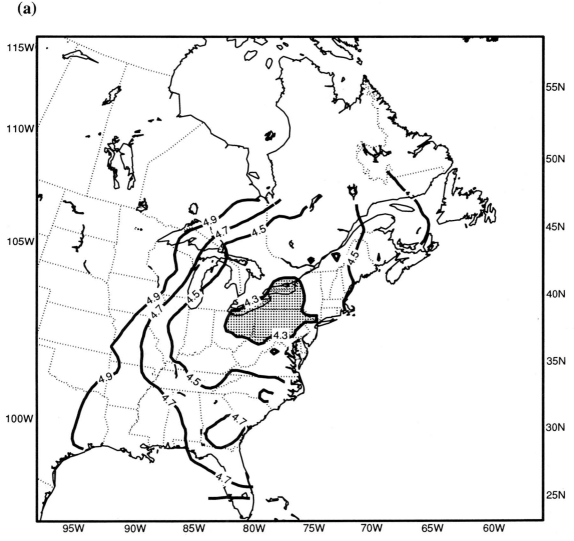

Fig. 4.1 Characteristics of wet-only precipitation in eastern North America in 1990. The isopleths enclose regions characterized by the indicated, mean-annual values: (a) pH, (b) excess sulfate deposition (i.e., corrected for sulfate of marine origin), kg/ha-year, and (c) nitrate deposition, kg/ha-year. Maps were provided by C. U. Ro of Environment Canada, using an integrated database of the U.S. National Atmospheric Deposition Program and Canadian federal and provincial agencies, as summarized in Anonymous (1993c) and Vet *et al.* (1993).

ated with storms from this sector. Air masses from the NNW–ESE sector had relatively small concentrations of H^+, NO_3^-, and SO_4^{2-}.

The chemistry of precipitation varies markedly on both regional (Fig. 4.1) and local scales. In part,

this reflects the patterns of emission of SO_2 and NO_x, their degree of oxidation to sulfate and nitrate, and the prevailing direction traveled by air masses containing these pollutants. Acid-neutralizing dusts in the atmosphere also influence precipitation

(b)

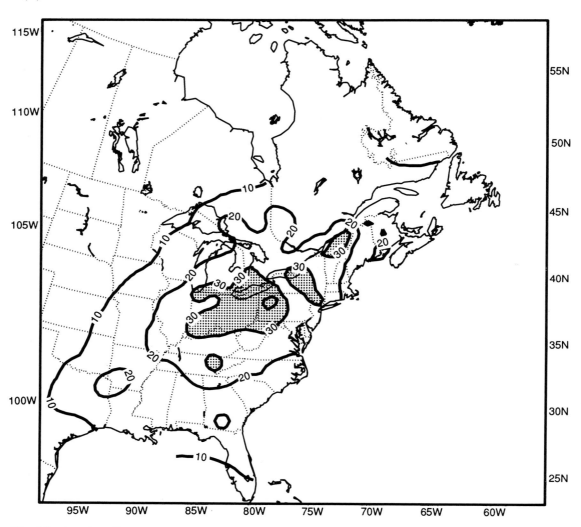

Fig. 4.1 (*continued*)

chemistry. This is especially true where there is little vegetation cover. For example, where agriculture is important, dusts are relatively easily entrained into the atmosphere by winds blowing over bare fields. Dusts entrained by vehicles from dry, unpaved roads can also be important (Barnard, 1986; Gatz *et al.*, 1986).

These influences are illustrated in Table 4.2, which summarizes data for precipitation chemistry at four sites where wet-only precipitation was collected (i.e., the precipitation sampler was only open to the atmosphere during precipitation events). The data for Dorset in southern Ontario are representative of a region where precipitation is quite acidic,

(c)

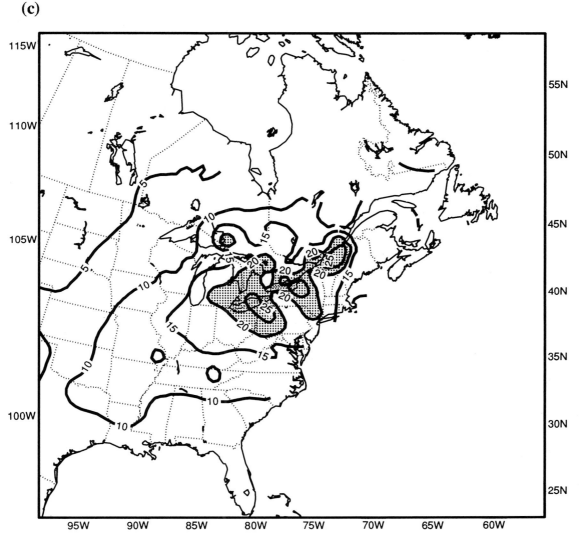

Fig. 4.1   (*continued*)

with an average pH of 4.1. Compared with the other stations, precipitation at Dorset had relatively large concentrations of H^+, SO_4^{2-}, and NO_3^-, implying that the acidity was largely composed of sulfuric and nitric acids. The large concentrations of sulfate and nitrate are believed to be partly anthropogenic, having been derived from oxidation in the atmo-

sphere of SO_2 and NO_x. These gases are emitted from power plants, metal smelters, and automobiles and are then dispersed over long distances before they are deposited to the surface as sulfate or nitrate (see also Chapter 2).

Like Dorset, the Experimental Lakes Area (ELA) site in northwestern Ontario is on a Precam-

Table 4.2 Comparison of the average volume-weighted concentrations of chemical constitutents in 1982–1983 precipitation collected at maritime, continental, and prairie sites in Canada[a]

Constituent		Maritime: Kejimkujik, N.S.	Industrial Continental: Dorset, Ont.	Remote Continental: ELA, Ont.	Prairie: Lethbridge, Alta.
Cations	H^+	25.1	73.6	18.6	1.0
	Ca^{2+}	4.3	10.0	12.0	112.8
	Mg^{2+}	2.9	2.4	2.4	25.5
	Na^+	26.1	3.9	4.3	9.6
	K^+	1.1	1.0	2.0	2.3
	NH_4^+	4.2	15.6	18.9	22.2
Anions	SO_4^{2-}	27.5	58.3	27.1	43.5
	NO_3^-	9.7	35.5	16.4	20.8
	Cl^-	29.5	4.2	5.4	9.9
pH		4.6	4.1	4.7	6.0

[a]Data are in μeq/liter. Calculated from CANSAP (1983, 1984).

brian Shield landscape, with bedrock and soil mainly composed of hard, plutonic minerals, such as granite, gneiss, and quartzite. However, the ELA site is located in a remote area that is infrequently affected by long-range transported air masses that are contaminated by anthropogenic emissions. Hence the precipitation at ELA is less acidic (mean pH 4.7) than at Dorset, and it has smaller concentrations of nitrate and sulfate.

The Kejimkujik, Nova Scotia, site is also fairly remote from sources of emission. It frequently receives air masses from emission-source areas in the northeastern United States and eastern Canada, but by the time these reach the Kejimkujik area, much of their acidic constituents have rained out, and hence precipitation is only moderately acidic (mean pH 4.6). The influence of maritime Atlantic weather systems at Kejimkujik is apparent from the relatively large concentrations of sodium and chloride in precipitation. In addition, marine-derived aerosols help to neutralize some of the acidity of precipitation, since the pH of ocean saltwater is about 8.

The fourth site at Lethbridge, Alberta, is located in a prairie landscape. The precipitation is nonacidic (mean pH 6.0) because of the acid-neutralizing influence of calcareous dusts that are entrained by wind into the atmosphere, where they affect the chemistry of precipitation. Note that the

concentrations of calcium and magnesium are almost one order of magnitude larger at Lethbridge than at the other sites.

There is a rather narrow transition between prairie and boreal-forest vegetation zones with respect to the neutralizing influence of calcareous dusts (Gorham, 1976; Eisenreich et al., 1980; Munger, 1982; Munger and Eisenreich, 1983). Eisenreich et al. (1980) collected snow and rain during precipitation events along a 560-km transect running from a calcareous, prairie-agricultural landscape in southwest North Dakota, through an area of forest transition, to granitic, conifer-forested terrain of northeastern Minnesota. The average, volume-weighted pH of snow and rainfall ranged from 5.3 in the agriculture prairie, to 4.6 in the conifer forest. This was paralleled by a change in the concentration of calcium in precipitation, from 1.6 mg/liter in the southwest, to 0.2 mg/liter in the northeastern forest.

Gorham (1976) sampled snow along a similar transect in 1975, after a windstorm that entrained large quantities of prairie dust into the atmosphere. The average pH of precipitation of 10 sites at the prairie-agricultural end of the transect was 8.0 and there was 274 mg/liter of particulates, while among 9 boreal forest sites in the northeastern part of his transect, the mean pH was 4.9 and there was only 17 mg/liter of particulates. Another study done in southern Ontario compared the precipitation chemistry of 8 sites in a largely forested, Precambrian Shield landscape with a thin mantle of hard plutonic minerals, to that of three sites in agricultural-forested terrain with calcareous soil, located just south of the Shield periphery (Dillon *et al.*, 1977). The average, volume-weighted pH of precipitation among the Shield sites ranged from 4.1 to 4.2, while for the sites south of the shield it was 4.8 to 5.8.

The above studies clearly show that calcareous dusts entrained into the atmosphere can strongly influence the chemistry of precipitation on both local and regional scales. This effect is strongest in terrain where agriculture is an important land use.

A few studies have been made of cloudwater, which can be sampled by aircraft-based instrumentation. Cloud moisture can sometimes be quite acidic, because: (a) ionic constituents are less "diluted" in cloud water than in precipitation, (b) the large surface areas of tiny water droplets make them highly efficient at scavenging sulfur dioxide from the atmosphere, and (c) the absorbed sulfur dioxide is converted to sulfate in the droplet. Cloudwater samples collected in Russia in 1961–1964 had a pH range of 3.4–5.9, while samples collected in the eastern United States during 1981–1983 had a pH range of 3.1–6.1 (Sisterton, 1990).

Fog moisture has been studied more frequently than cloudwater and has often been found to be acidic. This phenomenon was initially noted by Houghton (1955) in an analysis of fogwater samples collected at five locations in the northeastern United States. Fogwater from Mount Washington, New Hampshire, had a pH as acidic as 3.0, and samples from Brooklin, Maine, had a pH as low as 3.5. More recently, pH values as acidic as 3.0–3.5 have been reported in the same area of the eastern United States (Hileman, 1983; Shriner, 1990) and pH 4.0 at coastal locations in eastern Canada (S. Beauchamp, Environment Canada, personal communication). In Ohio, 10 fogwater events during 1985–1986 ranged in pH from 2.85 to 4.06 (median 3.8), while rainwater pH ranged from 3.94 to 4.26 (median 4.2) (Muir *et al.*, 1986). At a montane site in the Blue Ridge

Mountains of Virginia, fogwater during 1986–1987 had an average pH of 3.7 (205 μeq/liter), about four times as acidic as precipitation (pH 4.3; 52 μeq/liter) and six times as acidic as oak-influenced throughfall (pH 4.5; 33 μeq/liter) (Sigman *et al.*, 1989). Even in California, where rainwater pH is generally above 4.4, fogwater pHs of 1.7–2.3 have been recorded (Hileman, 1983; Shriner, 1990).

At relatively high-elevation sites where fog is frequent, windy conditions effectively impact small droplets of water vapor into the coniferous-forest canopy. Under these conditions, the direct depositions of cloud water and acidity can be considerable (Vogelmann *et al.*, 1968; Siccama, 1974; Lovett *et al.*, 1982; Sisterton, 1990). The total input of moisture to a balsam fir (*Abies balsamea*) stand at 1220 m in New Hampshire was estimated to be about 180 cm/year, while direct cloudwater deposition was 84 cm/year (Table 4.3; fog was present 40% of the time). Because many chemical constituents had relatively large concentrations in the cloudwater, their subsequent rates of deposition were also large. Fogwater deposition accounted for 62% of the total deposition of H^+ and 81% of the depositions of SO_4^{2-} and NO_3^- (Table 4.3). Clearly, for stands at high-altitude sites where fog is frequent, the direct cloudwater inputs of acidity and other atmospheric constituents can be substantial.

Spatial Patterns of Acidic Deposition

The area that is currently experiencing acidic precipitation is widespread in eastern North America (Fig. 4.1a), northwestern Europe (Fig. 4.2), and elsewhere (Cogbill, 1976; Likens and Butler, 1981; Persson, 1982; Rodhe and Granat, 1984; Galloway *et al.*, 1987b; Shriner, 1990; Sisterton, 1990). Acidic precipitation has been less well studied and monitored outside of North America and western Europe, but it is known to be a problem in other places where emissions of SO_2 and NO_x gases are substantial. In southern China, for example, at least 13 urban areas have mean-annual rainwater pHs <4.5 and experience events as acidic as pH 3.1 (Hongfa, 1989).

In eastern North America prior to 1955, the re-

Table 4.3 Chemical composition of cloud water and deposition of selected chemical constituents to a balsam fir (*Abies balsamea*) forest at 1220 m on Mt. Moosilauke, New Hampshire[a]

Constituent	Concentration (μeq/liter; mean $\pm$ SD)	Deposition (kg/ha-year)			Percentage of total from clouds
		Cloud	Bulk	Total	
H^+	288 $\pm$ 193	2.4	1.5	3.9	62
NH_4^+	108 $\pm$ 89	16.3	4.2	20.5	80
Na^+	30 $\pm$ 29	5.8	1.7	7.5	77
K^+	10 $\pm$ 4	3.3	2.1	5.4	61
SO_4^{2-}	342 $\pm$ 234	275.8	64.8	340.6	81
NO_3^-	195 $\pm$ 175	101.5	23.4	124.9	81
H_2O (cm/year)		84	180	264	32

[a]Modifed from Lovett *et al.* (1982).

gion that experienced precipitation with pH <4.6 was relatively small, localized, and centered over New York, Pennsylvania, and southern New England. This area expanded in the next few decades, as did the area receiving precipitation with pH less than 5.6, to the extent that most of the eastern United States and southeastern Canada is now receiving acidic precipitation (Cogbill and Likens, 1974; Cogbill, 1976; Shriner, 1990; Figure 4.1a). It seems that the general geographical pattern of acidic precipitation in eastern North America has existed since at least the early to middle 1950s. However, since that time, the phenomenon has become more widespread and to a degree its intensity has increased (severe acidic precipitation is characterized by a mean-annual pH of about ≤4.3). Clearly, an important aspect of acidic precipitation is its regional character; it affects large expanses of terrain.

Another notable characteristic of acidic precipitation is that it does not generally decrease in intensity with increasing distance from large, point-source emitters of acid-anion precursors, such as power plants or smelters that vent SO_2 to the atmosphere from tall smokestacks. This observation further reinforces the idea that acidic precipitation is regional in character. For example, several studies have reported an insignificant influence of distance from the world's largest point-source emitter of SO_2 near Sudbury, on the local acidity of precipitation (Dillon *et al.*, 1977; Freedman and Hutchinson, 1980a; Chan *et al.*, 1982, 1984b). Furthermore, the acidity of precipitation did not change when that smelter was closed by a strike in 1978–1979 (Scheider *et al.*, 1980). During the 7-month strike period,

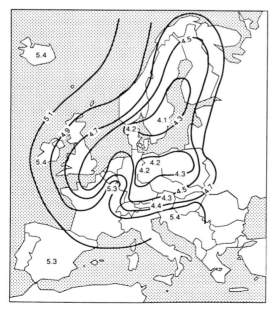

Fig. 4.2 The acidity of wet-only, volume-weighted precipitation in western Europe in 1985. Modified from Sisterton (1990).

the hydrogen ion concentration in precipitation averaged 33 μeq/liter or pH 4.49, compared with 30 μq/liter or pH 4.52 in a comparable period prior to the shutdown when there were large emissions of SO_2.

In part, this remarkable lack-of-a-distance effect at Sudbury is due to the fact that at any given time, only a relatively small sector of terrain (less than 10%) is directly beneath the smokestack plume. The rest of the area is solely under a regional influence. Millan *et al.* (1982) studied a rainfall event in 1980 and found that rain collected directly beneath the smelter plume had an average pH of 4.1, compared with 4.3 beyond the plume influence. Sulfate was more concentrated by 60% beneath the plume, and the metals Cu, Ni, and Fe were enhanced by factors of 5–20. Despite this measurable influence of the plume on the chemistry of precipitation that has passed through it, the average, ground-level influence in the Sudbury area is minor because the plume sector is small and spatially variable. As a result, the long-range transport of air pollutants appears to have a larger influence than local sources on the chemistry of precipitation, even in the vicinity of the large SO_2 sources at Sudbury (Tang *et al.*, 1987).

Qualitatively similar observations have been made in the vicinity of power plants that emit SO_2; that is, there is a relatively minor or no effect of distance from the point source on the acidity of local precipitation (Granat and Rodhe, 1973; Hutcheson and Hall, 1974; Enger and Hogström, 1979; Kelly, 1984; however, see Li and Landsberg, 1975).

Dry Deposition of Acidifying Substances

Dry deposition of atmospheric constituents is a continuous process, occurring in the intervals between precipitation events, as well as during those events. Dry deposition includes the direct uptake of gases such as SO_2 and NO_x by vegetation, soil, and water surfaces, plus the gravitational settling and impaction filtering of particulate aerosols.

These processes can result in substantial inputs of certain atmospheric constituents, including some

that can generate acidity when they undergo chemical transformations in the receiving ecosystem [Reuss, 1975, 1976; International Electric Research Exchange (IERE), 1981; Glass *et al.*, 1981; Hicks, 1990]. For example, gaseous SO_2 can dissolve directly into the surface water of a lake, or it can be absorbed at the moist, substomatal surfaces of the spongy mesophyll after entering plant foliage through stomata (behaving similarly to gaseous CO_2 in this respect). The dry-deposited SO_2 is oxidized to the sulfite (SO_3^{2-}) anion, which is then rapidly oxidized to the sulfate anion. The sulfate is electrochemically balanced by the same number of equivalents of hydrogen ion. Similarly, dry-deposited NO_x gas can be oxidized to the nitrate anion, which also generates an equivalent quantity of hydrogen ion.

Even certain atmospheric chemicals that are not in themselves acidic can generate acidity when they are chemically transformed in soil or water. For example, the gas ammonia and the cation ammonium can be deposited from the atmosphere to soil, where if conditions are suitable they will be oxidized to nitrate plus two equivalent units of hydrogen ion (see Fig. 4.3). Because inputs of these nonacidic substances can cause acidification when they are chemically transformed, the phrases "acid rain" and "acid deposition" are inaccurate (but shorter) replacements for "the deposition of acidifying substances from the atmosphere."

The dry depositions of sulfur and nitrogen compounds are largest in regions with large atmospheric concentrations of gaseous NO_x and SO_2. Unfortunately, processes of dry deposition to ecosystems are not well documented, and the quantitative estimation of dry fluxes is difficult. This is especially true of sulfur and nitrogen compounds. The major reasons for the uncertainty are the paucity of data concerning: (1) the spatial and temporal variations of concentrations of atmospheric gases and particulates and (2) the rates of their depositions to natural ecosystems of great structural complexity and under various weather conditions (Unsworth, 1979; Fowler, 1980a,b; Grennfelt *et al.*, 1980). In spite of these problems, several estimates of dry deposition to natural ecosystems have been made, and they

(a) Nitrogen Transformations

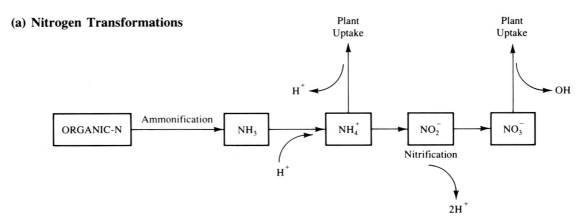

(b) Sulfur Transformations

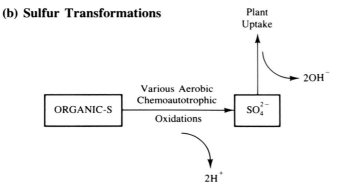

Fig. 4.3   A simplified flow diagram of the acidifying effect of some (a) nitrogen and (b) sulfur transformations in soil. See text for an explanation. Modified from Reuss (1975).

have helped to put into perspective the rates of dry deposition of sulfur and nitrogen compounds, and acidity, relative to their rates of wet input with precipitation.

A conceptually and mathematically simple model of dry deposition was computed by Fowler (1980a), using data from the literature on the average atmospheric concentrations of S- and N-containing gases and particulates and their likely deposition velocities to natural surfaces. These data were used to compute the dry-deposition fluxes to three landscapes located at different distances from sources of pollution. The computed dry inputs were then compared to the inputs via precipitation. Although this simple budget ignores a great deal of spatial and temporal variation by using average val-

ues for deposition velocities and for concentrations of pollutants, it is nevertheless useful for gaining an appreciation of the relative magnitudes of the various fluxes. Total (wet + dry) depositions of both S and N were much larger near the sources than further away (Table 4.4). This pattern was almost entirely due to a difference in dry deposition (especially of gaseous S and N), since wet deposition was much less affected by distance from the emission sources. Dry deposition dominated in the relatively polluted environment close to the sources, where it was 8.5 times larger than wet deposition for S and 4.9 times larger for N. Even at an intermediate distance, dry deposition was larger than wet, while at a relatively remote distance the two input sources were similar in magnitude. These simple calcu-

Table 4.4 Deposition budgets for several important sulfur and nitrogen compounds (expressed as S or N) to terrestrial landscapes at various distances from emissions sources[a]

	Deposition velocity (cm/sec)	Distance from emission sources					
		3–30 km		30–300 km		300–3000 km	
		Concentration (μg/m³)	Flux (kg/ha-year)	Concentration (μg/m³)	Flux (kg/ha-year)	Concentration (μg/m³)	Flux (kg/ha-year)
Dry deposition							
Gas SO_2	0.8	100	126.2	20	25.2	5	6.3
Gas NO_2	0.5	30	14.4	5	2.4	2	1.0
Particulate SO_4^{2-}	0.1	20	2.1	10	1.1	10	1.1
Particulate NO_3^-	0.1	5	0.4	2	0.1	2	0.1
Wet deposition (90 cm/year precipitation)							
SO_4^{2-}		5	15.0	4	12.0	3	9.0
NO_3^-		1.5	3.0	0.7	1.4	0.5	1.0
Total deposition, S			143.3		38.3		16.4
Total deposition, N			17.8		3.9		2.1
Dry/wet, S			8.5		2.2		0.8
Dry/wet, N			4.9		1.2		1.1

[a]Modified from Fowler (1980a).

105

lations indicate that the dry deposition of acidic substances and acid-forming precursors can be quantitatively important and that dry deposition dominates the total input in relatively contaminated environments close to sources of emission.

The determination of atmospheric deposition in the vicinity of large point sources also reveals this general pattern. Chan *et al.* (1984a) computed sulfur deposition within a 40-km radius of the largest smelter at Sudbury. They found that 55% of the total S deposition of 15.9×10^3 kg/day was due to dry deposition (Table 2.11, Chapter 2). Of this, 91% was dry deposition of SO_2, and the rest was dry input of sulfate aerosol. However, because the smokestack at Sudbury is so tall (380 m) and effective at dispersing emissions, the local, ground-level pollution is only moderate. As a result, the total S deposition within a 40-km radius is only 16% larger than the background S deposition that would be expected if the smelter was not present. In fact, <1% of the total emission of SO_2 from the smelter is deposited within 40 km of the source; virtually all of the SO_2 emission is transported over a longer distance.

Several other studies have estimated dry and wet inputs to forests. Grennfelt *et al.* (1980) calculated atmospheric deposition to a coniferous forest in southern Sweden and found that dry inputs accounted for 57% of the total input of 23 kg S/ha/year. Uptake of SO_2 accounted for 68% of the total dry input and impaction of sulfate aerosol the remainder. The dry input of nitrogen accounted for 54% of the total N input of 13 kg/ha/year. At Hubbard Brook, New Hampshire, total S deposition to a northern hardwood forest was 19 kg/ha/year, of which 31% was dry uptake of SO_2 and <2% was dry particulate deposition (Eaton *et al.*, 1980). Of the total deposition of S of 25.6 kg/ha/year to a hardwood forest in Tennessee, 56% was dry deposition, while for nitrogen, dry inputs accounted for 63% of the total input of 7.6 kg NO_3-N/ha/year and 33% of the 2.5 kg NH_3-N/ha/year (Lindberg *et al.*, 1986).

Dry inputs were also important for a beech (*Fagus sylvatica*) forest at Solling in central West Germany, where they accounted for at least 38% of the total S input, >17% of NO_3-N, and >56% of total hydrogen ion. The dry inputs resulted in a much larger sulfate flux in precipitation reaching the forest floor of the beech stand (measured as throughfall + stemflow) than in incident precipitation (i.e., 48 vs 24 kg/ha/year). Because of its more complex surface architecture, larger foliage biomass, and longer foliage persistence, this enhancement effect was even more pronounced in a nearby spruce (*Picea abies*) forest, where throughfall (TF) plus stemflow (SF) was 80 kg/ha/year (Mayer and Ulrich, 1974, 1977; Heinrichs and Mayer, 1977).

A large, industrialized region of eastern Europe has large and relatively uncontrolled emissions of sulfur, and the wet + dry rates of deposition are therefore extremely large, amounting to more than 100 kg S/ha-year (Stigliani and Shaw, 1990).

4.3
CHEMICAL CHANGES WITHIN THE WATERSHED

The chemistry of precipitation is greatly altered by interactions that take place within watersheds, especially in the terrestrial environment. Processes that are particularly important are ion exchange, uptake and leaching of nutrients by plants, and chemical transformations by microbial and inorganic reactions.

Canopy Effects

The first substrates that incoming precipitation encounters in a forest are the foliage and bark surfaces of trees. Most of the incoming water penetrates through the canopy and reaches the forest floor as a flux called throughfall, while another relatively minor fraction runs down the trunks of trees as stemflow. In addition, some of the incoming precipitation is intercepted by the canopy and evaporated back to the atmosphere (canopy evaporation, CE). During the growing season, the incoming precipitation to four conifer stands in Nova Scotia was partitioned into an average of 34% CE, 66% TF, and <1% SF, while for four angiosperm stands it was 22% CE, 75% TF, and 3% SF (Freedman and Prager, 1986). Averaged over an entire year, Eaton *et al.* (1973) reported 14% CE, 81% TF, and 5% SF for a hardwood forest at Hubbard Brook, NH. Because the evaporation of water back to the atmosphere as CE is essentially a distillation process, this should cause a proportionate concentration of chemical constituents in TF + SF, compared with ambient precipitation.

However, changes in the chemistry of throughfall and stemflow are much larger than this distillation effect. Several major ions, particularly potassium, are relatively easily leached from foliage. Freedman and Prager (1986) reported an average, volume-weighted concentration of potassium in TF + SF of 2.1 mg/liter (2.0 mg/liter in TF and 5.7 mg/liter in SF) among eight stands that they studied, compared with <0.2 mg/liter in the incident precipitation. The concentrations of calcium and magnesium were also enhanced in TF + SF (0.78 and 0.26 mg/liter, respectively) compared with precipitation (0.26 mg/liter and 0.06 mg/liter, respectively). Experimental studies have shown that treatment with acidic solutions increases the rates of leaching of K, Ca, and Mg from foliage, even at acidities that do not cause visible injury (Fairfax and Lepp, 1975; Lepp and Fairfax, 1976; Tukey, 1980; Tveite, 1980a). However, there is no evidence that increased leaching of these nutrients causes physiologically important decreases in the foliar concentrations of these nutrients (Tveite, 1980a,b; Morrison, 1984).

The average, volume-weighted concentration of hydrogen ion in the ambient precipitation sampled by Freedman and Prager (1986) was 0.041 mg/liter or pH 4.4. Acidity averaged slightly less in TF + SF of the four hardwood stands (0.018 mg/liter or pH 4.7) and in two of four conifer stands (these averaged pH 4.5; the other two were pH 4.4). The small decrease in acidity was probably caused by ion-exchange reactions taking place on foliage and bark surfaces, with H^+ largely exchanging for Ca^{2+},

Mg^{2+}, and K^+. Coupled with reduced water volume, the generally lesser acidity of TF + SF compared with incident precipitation resulted in an average canopy "consumption" of 66% of the precipitation input of hydrogen ion to the hardwood stands and 42% for the conifer stands.

A similar effect has been reported by others. Eaton *et al.* (1973) calculated an average annual H^+ consumption of 91% by a hardwood forest canopy at Hubbard Brook. Mahendrappa (1983) found a H^+ consumption of 21–80% among seven stands in New Brunswick. Mollitor and Raynal (1983) reported 43% consumption in an Adirondack, NY, hardwood stand and 14% in a conifer stand. Foster (1983) reported 38% consumption by a hardwood canopy in northern Ontario. Cronan and Reiners (1983) reported 31% for a New Hampshire hardwood stand, but an increase in H^+ flux of 16% under conifers. Finally, Abrahamsen *et al.* (1976) reported 47% consumption of H^+ flux (pH 4.5) under a *Betula* canopy in Norway compared with precipitation (pH 4.3), while flux increased by a factor of 2.2 times under *Pinus sylvestris* (pH 4.0) and was unchanged under *Picea abies*.

In an area where the atmospheric concentrations of SO_2 and aerosol sulfate are small, the sulfate flux in TF + SF tends to be only slightly larger than that in incident precipitation. Freedman and Prager (1986) reported an average growing season TF + SF flux of 4.2 kg SO_4-S/ha (equivalent to 1.3 mg SO_4-S/liter) in hardwood stands and 5.2 kg/ha (1.7 mg/liter) under conifers, compared with 3.2 kg/ha (0.8 mg/liter) in ambient precipitation, for an average enhancement of sulfate flux in TF + SF of 47%. In a more polluted environment, the enhancement of sulfate flux in TF + SF can be much larger, because of washoff of dry-deposited sulfur dioxide and sulfate. Eaton *et al.* (1973) observed an average-annual sulfate enhancement of 414% at Hubbard Brook, while in central West Germany Heinrichs and Mayer (1977) reported an enhancement of 200% in a beech forest, and 330% in a spruce forest. The study sites of Eaton *et al.* and Heinrichs and Mayer are located in relatively polluted regions, where there are enhanced atmospheric concentrations of aerosol sulfate, and to a lesser extent sulfur dioxide.

Chemical Changes in Soil

After precipitation reaches the forest floor, it begins to percolate through the soil. Important chemical changes take place as this water moves downward and interacts with mineral soil constituents, organic matter, microbes, and plant roots. Some examples of those changes are: (1) the biota can selectively take up or release chemical ions; (2) ions can be selectively exchanged at nonliving surfaces that have exchange capability, such as those of clay and organic matter; (3) water-insoluble minerals can be made soluble by weathering processes; and (4) secondary minerals such as certain clays and sesquioxides of iron and aluminum can be formed by crystallization/precipitation reactions involving particular ions.

These various chemical changes can contribute to acidification, to the leaching of basic cations such as calcium and magnesium, and to the mobilization of toxic cations of aluminum (especially Al^{3+}). These closely linked processes take place naturally in all sites where there is an excess of precipitation over evapotranspiration, and they may also be influenced by the nature of the vegetation. A potential influence of acidic deposition is to increase the rates of some of these processes and to enhance the rates of leaching of toxic Al^{3+} and H^+ to surface waters.

Some of these chemical changes have been examined experimentally by leaching soil monoliths isolated in plastic tubes with simulated rainwaters of various pH. These lysimeter experiments have shown that acidic leaching solutions can cause a number of effects on soil chemistry (Overrein, 1972; Abrahamsen *et al.*, 1976, 1977; Abrahamsen and Stuanes, 1980; Farrell *et al.*, 1980; Stuanes, 1980; Morrison, 1981, 1983; Tyler, 1981). These include: (1) increased rates of leaching of calcium, magnesium, and potassium, resulting in a decreased base saturation of the cation-exchange capacity; (2) increased concentrations of toxic, cationic forms of metals such as Al, Fe, Mn, Pb, Ni, etc.; (3) acidification of the soil; and (4) saturation

of sulfate adsorption capacity, after which sulfate leaches at about the rate of input, with a secondary effect on soil acidification if the mobile sulfate is accompanied by base cations and, more importantly, a secondary acidifying and toxic influence on surface waters if the leaching sulfate is accompanied by Al^{3+} and H^+ (Reuss *et al.*, 1987).

Some of these effects are illustrated in Tables 4.5 and 4.6. In the first experiment (Table 4.5), monolith profiles of two sandy soils were collected from jack pine (*Pinus banksiana*) stands in Ontario and leached with artificial "rainwaters" adjusted with mineral acid to pH 5.7, 4, 3, or 2. The percolates emerging from the bottom of the lysimeter were collected and analyzed chemically. The effects on soil acidity were minor. Even after 3 years of treatment with a solution as acidic as pH 2, the pH of the percolate was >6.5, compared with an initial pH of 7.3 (Morrison, 1981, 1983). Effects on the concentrations of calcium, magnesium, sum of basic cations (i.e., Ca + Mg + K + Na), and sulfate only occurred in the artificially acidic pH 2 treatment. For all of these chemical constituents, no effects were observed until a time (i.e., dose)-dependent threshold had passed, after which there were consis-

Table 4.5 Concentration of selected chemical constituents in percolate from lysimeters containing a profile of sandy soil from jack pine (*Pinus banksiana*) stands in northern Ontario[a]

	pH of leaching solution	Concentration in percolate (μeq/liter)			
		Ca	Mg	All bases	SO$_4$
Soil A	5.7	740	50	994	229
	4	670	49	896	224
	3	745	73	1098	152
	2	872	204	1348	527
Soil B	5.7	422	174	727	146
	4	446	183	765	152
	3	399	165	701	277
	2	3722	1474	5504	5427

[a]Data are for week 152; the leaching solution was added at 100 cm/year in equal weekly 1- to 2-hr events; average of three replicates per treatment. Modified from Morrison (1981, 1983).

tently larger concentrations in the percolate. For example, the sulfate concentration in the pH 5.7 treatment remained stable at less than 20 μeq/liter for both soil types. In the pH 2 treatment, sulfate concentration remained at about 20 μeq/liter for the first 100 and 150 weeks for the two soils, respectively, after which the concentrations in the percolate rapidly increased to 400 and 150 μeq/liter, respectively. These thresholds indicate that the sulfate adsorption capacity of the soils had become saturated, after which the added sulfate essentially flushed through the system, as a highly mobile anion.

The study of Farrell *et al.* (1980) involved the treatment of soil columns in the field with two rates of addition of sulfuric acid, which were equivalent to 15 and 33 kg S/ha/year (Table 4.6). These treatments simulated the rate of loading experienced with 100 cm/year of precipitation having a pH of 4.0 or 3.7, respectively. After 6 years of treatment, the effect on pH of the surface organic layer was minor. However, there was a large decrease in the base saturation of cation-exchange capacity, characterized by a decreased quantity of Ca + Mg + K + Na in the surface organic layer and in the top 5 cm of mineral soil, especially at the larger rate of acid loading. These effects are more marked than those observed by Morrison (1981, 1983) (Table 4.5),

who only found changes in the concentrations of these ions at pH 2. Farrell *et al.* (1980) found that the effect of acid treatment on base saturation could be largely overcome by the addition of NPK fertilizer; this was attributed to the presence of potassium in the fertilizer (Table 4.6). Farrell *et al.* (1980) also determined sulfur retention in their experiment. At their larger rate of acid loading, the total input to the columns was equivalent to 241 kg S/ha; 28% of this was ambient bulk deposition, and 72% was the experimental addition. Of the total input, 66% was recovered as sulfate leached from the lysimeter; the other 34% was retained in the columns by various sulfate-absorption mechanisms. The vertical distribution of the retained sulfur was investigated using a ^{35}S radiotracer technique. Three months after adding the radioactive sulfur, 24% of the retained ^{35}S was present in the organic strata (vegetation + forest floor). The rest was retained in the mineral soil, especially in the B horizon.

Soil Acidity

Factors Affecting Soil Acidity

The acidity of soil is influenced by a complex array of physical/biological chemical transformations and exchanges. Some of the more important chemi-

Table 4.6 Effect of irrigating a coniferous forest soil between 1972 and 1977 with several rates of acid ± NPK fertilizer: Average of eight replicate lysimeters per treatment[a]

Treatment[b]	Acid added (keq H^+/ha)	pH of A_0	% Base saturation of A_0	Quantity of base cations (keq/ha)			
					Mineral soil		
				A_0	0–5 cm	5–10 cm	Total
Control	0.0	4.0	16.0	5.2	1.8	0.7	7.7
Acid 1	6.1	3.9	12.2	3.2	1.9	0.6	5.7
Acid 2	13.3	3.8	9.0	2.1	0.9	0.5	3.5
NPK	0.0	4.2	20.1	5.7	—	—	—
NPK + acid 1	6.1	3.9	15.2	5.0	3.3	0.8	9.1
NPK + acid 2	13.3	3.8	11.5	3.5	2.7	0.8	7.0

[a]Modified from Farrell *et al.* (1980).

[b]Acid 1, annual application of H_2SO_4 at 15 kg S/ha-year: acid 2, annual application of H_2SO_4 at 33 kg S/ha-year: NPK, total application of 360 kg/N/ha as NH_4NO_3, plus 80 kg P/ha and 150 kg K/ha as a compound PK fertilizer.

cal processes that affect soil acidity are summarized below. All of these processes operate naturally, but their rates can potentially be affected by acidic deposition.

Carbonic Acid.   In the forest floor and in soil rich in organic matter, there can be large concentrations of CO_2 (frequently >1%) in the interstitial atmosphere because of respiration by saprophytes and plant roots. As a result, there is a relatively large equilibrium concentration of H_2CO_3 in the soil solution, and the pH can be more acidic than the pH 5.65 realized from a CO_2 equilibrium with the above-ground atmosphere. This carbonic-acid effect is most influential in soils having a pH >6 and with appreciable alkalinity—which usually means a calcareous substrate or a relatively young and unweathered, noncalcareous soil of glacial or other primary origin. In relatively acidic soils (i.e., pH <5.0–5.5) with less or no alkalinity, the influence of the partial pressure of CO_2 on acidity is relatively unimportant (Russell, 1973; Reuss, 1975, 1976; Hausenbuiller, 1978; Bache, 1980).

The Nitrogen Cycle.   Plant uptake/release and chemical transformations of nitrogen can have large and complex influences on soil acidity (Kirkby, 1969; Russell, 1973; Reuss, 1975, 1976). Most plants take up inorganic nitrogen from the soil solution as one or both of NH_4^+ or NO_3^-. However, in acidic soil with a pH less than about 5.0–5.5, the dominant form of inorganic N is NH_4^+. The NH_4^+ may have originated from ammonification of organic N to form ammonia (NH_3), which then consumes one H^+ to form an ion of NH_4^+ (Fig. 4.3). If the NH_4^+ is taken up by a plant root, an equivalent quantity of H^+ is excreted to the soil solution in order to maintain electrochemical neutrality. In this case, there is no net change in acidity of the soil. However, if there is a direct input of NH_4^+ to the soil, for example, from atmospheric deposition or by fertilization, then plant uptake of NH_4^+ with concomitant H^+ release will have a net acidifying effect.

In less acidic soil with pH >5.5, most of the inorganic nitrogen occurs as NO_3^-. Much of this nitrate is produced by microbial oxidation of NH_4^+ via a NO_2^- intermediate in the acid-sensitive process of nitrification, carried out by bacteria in the genera *Nitrosomonas* and *Nitrobacter*. Overall, the oxidation of NH_4^+ to NO_3^- generates two H^+ for every NO_3^- produced. If the ammonium substrate was derived from ammonification of organic-N, which consumes one H^+ per NH_4^+ produced, then the net effect is production of one H^+ per NO_3^- ion produced ultimately from organic N. However, if the nitrate is subsequently taken up by a plant root, then to maintain electrochemical neutrality one OH^- would be excreted, which is equivalent to the consumption of one H^+. Therefore, the net effect on soil acidity is zero.

If there is a direct input to the soil of NH_4^+, which then serves as a substrate for nitrification, then acidification will take place. Such a direct input could occur by the deposition of ammonia gas or ammonium ion from the atmosphere or by the addition of an inorganic fertilizer, such as ammonium nitrate, ammonium sulfate, or urea. The acidifying effect of the use of ammonium-containing fertilizers is a well-recognized phenomenon (Hausenbuiller, 1978; Bache, 1980; Jenny, 1980).

The Sulfur Cycle.   The transformation and uptake of sulfur compounds can also influence soil acidity (Russell, 1973; Reuss, 1975, 1976). Much of the sulfur in soil is present in the reduced form of organic sulfur. This can be transformed by various microbial processes through incompletely oxidized forms such as sulfides, elemental sulfur, and thiosulfates, but in an aerobic environment all of these are ultimately oxidized to sulfate by aerobic chemoautotrophs. The oxidation of organic S to SO_4^{2-} is accompanied by the release of an equivalent quantity of H^+ (i.e., one equivalent of H^+ per equivalent of SO_4^{2-} produced, or 2 moles H^+/mole SO_4^{2-} produced). If the SO_4^{2-} is taken up by plant roots, then an equivalent quantity of OH^- is excreted to the soil solution to conserve electrochemical neutrality, and there is no net effect on soil acidity. However, if there is a direct input of sulfate by atmospheric deposition, followed by plant uptake, then the net effect would be a reduction of soil acidity. If the soil is anaerobic, for example, because of a high water table, then the sulfate would be reduced to sulfide or another reduced-S compound by anaerobic chemoautotrophs, and this transformation would be accompanied by the consumption of an equivalent quantity of H^+ per equivalent of SO_4^{2-} reduced.

In most terrestrial ecosystems, the sulfur cycle has a much smaller effect on soil acidity than the nitrogen cycle. However, there are certain situations in which the sulfur cycle dominates. Most notably, the drainage of wetlands causes the previously anaerobic soil to become aerobic. Under this condition, a large quantity of acidity can be generated by the chemoautotrophic oxidation of reduced forms of sulfur, especially sulfides. In agriculture these are called "acid sulfate soils," and pH <3 can be present for as long as 10 years following the drainage of wetlands that are susceptible to this syndrome (Dost, 1973; Rorison, 1980).

In some situations, sulfide minerals, especially pyrites, can be exposed to atmospheric oxygen, and large quantities of acidity can be generated when the sulfides and reduced-iron species are oxidized. The net reaction occurring when pyrites are oxidized is as follows:

$$4FeS_2 + 15O_2 + 14H_2O \rightarrow 4Fe(OH)_3 + 16H^+ + 8SO_4^{2-}.$$

This phenomenon is known as "acid-mine drainage," and it can cause severe acidification, to an intensity of less than pH 2 in water, accompanied by large concentrations of sulfate and soluble iron and aluminum (Harrison, 1958; Hargreaves et al., 1975; Turner, 1990; J. P. Baker, 1990). Acid-mine drainage has degraded many surface waters in areas where coal mining has exposed mineral sulfides to the atmosphere. The U.S. Environmental Protection Agency sampled a large number of lakes and streams in acid-sensitive areas, mostly in the eastern United States, and of 1180 lakes and 4670 streams that were considered acidic, 3 and 26%, respectively, had been acidified by acid-mine drainage (L. A. Baker et al., 1991).

Leaching of Ions. In many well-drained terrestrial soils, sulfate, nitrate, and chloride leach readi-

ly to groundwater. For sulfate, this is especially true of the relatively young forest soils of formerly glaciated terrain of much of the northern hemisphere. In soils of this sort, sulfate adsorption capacity is small and easily saturated by enhanced atmospheric inputs caused by anthropogenic emissions. Older, highly weathered soils of more southerly latitudes tend to have considerably larger sulfate adsorption capacities (Reuss et al., 1987).

Soils have little capability to adsorb nitrate. This anion will leach readily if it is present in the soil solution in a large concentration, as often occurs when disturbance or fertilization of the site results in a large availability of nitrate in comparison with the biological demand (see also Chapter 9.3). In regions where the atmospheric depositions of nitrate and ammonium are large, the biological capacity for nitrate uptake can be exceeded within the watershed, and nitrate can leach at relatively large rates (Aber et al., 1989, 1991; Rosen et al., 1992).

In terrestrial, "nitrogen-saturated" systems, plant productivity is not limited by the availability of inorganic nitrogen. Potential, negative effects of nitrogen saturation could include an increased acidification of soil and water through aluminum and base-cation leaching (see below), nutrient loading to aquatic systems, and a predisposition of trees to suffering decline (discussed in Chapter 5). Sometimes, the relatively fertile conditions of nitrogen-saturated sites will favor the growth of certain species of plants, for example hairgrass (*Deschampsia flexuosa*) in Scandinavian forests (Rosen et al., 1992).

Sites vary in their predisposition to developing a nitrogen-saturated condition, depending on the character of their soils and vegetation. Aber et al. (1993) subjected three stand types to three years of chronic nitrogen additions at 146 kg/ha-year or 421 kg/ha-year, plus a reference treatment of 8 kg/ha-year. The ability of all of the stands and sites to retain the added nitrogen was high. A range of 75–92% of the added nitrogen was retained in soil, mostly through uptake by saprophytic microbes, which metabolize dead-biomass substrates having large ratios of carbon:nitrogen and therefore benefit from nitrogen fertilization. The only exception was for a red pine (*Pinus resinosa*) plantation, which began to leach nitrate in the third year of fertilization, possibly indicating the beginning of saturation of the system.

The phenomenon of nitrogen saturation can also occur in wetlands, as was demonstrated by a comparison of bogs in northern and southern Sweden, respectively characterized by relatively small (0.6–2 kg N/ha-year) and larger (7–9 kg N/ha-year) rates of nitrogen deposition from the atmosphere (Aerts *et al.*, 1992). In the low-input site, the growth of *Sphagnum* moss was increased by a factor of 2.0–2.5 by experimental nitrogen fertilization, but was unaffected by phosphorus fertilization, indicating that nitrogen was the primary limiting factor to productivity. In contrast, *Sphagnum* productivity at the high-deposition site was unaffected by nitrogen fertilization, but increased by a factor of 1.7–2.0 by phosphorus fertilization, indicating a phosphorus limitation in a nitrogen-saturated environment.

To preserve electrochemical neutrality, mobile SO_4^{2-} and NO_3^- anions that are leaching in soil must be accompanied by an equivalent quantity of cations. In calcareous soils and in relatively young and unweathered soils, bases such as Ca^{2+} and Mg^{2+} comprise a large fraction of the leaching cations. The Ca^{2+} and Mg^{2+} originate from either the solubilization of primary carbonates and silicates or by cation exchange with H^+. In soil with a small carbonate content, the loss of calcium and magnesium is important in reducing the acid-neutralizing capability of the soil and in acidification via a reduction of base saturation of cation-exchange capacity (Reuss, 1975, 1976; Bache, 1980; Tabatabai, 1987).

In addition, aluminum and hydrogen ions may leach with mobile anions from base-poor soils and into surface waters, where they can contribute to acidification and cause toxic stress to the aquatic biota (Cronan and Schofield, 1979; Reuss *et al.*, 1987). In many old, highly weathered soils of the humid tropics, intense leaching for a long period of time has removed much of the original content of silicate and base cations, leaving behind a matrix dominated by aluminum and iron hydroxides and silica-poor clays such as kaolinite (Ellenberg, 1986;

Lavelle, 1987). Such soils can be acidic, with pH values of about 4.0 or less, and there can be toxic concentrations of exchangeable aluminum (Uhl, 1983, 1987; Swift, 1984; dos Santos, 1987).

Soil Buffering Systems.   Soil is usually strongly buffered, in comparison to precipitation or freshwater. Therefore, soil is relatively resistant to pH changes, especially once it is already acidic. Different buffering systems come into play at particular ranges of soil pH. Carbonate minerals buffer soil within the pH range of >8 to 6.2; silicates from pH 6.2 to 5.0; cation-exchange capacity from pH 6.2 to 4.2; aluminum from pH 5.0 to 3.0; iron from pH 3.8 to 2.4; and humic substances from pH 5 to 3 (Hausenbuiller, 1978; Bache, 1980; Seip and Freedman, 1980; Ulrich, 1983; Stigliani and Shaw, 1990).

If they have small amounts of carbonate buffering associated with basic cations, particularly calcium and magnesium, then soils in the circumneutral pH range of 6–8 are generally sensitive to acidification. If such soils are subjected to large inputs of acidifying substances, they can acidify to pH of 3.5–4.5. In general, soils sensitive to acidification only produce about 200–500 eq/ha-year of new acid-neutralization capability through the weathering of basic cations (Stigliani and Shaw, 1990).

The role of aluminum buffering is especially important in strongly acidic soils (pH <5). In addition, exchangeable aluminum can contribute to total soil acidity through the ionization and hydrolysis of Al^{3+}, as follows (after Reuss, 1975; Bache, 1980; Reuss and Johnson, 1985):

$$Al^{3+} + H_2O \leftrightarrows AlOH^{2+} + H^+.$$

This reaction is most important at about pH 3.5. The $AlOH^{2+}$ can also hydrolyze to produce $Al(OH)_2^+$ + H^+ at about pH 5. However, since $Al(OH)_2^+$ is generally insoluble in acidic soil, it contributes little to the soil acidity.

Concentration of Soil Solutes.   Soil acidification is caused by the accumulation of soluble inorganic and organic acids, at a faster rate than they can be neutralized. Ionization of these acids produces free H^+ in the soil solution. Exchangeable H^+ from cation-exchange surfaces also contributes to acidity. The relative contribution of exchangeable H^+ is influenced by the total osmotic strength of the soil solution; the more concentrated the extracting solution, the greater the amount of H^+ that is exchanged.

This fact is important in the measurement of soil pH. If distilled water is used as an extracting solution, the measured pH value is typically 0.5–1.0 pH units higher than that obtained when using a standardized-salt solution as an extractant, usually 0.01 M $CaCl_2$ (Russell, 1973; Allen *et al.*, 1974).

Similarly, in natural ecosystems the acidity of soil may be influenced by temporal and spatial variations in the total concentration of solutes. In addition, locations with relatively large atmospheric inputs of total ions (e.g., a site near the ocean that is subject to a large rate of sea-salt deposition) may have a larger exchange acidity in soil than more continental sites (Rosenqvist, 1978a,b; Seip and Tollan, 1978).

Acidification of Soil

Soil acidification is a naturally occurring process. This has been illustrated by several classical studies of primary succession on newly exposed parent materials. Crocker and Major (1955) examined a chronosequence of deglaciated sites at Glacier Bay, Alaska. In this area the parent material that is initially available for biological colonization after the meltback of glacial ice is a fine morainic till with pH 8.0–8.4. The mineral composition is primarily of granite, gneiss, schist, and as much as 7–10% carbonate mineral. As this primary substrate is progressively leached by precipitation water and colonized and modified by the developing vegetation, the acidity of the mineral soil increases progressively. The pH decreases to ca. 5.0 after about 70 years, when a spruce–hemlock (*Picea–Tsuga*) forest begins to develop. Acidity of the mineral soil eventually stabilizes at pH 4.6–4.8 under a mature conifer forest, while the forest floor stabilizes at pH 4.4. This acidification is accompanied by a large reduction of calcium in the surface mineral soil,

from an initial average concentration of ca. 5–9% to less than 1%. The decreased concentration of calcium in soil is a result of both its leaching to below the rooting depth and its uptake by vegetation. Other soil changes include large successional accumulations of organic matter and nitrogen, due to the biological fixations of atmospheric CO_2 and N_2.

Other studies of primary succession examined chronosequences of sand dunes on Lakes Michigan and Huron (Olson, 1958; Morrison and Yarranton, 1973). The initial development of vegetation on newly exposed sand consists of a dunegrass community dominated by *Ammophila breviligulata* and *Calamovilfa longifolia*. With time, a tallgrass prairie develops, dominated by various grasses and dicotyledonous herbs. The prairie is invaded by shade-intolerant shrub and tree species, which form forest nuclei. Eventually an edaphic climax that is dominated by oaks (*Quercus* spp.) and tulip-tree (*Liriodendron tulipifera*) develops. Soil development in this sere is qualitatively similar to that observed at Glacier Bay (Olson, 1958). There are increases in acidity, organic matter, and fixed nitrogen and decreases in carbonate minerals. Initially, $CaCO_3$ comprises about 1.5–2.5% of the sandy substrate. This decreases to less than 0.15% in the upper 10 cm after about 400 years of succession, because of the uptake of Ca by vegetation and the solubilization and downward leaching of Ca, in conjunction with the development of a white, acidic, highly siliceous eluviated horizon through the process of podsolization. Concurrently, the soil acidifies from an initial pH of 7.7, to ca. pH 4.5 after about 1000 years.

An important cause of the acidification of soil by successionally aggrading vegetation is the uptake by plants of the bases Ca, Mg, and K, which is accompanied by an equivalent excretion of hydrogen ions into the soil solution. Over a long period of time (i.e., centuries) in natural ecosystems, there may not be a net acidifying effect via this process, because the bases are eventually returned to the soil surface by litterfall. If there is no net accumulation of organic matter on the soil surface, then the organically bound Ca, Mg, and K in litterfall are mineralized by microbes and returned to the soil. Alternatively, these cations may be returned to the soil in the form of base-rich ash following a wildfire. However, in the shorter term, soil acidification occurs if the quantity of organic matter is increasing on the site in the form of successively aggrading plant biomass and/or an increasing quantity of organic matter in the forest floor. This acidification occurs because the rate of incorporation of bases into organic matter exceeds their rates of input or recycling by such processes as mineralization, decomposition, weathering, and atmospheric deposition (Oden, 1976; Rosenqvist, 1978a, 1980; Rosenqvist *et al.*, 1980; Nilsson *et al.*, 1982; Ormerod *et al.*, 1989).

Oden (1976) calculated that a typical, aggrading pine forest could add 340 eq/ha/year of H^+ to the soil by the net uptake of basic cations and 55 eq/ha/year by the accumulation of organic matter in the forest floor. The total input of acid is similar in magnitude to the wet atmospheric inputs in many areas that are experiencing severe acidic precipitation. For example, the input of H^+ in 100 cm/year of precipitation with an average pH of 4.2 would be ca. 630 eq/ha/year. Therefore, an aggrading forest can cause an accumulation of a substantial quantity of acidity in the soil, and when this process combines with a large atmospheric deposition, the potential for acidification is great.

Additional acidification takes place if biomass is harvested from the site, since this removes large quantities of basic cations that are incorporated in the biomass (Russell, 1973; Rosenqvist, 1978a, 1980; Rosenqvist *et al.*, 1980; Nilsson *et al.*, 1982). Freedman *et al.* (1986) predicted the nutrient removals by stem-only and by whole-tree clearcuts of eight mature stands of forest in Nova Scotia (see also Chapter 9). On average, the stem-only clearcuts were predicted to remove 9300 eq Ca/ha, 1500 eq K/ha, and 1500 eq Mg/ha, for a total of 12,300 eq/ha. The whole-tree clearcuts were predicted to remove 18,000 eq Ca/ha, 3700 eq K/ha, and 3200 eq Mg/ha, for a total of 24,900 eq/ha. If a harvest rotation of 100 years is assumed, then acid-neutralizing capacity (associated with Ca, Mg, and K) of the site would be removed at an average rate of 123 eq/ha/year by a conventional, stem-only clearcut and 249 eq/ha/year by a whole-tree clearcut.

This removal of basic cations by harvesting would contribute to the acidification of the site, unless the cations were restored by the weathering of minerals, by atmospheric inputs, or by liming or fertilization as a management practice.

A number of studies have investigated whether the acidity of forest soils has increased as a result of atmospheric inputs of acidifying substances. Conceptually, the more important soil effects are likely to be: (1) a decrease in soil pH; (2) a decrease in base saturation of cation-exchange capacity; (3) an increase in the saturation of cation-exchange capacity by H^+ and aluminum ions, especially Al^{3+}; and (4) a saturation of sulfate-adsorption capacity, leading to the leaching of mobile sulfate anions, accompanied by base cations and toxic Al^{3+} and H^+ (Oden and Andersson, 1971; Wiklander, 1975, 1980; Frink and Voigt, 1977; McFee *et al.*, 1977; Bache, 1980; Morrison, 1984; Reuss *et al.*, 1987).

As described previously, these potential effects have been examined in experiments where soil-column lysimeters were treated with artificial rainwater solutions of various pH. In the shorter term, notable effects only occurred when highly acidic leaching solutions were used, that is, pH ≤ 3 (Abrahamsen *et al.*, 1976, 1977; Abrahamsen and Stuanes, 1980; Bjor and Tiegen, 1980; Stuanes, 1980; Morrison, 1981, 1983; however see Farrell *et al.*, 1980, for an effect at higher pH).

It could be argued that in terms of H^+ flux, 1 year of treatment with an artificial rain at pH 3 is equivalent to 10 years of treatment at pH 4. If this were true, then an effect observed with an artificially acidic pH treatment in a lysimeter experiment might be extrapolated to an effect that might occur over a longer time period at a less acidic pH, for example, in a field situation exposed to ambient acidic precipitation. However, this is a controversial point, and there is no direct evidence that such an extrapolation is reasonable. There is a need for longer-term experiments using realistic rates of acid loading, so that the potential effects of atmospheric depositions on soil acidity can be better understood.

In addition to the experimental approach, surveys of soil chemistry at the same location but conducted at different times can indicate whether acidification has taken place. Several studies have shown that the conversion of former agricultural land to forest can cause a substantial acidification of the soil, especially if conifers dominate the developing stand (Williams *et al.*, 1979; Alban, 1982; Brand *et al.*, 1986; Ormerod *et al.*, 1989). For example, the afforestation of abandoned farmland in Ontario with red pine (*Pinus resinosa*), Scotch pine (*P. sylvestris*), or white spruce (*Picea glauca*) caused an average acidification of about one pH unit (from pH 5.7 to 4.7) after about 46 years (Brand *et al.*, 1986). In general, it is well accepted that afforestation will cause an acidification of most sites.

Of more relevance to the atmospheric deposition of acidifying substances, is whether sites that are already forested will acidify further or more rapidly because of the enhanced atmospheric inputs. In part, this problem can be examined by resampling forest soils after widely spaced intervals of time. In one study, Linzon and Temple (1980) found no increase in soil acidity in a resurvey of six sites in Ontario after a 16-year interval, in a region where the mean annual pH of precipitation was 4.0–4.1. In another study, Troedsson (1980) compared analyses of forest floor samples collected in Sweden in 1961–1963 to those collected in 1971–1973. There was a general relationship between stand age and increasing soil acidity. A comparison of data from the two time intervals showed minor changes in pH and exchangeable aluminum, but decreasing concentrations of soil Ca, Mg, and K, implying that acidification may have taken place via a decrease in the base saturation of cation-exchange capacity. However, it is important to note that the designs of the above studies do not distinguish between the natural effects of increasing stand age on acidification and changes caused by atmospheric deposition.

In another study, Oden and Andersson (1971) mapped the distribution of soil pH and base saturation of cation-exchange capacity in Sweden. They found relatively acidic pHs and low base saturation in southern Sweden, where precipitation is more acidic than in the north. Because local sources of gaseous pollutants are relatively abundant in southern Sweden, the dry deposition of acidifying substances may also have influenced this spatial pattern, as various climatic and biological factors also could have.

Another study in Sweden combined aspects of several of the studies just described, that is, (1) resampling soil in permanent plots after a widely spaced time interval and (2) a comparison of areas receiving large or small depositions of acidifying substances from the atmosphere (Hallbacken and Tamm, 1985; Tamm and Hallbacken, 1986, 1988). These researchers resampled soil in 1982–1983 at sites that had originally been sampled in 1927. They used a permanently staked grid in an experimental forest in Sweden, a design that allowed them to dig their new soil pits less than 1 m from the originals. The 1927 pH measurements were made with a quinhydrone/calomel electrode, and to avoid analytical bias the same method was used in 1982–1983 instead of the then-conventional glass electrode. The original landscape of the site in southern Sweden was an open *Calluna vulgaris–Juniperus communis* heathland that had been planted to *Picea abies* beginning in the 1870s. In the mid-1920s there were middle-aged spruce plantations 37–50 years old, and there was also a natural hardwood forest dominated by beech, birch, and oak (*Fagus sylvatica*, *Betula* spp., and *Quercus robur*). In 1982–1983, there were mature spruce stands >100 years old and a mature beech–oak forest. A large increase in acidity occurred throughout the soil profile between 1927 and 1982–1983 (Table 4.7). This change was especially large in the surface humus layer, which in the beech forest acidified from an initial average pH of 4.5 to 3.8 in the early 1980s, and from 4.6 to 3.6 in the spruce stands. To a lesser degree the mineral soil also acidified, with the change being somewhat smaller under beech than under spruce. The changes in the B and C horizons were much larger than those observed by Tamm and Hallbacken (1988) in a similar study done in spruce stands in northern Sweden, where the rate of atmospheric deposition is considerably smaller. As a result, they interpreted their data to indicate that the atmospheric deposition of acidifying substances was the most important cause of acidification of the deeper soil horizons, whereas biological acidification was more important in the humus and upper A horizon.

Table 4.7 The change in average soil pH between 1927 and 1982–1983 in relocated pits in Swedish forests[a]

Horizon	Time	pH under *Fagus sylvatica*	pH under *Picea abies*
Humus	1927	4.5	4.6
	1982–1983	3.8	3.6
A₂	1927	4.5	4.7
	1982–1983	4.2	4.0
B	1927	4.9	4.9
	1982–1983	4.6	4.5
C	1927	5.3	5.2
	1982–1983	4.7	4.2

[a]Modified from Hallbacken and Tamm (1985).

These are important studies. However, because the data are variously complicated by succession, local spatial heterogeneity, harvesting, and regional variations of climate, in addition to differences in acidic deposition, they do not provide conclusive evidence of soil acidification via acidifying depositions—more data are still needed.

Surface Water Chemistry

Chemical Characteristics

Changes in water chemistry that occur when incoming precipitation interacts with foliage and bark surfaces (throughfall and stemflow) and with the forest floor and mineral soil were previously described. These interactions result in decreased concentrations of certain chemical constituents in the percolating solutions, and increased concentrations of others. The net effects of these various chemical interactions occurring in the terrestrial part of the watershed are also reflected in the chemistry of surface waters such as streams, rivers, and lakes.

A comparison of the chemistry of precipitation with that of two remote, headwater, oligotrophic lakes in Nova Scotia is presented in Table 4.8. Compared with precipitation, the lake waters are a relatively concentrated solution—their sum of cations plus anions averages 440 μeq/liter versus 135

A view of terrain in the LaCloche highlands of Ontario that is hypersensitive to acidification caused by atmospheric depositions. The terrestrial part of the watershed is characterized by thin, slowly weathering, oligotrophic glacial deposits. There are frequent exposures of whitish, quartzitic bedrock on hilltops and slopes. The lakes in this terrain had very little acid-neutralizing capacity, and they were quickly acidified by a combination of acidic precipitation and the dry deposition of sulfur dioxide (photo: B. Freedman).

μeq/liter in precipitation. This occurs in spite of the fact that these oligotrophic lakes are relatively dilute in comparison with most fresh waters, especially with respect to calcium concentration (Hutchinson, 1967; Bowen, 1979; Kerekes *et al.*, 1982, 1989).

In addition, the relative contributions of ionic constituents differ markedly—compared with precipitation, the lake waters are relatively enriched in calcium, magnesium, sodium, potassium, iron,

Table 4.8 Volume-weighted mean annual concentration of chemical constituents in wet-only precipitation, and in two headwater oligotrophic lakes in Nova Scotia, Canada: Beaverskin is a clearwater lake, and Pebbleloggitch is a brown-water lake[a]

Constituent	Wet-only precipitation (1981–1983)	Beaverskin Lake (1979–1980)	Pebbleloggitch Lake (1979–1980)
Ca	4.3	20.0	18.0
Mg	2.9	32.0	30.0
Na	26.1	126.0	126.0
K	1.1	8.0	6.0
Fe	<0.1	1.0	4.0
Al	<0.1	2.2	23.4
NH_4	4.2	1.0	2.4
H	29.9	5.0	33.0
SO_4	27.5	48.7	57.9
Cl	29.5	124.0	111.0
NO_3	9.7	1.0	0.9
Organic anions	<0.1	32.0	66.0
Alkalinity	0.0	0.0	0.0
Sum cations	68.5	195.2	242.8
Sum anions	66.7	205.7	235.8
Anions/cations	0.97	1.05	0.97
DOC (mg/liter)	0.0	3.9	10.0
TC (mg/liter)	0.0	4.5	13.8
Color (Hazen units)	0	6	87

[a]Data are in μeq/liter. Modified from Freedman and Clair (1987) and Kerekes and Freedman (1988).

aluminum, sulfate, chloride, organic anions, and dissolved and total organic carbon (Table 4.8). The increased concentrations of all of these constituents are largely due to their mobilization from the terrestrial part of the watershed. Because sodium, chloride, and sulfate have an important aerosol phase in the atmosphere, their dry depositions and subsequent solubilizations also contribute to their larger concentrations in lake water. In addition, evapotranspiration has the effect of physically concentrating surface waters in comparison with precipitation. However, this physical effect is fairly small, since evapotranspiration dissipates only about 33% of the annual atmospheric input of water to these watersheds (Kerekes and Freedman, 1988; Kerekes et al., 1989). In contrast, ammonium and nitrate have markedly smaller concentrations in the lakewaters than in precipitation, indicating that they have been "consumed" by biological uptake and

inorganic-exchange reactions within the terrestrial part of the watershed.

Beaverskin Lake is a typical, oligotrophic, clear water, slightly acidic lake. It is less acidic than incoming precipitation (average pH 5.3 versus 4.6, respectively), while the brown water Pebbleloggitch Lake (pH 4.5) has an acidity similar to that of precipitation. These two oligotrophic lakes are situated only 1 km apart on a similar geological substrate. However, they differ in acidity because of the strong influence of organic acids in tea-colored Pebbleloggitch Lake, which like most brown water lakes has a *Sphagnum*-heath bog covering a substantial part of its watershed (Kerekes and Freedman, 1988; Freedman et al., 1989). Water with a large concentration of dissolved organic substances is usually naturally acidic, with a pH range of about 4–5 (Oliver et al., 1983; Clymo, 1984; Gorham et al., 1984; Krug et al., 1985). In Pebbleloggitch

Airphoto of the watershed of Pebbleloggitch Lake, Nova Scotia. About one-third of the terrestrial watershed of this lake is characterized by an organic-rich boggy substrate, visible as the low-texture, whitish area at the bottom of the photo. The rest of the terrestrial watershed is covered with coniferous forest. Because of the large inputs of dissolved organic substances from the bog, the lake has tea-colored, dark-brown water and it is naturally acidic, with an average pH of about 4.5 (photo courtesy of J. Kerekes).

Lake, organic anions comprise 28% of the total anions, further suggesting that they influence its acidity.

Seasonal variations in the concentrations of chemical constituents are also important. In most places where a snowpack accumulates during the wintertime, there is a tendency for winter and especially spring meltwater flows to be relatively acidic, particularly in streams, rivers, and surface waters of lakes (Haapala *et al.*, 1975; Johannessen and Henriksen, 1978; Jeffries *et al.*, 1976; Goodison *et al.*, 1986; Freedman and Clair, 1987; Wigington, 1990). In part, this phenomenon is caused by soil that is saturated and/or frozen at the time of snowmelt, so that there is relatively little neutralization of the acidity of precipitation by interactions with soil constituents. "Acid-shock" events have also been linked to the initiation of snowmelt, since the

first melt fraction of the snowpack is more acidic than are later fractions (Johannesen and Henriksen, 1978; Cadle *et al.*, 1984). In the Adirondack Mountains of New York, the occurrence of relatively acidic streamwater in the springtime is believed to be due to: (1) the occurrence of relatively small concentrations of the base cations Ca^{2+}, Mg^{2+}, Na^+, and K^+ and their associated alkalinity; (2) an increase in the concentration of nitrate; and (3) the constant, large concentration of sulfate despite the smaller concentrations of bases (Galloway *et al.*, 1987a).

The concentrations of other chemical constituents also vary seasonally in surface waters. In streams in Nova Scotia, the concentrations of hydrogen ion, dissolved organic carbon, sulfate, nitrate, chloride, sodium, magnesium, calcium, and aluminum are all relatively large during the high-

A view of a set of tubular, clear polyethylene mesocosms in which columns of lake water were experimentally treated in order to examine the effects of acidification, liming, and fertilization on the open-water biota. The lake is located in a remote area of Nova Scotia, and it is oligotrophic and susceptible to acidification by atmospheric depositions of acidifying substances. The shallowness of the soil of its watershed can be appreciated from the many exposed boulders along the shoreline. The largest changes in phytoplankton and zooplankton were caused by fertilization, but stresses associated with acidification and liming also caused changes in species richness, standing crop, and productivity (Blouin *et al.*, 1984; photo courtesy of: P. Lane).

flow period of late autumn–winter–early spring (Fig. 4.4; Freedman and Clair, 1987). Overall, however, the seasonal variations in concentration are much smaller than the seasonal variations in water flow. As a result, water flow is by far the most significant determinant of the rate of export of chemicals from these watersheds; variations of concentration are relatively unimportant (see also Lewis and Grant, 1979; Hall and Likens, 1984; Kerekes and Freedman, 1989; Hall *et al.*, 1990).

Another important geochemical consideration is that of the flux of chemicals (for a particular chemical constituent, annual flux is calculated as the product of its mean-annual, volume-weighted concentration times water flux from the watershed). Because fluxes are standardized to time, area, and

water volume, they allow a comparison of the total input and the total output of material. The input and output fluxes of certain elements for a forested watershed at Hubbard Brook, New Hampshire, are summarized in Table 4.9. For elements with a net flux (i.e., inputs minus outputs) that is positive, there is an accumulation in the watershed; if net flux is negative there is a net depletion of the watershed "capital" of the chemical.

Elements with negative net fluxes include silicon, calcium, sodium, aluminum, magnesium, and potassium. The excess outputs of these derive from ion exchange and from the weathering of minerals in the watershed (Schindler *et al.*, 1976; Likens *et al.*, 1977; N. M. Johnson, 1979).

Hydrogen ion accumulates in the watershed at a

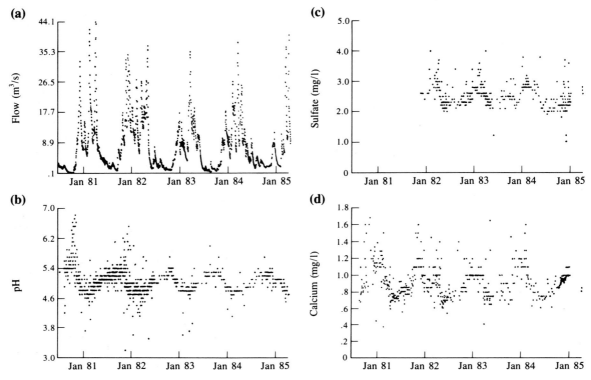

Fig. 4.4   Seasonal variations in water flow, pH, and the concentrations of sulfate and calcium in the Mersey River, Nova Scotia. Modified from Freedman and Clair (1987).

Table 4.9   Total inputs and outputs and net fluxes of chemical constituents for a forested watershed at Hubbard Brook, New Hampshire[a]

Constituent	Total inputs[b]	Total outputs	Net fluxes
Si	<0.1	23.8	−23.8
Ca	2.2	13.9	−11.7
Na	1.6	7.5	−5.9
Al	<0.1	3.4	−3.4
Mg	0.6	3.3	−2.7
K	0.9	2.4	−1.5
Organic C	1484	12.3	+1472
N	20.7	4.0	+16.7
Cl	6.2	4.6	+1.6
S	18.8	17.6	+1.2
H	0.96	0.10	+0.86
P	0.036	0.019	+0.017

[a]Data are in kg/ha-year. Modified from Likens *et al.* (1977).
[b]Based on bulk-collected precipitation, except for C, N, and S, where inputs include aerosol deposition and gaseous uptake.

rate that is equivalent to the consumption of 90% of the atmospheric input of H^+ via precipitation (Table 4.9). Other studies in regions receiving acidic deposition have also reported H^+ consumption by the terrestrial part of the watershed. Freedman and Clair (1987), for example, observed 75–90% H^+ consumption among four forested watersheds in Nova Scotia. Within the terrestrial part of the watershed, H^+ consumption is largely due to ion exchange, the dissolution of aluminum hydroxide compounds, and weathering reactions (Likens *et al.*, 1977; N. M. Johnson, 1979). In part, these processes cause the loss of basic cations such as Ca^{2+} and Mg^{2+}, and they result in an acidification of the terrestrial watershed.

The net flux of fixed nitrogen was also strongly positive at Hubbard Brook (+16.7 kg N/ha/year; Table 4.9). The nitrogen accumulates in the terrestrial part of the watershed, as organic N of the aggrading plant biomass and organic matter of the forest floor. The large, positive net flux of carbon is, of course, due to the net fixation of atmospheric CO_2 into organic carbon of the aggrading forest (Likens et l., 1977).

Acidification of Surface Water

Widespread acidification of weakly buffered surface waters has been attributed to the deposition of acidifying substances from the atmosphere. Some relatively well-studied, affected regions include: (1) Scandinavia (RMFA/RMA, 1971; Brakke, 1976; Wright and Gjessing, 1976; Seip and Tollan, 1978; Nilssen, 1980), (2) eastern Canada (Beamish and Harvey, 1972; Beamish et al., 1975; Thompson et al., 1980; Watt et al., 1979, 1983; Kelso et al., 1990), and (3) the eastern United States (Schofield, 1976a, 1982; Wright and Gjessing, 1976; A. H. Johnson, 1979; Glass et al., 1981; Schindler, 1988; L. A. Baker, 1990; L. A. Baker et al., 1991).

In a particularly intensive survey, the Environmental Protection Agency sampled many lakes and streams in the United States (L. A. Baker, 1990;

L. A. Baker et al., 1991). Out of a sample of 10,400 lakes, a total of 1180 were identified as acidic in this survey, mostly in the eastern United States (Table 4.10). The cause of acidification of 75% of the lakes was attributed to atmospheric-deposition processes, 3% to acid-mine drainage, and 22% to organic acids associated with bogs. Of the 4670 streams considered acidic, 47% were associated with atmospheric deposition, 26% with acid-mine drainage, and 27% with bogs. The largest frequency of acidic lakes is in Florida, but most of these are affected by natural, organic acids. Atmospheric deposition is more important in the northeastern United States, especially in the Adirondack region of New York state, where 10% of lakes had pH ≤ 5.0 and 20% had pH ≤ 5.5.

Some studies have been able to demonstrate actual changes in the acidity of particular lakes or groups of lakes. For example, the pH of 21 water bodies in central Norway decreased from an average of 7.5 in 1941 to 5.4–6.3 in the early 1970s (Brakke, 1976). In southwestern Sweden, the average pH of 14 surface waters decreased from 6.5–6.6 prior to 1950 to pH 5.4–5.6 in 1971 (Brakke, 1976). In Nova Scotia, the average pH of seven

Table 4.10 Frequency of acidic surface waters in the United States[a]

Region:	Northeast	Upper midwest	Interior southeast	Florida	West
No. lakes in sample ($\times 10^3$)	7.1	8.5	0.26	2.1	10.4
ANC frequency (%)					
0 eq/liter	6.2	2.9	0.0	22.7	0.1
<50	21.9	15.8	1.4	39.8	16.3
<100	37.1	26.1	11.6	51.1	42.9
<200	60.9	40.6	34.3	55.1	66.4
pH (%)					
<5.0	3.4	1.5	0.0	12.4	0.1
<5.5	8.6	3.6	0.0	20.6	0.1
<6.0	12.9	9.6	0.4	32.7	1.0
<6.5	26.3	23.6	9.5	48.9	5.6
Aluminum[b] (%)					
>200 g/l	1.4	0.1	0.0	0.0	0.7
>100	3.0	0.1	0.6	0.0	2.6
>50	5.7	0.5	2.9	0.0	6.5
>25	11.4	1.9	13.1	0.1	14.7

[a]Data are based on a large-scale, statistical sample of lakes in five geographic regions. Modified from J. P. Baker (1990).

[b]MIBK (methyl-isobutyl ketone) extraction, yielding total monomeric Al; Al^{3+}, $Al(OH)_n^{3-n}$, and simple organic complexes, but not polymeric forms of Al.

rivers was 4.9 in 1973, compared with 5.7 in 1954–1955 (Thompson *et al.*, 1980). In the Adirondack Mountains of the eastern United States, more than 51% of a sample of 217 high-altitude lakes had pH <5 in 1975 (90% of these acidic lakes were devoid of fish), compared with only 4% having pH <5 in a sample of 320 lakes from the same general area in the 1930s (note that the 1930s sampling was not restricted to high-altitude waterbodies; Schofield, 1976a).

It is important to note that some of the historical comparisons that suggest an acidification of surface waters may suffer from systematic biases that caused erroneously high values in the earlier pH and alkalinity determinations (Kramer and Tessier, 1982). These errors can result from the following:

1. The use of soft-glass containers, which may have contributed alkalinity to early samples.
2. A difference in analytical technique. (The pH meter with glass electrode came into common use in 1946–1960. Prior to that time the most frequent techniques for determining the concentration of H^+ involved the use of color-indicator comparators, which may have overestimated the pH of dilute, poorly buffered, oligotrophic waters.)
3. A difference in the timing and frequency of sampling between early and recent studies (on an annual basis, pH tends to be lower in winter, while on a diurnal basis it tends to be higher at mid-day due to alkalinity generated by primary production).

Therefore, pH determinations made prior to about 1960 should be regarded with some caution. However, in spite of these analytical problems, there is a broad consensus among researchers that there has been a recent acidification of many vulnerable surface waters. Even if there is some doubt about the accuracy of some of the older pH and alkalinity data, the evidence for biological changes caused by the acidification (described later) is incontrovertible (Kramer and Tessier, 1982; Havas *et al.*, 1984b).

In some regions, there have been large emissions of sulfur dioxide from coal burning since the beginning of the industrial revolution in the mid-1800s. In at least some cases, lakes in those regions may have begun to acidify at about that time, substantially sooner than the widespread recognition of this environmental problem beginning in the late 1960s. There are no reliable water-chemistry data that track this putative, early acidification, but the process can be inferred from paleoecological records of community changes, especially of diatoms. For example, Loch Laiden is a relatively large (473-ha), high-altitude lake in Scotland, with an average pH in 1988 of 5.3–5.5 (Flower and Battarbee, 1983; Flower *et al.*, 1987, 1988). Analysis of diatom remains in a sediment core from the lake indicates that, beginning in the mid-1800s, there was a decline of two then-common, acid-intolerant species, *Cyclotella kuetzingiana* and *Brachysira vitrea*, which were replaced as dominants by two acidophilous species, *Tabellaria flocculosa* and *Eunotia veneris*. These and changes in less prominent species of diatoms suggest that prior to the 1850s the pH of Loch Laiden was about 5.8–6.0, compared with about 5.6 at the turn of the century and 5.3–5.5 today. In a similar study, Cumming *et al.* (1992) determined that 25–35% of Adirondack lakes may have acidified during the 19th century.

Few data are available that illustrate the changes taking place in the concentrations of hydrogen ion and other chemicals as fresh waters acidify. The most important reason for this lack of information is that there are few accurate data on the characteristics of water bodies prior to their acidification. There have, however, been a few experimental studies of natural water bodies in which acidification was caused by the addition of concentrated mineral acid. A whole-lake experiment was performed in the Experimental Lakes Area (ELA) of northwestern Ontario (Schindler and Turner, 1982; Cook and Schindler, 1983; Mills, 1984; Schindler *et al.*, 1985; Rudd *et al.*, 1988; Schindler, 1990). The experimental water body is Lake 223, a 27-ha oligotrophic lake located on a Precambrian Shield landscape, with an average depth of 7.1 m and a volume of 19.5×10^5 m^3. The lake was studied for two years (1974, 1975) before its experimental manipulation, and then sulfuric acid was added to progressively acidify the system.

Prior to its acidification, Lake 223 had a mean-annual epilimnetic pH of 6.5, an average ice-free season pH of 6.8, and an alkalinity of 80 μeq/liter (the mean-annual pH of precipitation was 4.9–5.0). Beginning in 1976, sulfuric acid was added to Lake 223; by 1983, 27.4 thousand liters of 36 N H$_2$SO$_4$ had been added! The acidity of Lake 223 was increased progressively, from an initial, mean-annual epilimnetic pH of 6.49 in 1976, to 6.13 in 1977, 5.93 in 1978, 5.64 in 1979, 5.59 in 1980, and 5.02–5.13 between 1981 and 1983, when the objective was to maintain pH rather than to further acidify the lake. The pH of Lake 223 was then allowed to increase somewhat, to pH 5.5–5.6 during 1984–1986, and to pH 5.8 during 1987–1988.

The addition of sulfuric acid caused various direct and indirect chemical changes in Lake 223. As expected, both sulfate and hydrogen ions increased in concentration (sulfate averaged 35 μmol/liter in 1975, compared with 115 μmol/liter in 1979). There were also increases in the concentrations of manganese (a 980% increase in 1980, compared with the 1976 concentration), zinc (550%), aluminum (155%), and sodium (26%); these were probably solubilized from sediment by the acidifying fresh water. The wintertime concentration of ammonium also increased in Lake 223, after pH reached 5.4. The normal pattern under less acidic conditions is for a wintertime increase in the concentration of nitrate, but in Lake 223 ammonium increased instead because of the inhibition of bacterial nitrification. Decreased concentrations were observed throughout the year for alkalinity, dissolved nitrogen, iron, and chloride. An increase in the transparency of the water column was reflected by a change in the extinction coefficient from an initial 0.50–0.58 to 0.44–0.50 after acidification. The increased light penetration caused additional heating of deeper waters during the growing season and an increased depth of the thermocline. Many biological changes were also caused by the acidification of Lake 223; these are described later.

Interestingly, the observed pH decreases were substantially smaller than had been predicted by calculations assuming that Lake 223 would behave as an isolated, dilute, alkalinity-buffered system that was being titrated with sulfuric acid. In part, this was due to the consumption of H$^+$ by weathering and solubilization reactions at the water/sediment interface; these resulted in the increased concentrations of metals such as Al and Mn. However, the most important acid-consuming mechanism was associated with alkalinity produced by reactions that formed iron sulfide in the bottom 10% of the seasonally anoxic hypolimnion. Under this condition, Fe^{2+} is mobilized from the sediment. This ferric iron reacts with sulfide produced by the reduction of sulfate under anoxic conditions, to produce a FeS precipitate. This process is accompanied by the consumption of one equivalent of H$^+$ per equivalent of SO$_4^{2-}$ reduced. Overall, an estimated 66–81% of the added acid was neutralized by alkalinity, 75% of which was generated by sulfate reduction, iron reduction, and iron sulfide formation (Cook *et al.*, 1986). The exchange of H$^+$ for other cations in the sediment accounted for a further 19% of the alkalinity and inputs from the terrestrial watershed for about 5%. The remaining 19–34% of the H$^+$ that was not neutralized by alkalinity served to decrease the pH of the water column and to weather minerals, or it was exported from the lake. A sulfur budget for Lake 223 showed that streamwater outflow accounted for 33–35% of the total input of sulfate, while FeS sedimentation consumed 10–15%. The remaining 50–57% remained in the water column as increased sulfate concentration, as described earlier (Cook and Schindler, 1983).

Surface waters that are vulnerable to acidification tend to exhibit a syndrome of physical and chemical characteristics. The most important of these are discussed below (after Hendry *et al.*, 1980a; Harvey *et al.*, 1981; Henriksen, 1982; Kramer and Tessier, 1982; Schindler, 1988; T. J. Sullivan, 1990). One characteristic is that vulnerable waters tend to have a small alkalinity or acid-neutralizing capacity. Usually H$^+$ is absorbed until a buffering threshold is exceeded, after which there is a rapid decrease in pH until another buffering system comes into play (Fig. 4.5). In the

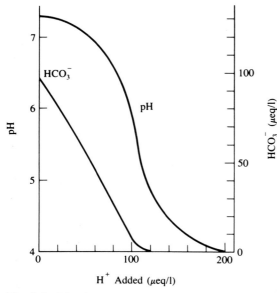

Fig. 4.5 Titration curve for a bicarbonate solution with an initial concentration of alkalinity (HCO_3^-) of 100 μeq/liter. This curve is qualitatively similar to that for a clear, oligotrophic surface water. Modified from Henriksen (1980).

Table 4.11 Percentage of inorganic carbon occurring as CO_2, HCO_3^-, and CO_3^{2-} in water at various pH levels[a]

pH	Free CO_2 (%)	HCO_3^- (%)	CO_3^{2-} (%)
4	99.6	0.4	1.3×10^{-7}
5	96.2	3.8	1.2×10^{-5}
6	72.5	27.5	0.0009
7	20.8	79.2	0.026
8	2.5	97.2	0.32
9	0.3	96.6	3.1
10	0.002	75.7	24.3

[a]Modified from Hutchinson (1957).

circumneutral-pH range (pH 6–8), bicarbonate alkalinity is the critical buffering system that can be depleted by acidic deposition, causing acidification of the water body. In this pH range, bicarbonate alkalinity (HCO_3^-) reacts with H^+ to form H_2O + CO_2; by this mechanism, added H^+ is neutralized, and there is little or no decrease in pH until the supply of alkalinity is exhausted. In this sense, the acidification of surface waters by the atmospheric deposition of acidifying substances can initially be viewed as large-scale, whole-lake titrations of alkalinity (Henriksen, 1980, 1982).

Even distilled water contains some bicarbonate as a result of the carbonic acid (H_2CO_3) equilibrium with atmospheric CO_2. At the equilibrium pH of 5.6, ca. 18% of the total inorganic carbon is present as HCO_3^-; the rest occurs as free CO_2 (Table 4.11). However, the concentration of bicarbonate in natural surface waters is influenced by other geochemical factors. When water comes in contact with min-

eral carbonates [especially calcite ($CaCO_3$) and dolomite ($Ca,MgCO_3$)] in the watershed soil, bedrock, or aquatic sediment, large quantities of bicarbonate alkalinity are generated, and the acid-neutralizing capacity of the water is large. Alkalinity is produced by the dissolution of these minerals, for example,

$$CaCO_3 + H_2O + CO_2 \leftrightarrows Ca^{+2} + 2HCO_3^-.$$

In most circumneutral, clear, dilute waters, the quantity of HCO_3^- (in equivalents) is strongly correlated with the sum of Ca^{2+} plus Mg^{2+}. The ratio of equivalents of HCO_3^- to (Ca^{2+} + Mg^{2+}) is close to 1:1, indicating that bicarbonate associated with these basic cations is the predominant source of alkalinity (Almer *et al.*, 1978; Harvey *et al.*, 1981; Schindler, 1988; J. P. Baker, 1990).

In a watershed with large quantities of calcium and magnesium carbonate in the terrestrial soil and/or aquatic sediment, there is enough alkalinity-generating capacity to effectively neutralize the acidifying inputs from the atmosphere. A water body in such a watershed will not acidify at ambient rates of acidifying inputs, even in a relatively polluted region with large rates of atmospheric deposition. However, the situation is different where the bedrock, soil, and sediment are composed of hard, slowly weathering, nutrient-poor minerals such as granite, gneiss, and quartzite, which do not contain appreciable quantities of carbonate minerals. In such a case, the alkalinity-generating capacity of the terrestrial watershed is small, and alkalinity generation reactions within the lake become relatively important. The *in situ* production of alkalinity is largely associated with the biological reduction of sulfate and nitrate, and the exchange of H^+ for Ca^{2+} in sediment (Schindler *et al.*, 1986). However, relatively small quantities of alkalinity are generated by these processes, so that the acid-neutralizing capacity is small, and acidification can therefore occur relatively easily as a result of atmospheric depositions of acidifying substances.

Vulnerable watersheds with minimal alkalinity are common in glaciated terrain of eastern Canada and Scandinavia, where thin soils of glacial till overlay hard, plutonic bedrocks of granite and gneiss. Vulnerable watersheds are also frequent at high elevation in older mountains where erosion has exposed crustal granite, for example, in the Adirondacks and other parts of the mountain chains of eastern North America.

Another important characteristic of soils of these vulnerable watersheds is that they have a small sulfate-adsorption capacity, which can be quickly overwhelmed by large sulfate inputs via atmospheric deposition. When this occurs, the additional inputs of sulfate flush from the terrestrial part of the watershed, electrochemically accompanied by H^+ and Al^{3+}, which can acidify surface waters and cause toxic stress to aquatic biota (Reuss *et al.*, 1987).

Even within these vulnerable regions, the occurrence of pockets of calcareous soil or till in the watershed can confer sufficient alkalinity for water bodies to be essentially nonsusceptible to acidification by atmospheric depositions. This spatial effect on vulnerability can even exist in areas where the till and bedrock are dominated by hard, noncalcareous minerals. Dillon *et al.* (1977) surveyed 15 lakes on a Precambrian Shield landscape in Ontario, in an area where the mean-annual pH of precipitation was 4.0–4.4. Fourteen of the lakes had relatively small alkalinities (range of 95–175 µeq/liter during the nonstratified period) and were slightly acidic (pH 5.8–6.7); these lakes were considered to be highly to moderately sensitive to acidification. However, one other lake had some calcareous glacial till in its watershed. As a result, it had an anomalously large alkalinity (1200 µeq/liter) and high pH (7.1), and because of its large acid-neutralizing capacity it would be unlikely to acidify.

Watershed area is another consideration with respect to the vulnerability of surface waters to acidification. In general, relatively high-altitude, headwater systems are more at risk, because they tend to have a small watershed area in comparison with their surface area of water and because the soil in their watershed is often relatively thin and oligotrophic. In a small watershed with little soil, acidic rainwater has less opportunity to interact with soil minerals and bedrock, and therefore relatively little of the acidity may be neutralized before the percolating groundwater reaches surface water. Water chemistry data for two small, headwater lakes in Nova Scotia was previously examined (Table 4.8). Both of these lakes have relatively small watersheds; their ratio of watershed area to lake surface area is less than 4. These headwater lakes have average calcium concentrations of only 18–20 µeq/liter and magnesium of 30–32 µeq/liter. Nearby, a larger downstream lake with a watershed : lake-area ratio of more than 30 has 38 µeq/liter of calcium and 39 µeq/liter of magnesium (Kerekes and Freedman, 1988).

Therefore, the acidification of typical, oligotrophic fresh waters is a process that is roughly analogous to the titration of a weak bicarbonate solution with sulfuric and nitric acids deposited from the atmosphere. Where there is a small supply of alkalinity in the watershed and where atmospheric depositions result in a large flux of sulfate from the terrestrial part of the watershed, accompanied by hydrogen and aluminum ions, the surface water is generally vulnerable to acidification. In general, surface waters with ≤50 µeq/liter of alkalinity are considered to be sensitive to acidification, while lakes with 100–200 µeq/liter may be sensitive after a longer-term exposure (Huckabee *et al.*, 1989; J. P. Baker, 1990).

4.4
BIOLOGICAL EFFECTS
OF ACIDIFICATION

Injury to Terrestrial Vegetation

Acidic deposition can potentially affect vegetation in several direct and indirect ways. As summarized by Tamm and Cowling (1976), the potential, direct effects include the following.

1. The integrity of the leaf cuticle could be damaged as a result of accelerated erosion of the waxy protective layer or by direct injury to surface cells by acidic droplets or acidic particles, causing micro- or macroscopic surface injuries.

2. There could be interference with the functioning of stomatal guard cells. To a degree, the turgidity of these cells is influenced by cytoplasmic pH, which may be affected by precipitation pH. An effect on guard cells could result in decreased control over stomatal aperture and thus over the rate of transpiration and the fluxes of CO_2, O_2, and other gases.

3. Acute injury could be caused to foliar cells after the penetration of acidic substances through the cuticle or stomata.

4. Hidden-injury metabolic effects could result from changes in the rates of photosynthesis, respiration, or some other metabolic function. Hidden injuries are not manifest by obvious damages such as necrotic tissues, but they may result in growth decrements, developmental dysfunctions, or premature senescence.

5. The chemistry or quantity of root or foliar exudates could be altered, with potential secondary effects on the microflora and microfauna of foliar and root surfaces, including organisms that are important in nutrient cycling and nutrient uptake and pathogens that cause disease.

6. There could be interference with plant reproduction, for example, by decreasing the viability of pollen, by interfering with stigmatic receptability, or by otherwise decreasing fruit set or viability.

In addition, Tamm and Cowling (1976) identified potential, indirect effects of acidic deposition on vegetation, including the following.

1. The rates of leaching of substances from foliage could be increased, especially mineral nutrients and organic chemicals. This could be an indirect effect of damage to the cuticle or guard cells, or it could be due to acid-leaching effects.

2. Damage to the cuticle or guard cells or other physiological disruptions could indirectly make plants more vulnerable to drought, air pollution, and other environmental stresses.

3. Metabolic disruptions of plants could affect their symbiotic associations with nitrogen-fixing microorganisms or mycorrhizal fungi, for example, via changes in the nature of root exudates.

4. Vulnerability to parasites, pathogens, and insect damage may be indirectly affected by damage to the protective cuticle, by changes in the nature of root or foliar exudates, by predisposing stresses associated with acidic inputs or metal toxicity in soil, or by changes in competitive or other interactions occurring among the microbial flora.

5. Root damage could be indirectly caused by increased concentrations of available metals in soil as a result of acidification. Aluminum toxicity would be most important in this respect.

6. There could be synergistic interactions with other environmental stressors such as gaseous air pollutants and drought, in such a way as to increase the damage caused to plants.

A review of the scientific literature indicates that cases where terrestrial vegetation has suffered acute injuries as a result of exposure to naturally occurring, acidic precipitation events have not yet been clearly demonstrated. In contrast, numerous laboratory and field-based experiments in which plants were treated with artificial acid-rain solutions have demonstrated injuries to natural and agricultural vegetation. However, the thresholds for acute toxicity in these experiments are usually at an intensity of acidity (less than about pH 3.6) that is more acidic than that observed in precipitation in nature, except during rare events (Tamm and Cowling, 1976; Abrahamsen, 1980; Jacobson, 1980; Cohen *et al.*, 1981; Evans, 1982; Inving, 1983; Amthor, 1984; Morrison, 1984; McLaughlin, 1985; Shriner, 1990).

Abrahamsen and Stuanes (1980) and Tveite (1980a,b) reported the results of field experiments in Norway in which artificial acid-rain solutions were sprayed on a young conifer plantation using an overhead apparatus. The effects on tree growth were small and sporadic. In lodgepole pine (*Pinus contorta*) watered for as long as 3 years at 50 mm H_2O/month, height growth was stimulated by 15–20% at pH values 4 and 3 compared with the pH 5.6–6.1 control treatment; at 25 mm/month there was no treatment effect at all. One year of treatment over the pH range 5.6 to 2.5 had no effect on the height growth of Norway spruce (*Picea abies*). The growth of Scotch pine (*Picea sylvestris*) was stimulated by as much as 15% by pH levels of 3.0 and 2.5 after 4 years of treatment, compared with pH 5.6–6.1, while pH 2.0 had a slightly negative effect. Birch (*Betula verrucosa*) was generally stimulated by the acid treatments. The feather mosses *Pleurozium schreberi* and *Hylocomium splendens* dominated the ground vegetation of these plantations. These bryophytes were sensitive to the acid treatments and were greatly damaged at pH ≤3.

Similar results were reported by Tamm and Wiklander (1980) in studies in Sweden. Six years of treatment of an 18-year-old Scotch pine stand with water containing the equivalent of 16, 33, or 49 kg H_2SO_4-S ha/year had a moderate effect on tree growth (there was a 30% increase in growth of basal area at the largest acid loading without NPK fertilization; with NPK there was a 10% decrease at the highest acidity). A dieback of the ground vegetation (dominated by feather mosses and the shrub, heather, *Calluna vulgaris*) occurred when the most acidic treatment was combined with fertilization. Note that negative effects on the ground vegetation of northern coniferous forests is usually observed after fertilization, as are increases of nitrophilous species such as raspberry (*Rubus* spp.), fireweed (*Epilobium angustifolium*), and certain grasses (e.g., *Agrostis capillaris* and *Deschampsia flexuosa*) (Kellner and Marshagen, 1991).

Because they are better controlled, laboratory experiments with artificial acid-rain solutions are more satisfactory than field experiments for the examination of dose–response effects on plants. In general, among the wide range of plant species that has been examined, growth reductions do not occur at laboratory treatment pHs >3.0, and in some cases growth stimulations have been observed at pH levels more acidic than this (Jacobson, 1980). In a greenhouse experiment, Wood and Bormann (1976) observed enhanced growth of white pine (*Pinus strobus*) seedlings that were treated with acidic solutions ranging from pH 2.3 to 4.0, compared with pH 5.6. They attributed this effect to nitrate fertilization, since their artificial rains contained that nutrient at a relative concentration of 24% of the total anion equivalents. The growth stimulation occurred in spite of an increase in leaching of base cations from the pine foliage caused by the acid treatments [see Fairfax and Lepp (1975), Wood and Bormann (1975), and Lepp and Fairfax (1976) for other studies of acid leaching of ions from foliage].

In another laboratory experiment, Percy (1983, 1986) exposed seedlings of 11 eastern North American tree species to acid-rain treatments that ranged from pH 2.6 to 5.6. Acute injuries in the form of macroscopic lesions on foliage of some species were only observed at pH 2.6. The lesions covered as much as 10% of the leaf surface, and 1 week of treatment with this very acidic pH was required to cause this effect. There were significant reductions of height growth, needle number, axillary meristems, and some other growth- or development-related variables of some species at pH ≤4.6. In general, conifer species were more sensitive than angiosperm species. Overall, however, these detrimental effects were not sufficient to cause significant decreases in the yield of dry weight of the seedlings (except at pH 2.6).

In general, it appears that trees and other vascular plants may not be at substantial risk of suffering direct, short-term, acute injuries from exposure to ambient acidic precipitation. However, there remains the possibility that hidden injuries or unidentified interactions with other environmental stresses could cause growth decreases (this is also discussed in Chapter 5, in the context of forest decline). Because acidic precipitation is regional in character, such yield decreases could occur over large areas and therefore would have great economic implications. This potential problem is most relevant to forests and other natural vegetation, since agricultural lands are regularly limed and have intrinsically larger rates of endogenous acid production, caused by cropping and fertilization, than those caused by acidifying depositions from the atmosphere.

Several studies carried out in western Europe and eastern North America have addressed the question of decreased forest productivity caused by acidifying depositions. These studies examined the temporal patterns of tree-ring width over large geographic areas, searching for recent trends of declining productivity that might be related to the regional patterns of acidifying deposition from the atmosphere (Johnsson and Sundberg, 1972; Abrahamsen *et al.*, 1976, 1977; Cogbill, 1977; Johnson *et al.*, 1981; Siccama *et al.*, 1982; Strand, 1983; Hornbeck *et al.*, 1987b). Many of these studies have demonstrated that productivity has recently declined in various tree species and in various regions. However, progressively decreasing ring widths are a natural phenomenon in aging forests because of: (1) the intensification of competitive stress when the forest canopy closes during the course of succession and (2) the immobilization of a large fraction of the site nutrient capital within aggrading biomass and organic matter. So far, it has not been possible in these tree-ring studies to clearly separate successional effects and other background influences, such as insect defoliation and climate change, from effects that might be related to acidifying depositions or other air pollution stresses (this is also discussed in Chapter 5).

Acidification and Freshwater Biota

Phytoplankton

The phytoplankton communities of lakes are generally species rich, but the taxa vary according to water chemistry. Nonacidic, oligotrophic lakes in the temperate zone are typically dominated by species of golden-brown algae and diatoms (Chrysophyceae and Bacilliarophyceae, respectively) (Schindler and Holmgren, 1971; Findlay and Kling, 1975; Ostrovsky and Duthie, 1975; Hendry *et al.*, 1980a). However, the dominant families of algae in acidic, clearwater, oligotrophic lakes are somewhat different. Yan and Stokes (1978) studied a pH 5.0 clearwater lake in Ontario and reported domination by dinoflagellates (Dinophyceae, especially *Peridinium limbatum*) and cryptomonads (Cryptophyceae, esp. *Cryptomonas ovata*). In a somewhat less acidic (pH 5.3) clearwater lake in Nova Scotia, Kerekes and Freedman (1988) reported domination by blue-green algae (Cyanophyceae, esp. *Agmenellum thermale*), green algae (Chlorophyceae, esp. *Sphaerocystis schroeteri*), and golden-brown algae. In two more acidic (pH 4.5), brown-water lakes nearby, the phytoplankton was dominated by golden-brown algae (esp. *Mallomonas caudata*), green algae (esp. *S. schroeteri*), diatoms (esp. *Asterionella formosa*), and yellow-green algae (Xanthophyceae); this general assemblage is typical of brown-water lakes (Ostrovsky and Duthie, 1975; Ilmavirta, 1980).

Species of diatoms and golden-brown algae can be highly responsive to variations of lakewater chemistry and can therefore be useful indicators of environmental conditions (Smol and Glew, 1992). In addition, because the silicious, outer cell wall of diatoms is persistent and diagnostic of species, fossil communities of these microalgae can be extracted from dated horizons of lake-sediment cores and used to infer historical conditions of water chemistry (as described earlier for Loch Laiden in Scotland). To calibrate the palaeoenvironmental conditions that can be inferred from diatom species, information must be gathered on their abundance in modern lakes with differing water-chemistry characteristics. For example, a study of 72 lakes in the larger vicinity of Sudbury found the following, acidification-related, indicator groups of diatoms (Dixit *et al.*, 1991):

1. indicators of low pH: *Eunotia pectinatus*, *Fragilaria acidobiontica*, *Pinnularia subcapitata*, *Tabellaria quadriseptata*;

2. indicators of low pH and high metals (Cu, Ni): *Eunotia exigua, E. tenella, Frustulina rhomboides saxonica, Pinnularia hilseana*;
3. indicators of high pH: *Achnanthes lewisiana, Cyclotella meneghiniana; Fragilaria construens, F. crotonensis.*

Therefore, because of the known, rather specific habitat requirements of these species in terms of water chemistry, their presence and relative abundance in dated sediment cores can be used to infer historical conditions of water quality.

In the experimental, whole-lake acidification described earlier, the phytoplankton of Lake 223 shifted from a preacidification community dominated by species of golden-brown algae, to one dominated by chlorophytes (especially *Chlorella* cf. *mucosa*). However, there were no substantial changes in gross community parameters such as species diversity and richness, because these are not sensitive to the presence of particular species (Findlay and Saesura, 1980; Findlay and Kasian, 1986). When the pH of Lake 223 was allowed to increase again, from pH 5.0–5.1 during 1981–1983 to pH 5.5–5.6 during 1984–1986, species typical of the preacidification phytoplankton community rapidly reappeared (Schindler, 1990).

Unlike species composition, the standing crop and productivity of the phytoplankton of oligotrophic lakes are rather unresponsive to changes in acidity (Hendry *et al.*, 1980a). In fact, if acidification is accompanied by a clarification of the upper water column, then standing crop and production can increase due to a greater depth of the euphotic zone. In Lake 223, the preacidification, whole-lake phytoplankton biomass was estimated as 1465 kg, compared with 2410 kg following acidification (Findlay and Saesura, 1980). The average, preacidification, annual production of Lake 223 ranged from 16 to 26 mg C/m^2-year over 3 years when the lakewater pH was 6.5 to 6.7, but it ranged from 28 to 47 mg C/m^2-year during the 4 years that the pH was 5.6 to 6.1, and from 31 to 60 mg C/m^2-year during the 3 years that the pH was 5.0 to 5.1 (Shearer *et al.*, 1987b). These data indicate relatively small effects of the acidification of Lake 223 on primary production, in contrast to the large changes on species composition.

Changes in the trophic status of water bodies have much larger effects on their standing crop and productivity of phytoplankton. To specifically illustrate this for acidic water, consider the case of a pair of small, adjacent lakes in Nova Scotia (Kerekes *et al.*, 1984). These lakes were acidified by the oxidation of pyrites in slate within their watershed, after bedrock was exposed to the atmosphere during the construction of a highway. Little Springfield Lake (mean annual pH 3.7) is oligotrophic, while Drain Lake (pH 4.0) receives an input of sewage and is eutrophic in spite of its extreme acidity. Drain Lake had a midsummer phytoplankton volume of 15.5 mm^3/liter and a chlorophyll-a concentration of 10.3 mg/liter. These are much larger than in (1) oligotrophic Little Springfield Lake with 0.85 mm^3/liter and 0.49 mg/liter, respectively or (2) the range of values among 15 nearby, less acidic (pH >4.5), oligotrophic lakes (0.21–3.31 mm^3/liter, 0.23–2.76 mg/liter, respectively) (Blouin, 1985).

Yan and Lafrance (1984) experimentally fertilized an acidic lake (pH 4.6) in Ontario and observed large increases in primary production. In the first year of their study, there was also a substantial increase in herbivorous zooplankton, which grazed the phytoplankton to a relatively small standing crop by the end of that growing season. However, in the second and third years of their study, the development of large populations of a predacious zooplankter, *Chaoborus* spp. (primarily *Chaoborus americana*), prevented the development of large populations of herbivores, and larger standing crops of phytoplankton were sustained.

Periphyton

Periphyton are microalgae found on solid substrates, such as sediment, rocks, sunken woody debris, or the foliage of macrophytes. Periphyton can be species rich, even in acidic lakes. Nakatsu (1983) studied three brown-water lakes in Ontario and identified 460 taxa of algae, most of which were littoral periphyton.

In some acidic lakes with transparent water, periphyton can be quite prominent and may even form benthic clouds or felt-like mats. The dominant taxa of periphyton in such oligotrophic water bodies tend to be *Mougeotia* spp., *Spirogyra* spp., and *Binuclearia tatrana* (all Chlorophyceae), and *Tabellaria flocculosa* and *Eunotia lunaris* (Bacillariophyceae) (Hendry *et al.*, 1980a; Stokes, 1980).

In the Lake 223 whole-lake acidification, a mat of the filamentous green alga *Mougeotia* sp. began to develop in the littoral zone after the pH had decreased to less than 5.6 (Mills, 1984; Schindler *et al.*, 1985). Similar observations of increased abundance of *Mougeotia* plus other Chlorophyceae of the periphyton have been made in other experimentally acidified lakes in northwestern Ontario (Turner *et al.*, 1987). These filamentous green algae (1) are probably relatively tolerant of physiological stresses associated with acidity; (2) are probably relatively efficient at obtaining dissolved inorganic carbon for photosynthesis (in acidic water, DIC is present in a relatively small concentration, and it largely occurs as CO_2, which most algae are not well adapted to utilize); and (3) they may experience less grazing pressure from invertebrates at low pH (Turner *et al.*, 1987).

Muller (1980) performed an acidification experiment in cylindrical mesocosms in Lake 223 and found no effects of acidification to pH 3.7–4.7 on the biomass or productivity of periphyton. At pH ≥ 6.3 diatoms dominated the periphyton biomass; at pH <6 chlorophytes were dominant. At pH 4, *Mougeotia* sp. was the only dominant, similar to what occurred in the whole-lake acidification.

Macrophytes

Certain species of macrophytic plants have apparently decreased in abundance in some acidified lakes [e.g., *Phragmites communis* (Almer *et al.*, 1974); *Lobelia dortmanna*, *Litorella uniflora*, and *Isoetes* spp. (Grahn *et al.*, 1974; Grahn, 1977, 1986)]. In some cases in Sweden, declines of particular species of vascular plants were accompanied and possibly caused by increases in abundance of acidophilous peatmoss species, especially *Sphagnum subsecundum*, and benthic filamentous fungi and algae (Grahn *et al.*, 1974; Hultberg and Grahn, 1975a; Grahn, 1977, 1986).

Benthic mats of *Sphagnum* have also been reported in some acidic lakes in North America, usually where there is extremely clear water (Hendry and Vertucci, 1980; Stewart and Freedman, 1989). Because of its apparent restriction to clearwater lakes, benthic *Sphagnum* is not characteristic of acidic water bodies (Singer *et al.*, 1983; Wile and Miller, 1983; Wile *et al.*, 1985). *Sphagnum* and other mosses have also increased in abundance in water bodies that have been affected by acid-mine drainage (Harrison, 1958; Hargreaves *et al.*, 1975). Acidophilous mosses are also abundant in acidic, volcanic lakes in Japan (*Leptodictyum* sp.; Yoshimura, 1935) and in acidic lakes near Sudbury and at the Smoking Hills (Chapter 2).

It has been hypothesized that the invasion of acidified lakes by bryophytes, particularly *Sphagnum* spp., may contribute to a "self-accelerating oligotrophication" (Grahn *et al.*, 1974). This process has been attributed to: (1) the acid-generating potential of acidophilous *Sphagnum* spp., which have an efficient cation-exchange ability that allows them to remove calcium, magnesium, and other basic cations from water, in exchange for hydrogen ion (Skene, 1915; Clymo, 1963, 1964, 1984; Hutchinson, 1975; Grahn, 1977), and (2) the physical isolation of sediments from the water column by the intervening benthic mat of *Sphagnum*, thereby interfering with acid-neutralization, alkalinity-generating, and nutrient-cycling processes that would otherwise take place at the sediment/water interface (Hultberg and Grahn, 1975a).

The macrophyte communities of brown-water and clearwater acidic lakes are quite different. This can be illustrated by comparison of the macrophytes of two small, oligotrophic lakes in Nova Scotia (Stewart and Freedman, 1989). Beaverskin Lake has a mean-annual pH of 5.3, clear water, and a euphotic zone that extends to all of its bottom. Macrophytes are found as deep as 6.5 m, and 96% of the lake bottom is vegetated. The most important littoral community is dominated by *Eriocaulon sep-*

tangulare, *Eleocharis acicularis*, and *Lobelia dort-manna*, and this type covers 25% of the lake bottom. A deeper community is dominated by *Sphagnum macrophyllum* and *Utricularia vulgaris*, and it covers 71% of the bottom. The average standing crop of macrophytes in Beaverskin Lake is 61 g dry weight (d.w.)/m^2.

In contrast, the nearby, relatively acidic (pH 4.5), brown-water Pebbleloggitch Lake has a much more restricted distribution of macrophytes. Plants are only found in a shallow, littoral fringe, and only 15% of the bottom of the lake is vegetated. The most extensive macrophyte community is dominated by the floating-leaved *Nuphar variegatum*. A second community has relatively sparse *Eriocaulon*, *Lobelia*, and *Eleocharis* and only occurs at depths <0.7 m. The average standing crop of macrophytes is only 2.6 g/m^2, but 17 g/m^2 within the vegetated zone.

Clearly, water clarity is a critical factor in determining the distribution and abundance of macrophytes in oligotrophic lakes; acidity per se is relatively unimportant. However, if a sample of lakes that covers a wider range of acidity is considered, rather different conclusions can be reached—acidity and associated chemical variables, especially calcium and alkalinity, appear to have important influences on macrophytes (Seddon, 1972; Hutchinson, 1975; Kadono, 1982; Catling *et al.*, 1986).

Aquatic macrophytes respond vigorously to fertilization (Lind and Cottam, 1969; Seddon, 1972; Hutchinson, 1975; Porcella, 1978). This is also true of acidic lakes. The previously described, eutrophic Drain Lake (pH 4.0) has a lush growth and large productivity of macrophytes (Kerekes *et al.*, 1984). There are also some anomalous species present. Two species of *Potamogeton* occur in Drain Lake (*P. pusillus* and *P. oakesianus*); the next-most acidic waters from which that genus has been reported in northeastern North America are pH 5.0 (Hellquist and Crow, 1980; Catling *et al.*, 1986). Therefore, trophic status can have more substantial, direct effects than acidity on the distribution and productivity of aquatic macrophytes.

Crustacean Zooplankton

The response of zooplankton to the acidification of oligotrophic water bodies is a complex function of: (1) the toxicities of hydrogen ion and associated metals such as aluminum, (2) the effects of any changes in primary production and the standing crops of the phytoplankton food of zooplankton, and (3) the effects of any changes in the nature and intensity of predation, particularly if acidification causes the extirpation of planktivorous fish. The toxic influences can act directly to eliminate relatively vulnerable taxa of zooplankton, while all three of the above factors can act indirectly, especially by changing competitive interactions within the zooplankton community, and by affecting the nature of predation (Sprules, 1975a; Hendry *et al.*, 1980a; Hobaek and Raddum, 1980).

Regional surveys have documented the typical, crustacean zooplankton communities of oligotrophic lakes, including acidic ones. Sprules (1975b) surveyed the midsummer zooplankton community of 47 lakes in Ontario that covered a pH range of 3.8 to 7.0. Species occurring primarily in lakes with pH <5 and that were considered to be indicators of acidity were *Polyphemus pediculus*, *Daphnia catawba*, and *D. pulicaria*. Acid-intolerant species that occurred only at pH >5 included *Tropocyclops prasinus mexicanus*, *Epischura lacustris*, *Diaptomus oregonensis*, *Leptodora kindtii*, *Daphnia galeata mendotae*, *D. retrocurva*, *D. ambigua*, and *D. longiremis*. Apparently acid-indifferent species that were present over the entire pH range were *Mesocyclops edax*, *Cyclops bicuspidatus thomasi*, *Diaptomus minutus*, *Holopedium gibberum*, *Diaphanosoma leuchtenbergianum*, and *Bosmina* sp. The most frequent species was *Diaptomus minutus*; this was also the most acid-tolerant species and was the sole zooplankter in the most acidic (pH 3.8) lake that was sampled. In lakes with high pH, the zooplankton community was relatively species rich and had a relatively even distribution of dominant species. Above pH 5 there were 9–16 species with three or four dominants; below this pH there were 1–7 species with one or two dominants.

Sprules (1977) performed a multivariate, principal components analysis (PCA) on his data matrix of lakes × zooplankton species (Table 4.12). The first PCA axis accounted for 50% of the variance and separated small, acidic, clearwater lakes dominated by *Diaptomus minutus*, from relatively large, neutral, low-clarity lakes dominated by four acid-indifferent species. The second PCA axis accounted for 15% of the variance, and separated acidic, high-sulfate, clear lakes from neutral, low-sulfate, low-clarity lakes. Of the various physical and chemical limnological variables that Sprules correlated with the PCA axes, pH showed the strongest relationship. The correlation of pH with axis 1 was 0.63, and it was −0.54 for axis 2 (both $P < 0.001$). Note,

however, that these are not strong correlations, since they only account for 40 and 29% of the variance, respectively. The results of this study indicate that a complex of environmental factors that includes, but is not necessarily overwhelmed by, pH is important in structuring the zooplankton community. Other important (but unmeasured in this study) environmental factors that may be affected by acidity and that are known to influence the zooplankton community include biological interactions, such as predation and competition.

In another survey, of 27 lakes in Norway, Hobaek and Raddum (1980) found that acidic lakes had a zooplankton fauna that was dominated by *Bosmina longispina*, *Eudiaptomus gracilis*, and

Table 4.12 Principal components analysis of the distribution matrix of 23 crustacean zooplankton species in 60 lakes in the Killarney area of Ontario[a]

Species	Principal component 1 (50.1%)		Principal component 2 (14.9%)	
	Eigenvector	Correlation	Eigenvector	Correlation
Diaphanosoma leuchtenbergianum W	0.215	0.66[b]	−0.142	−0.24
Bosmina longirostris W	0.346	0.64[b]	0.605	0.84[b]
Mesocyclops edax W	0.193	0.60[b]	−0.249	−0.27[c]
Cyclops bicuspidatus thomasi W	0.275	0.54[b]	−0.593	−0.63[b]
Daphnia retrocurva I	0.067	0.45[b]	−0.094	−0.35[d]
Tropocyclops prasinus mexicanus I	0.077	0.37[d]	−0.079	−0.20
Diaptomus oregonensis I	0.086	0.37[d]	−0.190	−0.45[b]
Leptodora kindtii I	0.007	0.34[d]	−0.013	−0.34[d]
Daphnia galeata mendotae I	0.064	0.34[d]	−0.151	−0.44[b]
Diaptomus reinhardi	0.057	0.33[d]	0.017	0.05
Ceriodaphnia reticulata	0.021	0.30[c]	0.008	0.06
Holopedium gibberum W	0.119	0.28[c]	0.330	0.42[b]
Daphnia longiremus I	0.039	0.27[c]	−0.077	−0.29[c]
Daphnia ambigua I	0.021	0.26[c]	−0.022	−0.15
Senecella calanoides	0.005	0.25	−0.008	−0.22
Cyclops scutifer	0.012	0.17	−0.021	−0.16
Epischura lacustris I	0.030	0.13	−0.062	−0.14
Polyphemus pediculus A	0.003	0.08	0.033	0.50[b]
Daphnia sp.	0.001	0.04	−0.008	−0.17
Daphnia catawba A	−0.002	−0.01	−0.007	−0.02
Daphnia pulicaria A	−0.007	−0.01	0.055	0.19
Orthocyclops modestus	−0.017	−0.10	−0.020	−0.07
Diaptomus minutus W	−0.823	−0.99[b]	−0.044	−0.03

[a] A, acid indicator; I, acid intolerant; W, acid independent (see text). Modified from Sprules (1977).
[b] Probability (P) < 0.001 of obtaining a product–moment correlation if true correlation is zero (two-tailed test).
[c] $P < 0.05$.
[d] $P < 0.01$.

Kellicottia longispina. These, along with the less abundant species *Heterocope saliens*, *Holopedium gibberum*, *Keratella serrulata*, and *Polyarthra* spp., were regularly present in acidic lakes, but were not necessarily restricted to them. Hobaek and Raddum noted that *Daphnia* spp. and cyclopoid copepods were generally absent from acidic lakes, in contrast to Sprules's study, where two of the three acid-indicator taxa were *Daphnia* species. The acid-lake community in Norway was characterized by a small species richness and diversity and by a tendency to have only a few, strongly dominant taxa.

Another study compared the crustacean zooplankton of a pH 5.3 clearwater lake and a nearby, pH 4.5 brown-water lake (Blouin, 1985; Kerekes and Freedman, 1988). The clearwater lake was dominated by the copepod *Diaptomus minutus* and the rotifer *Keratella cochlearis*. The brown-water lake was dominated by the rotifers *Keratella cochlearis* and *Conochilis unicornis* and the copepod *Diaptomus minutus*. Zooplankton were somewhat more abundant in the brown-water lake, with a biomass of 330 mg/m^3 and a density of 222/m^3, compared with 200 and 124 mg/m^3, respectively, in the clearwater lake. The greater abundance of zooplankton in the brown-water lake could have been due to several factors. In brown waters, large concentrations of dissolved and suspended organic matter from allochthonous sources could be partly responsible for sustaining a relatively large productivity of zooplankton (Nauwerck, 1963; Schindler and Noven, 1971; Ostrovsky and Duthie, 1975; Petersen *et al.*, 1987). Predation could also be important. Although fish were equally abundant in the two lakes (Kerekes and Freedman, 1988), the feeding efficiency of visual predators is undoubtedly less in a brown-water lake.

The above studies describe the crustacean zooplankton communities of acidic lakes, but they do not directly shed light on the changes that take place while acidification is proceeding. For such information, we can consider the whole-lake acidification of Lake 223, because the dynamics of zooplankton were monitored in this experiment (Malley *et al.*, 1982; Nero and Schindler, 1983). Overall, acidifi-

cation caused an increase in the abundance of zooplankton. The total density of cladocerans was larger by 66% in 1980 (pH 5.4) compared with 1974 (pH 6.6), while the density of copepods was 93% greater. Throughout the study, the numerically dominant zooplankters were the copepods *Diaptomus minutus* and *Cyclops bicuspidatus*. The cladocerans *Bosmina longirostris*, *Daphnia galeata*, *Holopedium gibberum*, and *Diaphanosoma brachyurum* were also present throughout, but at a relatively small density. An important extirpation was of the nocturnal predator *Mysis relicta*, which declined from a whole-lake abundance of 6.78×10^6 in August 1978 (pH 5.9), to 0.27×10^6 in August 1979 (pH 5.6), and then to zero in the following year. Two other, relatively minor zooplankters (*Epischura lacustris* and *Diaptomus silicilis*) also became extirpated. *Daphnia catawba* × *schroederi*, a species that was not recorded prior to acidification, appeared suddenly, and by 1980 it comprised 12% of the total cladoceran abundance. The overall increase in abundance of the zooplankton was attributed to the previously described increases in phytoplankton productivity and biomass. Toxicity of acidic water and changes in the nature and/or intensity of predation and competition are also believed to have had important influences on the zooplankton of Lake 223.

Benthic Invertebrates

Various surveys have demonstrated that benthic invertebrates are generally less species rich in acidic oligotrophic waters, compared with less acidic waters (Sutcliffe and Carrick, 1973; Conroy *et al.*, 1976; Hendry and Wright, 1976; Leivestad *et al.*, 1976; Almer *et al.*, 1978; Hendry *et al.*, 1980a). However, benthic invertebrates can be abundant in acidic lakes, especially if predatory fish have disappeared or have been reduced in numbers. The benthos of acidic lakes is generally dominated by (1) crustaceans—various Copepoda, Cladocera, Amphipoda, and Isopoda, and (2) insects—especially Notonectidae, Corixidae, Chrironomidae, and Megaloptera; Trichoptera, Ephemeroptera, and Plecoptera may also be present.

All of the above groups, however, also have species that are intolerant of acidity. For example, Raddum (1978) found that while many Plecoptera species were indifferent to pH in Scandinavian fresh waters, there were also several sensitive taxa, including *Amphinemura sulcicollis*, *Brachyptera risi*, and *Leuctra hippopus*. The Ephemeropteran *Baetis rhodani* tends to dominate its order in circumneutral streams, but it disappears with acidification. In northern Europe, acidification-indicating mayflies include *Baetis rhodani*, *B. lapponicus*, and *B. macani* (Raddum and Fjellheim, 1984). The benthic amphipods *Gammarus lacustris* and *Hyallela azteca* and the snails *Valvata macrostoma* and *Ancyclus fluviatilis* are also sensitive to acidification (Raddum and Fjellheim, 1984; Stephenson and Mackie, 1986).

Because of their need to deposit shells of calcium carbonate and the apparent physiological and physical/chemical difficulties in accomplishing this function in acidic waters, mollusks as a group are sensitive to acidification. Okland and Okland (1986) studied the benthos of more than 1000 lakes in Norway and reported that mollusks were especially sensitive to acidification, with 17 of 22 species of snail being absent at pH <6.0, no snails present at pH <5.2, and no clams at pH <6.0.

Hall and Ide (1987) resampled two oligotrophic, low-alkalinity streams in Ontario in which benthic insects had been thoroughly studied 48 years previously. In one stream without severe acid pulses in the spring (pH 6.4–6.1), there were few differences in insect taxa between the two samplings. The other stream currently experiences severe acid pulses (pH 6.4–4.9), and there were some notable changes in the community of benthic insects between the two samplings. Four species of Ephemeroptera (mayflies) disappeared since 1942, but seven acid-tolerant species were newly recorded, while two species of Plecoptera (stoneflies) disappeared and two new species were added.

As an example of the benthos of typical, acidic water bodies, consider the fauna of two previously described oligotrophic lakes in Nova Scotia (Kerekes and Freedman, 1988). Insects were the most abundant group in both the clearwater pH 5.3 and

the brown-water pH 4.5 lake, comprising 56 and 78% of the total density of benthic invertebrates, respectively. The insects were dominated by Diptera (45 and 68% of total insect density, respectively), especially Chironomidae (96 and 92% of dipteran density). Dominance by Chironomidae appears to be a general characteristic of the benthos of acidic lakes (Roff and Kwiatkowski, 1977; Raddum, 1978; Havas and Hutchinson, 1982; Kerekes *et al.*, 1984). Other important insects in the Nova Scotian lakes were Trichoptera (5% of total invertebrate abundance in both lakes), Ephemeroptera (1 and 3%, respectively), and Odonata (both 2%). Crustacea comprised 18–19% of the benthic invertebrate density. The most abundant orders were Copepoda, Cladocera, Amphipoda, and Isopoda. The oligochaete families Naididae and Enchytraeidae accounted for 13% of the total invertebrate abundance in the clearwater lake and 3% in the brown-water lake. Overall, there was a rather minor difference in the benthic invertebrates of these two lakes, in spite of large differences in their water chemistry.

It is notable that the acidity of the surface sediment did not change during the Lake 223 experiment, even after 8 years of acidification (Kelly *et al.*, 1984). Microbial processes such as sulfate reduction (described earlier) kept pH >6 within just a few millimeters of the sediment–water interface. In 1981, the pH at and just above the sediment surface was 5.3–5.4, while at a depth of 0.5 cm in the sediment it was >6.0 and at 2 cm it was 6.7–6.8. After the acidification of Lake 223, there was an increased emergence of adult dipterans, especially of chironomids, which peaked in abundance at pH 5.6 (Mills, 1984). In contrast, the previously abundant larvae of mayflies (*Hexagenia* spp.) disappeared at about pH 5 (L.A. Baker, 1990).

The population of the crayfish *Orconectes virilis* in Lake 223 suffered a demise as a result of the acidification (Schindler and Turner, 1982; Mills, 1984; France and Graham, 1985; France, 1987). This was largely caused by reproductive failure at pH ≤5.6, along with an inhibition of carapace hardening following moult (which makes the animals vulnerable to predation, including cannibalism), and the effects of a microsporidian parasite. Even though the crayfish population was extirpated after 3 years of reproductive failure, direct toxicity to mature animals did not seem to be an important problem, even at pH 5.1. Interestingly, the extirpation of another population of *Orconectes virilis* was observed in a remote lake in south-central Ontario, as it acidified over a 6-year period from pH 5.8 to pH 5.6 under the influence of atmospheric deposition (France and Collins, 1993).

In another whole-ecosystem experiment, Norris Brook, NH, was rapidly acidified from an initial pH 5.7–6.4 to pH 4.0 (Hall and Likens, 1980). The drift of dead invertebrates increased by a factor of 13 during the first week of acidification, followed by a return to reference values. The total density of benthic invertebrates decreased by 75% overall, with the largest decreases being exhibited by Ephemeroptera and some species of Plecoptera and Diptera. Because of the decreased grazing by herbivorous invertebrates in Norris Brook, the biomass of periphyton algae increased significantly.

Fish

Arguably the most important and highest-profile victims of acidification are populations of susceptible species of fish. Extirpations of fish populations, particularly of commercially important Salmonidae, have been reported from various locations in the northern hemisphere where there have been severe acidifications of surface waters.

In Scandinavia, for example, many surface waters have lost sport and commercial salmonid fisheries, especially in southern areas where acidifying depositions are relatively severe (Fig. 4.6) (Jensen and Snekvik, 1972; Wright and Snekvik, 1978; Rosseland *et al.*, 1986). A survey of 700 small Norwegian lakes done in 1974–1975 showed that fish populations, primarily brown trout (*Salmo trutta*), were absent from 40% of the lakes, and that fish were sparse in another 40%. Prior to the 1950s most of these lakes had sustained healthy fish populations. Another survey of more than 2000 lakes in southern Norway showed that one-third of them had lost their fish population since the 1940s. Most of the fishless lakes had a pH <5.0. Another Nor-

wegian survey, of 890 lakes in 1986, found that fish were not present or were extirpated in 27% of the lakes (with an average pH of 4.8), while fish populations were not reproducing well in 25% of the lakes (mean pH 5.4), and 48% of the lakes had healthy populations (mean pH 6.1) (Henriksen *et al.*, 1989). The damages were most important in southern Norway, where acidifying depositions are most intense.

Salmonid fish species differ in their sensitivity to acidification in Norway. Atlantic salmon (*Salmo salar*) requires a pH greater than 5.0–5.5 for successful hatching and development of fry; sea trout (a distinct, anadromous form of brown trout) also require a pH >5.0–5.5; stationary brown trout are more tolerant, and require a pH >4.5. In general, younger life-history stages are more sensitive than are adult fish. However sporadic kills of adults have been documented in the springtime, when water pH is the most acidic.

In the Adirondack Mountains of New York state, many high-altitude, oligotrophic lakes have lost their fish populations as a result of acidification (Schofield, 1976b, 1982; Haines and Baker, 1986). In sport-fishery surveys done in the 1930s, brook trout (*Salvelinus fontinalis*) was the dominant species, and it was present in 82% of the lakes that had fish. In the same area in the mid-1970s, brook trout were extirpated from at least 26 of the previously surveyed lakes. In nearby Ontario, brook trout become extirpated when the average pH decreases below about 5.0 (Beggs and Gunn, 1986). In total, fish were not present in 93 of the 215 New York lakes surveyed by Schofield (1982) in the mid-1970s.

In another survey of fish status in Adirondack lakes, Haines and Baker (1986) reported the following losses of fish populations: (a) 42% of 707 brook trout populations, (b) 39% of 111 lake trout (*Salvelinus namaycush*) populations, (c) 36% of 90 rainbow trout (*Salmo gairdneri*) populations, (d) 18% of 412 white sucker (*Catostomus commersoni*) populations, (e) 19% of 520 brown bullhead (*Ictalurus nebulosus*) populations, (f) 32% of 351 pumpkinseed sunfish (*Lepomis gibbosus*) populations, (g) 34% of 326 golden shiner (*Notemigonus crysoleucas*) populations, (h) 42% of 254 creek

Fig. 4.6 The declining yield of Atlantic salmon from seven acidified rivers in southernmost Norway with an average pH of 5.1, compared with 68 rivers in the rest of the country with an average pH of 6.6 (upper curve, left vertical axis). Modified from Leivestad *et al.* (1976).

chub (*Semotilus atromaculatus*) populations. These dramatic effects on previously vigorous sport fisheries, and on the fish community in general, have paralleled the documented, concurrent acidification of lakes in that area.

Losses of sport-fish populations have also occurred in acidified lakes and rivers in Canada. Watt *et al.* (1983) reported the extirpation of Atlantic salmon from seven acidic (pH <4.7) rivers in Nova Scotia that had previously supported that species. Atlantic salmon were declining in other rivers with pH 4.7–5.0, but were stable at pH >5.0 (Fig. 4.7). Lacroix and Townsend (1987) penned juvenile Atlantic salmon in four acidic streams in Nova Scotia and reported no survival in the two streams with pH <4.7 but complete survival where pH stayed >4.8.

Other studies documented the loss of fish popu-

lations from the Killarney region of Ontario (Beamish and Harvey, 1972; Beamish, 1974; Beamish *et al.*, 1975; Harvey and Lee, 1982). This area receives acidic precipitation (pH 4.0–4.5), and because it is fairly close to the Sudbury smelters, it has episodes of dry deposition of acidifying SO_2. In a survey done in the early 1970s, 33 of 150 lakes in the area had a pH <4.5. The extirpations of several species of fish were actually monitored for two lakes (Lumsden and George Lakes). There was also circumstantial evidence for the extirpation of fish in other lakes, in the form of eyewitness accounts of historical sport fisheries in presently fishless lakes. In total, there are known extirpations of lake trout in 17 lakes in the Killarney area. Lake trout is the most important sport fish in this region. In general, this species fails to recruit at pH <5.5, and it is extir-

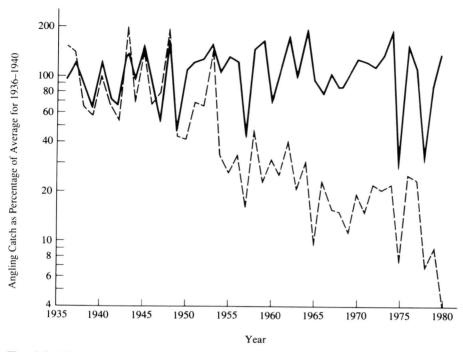

Fig. 4.7   The catch of Atlantic salmon by sport fishing in Nova Scotia rivers. The data were standardized to facilitate the comparison of acidic and less acidic rivers. (—) Mean for 12 rivers with pH >5.0 (1980). (---) Mean for 10 rivers with pH ≤ 5.0 (1980). From Watt *et al.* (1983).

pated from Ontario lakes with an average pH of about <5.2 (Beggs and Gunn, 1986). In addition, smallmouth bass (*Micropterus dolomieiu*) have disappeared from 12 lakes in the Killarney area, largemouth bass (*M. salmoides*) and walleye (*Stizostedion vitreum*) from four, and yellow perch (*Perca flavescens*) and rock bass (*Ambloplites rupestris*) from two (Harvey and Lee, 1982). Therefore, it seems that yellow perch and rock bass are among the most acid-tolerant fish species in the Killarney lakes.

Yellow perch and rock bass are also relatively tolerant of acidification elsewhere in eastern North America, as are the central mudminnow (*Umbra limni*), largemouth bass, bluegill (*Lepomis macrochirus*), black bullhead (*Ictalurus melas*), brown bullhead, golden shiner, and American eel (*Anguilla rostrata*), all of which are known to be present in some water bodies in North America with pH ≤4.5 (Rahel and Magnuson, 1983; Kerekes and

Freedman, 1988; L. A. Baker, 1990). In western Europe, *Umbra pygmaea* can persist in lakes to pH 4.0 and can tolerate laboratory bioassays of pH 3.5 (Dederen *et al.*, 1987).

In Lumsden Lake, the water pH decreased from 6.8 in 1961 and to 4.4 by 1971 (Beamish and Harvey, 1972). This acidification was accompanied by reproductive failures and losses of the populations of several fish species, including lake trout, lake herring (*Coregonus artedii*), and white sucker. In George Lake with a pH of 4.8–5.3, lake trout, walleye, burbot (*Lota lota*), and smallmouth bass disappeared. In 1973, unsuccessful reproduction was observed in other species that were not-yet extirpated. These were northern pike (*Esox lucius*), rock bass, pumpkinseed sunfish, brown bullhead, and white sucker. Clearly, an important cause of the extirpations of fish in these lakes was a persistent failure to reproduce.

Fish populations were carefully monitored during the Lake 223 acidification, and consideration of that study gives insight into the dynamics of the response of the fish community (Schindler and Turner, 1982; Mills, 1984; Mills *et al.*, 1987). Initially, there were five species of fish in Lake 223: lake trout, white sucker, fathead minnow (*Pimephales promelas*), pearl dace (*Semotilus margarita*), and slimy sculpin (*Cottus cognatus*). These species were all abundant, except for the relatively uncommon pearl dace.

As acidification progressed, there were marked changes in the fish community of Lake 223. The most sensitive species was the fathead minnow, which declined precipitously in 1979 when the lake pH reached 5.6 and became extirpated in 1980. The first of many year-class failures of lake trout was in 1980 (pH 5.4) and of white sucker in 1981 (pH 5.1). Slimy sculpin declined throughout the experiment. The pearl dace began to increase markedly in abundance in 1980, and it quickly became the most abundant small-fish species. This species may have experienced a competitive release following the demise of the ecologically similar fathead minnow. However, the pearl dace also eventually declined, in 1982 when the pH of Lake 223 reached 5.1. In fact, by 1982 and pH 5.0–5.1, no fish species were reproducing in Lake 223. Although lake trout and white sucker were still quite abundant in 1983, in the absence of successful reproduction in water at ca. pH 5.0–5.5, they will of course become extirpated.

The Lake 223 experience indicates a general sensitivity of the fish community to lake-water acidification. However, within limits set by the ultimate pH that is reached, there can be a replacement of sensitive species by relatively tolerant ones.

Many studies have examined the physiological effects of acidity and associated chemical stresses on fish. An important generalization is that early life-history stages are more susceptible to acidity than are adults. As was described above, most extirpations of fish populations have been attributed to reproductive failure, rather than to the death of adult fish (although episodic mortality of adult fish has been caused by events of acid shock; Leivestad and Muniz, 1976, Leivestad *et al.*, 1976).

Laboratory studies of egg and embryonic development of Atlantic salmon have indicated a toxic threshold (measured as LC_{50}—the concentration at which 50% of the exposed eggs fail to develop successfully) at pH 3.9, while alevins were affected at pH 4.3 (Daye and Garside, 1979). In another laboratory study, Peterson *et al.* (1980) found that the hatching of eggs of Atlantic salmon was inhibited at pH 4.0. This effect was eliminated by transferring the eggs to pH 6.8 water, providing that the pH 4.0 exposure occurred for less than 7–10 days.

Lacroix (1985) incubated eggs of Atlantic salmon in five streams in Nova Scotia with pH 4.6–6.5 and found an LC_{50} of pH 4.7 in the interstitial water of gravel in spawning beds. The toxic threshold reported in this field study was at a higher pH than would have been predicted from the results of the two previously described laboratory experiments. The discrepancy could be related to some unknown influences of environmental variables that are controlled in the laboratory, but are important in the field, such as oxygen tension or water temperature. For example, Kwain (1975) examined the effects of temperature and acidity on embryonic development of rainbow trout and found greater vulnerability to acidity at lower temperatures. At 15°C there was a toxic effect at pH 4.5; at 10°C the toxic threshold was pH 4.8; and at 5°C it was at pH 5.5. The pattern of toxicity differed between yearling and fingerling trout (for fingerlings, pH 4.5 was the toxic threshold at 20°C; pH 4.2 at 15°C; and pH 4.1 at 10°C), but both of these stages were more tolerant of acidity than were embryos.

As has been previously described, certain chemical constituents increase in concentration at low pH. From the perspective of toxicity to fish and other aquatic biota, the most notable example of this effect is aluminum (although the concentrations of most other metals also increase in acidic water). In fact, aluminum concentrations in many acidic waters can be sufficient to cause fish mortality, irrespective of any direct effects of hydrogen ion. J. Baker and Schofield (1982) reported that over the pH range 4.2 to 5.6, there were reduced survival and growth of larvae and older life-history stages of white sucker at 0.1 mg/liter of aluminum and at 0.2 mg/liter for brook trout. Aluminum concentrations of this magnitude are regularly exceeded in acidic waters. The most toxic forms of aluminum in acidic clearwaters are the ions Al^{3+} and $AlOH^{2+}$ (Havas, 1986).

It is important to note that in brown waters with large concentrations of dissolved organic matter, most of the soluble aluminum and other toxic metals is present in complexed forms. In this state, metals are relatively nonexchangeable and less bioavailable than are free-ionic forms, and aluminum in this bound form is much less toxic than an equivalent concentration in clear water (Driscoll *et al.*, 1980; J. Baker and Schofield, 1982; Campbell *et al.*, 1983; Lazerte, 1984; Petersen *et al.*, 1987). In four acidic brown-water streams in Nova Scotia (pH 4.6–5.6; dissolved organic carbon 6.6–18 mg/liter), virtually all of the aluminum dissolved in the water (0.08–0.19 mg/liter) was organically bound (Clair and Komadina, 1984). The lowest pH at which *Perca fluviatilis* occurred in a survey of Norwegian lakes was strongly influenced by the concentration of dissolved-organic matter (TOC), with only pH ≥4.9 being tolerated at TOC < 3.0 mg/liter, but pH ≥4.45 at TOC > 5.0 mg/liter (Henriksen *et al.*, 1989).

Amphibians

Most species of amphibians are either completely aquatic or they enter the water seasonally to breed. Several studies have suggested that acidification of their aquatic habitats could reduce the population sizes or restrict the distributions of amphibians (reviewed in Freda, 1986). Gosner and Black (1957), for example, found that the distributions of frog species in the New Jersey pine barrens are strongly influenced by acidity (pH range 3.6–5.2; the acidity is due to drainage from *Sphagnum*-dominated peatlands). The carpenter frog (*Rana virgatipes*) and the pine barrens tree frog (*Hyla andersoni*) are restricted to this area and are the most acid tolerant of the species that were tested, surviving pH 3.8 in a laboratory bioassay. Other species with more widespread distributions are less tolerant of acidity.

Saber and Dunson (1978) showed that the species richness of amphibians was smaller in an acidic bog in Pennsylvania than in less acidic water. For five of the eight species that were present in the acidic bog, reproductive and adult stages were both present. For the other three species only adults were present, possibly indicating a reproductive failure. In England, Cooke and Frazer (1976) found that the smooth newt (*Triturus vulgaris*) rarely breeds in ponds with pH <6.0, whereas the palmate newt (*T. helveticus*) was present in water as acidic as pH 3.9, but not at greater acidity. In the Netherlands, Strijbosch (1979) found that six frog species avoided water with pH <4.5 and that there was a relatively large frequency of dead egg masses in acidic habitats.

In Nova Scotia, the distributions of 11 species of amphibians were not clearly related to acidity among 159 sites encompassing a pH range of 3.9 to 9.0 (Dale *et al.*, 1985). The bullfrog (*Rana catesbeiana*) was present between pH 4.0 and 9.0; spring peeper (*Hyla crucifer*) and yellow-spotted salamander (*Ambystoma maculatum*) between pH 3.9 and 7.8; green frog (*R. clamitans*) between pH 3.9 and 7.3; and wood frog (*R. sylvatica*) between pH 4.3 and 7.8. Habitat structure, predation, and competition may have had a more important influence than acidity in the distribution of these amphibians. Successful reproduction was indicated at pH 4 by the presence of eggs, developing larvae, and adults of five species: yellow-spotted salamander, and green, bull, wood, and pickerel (*R. palustris*) frogs. In contrast, the experimental acidification (to pH 4) of a small stream in New Hampshire caused salamanders to leave the treatment area (Hall and Likens, 1980).

In addition to the above field studies, laboratory experiments have revealed that among 14 species of amphibians, exposure to a pH of 3.7–3.9 during embryonic development caused mortality that exceeded 85%, and a prolonged exposure to pH 4.0 caused mortality that exceeded 50% (for reviews, see Tome and Pough, 1982; Dale *et al.*, 1986; Freda, 1986; Freda *et al.*, 1991). However, pH levels this acidic are uncommon in nature, and are not generally associated with an acidification of fresh water that was caused by acidifying depositions from the atmosphere (which has a threshold of about pH 4.5). It seems that at least some amphibian species may be at lesser risk than fish for population changes from this source of acidification.

Waterfowl

Direct, toxic effects of acidification to aquatic birds are unlikely and have not been documented. However, if acidification were to cause important changes in the habitats of waterfowl, then indirect effects on bird populations would be anticipated (Eriksson, 1984). For example, any reduction or extirpation of fish populations would be detrimental to piscivorous water birds. In contrast, an increased abundance of aquatic insects or zooplankton, possibly resulting from decreased predation caused by an extirpation of predatory fish, could be beneficial to birds that eat these arthropods (M. L. Hunter *et al.*, 1985; McNeil *et al.*, 1987; DesGranges and Houde, 1989; Parker *et al.*, 1992).

Piscivorous waterfowl that breed on oligotrophic lakes of the northern hemisphere include common and red-throated loons (*Gavia immer* and *G. stellata*, respectively) and mergansers (*Mergus merganser* and *M. serrator*); osprey (*Pandion haliaetus*) may also be at risk in certain locations. Arthropod-eating birds that might benefit from an increased abundance of invertebrates include diving ducks such as the common goldeneye (*Bucephala clangula*), ring-necked duck (*Aythya collaris*), and hooded merganser (*Lophodytes cucullatus*), and dabbling ducks such as mallard and black duck (*Anas platyrhynchos* and *A. rubripes*).

Alvo *et al.* (1988) studied the breeding success of common loons on 84 lakes in central Ontario. Only 9% of breeding attempts were successful on low-alkalinity lakes (<40 μeq/liter), compared with 57% for lakes with alkalinity of 40–200 μeq/liter and 59% for lakes with >200 μeq/liter. These observations presumably reflect the status of fish populations in the lakes.

Parker *et al.* (1992) studied the abundance of invertebrates in the littoral zone of 79 small lakes and ponds in New Brunswick. In water bodies with pH 4.5–4.9, there was about two times as large a standing crop as in those with pH >5.5. Odonata were more abundant in the acidic water bodies, while Amphipoda, Ephemeroptera, and Pelycopoda were less abundant, and there was little difference in Hemiptera, Coleoptera, Trichoptera, Diptera, or Acarina. Five species of arthropod-eating ducks averaged about 3.5 broods/ha on water bodies with pH <5.5, compared with 0.65 broods/ha at pH ≥5.5.

Interestingly, the previously described, acidic (pH 4.0), eutrophic Drain Lake is relatively productive of black ducks, sustaining a productivity of 0.5 broods/ha (Kerekes *et al.*, 1984). Comparable productivities of this species are only observed on other relatively fertile but nonacidic lakes in Nova Scotia (Staicer *et al.*, 1993). In acidic Drain Lake, the large duck productivity is sustained by vigorous macrophyte and invertebrate communities. DesGranges and Hunter (1987) observed the feeding behavior of 49 semicaptive, imprinted ducklings on seven lakes in Quebec and Maine that varied in acidity and productivity. They also observed that food availability was the key influence on duckling productivity, irrespective of acidity (see also Staicer *et al.*, 1993).

4.5
RECLAMATION OF ACIDIFIED WATERBODIES

Since at least the 1950s, limnologists have been interested in the treatment of brown-water lakes by liming, as a potential management option to reduce acidity, clarify the water, improve productivity, and create sport-fish habitat (Hasler *et al.*, 1951; Waters, 1956; Waters and Ball, 1957; Stross and Hasler, 1960; Stross *et al.*, 1961). More recently, much work has been done on the neutralization of acidified clearwaters by the addition of bases, usually limestone ($CaCO_3$) or lime [$Ca(OH)_2$] (Flick *et al.*, 1982). These treatments can be viewed as whole-lake titrations to raise pH.

In some parts of Scandinavia, liming is now used in wide-scale remedial programs to mitigate the biological symptoms of acidification. For example, by 1988 about 5000 surface waters had been experimentally or operationally limed in Sweden, mostly with limestone, along with another several hundred lakes in southern Norway [Ministry of Agriculture and Environment (MAEC), 1983; Lessmark and Thornelof, 1986; Olem, 1990]. In the early 1980s there was also a program to lime 777 acidic lakes in the Adirondack region of New York, at a cost of $40–75/ha for accessible lakes (1982 dollars) and $450–675/ha for remote lakes (Cooke *et al.*, 1986).

The effects of base addition on lake-water pH are illustrated in Fig. 4.8 for three limed and one reference lake in Ontario. Prior to treatment, the lake waters had a pH of 4–5, which after liming increased to pH 7–8. The pH of Middle and Hannah Lakes remained fairly stable after treatment, while that of Lohi Lake quickly drifted back toward an acidic condition. This difference largely reflects the sizes of the drainage basins of these lakes; that of Lohi is relatively large, the lake flushes rapidly because of large water inputs, and therefore neutralization was relatively short lived (Dillon *et al.*,

1979). Similar liming experiments have been conducted in Scandinavia (Wilander and Ahl, 1972; Hultberg and Grahn, 1975b; Bengtsson *et al.*, 1980; Wright, 1985; Hultberg and Grennfelt, 1986) and the northeastern United States (Cooke *et al.*, 1986).

Initially in the Ontario liming experiment, there was a drastic, postliming decline in the abundance and productivity of phytoplankton and zooplankton (Scheider *et al.*, 1975, 1976; Dillon *et al.*, 1979). Phytoplankton abundance returned to the pre-neutralization condition fairly quickly, but there was a shift in dominance from Dinophyceae to Chrysophyceae. Zooplankton were slower to recover, and 3 years after the first addition of lime to Lohi Lake, the zooplankton had not returned to its pre-neutralization abundance. In addition, fish caged in the neutralized lakes suffered a high rate of mortality (Yan *et al.*, 1979), contrary to the expectation

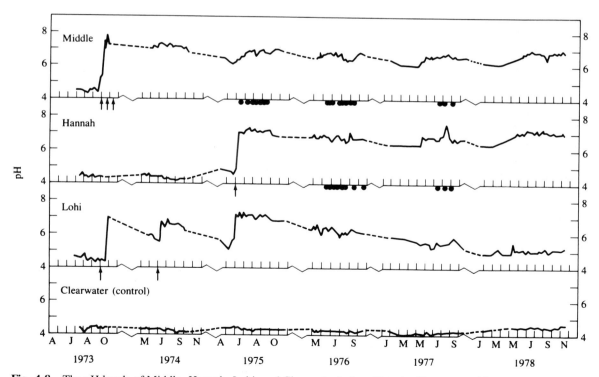

Fig. 4.8   The pH levels of Middle, Hannah, Lohi, and Clearwater Lakes, Ontario. Treatment with neutralizing agents [$CaCO_3$, $Ca(OH)_2$] is indicated by arrows and addition of phosphorus by solid dots. From Dillon *et al.* (1979).

based on liming experiments elsewhere (Hultberg and Grahn, 1975b). The results in Ontario were influenced to some degree by metal toxicity, since the experimental lakes are close to the Sudbury smelters. Although the concentrations of Cu, Ni, Zn, and Al all decreased after the lakes were limed, they still remained large and probably continued to exert a toxic stress; this was especially true of copper (Yan et al., 1979).

The biological effects of liming were also studied in Lake Gardsjon, in southern Sweden (Henricksen et al., 1984; Grahn and Sangfors, 1988; Olem, 1990). Prior to liming in 1982, the lake had an average pH of 4.7, while after liming the pH was 5.5–7.9. Prior to liming the annual production of planktonic algae was 8.3 gC/m^2-year, while epiphyton and benthic algae contributed 1.8 gC/m^2-year and macrophytes another 3.6 gC/m^2-year, for a total primary production of 14 gC/m^2-year. Liming produced initial decreases in algal species richness and biomass. However, by 1985 the phytoplankton diversity and production had increased (to 20–40 gC/m^2-year), with much of the increase in productivity due to a bloom of a particular species (*Cosmocladium perissum*), while the productivity of epiphytic and benthic algae and macrophytes decreased. The macrophyte decrease was caused by a 50% reduction in biomass of *Sphagnum subsecundum*, as this moss is intolerant of high-pH and high-calcium conditions (Clymo, 1973). At the same time, the angiosperm macrophyte, *Potamogeton alpinus*, increased in abundance. The cladocerans *Bosmina* spp. were replaced by another cladoceran, *Diaphanosoma brachyurum*, after liming. Benthic invertebrates also increase significantly after liming lakes in Scandinavia (Eriksson et al., 1983), as do fish if they can invade the waterbody or are introduced (Olem, 1990).

These experiments clearly show that acidified, oligotrophic water bodies can be neutralized. Not unexpectedly, liming causes severe shock to the acid-adapted biota, resulting in shifts in species dominance until new, steady-state communities are achieved. However, it is important to recognize that liming is not a longer-term, permanent solution to the acidification of freshwaters, because the acidic

waters must be re-treated periodically, as the neutralizing substances are exhausted or flushed from the system.

Another possible way to mitigate acidified water is by fertilization. Even in acidic waters, there are strong trophic responses to fertilization (Dillon et al., 1979; DeCosta et al., 1983; Kerekes et al., 1984; Yan and Lafrance, 1984), as was described earlier for Drain Lake. Fertilized but still-acidic lakes can potentially sustain large rates of primary and secondary productivity, nurture waterfowl and other wildlife, and have other positive ecological attributes. However, the creation of large numbers of mesotrophic or eutrophic lakes may not be an appropriate management objective in many situations, especially where nonconsumptive recreation is an important use of surface waters. Nevertheless, the fertilization of acidic lakes may be an effective management practice for many acidified water bodies.

4.6
ABATEMENT OF EMISSIONS OF ACIDIFYING SUBSTANCES AND THEIR PRECURSORS

It must be stressed that the liming of acidified ecosystems treats the symptoms but not the causes of acidification. Moreover, liming in a sense transforms the water body from one polluted state to another condition that is still polluted but less toxic.

Clearly, large reductions in the anthropogenic emissions of acid-forming gases will be the ultimate solution to the widespread environmental problems associated with acidification caused by atmospheric depositions. However, there is a great deal of controversy about: (a) the amounts of reduction of emissions of SO_2, NO_x, and other gases that will be required to reduce the effects of acidic deposition and (b) the various emissions–reduction strategies to be pursued. For example, is it better to target large, point sources such as power plants and smelters, while paying less attention to smaller, individual sources such as automobiles and oil- or coal-burning furnaces in homes?

Not surprisingly, industries and political jurisdictions that are large emitters of acid precursors have lobbied forcefully against substantial reductions of point-source emissions, for which they argue the scientific justifications are not yet adequate, while the economic costs of abatement are known to be large.

There is also controversy about how small the rates of sulfur and nitrogen deposition should be to avoid further acidification of sensitive fresh waters or to allow their recovery. For example, for the purposes of bilateral negotiations about the transboundary transport of acidic precipitation precursors between the United States and Canada, the Canadian government has advocated a desirable rate of sulfate deposition of 20 kg/ha-year. It is awkward or impossible, however, to rigidly justify this desired level of sulfate deposition using the available scientific data. In fact, it appears that many sensitive freshwaters could acidify and suffer biological damage at sulfate loadings as small as 10 kg/ha-year (Schindler, 1988). Similarly, so-called critical loads of sulfate have been estimated for Scandinavia, where they are generally believed to be about 15 kg/ha-year, except for relatively vulnerable landscapes in northern Norway and central and northern Finland, where they range from 6 to 12 kg/ha-year (Brodin and Kuylenstierna, 1992).

Critical loads for nitrogen deposition depend on many factors, including the rate of sulfur deposition and the nature of the soil and vegetation on the site. Schultze *et al.* (1989) suggested that critical nitrogen depositions in northern Europe range from only 3–14 kg/ha-year on shallow, nutrient-poor silicate soils, to as much as 48 kg/ha-year on productive sites with calcareous soils.

In spite of all of the uncertainties about the specific causes and magnitudes of the damage caused by the deposition of acidifying substances from the atmosphere, it seems intuitively clear that what goes up (i.e., the acid-precursor gases) must come down (i.e., as acidifying depositions). Much scientific evidence supports this common-sense notion,

and because of widespread public awareness and concern about acid rain in many countries, politicians have began to act effectively, and emissions of sulfur dioxide and oxides of nitrogen are being reduced somewhat, especially in western Europe and North America. In the United States, for example, emissions of sulfur dioxide decreased from 23.1 million tonnes in 1980 to 21.0 million tonnes in 1990, a reduction of about 9% (Anonymous, 1993b). During the same period, emissions of nitrogen oxides decreased by 6%, from 20.7 million to 19.4 million tonnes (Anonymous, 1993b).

In 1992, the governments of the United States and Canada culminated years of scientific research and political negotiations by signing a binational air-quality agreement aimed at reducing the intensity of acidifying depositions in both countries, including the transboundary transportation of air pollutants. This agreement calls for large expenditures by governments and industries to achieve large reductions in the emissions of air pollutants during the 1990s (Anonymous, 1993b).

Although this air-pollution agreement is a positive accomplishment, it remains to be seen whether (a) the reductions of emissions will actually be accomplished in the face of emerging political and economic challenges and (b) the actions will be sufficient to achieve substantial improvements in environmental quality related to acidification.

So far, actions to reduce the emissions of the precursor gases of acidifying deposition have only been relatively vigorous and effective in western Europe and North America. Actions are also needed in other, less wealthy regions of the world where the political focus is on industrial growth, and not on control of the many types of air pollution and other environmental degradations that can be used to subsidize that growth. In the coming decades, much more attention will have to be paid to acid rain and other pollution problems in eastern Europe and the former USSR, China, India, southeast Asia, Mexico, and other so-called "developing" regions, where emissions of sulfur dioxide, oxides of nitrogen, and other important air pollutants are still rampant, and are increasing rapidly.

5

FOREST

FOREST

DECLINES

5.1

INTRODUCTION

Around large point sources of SO_2 pollution, predictable zonations of vegetation damage can often be observed. In parallel with the patterns of pollution-related stresses, plant damage decreases geometrically with increasing distance from the sources of emission. In contrast to these spatial patterns near point sources, the effects of air pollution on regional scales are not usually obvious, and they are always difficult to measure.

In recent years, there have been concerns over declines in vigor of mature forests in many parts of the world, often leading to mass, stand-level mortality. These ecological phenomena have been well documented, and their occurrences can be considered to be facts. However, conclusive statements cannot yet be made about the causal factor(s) of the forest damage. Many of the forest declines are occurring in regions where the airshed is contaminated by various combinations of acidic deposition, ozone, sulfur dioxide, nitrogen compounds, and toxic elements, and it has been suggested that pollution is an important contributing factor in the forest damage. However, forest declines also occur in some areas where pollution is unimportant. In these cases, the decline may be a natural process that

variously involves climatic change, the synchronous senescence of a cohort of mature trees, or the effects of unrecognized pathogens or other environmental stresses.

In this chapter, the phenomenon of forest decline is examined, and some of the current hypotheses are discussed for its occurrence in landscapes subject to and free from the effects of pollution.

5.2

THE NATURE OF FOREST DECLINES

Forest decline is characterized by progressive, often rapid deteriorations in vigor of one or several species of tree, caused by an unknown etiology, and often resulting in a progressive dieback of branches. Ultimately, these can result in an event of synchronous, mass mortality affecting stands over a large area.

The decline syndrome often preferentially affects mature and older individuals and is thought to be triggered by one or a combination of stresses, such as severe weather, nutrient deficiency, toxic substances in soil, air pollution, or pathogens. According to this scenario, stressed trees suffer marked declines in vigor and dieback, and in this

144

weakened condition they are vulnerable to attack by insects and pathogens. Normally, secondary agents such as these are not harmful to vigorous individuals, but they may cause the death of severely stressed trees. It is important to note that although the occurrence and characteristics of forest declines can be well documented, the primary environmental variables that trigger the disease are not known. As a result, the etiology of the decline syndrome is often attributed to a vague, undefined concatenation of biotic and abiotic factors (Manion, 1985; Wargo, 1985; Mueller-Dombois, 1986, 1987a, 1992; Klein and Perkins, 1987; Last, 1987; Barnard, 1990).

Decline symptoms vary markedly among tree species. Frequently observed effects include: (1) decreased net production of biomass, (2) chlorosis, abnormal size or shape, or premature abscission of foliage, (3) dieback of branches, beginning at the extremities and often causing a "stag-headed" appearance, (4) root dieback, (5) a high frequency of attack by secondary agents such as fungal pathogens and insects, and ultimately (6) mortality, often as group- or stand-level mortality (Hepting, 1971; Cowling, 1985; Mueller-Dombois, 1986, 1987a; Klein and Perkins, 1987; Barnard, 1990).

It must be noted, however, that not all forest declines are of unknown etiology. The causes of some synchronous, mass mortalities of trees are well understood, as in the cases of declines associated with: (1) certain fungal diseases of trees, such as Dutch elm disease and chestnut blight (Chapter 1); (2) damage associated with epidemic insects, such as spruce budworm, gypsy moth, and bark beetles (Chapter 8); and (3) the effects of severe windstorms. Sometimes, however, the causes of forest mortality are not known, and the phenomenon may be the culmination of rapid or prolonged periods of decline.

5.3
CASE STUDIES

Naturally Occurring Declines

Periodic occurrences of local or widespread declines of various tree species have been recognized

for more than a century, and in many of these cases the phenomena may be unrelated to human activities. Some of these cases are described below.

Ancient Declines

Davis (1978, 1981) described what appears to have been a palaeodecline of eastern hemlock (*Tsuga canadensis*) in North America. She documented a decrease of about 90% in the influx of hemlock pollen to lake sediment, occurring synchronously at many sites throughout eastern North America during a 50-year period about 4800 years ago. Similarly widespread declines of eastern hemlock have not been observed in modern times. The hemlock decline was probably not caused by climatic change, since other tree species that presently co-occur with hemlock, and that presumably have similar habitat requirements, were not affected. In fact, during and after the hemlock decline these other species apparently increased in abundance to compensate for the decreased prominence of hemlock in the forest. The reason for the putative hemlock decline is unknown, but the triggering stress could have been a fungal-disease pandemic, an outbreak of a defoliating insect such as hemlock looper (*Lambdina fiscellaria*), or some other environmental factor.

There is similar palynological evidence for a widespread decline of elm (*Ulmus* spp.) in northwestern Europe about 5000 years ago (Smith and Pilcher, 1973). It has been speculated that this phenomenon could have been caused by a widespread clearing of the forest by Neolithic humans, but an unknown disease pandemic is an alternate, causal hypothesis (Perry and Moore, 1987).

Birch Decline

Because they occurred so long ago, little is known about the hemlock and elm declines. There are other, more recent examples of forest declines that do not appear to have been caused by an anthropogenic influence.

In North America, the best example is the widespread decline of birches that occurred throughout the northeastern United States and eastern Canada, especially from the 1930s to the 1950s (Fowells, 1965; Hepting, 1971; Auclair, 1987). The most susceptible species were yellow birch (*Betula alleghaniensis*), paper birch (*B. papyrifera*), and gray birch (*B. populifolia*). Birches over a vast area were afflicted by this condition, and there was extensive mortality. For example, in 1951 at about the time of the greatest decline damage in Maine, an estimated 67% of the birch trees had been killed. Birch decline is less important today, but it still causes some mortality.

In spite of a considerable research effort, a single primary cause was not determined for birch decline. It is known that a heavy mortality of fine roots often preceded the deterioration of the above-ground tree, but the environmental causes of this below-ground effect are unknown. No biological agent was identified as a primary predisposing factor, although a number of secondary fungal pathogens were observed to attack weakened trees and to cause their death, as did the bronze birch borer beetle (*Agrilis anxius*).

One causal hypothesis for the extensive birch decline involved stresses associated with severe weather during the late 1930s and early 1940s (Pomerleau, 1991). In Quebec, for example, the winters of 1938 and 1942–1944 were characterized by prolonged periods during which there was little or no snow cover. Under such conditions, frost can penetrate the ground relatively effectively, and the depth of soil freezing can extend to more than 1 m, causing stress to shallow-rooted trees like birches, maples, and some other species. Decline of birch was severe in Quebec in the late 1930s and especially during 1941 to 1944, followed by recovery after 1950.

In a simple yet telling experiment performed in the winter of 1953 (but not published until 1991), 9.1-m-diameter plots in a birch–maple stand in Quebec were kept snow free by shovelling after each snowfall (Pomerleau, 1991). As a result, soil freezing extended to 1.3 m into the soil, compared with the unfrozen condition of soil in control plots (where temperature at a soil depth of 2.5 cm was no less than 0.3°C). The penetration of frost resulted in

ground heaving and root breakage, a late thaw of soil, the occurrence of massive ice in the soil and around tree roots, and signs of dieback in birch and maple trees in the following growing season.

By definition, the birch-decline syndrome only affects birch trees in a natural forest environment. However, the symptoms of the disease are virtually identical to another decline called "postlogging decadence" that affects the same species of birch on logged sites. The primary cause of this latter syndrome is also unknown, but it is suspected that stresses associated with the large microclimatic changes that occur after removal of much of the stand by logging are involved. Such environmental changes could predispose birch trees to damage or death caused by disease or defoliating insects that are normally tolerated by more vigorous individuals.

Some other possibly "natural" tree declines in North America have affected black willow (*Salix nigra*), oak (*Quercus* spp.), ash (*Fraxinus* spp.), balsam fir (*Abies balsamea*), pine (*Pinus* spp.), and others (Staley, 1965; Hepting, 1971; Sprugel, 1976; McCracken, 1985a,b; Auclair, 1987; Mueller-Dombois, 1987a; Weiss and Rizzo, 1987; Zahner *et al.*, 1989; Barnard, 1990).

Ohia Decline

Other forest declines have been observed on Pacific islands. One of the best known is that of the ohia tree (*Metrosideros polymorpha*), an endemic species of the Hawaiian archipelago (see below). Similar declines have occurred elsewhere in the Pacific, for example, in New Zealand forests of *Metrosideros umbellata–Weinmannia racemosa* and of *Nothofagus* spp. (Batcheler, 1983; Stewart and Veblen, 1983; Wardle and Allen, 1983; Mueller-Dombois, 1987b, 1992; Newhook, 1989; Stewart, 1989).

In the Hawaiian Islands, *Metrosideros polymorpha* is the dominant tree species of the native forest (Mueller-Dombois *et al.*, 1983; Hodges *et al.*, 1986; Mueller-Dombois, 1980, 1986, 1987b). It usually occurs in pure stands, which comprise about 62% of the total forest area of the islands.

Stand-level mortality of *Metrosideros polymorpha* on a dry upland site in Hawaii. Note the dense advance regeneration of younger ohia trees in the understory (photo courtesy of D. Mueller-Dombois).

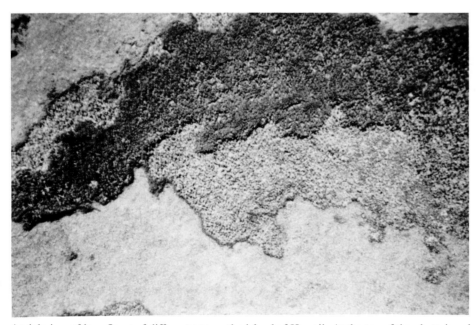

Aerial view of lava flows of different age on the island of Hawaii. At the top of the photo is a dark, healthy, even-aged forest of *Metrosideros polymorpha*. Below is a lighter-colored, older forest that is suffering dieback. These are surrounded by an immature forest on a younger lava flow (photo courtesy of J. Jacobi).

From anecdotal accounts, events of widespread mortality of ohia are known to have occurred for at least a century, but the phenomenon is probably more ancient than this. The most recent decline began in the late 1960s. A survey of 76,900 ha of the island of Hawaii in 1982 found that about 50,000 ha exhibited symptoms of ohia decline. On about 35% of that area the damage was slight (i.e., less than 10% of the trees were declining or were dead), while 24% exhibited moderate damage (11–50%), and in 41% the damage was severe (>50%). It is important to note that in most declining stands, only the canopy individuals are affected. Understory saplings and seedlings do not decline, and in fact they are released from competitive stress by the deterioration of the overstory.

As with all forest declines, the specific etiology of the ohia decline is unknown. Soil waterlogging may be an important, predisposing environmental stress. An important role of nutrient deficiency has been ruled out by fertilization experiments, which did not reduce the mortality rate in declining stands (Gerrish *et al.*, 1988). A number of secondary biological stresses become important in many declining stands of ohia. These include the fungal root pathogens *Phytophthora cinnamomi* and *Armillaria mellea*, and the endemic borer beetle *Plagithmysus bilineatus*. Often it is these secondary agents that kill the weakened trees.

An interesting hypothesis has been developed to explain the etiology of ohia decline in Hawaii, and the extension of the theory to forest declines elsewhere has been suggested (Mueller-Dombois, 1980, 1986; Mueller-Dombois *et al.*, 1983). Stand-level decline may be envisaged as due to the phenomenon of "cohort senescence." This is a stage of the tree life history that is characterized by a simultaneously decreasing vigor of many individuals, occurring in an even-aged population of the same generation (i.e., a cohort). The development of senescence in individuals is primarily governed by intrinsic genetic factors, but the timing of its onset can be influenced by extrinsic stresses. The decline-susceptible, life-history stage follows a more vigorous, younger, mature stage of the cohort, initially establishing on the site following a severe disturbance.

In Hawaii, lava flows, volcanic-ash deposition, and hurricanes are catastrophic disturbances that initiate primary succession. Sites disturbed in this way are colonized by a cohort of ohia individuals, resulting in an even-aged stand. In the absence of an intervening catastrophic disturbance, the ohia stand eventually matures and then becomes senescent and enters a decline phase. The original stand is then replaced by another ohia forest composed of an advance regeneration of small individuals that are already established in the understory.

Therefore, according to the cohort senescence theory, the ohia decline should be considered to be a characteristic of the natural population dynamics of the species *Metrosideros polymorpha*. The extension of this theory to other tree species appears to be most reasonable for cases of forest decline that involve relatively short-lived species that are intolerant of the stressful conditions of mature, closed forest. As such, it may be useful in explaining the declines that have occurred in species of birch, willow, ash, oak, and some other trees. However, the cohort senescence model may be less tenable as an explanation for declines that affect relatively tolerant, longer-lived species such as sugar maple and red and Norway spruce (described below). Declines of these latter species appear to involve stands that are younger than the potential longevity of these trees, which can exceed several centuries.

Fir Waves

Ecologists working in montane forests dominated by species of fir (*Abies*) in various locations have noted occurrences of a periodic, travelling decline that appears as distinct strips that advance upward along slopes. Above the current wave of decline is intact, mature forest, and immediately below are progressively older waves with a vigorous, fir-dominated regeneration beneath standing, dead snags. Overall, the sere comprises a longer-term, self-perpetuating, sequence of stand decline, establishment, and growth (Sprugel, 1976; Reiners and Lang, 1979; Sprugel and Bormann, 1981). Fir waves are best described for balsam fir in the northeastern United States and Newfoundland, but they are also known for other fir species, for example,

Abies veitchii and *A. mariesii* in Japan, where the phenomenon is known as Shigamare, or "the mountain with dead stripes of trees" (Boyce, 1988; Mueller-Dombois, 1992).

Boyce (1988) studied fir waves at 11 sites in the White Mountains of New Hampshire, using a time series of aerial photographs. The wave-front velocity was 1–3 m/year. A common symptom of damage was "flagging" of emergent trees, characterized by branches that only extend in the downwind direction, producing an asymmetrical crown (in some alpine and exposed, coastal situations, this conifer-growth form is called "krummholz"). The flagging is apparently caused by a wintertime buildup of rime ice, which is most severe at the leading edge of the wave. The ice damage, coupled with root damage caused by the wind-induced swaying of exposed trees, are believed to be the causes of the dieback and death of trees in the active-wave zone (Marchand, 1984; Boyce, 1988; Rizzo and Harrington, 1988).

Wave declines are also known for Fraser fir (*Abies fraseri*) in the southern Appalachian Mountains of the eastern United States. In the case of Fraser fir at Mount Mitchell, North Carolina, the wave decline has been exacerbated by damage caused by an introduced insect, the balsam wooly adelgid (*Adelges picea*). This insect can kill trees after 2–9 years of infestation, although good regeneration follows the death of the initial stand (Witter and Ragenovich, 1986; Nicholas and Zedaker, 1989). Mount Mitchell is also known to have high ozone concentrations and acidic rain, fogs, and soils (Bruck, 1989). However, the roles of these potentially toxic factors in the fir decline are not understood and are difficult to reconcile with a spatial pattern characterized by waves.

Forest Declines Possibly Triggered by Air Pollution

In recent years, declines of uncertain etiology have taken place in forests located in regions with polluted air. In western and central Europe, the tree species that have been most severely affected are Norway spruce (*Picea abies*) and, to a lesser extent, beech (*Fagus sylvatica*). In North America, the

most prominent declines have occurred in red spruce (*Picea rubens*) and sugar maple (*Acer saccharum*). Declines of conifers have also been described for forests in southern California, but in those cases the damage has been clearly ascribed to toxicity caused by photooxidants (see Chapter 2). Some of these forest declines are described below.

The "New" Type of Forest Decline in Europe

Some of the recent, widespread forest damage in Europe has been described as a "new" decline syndrome that may be triggered by stresses associated with air pollution. Although the symptoms appear to be similar, the modern declines are believed to be different from declines that are known to have occurred historically and that are generally believed to have been natural, apart from some cases of local, industrial pollution (Prinz, 1984; Blank, 1985; Hinrichsen, 1986; Huettl, 1986a, 1989a; Krause *et al.*, 1986; Blank *et al.*, 1988; Barnard, 1990; Kandler, 1992). In Germany, the modern forest declines have been called *waldsterben*, or "forest death."

A modern decline syndrome was first noted in fir (*Abies alba*) in Germany in the early 1970s. In the early 1980s, a larger-scale decline was apparent in Norway spruce (*Picea abies*), the economically most important tree species in the region. In the mid-1980s, decline also became apparent in beech (*Fagus sylvatica*) and, to a lesser extent, oak (*Quercus* spp.).

Decline symptoms in the conifers vary regionally and with tree species, but usually include: (1) chlorosis or yellowing of needles, especially of older foliage; (2) premature shedding of foliage, beginning at the base of the crown and the inner parts of branches; (3) a decrease or elimination of the net production of woody tissue, usually beginning after the loss of older foliage biomass reaches 30–50%; (4) dieback of branches, often from the top of the tree and causing a "stag-headed" appearance; (5) a diagnostic nutrient deficiency in foliage, especially of magnesium, but sometimes of calcium, potassium, or zinc; (6) root dieback; and ultimately (7) the death of trees, often as a synchronous, stand-level mortality. Symptoms of decline in beech include: (1) abnormal coloration of foliage; (2) premature leaf drop; (3) dieback of lateral branches; (4) reduced net production; and (5) death.

Declines of this type have been observed in many countries of Europe, including Austria, Britain, Czechoslovakia, France, Germany, Italy, Yugoslavia, Poland, Scandinavia, and Switzerland. In 1986, more than 19 million ha or 14% of the total forest area was damaged in Europe west of Russia, Belarus, and Ukraine, comprising about 15% of the coniferous-forest volume and 17% of the angiosperm-forest volume (Nilsson and Duinker, 1987; Postel, 1987).

The spatial extent of the decline has been relatively well documented in western Germany (Blank, 1985; Krause *et al.*, 1986). In 1984, some degree of decline was observed in 3.7 million ha, or about 50% of the western German forest area (33% of the forest was "lightly damaged" with 11–25% foliage loss; 16% was "moderately damaged" with 26–60% loss; and 1.5% was "severely damaged" with >60% loss). The area of damage had increased substantially from the condition in 1983, when 34% of the total forest area was classified as suffering damage, and from 1982 when the total was only 8%. [Note, however, that these apparently large increases in the area of damaged forest may in part be an artifact of different methodologies that were used to assess the damage at different times (Blank *et al.*, 1988).]

By the late 1980s, a widespread forest decline was not apparent in western Europe, although there were some severely damaged stands. The intensity of the relatively severe forest damage in western Germany decreased in the late 1980s, from affecting about 24% of the forest area in 1985, to 15% in 1988 (Barnard, 1990).

Decline symptoms were variable in the German stands. However, in general: (1) mature stands older than about 60 years tended to be more severely affected; (2) dominant individuals within the stand seemed to be relatively vulnerable; and (3) more exposed individuals located at or near the edge of the stand were more severely affected by decline symptoms. Interestingly, epiphytic lichens often flourish in badly damaged stands, probably because of a greater availability of light and other resources caused by the diminished cover of tree foliage. This is a paradoxical observation (at least if air pollution is believed to be the cause of forest decline), because lichens as a group are usually hypersensitive to toxic gases (see Chapter 2).

It has been difficult to generalize about the role of site characteristics in predisposing stands to decline. Declines have occurred in stands growing on both calcareous and acidic soils. However, it appears that stands on sites with oligotrophic soil, and on sites that experience drought during the growing season, are relatively vulnerable to decline damages.

From the information that is now available, the "new" types of forest decline in Europe seem to be triggered by a variable concatenation of environmental stresses. The weakened trees then start to decline, sometimes rapidly, and may die as a result of attack by secondary agents such as fungal disease, or defoliating or wood-boring insects. A number of suggestions regarding the identity of the primary, inducing factor(s) have been made. These include gaseous air pollution, acidification, and toxic metals in soil. Other proposals suggest a natural climatic effect, in particular, drought. However, there is not yet a consensus as to which of these interacting factors is the primary trigger that induces forest declines in Europe. It is quite possible that no single natural or air pollution-related stress will prove to be the primary cause of the new forest declines (Hinrichsen, 1986, 1987; Last, 1987). In fact, there may be several different declines occurring simultaneously, but in different areas (Blank *et al.*, 1988; Huettl, 1989b; Kandler, 1992).

To prove that a specific environmental factor is the primary, predisposing mechanism of forest decline, several or all of the following items of evidence would have to be demonstrated: (1) the environmental stress would have to be shown to be present at an intensity that is known to be sufficient to cause acute or at least chronic toxicity; (2) experimental amelioration of the environmental factor should arrest or reverse decline symptoms; (3) experimental intensification of the environmental factor should predispose stands to decline; and (4) in the case of a pathogenic, biological agent, Koch's postulates must be demonstrated; that is, the pathogen must be isolated from declining trees, and it must be shown to induce decline in healthy trees that are injected or otherwise treated with an axenic culture or suspension of the putative pathogen (Agrios, 1969; Last, 1987).

So far, these requirements have not been met by any of the proposed agents of the new forest declines. This indeterminate state of affairs with respect to the causes of forest declines is not unlike that for some human diseases, for example, many types of cancer for which the predisposing or causal agent(s) have not been identified.

The major hypotheses put forward as mechanisms for the initiation of the European forest declines are summarized briefly below. Note that they are not listed in any particular order.

1. *Aluminum toxicity in acidified soil* was one of the first suggestions for the triggering stress in the new forest declines. In this scenario, aluminum is thought to be mobilized from naturally occurring minerals as a consequence of the acidification of forest soil, with the latter possibly caused by the deposition of acidifying substances from the atmosphere (Ulrich *et al.*, 1980). The aluminum hypothesis is consistent with: (1) the presence of large amounts of extractable aluminum in soils at many sites where decline has occurred and (2) frequent observations of damaged and dead roots, exhibiting symptoms that resemble those of typical aluminum toxicity. In an experiment carried out in a mature *Picea abies* stand in Germany, Matzner *et al.* (1986) limed and fertilized soil and then measured the *in situ*, net production of root biomass after 7 months. At the initial soil pH of 3.9–4.0, there was a total root production of 180 kg/ha. However, liming to pH 4.3 allowed the production of 630 kg/ha, while a fertilized, pH 4.5 soil produced 1820 kg/ha. This experiment suggests that root productivity was limited by acidity and associated aluminum toxicity, along with nutrient deficiency. As noted previously, however, stands are also declining on calcareous sites where aluminum availability in

soil is usually low. Furthermore, laboratory bioassays have shown that many tree species have a considerable tolerance to plant-available aluminum (McCormick and Steiner, 1978; Ogner and Tiegen, 1980; Steiner *et al.*, 1980; Hutchinson *et al.*, 1986; Thornton *et al.*, 1986). This tolerance is a result of selection by the naturally large concentrations of soluble aluminum in acidic forest soils.

2. *Ozone pollution* has also been implicated as a potential contributor to the European forest declines. During sunny periods in the growing season, peak ozone concentrations of up to 400 $\mu g/m^3$ have been measured in western Germany. This is well above the toxic thresholds of sensitive plants (100–200 $\mu g/m^3$; Hinrichsen, 1986, 1987), and ozone damage has occasionally been observed in declining stands. However, the foliar damage that most often occurs in such stands does not resemble the diagnostic, acute injuries caused by ozone. This may mean that acute damage by ozone is not a primary cause of the forest damage, but it does not rule out secondary, predisposing effects, especially subacute or chronic damage (Blank, 1985; Krause *et al.*, 1986; Last, 1987; Huettl, 1989a).

3. *Sulfur dioxide and oxides of nitrogen* are present in large concentrations throughout the regions where forest declines are occurring (Hinrichsen, 1986). There have, however, been only infrequent observations of the characteristic, acute foliar injuries caused by exposure to these air pollutants in the European forests where decline is occurring. It is possible that these gases are causing chronic damage (i.e., "hidden injuries" occurring without obvious symptoms; see Chapter 2), and the resulting stresses could be contributing to the decline. On the other hand, the previously noted, flourishing communities of epiphytic lichens in the badly damaged forests suggest that SO_2 and other gaseous pollutants are not present in phytotoxic concentrations, since lichens are well known for their sensitivity to these toxic gases.

4. *A climatic effect*, or extremes of weather during the growing season, have also been con-

sidered as possible agents in the etiology of the "new" forest declines. The declines were first noted in the mid-1970s, following several droughty growing seasons. The intensification of symptoms in the early 1980s may also be related to relatively dry growing conditions at that time (Blank, 1985; Andersson, 1986; Hauhs and Wright, 1986; Becker, 1989; Auclair *et al.*, 1992). Drought could have caused severe stress to trees, predisposing them to damage from other agents.

5. *Nutrient imbalance* has also been implicated in the decline disease in Europe. Two primary mechanisms have been proposed: (1) nitrogen fertilization, especially by the deposition of ammonium from the atmosphere, and (2) nutrient deficiency, especially of magnesium:

(a) *Nitrogen fertilization*. Many of the declining stands are subject to large rates of atmospheric deposition of inorganic nitrogen compounds, largely originating from anthropogenic emissions. A typical, background deposition of fixed nitrogen from the atmosphere is about 2–4 kg N/ha/year. However, in central Europe the rate of wet plus dry deposition is much larger than this, typically 10–25 kg/ha/year, and reaching more than 100 kg/ha/year at some sites (Nihlgard, 1985; Grennfelt and Hultberg, 1986; Huettl, 1989a). Anthropogenic sources of emission of inorganic nitrogen include: (1) discharges of NO_x from automobiles, power plants, and home heating, (2) ammonia volatilization from fertilized agricultural fields and animal manure, and (3) emission of N_2O by microbial denitrification in fertilized agricultural fields. These various sources can cause regionally elevated rates of wet and dry depositions of inorganic nitrogen. This is especially true of forests, which because of their complex architecture are relatively efficient at filtering gases and particulates from the atmosphere. Relatively large rates of fertilization of forests with nitrogen can have various effects (after Nihlgard, 1985),

including: (1) an increased net productivity of forest biomass, although this is often accompanied by relatively little allocation of resources to the root system, causing large shoot:root ratios; (2) a possible accumulation of toxic concentrations of nitrogen-containing metabolites in foliage, leading to premature abscission as a detoxification mechanism; (3) decreased or delayed frost hardiness, a frequent observation after the fertilization of conifer forest (Soikkeli and Karenlampi, 1984); (4) an increased susceptibility to pathogens and insect attack; and (5) nutritional imbalances caused by the abnormally large supply of fixed nitrogen (see below).

(b) *Nutrient deficiency*. In addition to a nutritional imbalance, trees may become deficient in magnesium, calcium, or potassium. Such deficiencies could be caused by increased leaching of these basic cations from foliage or soil by acidic precipitation (see Chapter 4) or by past forest harvesting, which can remove substantial quantities of nutrients from the site (see Chapter 9). Some German researchers have claimed that nutrient deficiency is the dominant predisposing stress in at least some of the "new" types of forest decline, particularly in a particular area in southwestern Germany. To support this hypothesis, they have presented data from field trials indicating that declining stands can be revitalized with an appropriate fertilization regime (Huettl, 1986a,b, 1989a,b; Huettl and Wisniewski, 1987; Zoettl and Huettl, 1986). This is illustrated in Table 5.1 for a young stand of *Picea abies*, originally having deficient foliar concentrations of K and Mg. This stand responded favorably to fertilization with these nutrients, the symptoms of deficiency disappeared, and the stand appeared to be revitalized.

As with most cases of forest decline, those of western and central Europe are rather well-

Magnesium deficiency in foliage of Norway spruce (*Picea abies*) in western Germany. A diagnostic, yellowish coloration of spruce foliage, particularly of needle tips, often occurs in declining stands at high altitude. Usually, older foliage is affected first. If the damage is not too severe, magnesium deficiency can be alleviated by fertilization (photo courtesy of Kali und Salz AG, Germany).

documented ecological phenomena. The forest damage is fact, but at the present time only hypotheses can be suggested about their causes, and no conclusive statements can yet be made about the causal factor(s). According to Huettl (1989a, p. 55):

> The new forest damages are characterized by many symptoms, of which only a few are truly new. The recent changes cannot be related to one specific, causal mechanism. The problem is complex and probably caused by regionally varying sets of stress factors, both natural and unnatural in origin. However, most of

the scientists involved in this research in central Europe consider anthropogenic air pollution an important factor in the cause of this serious phenomenon.

Decline of Sugar Maple in Eastern North America

Sugar maple (*Acer saccharum*) is a long-lived, shade-tolerant tree of eastern North America, but it is shallow-rooted and susceptible to drought (Fowells, 1965).

The modern epidemic of decline in sugar maple began in the late 1970s and early 1980s. It has been most prominent in Quebec, Ontario, New York, and parts of New England (Bordeleau, 1986; Carrier, 1986; McIlveen *et al.*, 1986; Houston *et al.*, 1990). The modern symptoms are similar to those described for earlier declines (Westing, 1966; Barnard, 1990) and include abnormal coloration, size, and shape of foliage; premature abscission; dieback of branches from the top of the tree downward; reduced productivity; and death of trees. Some declining stands have symptoms of nutrient deficiency, variously of potassium, magnesium, and phosphorus (Bernier and Brazeau, 1988a,b; Ouimet and Fortin, 1992).

An aerial survey in 1985 of the region of southern Quebec with the most severe damage to sugar maple found that about 35% of the maple stands had light damage (11–25% foliage loss), 4% had moderate damage (25–50%), and 1% had severe damage (>51%) (Barnard, 1990). In 1988, permanent plots were established in 165 sugar maple stands in four eastern Canadian provinces and seven northeastern U.S. states. Data from that network indicate that by 1990 the severity and incidence of the decline had decreased substantially. In 1988, 7.3% of natural stands of sugar maple had ≥16% crown dieback (indicative of severe damage), but by 1990 this had decreased to 4.1%. Comparable values for stands managed for the production of maple sugar were 10.7% in 1988 and 5.6% in 1990 (Millers *et al.*, 1992).

There has been a frequent association of the pathogen *Armillaria mellea* with declining trees, but this fungus is believed to be a secondary agent that attacks weakened trees (Hibber, 1964; Wargo

Table 5.1 Effects of fertilizing a 12-year-old *Picea abies* stand in southern Germany on foliar nutrient concentration: The stand had symptoms of deficiency and decline[a]

| Nutrient | Foliar nutrient concentration | | |
	May 1984 (before fertilization)	Oct. 1984 (after fertilization)	Optimal foliar concentration
N (mg/g, d.w.)	15.0	14.5	13.0
P	1.4	1.8	1.3
K	2.6	5.6	4.5
Mg	0.4	0.9	0.7
Ca	1.3	2.6	1.0
Zn (μg/g, d.w.)	11	17	15
Mn	80	70	15

[a] The ferilizer treatment was 800 kg/ha of $MgSO_4 + K_2SO_4$. Modified from Huettl and Wisniewski (1987).

and Houston, 1974; Bauce and Allen, 1992). Declining trees are also more susceptible to infestation by an insect pest, the sugar maple borer (*Glycobius speciosus*) (Bauce and Allen, 1992).

Some declining stands had recently been severely defoliated by the forest tent caterpillar (*Malacosoma disstria*), for as many as 3 consecutive years. This insect causes extensive defoliation early in the growing season, from which trees usually refoliate within 5–6 weeks. Productivity of the trees is affected, however, and affected trees may be less tolerant of other environmental stresses, such as drought, air pollution, and pathogens (Houston, 1989; Barnard, 1990). In addition, some declining stands of sugar maple were severely damaged by another insect, the pear thrips (*Taeniothrips inconsequens*) (Allen *et al.*, 1992).

Many of the declining stands of sugar maple were commercial sugar bushes, in which trees are tapped each spring for sap to produce maple sugar. Management for sugar production is not, however, believed to be a predisposing environmental factor for decline, as long as simple precautions are taken to avoid fungal infections of tap holes and pruning wounds (Houston *et al.*, 1990). Moreover, the damage between natural stands of sugar maple and those exploited for the production of maple sugar has not been significantly different (Barnard, 1990; Houston *et al.*, 1990; Allen *et al.*, 1992).

An increasingly popular hypothesis with which to explain the onset of the sugar maple decline of the mid-1980s relates to several years of severe weather in the late 1970s and early 1980s (Barnard, 1990; Bauce and Allen, 1991; Allen *et al.*, 1992). In 1980, a severe late-spring frost killed more than 75% of sugar maple foliage in parts of Appalachia and Quebec, including more than 50,000 ha in Vermont (Barnard, 1990). In 1981, a midwinter thaw caused a premature dehardening of buds and woody tissues, which were then damaged by a return of cold weather, while at the same time soil froze to a relatively great depth because of the absence of an insulating cover of snow (Barnard, 1990; Auclair *et al.*, 1992). In addition, the summers of 1982 and 1983 were characterized by severe drought in parts of Quebec, where maple decline later became prominent (Barnard, 1990).

As with red spruce (below), the declining maple stands are located in a region that is subject to large rates of atmospheric deposition of acidifying substances, and this has also been suggested as a possible predisposing factor, along with soil acidification and mobilization of available aluminum. However, in the extensive, 165-stand, permanent-plot network of Allen *et al.* (1992), there was no apparent relationship between the spatial patterns of the intensity of sugar maple decline and that of sulfate deposition from the atmosphere.

Decline of Red Spruce in Eastern North America

Red spruce (*Picea rubens*) can potentially live to be 300–400 years old, and it is tolerant of shaded understory conditions, but shallow-rooted and susceptible to drought (Fowells, 1965).

Stand-level declines of red spruce have been most frequent in high-elevation sites of the northeastern United States, especially in upstate New York, New England, and the middle and southern Appalachian states (Siccama *et al.*, 1982; Johnson, 1983; Johnson and Siccama, 1984, 1989; Johnson *et al.*, 1986; Scott *et al.*, 1984; Adams *et al.*, 1985; Bruck, 1985; Friedland and Battles, 1987; Barnard, 1990; McLaughlin *et al.*, 1990; Pitelka and Raynal, 1990). These sites are variously subject to acidic precipitation (mean-annual pH about 4.0–4.1), to acidic fog waters (pH as low as 3.2–3.5), to large rates of deposition of sulfur and nitrogen from the atmosphere, and to stress from metal toxicity in acidic soils, particularly by aluminum (Lovett *et al.*, 1982; Barnard, 1990; McLaughlin *et al.*, 1990; see also Chapter 4).

The modern decline of red spruce can involve all mature size classes. Symptoms include (after Bruck, 1985; Johnson *et al.*, 1986): (1) discoloration and premature death and abscission of foliage, starting from the top of the tree and working downward and from the outside of the crown and progressing inward; (2) dieback of branches, starting from the top of the tree; (3) a general thinning of the crown; (4) decreased or zero production of radial growth; and ultimately (5) death of trees, singly or in patches of various size. The pathogenic root fungus *Armillaria mellea* frequently plays a secondary role and may kill weakened trees, especially at relatively low-altitude sites (Carey *et al.*, 1984). Secondary attacks by the bark beetle *Dendroctonus rufipennis* and other damaging insects are also frequent (Johnson *et al.*, 1986).

Where red spruce decline is severe, there may be large changes in the species composition of stands. Siccama *et al.* (1982) observed such a change between 1964 and 1979 in permanent plots in high-elevation, mixed-conifer stands at Camel's Hump in the Green Mountains of Vermont. The live basal

Dieback of red spruce (*Picea rubens*) at high altitude on Camel's Hump Mountain in Vermont. Between 1965 and the early 1980s, about one-half of the spruce individuals died. Although the etiology of the dieback has not yet been demonstrated, it is suspected that the deposition of acidifying substances from the atmosphere may play a key role (photo courtesy of H. W. Vogelmann).

Another view of damage in a red spruce dieback stand on Camel's Hump (photo courtesy of H. W. Vogelmann).

Table 5.2   Tree regeneration at Camels Hump, Vermont, in 1989. Data were gathered spearately in gaps created by dieback of red spruce and beneath a relatively intact canopy[a]

	Gap	Canopy
Average seedling density (m^{-2})		
Red spruce	0.3	0.3
Balsam fir	9.4	11.6
Birch	6.5	4.4
Total	16.2	16.3
Average seedling height (cm)		
Red spruce	52	29
Balsam fir	19	12
Birch	48	15
Average length of youngest leader (cm)		
Red spruce	7.5	1.8
Balsam fir	3.9	2.3
Birch	6.4	6.5

[a]Modified from Perkins *et al.* (1992).

area of the stand was relatively unaffected by mortality caused by spruce decline (29.6 m²/ha in 1965 versus 26.5 m²/ha in 1979), but over that period the basal area of red spruce decreased by about 45% (from 6.7 m²/ha to 3.7 m²/ha), and its relative dominance decreased from 23 to 14%.

In 1989, 40% of another study area on Camel's Hump was occupied by canopy gaps associated with spruce decline, and these gaps were increasing at 0.8–1.2 m/year, following the prevailing wind direction (Perkins *et al.*, 1992). The density of regeneration of red spruce was poor in the gaps, comprising only 2% of tree seedlings, compared with 48% for balsam fir and 40% for birch (Table 5.2). Growth of the spruce regeneration was good, however, with seedling height and leader length indica-

tive of relatively free growth, compared with individuals growing beneath an intact canopy.

Scott *et al.* (1984) remeasured permanent plots on Whiteface Mountain in New York. Between 1964–1969 and 1982, stands of red spruce at elevations less than 900 m suffered reductions of basal area of 40–60%, while higher-elevation stands decreased by 60–70%. These changes were due to a high rate of mortality of red spruce, along with decreased productivity of surviving trees. Data from several other studies also illustrate the latter effect. Siccama *et al.* (1982) found that the average ring width of red spruce at Camel's Hump had decreased from 1.33 mm/year between 1965 and 1969, to 0.92 mm/year between 1967 and 1971, and 0.63 mm/year between 1975 and 1979. In a larger-scale survey of 3000 dominant and codominant red spruce trees from New England and upstate New York, the average ring width was 13–40% smaller in 1980 than in 1960 (note that 1960 was a time of peak productivity of red spruce in the study region) (Hornbeck and Smith, 1985; Hornbeck *et al.*, 1986, 1987b).

Hornbeck *et al.* (1986, 1987b) also noted a de-

A view of forest damage near the summit of Mount Mitchell, North Carolina. This forest dominated by Fraser fir (*Abies fraseri*) has been affected by wave declines, but this damage has been exacerbated by the effects of an introduced insect, the balsam wooly adelgid (*Adelges picea*), which can kill trees after 2–9 years of infestation. Mount Mitchell is also known to have high ozone concentrations and acidic rain, fogs, and soils, but the roles of these factors in the fir decline are not understood. Note the vigorous regeneration of fir and red spruce (*Picea rubens*) in the understory of this damaged stand.

crease of 7–37% in the average ring width of balsam fir over the same time period, even though that species was not apparently suffering from decline. These observations suggest that a climatic effect or a natural successional influence, related to the closure of the canopy of developing stands, could have been a cause of the decreased net production of both species (see also below). However, note that in more southern mature forests of red spruce and Fraser fir in the Appalachian Mountains, reduced productivity and decline of both species is occurring (Weiss and Rizzo, 1987; McLaughlin *et al.*, 1990).

Declines of red spruce are known from anecdotal accounts to have occurred in the past. A major episode of decline occurred in the 1870s and 1880s in the same general area where the "new" decline is occurring. An estimated one-third to one-half of the mature red spruce in parts of the Adirondacks of New York was lost during that early decline epi-

sode, and there was also extensive damage in New England (Johnson *et al.*, 1986; Hamburg and Cogbill, 1988; Barnard, 1990). As in the case of the European forest declines, the old and new episodes appear to have similar symptoms, and it is possible that both occurrences are examples of the same disease.

Most of the hypotheses that have been suggested to explain the initiation of red-spruce decline are similar to those proposed for the European forest declines. These causal hypotheses include: (1) acidification of soil and concomitant aluminum toxicity, (2) acidic deposition, (3) gaseous air pollution, (4) drought, (5) winter injury exacerbated by insufficient hardiness due to nitrogen fertilization, (6) heavy metals in soil, and (7) calcium deficiency, possibly caused secondarily by large concentrations of aluminum in soil, which may interfere with calcium uptake by tree roots (Friedland *et al.*, 1985,

1988; Hornbeck and Smith, 1985; Vogelmann *et al.*, 1985; Evans, 1986; Hornbeck *et al.*, 1986; Johnson *et al.*, 1986, 1988; Shortle and Smith, 1988; Robarge *et al.*, 1989; McLaughlin *et al.*, 1990, 1991; Hadley *et al.*, 1991; Amundson *et al.*, 1992; Schlegel *et al.*, 1992; Miegroet *et al.*, 1993).

Additionally, more novel hypotheses include the following: (8) decline of spruces, including red spruce, may be related to a nitrogen deficiency that occurs in late-successional stands, caused by the immobilization of this primary, limiting nutrient in aggrading biomass and spruce litter (Pastor *et al.*, 1987); (9) climatic change, in particular a longer-term warming that has occurred subsequent to the end of the Little Ice Age in the mid-1800s, may be important in the decreased prominence of red spruce in the northeastern United States since that period (Hamburg and Cogbill, 1988); and (10) the modern growth decline of red spruce may be a natu-ral consequence of stand dynamics, possibly following from a period of ecological release since about 1945–1955, associated with the death of birches in mixed spruce–birch stands, caused by birch decline (Reams and Huso, 1990). In this latter scenario, the reduction of red spruce growth is a natural consequence of an intensification of intraspecific competition, occurring as mature red spruce trees fully occupied the canopy space left by dying birch trees.

At present, so little is known about the etiology of the red spruce decline that the possible roles of anthropogenic air pollution and of natural environmental factors remain quite uncertain. This does not necessarily mean that air pollution is not involved. Rather, it suggests that more information is required before any conclusive statements can be made regarding the causes of the phenomenon of red spruce decline.

6

OIL POLLUTION

6.1

INTRODUCTION

Petroleum is a critically important but nonrenewable natural resource. In its refined forms, petroleum is used for the production of energy and for the manufacture of synthetic materials such as plastics, while its asphaltic residues are used for heating, construction, and roads. Energy production is by far the largest of the uses of petroleum, globally accounting for an energy equivalent of about 143×10^{18} J in 1991, or 39% of the total production of energy by all means (Anonymous, 1992a).

In aggregate, about 49% of the 1990 global use of petroleum was in the world's most developed regions, i.e., North America (18%), Western Europe (20%), and Japan (11%), even though these regions only comprise about 14% of the human population (WRI, 1990). Global petroleum use increased by about 3% per year during the late 1980s and early 1990s, but the fastest increases were in the world's most rapidly growing economies in southeast Asia (14%/year in 1988), led by Taiwan (21%/year in 1988) (WRI, 1990; Anonymous, 1992a).

The known, recoverable, reserves of petroleum were about 139 billion tonnes in 1991, of which 87.5×10^9 tonnes or 63% of the total were in the Middle East, 11.7×10^9 tonnes (8%) were in North America, and 8.3×10^9 tonnes (6%) in the former USSR (Anonymous, 1992a).

Because most petroleum is extracted in locations that are remote from the places where most consumption occurs, it is a commodity that must be transported in large quantities. For example, in 1988 western Europe produced about 8.3×10^{18} J of oil equivalent but consumed 24.9×10^{18} J, while North America (the United States and Canada) produced 22.9×10^{18} J but consumed 36.2×10^{18} J, and Asia plus Australasia produced 6.8×10^{18} J but consumed 19.5×10^{18} J (WRI, 1990). In contrast to these large net consumers, the Middle East produced the oil equivalent of 30.9×10^{18} J, but consumed 5.7×10^{18} J (WRI, 1990).

The most important methods of transportation of petroleum are by oceanic tankers and overland pipelines. These transportation methods can pollute the environment through accidental oil spills and by operational discharges (e.g., the cleaning of storage and ballast tanks of tankers). There have been some spectacular, accidental spills, involving the loss of large quantities of crude oil from disabled super-tankers and offshore platforms. In addition, the large facilities that are used for the refining of petroleum cause chronic contamination and pollution by the discharge of hydrocarbon-laden waste waters and by frequent small spills. The purpose of this chapter is to examine the ecological effects of oil pollution from these various sources.

6.2

CHARACTERISTICS OF PETROLEUM AND ITS REFINED PRODUCTS

Petroleum is a complex, naturally occurring mixture of organic compounds, mostly hydrocarbons. Petroleum is produced from biomass over geologically long periods of time, by complex reactions occurring under conditions of high pressure and temperature deep in sedimentary formations. Petroleum compounds can occur in a gaseous form that is often called natural gas, as a liquid called crude oil, and as a solid or semisolid asphalt or tar found in oil sands and oil shales. These materials are chemically complex and can be composed of hundreds of molecular species. The molecules range in size and complexity from methane, a gaseous hydrocarbon with a molecular weight of only 16 g/mol, to solid substances having molecular weights greater than 20,000 g/mol [Clark and Brown, 1977; Kornberg, 1981; National Resource Council (NRC), 1985b].

Hydrocarbons are quantitatively the most important constituent of petroleum. Hydrocarbons can be classified into three broad groups, each with various subclasses (Clark and Brown, 1977; Kornberg, 1981):

1. *Aliphatic hydrocarbons* are open-chain compounds. If there is only a single bond between all adjacent carbon atoms, the molecule is said to be saturated. Unsaturated molecules have at least one double or triple bond. This is illustrated by the following series of two-carbon aliphatics: ethane, $H_3C—CH_3$; ethylene,

$H_2C{=}CH_2$; and acetylene, $HC{\equiv}CH$. Saturated aliphatics are known as paraffins or alkanes, and they are chemically more stable than unsaturated aliphatics. The latter are not present in crude oil, but they can be produced secondarily during industrial refining processes or photochemically after crude oil is spilled and exposed to environmental influences.

2. *Alicyclic hydrocarbons* have some or all of their carbon atoms arranged in a ring structure, and they can be saturated or unsaturated.

3. *Aromatics* are hydrocarbons that contain at least one six-carbon ring in the molecular structure. The basic C_6H_6 ring is known as benzene.

Crude oils from different locations vary greatly in their hydrocarbon composition (Table 6.1). On average, the three most important groups of hydrocarbons in petroleum are paraffin molecules, ranging from 1 to >78 carbons, saturated and unsaturated five- and six-carbon alicyclics or naphthenes, and a great variety of aromatics. Other elements that are present in crude oil include sulfur at a concentra-

Table 6.1   Chemical characteristics of several crude oils[a]

Component	Source of crude oil		
	Prudhoe Bay	South Louisiana	Kuwait
Sulfur (wt%)	0.94	0.25	2.44
Nitrogen (wt%)	0.23	0.69	0.14
Nickel (ppm)	10	2.2	7.7
Vanadium (ppm)	20	1.9	28
Naphtha (20–205°C) (wt%)	23.2	18.6	22.7
Paraffins	12.5	8.8	16.2
Naphthenes	7.4	7.7	4.1
Aromatics	3.2	2.1	2.4
High-boiling fraction (>205°C)	76.8	81.4	77.3
Saturates	14.4	56.3	34.0
Aromatics	25.0	16.5	21.9
Polar materials	2.9	8.4	17.9
Insolubles	1.2	0.2	3.5

[a]Modified from Clark and Brown (1977).

tion of from <0.1% to 5–6% by weight and nitrogen at <0.1% to 1%. Both of these are typically present in an organically bound form. Oxygen is also present, at up to 2%. The most important trace elements in petroleum are vanadium and nickel, both at concentrations of up to 1400 ppm and present as organometallic complexes (Clark and Brown, 1977; Freedman, 1991).

During the refining of crude oil, various hydrocarbon fractions are separated by fractional distillation at specific temperatures (see the boiling-point range for major refinery products in Table 6.2). A typical yield of products by the refining of petroleum from the Prudhoe Bay, Alaska, field is natural gas, 3%; gasoline, 18%; kerosene, 2%; middle distillates, 25% (including heating oil and jet and rocket fuels); wide-cut gas oil, 35% (lubricating oils, waxes, feedstock for catalytic cracking); and residual fuel oil, 18% (bunker fuel for ships and fuel for oil-fired electrical utilities) (Table 6.2). In addition, the process of catalytic cracking is often used to convert some of the heavier fractions to lighter, more valuable products. This secondary process can increase the yield of gasoline to more than 50% of the original quantity of raw petroleum.

6.3
OIL POLLUTION

Oil Spillage

Oil pollution can be caused by any spillage of crude oil or its refined products. However, the largest and most damaging pollution events usually involve spills of petroleum or heavy bunker fuel from disabled tankers or drill platforms at sea, from barges or ships on major inland waterways, or from blowouts of wells or broken pipelines on land.

A spill on land can occur in many ways, but the largest events generally involve a pipeline rupture or a well blowout. Globally, in 1982 there were 64.5 thousand km of pipeline for the transportation of liquid petroleum and another 136 thousand km of natural gas pipeline (Gilroy, 1983). The causes of pipeline ruptures are diverse. They include faulty pumping equipment and pipe seam welds, earthquakes, sabotage, deliberate spillage as in the Gulf War, and sometimes hunters using aboveground pipelines for target practice. The total quantity of oil spilled from pipelines is not well quantified in many parts of the world. However, because of the wide-

Table 6.2 Typical analysis of the components of a crude oil from Prudhoe Bay, Alaska[a]

	Crude oil	Natural gas	Naphtha		Middle distillate	Wide-cut gas oils	Residuum
			Gasoline	**Kerosene**			
Boiling-point range (°C)	—	<20	20–190	190–205	205–343	343–565	565+
Yield: crude oil (vol %)	100	3.1	18.0	2.1	24.6	35.0	17.6
Paraffins	27.3	100	47.3	41.9	8.9	9.3	9.3
Naphthenes	36.8	0	36.8	38.1	14.4	22.8	22.8
Aromatics	25.3	0	15.9	20.0			
Others	10.6	0	0	0	76.7[b]	67.9[b]	67.9[b]
Composition: sulfur (wt%)	0.94	—	0.011	0.04	0.34	1.05	2.30
Nitrogen (wt%)	0.23	—	0.02	0.02	0.04	0.16	0.68
Vanadium (ppm)	18	0	0	0	0	<1	93
Nickel (ppm)	10	0	0	0	0	<1	46
Iron (ppm)	4	0	0	0	0	<1	25

[a]Modified from Clark and Brown (1977).
[b]Numbers indicate a percentage derived from a mixture of aromatics and others.

spread use of spill sensors and mechanisms for shutting down sections of pipeline, individual events are usually much smaller than can potentially be spilled by oceanic supertankers or by blowouts of offshore platforms. Because the spread of spilled oil is much more restricted on land than on water, terrestrial spills usually affect relatively localized areas (unless the spilled oil reaches a watercourse). The characteristics and ecological effects of terrestrial spills are described in more detail later in this chapter, in the context of oil spills in the Arctic.

Much information is available about the magnitude, behavior, and effects of oil spilled at sea. Recent estimates of the input of petroleum hydrocarbons to the world's oceans during the 1970s and early 1980s ranged from about 3.2 to 6.1 million tonnes/year (Table 6.3). This was equivalent to about 0.2–0.3% of the total quantity of petroleum transported by tankers in 1980 and 0.1–0.2% of global petroleum production (Cormack, 1983). More recent estimates of global inputs of oil to the marine environment suggest that there was a substantial reduction during the 1980s, from 1.47 million tonnes in 1981 to 0.57 million tonnes in 1989 [International Marine Organization (IMO), 1990; GESAMP, 1993].

In comparison to these anthropogenic inputs of petroleum to the oceans, the natural production of nonpetroleum hydrocarbons by marine plankton has been estimated at about 26 million tonnes/year, or about four to eight times the total input of petroleum hydrocarbons. These biogenic hydrocarbons are an important component of the background concentration of hydrocarbons in the marine environment, but they are, of course, well dispersed and should not be considered to be an important source of marine pollution. An estimate of the emission of petroleum hydrocarbons from natural submarine and coastal oil seeps is about $200–600 \times 10^3$ tonnes/year, or about 6–13% of the total petroleum input to the oceans.

The remaining inputs are anthropogenic, except for an unknown fraction of the atmospheric deposition of hydrocarbons, which might have originated with emissions from terrestrial vegetation and from other natural sources. Of the anthropogenic sources, point discharges contaminated by urban runoff, refineries, and other coastal effluents are in aggregate substantial and are important in causing local, chronic pollution in the vicinity of harbors and coastal cities around the world (GESAMP, 1993). The discharges from tankers, other ships, and offshore exploration and production platforms are essentially episodic, and they occur as accidental

Table 6.3   Estimates of inputs of petroleum hydrocarbons to the world's oceans[a]

Source	1973[b]	1979[a]	1981[d]	1983[e]
Natural seeps	600	600	300 (30–2600)	200 (20–2000)
Atmospheric deposition	600	600	300 (50–500)	300 (50–500)
Urban runoff and discharges	2500	2100	1430 (700–2800)	1080 (500–2500)
Coastal refineries	200	60	—	100 (60–600)
Other coastal effluents	—	150	50 (30–80)	50 (50–200)
Accidents from tankers at sea	300	300	390 (350–430)	400 (300–400)
Operational discharges from tankers	1080	600	710 (440–1450)	700 (400–1500)
Losses from nontanker shipping	750	200	340 (160–640)	320 (200–600)
Offshore production losses	80	60	50 (40–70)	50 (40–60)
Total discharges	6110	4670	3570	3200

[a]Data are in units of 10^3 tonnes/year.
[b]National Academy of Sciences (NAS) (1975d).
[c]Kornberg (1981).
[d]Baker (1983), best estimate, range in parentheses.
[e]Koons (1984), best estimate, range in parentheses.

spills and deliberate discharges of various size.

It is impossible, of course, to predict the location or magnitude of any accidental spill of petroleum. As might be expected, spills from tankers are most frequent in coastal areas in the most heavily traveled sea lanes (Fig. 6.1). In magnitude, marine spills range from frequent losses of minor quantities to infrequent but massive, accidental events. Some examples of the latter, disastrous spills include supertanker accidents such as (Smith, 1968; Grose and Matton, 1977; Thomas, 1977; Anonymous, 1982; Koons and Jahns, 1992): (1) the *Torrey Canyon* in 1967 off southern England with about 117 thousand tonnes spilled, (2) the *Arrow* incident in 1970 off Nova Scotia (11 thousand tonnes), (3) the *Metula* in Estrecho de Magallanes (Strait of Magellan) in 1973 (53 thousand tonnes), (4) the *Argo Merchant* in 1976 off Massachusetts (26 thousand tonnes), (5) the *Amoco Cadiz* in 1978 in the English Channel (230 thousand tonnes), (6) the *Exxon Valdez* in 1989 in Prince William Sound in southern Alaska (35 thousand tonnes), and (7) the *Braer* spill in 1993 off the Shetland Islands of Scotland (84 thousand tonnes).

Massive, accidental spills have also occurred from offshore platforms. The 1979 blowout of the IXTOC-I exploration well in the Gulf of Mexico spilled more than 500 thousand tonnes of petroleum, making it the largest, accidental spill to date [Advisory Committee on Oil Pollution of the Sea (ACOPS), 1980; Koons and Jahns, 1992)]. Other examples are the 1969 Santa Barbara blowout off the coast of southern California (about 10 thousand tonnes; Foster and Holmes, 1977) and the 1977 Ekofisk blowout in the North Sea off Norway (30 thousand tonnes; Cormack, 1983).

Petroleum has also been spilled during warfare, as a result of the deliberate targeting of tankers or offshore production platforms, or as a tactical strategy of war. During the Second World War, German submarines sank 42 tankers off the east coast of the United States, resulting in an aggregate spillage of about 417 thousand tonnes of petroleum and refined products (Koons and Jahns, 1992). During the Iran–Iraq War of 1981–1987, there were 314 attacks on oil tankers, 70% of them carried out by Iraqi forces and 30% by Iranian. The largest spill event of that war began in March 1983 when Iraq attacked five

Fig. 6.1 Major oceanic transportation routes for petroleum, from National Academy of Sciences (NAS) (1985b).

tankers and damaged three producing wells at the offshore *Nowruz* facility, causing a massive spill into the Persian Gulf, estimated at more than 260 thousand tonnes (Holloway and Horgan, 1991; Horgan, 1991).

The world's largest-ever marine spill occurred during the Gulf War of 1991, when Iraqi forces deliberately released about 0.8 million tonnes of crude oil (the range of estimates is $0.5-2 \times 10^6$ tonnes) from several tankers and an offshore loading facility into the Persian Gulf at Kuwait's *Sea Island Terminal* (Holloway and Horgan, 1991; Horgan, 1991; Jones, 1991; Hopner *et al.*, 1992; Sorkhoh *et al.*, 1992).

Another, extraordinarily large spillage occurred on land during the Gulf War, beginning in January 1991 when more than 700 Kuwaiti production wells were sabotaged and ignited, resulting in tremendous releases of petroleum to land and the atmosphere. Estimates of the daily emission of petroleum from the burning oil wells ranged from $2-6 \times 10^6$ tonnes/day, with about one-half of the wells being capped within 6 months, and the last one in early November 1991 (Bakan *et al.*, 1991; Johnson *et al.*, 1991; Popkin, 1991; Warner, 1991; Earle, 1992). If a total spilltime of 150 days is assumed, then the aggregate spillage from the blowout wells in Kuwait might have been $42-126 \times 10^6$ tonnes. Much of the oil burned in the atmosphere, but a large amount of petroleum accumulated locally in oil lakes, which in November 1991 contained an estimated $5-21 \times 10^6$ tonnes of crude oil (Hoffman, 1991; Earle, 1992).

Most of these spillages caused considerable, although not necessarily well documented, ecological damage, as described later in the case studies of large marine oil spills.

It is important to emphasize that in itself, the size of the spill does not necessarily tell much about its potential to cause damage. Even a small spill can wreak havoc in an ecologically sensitive environment. For example, near Norway in 1981, a small, operational discharge of oily bilge washings from the tanker *Stylis* killed an estimated 30 thousand seabirds, because the oil affected a location where birds were seasonally abundant (Kornberg, 1981).

In another case, more than 16.5 thousand oiled Magellanic penguins (*Spheniscus magellanicus*) were beached along an Argentinean coast, without any direct signs of an oil slick (Fry, 1992). In addition, the type of petroleum product can influence the severity of the ecological damage. The most important considerations in this respect are the relative toxicities and environmental persistence of the spilled materials (these topics are discussed later).

Operational discharges of oily tank washings from tankers are also a frequent source of marine spills, although the importance of this source has been decreasing (GESAMP, 1993). This source of pollution is associated with the practice of filling the holding tanks with seawater ballast after the delivery of a load of petroleum or a refined product, and later discharging the oily bilge water into the sea as the ship travels to pick up its next load (Clark and MacLeod, 1977; Koons, 1984). The discharge typically contains a hydrocarbon residue that is equivalent to about 0.35% of the tanker's capacity (range of 0.1% for light refined oil, to 1.5% for heavy bunker fuel).

Beginning in the early 1970s, this operational source of oil pollution was decreased substantially by a simple change in the tank-cleaning procedure called the load-on-top (LOT) process. In LOT, the oily wash water is retained on board for some time as ballast. This allows most of the oil to separate from the seawater. The oily residue is combined with the next cargo, while the relatively clean seawater is discharged. The LOT technique has a potential recovery efficiency of 99%, but in operational use it is usually 90% or less, depending on how turbulent the sea is during the separation phase (Clark and MacLeod, 1977). As a result of the widespread (but not universal) adoption of LOT by tankers, oil pollution from this source was reduced from about 1.1 million tonnes/year in 1973, to 0.7×10^6 tonnes/year in 1983, and 0.25 million tonnes in 1989 (Table 6.3; GESAMP, 1993).

Partitioning of the Spilled Oil

After oil is spilled, it undergoes partitioning within the environment in a number of ways (Fig. 6.2).

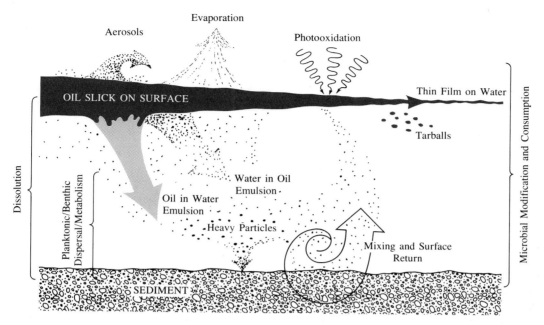

Fig. 6.2 Diagrammatic representation of the processes involved in the dissipation of spilled petroleum at sea, from Clark and MacLeod (1977).

1. *Spreading* is the process by which spilled oil physically moves and dilutes itself over the surface of the water. The surface slick can then be transported by a water current, or it can be moved by wind at a rate of about 3–4% of the wind speed (Kornberg, 1981). The degree of spreading is directly influenced by the viscosity of the spilled oil and by such environmental conditions as wind strength, turbulence, and the presence of ice on the water surface. A surface oil slick can be visually detected as a sheen on calm water when only 0.0003 mm thick. After an experimental spill of 1 m³ of a Middle East crude oil, a slick 0.5 mm thick and 48 m in diameter formed in 10 min, and after 100 min it was 0.1 mm thick and 100 m wide (Kornberg, 1981).

2. *Evaporation* is initially important in reducing the volume of spillage that remains in the aqueous environment (and also in terrestrial environments). Evaporation is most important in the dissipation of relatively light and volatile hydro-carbon fractions, and it is aided by high ambient temperatures and wind speed, and in the marine environment, by rough seas that move spilled oil into the atmosphere by the formation of a fine aerosol at wave crests. At sea, evaporation typically accounts for as much as 100% of spilled gasoline, 75% of number 2 fuel oil, 30–50% of spilled crude oil, and 10% of bunker C (Clark and MacLeod, 1977; Kornberg, 1981). Since the lower-molecular-weight fractions of petroleum are evaporated preferentially, the relative concentration of heavier molecules increases greatly in the residual spill volume. For example, after a spill of an Alaskan oil, there was a 15–20% loss of mass by evaporation. This caused the relative concentration of the heavy, nondistillable molecular fraction to increase from an initial 34% of the mass to more than 50% (Payne and McNabb, 1984).

3. *Solubilization* is the process by which oil fractions dissolve into the water column. This causes contamination of water in the vicinity of

the oil spill. For example, the concentration of total hydrocarbons dissolved in water directly beneath (at a 2-m depth) a petroleum slick in the North Sea was as large as 4 g/m^3 (ppm), compared with about 1 mg/m^3 (ppb) or less in uncontaminated seawater (Cormack, 1983). In general, lighter fractions are more soluble in water than are heavier ones, and aromatics are much more soluble than alkanes. Among the alkanes, the solubility of the gases C_1–C_4 ranges from 24 to 62 g/m^3 in freshwater, while the light liquids C_5–C_9 are 0.05–39 g/m^3, species in the kerosene fraction C_{10}–C_{17} are 1–2 $\times$ 10^{-4} g/m^3, gas oils C_{16}–C_{25} are 6 $\times$ 10^{-4} to 10^{-8} g/m^3, lubricating oils C_{23}–C_{37} are 10^{-7} to 10^{-14} g/m^3, and residual hydrocarbons >C_{37} are less than 10^{-14} g/m^3. Benzene, the lightest aromatic, has a solubility of 1780 g/m^3, while toluene is 515 g/m^3 and naphthalene is 31 g/m^3 (C. A. Parker et al., 1971; Clark and MacLeod, 1977).

4. *Residual material* is the fraction that remains after most of the evaporation and solubilization of lighter fractions has occurred. This residuum forms a rather stable and gelatinous water-in-oil emulsion (although it can contain up to 70–80% water), known as "mousse" because of its superficial resemblance to the whipped desert with that name. It is in this partially weathered emulsion that most oil that has been spilled offshore actually impacts shorelines. As they wash ashore, these emulsions can combine with sediment particles to form patties of tar-like oil and sand, which subsequently get buried in the beach or are washed back to sea. At sea, further degradation and weathering of the emulsions by biological oxidation and photooxidation of relatively light components creates lumps of a dense, semisolid, asphaltic residuum, known as "tar balls."

Tar balls can be important in the chronic pollution of beaches and some pelagic environments. In most cases, the primary source of tarballs is thought to be tank washings, rather than the weathered residue of accidental spills. The presence of a large quantity of floating tar was first reported in the mid-Atlantic gyre known as the Sargasso Sea by Horn et al. (1970). However, the phenomenon did not receive much attention from the popular media until a dramatic report by the anthropologist Thor Heyerdahl (1971) of "shocking" and "terrible" pollution by "floating asphalt-like material" along the edge of the Sargasso Sea. He made this observation at sea level during his two east-to-west crossings of the Atlantic Ocean in 1969 and 1970, in small papyrus raft-boats.

Other studies of pelagic tar in the northwest Atlantic Ocean estimated a total quantity of >86 thousand tonnes, 75% of which was in the Sargasso Sea (Butler et al., 1973). The tar balls accumulate quite efficiently in the Sargasso Sea, because of the natural ability of this gyre to collect floating material, including the locally dominant seaweed *Sargassum* spp., as well as human detritus. Because the tar balls are typically composed of C_{30} to C_{40} hydrocarbons, their surfaces are highly weathered of the relatively toxic, light-hydrocarbon fractions, and they can serve as an ecological substrate for some marine life. Typical species of these tiny pelagic ecosystems include the isopod *Idothea metallica*, the goose barnacle *Lepas pectinata*, and the epipelagic fish *Scomberesox saurus* (Butler et al., 1973).

6.4
BIOLOGICAL EFFECTS
OF HYDROCARBONS

Petroleum and hydrocarbon toxicities are well-known and well-studied phenomena, with a large amount of both field- and laboratory-based bioassay data. Comprehensive reviews of specific toxicity data for a wide variety of organisms are available in various sources (e.g., Currier and Peoples, 1954; Baker, 1971f, 1983; O'Brien and Dixon, 1976; Malins, 1977; Corner, 1978; Boyles, 1980; Nounou, 1980; Neff and Anderson, 1981; Wells, 1984; NRC, 1985b; Wells and Percy, 1985; Capuzzo, 1987; Vandermeulen, 1987; GESAMP, 1993). This subject is not detailed here.

Several recent syntheses have attempted to derive empirical relationships between the physical/chemical properties of hydrocarbons and their toxicity. The value of such relationships lies in their potential ability for prediction of the likely toxicities of the multitude of hydrocarbon species and mixtures that have not been directly bioassayed (Birge, 1983; Miller, 1984). An important observation that has emerged from these studies is that the toxicity of particular hydrocarbons is strongly related (in a statistical sense) to their chemical structure and hydrophobicity (Hutchinson *et al.*, 1979; Veith *et al.*, 1979; Bobra *et al.*, 1985; Hermens *et al.*, 1985). To state the relationship in another way, those hydrocarbons that are most soluble in water are least toxic (on a molar basis). This statistical relationship has been found to be remarkably strong and consistent across a wide range of hydrocarbons and organisms.

The biophysical mechanism of the hydrophobicity effect is that the rate of transport of hydrocarbons into organisms depends on their solubility in the lipid phase of cellular membranes. Therefore, lipid solubility is a major controlling factor for the rate and degree of bioconcentration of specific hydrocarbons from the aqueous environment (i.e., organism–water partitioning). Furthermore, in cases of acute exposure, lipid solubility influences the degree of membrane disruption that is caused (loss of integrity of the plasma membrane is a frequently observed toxic effect of an acute exposure to hydrocarbons).

From these observations, a toxicity index based on the partitioning of particular hydrocarbons between water and *n*-octanol phases has been proposed (octanol chemically mimics partitioning into a generalized lipid phase) (Veith *et al.*, 1979; Bobra *et al.*, 1985; Hermens *et al.*, 1985; M. Miller *et al.*, 1985; Abernethy *et al.*, 1986). This index is also useful for modeling the toxicity and food-web accumulation of chlorinated hydrocarbons (described in Chapter 8).

However, it should be stressed that even though large, water-insoluble hydrocarbon species have a greater toxicity *on a per mole basis* than do relatively light and water-soluble ones, the greater part of the ecological damages after an actual petroleum

spill is often attributed to the lighter fractions. The reason for this apparent contradiction is that the light hydrocarbons typically make up a large fraction of the spill volume, and since they are also of relatively small molecular weight, they comprise a much larger fraction of the total number of moles of hydrocarbons in the spill that contact organisms than do the heavier hydrocarbons.

The bioconcentration of hydrocarbons from the aqueous environment has been determined in many situations of chronic exposure and after spill events [e.g., Clark and MacLeod, 1977; Lee, 1977; Kornberg, 1981; National Oceanic and Atmospheric Administration (NOAA), 1982; NRC, 1985b]. A few selected examples of hydrocarbon concentration in organisms of polluted marine environments are as follows (from Clark and MacLeod, 1977; Ohlendorf *et al.*, 1978).

1. Plants. The littoral macroalga *Enteromorpha clathrata*, 429 ppm after a fuel oil spill; the salt marsh grass *Spartina alterniflora*, 15 ppm after a fuel oil spill; the estuarine dicot *Zostera marina*, 17 ppm after a fuel oil spill.

2. Invertebrates. The snail *Littorina littorea*, 27–604 ppm after a bunker C spill; the mussel *Modiolus edulus*, 21–372 ppm after a bunker C spill; the mussel *Mytilus edulis*, 77–103 ppm after a bunker C spill; the oyster *Crassostrea virginica*, 38–126 ppm after a fuel oil spill, 236 ppm in a chronically polluted estuary, 160 ppm in a chronically polluted harbor; the starfish *Asterias vulgaris*, 14–400 ppm after a bunker C spill; the lobster *Homarus americanus*, 103–130 ppm after a bunker C spill.

3. Birds. Herring gull *Larus argentatus*, 584 ppm in brain tissue after a fuel oil spill; common murre *Uria aalge*, 8820 ppm in a composite organ sample after a fuel oil spill; western grebe *Aechmophorus occidentalis*, 9100 ppm in liver after a fuel oil spill.

Of course, the toxic effects of an exposure to a particular concentration of hydrocarbons in the environment are influenced by many variables. Some of the more important of these are (after Mackay and Wells, 1981; Baker, 1983; NRC, 1989): (1) the amount of oil; (2) the type of oil and the relative toxicities of its component hydrocarbons; (3) the frequency and timing of the exposure event (i.e., chronic versus episodic pollution), including the persistence of residues under particular environmental conditions; (4) the condition of the oil, for example, thickness of the slick, nature of the emulsion, degree of weathering, etc.; (5) environmental variables that affect exposure and toxicity, such as weather and climatic conditions, oxygen status, and presence of other pollutants; (6) toxicity associated with chemical dispersants that may be used to create an oil-in-water emulsion for purposes of cleanup; and (7) the sensitivity of the specific biota of the affected ecosystem to the toxic effects of hydrocarbons.

6.5

ECOLOGICAL EFFECTS OF OIL POLLUTION

A number of case studies are described to illustrate the ecological effects of oil pollution in various environments. The marine cases describe the effects of (1) massive spills from wrecked supertankers, (2) spills from offshore drilling platforms, (3) experimental oiling of salt marsh vegetation, and (4) chronic oil pollution. To illustrate the effects of oil spilled on terrestrial ecosystems, the effects of petroleum spilled on arctic terrain are examined.

Oil Spills from Wrecked Tankers

The Torrey Canyon

One of the largest and better-studied cases of oil pollution caused by the wreckage of a tanker is the *Torrey Canyon* incident of 1967 (Smith, 1968; Michael, 1977; Nelson-Smith, 1977). The *Torrey Canyon* was a supertanker bound for Milford Haven in southwestern Wales when it ran aground on the Seven Stones Rocks. This caused the eventual loss of its entire cargo of 117 thousand tonnes of Kuwait crude oil. Pollution from this wreck contaminated about 225 km of the Cornish coastline, and to a

lesser degree the Brittany coast of France. Most of the oil that washed on shore was a viscous mousse of about 80% water-in-oil.

Marine oil spills have killed millions of seabirds in the twentieth century (Piatt *et al.*, 1991), and in the perceptions of the popular press and many people, birds are among the most tragic victims of oil pollution. Birds were also among the most obvious victims of the *Torrey Canyon* incident. The *Torrey Canyon* spill was estimated to have caused the deaths of at least 30 thousand birds, of which 97% were common murres (*Uria aalge*) and razorbills (*Alca torda*) (Ohlendorf *et al.*, 1978). About 7850 oiled birds were captured and cleaned. However, partly because of inexperience in the cleaning and handling of a large number of oiled seabirds, the rehabilitation program was not successful. Only 6% of the cleaned birds survived for at least 1 month.

The total avian mortality was sufficient to cause population decreases of 80–88% among breeding puffins (*Fratercula arctica*), murres, and razorbills on the Brittany coast in the year following the *Torrey Canyon* spill (Michael, 1977; Evans and Nettleship, 1985). Puffins, for example, maintained a population of about 2500 pairs on their breeding islands of Les Sept Isles in the English Channel, but only 400–500 pairs after the *Torrey Canyon* oil spill and only 135–170 pairs in 1981 after the *Amoco Cadiz* spill of 1978 (Piatt *et al.*, 1991).

Seabirds that spend much of their time on the surface of the ocean are especially vulnerable to oil, and pelagic species that seasonally aggregate in large numbers as surface "rafts," such as sea ducks, alcids, and penguins, can suffer tremendous mortality (R. G. B. Brown *et al.*, 1973; Bourne, 1976; Holmes and Cronshaw, 1977; Ohlendorf *et al.*, 1978; Clark, 1984; Evans and Nettleship, 1985; Leighton *et al.*, 1985; Cairns and Elliot, 1987; Fry, 1992). In addition, the alcids and penguins have relatively small reproductive potentials, and as a result it can take a considerable time for their populations to recover from an event of mass mortality. For example, most murres (*Uria* spp.) do not breed until they are at least 5 years old, not all adults breed in a given year, they lay only one egg per clutch, and they fledge an average of only about one young per

Aquatic birds are among the most evocative victims of oil spills. This blue-winged teal (*Anas discors*) was impacted by a spill of heavy fuel oil in the St. Lawrence River near Alexandria Bay, New York (photo: B. Freedman).

five breeding adults per year (Birkhead and Harris, 1985; Harris and Birkhead, 1985; Piatt *et al.*, 1991).

The most important lethal effect of petroleum on birds occurs through the fouling of their feathers. This causes a loss of the critical feather properties of insulation and buoyancy, so that animals die from drowning or from an excessive heat loss leading to hypothermia. There are also direct toxic effects of oil ingested during attempts to clean feathers by preening, and developing embryos may suffer mortality from even a light oiling of their egg from the feathers of a contaminated adult (Bourne, 1976; Ohlendorf *et al.*, 1978; Clark, 1984; Holmes, 1984; Leighton *et al.*, 1985; Vandermeulen, 1987; Fry, 1992; Nisbet, 1993).

An important aspect of the *Torrey Canyon* case is that an intensive effort at beach clean up was mounted, that emphasized the use of detergent and first-generation dispersants to create milky, oil-in-water

emulsions that washed out to sea. In total, about 10 thousand tonnes of detergent and dispersant was used to clean up the 14 thousand tonnes of crude-oil residue that had washed up on the Cornish beaches. These chemicals are highly toxic in their own right, and their use greatly exacerbated the toxic effects of the oil on the shoreline and nearshore ecosystems.

The ecological effects of the *Torrey Canyon* spill were relatively well described (Smith, 1968). A rocky intertidal habitat was common in the area and it was heavily damaged, especially where detergent and dispersants were used in the cleanup. In the absence of such cleaning, algae were damaged by oil but they generally survived (including species that proved to be quite sensitive to detergent; see below). In addition, limpets (*Patella* spp.) often survived a complete covering by oil, and they were even observed grazing on oil-covered rocks. There was much more intensive damage to seaweeds on oiled-and-cleaned shores. Relatively sensitive species of macroalgae included *Fucus serratus*, *F. spiralis*, *F. vesiculosus*, *Pelvetia canaliculata*, and *Ulva lactuca*. *Laminaria digitata* suffered widespread damage to its fronds, but it was able to regenerate vigorously from this acute injury. The algae that were most resistant to oil plus detergent were *Chondrus crispus* and *Ascophyllum nodosum*.

The most resistant invertebrates were the beadlet *Actinia equina* and the dahlia anemone *Tealia felina*. These survived in tidal pools that were devoid of all other animals as a result of the combined toxicities of oil and detergent/dispersant. Among the more notable crustacea that were devastated on rocky subtidal environments were the lobster *Homarus vulgaris* and the crabs *Cancer pagurus*, *Carcinus maenas*, *Pilumnus hirtellus*, *Porcellana platycheles*, *Portunus puber*, and *Xantho incisus*. The limpets *Patella vulgata*, *P. intermedia*, and *P. asper* were also devastated, as were other mollusks, such as the previously abundant saddle oyster *Anomia ephippium*. There was also a demise of echinoderms, including the starfish *Marthasterias glacialis*, the cushion-star *Asterina gibbosa*, and the urchin *Psammechinus miliaris*. A widespread mortality occurred among fish of the rocky nearshore, especially the blenny *Blennius pholis*, the Cornish sucker-fish *Lepadogaster lepadogaster*, the eel *Conger conger*, and others. Virtually no fish were observed in the vicinity of oiled and cleaned beaches.

Because such a large fraction of the invertebrate herbivore population was killed on the oiled and cleaned rocky beaches, the regenerating intertidal habitat was characterized by an early successional plant community dominated by the green alga *Enteromorpha* spp. As herbivores gradually recovered in abundance, the *Enteromorpha* was replaced by species of brown algae that are more typical of the rocky intertidal. However, during the initial years of the post oiling succession, the zonation of algal species was somewhat abnormal, an artifact probably of the disruption of grazing by limpets on oiled and cleaned shores. For example, the upper limit of the distribution of *Laminaria digitata* and *Himanthalia elongata* was higher by as much as 2 m during the first few years of succession. As limpets

progressively recolonized, the distributions of seaweed species returned to their more typical limits (Southward and Southward, 1978).

Another important consequence of the massive use of detergent and dispersant during the *Torrey Canyon* cleanup was the damage caused in submersed habitat as far as 0.4 km offshore. Often, a marine oil spill causes severe damage only to biota that are directly contacted by petroleum at the surface or in the intertidal zone. However, the exposure of submersed habitats to water having large concentrations of dissolved detergent, dispersant, and emulsified petroleum can also cause mortality of many marine species. In the case of the *Torrey Canyon*, especially susceptible subtidal decapods were the squat lobster *Galathea strigosa*, the swimming crab *Portunus puber*, and the edible crab *Cancer pagurus*. Serious mortality occurred in the bivalve molluscs *Ensis siliqua* and *Mactra corallina* at depths as great as 14.5 m. Susceptible echinoderms included the starfish *Martasterias glacialis*, the heart urchin *Echinocardium cordatum*, and the sea urchin *Echinus esculentus*. However, some invertebrates tolerated an exposure that was fatal to other species in their phylum, notably the spider crab *Maia squinado*, the starfish *Asterias rubens*, and the brittle star *Acrocnida brachiata*. In addition, there was relatively little mortality of macroalgae in subtidal habitats (Smith, 1968).

The unexpectedly great ecological damage caused by the oil and detergent/dispersant during the *Torrey Canyon* cleanup proved to be an important lesson. These effects resulted in a much more judicial use of detergent and dispersants during subsequent oil-spill cleanups, with the practice largely being restricted to offshore locations and to sites that are of high value for industrial or recreational purposes. In addition, less toxic dispersants were developed for dealing with oil-spill emergencies (NRC, 1989). For the latter purpose, small concentrations of dispersant are often used in a seawater diluent to remove oil residues. Except for birds, the ecological damage caused by the *Torrey Canyon* oil alone proved to be relatively small, and recovery was fairly vigorous. In contrast, some sites treated with detergent and/or dispersants took as long as a decade to reestablish communities similar to those present before the spill (Southward and Southward, 1978).

The Amoco Cadiz

In late winter of 1978, another supertanker broke up in the same general area as the *Torrey Canyon*, but closer to France. This was the *Amoco Cadiz* accident, in which about 233 thousand tonnes of crude oil were spilled off the Brittany Coast. The accident was caused by a steering failure of the fully laden ship during a storm. The vessel drifted helplessly for several critical hours while its crew was attempting repairs. At the same time, the vessel's captain and a marine salvage company negotiated assistance terms under conditions complicated by different native languages, long-distance communications between the captain and owners of the vessel, and other factors. In any event, the *Amoco Cadiz*

went aground on a nearshore reef before any effective rescue procedures could be implemented.

Of the total oil spilled, about 32% was estimated to have evaporated, 11% sank to the bottom, 11% was dispersed at sea, 10% was recovered and disposed, and the remaining 36% contaminated beaches, rocky shores, and other parts of the marine environment (Ganning *et al.*, 1984). About 62 thousand tonnes of oil impacted more than 360 km of shoreline with some degree of oiling. This included about 140 km of heavily oiled Brittany coast, which was polluted by an approximately 50% water-in-oil mousse (Berne *et al.*, 1980; Finkelstein and Gundlach, 1981; Anonymous, 1982).

There was a large-scale, intensive cleanup of much of the contaminated shoreline, an effort that cost an estimated $117 million (in 1978 U.S. dollars; Berne and Bodennec, 1984). The cleanup largely involved the physical removal of oily residues and contaminated sand and sediment. The use

A rocky shore on the Brittany coast of France that was heavily contaminated by an oil-in-water emulsion after the wreck of the *Amoco Cadiz*. The dominant macroalgae are *Fucus serratus* and *Ascophyllum nodosum*. When these were directly oiled, their fronds were killed, but most of their holdfasts survived, so that regeneration was fairly rapid (photo courtesy of J. Vandermeulen).

of dispersants and detergents was restricted to sites of high economic value, especially harbors, and to some deeper-water environments where dispersants were applied to pelagic windrows of oil. In addition, the specific dispersants used were much less toxic than those used in the *Torrey Canyon* incident, and they were mostly applied as a dilute fraction of the wash water.

Much effort went into monitoring the degree of environmental pollution by hydrocarbons and their subsequent dissipation by biodegradation and physical weathering. For example, 14–28 days after the spill, a polluted nearshore site had an average hydrocarbon concentration in water of 122 ppb. This decreased to 39 ppb after 28–39 days, 10 ppb after 54–63 days, 2–5 ppb after 87–112 days, and 1 ppb after 1 year (the latter is the approximate background concentration of hydrocarbons in seawater) (Berne *et al.*, 1980). Nearshore sediment contained as much as 1100 ppm of hydrocarbons, but the degree of contamination decreased rapidly in seaward transects. Anaerobic sediments of an estuary had a concentration as large as 50 thousand ppm (5%) of hydrocarbons (Berne *et al.*, 1980). The oyster *Crassostrea gigas* accumulated as much as 2700 ppm d.w. of total petroleum hydrocarbons in its flesh, and even after 2 years some individuals contained as much as 100 ppm (Boehm, 1982). The initial contamination of some commercial fish species in the vicinity of oiled areas was as much as 154–170 ppm (Boehm, 1982).

The presence of large concentrations of hydrocarbons in the contaminated sediment of various habitats stimulated a strong numerical response by microorganisms (Ward *et al.*, 1980; Atlas, 1982). Interestingly, the general, indigenous microbial community proved to be quite capable of utilizing hydrocarbons as a metabolic substrate; the response of taxa that are metabolically specific to hydrocarbons comprised only a small proportion of the total microbial response to the oil pollution (Table 6.4). The average rate of biodegradation was about 0.5 μg hydrocarbon/g d.w. sediment-day, and in an aerobic, nutrient-rich environment as much as 80% of the mass of aliphatic and aromatic hydrocarbon residues was metabolized by microorganisms in the first 7 postspill months (Ward *et al.*, 1980).

Table 6.4   Average abundance of bacteria in various Brittany sediments 1 month after the *Amoco Cadiz* oil spill[a]

	Total number of bacteria (10^3/g sediment)	Hydrocarbon-utilizing bacteria (10^3/g sediment)
Oiled sites		
Salt marsh	630,000	14,000
Estuary—high intertidal	190,000	18
Estuary—low intertidal	690,000	390
Sandy beach	100,000	350
Reference sites		
Salt marsh	96,000	0.52
Beach	45,000	0.38

[a] Modified from Ward *et al.* (1980).

Many of the ecological consequences of both oil pollution and the subsequent cleanup after the *Amoco Cadiz* incident were less severe than those of the *Torrey Canyon* spill. In addition, recovery was more rapid and was largely complete within several years (NRC, 1985b). The principal reason for this smaller ecological effect was the effective use of relatively small quantities of low-toxicity dispersants, rather than the industrial detergent and first-generation dispersants used to clean up the *Torrey Canyon* residues.

Seaweeds of the rocky coast were most affected in the intertidal zone (Floc'h and Diouris, 1980). The red alga *Rhodothamniella floridula* is present on rock surfaces that are exposed except at high tide. Where this species was directly exposed to oily residues, it died within a few days. The brown alga *Pelvetia canaliculata* occupies rocks that are only exposed at low tide. Where this species was directly oiled it suffered initial damage, and then for as long as 8 months it experienced a progressive decline that resulted in a virtual denudation of previously vegetated substrates. About 18 months after the spill regeneration of *Pelvetia* began by the establishment of germlings, and this led to a progressive and rapid recovery. Lower down in the intertidal zone, the dominant alga *Fucus spiralis* suffered some damage, but it was not as intensively damaged as were the preceding two algae of more xeric habitat. However, the encrusting intertidal alga *Hildenbrandia* was drastically affected. In the mid- and low-tide zones, the dominant algae *Ascophyllum nodosum* and *Fucus serratus* were little affected, except for relatively minor injuries occurring where there was direct contact with oil. As in the *Torrey Canyon* incident, the opportunistic green alga *Enteromorpha* spp. formed a thick, but ephemeral carpet on denuded rock substrates in the intertidal zone, wherever herbivores were eliminated or reduced greatly in density by oil pollution. Overall, injury to seaweeds was largely restricted to situations where there was direct contact with oil. Compared with the *Torrey Canyon* incident, much less longer-term damage was caused to intertidal habitat by the *Amoco Cadiz* oil spill.

Similarly, the overall effect was not great in a small eelgrass (*Zostera marina*) community that was affected by *Amoco Cadiz* oil (Zieman *et al.*, 1984). The eelgrass (an angiosperm plant) was only damaged by direct contact of foliage with oil. However, there was little plant mortality since the perennating tissues of this plant are in the sediment, where they were not directly oiled. There was great mortality of some of the fauna of this ecosystem, particularly among the amphipods. Prior to the spill, there were 26 species of amphipods present, but even 1 year after the spill only five species were found in the oiled area. In contrast, gastropod molluscs were little affected in the seagrass ecosystem.

There was some damage to salt-marsh vegetation, but most of this was caused by the physical removal and trampling of vegetation during cleanup of the thick deposits of mousse from the salt marsh. The subsequent plant regeneration was fairly vigorous and involved the resprouting of surviving perennials and the establishment of seedlings of salt-marsh annuals (Levasseur and Jory, 1982; Seneca and Broome, 1982).

Biological effects in the sublittoral zone were important, but less than after the *Torrey Canyon* spill (Cabioch, 1980; Glemarec and Hussenot, 1982). In this zone the biota was affected by water contaminated by dissolved and finely dispersed hy-drocarbons. In general, severe damage was restricted to nearshore locations, usually less than 30 m from shore. The most heavily damaged community was dominated by the bivalve *Abra alba* and the amphipod *Amplelisca tenuicornis*. This community suffered mortality equivalent to about 40% of its total invertebrate biomass. Overall, the most sensitive taxa of the nearshore benthos were molluscs, amphipods, and the heart-urchin *Echinocardium cordatum* (millions of dead urchins washed up along a polluted beach).

Commercially important oyster (*Crassostrea gigas*) beds were contaminated by oil residues, which caused a discernible "tainting" of the animals. However, the oysters themselves did not exhibit any histopathological injuries, and they were not obviously affected (Neff and Haensley, 1982). In contrast, a flatfish (plaice, *Pleuronectes platessa*) was affected by serious and progressive tissue and biochemical injuries, even though it was much less contaminated by petroleum residues than were the oysters (Neff and Haensley, 1982).

At least 4600 seabirds died as a result of oiling by the *Amoco Cadiz* spill (Clark, 1984). A larger number of deaths was prevented by the fact that the avian population was at a seasonally small abundance at the time of the oil spill.

The Exxon Valdez

About one-quarter of the crude oil produced in the United States, averaging about 2.9×10^5 tonnes/day, is recovered in fields near Prudhoe Bay on the North Slope of Alaska. This petroleum is transported through the 1280-km Trans-Alaska Pipeline to a massive, 18-unit, storage-tank farm at the port of Valdez in southern Alaska and then loaded onto tankers for shipment to the continental west-coast states (Baker *et al.*, 1989; McCoy, 1989; Anonymous, 1992c). In the initial stage of the oceanic trip, the loaded tankers pass through a narrow, but well-defined shipping channel in Prince William Sound, prior to entering the open sea. Before the accidental grounding of the *Exxon Valdez*, the passage of tankers through Prince William Sound had been successfully made about 16,000 times, transporting about 1.25 billion tonnes of petroleum.

(**Top**) An initial, 1989 view of a heavily oiled cobble beach on Green Island, Prince William Sound, Alaska. This site was treated with a warm water wash in 1989 and by manual cleaning in 1990. The site was also fertilized in 1989 and 1990, to enhance microbial oxidation of petroleum residues. There were no treatments in 1991 or 1992. (**Bottom**) Condition of the same beach in 1992. Although there may be lingering hydrocarbon residues below the surface, the beach has substantially recovered through the combined actions of cleanup and natural processes (photos courtesy of Exxon Co., Houston, TX).

Prince William Sound is not only a route for the transport of crude oil. It also supports an important commercial fishery for salmon and herring and a number of salmon hatcheries. The region is also renowned for its populations of marine mammals and birds, as well as its spectacular scenery.

Just after midnight on March 23, 1989, a large tanker, the *Exxon Valdez*, ran aground onto Bligh Reef, a submerged, rocky ledge well outside of the shipping channel through Prince William Sound. The 300-m-long *Exxon Valdez* was only 3 years old and was the newest, best-equipped, and largest ship in the Exxon fleet. At the time that it ran aground, the *Exxon Valdez* contained a full load of 175.6 thousand tonnes of crude oil. Of this, about 36 thousand tonnes were spilled, and the rest was off-loaded to another tanker (Baker *et al.*, 1989; Holloway and Horgan, 1991; Koons and Jahns, 1992).

Of the oil that was spilled into the waters of Prince William Sound, about 35% evaporated, 1% dissolved into the water column, 40% washed onto shores, 25% left Prince William Sound as floating oil, and less than 10% was recovered or burned (Galt *et al.*, 1991; Kelso and Kendziorek, 1991).

The captain of the *Exxon Valdez* was Joseph Hazelwood, who had sailed for Exxon Shipping for 19 years and was a well-respected seaman (Davidson, 1990). Hazelwood had trained for the merchant marine at New York Maritime College, where he was considered to be an excellent student. In the yearbook for his graduating class, the following motto underscored Hazelwood's picture: "It can't happen to me." Hazelwood was well known as being an excellent ship's captain, but it also appears that he had a drinking problem. At the time of the grounding of the *Exxon Valdez*, Hazelwood's automobile-driving license had been revoked because of an intoxicated-driving incident, and he had three driving-license suspensions since 1984 (Davidson, 1990). According to a former crewmate: "There's a bad joke in the [Exxon] fleet that it's Captain Hazelwood and his Chief Mate, Jack Daniels, that ran the ship [the *Exxon Valdez*]" (Davidson, 1990).

The grounding of the *Exxon Valdez* was certainly accidental, but it was an accident of the worst kind, because it could have been prevented by sensible tanker operations and personnel management (Davidson, 1990). The ship had been taken out of Valdez at 21:12 by a Pilot, Ed Murphy, who handed the bridge back to Captain Hazelwood at 23:20. Hazelwood did not stay at the helm for long, however, and by 23:53 he had left the bridge to go to his cabin, having turned over control of the vessel to his Third Mate, who was not, as it turned out, well versed in the geographic details of the shipping channel through Prince William Sound. Soon after the Third Mate assumed control of the *Exxon Valdez*, at 00:04, the ship was aground on Bligh Reef, and at 00:20 Hazelwood reportedly remarked to his First Mate: "I guess this is one way to end your career" (Davidson, 1990).

Six hours after the grounding, Captain Hazelwood had a blood-alcohol concentration of 0.06%, greater that the legal limit for an on-duty vessel master of 0.04% (Davidson, 1990). It has been estimated that at the actual time of the grounding, Hazelwood's blood–alcohol concentration may have been as large as 0.2% (once ingested, ethanol is progressively metabolized by dehydrogenase enzymes, particularly in the liver, and this accounts for the diminishing concentrations in blood as time passes). These observations do not necessarily mean that Captain Hazelwood was incapacitated by alcohol at the time of the grounding. He was an experienced drinker and probably had a higher tolerance of alcohol than most people. Moreover, eye-witness accounts from the time of the grounding and shortly afterward attested to Hazelwood not being obviously intoxicated (Davidson, 1990). Nevertheless, it is obvious that alcohol must have been a substantial, and inexcusable, factor in the grounding of the *Exxon Valdez*.

In terms of the oil-spill contingency response, the immediate reaction to the oil spill from the *Exxon Valdez* also left much to be desired, largely because personnel, equipment, and overall capabilities were deficient. In testimony to the U.S. Congress, Dennis Kelso, Commissioner of the Alaska Department of Environmental Conservation, stated that "The industry's response during the first, critical 72 hours of the spill was ineffective, in part because of Alyeska Pipeline Service's [the owner and operator of the Trans-Alaska Pipeline and its associated facilities at Valdez] decade-long efforts to scuttle any meaningful oil-spill contingency plan" (cited in Baker *et al.*, 1989).

Because the transport of oil from Valdez had been successfully accomplished for years without any major incidents, the petroleum industry had been progressively decreasing its capabilities to prevent tanker accidents and to deal with oil spills once they happened (Baker *et al.*, 1989; McCoy, 1989). To decrease operating costs, the staffing and training of spill-response crews had been progressively deemphasized over time, and equipment was not always in repair and/or available on short notice to deal with an emergency. All of these factors came into play on the night that the *Exxon Valdez* ran aground, and they greatly exacerbated the damage associated with the incident by diminishing the immediate response to the spill.

For example, on the night of the grounding (after McCoy, 1989):

1. It took 14 hr to locate ship fenders so that a second tanker could come alongside the *Exxon Valdez* to lighten its remaining, unspilled cargo of crude oil. The fenders were eventually found, buried under more than 4 m of snow.
2. The boom-deployment barge that carries the offshore containment and cleanup equipment, most importantly a 2100-m containment boom, was damaged, unloaded, and out of action. It took a critical 14 hr to get the barge ready to deal with the spill from the *Exxon Valdez*.
3. Initially, only one person who knew how to operate a forklift and a crane was available, and for a while this fellow ran from machine to machine to deploy containment equipment.
4. In 1982, to save costs, Alyeska's 12-person, task-dedicated, spill-response team was dissolved, and their responsibilities assigned to other workers to execute along with their existing duties.
5. Most of the tankers on the Alaska route, including the *Exxon Valdez*, were not double hulled. This safety feature can greatly reduce the risk of oil spillage from ship groundings and had earlier

been promised during the environmental hearings associated with the initial proposal to develop a trans-Alaska pipeline and its associated shipping.

It was not until the second day of the spill that oil began to be lightened from the *Exxon Valdez* to other tankers, and it was not until the third day that sea booms were deployed to contain the spilled oil. On the fourth day a storm developed, with winds as strong as 110 km/hr, making it impossible to contain the spilled oil in the vicinity of the tankers. Consequently, the spilled oil dispersed widely, much of it eventually fouling great lengths of shoreline (McCoy, 1989; Anonymous, 1992c).

In the aftermath of the *Exxon Valdez* spill, some Alaskan environmental regulators, familiar with conditions around Valdez, made some telling statements about the capabilities for environmental protection and emergency measures for the cleanup of oil spills. Dennis Kelso, Commissioner of the Alaska Department of Environmental Conservation, stated that "Alyeska stands as a monument to a powerful and rich industry's fundamental failure to keep its commitments. They have operated as if they were a sovereign state, with terrible consequences" (cited in McCoy, 1989). James Woodie, who had served as the Coast Guard Commander for the port of Valdez and as a Marine Superintendent for Alyeska, stated that "Based on my experience with Alyeska, the only surprise is that disaster didn't strike sooner" (cited in McCoy, 1989). Clearly, the accidental grounding of the *Exxon Valdez* and the resulting oil spill were, more than in most cases, a most preventable accident.

About 1900 km of shoreline in Prince William Sound and on the nearby Kenai Peninsula and Kodiak Island were eventually oiled to some degree (Hodgson, 1990; Maki, 1991). The intensity of oiling varied greatly. A survey in 1989 of 1437 km of shoreline in Prince William Sound found that: (a) 140 km of beach were "heavily" oiled, with a greater than 6-m width of oiled substrate; (b) 93 km were "moderately" oiled, with a 3- to 6-m width of oiled beach; (c) 323 km were "lightly" oiled, with a less than 3-m width of oiled substrate; (d) and 222 km were "very-lightly" oiled, with less than 10% oil cover (Owens, 1991). Overall, about 54% of the surveyed area was oiled. Because this survey focused on places that actually had oil, it overstated the actual percentage of the shoreline of Prince William Sound that was oiled, a figure that was estimated to be closer to 20% (Owens, 1991). An additional 14% of the shoreline of the Kenai Peninsula and Kodiak Island was oiled (Owens, 1991).

Once industry and government mobilized their response teams, a massive effort was dedicated to cleanup of the oil spilled from the *Exxon Valdez*. As noted previously, about 80% of the load of the grounded ship was safely transferred to another tanker. Of the 36 thousand tonnes that were spilled into Prince William Sound, about 30% evaporated within days. Most of the rest eventually washed onto beaches, and much of this residue was removed during a large, and expensive, cleanup program. In total, about 11 thousand people were involved in the cleanup, about $2.5 billion was spent by Exxon, and another $154 million was spent by the U.S. federal government (Hodgson, 1990; Holloway and Horgan, 1991; Maki, 1991).

A variety of cleanup procedures was used (Hodgson, 1990; Holloway and Horgan, 1991). On some heavily oiled beaches, most of the petroleum residues were removed mechanically by backhoes and by people with shovels and bags, while other beaches were cleaned by washing with pressurized hot and/or cold water. On some beaches, individual rocks were literally hand wiped with absorbent cloths by people earning $16.69 per hour, an activity that was euphemistically known as "rock polishing."

A semioperational attempt to enhance microbial degradation of petroleum residues was made. In 1989, about 118 km of cleaned or relatively lightly oiled beach were treated with an oleophilic, nitrogen- and phosphorus-containing fertilizer. This material adhered to the oil residues and enhanced degradation by naturally occurring hydrocarbon-degrading microorganisms by reducing the otherwise unfavorable C:N and C:P ratios of the residues. In 1990, about 400 patches of residual oil were similarly treated. No attempts were made to enhance microbial populations by "seeding" with strains or species that are specifically adapted to utilizing hydrocarbon substrates. It was believed that these specific microbes were naturally present, and their activity and that of microbes with a broader substrate tolerance only had to be enhanced by making the ecological conditions more favorable, i.e., by fertilizing. The treatment with oleophilic fertilizer increased the rate of microbial oxidation of residues by about 50% (Exxon, 1990; Chianelli *et al.*, 1991).

As a composite result of the cleanup, natural cleansing by wave action on high-energy beaches, and microbial oxidation, the amount of oil on the surface of rocks and beaches was much smaller in years subsequent to the spill. One survey of 28 oiled sites in Prince William Sound, the Kenai Peninsula, and Kodiak Island found an average of less than 2% oil cover in late summer of 1990, one winter and two summers after the spill, compared with 37% oil cover in the summer of 1989 (Owens, 1991). Another survey conducted in late summer of 1991, two winters and three summers after the spill, found that less than 2% of the beaches of Prince William Sound were still visibly oiled at the surface, compared with about 20% in 1989, while less than 1% of the nearby Kenai–Kodiak beaches remained visibly oiled (J. M. Baker *et al.*, 1991; Anonymous, 1992b).

The intertidal community was severely affected by the short-term effects of oiling, and these effects were greatly exacerbated by some of the cleanup practices, especially the use of pressurized hot-water washings. However, the recovery from these stresses was rapid (J. M. Baker *et al.*, 1991; Maki, 1991). In 1990, at the end of two growing seasons after the spill, ephemeral green algae were abundant, and there was a widespread, initial recruitment of fucoid algae and intertidal invertebrates. At the end of the summer of 1991, ecological recovery was more advanced, including communities on shores that had been heavily oiled and further stressed by intensive cleaning. According to J. M. Baker *et al.* (1991), based on visual impressions made in late summer of 1991, "After the largest oil-spill cleanup program in history, the recovery of the shore of Prince William Sound is essentially complete."

Prince William Sound and its surrounding area support large numbers of sea mammals, and substantial mortality was caused to some species. Hardest hit were sea otters (*Enhydra lutris*), of which about 5–10 thousand occur in Prince William Sound, and at least 1 thousand were killed by oiling (J. M. Baker *et al.*, 1989; Hodgson, 1990; Maki, 1991). Exxon spent about $18 million to employ 320 people to recover and treat 357 sea otters, of which 223 were rehabilitated to the point where they could be released or placed in zoos (Maki, 1991).

Seabirds are also abundant in and around Prince William Sound. The abundance of birds and the dominant species vary greatly with season. During both the northern and southern migrations, millions of birds stage in the area. Others spend the winter there, and yet other seabirds breed. The peak, seasonal abundance of seabirds in Prince William Sound is about 10 million individuals (Maki, 1991).

At the actual time of the spill, an estimated 600 thousand wintering seabirds were present in Prince William Sound. The body count of dead birds was about 36 thousand. However, most dead birds do not wash up onto the shore, and the actual number killed has been estimated to be at least 100 thousand, and possibly more than 300 thousand (Hodgson, 1990; Piatt *et al.*, 1989, 1990, 1991). In addition, at least 153 bald eagles (*Haliaeetus leucocephalus*) were poisoned because they scavenged oiled carcasses of seabirds (Hodgson, 1990; Maki, 1991).

Exxon spent about $25 million to engage 400 people, 140 boats, and 5 aircraft to recover and treat more than 1600 birds of 71 species, with an eventual release rate of 50% (Maki, 1991).

Prince William Sound also supports an important fishery, with the value of the prespill 1988 catch estimated at more than $90 million (J. M. Baker *et al.*, 1989). Because of the oil spill, much of the fishery was closed for the 1989 season, and Exxon paid compensation of $169 million to fishers and another $133 million to processors and allied industries, in addition to the $105 million that was paid to many of these same claimants for wages and vessel charters during the cleanup operations (Royce *et al.*, 1991).

The effects of the spill on the fishery two summers afterward in 1990 were apparently not substantial (J. M. Baker *et al.*, 1991; Royce *et al.*, 1991; Schroeder, 1991). The 1990 catch of pink salmon (*Oncorhynchus gorbuscha*) in Prince William Sound was 44.2 million fish, substantially larger than the previous record-high catch of 29.2 million fish. These 2-year-old fish would have passed through Prince William Sound on their seaward migration in 1989, the year of the spill. The catch of pink salmon in 1991 was also large, more than 37 million fish. These 1991 salmon were fry in 1989, and about one-quarter were the progeny of river-spawning adults that passed through Prince William Sound on their landward, breeding migration about 2 weeks after the spill (the other three-quarters were reared in commercial hatcheries). The catch of Pacific herring (*Clupea harengus pallasi*) was also large in 1990, when about 7500 tonnes were harvested. In 1991 about 10,800 tonnes were taken, the largest catch in a decade.

Pelagic Oil—The Argo Merchant

Another case of oil pollution from a stricken tanker concerns the *Argo Merchant* wreck in 1976 off the coast of Nantucket Island, Massachusetts (Grose and Matton, 1977; NRC, 1985b). About 26 thousand tonnes of fuel oil was lost at sea in this incident. However, because the spilled oil was mainly composed of light hydrocarbons and there was a major storm at the time, virtually all of the spill evaporated to the atmosphere or dispersed into the water column. Furthermore, the prevailing wind kept the oil slicks at sea until they dissipated, so that there was no contamination of relatively susceptible, littoral and other shallow-water ecosystems.

Relatively minor effects on the pelagic biota were documented. For example, up to 55% of sampled individuals of the zooplankter *Calanus finmarchicus* were visibly contaminated by oil, 61% of *Centropages typicus*, and 34% of *Pseudocalanus* sp., largely in the form of oil-containing fecal pellets. However, this contamination did not cause a measurable change in the zooplankton community. There was no demonstrable effect on the phytoplankton of oiled compared with reference areas. Of the six fish species that were common in the oiled area, only the sand lance *Ammodytes americanus* was apparently reduced in abundance. Although the mortality of seabirds was not determined, it likely was not large since bird density in the spill area was small. The most frequently oiled species were the gulls *Larus argentatus* and *L. marinus*.

Therefore, the *Argo Merchant* oil spill caused relatively minor ecological damages. This was largely because the oil did not wash ashore. The spill dispersed rapidly offshore, and seabirds were not abundant in the vicinity of the spill.

Oil Spills from Offshore Drill Platforms

The world's largest accidental oil spill occurred in 1979 from the *IXTOC-I*, an exploration, semisubmersible, platform well located 80 km off the eastern coast of Mexico (ACOPS, 1980; Kornberg, 1981). The rate of discharge was as large as 6.4 thousand m^3/day, and over the more than 9-month blowout period, an estimated 476 thousand tonnes of crude oil were spilled. Within a week of the beginning of the spill, a slick 180 km long and as wide as 80 km had formed. An estimated 50% of the spilled oil evaporated to the atmosphere, 25% sank to the bottom of the Gulf of Mexico as weathered residue, 12% was degraded by microorganisms and photochemical processes, 6% was mechanically removed or burned at the well site, 6% reached and contaminated about 600 km of Mexican shoreline, and another 1% landed on beaches in Texas (Ganning *et al.*, 1984).

This massive spill caused great disruptions of tourism in the Gulf of Mexico and affected the fishing industry by fouling boats and gear, by tainting commercial species, and by eliminating fishing in the large areas where slicks were present. There was relatively little documentation, however, of the effects of this spill on offshore and coastal ecosystems (NRC, 1985b).

Better ecological information about a blowout that took place in 1969 in the Santa Barbara Channel of southern California is available (Straughan and Abbott, 1971; Steinhart and Steinhart, 1972; Foster and Holmes, 1977). In total, approximately 10 thousand tonnes of crude oil were spilled, resulting in contamination of the entire channel and more than 230 km of coastline. The average pollution of beaches by oil residues was 115 tonnes/km, compared with 10.5 tonnes/km in a nearby area that was only affected by natural petroleum seepages (see below) and 0.03 tonnes/km for all California beaches.

The Santa Barbara area has long been known for its naturally occurring underwater and inland petroleum seepages. These have caused slicks on the ocean surface, deposits of oil and tar on beaches, and tar deposits inland (the latter include the fossil-rich La Brea tar pits). The seepages come from shallow, underground reservoirs of petroleum, which leak to the surface through fractured or porous bedrock. One estimate of the rate of marine seepage is 3–4 thousand tonnes/year (Allan *et al.*, 1970). Because of the longer-term occurrence of natural oil seepages in the Santa Barbara area, it has been speculated that some of the local biota may have evolved a physiological tolerance to the toxic effects of hydrocarbons. However, studies of marine invertebrates have generally not shown this effect (NRC, 1985b).

As in many cases of marine pollution by oil spills, the most evocative victims of the Santa Barbara spill were birds (Steinhart and Steinhart, 1972; Foster and Holmes, 1977). An estimated 9 thousand birds were killed, or about 45% of the estimated population that was present at the time of the spill (the area is a wintering ground for several species of migratory aquatic birds). About 60% of the deaths comprised three species of loons (*Gavia immer*, *G. stellata*, and *G. arctica*) and western grebes (*Aechmophorus occidentalis*). However, in spite of this mortality the number of birds in the channel in subsequent winters was not measurably reduced. An attempt was made to clean and rehabilitate many of the oiled birds that reached the beaches alive, but the techniques were not as advanced as those used today, and the effort was not successful. Of more than 1600 individuals that were treated, only 246 were still alive after 1 month.

The habitat type that was most affected by the Santa Barbara spill was the rocky intertidal, which comprised about 13% of the affected coastline (Foster and Holmes, 1977). There was a great mortality of barnacles, especially *Chthamalus fissus* and *Balanus glandula*. Mortality ranged from 60–90% under "very heavily" oiled conditions to 20% in "moderately" oiled situations and 1–10% after "light" oiling. Relatively fresh oil was especially toxic to barnacles. Weathered residues caused much less mortality; in some cases, *Chthamalus fissus* survived a complete covering by weathered residues. The rocky intertidal surfgrasses *Phyllospadix souleri* and *P. torreyi* were also badly damaged, suffering 50–100% damage after heavy oiling and 30–50% after light exposure. These were the most prominent of the short-term effects of the Santa Barbara oil spill on the rocky intertidal ecosystem. However, most other species of this habitat were also damaged. The recovery of most species was, however, fairly rapid. Within a year of the spill, barnacles began resettling, even on rocks that were still covered with asphaltic residues.

In recreationally important areas of the Santa Barbara Channel, there was a cleanup of the oil that washed onto beaches. Various techniques, including the physical removal of oil and oily sand, blast-cleaning with steam, sand, or water, and spraying with a light naphtha-like mixture to dissolve the residues and wash them to sea, were used. When these cleanup techniques were used in natural habitats they usually exacerbated the ecological damage, because they physically removed the biota along with the oil residues or they engendered additional toxic effects.

Effects of Oil on Salt Marshes

Salt marshes occur in coastal situations that may be frequently impacted by oil from coastal refineries or terminals or by oil spilled offshore. Because salt marshes are a relatively accessible natural habitat for experimental research, they have been better studied with respect to the effects of oil pollution than any other type of marine ecosystem. In this section, a series of experiments done in salt marshes in southern Britain will be briefly examined (similar research has also been done in North America; e.g., DeLaune *et al.*, 1979; Getter *et al.*, 1984; Webb *et al.*, 1985).

Baker (1971a) reported the effects of a single experimental treatment with crude oil on three graminoid-dominated communities along a salt-marsh gradient in southern England (Table 6.5). Initially, the oil caused damage to all species. However, since the regenerative meristematic tissues of most plants were not killed, there was a progressive refoliation during the growing season. The five most prominent plant species of these habitats displayed a wide range of tolerance to crude oil (Table 6.5). The rush *Juncus maritimus* was sensitive and suffered a large decrease in cover even in the "lightly oiled" treatment. The grass *Agrostis stolonifera* was most tolerant, and it actually increased in abundance in the lightly oiled treatment. Apparent growth stimulations after a light oiling were also observed with the grasses *Puccinellia maritima* and *Festuca rubra* in another field experiment (Baker, 1971e). The reasons for such positive responses are not clear, but the stimulations could have been due to (1) indirect effects on nutrient cycling associated with the vigorous microbial activity that often occurs in oiled soil, (2) a microclimatic effect related to relatively dark and warm oiled soils, and/or (3) a hormone-like action of particular hydrocarbons.

Baker (1971d) also carried out an experiment designed to examine the effects of cleaning oiled salt marsh vegetation by (1) cutting and removing oiled biomass, (2) burning, and (3) spraying with detergent to emulsify the oil residues. At best, none of these treatments provided a better recovery than was achieved if the oiled vegetation was simply left alone. Moreover, some of the cleanup procedures greatly exacerbated the initial damage caused by the oil, especially the use of detergent or burning. The conclusion was that it was best to leave salt-marsh vegetation to recover naturally after relatively light oilings. However, in situations contaminated by thick deposits of mousse or other residues, an active cleanup may be required. In addition, it should be noted that the detergent used in this study was relatively toxic, compared with the detergents and emulsifiers that are available today.

In another set of experiments, Baker (1971c,g) investigated the seasonality of the effects of crude oil spills on salt-marsh vegetation. She found that most plant species were relatively sensitive to oiling

Table 6.5   Effects of a single experimental treatment of salt-marsh vegetation with crude oil: The plots were oiled in June and then sampled 3 months later in September[a]

Treatment	Low marsh: *Spartina anglica* density (number/0.025 m²)	Middle marsh (% cover)		High marsh (% cover)	
		Puccinellia martima	*Festuca rubra*	*Agrostis stolonifera*	*Juncus maritimus*
Unoiled	19.9	84	60	10	17
Lightly oiled (9 liters/36 m²)	8.1	47	41	54	3
Heavily oiled (54 liters/36 m²)	1.4	10	9	10	5

[a]Modified from Baker (1971a).

when their buds were expanding early in the growing season. Plants were less sensitive when they were growing less actively or were senescent or when their foliar stomata were closed because of drought. The annual *Suaeda maritima* was sensitive to oil treatment at any time in the growing season, and once oiled it was unable to recover. In contrast, although the foliar tissues of perennial plants were killed by contact with oil, meristematic tissues did not always die, and therefore regrowth of new foliage could occur.

Baker (1971b, 1973) also investigated the effects of successive oilings of salt-marsh vegetation. This was done to model the effects of frequent exposures in chronically polluted situations, for example, near a refinery or a marine oil terminal. At the higher frequencies of oiling, all of the important graminoid species suffered relatively large, initial damage (Table 6.6). Afterward, most species exhibited progressive recoveries of abundance on the oiled plots. The recoveries were largely by vegetative incursions from outside the experimental plots, rather than by the establishment of seedlings. Of course, there were large differences among species in their relative tolerance to the crude-oil treatments. The most sensitive taxa were the annuals *Suaeda maritima* and *Salicornia* spp., while the most tolerant species was the umbellifer *Oenanthe lachenalii*, which survived 12 oilings in one growing season. Among the dominant graminoids, *Juncus maritimus* and *Puccinellia maritima* were initially most

sensitive and were slow or unable to recover after four post-oiling years. The most tolerant grass was *Agrostis stolonifera*, the growth of which was initially stimulated by as many as 8 oiling treatments, and even by up to 12 oilings after several years of regeneration.

Overall, it appears that some species of the salt marsh are sensitive to oiling, but the ecosystem itself is resilient. The plant community responds to the experimental pollution stress by a change in species composition, but not necessarily by decreases in productivity or biomass.

Effects of Chronic Oil Pollution

The environment near petroleum refineries or tanker terminals can be subject to chronic oil pollution from frequent spills, and from the continuous discharge of contaminated process effluents. Chronic oil pollution of the coastal environment is also associated with cities, where hydrocarbons are often discharged into storm or sanitary sewers. In some cases these can cause longer-term ecological damages.

Dicks (1977) studied changes in a *Spartina anglica*-dominated salt marsh in the vicinity of a large oil refinery at Southampton Water, England. He observed a progressive deterioration of the salt marsh, which began when the refinery began operating in 1951. After 1970, the ecosystem started to recover as a result of a decreased frequency of oiling

Table 6.6 Response of salt-marsh graminoids to as many as 12 treatments of crude oil at 4.5 liters/10 m², followed by 5 years of regeneration[a]

Species	Number of oil treatments	Prespray	Postspray years of recovery			
			1	2	3	4
Spartina anglica	0	400	240	256	336	256
(shoots/m²)	2	352	160	256	240	288
	4	288	80	240	208	272
	8	336	32	112	160	272
	12	288	0	0	48	160
Puccinellia	0	86	85	90	86	78
martima	2	95	78	84	91	74
(% cover)	4	91	71	80	75	65
	8	90	0	0	0	6
	12	92	0	0	0	0
Juncus martimus	0	22	16	20	25	18
(% cover)	2	28	2	4	8	7
	4	21	0	0	2	5
	8	28	0	0	0	0
	12	30	0	0	0	0
Festuca rubra	0	72	30	40	35	41
(% cover)	2	75	30	52	45	40
	4	73	35	40	50	32
	8	68	0	0	5	8
	12	62	0	0	10	8
Agrostis stolonifera	0	10	13	13	15	12
(% cover)	2	16	22	40	30	12
	4	14	18	32	15	8
	8	18	17	43	36	46
	12	16	0	15	35	42

[a]Modifed from Baker (1973).

and an improved chemical quality of the refinery effluents. In 1970, the ecological damage was severe but local and was characterized by death of the dominant *Spartina* grass over an area of about 60 ha, resulting in a devegetated area of bare mud. In the primary succession that began in the denuded area in 1970, the initial colonists were the annuals *Salicornia* spp. and *Suaeda maritima*. These were followed by the perennial dicot herbs *Aster tripolium* and *Halimione portulacoides* and ultimately by *Spartina anglica*.

Baker (1976) reported qualitatively similar changes in a *Spartina anglica* salt marsh subjected to chronic hydrocarbon pollution in the vicinity of a large terminal and refinery complex at Milford Ha-ven in southwestern Wales. However, the damage was restricted to a smaller area. She also described short-term effects of small oil spills on the intertidal biota and observed an increased abundance of the opportunistic green alga *Enteromorpha* in a small area of frequently impacted rocky shore. Overall, Baker concluded that the petroleum industry at Milford Haven had not caused important damage to the salt marsh or the marine biota.

Chronic exposures to diverse arrays of hydrocarbons and other pollutants occur in many situations where effluents from industrial and municipal sources are discharged into natural waters. In some cases, high frequencies of both cancerous and non-cancerous diseases of fish and/or shellfish have been observed in chronically contaminated habitats (e.g., E. R. Brown *et al.*, 1977; Mearns and Sherwood, 1977; Sherwood and Mearns, 1977; Sonstegard, 1977; Malins *et al.*, 1983; Fabacher *et al.*, 1986). Although the precise etiology of these diseases is unknown, it is suspected that they may be somehow caused by chronic pollution and that they could represent an important bellwether of environmental degradation.

For example, Sonstegard (1977) found a large incidence of gonadal tumors (up to 100% among older male fish) in goldfish × carp hybrids (*Carassius auratus × Cyprinus carpio*) collected in 1977 from the River Rouge, a polluted river in Detroit, Michigan. Examination of a large museum sample of hybrids collected in the same location in 1952 indicated that there were no tumors at that time. More generally, however, it has proven difficult to demonstrate consistent linkages of animal diseases with chronic water pollution. Yevich and Barszcz (1977) examined biota that were chronically exposed to hydrocarbons at 16 oil-spill sites on the east coast of the United States. They found cancerous neoplasia in one of the 18 species of invertebrates that they surveyed (the soft-shell clam, *Mya arenaria*), but only in apparent conjunction with two of the 16 spill incidents. R. S. Brown *et al.* (1977) also investigated the frequency of neoplasia in the soft-shell clam in a survey of the New England coast. They found that a high incidence of this cancerous disease was widespread in their study area, but the neoplasia was not obviously linked to pollution (some polluted sites had a large frequency of neoplasia, while other polluted sites did not, and some nonpolluted sites also had animals with neoplasia). Clearly, the role of the environment (including pollution) in the etiology of neoplasia and other diseases of fish and invertebrates is complex and uncertain [see Mix (1986) and GESAMP (1991) for recent reviews].

Oil Spills in the Arctic

In this section, some of the effects of oil spills on terrestrial arctic ecosystems are described. In a general way, this section also serves to illustrate the effects of terrestrial oil spills in other biomes. For reviews of the effects of oil in arctic marine environments, see Percy and Wells (1984) and Engelhardt (1985).

Natural Oil Seeps

Some insights into the effects of petroleum and its residues can be gained from consideration of natural oil seeps on the North Slope of Alaska. Fresh seepages range in quality from a heavy fluid oil to tar. The seepage weathers progressively to a thick tar, and ultimately to a dry, oxidized, crumbly asphalt (McCown *et al.*, 1973a). Associated with the fresh seepage materials is a vigorous bacterial community that is adapted to the oxidation of petroleum hydrocarbons, with the most prominent genera being *Pseudomonas*, *Serratio*, *Vibrio*, *Cytophaga*, and *Flavobacterium* (Agosti and Agosti, 1973). Bacteria are also abundant in shallow tundra ponds that are affected by oil seepage (Barsdate *et al.*, 1973). In both the terrestrial and aquatic environments, these microorganisms are responsible for much of the oxidative weathering of seepage materials. In laboratory studies, these microbes are capable of oxidizing hydrocarbons at temperatures as low as 4°C (Agosti and Agosti, 1973).

Plants of tundra wet meadows can grow in intimate contact with the seepages. Plants suffer foliar damage when they contact a fresh seepage, but they can tolerate weathered seepage without obvious injury. In fact, some species are relatively lush and vigorous when growing in tar, apparently because of the warm microclimate that is associated with the dark surface. This is especially true of the sedge *Carex aquatilis*, the cottongrass *Eriophorum scheuchzeri*, and the grass *Arctagrostis latifolia* (McCown *et al.*, 1973b; Deneke *et al.*, 1975).

Experimental Oil Spills on Land

Since the early 1970s, several research projects have examined the environmental effects of oil and gas development in the Arctic of North America. This research has paralleled the hydrocarbon exploration and resource development activities in the

A 1-year-old spill of crude oil in a freshwater marsh near Norman Wells in the Canadian subarctic. The dominant plant is the sedge *Carex aquatilis*. The foliage of this plant was killed by oiling, but its perennating rhizomes in the sediment were little affected. As a result, after only 1 year there was a vigorous regeneration, which at this location has penetrated a weathered residue of crude oil without suffering much damage (photo: B. Freedman).

north, although these activities have since slowed down due to unfavorable crude-oil pricing.

Most of the research on the potential effects of oil spills on terrestrial vegetation involved the perturbation of permanent plots, which were then monitored over time (e.g., McCown *et al.*, 1973b; Freedman and Hutchinson, 1976; Hutchinson and Freedman, 1978). As in other experimental studies of the effects of oil on vegetation, there was initial contact damage that killed foliage and some exposed woody tissues. However, in many species, not all of the perennating tissues were killed. Vegetative growth that originated from surviving meristematic tissues was the most important mechanism of progressive, postoiling regeneration of plants.

This general pattern of effect can be illustrated by the results of field experiments done in the western Canadian Arctic (Table 6.7). In two study sites in black spruce (*Picea mariana*) boreal forest, the treatment of vegetation with crude oil caused a rapid defoliation of the ground vegetation to 21–37% of the prespray plant cover after 1 month. Subsequently, most of the plants that survived the initial oiling succumbed to normally tolerated winter stress. As a result, plant cover 1 year after the spill was only about 5% of the prespray value. A protracted period of postspray mortality was especially marked in black spruce trees, which dominate that boreal ecosystem. Individual spruce trees took as long as 4 years to die after being oiled near ground level. After the initial period of mortality, a slow recovery of many species began. An exception was black spruce, for which no seedlings were observed during the 5 years of the study.

There were similar patterns of vegetation dam-

A 1-year-old experimental spill of crude oil in dwarf shrub tundra near Tuktoyaktuk in the Canadian low arctic. Initially, the crude oil acted as an herbicide and killed all foliage and active woody stem buds that were directly contacted. However, not all individuals of the dominant shrubs [willow (*Salix glauca*) and birch (*Betula glandulosa*)] were killed. In this photo, regeneration can be seen issuing from lateral stem buds (which were stimulated to break dormancy by the death of the terminal stem bud), or by suckering from the root crown. Other plant species of this ecosystem were less resilient to the effects of the oil spill (photo: B. Freedman).

Table 6.7 Effect of experimental crude oil spills (9 liters/m²) on live vegetation cover of four arctic plant communities: (1) mature *Picea mariana* boreal forest, (2) 40-year-old *P. mariana* boreal forest, (3) cottongrass wet meadow tundra, and (4) dwarf shrub tundra[a]

			Total foliage cover (%) of green plants				
			Postspill				
	Treatment	**Prespill**	**1 month**	**1 year**	**2 years**	**3 years**	**5 years**
Mature forest	Reference	195	198	215	255	210	240
	Oil spill	350	130	18	10	35	20
40-year-old forest	Reference	355	350	360	260	210	235
	Oil spill	420	90	20	20	23	95
Cottongrass tundra	Reference	339	342	284	268	—	—
	Oil spill	358	56	26	34	—	—
Dwarf shrub tundra	Reference	339	342	338	292	—	—
	Oil spill	322	57	55	82	—	—

[a]Modified from Freedman and Hutchinson (1976) and Hutchinson and Freedman (1978).

age in tundra communities that were experimentally treated with crude oil. Both a cottongrass-dominated (*Eriophorum vaginatum*) wet meadow and a mesic, dwarf-shrub (*Salix glauca* and *Betula glandulosa*) community showed large initial defoliations. There was a further winter kill in the wet meadow, and then both vegetation types began to recover from the perturbation. In both of these tundra communities, the dominant plant species were able to recover rather vigorously from the oiling treatment. The regeneration emerged from surviving basal meristematic tissue of the cottongrass and from previously dormant lateral stem buds of the dwarf willow and birch. As a result, the initial community dominants were also prominent in the postdisturbance secondary succession.

Lichens and bryophytes were notably susceptible to the experimental oiling of tundra and boreal forest vegetation. These nonvascular plants suffered almost complete mortality and had little ability to recolonize the oiled plots in the first few postoiling years.

Several studies in Alaska examined the ability of the natural, terrestrial microflora to oxidize petroleum residues. The most frequent genera of oil-degrading bacteria were *Arthrobacter*, *Brevibacterium*, *Pseudomonas*, *Spirillum*, and *Xanthomonas*, while the most frequent microfungi were *Beauveria bassiana*, *Mortierella*, *Penicillium*, *Phoma*, and *Verticillium*. These and many other taxa are more-or-less ubiquitous in northern soils. After oiling they respond variously, depending on their relative competitive abilities in the presence of large quantities of hydrocarbon substrate (Scarborough and Flanagan, 1973; Linkins *et al.*, 1984). Other studies on the North Slope of Alaska showed that experimental oiling elicits vigorous but short-term numerical responses by heterotrophic microbes. For example, Scarborough and Flanagan (1973) found that microfungal propagules were 17 times more abundant in an oiled soil than in a reference soil at Prudhoe Bay, while yeasts were 20 times as abundant. Similarly, Campbell *et al.* (1973) found that the respiratory activity (i.e., efflux of CO_2 from soil) of experimentally oiled soil near Barrow was

about two times larger than that of reference soil, while bacteria were about five times as abundant.

In general, it appears that hydrocarbon-degrading microbes are present widely in natural soils and waters, and they can exhibit strong, numerical and functional responses to the availability of hydrocarbon substrates after spills. The speed with which the microbial community can degrade hydrocarbon residues mostly depends on the availabilities of oxygen and of nutrients such as nitrogen and phosphorus, both of which are presently in small concentrations in petroleum and its refined products and residues. The influences of these limiting factors on the rate of hydrocarbon degradation can be effectively managed by occasionally tilling oiled soils to increase oxygen concentration and by fertilizing to create more favorable ratios of C:N and C:P.

Overview of Arctic Oil Spills

Overall, the potential effects of oil spills from pipelines onto northern terrain appear to be relatively moderate. Damage would be caused to vegetation, but its spatial extent would be relatively restricted. Except for a large spill, the area of land that would be affected is fairly small because of the great absorptive capacity of the terrain and because much of the spilled oil would accumulate in low spots and therefore not spread widely. This is an important difference between terrestrial and aquatic oil spills, because the latter affect much more extensive areas.

Furthermore, many arctic plants have the ability to survive an exposure to oil and to invade habitat damaged by an oil spill. In addition, even at relatively cool ambient temperatures, the microflora is capable of oxidizing spilled hydrocarbons. Along with the evaporation that takes place after a spill, this microbial activity would progressively remove most of the spilled oil from the environment. However, because of the relatively short growing season and cold ambient temperatures during much of the year, this process could take several decades in the Arctic (Atlas, 1985). The postspill recovery of terrestrial ecosystems can be enhanced by fertilization and by artificial revegetation with opportunistic arctic grasses (Linkins *et al.*, 1984). Of course, oil that reaches watercourses would cause damage there, especially in relatively stagnant waters. (Depending on the volume of oil that is spilled, some rivers and streams might be less affected because of rapid dilution and dissipation of the oil.)

Overall, the ecological damage caused by isolated, accidental terrestrial spills of oil in the Arctic may be considered to be environmentally acceptable by resource managers, in view of the economic and strategic importance of hydrocarbon resource development in the north (Hanley *et al.*, 1981; Alexander and Van Cleve, 1983).

Of course, the environmental consequences of northern oil development are much broader than oil spills per se. They include a socioeconomic impact (Freeman, 1985), the various environmental consequences of building roads and pipelines in remote and inhospitable terrain, the effects on wildlife, and many other problems that are not considered here. However, some ecologists have concluded that hydrocarbon-resource development in the North Slope of Alaska has been an ecological success story, even after considering the environmental effects of exploration, oil extraction, the building and operating of an oil pipeline across the state, and the operation of a large tanker terminal at Valdez (Alexander and Van Cleve, 1983; note that their review was written prior to the grounding of the *Exxon Valdez*, and the subsequent oil spill into Prince William Sound).

For example, it has been shown that movements of caribou (*Rangifer tarandus*) and moose (*Alces alces*) have been relatively little affected by the presence of the Trans-Alaska Pipeline and its associated road (Eide *et al.*, 1986). In most years, the Nelchina herd of caribou cross this transportation corridor twice each year during their spring and autumn migrations. This herd increased in abundance from about 14 thousand animals at the beginning of construction of the pipeline in 1977 to 25 thousand in 1983 (Eide *et al.*, 1986). Perhaps in response to this study of caribou, in 1988 the U.S. Republican presidential candidate George Bush

(subsequently elected) was quoted as saying: "The caribou love the pipeline, they rub up against it and have babies, so now there are more caribou in Alaska than you can shake a stick at" (cited in Anonymous, 1988).

In contrast to the terrestrial situation, the potential environmental consequences of oil spills resulting from offshore exploration and production activity in the Arctic are much more problematic, and potentially catastrophic. Some of the more prominent reasons for this conclusion are as follows (after Pimlott *et al.*, 1976; Milne and Herlinveaux, 1977; Ross *et al.*, 1977; Percy and Wells, 1984; Engelhardt, 1985):

1. Because of the severe physical and climatic operating conditions in the Arctic, there are relatively large risks of oil spills caused by equipment failure or by human error.

2. A blowout from a drilling or production platform in the Arctic Ocean could potentially remain uncontained for as long as several years, because ice cover causes great difficulties in quickly drilling a relief well.

3. The cleanup of spilled oil would be extraordinarily difficult in the rigorous Arctic marine environment.

4. Oil trapped under sea ice would not weather appreciably by evaporation and dissolution, and biodegradation would be slow. As a result, the quantity of spilled petroleum would not decrease much, and the oil would retain most of its initial toxicity for a long time.

5. When they return to their northern breeding habitat in the spring, arctic marine seabirds and mammals often congregate in large numbers in ice-free water, such as leads and polynyas. Spilled oil would accumulate in these open-water sites and cause a substantial mortality of these animals.

6. Arctic marine foodwebs are relatively simple, and therefore the elimination or great reduction in abundance of a few species as a result of an oil spill could cause relatively great ecological damage.

7. The successional recovery of ecosystems

devastated by an offshore oil spill in the Arctic would be slow.

To date, there have been no large oil spills caused by offshore drilling activities in the Arctic Ocean of North America. However, it appears likely that when a severe spill incident does occur, it might cause great ecological damage, from which recovery would be slow. (Note that in the sense used here, the southern-Alaskan ecosystems affected by spillage from the *Exxon Valdez* are subarctic in character, rather than arctic.)

7

EUTROPHICATION OF FRESHWATER

7.1

INTRODUCTION

Although there is continuous variation among water bodies in their biological production, they are often grouped into three or more classes. Eutrophic water bodies are characterized by large rates of annual, biological production, as a result of an enriched supply of nutrients. This contrasts with oligotrophic waters, which are relatively unproductive because of restricted availabilities of nutrients, and mesotrophic waters, which are intermediate between these two conditions.

The most conspicuous symptoms of increasing eutrophication of a water body are large increases in primary production and in the standing crop of phytoplankton, which in severe cases is known as an algal "bloom." Usually, there is also a change in algal species composition. In shallow water bodies, there may also be a vigorous growth of vascular plants. These primary responses are generally accompanied by secondary changes at higher trophic levels, in response to greater food availability and other habitat changes, sometimes including the development of anoxia in deeper water.

So-called "cultural eutrophication" is caused by anthropogenic influences, especially activities that increase the inputs of phosphorus and other nutrients into water bodies. Usually the nutrient loading is associated with the dumping of untreated or incompletely treated sewage wastes of human or animal origin or with agricultural practices such as the excessive use of fertilizers. Eutrophication is an important environmental problem, because of the degraded quality of water chemistry and of aquatic communities.

Shallow lakes are relatively vulnerable to the development of an extremely productive or hypertrophic condition. Hypertrophic water bodies can be characterized by noxious blooms of blue-green algae (also known as cyanobacteria), an off flavor of drinking water, the production of toxic substances by algae and other microorganisms, periods of hypolimnetic oxygen depletion causing mortality of fish and other biota, and emissions of noxious gases such as hydrogen sulfide (Vallentyne, 1974; Wetzel, 1975; Barica, 1980; Vollenweider and Kerekes, 1982; Harper, 1992). These conditions are not necessarily restricted to highly productive waters, but they are a common and often intense response to eutrophication, sometimes occurring together as a degraded-water syndrome. These characteristics adversely affect the possibilities for multiple uses of hypertrophic water bodies for such purposes as drinking water, a fishery, recreation, aesthetics, and natural ecological values.

In this chapter, the causes, ecological consequences, and control of the eutrophication of freshwater are examined.

7.2
CAUSES OF EUTROPHICATION

Background

Anthropogenic or cultural eutrophication is most frequently caused by the fertilization of water with nutrients (especially phosphorus) in: (1) sewage that contains phosphorus-rich detergents, human wastes, and/or animal wastes and (2) agricultural runoff contaminated by fertilizers. These anthropogenic influences have affected lakes and other surface waters wherever the human population density is large or agricultural land use is intensive.

The term eutrophication has also been used to describe the slow, natural process by which geologically young and unproductive water bodies (such as lakes that have recently developed through deglaciation) gradually increase in production as nutrients are accumulated over time and as the lake basin becomes shallow due to sedimentation (Hutchinson, 1969; Vallentyne, 1974; Wetzel, 1975; Harper, 1992). In addition, some surface waters are naturally eutrophic, for example, many shallow, prairie lakes that have large rates of recycling of nutrients (this is discussed in the next section).

In contrast, a condition of decreasing production or oligotrophication often accompanies the natural paludification of landscapes in cool, wet climates. The decreased production is caused by the development of ombrotrophic, raised bogs on initially minerotrophic, relatively productive landscapes. Because ombrotrophic systems rely entirely on atmospheric inputs for their supply of nutrients, the process of paludification results in decreased biological production (Hutchinson, 1969). Oligotro-

phication may also be caused by the acidification of surface waters (Chapter 4). However, more to the point of this chapter, the biological production of eutrophic water bodies can often be decreased by abatements of the inputs of phosphorus or by decreasing the rates of internal cycling of this critical nutrient.

The most widely acknowledged cause of cultural eutrophication of freshwaters is excessive nutrient loading, particularly with phosphorus. This process of fertilization stimulates primary production, which results in secondary changes at higher trophic levels. In some situations, changes in trophic structure may be another factor that affects primary production and, especially, algal standing crop. There is experimental evidence that increases in the population of planktivorous fish in small water bodies can cause a decreased abundance of herbivorous zooplankton. Decreased grazing can then result in an increased standing crop of phytoplankton, assuming that the nutrient supply is sufficient to sustain an increased abundance of these autotrophs. It must be stressed, however, that nutrient addition, especially of phosphorus, is the most widely accepted theory for the cause of eutrophication of freshwaters. Environmental factors that influence the trophic status of freshwaters are described in more detail below.

Nutrient Loading

According to the Principle of Limiting Factors, the rates of ecological processes are controlled by the metabolically essential environmental factor that is present in least supply relative to demand. If it is assumed that environmental factors other than nutrients (e.g., temperature, light, moisture, and oxygen supply) are adequate, then primary productivity is limited by whichever nutrient is present in least supply relative to demand.

It is important to distinguish between primary production, which is a longer-term function (often expressed on an annual basis), and productivity, which is a shorter-term, even instantaneous function. As is repeatedly noted in this chapter, the primary production of most freshwaters is limited by the availability of phosphorus. This does not, however, mean that at particular times productivity cannot be limited by the availability of other nutrients, at least on the short term. In general, the continuous or steady-state rate of supply of a limiting nutrient is of particular importance to longer-term, biological production; pulses of availability of nutrients tend only to result in short-term changes in productivity (Wetzel, 1975; Odum, 1983; Hecky and Kilham, 1988).

Of the various nutrients that can potentially affect the rate of primary production in freshwater, phosphorus is the one that is most frequently limiting, especially in the form of ionic orthophosphate (PO_4^{-3}). The typical concentration of phosphorus in freshwater (an index of "supply") is small compared with the concentration of P in plants (an index of "demand"). Moreover, for P, the ratio of supply to demand is much smaller than is observed for other important inorganic nutrients (Table 7.1). Following from Table 7.1, the next-most frequent chemical limiting factor for primary production in freshwater would be inorganic nitrogen (i.e., in the form of ammonium or nitrate).

Large amounts of plant nutrients can be delivered to water bodies by natural processes and by the activities of humans. In the late 1960s, the average rate of anthropogenic supply of P was about 2 kg/person-year in the United States, or about one-

Table 7.1 Demand and supply of selected essential elements in freshwater[a]

Element	Concentration in plants (%)	Concentration in water (%)	Ratio of plants to water (approx.)
C	6.5	0.0012	5,000
Si	1.3	0.00065	2,000
N	0.7	0.000023	30,000
K	0.3	0.00023	1,300
P	0.08	0.000001	80,000

[a]Demand is indexed as the typical concentration in aquatic plants, while supply is average river-water concentration. Modified from Vallentyne (1974).

Table 7.2 Per capita output of N and P in the United States, 1965–1970[a]

Source	Output (kg/person-year) N	P
Sewage		
Physiological waste	4.5	0.6
Detergents	0.0	1.1
Industry	0.5	0.1
Total	5.0	1.8
Delivered to water	4.5	1.6
Agriculture		
Animal wastes	45	6
Fertilizers	20	8
Total	65	14
Delivered to water	8	0.3

[a]From Vallentyne (1974).

sixth the rate of supply of N (Table 7.2). The relatively smaller rates of anthropogenic supply of P compared with N and C are an important reason why P is the primary limiting nutrient in the cultural eutrophication of freshwaters. Moreover, atmospheric inputs of nitrogen and carbon to surface waters are much more substantial than those of phosphorus. Inputs of nitrogen occur through dinitrogen fixation, wet depositions of nitrate and ammonium, and dry depositions of nitrate, ammonium, and NO_x gases, while carbon inputs occur through the solubilization of gaseous carbon dioxide, which then enters the aqueous carbonate cycle (Chapters 2 and 4).

In many areas, the rates of input of anthropogenic phosphorus and nitrogen into surface waters have been substantially reduced by the construction of sewage-treatment facilities, many having a capability of removing nutrients from their effluents. For example, during the 1970s and 1980s, the United States invested more than $75 billion in the construction of municipal sewage-treatment facilities, and the number of people served by facilities with secondary treatment or better increased from 85 million in 1972 to 144 million in 1988 [Council on Environmental Quality (CEQ), 1992a].

This has resulted in substantially smaller inputs of phosphorus to water bodies than the average of 2 kg/person-year cited above for the late 1960s.

Studies that have analyzed data for surface waters in many parts of the world have provided statistical evidence that demonstrates the importance of phosphorus as the most typically limiting nutrient for primary production in freshwaters. The intent of these studies was to identify the nutrient that most strongly correlated with primary production and algal standing crop, thereby indicating a potential controlling mechanism for the rates or amounts of these biological variables (e.g., Sakamoto, 1966; Dillon and Rigler, 1974; Schindler, 1978; Fee, 1979; Smith, 1980; Smith and Shapiro, 1981; Canfield and Bachman, 1981; Vollenweider and Kerekes, 1981, 1982; Hecky and Kilham, 1988). In all cases, the correlations with total-phosphorus concentration are strongest, indicating that it is the most likely limiting nutrient.

For example, Vollenweider and Kerekes (1982) examined a data matrix of the aqueous concentrations of chlorophyll (an index of the standing crop of phytoplankton) and nutrients among a large number of lakes in the northern hemisphere. A log–log plot of the mean-annual total-P concentration (this includes phosphate and organically bound P) and the mean-annual concentration of chlorophyll yielded a highly significant correlation coefficient of 0.88 ($n = 77$ lakes), while the correlation of total P with peak chlorophyll concentration (which oc-curs during the summer "bloom" of phytoplankton) was $r = 0.90$ ($n = 50$ lakes). The log–log correlations with total N were also significant, but much weaker (for mean-annual chlorophyll, $r = 0.61$ among $n = 41$ lakes; for peak chlorophyll, $r = 0.66$ among 40 lakes).

Largely because of these statistical relationships, Vollenweider and Kerekes concluded that phosphorus was the most generally limiting nutrient for primary production in freshwater, and that nitrogen did not play an important role except in situations of exceptional supply of phosphorus. They were so confident in the predictive capability of phosphorus that they used its concentration in a scheme of boundary values for the trophic status of lakes (Table 7.3). They also proposed a probability distribution for these trophic categories around mean-annual concentrations of phosphorus (Fig. 7.1). Although these boundary values and frequency distributions are interesting in a descriptive sense, they may be of little relevance to the management of particular lakes, because primary production varies continuously among surface waters.

Trophic status can, of course, be measured directly via the biological characteristics of the water body, for example, algal standing crop or chlorophyll in Table 7.3 and Fig. 7.1. The importance of the strong statistical relationship between phosphorus and trophic status is that it suggests a mechanism for eutrophication via P fertilization. A corollary of this relationship is that cultural eutrophication

Table 7.3   Proposed boundary values for trophic categories of inland lakes and reservoirs[a]

Trophic category	Mean annual total P (mg/m³)	Mean annual chlorophyll (mg/m³)	Maximum chlorophyll (mg/m³)	Mean annual secchi transparency (m)	Minimum annual transparency (m)
Ultraoligotrophic	<4.0	<1.0	<2.5	>12	>6
Oligotrophic	<10	<2.5	<8	>6	>3
Mesotrophic	10–35	2.5–8	8–25	3–6	1.5–3
Eutrophic	35–100	8–25	25–75	1.5–3	0.7–1.5
Hypertrophic	>100	>25	>75	<1.5	<0.7

[a]Note that the concentration of chlorophyll and secchi transparency are indices of the standing crop of phytoplankton and that maximum values correspond to the summer "bloom" condition. Modified from Vollenweider and Kerekes (1982).

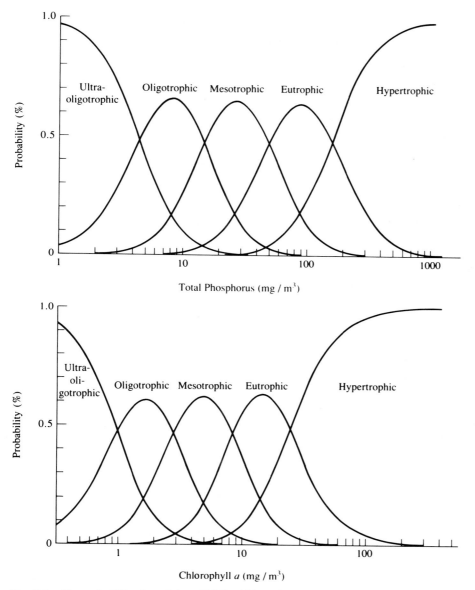

Fig. 7.1 The probability that a lake will fall within a given trophic category at a particular average annual concentration of total P or chlorophyll. Modified from Vollenweider and Kerekes (1982).

should be controllable or reversible by reductions of the anthropogenic inputs of phosphorus.

A number of studies of freshwaters have demonstrated limitations to primary productivity by nutrients other than phosphorus. These cases most frequently involve limitations by inorganic nitrogen. For example, a bioassay-based survey of 238 Scandinavian lakes indicated that 22 of them were primarily limited by nitrogen and 216 by phosphorus (Vollenweider and Kerekes, 1982). It is notable that in cases where surface waters are well supplied with phosphorus, the rates of supply of biologically useful nitrogen and carbon can, within limits, be increased by natural, compensatory changes in the rates of fixation of atmospheric N_2 by blue-green algae and of physical diffusion of atmospheric CO_2 into the water column (Schindler *et al.*, 1972; Schindler, 1977; Smith, 1983; see also the section on whole-lake fertilization experiments later in this chapter).

Rarely, micronutrient deficiencies can limit the productivity of particular species or groups of phytoplankton. Goldman (1961, 1965, 1972) demonstrated a stimulation of phytoplankton productivity in Castle Lake, California, after fertilization with molybdenum. The chrysophycean alga *Dinobryon sertularia* was especially limited by molybdenum. In a laboratory bioassay this alga responded to fertilization with 100 ppb Mo by a 2.6-fold increase in net productivity. In plants, molybdenum is necessary for synthesis of the enzyme nitrate reductase, which reduces absorbed nitrate to ammonium, and as a cofactor in the enzymatic fixation of atmospheric dinitrogen to ammonium. The molybdenum deficiency in Castle Lake was apparently caused by the development of a stand of the alder *Alnus tenuifolia* around the fringe of the water body. This N_2-fixing shrub presumably tied up so much of the watershed's molybdenum that this element became limiting to the phytoplankton.

Other studies have shown that silica can limit the productivity of diatoms, a family of phytoplankton that require this element in relatively large quantities for the construction of their siliceous cell wall, or frustule (Lund, 1950; Lund *et al.*, 1963; Heron, 1961; Schelske and Stoermer, 1972).

Because algal species are variously affected by particular micronutrient limitations, variations of the supply and ratios of particular nutrients can influence the species composition and seasonal dynamics of the phytoplankton community. However, the influence of micronutrients on the longer-term production of the phytoplankton community considered as a whole is much smaller than that observed with a macronutrient such as phosphorus.

Some productive, hypertrophic water bodies can have such large concentrations of phosphorus and nitrogen that it can be difficult to demonstrate relationships between primary productivity and variations of the concentrations of these nutrients. This situation exists in some prairie lakes of central North America, which are shallow and saucer-shaped and located in watersheds with agriculturally fertilized or naturally fertile soils (Hammer, 1969; Haertl, 1976; Fee, 1979; Bierhuizen and Prepas, 1985). Light, temperature, and winds that cause mixing of the water column appear to the important factors in the initiation of conditions appropriate for algal blooms in these lakes (Haertl, 1976). Because the shallow water column is well mixed by summer winds, it does not stratify and the productive surface waters are not isolated from nutrient-rich deeper waters, as occurs in lakes that do stratify during the growing season. Nutrient supplies in these hypertrophic systems are sustained at high rates throughout the year. This allows the development of protracted blooms of blue-green algae, which persist from about mid-June to autumn. Because this situation is so fertile, the eutrophic condition does not usually diminish quickly if anthropogenic nutrient loadings are decreased. This is largely because internal phosphorus loadings occur at such high rates in these lakes (i.e., P release from anaerobic sediment; Jacoby *et al.*, 1982; Ryding, 1985).

Sources and Control of Phosphorus Loading

There are many natural sources of phosphorus inputs to fresh water. These are mostly relatively diffuse, nonpoint sources, such as wet and dry depositions from the atmosphere, leaching and surface runoff from the watershed, and even biological transport (e.g., inputs of marine phosphorus to the upper reaches of water courses via the bodies of anadromous fish, such as salmon, which may die after spawning and release nutrients via decomposition). Typically, the natural inputs of P total considerably less than 1 kg/ha-year, averaged over the entire watershed (Freedman, 1981a).

Anthropogenic inputs of phosphorus can be

much larger than this in watersheds with large populations of humans, domestic animals, and/or livestock. In the United States, the average per-capita discharge of phosphorus to water was about 1.9 kg/person-year between 1965 and 1970 (Table 7.2). About 84% of this total came from municipal sources and the rest from agriculture (the total nitrogen discharge was 12.5 kg/person-year: 36% from municipal sources and 64% from agriculture). For Lake Erie, the most productive of the Great Lakes, the anthropogenic inputs of phosphorus comprised at least 87% of the total P input in 1976 (about 2 kg/ha-year), compared with 11% for oligotrophic Lake Superior (0.2 kg/ha-year) (Gregor and Johnson, 1980).

Because phosphorus is so clearly the limiting nutrient in most freshwater systems, there has been a great effort to reduce its anthropogenic inputs so as to alleviate eutrophication in water bodies that receive waste waters.

Detergents were an important contributor to municipal phosphorus loading in the 1960s and early 1970s. A typical chemical formulation of domestic detergents used at that time included about 50–65% by weight (12–16% as P) of complex phosphates, especially sodium tripolyphosphate. These so-called "builders" were used to chelate ionic Ca, Mg, Fe, and Mn in the wash water, so that the chemical surfactants, the actual cleaning agents in the detergent, were not immobilized by complexation (Duthie, 1972a,b; Perry *et al.*, 1984; Harper, 1992). In the early 1970s, the total production and use of detergents in North America was about 3 million kg/year, and much of this quantity was eventually flushed into freshwaters with municipal sewage effluents (most of the rest was discharged into the oceans by coastal cities). Detergent phosphorus comprised about one-third to one-half of the total P in waste water discharges at that time (Duthie, 1972a; Rohlich and Uttormark, 1972; Vallentyne, 1974; Harper, 1992).

Because detergent use is a discrete activity sector and there were substitutes available for phosphorus in the builder function, detergents were a relatively easy and inexpensive target through which rapid reductions of phosphorus loading to freshwater

could be achieved. As a result, during 1970 and 1973, the maximum permissible concentration of phosphorus in detergents in Canada was reduced from 16% to the present maximum concentration of 2.2%. Larger reductions have been legislated in other places, for example, to less than 0.5% in Dade County, Florida, and in most of the U.S. Great Lakes basin. In Switzerland, no phosphate-containing detergents are allowed (Duthie, 1972b; Prakash, 1976; Ryding and Rast, 1989).

Another strategy for reducing the rate of phosphorus loading to particular water bodies is the diversion of sewage effluents to some other place (see the discussion of Lake Washington, later in this chapter). If diversion is not possible, then sewage can be collected and treated to reduce the concentrations of phosphorus in the effluent. Unfortunately, this latter practice has not been actively pursued in many places, because it can involve expensive investments in facilities, technology, and operating costs. Sewage treatment is often considered to be especially impractical in the cases of agricultural animals and many low-density, residential populations of humans. In these cases, the sewage inputs are nonpoint and/or diffuse and are difficult to control or collect for treatment.

The concentrations of nutrients and other materials in effluent waters from sewage-treatment plants can be greatly reduced by the treatment of sewage wastes (Rohlich and Uttormark, 1972; Anonymous, 1986), as follows:

1. Primary sewage treatment involves the screening or settling of raw sewage to remove larger materials. The effluent may also be disinfected to kill pathogenic microorganisms. This process will remove about 25% of biological oxygen demand (BOD), 60% of suspended solids, and 5–15% of phosphorus.

2. Primary treatment may be followed by a secondary treatment process intended to reduce BOD and create soluble forms of nutrients and colloidal organic material. This is usually done by the use of an activated sludge or a trickling filter, in either of which microorganisms are used to aerobically degrade the organic wastes. A waste

product of this process is a humus-like sewage sludge that is often disposed by application to agricultural land (see also Chapter 3), by solid waste disposal, or by incineration. Secondary treatment can remove about 80% of BOD, 90% of suspended solids, and 30–50% of phosphorus.

3. Sometimes artificial lagoons are constructed for the treatment of sewage wastes. These are biologically productive systems, which rely on microbial processes to decompose organic wastes, and may use uptake by green plants to reduce nutrient concentrations. The efficiencies of lagoons vary greatly depending on climate, water-retention time, and character of the ecosystem, but they may achieve removal efficiencies equivalent to 90% of BOD, 75% of suspended solids, and 30% of phosphorus.

4. Tertiary treatment usually involves the use of a suite of chemical processes designed to remove nutrients from the sewage effluent. Phosphorus removal can be achieved by the flocculation or precipitation of phosphate compounds by aluminum, iron, calcium, or other chemical agents, or by the use of algae or macrophytes to incorporate phosphate into aggrading biomass that is later harvested. The removal efficiency of these processes is typically 90% or better. Tertiary treatment can also include nitrogen removal, using biological uptake and harvesting, denitrification, ammonia stripping, ion exchange, or some other process.

Typical data for U.S. effluent waters of sewage-treatment plants that receive no more than primary or secondary treatment are 5–15 mg/liter of total P, 25–35 mg/liter total N, 150–250 mg/liter of BOD, and 200–300 mg/liter of suspended solids (Rohlich and Uttormark, 1972; Ryding and Rast, 1989). Effluents with this sort of chemical quality are rich growth media for algae and microorganisms, but are suboptimal nutritionally because the P:N ratio (about 1:2) is large relative to the optimal for the growth of freshwater plants (about 1:9; Vallentyne, 1974). Effluents from sewage-treatment facilities with tertiary treatment typically have phosphorus concentrations of 0.2–0.5 mg/liter and can be as small as 0.1 mg/liter.

Effects of Changes in Trophic Structure on Symptoms of Eutrophication

Studies have shown that in some circumstances, changes in trophic structure can affect the standing crop of phytoplankton. Most of these studies have been experimental and have examined the effects of manipulation of trophic structure in mesocosms or enclosures or in small ponds or lakes in which large changes in fish-community structure were used to affect the herbivorous zooplankton community, and thereby the phytoplankton. These studies have demonstrated that large changes in trophic structure can have a strong influence on the symptoms of eutrophication, at least over the shorter term.

In an experiment involving 2-m^3 plastic enclosures in a pond, Shapiro (1980) found that the addition of planktivorous bluegill sunfish (*Lepomis macrochirus*) caused a large increase in the standing crop of algae by reducing the abundance of cladoceran zooplankton that filter-feed on phytoplankton. In two enclosures without fish, the concentration of chlorophyll was <5 μg/liter and secchi depth extended 2.3 m to the bottom of the pond, while in enclosures with four or five fish, chlorophyll was 42–50 μg/liter and secchi depth 0.9–1.0 m. The concentration of total phosphorus in the water was unaffected by the experimental treatment. In another experiment, five enclosures with planktivorous fathead minnows (*Pimephales promelas*) had a relatively small abundance of cladoceran zooplankton and an average late-summer phytoplankton biomass of 60×10^5 μm^3/ml, while five enclosures without fish averaged 24×10^5 μm^3/ml (Lynch and Shapiro, 1981).

In a larger-scale experiment, Spencer and King (1984) examined four eutrophic ponds that varied in trophic structure (Table 7.4). One of the ponds was fishless and had a relatively large population of zooplankton, a small standing crop of phytoplankton because of intense grazing, and since the water was transparent, a great abundance of macrophytes. Another pond had piscivorous largemouth bass (*Micropterus salmoides*), which caused a small density of planktivorous minnows and a consequent large density of zooplankton, a small standing crop of

Table 7.4 Role of predation and grazing in some small (3.3–5.0 ha), experimental, eutrophic ponds in Michigan: Pond 1 had no fish, pond 2 only had piscivorous largemouth bass, ponds 3 and 4 had planktivorous minnows but no bass[a]

	Pond 1	Pond 2	Pond 3	Pond 4
Fish community	No fish	Bass, 3000/ha	Minnows, 25,000/ha Sticklebacks, 1,000/ha	Minnows, 13,000/ha Sticklebacks, 58,000/ha
Zooplankton (number/liter)	24.7	35.0	0.06	0.03
Phytoplankton (mm³/liter)	2.6	7.4	20.1	36.2
Light penetration	17.4	10.9	0.45	0.25
(% of surface at bottom)	196	264	3	75
Macrophytes (g d.w./m²)				

[a]See text for discussion. Modified from Spencer and King (1984).

phytoplankton, and a large standing crop of macrophytes. Two other experimental ponds had a large population of planktivorous fathead minnows and brook stickleback (*Culaea inconstans*), a small density of zooplankton, a large standing crop of phytoplankton, and a small standing crop of macrophytes.

Predation may also play a role in relieving the symptoms of eutrophication in much larger waterbodies. It has been suggested that predation by the burgeoning population of introduced salmon species in Lake Michigan has caused a large decrease in abundance of the planktivorous alewife (*Alosa pseudoharengus*), resulting in an increased abundance of herbivorous zooplankters such as *Daphnia* spp., which then caused a decreased standing crop of phytoplankton and an increased clarity of the water column (Scavia *et al.*, 1986). If the above speculations are correct, then trophic effects may have helped to relieve the symptoms of eutrophication in this large lake, in conjunction with decreases in nutrient loading since the mid-1970s.

Some other researchers have disagreed about this hypothesized, top-down trophic effect in Lake Michigan (Lehman, 1988; Evans, 1992). Recently, an introduced Eurasian cladoceran, the spiny water flea *Bythotrephes cederstroemii*, has reduced the abundance of the larger herbivorous cladocerans, such as *Daphnia* spp. in Lake Michigan, without causing an increased standing crop of phytoplank-

ton (Lehman, 1988). This latter observation suggests that the effect of changes in trophic structure in Lake Michigan may not be as important as was earlier suggested. Studies that have examined the statistical relationships within tropic webs in other lakes have also tended to not show strong or consistent relationships among the standing crops of chlorophyll and the abundance of zooplankton and/or planktivores and/or fish that feed on planktivores, suggesting that nutrients have the major influence on eutrophication (Benndorf, 1987; McQueen, 1990; McQueen *et al.*, 1989, 1990, 1992).

Clearly, in some situations, trophic structure can influence the characteristics of a eutrophic water body. This effect can potentially be manipulated to ameliorate some of the symptoms of eutrophication, particularly the standing crop of phytoplankton (Uhlmann, 1980; Shapiro and Wright, 1984). However, it is likely that this technique would be especially useful in eutrophic ponds or lakes that are relatively small and shallow. Because these smaller water bodies have large internal rates of phosphorus loading from their sediment, they do not respond quickly to decreases in the external supply of nutrients. However, the symptoms of eutrophication in larger lakes are unlikely to be reliably alleviated by top-down manipulations of trophic structure. In such cases, it is more important to focus on the control of phosphorus inputs, which are the actual cause of eutrophication.

7.3
CASE STUDIES OF EUTROPHICATION

The first case of eutrophication to be considered involves several whole-lake experiments that investigated the role of nutrients in limiting the primary production of remote oligotrophic lakes. The next example describes the response of an ecologically simple lake in the Arctic to nutrient loading via sewage. The relatively complex case of Lake Erie, the most productive of the Great Lakes of North America, is then considered. This lake has been affected by nutrient loading, extensive disturbance and conversion of its watershed that caused severe siltation, a vigorous commercial fishery, toxic chemicals, and other anthropogenic stresses. Finally, the recovery of eutrophied waterbodies is discussed, by reference to the case of Lake Washington near Seattle.

Whole-Lake Fertilization Experiments

Much has been learned about the mechanisms and dynamics of the eutrophication of oligotrophic lakes from a series of whole-lake experiments done in the Experimental Lakes Area (ELA) of northwestern Ontario by D. Schindler and his co-workers.

In a relatively longer-term, ongoing experiment, Lake 227 has been fertilized since 1969 at an average rate of 5.7 kg PO_4–P/ha-year (from 1969 to 1974 it also received 7.2 kg NO_3–N/ha-year, but since 1975 it received 3.3 kg NO_3–N/ha-year; Schindler, 1985, 1990). Observations were made of phosphorus dynamics in Lake 227. Typically, 1–5% of the mass of phosphorus in the euphotic zone sedimented to deeper water each day. If the lake was thermally stratified, the sedimented phosphorus became unavailable to sustain primary production in surface waters. (Note that this requires that the hypolimnetic waters are not anoxic, as was the case for this lake.) The residence times of phosphorus (0.6 years) and nitrogen (2.5 years) within the water column of Lake 227 were considerably shorter than

that of water (6.1 years), because of the rapid rates of sedimentation of these nutrients.

Studies using radioactive ^{32}P as a tracer in Lake 227 showed that within minutes of its addition to the surface water, more than 90% of the radiophosphorus was incorporated into the smallest plankton fraction, known as nanoplankton (composed of bacteria and microphytoplankton <10 μm in diameter) (Levine et al., 1986). The ^{32}P tended to remain in the nanoplankton, and less than 5% of the added ^{32}P occurred in larger phytoplankton and zooplankton.

The ELA observations of the dynamics and size fractionation of phosphorus have been generally confirmed in research on P-limited lakes elsewhere (Currie and Kalff, 1984; Currie et al., 1986; Heath, 1986). Collectively, these studies indicate that relatively large phytoplankton are inefficient competitors for phosphate, and they raise questions about the mechanisms by which the larger species of phytoplankton acquire this critical nutrient.

Because phosphate is so rapidly immobilized by bacteria and microphytoplankton, its dissolved concentrations in surface water are small during the growing season. However, it is important to note that the tiny pool of dissolved phosphorus, especially phosphate, is quite labile. For example, the turnover time of phosphorus in the epilimnetic water of several ELA lakes is only 8–14 min and 15–77 min in the hypolimnion (Planas and Hecky, 1984). Because of the dynamic character of the pool of dissolved phosphorus, relatively small concentrations of phosphate can drive large rates of planktonic primary production.

As noted earlier, Lake 227 was fertilized with both phosphate and nitrate. It subsequently responded with a large increase in primary production, but because of the experimental design (which was intended to test carbon limitations to primary production) it was not possible to determine which of these two key nutrients had acted as the primary limiting factor. However, observations from other experimental treatments in ELA lakes clearly indicate that phosphorus is the primary limiting nutrient in these oligotrophic water bodies (Schindler and Fee, 1974; Findlay and Kasian, 1987; Shearer *et*

Aerial view of Lake 226 in the Experimental Lakes Area of northwestern Ontario. This remote lake was partitioned into two experimental basins using a heavy vinyl curtain. The upper basin in the photograph received fertilizer containing phosphorus, nitrogen, and carbon, while the lower basin received nitrogen and carbon. The role of phosphorus as the limiting nutrient for primary productivity was clearly demonstrated, since only the basin that was fertilized with phosphorus became eutrophic and developed a large standing crop of phytoplankton. Because of the bloom of phytoplankton, the water of the eutrophic basin developed a greenish color, represented by a whitish hue in this photograph. (Photo courtesy of D. Schindler.)

al., 1987b; Levine and Schindler, 1989; Schindler, 1990), as follows:

1. Lake 304 was fertilized for 2 years with phosphorus, nitrogen, and carbon, and it became eutrophic. It recovered rapidly and became oligotrophic again when the P fertilization was stopped, even though N and C fertilization was continued.

2. Lake 226 is an hourglass-shaped lake that was partitioned with a heavy vinyl curtain into two experimental basins. One basin was fertilized with C + N + P at a ratio of 10:5:1, and the other with C + N at 10:5. Only the former treatment increased the production of phytoplankton and caused algal blooms. When the nutrient additions were stopped after 5 years, the phytoplankton biomass and species composition returned to the reference condition within only 1 year. Another lake, number 303, exhibited a similarly rapid recovery of reference conditions upon the cessation of fertilization with phosphorus.

3. Lake 302N received an injection of P + C + N directly into its hypolimnion during the summer. Because the lake was thermally stratified at that time, the hypolimnetic nutrients were not available to fertilize phytoplankton in the epilimnetic euphotic zone, and no algal bloom resulted from the fertilization. This example does not directly address the specific role of phosphorus, but it does demonstrate that in terms of nutrient cycling, the hypolimnion and epilimnion are relatively independent compartments while lakes are seasonally stratified.

It is notable that even though N and C were not directly added to the lakes that were only fertilized with P, the rates of supply of these nutrients nevertheless increased markedly, in apparent response to decreases in the ratios of N:P and C:P. In the case of N, the response involved an increase in the rate of fixation of atmospheric dinitrogen by blue-green algae, while C exhibited an enhanced rate of diffusion of atmospheric CO_2 into the water column (Schindler *et al.*, 1972; Schindler, 1977; Smith, 1983). To further explore the importance of the ratios of available nitrogen and phosphorus, the fertilization of

Lake 227 with N was stopped in 1990, while P additions continued (Findlay *et al.*, 1993; Hendzel *et al.*, 1993). In response to decreased ratios of N:P in the euphotic zone, there was a substantial increase in the rate of fixation of atmospheric dinitrogen. In the 2 years prior to cessation of N fertilization, the whole-lake rate of dinitrogen fixation averaged 7.2 kg N/year, while during the next 3 years when only phosphorus was added the rate of fixation averaged 63.9 kg N/year. It appears that the secondary emergence of N and C limitations after fertilization with phosphorus was compensated by natural mechanisms that increased the rates of supply of these nutrients.

Detailed observations of the phytoplankton community's response in the Lake 227 eutrophication experiment were made. The initial responses are briefly summarized below (Schindler *et al.*, 1973).

1. Prior to fertilization, the phytoplankton biomass averaged about 1 g/m³ in winter and 2 g/m³ during the ice-free season, with a maximal summer value of 4–5 g/m³ and a chlorophyll concentration of 1–5 mg/m³. The dominant taxa during the spring and summer were *Dinobryon* spp., *Chrysochromulina parva*, *Mallomonas pumilo*, *Chrysoikos skujai*, *Pseudokepherion* spp., and *Kepherion* spp., and the overall dominant group within the phytoplankton was the Chrysophyta (golden-brown algae).

2. In the first year of fertilization (1969), the peak summer biomass increased more than threefold to 16 g/m³, while chlorophyll increased to 50 mg/m³. The dominant group of phytoplankton in summer changed to the Chlorophyta (green algae).

3. In the second year of fertilization, the peak summer biomass increased further to 35 g/m³, and chlorophyll increased to 90 mg/m³. The peak summer phytoplankton in this and subsequent years of the experiment was dominated by blue-green algae (Cyanobacteria), especially *Oscillatoria geminata*, *O. amphigranulata*, *Pseudoanabaena articulata*, and *Lyngbya lauterbornii*.

4. In the third year of fertilization, the peak biomass was only 12.0 g/m³ and chlorophyll 30

mg/m³. The decreased standing crop of phytoplankton compared with the previous year was attributed to grazing pressures from an exceptionally great abundance of rotifer zooplankton.

5. In the fourth year of fertilization, the peak biomass increased again to 30.0 g/m³ and chlorophyll to 180 mg/m³.

Observations of the responses of other trophic levels to some of the experimental fertilizations in the Experimental Lakes Area were also made. In the hourglass-shaped Lake 226, the abundance and growth rate of whitefish (*Coregonus clupeaformis*) were measured in each of the oligotrophic and eutrophic basins (Mills, 1985; Mills and Chalanchuk, 1987). In the first year of fertilization, the growth rate of fish in the basin receiving phosphorus averaged 21% greater than in the oligotrophic basin. In the second year of fertilization, this increased to 106% greater, then to 141% greater in year three and 171% in year four. Therefore, the increased primary production that was caused by experimental fertilization with phosphorus resulted in increased production and biomass at a higher trophic level.

Eutrophication of a Lake in the Arctic

Meretta and Char Lakes are small water bodies located in the Canadian High Arctic at about 75°N latitude (Kalff and Welch, 1974; Kalff *et al.*, 1975; Schindler *et al.*, 1974). These lakes are relatively simple ecosystems, because of severe climatic limitations on ecological processes. Char Lake is an ultraoligotrophic polar lake, while Meretta Lake receives sewage from a small community and is moderately eutrophic.

The loading rate of phosphorus was about 13 times larger in Meretta Lake than in Char Lake, and that of nitrogen was 19 times larger (Table 7.5). The fertilized lake had a summer chlorophyll concentration that averaged about 12 times larger than the reference lake, while the maximum summer biomass was 40 times greater and annual primary production 2.8 times greater. Both lakes had a similar richness of phytoplankton species, and the phytoplankton biomass of each was dominated by only a few species, but these were generally different in the two water bodies (Table 7.5). Because of the large demand for oxygen to sustain the decomposition of organic materials from sewage and the pri-

Table 7.5   A comparison of two High Arctic lakes at 75°N on Cornwallis Island, Canada: Char Lake is a typical, ultraoligotrophic polar lake, while Meretta Lake receives sewage and is eutrophic[a]

Characteristics	Char Lake	Meretta Lake
Nutrient inputs (kg/ha/year)		
-P	0.16	2.0
-N	0.32	6.0
Phytoplankton		
Summer chlorophyll (mg/m³)	0.13–0.69	3–5
Maximum summer biomass (g/m³)	0.20–0.55	14–16
Primary production (gC/m²/year)	4.1	11.3
Dominant phytoplankton species		
	Gymnodinium helveticum	*Gymnodinium mirable*
	Rhodomonas minuta	*G. veris*
	Chromulina sp.	*G. lacustre*
	Two Chrysophycean spp.	Two Chrysophycean spp.
	Chyrsococcus sp. (episodic)	*Cyclotella glomerata* (episodic)
	Cyclotella comensis (episodic)	

[a]Data are from Kalff and Welch (1974), Kalff *et al.* (1975), and Schindler *et al.* (1974).

Gull Lake, and part of its watershed, in southwestern Michigan. The lake area is 827 ha and, as can be clearly seen in the airphoto, the predominant land-use in its 6960-ha terrestrial watershed is agriculture [mainly cultivation of corn (*Zea mays*)], which results in a large input of fertilizer nutrients to the lake. The lake itself is used for recreation (the many white spots on the water are sailboats and motorboats), and there are more than 500 cottages and other recreational facilities along the shore, which also contribute nutrients via sewage, lawn fertilizers, and other sources. As a result of the large inputs of nutrients, Gull Lake has changed from an initially oligotrophic condition to a meso-trophic/eutrophic state. The consequent biological changes have included (after P. Lane, unpublished observations): (1) a change in phytoplankton from an assemblage dominated by diatoms of the genera *Asterionella*, *Fragilaria*, and *Melosira*, to a community dominated by blue-green algal genera such as *Aphanizomenon*, *Anabaena*, and *Oscillatoria*; (2) the development of nuisance growths in the littoral zone of the macrophytic alga *Cladophora* and various aquatic angiosperms; (3) a change in zoo-plankton from an assemblage dominated by *Daphnia galeata*, *D. retrocurva*, and *Diaptomus ore-gonensis* to one dominated by relatively small species such as *Bosmina longirostris*, *Cyclops bicus-pidatus*, and *Tropocyclops parsinus*; and (4) a change from a relatively species-rich fish community dominated by piscivorous lake trout (*Salvelinus namaycush*) and planktivorous lake herring (*Core-gonus artedii*) to one dominated by an introduced planktivore, the rainbow smelt (*Osmerus mordax*). (Photo courtesy of W. K. Kellogg Biological Station, Michigan State University.)

mary production in Meretta Lake, there was a hypo-limnetic oxygen deficit. The anoxia may have been responsible for the disappearance of the zooplankter *Limnocalanus macrurus*, and it caused recruitment problems in arctic char (*Salvelinus alpinus*), a salmonid fish.

Consideration of the case of these polar lakes shows that even simple aquatic ecosystems that are under severe climatic stress can exhibit a profound eutrophication response to fertilization.

Lake Erie—Effects of Eutrophication in Combination with Other Stressors

Lake Erie is the most productive of the Great Lakes of North America. Lake Erie has been affected by a variety of anthropogenic stressors in addition to nutrient loading. These include conversion of the natural ecosystems of its watershed, other disturbances in its watershed, an intensive fishery, pollution by toxic chemicals, and numerous introductions of exotic species of plants and animals. In spite of the complexity of its stressor regime, Lake Erie is an important case study of eutrophication because of the scale of the changes that have been caused by fertilization of this large lake.

The watershed of Lake Erie is much more agricultural and urban in character than are those of the other Great Lakes (Table 7.6). Consequently, the dominant sources of phosphorus to Lake Erie are agricultural runoff and municipal point sources (Table 7.7). The total inputs of phosphorus to Lake Erie

in 1978 (standardized to watershed area) were about 1.3 times larger than to Lake Ontario and more than five times larger than to the other Great Lakes (Tables 7.6 and 7.7).

The eutrophication of Lake Erie was most intense during the late 1960s and early 1970s. At that time, the concentrations of total P and inorganic N (i.e., nitrate + ammonium) were larger than in any of the other Great Lakes (Table 7.8). The supply of both of these nutrients, but primarily P, are critical in the limitation of primary production in the Great Lakes, along with silica for diatoms under a condition of P fertilization (Thomas *et al.*, 1980; Schelske *et al.*, 1986). During the late 1960s and early 1970s, the eutrophic western basin of Lake Erie (which is relatively shallow and warm and directly receives large inputs of sewage and agricultural runoff) had spring phosphorus concentrations that averaged about eight times larger than those of oligotrophic Lake Superior and concentrations of inorganic nitrogen that were almost three times as large (Table 7.8).

As a consequence of the large loading rates and concentrations of nutrients in Lake Erie, it is more productive and has a larger standing crop of phytoplankton than do the other Great Lakes (Table 7.8). During the late 1960s and early 1970s, the eutrophic western basin of Lake Erie had summer chlorophyll concentrations that averaged about twice as large as in Lake Ontario and 11 times larger than those in oligotrophic Lake Superior. Transparency showed a similar pattern.

Table 7.6 Size and watershed characteristics of the Great Lakes[a]

	Superior	Michigan	Huron	Erie	Ontario
Lake surface area (km²)	82,414	58,068	59,596	25,719	19,477
Watershed area (km²)	138,586	117,408	128,863	78,769	75,272
Land use in watershed (%)					
Forest	95	50	66	17	56
Agriculture	1	23	22	59	32
Urban	<1	4	2	9	4
Brush, wetland, other	4	23	10	15	8
Average P export (kg P/km²-year)	19	40	24	198	86

[a]Modified from Gregor and Johnson (1980).

Table 7.7 Importance of various source categories to the total 1978 phosphorus load to the Great Lakes[a]

Lake	Inputs of P (tonnes/year)			Relative contribution of source categories (% of total)					
	Total[b]	Point sources	Nonpoint sources	Industrial point	Municipal point	Atmospheric inputs	Nonpoint inputs via tributaries		
							Agriculture	Urban	Forest
Superior	4,200	161	4,039	2	1	40	4	4	49
Huron	6,350	155	6,195	1	2	64	23	4	6
Michigan	4,850	1,072	3,778	1	21	28	34	6	10
Erie	17,450	6,010	11,440	2	33	5	40	12	8
Ontario	11,750	2,204	9,546	1	25	41	23	6	5

[a]Data modified from Zar (1980) and Gregor and Johnson (1980).
[b]For comparison, the total phosphorus loads in 1989 (in tonnes) were: Superior, 2320; Huron, 3230; Michigan, 4360; Erie, 8570; Ontario, 6720 (CEQ, 1992a). These compare favorably with the target loadings under the *Great Lakes Water Quality Agreement,* negotiated between the United States and Canada, of (in tonnes/year): Superior, 3400; Huron, 4360; Michigan, 5600; Erie, 11,000; Ontario, 7000 [Council on Environmental Quality (CEQ) 1992a].

As discussed later in this section, the eutrophication of Lake Erie has been alleviated somewhat since the late 1960s and early 1970s, in direct response to decreases in phosphorus loading (Table 7.9).

A consequence of the eutrophic state of Lake Erie was the development of anoxia in its hypolimnion during the summer stratification (Hartman, 1972; Reynoldson and Hamilton, 1993). During the summers of 1953 and 1955 there were extended periods of hot, calm weather, which caused a stable thermal stratification of the water column. Because of large demands for oxygen for the decomposition of organic materials in the hypolimnion, wide-

Table 7.8 Average values for nutrients and other water quality variables in the Great Lakes during the late 1960s and early 1970s[a]

Lake	Spring values			Summer values		
	Total P (μg P/liter)	Inorganic N (μg N/liter)	Reactive Si (mg SiO_2/liter)	Chlorophyll-*a* (μg/liter)	Secchi depth (m)	Trophic status[b]
Superior	4.6	275	2.25	1.0	8.8	0
Huron	5.2	259	1.36	1.2	8.3	0
Michigan	9.0	200	1.50	2.0	6.0	0–M
Erie						
Western	39.5	631	1.32	11.1	1.5	E
Central	21.2	133	0.33	3.9	4.4	E–M
Eastern	23.8	180	0.30	4.3	4.5	E–M
Ontario	24.0	279	0.42	5.3	2.5	E–M

[a]Modified from Thomas *et al.* (1980).
[b]O, oligotrophic; M, mesotrophic; E, eutrophic.

Table 7.9 Recent reductions in algal standing crop and phosphorus concentration in Lake Erie[a]

Time period	Western basin	Central basin	Eastern basin
Total phosphorus in water (μg/liter; spring maximum)			
1970–1975	39 ± 5	19 ± 2	24 ± 6
1978–1980	31 ± 4	14 ± 1	12 ± 2
Chlorophyll a (μg/liter; summer values)			
1970–1975	11.6 ± 2.4	4.8 ± 0.8	4.3 ± 1.0
1979–1980	10.0 ± 2.2	4.1 ± 1.4	2.7 ± 0.9

[a]Modified from Rapport (1983).

spread anoxic conditions developed in the deeper waters of Lake Erie, especially in the western end of the lake.

This deoxygenation had a great effect on benthic animals (Beeton and Edmondson, 1972; Hartman, 1972; Wetzel, 1975). Prior to this time, the benthos was dominated by mayfly larvae, especially *Hexagenia rigida* and *H. limbata*. In 1929 their density was about 397/m², and in 1942–1943 they averaged 422/m². Just prior to the severe stratification in 1953, the density of *Hexagenia* was 300/m², but this collapsed to only 44/m² in September. Density remained small until 1957 when it averaged 37/m², but by 1961 these insects had virtually disappeared, as density was then less than 1/m². The collapse of the population of benthic mayflies was widely reported in the popular press, which interpreted the phenomenon to indicate that Lake Erie was "dead." After this effect on the initially mayfly-dominated benthos, a low-oxygen tolerant benthic community was established. This was dominated by the tubificid worms *Limnodrilus hoffmeisteri* and *L. cervix*, by chironomid midges, and by gastropod and sphaeriid molluscs.

It has been suggested that the western and central basins of Lake Erie have long been subject to periodic incidents of oxygen depletion, even prior to their cultural eutrophication (Charlton, 1979, 1980; Delorme, 1982; Reynoldson and Hamilton, 1993). This effect apparently resulted from the morphometry of the basins, which causes them to be relatively

vulnerable to the formation of a stable thermal stratification. In addition, the deeper waters of western Lake Erie have always been subject to large inputs of oxygen-consuming organic matter from biological production. Nevertheless, it is likely that the frequency and intensity of the events of deoxygenation in Lake Erie were exacerbated by the modern increases in biological production and by the inputs of organic matter from sewage outfalls and other sources (Charlton, 1979, 1980).

A conspicuous change caused by the eutrophication of Lake Erie has been increased production and standing crops of phytoplankton and macrophytes. The standing crop of phytoplankton (indicated by the concentration of chlorophyll-*a* and secchi depth) was previously shown to be larger than that in the other Great Lakes (Table 7.8). Some studies of phytoplankton in Lake Erie have suggested that there may not have been a large change in species composition in recent decades compared with the previous century, in spite of the apparently more severe eutrophication of the lake. Harris and Vollenweider (1982) reviewed historical data on algal species composition, using the records of a water-treatment facility at Buffalo, New York. They also examined siliceous diatom frustules in a sediment core taken from the central basin of Lake Erie. These authors concluded that the lake had been characterized by a meso-to-eutrophic condition since at least 1850 (p. 624): "The phytoplankton communities already present in the last century (from historical data) resemble those of the present time." In the earliest years of their sediment core, the most frequent diatoms were *Melosira distans* and *Stephanodiscus niagarae*, while more recently *Fragilaria capucina* was most frequent. However, large fluctuations in the abundance of these dominants made it impossible to generalize a clear trend in longer-term dominance. In the data from the Buffalo water works, the earliest appearance of a phytoplankton species considered to be a reliable indicator of a eutrophic condition was in the late 1880s, when *Cyclotella bodanica* was present. By 1900 there were other indicators of eutrophication, including the blue-green algae *Aphanizomenon*, *Anabaena*, and *Oscillatoria*.

The historical, mesotrophic to eutrophic condition of Lake Erie was likely partly natural, especially in the western basin, which had long been well known for its lush growth of macrophytes and its large fish populations. In addition, events in the mid-1800s, such as the draining at the southwestern end of the lake of a million-hectare wetland known as the Great Black Swamp, the general conversion of forest in the Lake Erie watershed to agriculture, and sewage discharges from developing cities, must have caused a degree of cultural eutrophication in the 19th century (Regier and Hartman, 1973). Nevertheless, the rate and intensity of eutrophication have undoubtedly increased in modern times, in response to accelerated nutrient loadings caused by the rapidly increasing human and livestock populations in the Lake Erie basin, and the introduction of phosphorus-containing detergents in the late 1940s.

The phytoplankton of Lake Erie differ markedly among its three basins and between its nearshore and offshore water. The eastern and central basins are considered to be mesotrophic to eutrophic, while the western basin is eutrophic (Table 7.8), and in all basins the relatively shallow nearshore waters are more productive than the offshore. In general, the spring algal bloom in relatively eutrophic situations is dominated by the diatom *Melosira*, while the late summer–autumn bloom is dominated by the blue-greens *Anabaena*, *Microcycstis*, and *Aphanizomenon*, along with the diatom *Fragilaria* and the green alga *Pediastrum* (Hartman, 1972, 1973).

A notable development that occurred after about 1940 in rocky, nearshore habitats was a profuse growth of the filamentous green alga, *Cladophora glomerata*. Mats of this plant frequently separated from their rocky substrate and drifted on the surface of the lake, eventually washing ashore as a rotting, smelly nuisance or sinking to the bottom where they created a large oxygen demand and contributed to the anaerobic condition during thermal stratification (Hartman, 1972, 1973). In Lake Erie and in localized eutrophic situations in some of the other Great Lakes, *Cladophora* appears to become established when the total P concentration in the spring exceeds 15 μg/liter (Thomas *et al.*, 1980).

There have also been changes in the standing crop, species composition, and size spectrum of the zooplankton of Lake Erie. In 1939–1940 the zooplankton density of the western basin never exceeded about 7 thousand/m^3 during July and August, compared with 10–22 thousand/m^3 in 1949 and 26–110 thousand/m^3 in 1959 (Brooks, 1969; Hartman, 1972, 1973). Historical collections of zooplankton were dominated by relatively large species, such as *Limnocalanus macrurus* and *Daphnia* spp. By the late 1960s and early 1970s, these taxa had declined or disappeared and were replaced by previously rare species, especially the eutrophic indicator *Diaptomus siciloides*. This change in the zooplankton was probably caused in part by the increasing productivity of Lake Erie. However, at about the same time that Lake Erie was undergoing cultural eutrophication, a vigorous commercial fishery caused an increasing dominance of the fish community by relatively small planktivorous species, which replaced the largely piscivorous taxa that were dominant earlier (described below). This change in predation must also have influenced the size spectrum and species composition of the zooplankton of Lake Erie (Brooks, 1969; Hartman, 1972, 1973).

The changes in zooplankton were most marked in the shallow, western basin of Lake Erie. In the deeper eastern basin, some taxa characteristic of an oligotrophic condition managed to survive. The most notable examples of these oligotrophic taxa are the opossum shrimp *Mysis relicta* and the amphipod *Pontoporeia affinis* (Regier and Hartman, 1973).

Tremendous changes also occurred in the fish community of Lake Erie during the time that it was becoming more eutrophic. It is possible that these effects were partly due to habitat changes caused by eutrophication. However, it is more likely that the changes in the fish community were caused by its intensive exploitation by a vigorous fishery and by other habitat effects, such as the damming of streams required by anadromous fish for spawning and the sedimentation of shallow water habitats by silt eroded from deforested parts of the watershed (Hartman, 1972, 1973; Regier and Hartman, 1973).

In terms of fishery landings, Lake Erie has always been the most productive of the Great Lakes. Over the past 150 years, the average yield of Lake Erie's commercial fishery has exceeded the combined landings of all the other Great Lakes (Regier and Hartman, 1973; Hartman, 1988). The peak years of the commercial fishery in Lake Erie were in 1935 (28.5 million kg) and 1956 (28.3 million kg), while the minima were in 1929 (11.2 million kg) and 1941 (11.6 million kg). Overall, the total catch by the commercial fishery has been remarkably stable, in spite of large changes in fish species, in fishery effort, in the degree of eutrophication and pollution by toxic chemicals, and other habitat effects.

There has been a dramatic change in the species composition of the fish community of Lake Erie during the past 150-or-so years, and this reflects a deterioration of the quality of that natural resource. The historical pattern of development of the Lake Erie fishery is similar to that of most previously unexploited, biological resources. The most desirable and valuable species were exploited first. As these species declined in abundance because of unsustainable fishing pressure, coupled with deterioration of their habitat, the industry diverted to a progression of less desirable species of fish. This pattern of fishery resource development was labeled "fishing up" by Regier and Loftus (1972).

The history of the Lake Erie fishery and its fish species has been described by various researchers (Hartman, 1972, 1973, 1988; Regier and Hartman, 1973; Smith, 1972a,b; Mills *et al.*, 1993; Sonzogni *et al.*, 1983). The initial fishery exploited nearshore, and then offshore, populations of lake whitefish (*Coregonus clupeaformis*) and lake trout (*Salvelinus namaycush*; this species was almost exclusively in the relatively deep eastern basin), along with lake herring (*Leucichthys artedi*). Lake sturgeon (*Acipenser fulvescens*) were also fished at this time, but not for food; they were killed because they damaged nets. The species of the initial resource were rapidly overfished, and they declined to a small abundance or extirpation (Fig. 7.2).

The initially most-desired species were then replaced as targets of the fishery by "second choice" percid species, such as blue pike (*Stizostedion vit-*

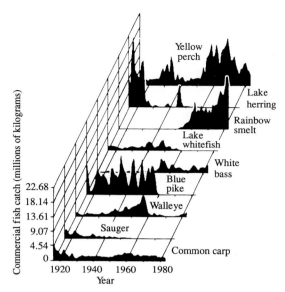

Fig. 7.2   Changes over time in the commercial fishery of Lake Erie. Although the total catch of fish has not changed substantially, there has been an important change in the species composition and size spectrum of the fishery. In the earlier years, the fishery was focused on relative large individuals of desirable species such as lake trout (*Salvelinus namaycush*), lake herring (*Leucichthys artedi*), lake whitefish (*Coregonus clupeaformis*), blue pike (*Stizostedion vitreum glaucum*), walleye (*S. v. vitreum*), and sauger (*S. canadense*). This valuable resource was degraded by overexploitation and habitat changes, including eutrophication. The fishery now relies on small fish, mostly of introduced species such as rainbow smelt (*Osmerus mordax*) and common carp (*Cyprinus carpio*), but also previously undesirable native species, such as yellow perch (*Perca flavescens*). Modified from Hartman (1988), Copyright John Wiley & Sons, Inc.

reum glaucum), walleye (*S. v. vitreum*), sauger (*S. canadense*), and yellow perch (*Perca flavescens*). By the early 1970s, the *Stizostedion* species were extinct or rare. The fishery is now dominated by species that were initially of lowest value, such as yellow perch, and by introduced species such as rainbow smelt (*Osmerus mordax*), freshwater drum (*Aplodinotus grunniens*), and carp (*Cyprinus carpio*).

The ecological condition of Lake Erie has substantially improved since the mid-1970s (Makare-

Even very acidic lakes can exhibit a strong trophic response to nutrient input. Drain Lake in Nova Scotia became very acidic (pH 4.0) when pyritic minerals in its watershed were oxidized after their exposure to the atmosphere by construction activity. However, the lake also receives input of sewage, and the nutrients in this waste caused Drain Lake to become eutrophic. For such an acidic body of water, there are remarkably large standing crops and productivity of phytoplankton and macrophytes, a great abundance of crustacean and insect invertebrates, and a large population of breeding waterfowl and muskrats (*Ondatra zibethicus*). Obvious in the photo are a productive fringing community dominated by the grass *Calamagrostis canadensis* and dense beds of the floating-leaved macrophyte *Nuphar variegatum*, which are visible as a stippling of the water surface (see also Chapter 4). (Photo courtesy of B. Freedman.)

wicz and Bertram, 1991). In large part, this has been accomplished by the investment of more than $7.5 billion since 1972 to improve sewage-treatment capabilities for cities and towns fringing Lake Erie in the United States and Canada, with particular attention paid to the abatement of phosphorus inputs. In 1972, the total municipal discharge of phosphorus to Lake Erie was 15.3 thousand tonnes and up-watershed loading from the Detroit River contributed another 12.0 thousand tonnes. By 1985, these inputs had been reduced to 2.45 thousand and 3.80 thousand tonnes, respectively (International Joint Commission, 1987). In 1990, these inputs were 2.29 thousand and 1.08 thousand tonnes, respectively (Dolan, 1993).

The reductions of phosphorus input have resulted in substantial improvements in water quality and trophic conditions in Lake Erie. Changes in the biomass of phytoplankton have been particularly substantial (Makarewicz and Bertram, 1991; Makarewicz, 1993a). The average standing crop of phytoplankton in Lake Erie has decreased, from 3.4 g/m^3 in 1970 to 1.2 g/m^3 between 1983 and 1985. The western basin still has the largest standing crops of phytoplankton, averaging 1.9 g/m^3 during 1983 to 1987, compared with 1.0 g/m^3 in the central basin, and 0.6 g/m^3 in the relatively deep, eastern basin. Overall, depending on the basin, algal biomass during 1983 to 1987 was 52–89% smaller than in 1970.

There have also been large changes in species composition of the phytoplankton community (Makarewicz and Bertram, 1991; Makarewicz, 1993a). In 1970 *Aphanizomenon flos-aquae*, a nuisance blue-green alga, had a standing crop as large as 2.0 g/m³, but its blooms averaged 0.22 g/m³ during 1983–1985. The biomass of eutrophic-indicator species of diatoms has also decreased greatly. In the early 1970s, these species were widespread throughout Lake Erie, but most species now only occur in the western basin. Even in the western basin, their abundance is greatly reduced, *Stephanodiscus binderanus* by 85% and *Fragilaria capucina* by 94%. At the same time, diatoms that are indicators of mesotrophic or oligotrophic conditions have become more abundant, notably *Asterionella formosa* and *Rhizosolenia eriensis*. The previously eutrophic, western basin of Lake Erie is now considered to be in a mesotrophic condition, while the eastern basin, considered to be meso-eutrophic in 1970, is now oligotrophic.

Complex changes have also occurred in the animal communities of Lake Erie since the 1970s. Among the zooplankton, oligotrophic indicator species of Calanoida have become relatively abundant, while eutrophic indicators of Cyclopoida and Cladocera are less so and are largely restricted to the western basin of the lake (Makarewicz and Bertram, 1991; Makarewicz, 1993b). *Daphnia pulicaria*, a relatively large herbivorous Cladoceran, was first observed in Lake Erie in 1983, and it was the most abundant zooplankter in 1984 at a density of 492/m³ and 15% of zooplankton biomass, but by 1985 it had declined to 44/m³. The decline of *D. pulicaria* may have been caused by an irruption of a large, newly introduced, predacious Cladoceran, the spiny water flea (*Bythothrepes cederstroemii*), which reached a maximum abundance of 72/m³ in 1985.

Since 1972, there has been a large increase in the abundance of piscivorous species of fish in Lake Erie, especially of walleye and introduced Pacific salmon (*Oncorhynchus* spp.) (Makarewicz and Bertram, 1991). This has resulted in substantial decreases of smaller, planktivorous fish such as alewife, smelt, spottail shiner (*Notropis hudsonius*), and emerald shiner (*N. atherinoides*). The de-

creases in these fish species may have allowed secondary increases of larger-bodied zooplankters such as *Daphnia pulicaria*, which then made possible the successful establishment of the predatory *Bythothrepes cederstroemii*.

Another ecologically important introduction to Lake Erie has been the zebra mussel (*Dreissena polymorpha*). This species was first noted in 1988 in Lake St. Clair, up river of Lake Erie, to which it probably immigrated from Europe via ship-ballast water (Hebert *et al.*, 1989; Mackie, 1991; Strayer, 1991). This fecund and vagile mollusk has now spread throughout the Great Lakes drainage and to many lakes beyond, where it can develop extremely large population densities (up to 10–50 thousand/m²) on hard substrates, such as rock, concrete, wood, or metal (Griffiths *et al.*, 1988; Hebert *et al.*, 1991; Mackie, 1991; Neary and Leach, 1992; Ludyanskiy *et al.*, 1993). The zebra mussel can filter a wide range of particle sizes, from 10 to 450 μm. The tremendous filtering capacity of the huge populations of zebra mussels is believed to have contributed to the clarification of water in several of the more productive areas of the Great Lakes, including Lake Erie (Reeders *et al.*, 1989; Reeders and Bij der Vaate, 1990; Wisniewski, 1990; Ludyanskiy *et al.*, 1993). The shoals of zebra mussel have been detrimental to some native species of mollusk and to industry through the clogging of water-intake pipes, but some species of wintering, diving ducks have become considerably more abundant on Lake Erie and elsewhere on the Great Lakes, because they have apparently benefitted from the mussel-food resource (Bij der Vaate, 1991; Wormington and Leach, 1992).

In summary, Lake Erie is an important example of the effects of cultural eutrophication on the ecological structure and function of a large lake. Its case is also representative of the detrimental effects of other anthropogenic stressors, especially overexploitation of a potentially renewable natural resource (the fishery), deterioration of aquatic habitat caused by the clearing of forest for agricultural purposes, pollution by oxygen-demanding sewage and toxic chemicals, and the introductions of exotic species. Lake Erie is also, however, an example of how a degraded ecological condition can be substantially improved by attacking the anthropogenic causes of the changes. In the case of Lake Erie, this has been accomplished largely by reducing inputs of phosphorus with sewage.

Prevention and Mitigation of Cultural Eutrophication

In certain situations, some of the symptoms of eutrophication have been operationally managed by biomanipulations of various sorts. For example, the herbivorous grass carp (*Ctenopharyngodon idella*), a species native to Indonesia, has been introduced to some shallow, warm, eutrophic water bodies that had become overgrown with macrophytes (Cooke *et al.*, 1986). This fish was introduced to Red Hawk Lake, Iowa, a 29-ha impoundment with growths of the macrophytes *Potamogeton*, *Najas*, *Ceratophyllum*, and *Elodea* that were sufficiently dense to be considered a weed problem. In 1973–1974, 780 grass carp were introduced, and within 3 years they had reduced the macrophyte biomass by 91%, allowing an increased use of the water body for sport fishing. In another case, Lake Baldwin in Florida had a nuisance growth of *Hydrilla verticillata*. Grass carp weighing an average of 0.8 kg were released in 1979, when standing crop of the macrophyte was an estimated 780 tonnes. After 3 years, the fish averaged 9.2 kg each, and the *Hydrilla* biomass was reduced to 32 tonnes. Of course, the ecological changes in these water bodies must have been rather complex, and important damages may have been caused to native species. It is not generally desirable to introduce nonnative species to manage the symptoms of aquatic problems such as eutrophication.

It is much more desirable to manage the causes of eutrophication than to treat the symptoms. In almost all situations where freshwaters have become culturally eutrophic, the primary cause has been fertilization with phosphorus. It is therefore reasonable to expect that reductions of the anthropogenic inputs of P and other nutrients would result in a reversal of eutrophication.

The best-known case of the reversal of cultural eutrophication is the recovery of Lake Washington (Edmondson, 1972, 1977; Edmondson and Lehman, 1981; Edmondson and Litt, 1982; Cooke *et al.*, 1986). In the early 1900s, Lake Washington received raw sewage from Seattle, then a town of 50 thousand people. Because of the ensuing eutrophication and public-health problems caused by sewage dumping, the city diverted its effluents to Puget Sound, and Lake Washington quickly recovered its oligotrophic character.

A second and more intense episode of eutrophication of Lake Washington took place between 1941 and 1963, when there were as many as 10 secondary sewage treatment plants with outfalls into the lake, serving a population of more than 64 thousand. In addition, private septic-field systems were used by about 12 thousand people, and there was a storm-flow input that was equivalent to the annual sewage output of another 4.5 thousand people. In 1964, phosphorus loading to Lake Washington was at its maximum, and sewage accounted for 72% of the total P input of about 204 tonnes P/year (the input of N was 1400 tonnes/year).

This large nutrient loading caused summer blooms of phytoplankton. When the nutrient loading was at its maximum in the mid-1960s, about 98% of the phytoplankton volume during the summer bloom was composed of colonial blue-green algae, especially *Oscillatoria rubescens* and *O. agardhii*.

In 1968, a system to divert most of the sewage from Lake Washington was completed, and the total input of P was decreased to 21% of its peak loading in 1964 (the loading of N was decreased to 52% of the 1964 rate). This large decrease in nutrient loading led to: (1) smaller nutrient concentrations in the euphotic zone, from about 64 μg/liter during 1961–1964 to <20 μg/liter after the diversion; (2) decreases in the summer standing crop of algae, with chlorophyll-*a* decreasing from about 36 μg/liter during 1961–1964 to 6 μg/liter during the first 7 years after the diversion and 3 μg/liter in years 8–12; (3) a diminished dominance of *Oscillatoria* in the summer bloom (Fig. 7.3); (4) increases in transparency of the water column, with the average secchi transparency in summer increasing from only 1.0–1.1 m in the mid-1960s to about 3 m during the

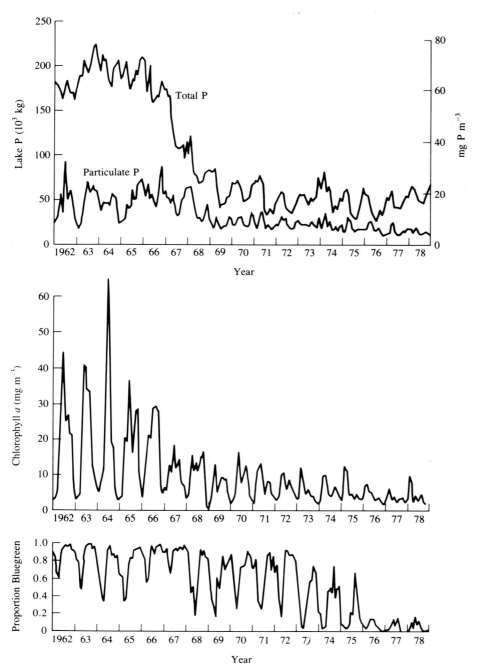

Fig. 7.3 The oligotrophication of Lake Washington as the result of a diversion of sewage effluents that was completed in 1968, from Edmondson and Lehman (1981).

first 7 years after the diversion and 6–8 m in years 8–12; and (5) the reappearance in 1976 of oligotrophic-indicator species of *Daphnia* in the zooplankton.

It is believed that the oligotrophication of Lake Washington was primarily achieved by reductions of the loading of phosphorus and that reductions of nitrogen input had a smaller influence. This conclusion was based in part on the observation that the summer biomass of phytoplankton was most strongly correlated with the maximum concentration of phosphorus in water (which occurs in winter). Over the 12-year period of 1950 to 1972: (1) the correlation of summer chlorophyll concentration × maximum (winter) phosphate concentration yielded a correlation coefficient (r) of 0.91, while the correlation with total P was 0.93, and (2) maximum phytoplankton biomass × maximum PO_4 gave $r = 0.93$, while (3) summer chlorophyll × maximum nitrate concentration only gave $r = 0.60$.

In summary, the cultural eutrophication of Lake Washington was rapidly alleviated by reductions of the loading of phosphorus. From this and other observations of oligotrophication following reductions of nutrient loading, it is clear that in many cases the process of cultural eutrophication of freshwaters can be reduced or avoided by actions that reduce the supply of phosphorus (Bjork, 1977; Ahlgren, 1978; Loehr *et al.*, 1980; Vollenweider and Kerekes, 1982; Ryding and Rast, 1989; Harper, 1992). This can be accomplished by the diversion of sewage effluents (as in Lake Washington), by the replacement of phosphorus-containing detergents with nonfertilizing alternatives, and by the installation of tertiary sewage-treatment processes to remove phosphorus from sewage effluents.

There are, however, exceptions to this generalization about the mitigation of eutrophication through phosphorus control. This tactic may not work with shallow, hypertrophic lakes that develop hypoxic bottom waters and have large quantities of phosphorus in sediment. In these cases, the internal cycling of phosphorus may sustain a eutrophic condition, even if further inputs of phosphorus are substantially reduced. Shallow, hypertrophic lakes might have to be treated with more drastic measures, such as the dredging of sediment, aeration of the hypolimnion, or *in situ*, chemical precipitation of phosphate, for example, by the addition of aluminum (Cooke *et al.*, 1986; Ryding and Rast, 1989; Olem and Flock, 1990; Harper, 1992).

8

PESTICIDES

8.1

INTRODUCTION

Pesticides are substances that are used to protect humans against the insect vectors of disease-causing pathogens, to protect crop plants from competition from abundant but unwanted plants (i.e., "weeds"), and to protect crop plants and livestock from diseases and depredations by fungi, insects, mites, and rodents.

Pesticide use is not just a modern practice (Hayes, 1991). Perhaps the first recorded use of pesticide was around 1550 B.C., when Egyptians used unspecified chemicals to drive fleas from homes. The Greek poet, Homer (ca. 800 B.C.) wrote of the mythological hero, Odysseus, during his wanderings in the Odyssey, burning sulfur "to purge the hall and the house and the court." Around 900 B.C., arsenic was used as an insecticide in China, and by 1870 A.D., many inorganic chemicals were used as pesticides.

In modern times, however, pesticide use has been much more prevalent, and by 1990, about 300 insecticides were in use, as were 290 herbicides, 165 fungicides, and other pesticidal chemicals, with a grand total of more than 3000 formulations (actually, there is an even larger number of separately registered commercial products, many based on the same or similar formulations) (Hayes, 1991).

For diverse, pesticidal purposes, use has been made of a great array of natural biochemicals extracted from plants and of inorganic and synthetic organic chemicals. For the newer, synthetic chemicals, the costs of development and of toxicological and environmental testing can be enormous, about $20–30 million per chemical in 1983 (Hayes, 1991). The profits are also potentially large, however, if a new pesticide is adopted for common usage.

There are many tangible benefits to humanity of the use of pesticides. The most important of these have been: (1) an increased production of food and fiber because of the protection of crop plants from pathogens, competition from weeds, defoliation by insects, and parasitism by nematodes; (2) the prevention of spoilage of harvested, stored foods; and (3) the saving of many millions of human lives by the prevention of certain diseases.

In North America, it has been estimated that pests destroy 37% of the potential production of food and fiber crops (Pimentel *et al.*, 1992). There is no doubt that the agricultural production of some crops that are especially vulnerable to pest problems would not be economically viable without the use of pesticides. Some studies of the use of pesticides in agriculture have estimated that, for every dollar spent on pesticides, there is a $4.00 gain of crop yield or prevented depredations of stored products (Pimentel *et al.*, 1992). (Note that such calculations only involve "conventional" economic costs and benefits; the costs associated with environmental damages caused by pesticide use are not taken into account—see Chapter 12.)

Because of the important, tangible benefits of pesticide use, reliance on these chemicals has increased greatly in recent decades. For example, between 1945 and 1989, the use of pesticides increased by a factor of about 10 in the United States (Pimentel *et al.*, 1992). The practice of chemical pest control is now a firmly entrenched component of the technological culture of modern humanity.

Unfortunately, the considerable benefits of the use of pesticides are partly offset by some important human-health and environmental damages. For example, each year, there is an estimated global total of 1 million pesticide poisonings, resulting in 20 thousand fatalities (Pimentel *et al.*, 1992). Although developed countries account for about 80% of global pesticide use, they only sustain about half of the associated poisonings (Pimentel *et al.*, 1992). The frequency of poisoning incidents is much greater in developing countries because of (1) illiteracy, (2) relatively lax regulations, standards, and (especially) enforcement, and (3) an inadequate availability of protective equipment and clothing and washing facilities for workers.

There have been some rare but spectacular incidents of pesticide-related toxicity to humans. The most widely known case occurred in 1984 at Bhopal, India, where more than 2.8 thousand people were killed and more than 20 thousand seriously injured by a large emission (about 40 tonnes) of poisonous methyl isocyanate vapor, a chemical intermediate in the production of an agricultural insecticide (Rozencranz, 1988).

Another important problem with most pesticide applications is that they kill many organisms that are not the pests that are the target of the treatment. This is an important consideration whenever pesticides with a wide spectrum of toxicity (i.e., not specifically toxic only to the pest) are broadcast sprayed over large areas, such as entire agricultural fields or stands of forest. Many nonpest organisms are exposed to these sorts of treatments, in addition to the intended target of pests. Depending on the pesticide and the susceptibility of the nonpest species, this

exposure can result in a substantial, unintended, but unavoidable nontarget mortality.

For example, in a typical agricultural field or forestry plantation, only a few species of plants would be sufficiently abundant to significantly interfere with the growth of crop plants. These noncrop, competitive plants are the "weeds" that might be targeted by an herbicide application. However, there would be many other species of plants in the same community that do not interfere significantly with the growth of the crop plants. These nontarget plants are also affected by the herbicide application, but to no beneficial purpose in terms of pest management. In fact, the nontarget plants may have beneficial roles to play in their ecosystem, by helping to prevent erosion and nutrient leaching, or by providing food and habitat for animal wildlife. Similar stories could be developed about nontarget arthropods, birds, and other species that are exposed to insecticide during a spray directed against a particular, pest insect. In general, any broadcast spray of a broad-spectrum pesticide causes a substantial mortality of nontarget species.

An ecologically more pervasive problem is a widespread, environmental contamination by persistent pesticides, including the presence of chemical residues in wildlife, in drinking water, and in humans. Ecological damage has included the poisoning of wild life by some pesticides and the disruption of ecological functions such as productivity and nutrient cycling. Many of the worst cases of environmental damage caused by pesticides have been associated with the use of relatively persistent chemicals, such as DDT. Modern pesticide usage mostly involves less persistent chemicals, although these can be very toxic.

A polarization of society has been caused by wide dichotomies in the perceptions of the benefits and risks of pesticide use. This state of affairs can be illustrated by quoting two influential persons (from McEwen and Stephenson, 1979). First, consider a phrase from a speech by Winston Churchill after the successful use of DDT to control a potentially deadly plague of typhus among Allied troops in Naples during the Second World War: "that miraculous DDT powder." In contrast, in her seminal book "Silent Spring," which was the first high-profile chronicle of environmental consequences of the widespread use of persistent pesticides, especially DDT, Rachael Carson (1962) referred to that same chemical as the "elixir of death," because of her concerns about the ecological damage that was associated with its use.

The global usage of pesticides is expanding in scale and intensity. Although we know a great deal about some of the environmental consequences of this technological practice, not all of the potential effects are understood. This chapter describes some of the ecological effects of the use of pesticides. The most prominent uses of pesticides are considered first, and then the characteristics of selected groups of these chemicals are described. This background material is followed by a consideration of some notable examples of the detrimental effects of pesticides on off-site wildlife, especially birds. Then several cases of nontarget damage associated with pesticide use in agriculture are described. Then, to give an appreciation of the ecological effects of large-scale pesticide spraying in complex natural ecosystems, two additional case studies are examined. The first deals with the extensive spraying of conifer forests infested with spruce budworm. The second case examines the use of herbicides to manage plant regeneration on forest clear-cuts.

8.2
CLASSIFICATION OF PESTICIDES BY THEIR USE AND CHEMICAL CHARACTERISTICS

Pesticides comprise a diverse group of chemicals, which can be classified according to: (1) the pest organisms against which they are targeted, (2) the use sector, such as in agriculture, around the home, or in forestry, and (3) their similarities of chemical structure.

Classification of Pesticides by Their Biological Target

Pesticides can be classified on the basis of their intended pest targets. These categories are de-scribed below, and some examples of the more prominent chemicals are given [after Metcalf, 1971; Eto, 1974; McEwen and Stephenson, 1979; Entomological Society of America (ESA), 1980].

1. *Fungicides* are used to protect crop plants and animals from fungal pathogens. Fungicides include: (a) inorganic chemicals, such as elemental sulfur, and copper compounds, such as Bordeaux mixture (more detailed descriptions of the pesticides used as examples in this section are given later in this chapter); (b) organometallic compounds of mercury and tin; (c) chlorophenolics such as tri-, tetra-, and penta-chlorophenol; and (d) synthetic organics such as the dithiocarbamates and captans.

2. *Herbicides* are used to kill weedy plants, so as to release desired crop plants from competition. Herbicides include: (a) amides such as alachlor and metolachlor; (b) triazines such as atrazine, hexazinone, and simazine; (c) thiocarbamates such as butylate; (d) dinitroanilines such as trifuralin; (e) chlorophenoxy acids such as 2,4-D and 2,4,5-T; (f) chloroaliphatics such as dalapon and trichloroacetate; (g) organic phosphorus chemicals such as glyphosate; and (h) inorganics such as various arsenicals, cyanates, and chlorates.

3. *Insecticides* are used to kill insect pests and vectors of deadly human diseases, such as malaria, yellow fever, trypanosomiasis, plague, and typhus. Some prominent insecticides are: (a) inorganic arsenicals and fluorides; (b) "natural" plant-derived chemicals and their synthetic analogs, such as nicotine, pyrethroids, and rotenoids; (c) the DDT group of chlorinated hydrocarbons, including DDT, DDD, and methoxychlor; (d) lindane, an insecticidal isomer of benzene hexachloride; (e) highly chlorinated cyclodienes such as chlordane, heptachlor, mirex, aldrin, and dieldrin; (f) chlorinated terpenes such as toxaphene; (g) organophosphorus esters such as parathion, diazinon, fenitrothion, malathion, and phosphamidon; (h) carbamates such as carbaryl and aminocarb; and (i) microbial agents such as *Bacillus thuringiensis* (or *B.t.*) and nuclear polyhedrosis virus.

4. *Acaricides* are used to kill mites, which are pests in agriculture, and ticks, which can carry encephalitis of humans and domestic animals. Most insecticides are effective against these arachnid arthropods, but there are also some specific acaricides with a relatively specialized use.

5. *Molluscicides* are used against snails and slugs, which can be important pests of citrus groves and vegetable and flower gardens. In addition, aquatic snails are the vector of certain human diseases, most prominently schistosomiasis. Important molluscicides include copper sulfate; the carbamates methiocarb, isolan, and zectran; the organophosphate guthion; and the acetaldehyde polymer, metaldehyde.

6. *Nematicides* are used to kill nematodes, which can be important parasites of the roots of crop plants. Chemical control involves fumigation of the soil with halogenated organics such as ethylene dibromide, dichloropropane, dichloropropene, dibromo-chloropropane, or with certain carbamates or organophosphorus chemicals.

7. *Rodenticides* are used to control rats, mice, gophers, and other rodent pests of human habitation and agriculture. Important rodenticides include plant-derived chemicals such as the alkaloid strychnine and the cardiac glycoside red squill, the hydroxycoumarin compound warfarin, and others.

8. *Avicides* are used to control birds, which are sometimes considered pests in agriculture. The world's most important avian pest is probably an African weaver finch, the quelea (*Quelea quelea*), which is aerially sprayed at communal roosts with organophosphates, with as many as one billion being killed each year (Feare, 1991). In North America, red-winged blackbirds (*Agelaius phoeniceus*) and starlings (*Sturnus vulgaris*) are also killed in the millions, because when they aggregate in large numbers in winter they can be agricultural pests and sources of histoplasmosis, a fungal disease of the lungs that can be spread from avian fecal deposits beneath communal roosts (Feare, 1991).

9. *Antibiotics* are used to control bacterial-caused infections and diseases, which can be important considerations in the health of humans and agricultural animals. Important antibiotics include antimycin, blasticidin, penicillin, and streptomycin.

Production and Use of Pesticides

It has been estimated that the global use of pesticidal chemicals is about 2.5–2.9 billion kg/year (this total includes "conventional" pesticides such as insecticides, herbicides, and fungicides, as well as preservatives, disinfectants, and sulfur compounds) (Briggs, 1992; Pimentel *et al.*, 1992). The value of these chemicals in the early 1990s was about U.S. $20 billion (Pimentel *et al.*, 1992).

Among major countries or regions, the patterns of manufacture and use of pesticides are best documented for the United States, a country that accounts for about one-third of the global use of pesticides (Briggs, 1992).

The total production of pesticides in the United States in 1977 was 631 million kg of active ingredient (a.i.)., while total imports were 23.5 million kg and exports 274 million kg, leaving 381 million kg that were used domestically (ESA, 1980). These numbers were significantly different in 1989, when the production was 591 million kg, imports were 90 million kg, exports 180 million kg, and domestic-U.S. usage 500 million kg (Briggs, 1992). Note that these data refer only to conventional pesticides—another 455 million kg of preservatives and disinfectants were used in 1989 (Briggs, 1992).

Herbicides accounted for about 61% of the total quantity of pesticides used in the United States in 1989, while insecticides comprised 21%, fungicides 10%, and all others 7% (Briggs, 1992). The comparable data for 1979 were herbicides, 46%; insecticides, 43%, and fungicides and others, 11% (ESA, 1980).

The total expenditures for the use of pesticides in the United States in 1989 were $7.6 billion (Briggs, 1992). About 10% of the land area of the continental United States is treated annually with pesticides (ESA, 1980).

Eight of the top 10 pesticides used in the United

States in 1989 were herbicides, and two were insecticides, as follows (Briggs, 1992):

1. alachlor, an amide herbicide, 45 million kg
2. atrazine, a triazine herbicide, 45 million kg
3. 2,4-D, a phenoxy herbicide, 24 million kg
4. butylate, a thiocarbamate herbicide, 20 million kg
5. metolachlor, an amide herbicide, 20 million kg
6. trifluralin, a dinitroaniline herbicide, 14 million kg
7. cynazine, a triazine herbicide, 9.2 million kg
8. malathion, an organophosphate insecticide, 6.9 million kg
9. metribuzin, a triazine herbicide, 6.0 million kg
10. carbaryl, a carbamate insecticide, 5.6 million kg.

Classification of Pesticides by Their Use Category

Pesticides can also be considered from the perspective of use category. The most important of these are human health, agriculture, and forestry, as described below (after McEwen and Stephenson, 1979).

Human Health and Sanitation

In various parts of the world, species of insects and ticks play critical roles as vectors in the transmission of certain disease-causing pathogens of humans, livestock, and wild animals. Worldwide, the most important of these human diseases and their vectors are: (1) malaria, caused by the protozoan *Plasmodium* and spread to humans by a mosquito vector, especially *Anopheles* spp.; (2) yellow fever and related viral diseases such as encephalitis, which are spread by mosquito vectors, especially *Aedes aegypti* and *Culex* spp. in the case of yellow fever; (3) trypanosomiasis or sleeping sickness, caused by the flagellated protozoans *Trypanosoma gambiense* and *T. rhodesiense* and spread by the tsetse fly *Glossina* spp.; (4) plague or black death, caused by the bacterium *Pasteurella pestis* and transmitted to people by the oriental rat flea *Xenopsylla cheops*, a parasite of various species of rat that live in association with humans; and (5) typhoid fever, caused by the bacterium *Rickettsia prowazeki* and transmitted to humans by the body louse *Pediculus humanus*.

The incidences of all of these diseases can be greatly reduced by the judicious use of pesticides to control the abundance of their arthropod vectors (McEwen and Stephenson, 1979). For example, (1) there are many cases where the local abundance of mosquito vectors has been reduced by the application of insecticide to their aquatic breeding habitats or by the application of a persistent insecticide to walls and ceilings, which serve as resting places for these insects; (2) infestations of the human body louse have been treated by dusting people with insecticide; and (3) plague has been controlled by reductions of rat populations, using rodenticides in conjunction with manipulation of ratty habitats through sanitation programs.

The use of insecticides to reduce the abundance of the mosquito vectors of malaria has been especially successful, although in many areas this disease is now reemerging because of the evolution of tolerance by mosquitoes to various insecticides. The use of DDT to control mosquitoes during and just after the Second World War was so successful that hopeful predictions were made of the eradication of this debilitating and often-fatal disease.

Malaria has always been an important disease in the tropics and subtropics. During the 1950s, it was estimated that each year more than 5% of the world's population was infected with malaria. For example, an epidemic in Sri Lanka in 1934–1935 affected half of the population, and 80 thousand people or 1.4% of the population died as a result (Hayes, 1991). In the early 1960s, an estimated 2–5 million children died of malaria each year in Africa (Hayes, 1991).

Large reductions in the incidence of malaria were rapidly achieved through the use of insecticides, by drastically reducing the abundance of anopheline mosquitoes. For example, between 1933 and 1935, India annually recorded about 100 million cases of malaria and 0.75 million deaths. However, there were only 0.15 million cases and 1500 deaths in 1966, largely because of the vigorous use of DDT (McEwen and Stephenson, 1979; Hayes, 1991). [In 1985, India utilized about 40–50% of its public health budget for anti-malarial insecticide spraying, and about 15×10^6 kg of insecticide was used for that purpose (Hubendick, 1987)].

Similarly, in Sri Lanka there were 2.9 million cases of malaria in 1934 and 2.8 million in 1946, but mosquito suppression based on the use of DDT reduced this incidence to only 17 cases in 1963, and mosquito control was discontinued in 1964 (Hayes, 1991). However, malaria has resurged in many tropical countries, largely because of the development of insecticide resistance in mosquitoes. In 1970, 31% of the population of Sri Lanka tested positive in an immunomalarial assay (Hayes, 1991). This indicated an extensive exposure to the malarial parasite by mosquito bites, although the disease itself may have been controlled by the use of prophylactic or other drugs.

In 1962, during a vigorous campaign to control malaria in the tropics, about 59.1×10^6 kg of DDT was used, as was 3.6×10^6 kg of dieldrin and 0.45×10^6 kg of lindane. Most of the insecticide was sprayed inside homes and on other likely resting sites for mosquitoes, rather than in their breeding habitat (McEwen and Stephenson, 1979).

Agriculture

Modern, technological agriculture uses pesticides for the control of weeds, arthropods, and plant diseases. Worldwide, pests and diseases cause losses that are equivalent to about 24% of the potential crop of wheat (*Triticum aestivum*), 46% of rice (*Oryza sativa*), 35% of corn or maize (*Zea mays*), 55% of sugar cane (*Saccharum officinale*), 37% of grapes (*Vitis vinifera*), and 28% of vegetables (McEwen and Stephenson, 1979). Typically in North America, pests destroy 37% of the potential production of food and fiber crops (Pimentel *et al.*, 1992).

There is no doubt that intensifications of management practices in agriculture have resulted in substantial increases in productivity. An increased reliance on pesticides has been an important contributor to these increases in yield, along with mechanization, improved crop varieties grown in monoculture, and the use of fertilizers. For example, in the United States, the typical yield of corn was 1400 kg/ha-year in 1933, compared with 4240 kg/ha-yr in 1963 and 5080–7100 kg/ha-year during 1978–1984. In Mexico, the yield of wheat increased from 750 kg/ha-year in 1945 to 2600 kg/ha-year in 1964. In Japan, the yield of rice has increased from historical levels of 1800 kg/ha-year, to 4000 kg/ha-year in 1963, while in the United States, rice yielded 4940–5520 kg/ha-year during 1978–1984 (Hayes, 1991).

Most intensively managed systems in agriculture have an intrinsic reliance on the use of pesticides. For example, high-yield crop species may be intolerant of competition from weeds, so herbicides must be used if those crops are to be successfully grown. Similarly, the high-yield varieties may be vulnerable to infestations by certain insect pests, or to diseases, so that pesticides must be used to manage those problems. Furthermore, extensive use of monocultural agroecosystems without crop rotation may result in reduced populations of natural predators or parasites, causing an exacerbation of existing pest problems or new ones to emerge. Because intensively managed agricultural systems have an inherent reliance on the use of pesticides, often causing new and unanticipated pest problems to emerge as ecological "surprises," these systems have been described as pesticidal "treadmills."

Nevertheless, because of the perceived benefits of pesticide use in agriculture, their use has increased tremendously, by about 10 times in the United States between 1945 and 1989 (Pimentel *et al.*, 1992). Interestingly, over that same time span the total crop losses (to insects only) increased from an estimated 7% during 1941–1951 to 13% during 1951–1960 and 13% in 1974 (Hayes, 1991). The reasons for these paradoxically opposite trends are not totally clear, but they are partly related to: (1) recent increases in the tolerance of some important pests to certain, previously effective pesticides and

(2) the emergence of new pests, because of accidental introductions, because pesticide use and other management practices have caused changes in the trophic structure of agricultural ecosystems, especially upsets of predator–prey relationships, and because of the cultivation of crop species and varieties that are relatively vulnerable to pests.

In agriculture, arthropod pests compete with humans for a common food resource. From the human perspective, that competition is direct when insects cause large reductions in agricultural yield by the defoliation of crops in fields or when stored foods are attacked. In a few exceptional cases, defoliation can cause total losses of the economically harvestable agricultural yield, as in the case of acute infestations by "locusts" (in North America, irruptive locusts are composed of four species of spurthroated grasshopper, *Melanoplus* spp.; in Eurasia, various genera are important, including the desert locust, *Schistocerca gregaria*).

More usually, however, insect defoliation causes reductions in the yield of crops. For example, in the United States between 1963 and 1973, defoliation by the European corn borer (*Ostrinia nubilalis*) resulted in an average loss of yield of maize of 9% (range of 3 to 17%; McEwen and Stephenson, 1979).

In many cases, pests may only cause trivial damage in terms of the quantity of biomass that they consume, but by causing cosmetic damage they can still greatly reduce the economic value of the crop. For example, in unsprayed orchards of apples (*Malus pumila*) the frequency of infestation of fruit by codling moth (*Carpocapsa pomonella*) can be 20–90% (McEwen and Stephenson, 1979). This can render sale of the crop economically impractical. Although codling-moth larvae do not consume much of the fruit that they infest, they cause great psychological damage to any consumer who finds a half worm in a just-bitten apple. Even seemingly "minor" deficiencies in the appearance of fruits or vegetables, for example, scabbing of apples or russeting of oranges, may be interpreted as making the product less desirable to consumers in some markets. This can be an important economic consideration for farmers and the food industry. However, from the ecological and environmental perspectives, the use of pesticides to deal with damage that is trivial in terms of crop productivity or nutritional quality can only be viewed as an unwarranted practice. Pesticides, like many prescription drugs, can be overprescribed in modern societies.

In agriculture, a weed can be considered to be any plant that significantly interferes with the productivity of a crop plant (even though in other contexts weed species may have ecological and economic values that are positive). Weeds exert this effect by competing with the crop for light, water, and nutrients. Studies in Illinois demonstrated an average reduction of yield of corn of 81% in unweeded plots, while an average 51% reduction was reported in Minnesota. Competition from weeds can reduce the yield of small grains such as wheat and barley by 25–50% (McEwen and Stephenson, 1979). To reduce the influence of weeds on agricultural productivity, fields may be sprayed with a herbicide that is toxic to the weeds, but to which the crop plant is insensitive.

There are several herbicides that are toxic to dicotyledonous weeds, but not to grasses. As a result, herbicides are intensively used in grain crops of the Gramineae. In North America, more than 83% of the acreage of maize is treated with herbicides (McEwen and Stephenson, 1979). This practice is especially important in maize agriculture because of the widespread use of no-tillage cultivation, a system that reduces erosion and saves fuel. Since an important purpose of plowing is to reduce the abundance of weeds, the no-tillage system would be impracticable if it were not accompanied by the use of herbicides. Prominent herbicides used in corn cultivation in the United States include atrazine, propachlor, alachlor, 2,4-D, and butylate. In aggregate, the use of these amounted to about 39×10^6 kg/year in the early 1970s (NAS, 1975c). Most of the area planted with other agricultural grass crops, such as wheat, rice, and barley, is also treated with herbicide. About 50–80% of the small-grain area in North America was treated with the phenoxy herbicides 2,4-D or MCPA in the mid-1970s. In 1975, about 28×10^6 kg of these herbicides was used for that purpose (McEwen and Stephenson, 1979). More recently, the use of these chemicals has been substantially supplanted by other herbicides.

In the mid-1980s, amide herbicides accounted for 30% of the quantity of herbicides used in the United States, while triazines comprised 22%, carbamates 13%, *N*-anilines 11%, and phenoxys only 5% (Stevens and Sumner, 1991).

There are many diseases of agricultural plants that can be managed by the use of pesticides. In some cases, insecticides can be used to control the insect vectors of viral, bacterial, and fungal diseases of plants (Borror *et al.*, 1976). More importantly, fungicides are used to control diseases caused by fungal pathogens that could otherwise reduce or totally destroy the production of crops. Examples of important fungal diseases of crop plants include (after Agrios, 1969): (1) late blight of potato (*Phytophthora infestans* on *Solanum tuberosum*), (2) apple scab (*Venturia inequalis* on *Malus pumila*), (3) powdery mildew of peach and other rosaceous crops (*Sphaerotheca pannosa* on *Prunus persica*), and (4) Pythium seedrot, damping-off, and root rot of many agricultural species (caused by *Pythium* spp.). These diseases can be controlled by the use of fungicides, usually in conjunction with the cultivation of resistant plant varieties and with particular cultural practices that help to reduce the incidence and severity of the disease. Another important use of fungicides in agriculture is to help prevent the spoilage of stored crops and the development of mycotoxins that would make the crop unfit for consumption. For example, toxic aflatoxins can be synthesized by certain strains of *Aspergillus flavus* in stored grains, legumes, and nuts. The aflatoxins can make the produce poisonous, in spite of minimal change in taste or overall appearance.

Forestry

In forestry, the most important uses of pesticides are for the control of defoliation by epidemic insects and the reduction of weeds in plantations. If left uncontrolled, these pest problems could result in large decreases in the yield of merchantable timber. In the case of some insect infestations, particularly eastern spruce budworm (*Choristoneura fumiferana*), repeated defoliations can cause the death of large areas of forest (see the case study of spruce budworm later in this chapter).

The quantities of pesticide that are used in forestry are much smaller than those used in agriculture. In spite of this fact, in some regions pesticide use in forestry has attracted a disproportionate amount of high-profile attention from environmental advocates and the media. Some of the reasons for this phenomenon include the following (Freedman, 1991):

1. In forestry, large tracts of natural and semi-natural ecosystems are sprayed with pesticides. The broader public has a higher regard for these types of habitats than for the intensely and frequently disturbed, technological agroecosystems that are treated with pesticides in agriculture.

2. In intensively managed agroecosystems, populations of wildlife are often relatively sparse and may not be directly affected by pesticide spraying. In contrast, silvicultural habitats support a great richness of native plant and animal species. Therefore, there are legitimately disproportionate concerns about toxicity caused to wildlife by the use of pesticides in forestry, compared with agriculture. (There are, however, some important exceptions to the above generalization about toxic exposures to wildlife in agriculture. Several of these are discussed later in this chapter, in the case studies dealing with carbofuran and DDT.).

3. In forestry, spraying is often conducted using aircraft flying relatively high, above the height of the tallest trees. Such applications greatly increase the risks of drift of the spray, and therefore there is a much greater risk of off-target ecological exposures and of "involuntary" exposures of humans.

4. Some earlier forestry programs caused widely publicized toxicity to high-profile, nontarget wildlife, particularly birds and sportfish.

5. In the case of herbicide use in forestry, there are popular concerns about TCDD, a very toxic dioxin isomer that is a trace contaminant of 2,4,5-T. Until 1979 in the United States and the mid-1980s in Canada, 2,4,5-T was the most commonly used silvicultural herbicide, usually in a

1:1 mixture with 2,4-D. These chemicals had also been used at about 10 times the normal forestry spray rates to defoliate large areas of Vietnam during the Second Indochina War (see Chapter 11). Some of the 2,4,5-T used in Vietnam was badly contaminated (to 45 μg/g) with TCDD, but using postwar manufacturing technology, the contamination of 2,4,5-T with TCDD was kept to <0.1 μg/g for silvicultural usage, in accordance with regulations of the U.S. Environmental Protection Agency. Nevertheless, severe damage had been rendered to the collective public and political psyches, and the continued use of 2,4,5-T contaminated to any degree with TCDD became unacceptable, so this herbicide is no longer used in forestry. Interestingly, in 1989 it was discovered that another commonly utilized silvicultural herbicide, glyphosate, was used in a formulation containing a surfactant that was contaminated to about 2000 mg/liter by 1,4-dioxane. Because the word "dioxane" is visually and phonetically similar to the chemically different "dioxins," the family to which TCDD belongs, there were some misdirected concerns about the presence of dioxane in glyphosate formulations.

6. There is a widespread perception among much of the public, media, and politicians that forest management in general may be becoming dangerously intensive. In North America and elsewhere, concerns are frequently expressed about: (1) the conversion of natural forests into plantation forests, (2) the favoring of conifer growth over that of angiosperm hardwoods, (3) changes in landscape character in terms of vegetation and wildlife habitat, (4) the widespread use of intensive harvesting methods, especially clearcutting, (5) incompatibility of intensive forest management with other potential options in multiple land-use, and (6) broadcast applications of insecticides and herbicides. Although pesticide uses are only components of the intensive harvesting and management systems that are increasingly being used in forestry, they have become somewhat of a focus of opposition. Intensive management systems can clearly be beneficial in terms of short-term, conventional economics and productivity of the forest resource, but industrial forest managers have not yet convincingly justified these practices and their associated ecological detriments to the general public (see also Chapter 9).

7. The government agencies and multinational forestry companies that are most involved in silvicultural pesticide use are large, immobile, and impersonal targets for environmental activists. In contrast, farmers are a pesticide-using group for which the public is generally more empathetic and supportive.

In North American forestry, by far the largest insecticide spray programs are carried out against several species of budworm (*Choristoneura* spp., and *Acleris variana*), which defoliate several conifer species (see the case study on eastern spruce budworm, later in this chapter). In the United States between 1945 and 1974, about 6.4 million hectares of budworm-infested forest were sprayed with insecticide (including repeatedly sprayed areas), accounting for 52% of all forest insecticide spraying (NAS, 1975a). The largest spray programs in North America have been in New Brunswick, where between 1952 and 1992 an aggregate area of about 48.9 million ha was sprayed (see Table 8.7).

DDT was the most important chemical used during the earlier forest insecticide spraying in North America, mostly against spruce budworm. In the United States, DDT was used on 84% of the total area of forest sprayed with insecticides between 1952 and 1974. The peak year of DDT use was 1957, when 2.3 million kg was used; however, its use for this purpose was prohibited in 1972. Also important in the United States have been carbaryl (used over 7.5% of the sprayed area), mexacarbate (5.0%), and malathion (1.2%) (NAS, 1975a). DDT was also used in forest spray programs in Canada against eastern spruce budworm, along with fenitrothion, phosphamidon, and aminocarb, and since the mid-1980s, *Bacillus thuringienis* (see the case study under "Spraying Forests Infested with Eastern Spruce Budworm").

The secondmost important forest insecticide program in the United States has been targeted against the gypsy moth (*Lymantria dispar*), a defoliator of many tree species. Between 1945 and 1974, spraying against this pest totalled about 5.1 million ha and accounted for 41% of total insecticide spraying in forestry (NAS, 1975a). The gypsy moth was accidentally introduced to North America in 1869 in Medford, Massachusetts, by Leopold Trouvelot, who had hoped to develop a commercial silkworm industry by crossing the gypsy moth with the silkworm moth (*Bombyx mori*). Instead, North America's secondmost important forest defoliator was released one day, when a cage fell over and some moths escaped. The introduced moths remained at a low level of abundance, until 1889 when there was a local population explosion, and the species has since been expanding its area of infestation. The continental area of infestation is now more than 1 million ha, almost entirely in the eastern United States and southeastern Canada (Gerardi and Grimm, 1979; Doane and McManus, 1981). The gypsy moth attacks many species of tree. Although oaks (*Quercus* spp.) on dry sites can tolerate many defoliations, a single defoliation is sufficient to kill sensitive conifers, such as eastern hemlock (*Tsuga canadensis*), while many angiosperm trees will die after three or more defoliations. Some stands ultimately suffer 80–100% overstory mortality, but often the damage is less (Benoit and Lachance, 1990; Twery, 1990).

Other insect infestations that have been treated by the broadcast spraying of forests include those of Douglas-fir tussock moth (*Hemerocampa pseudotsugata*), hemlock looper (*Lambdina fiscellaria*), tent caterpillars (*Malacosoma disstria* and *M. americanum*), and bark beetles (especially *Ambrosia* spp.) (Johnson and Lawrence, 1977). In addition, individual-tree spraying has been used to control infestations of urban elm trees (mainly *Ulmus americana*) by Dutch elm disease, caused by an introduced fungal pathogen (*Ceratocystis ulmi*) that is spread by two species of elm bark beetle (the introduced *Scolytus multistriatus* and the native *Hylurgopinus rufipes*) (Agrios, 1969). The practice of spraying urban elm trees has largely been discontinued, mostly because of controversy caused by the broadcast spraying of insecticides close to homes. Today, elm trees can be treated using root injections

of a systemic fungicide, but because this is a relatively expensive treatment it is only administered to particularly valuable, urban trees.

A widespread use of herbicides in forestry began in the 1950s, and it is progressively becoming a routine silvicultural practice in intensively managed systems. Most herbicide use in forestry is intended to release desired conifer species from the effects of competition with herbaceous and shrub-sized angiosperm plants. The topic of silvicultural herbicide use is discussed in detail later in this chapter.

Chemical Classification of Pesticides

Pesticides can be classified according to their similarity of chemical structure, as outlined in the following section (after Metcalf, 1971; Eto, 1974; McEwen and Stephenson, 1979; ESA, 1980; Newton and Knight, 1981; Briggs, 1992) (see Table 8.1 for chemical names). Note that this brief treatment only gives a sampling of the great variety of pesticides that is currently available for use. Note also that pesticides based on microbial formulations, such as the bacterium *Bacillus thuringiensis* (abbreviated as *B.t.*) and certain viruses such as nuclear polyhedrosis virus, are not discussed in this section (see the case study on spruce budworm).

Inorganic Pesticides

This group of pesticides is composed of compounds of various toxic elements, predominantly arsenic, copper, lead, and mercury. Compounds of these elements do not degrade in the conventional sense, and when used as a pesticide they can have a long persistence as toxic substances. However, a degree of environmental detoxification may take place by changes in molecular structure caused by organic and inorganic chemical reactions. In addition, the persistence of inorganic chemicals in soil is affected by dissipative processes that physically remove residues, for example, leaching, or soil erosion by wind and water. Chapter 3 gave some examples of environmental contamination and damage to wildlife caused by the use of inorganic pesticides. Some

Table 8.1 Common and chemical names of selected pesticides mentioned in the text[a]

Class	Common name	Chemical name
1. Inorganic pesticides		
a. Bordeaux mixture	Tetracupric sulfate + pentacupric sulfate	$4CuO \cdot SO_3 + 5CuO \cdot SO_3$
b. Arsenicals	Arsenic trioxide	As_2O_3
	Sodium arsenite	$NaAsO_2$ and Na_2HAsO_3
	Calcium arsenate	$Ca_3(AsO_4)_2$
	Paris green	$Cu(C_2H_3O_2)_2 \cdot 3Cu(AsO_2)_2$
	Lead arsenate	$PbHAsO_4$
2. Organic pesticides		
a. Natural organics	Nicotine	l-l-Methyl-2(3′-pyridyl)-pyrrolidine
	Nicotine sulfate	$(C_{10}H_{14}N_2)_2 \cdot H_2SO_4$
	Pyrethroids	Pyrethrins I,II, cinerins I,II, jasmoline II
	Rotenone	1,2,12,12a-Tetrahydro-2-isopropenyl-8,9-dmethoxy-[1]benzo-pyrano[3,4-b]furo-[2,3-h][1]-benzopyran-6(6aH)-one
	Red squill	Various cardiac glycosides
	Strichnine	Complex alkaloid
b. Organomercurials	Phenyl Hg acetate	$C_6H_5HgOCOCH_3$
	Methylmercury	CH_3Hg
	Methoxyethyl Hg chloride	$CH_3OCH_2CH_2HgCl$
c. Phenols		2,4,5-Trichlorophenol
		Pentachlorophenol
d. Chlorinated hydrocarbons		
1. DDT and relatives	DDT	2,2-Bis-(p-chorophenyl)-1,1,1-trichloroethane
	DDD or TDE	2,2-Bis-(p-chorophenyl)-1,1-dichloroethane
	Methoxychlor	2,2-Bis-(p-methoxphenyl)-1,1,1-trichloroethane
	DDE	2,2-Bis-(p-chorophenyl)-1,1-dichloroethylene
2. Lindane	Lindane	1,2,3,4,5,6-Hexachlorocyclohexane
3. Cyclodienes	Chlordane	2,3,4,5,6,7,8,8-octochloro-2,3,3a,4,7,7a-hexahydro-4,7-methanodiene
	Heptachlor	1,4,5,6,7,8,8-Heptachloro-3a,4,7,7a-tetrahydro-4,7-methanoindene
	Aldrin	1,2,3,4,10,10-Hexachloro-1,4,4a,5,8,8a-hexahydro-1,4-$endo,exo$-5,8-dimethanonaphthalene
	Dieldrin	1,2,3,4,10,10-Hexachloro-6,7-epoxy-1,4,4a,5,6,7,8,8a-octahydro-1,4-$endo,exo$-5,8-dimethanonaphthalene
4. Chlorophenoxy acids	2,4-D	2,4-Dichlorophenoxyacetic acid
	2,4,5-T	2,4,5-Trichlorophenoxyacetic acid
	MCPA	2-Methyl-4-chlorophenoxyacetic acid
	Silvex	2-(2,4,5-Trichlorophenoxy)-propionic acid
e. Organophosphorus compounds	Parathion	O,O-Diethyl O-p-nitrophenylphosphorothionate
	Methyl parathion	O,O-Dimethyl O-p-nitrophenylphosphorothionate
	Fenitrothion	O,O-Dimethyl O-3-methyl-4-nitrophenylphosphorothionate
	Malathion	O,O-Dimethyl S-(1,2-dicarboxyethyl)-phosphorodithioate
	Phosphamidon	Dimethyl 2-chloro-2-diethylcarbamyl-1-methyl vinyl phosphate
	Glyphosate	N-Phosphonomethylglycine
f. Carbamate insecticides	Carbaryl	1-Naphthyl N-methylcarbamate
	Aminocarb	4-Dimethylamino 3-tolyl N-methylcarbamate

(continued)

Table 8.1   (*Continued*)

Class	Common name	Chemical name
	Carbofuran	2,2-Dimethylbenzofuran-7-yl *N*-methylcarbamate
	Aldicarb	2-Methyl-2-(methylthio)propionaldehyde *O*-(methylcarbamoyl)oxime
	Carbofuran	2,3-Dihydro-2,2-dimethyl-7-benzo-furanol Methylcarbamate
	Methiocarb	3,5-Dimethyl-4-(methylthio)phenol methyl carbamate
g. Triazine herbicides	Simazine	2-Chloro-4,6-bis-(ethylamino)-*s*-triazine
	Atrazine	2-Chloro-4-(ethylamino)-6-(isopropylamino)-*s*-triazine
	Hexazinone	3-Cyclohexyl-6-(dimethylamino)-1-methyl-1,3,5-triazine-2,4(1*H*,3*H*)-dione
	Cynazine	2-(4-Chloro-6-ethyamino-5-triazin-2-ylamino)-2-methylpropionitrile
	Metribuzin	4-Amino-6-tert-butyl-3-(methylthio)-as-triazin-5(4H)-one
h. Amide herbicides	Alachlor	2-Chloro-2′,6′-diethyl-*N*-(methoxymethyl)-acetanilide
	Metolachlor	2-Chloro-6′-ethyl-*N*-(2-methoxy-1-methylethyl)acet-*o*-toluidide
i. Thiocarbamate herbicide	Butylate	*S*-ethyl diisobutylthiocarbamate
j. Dinitroaniline herbicide	Trifluralin	*a,a,a*-trifluoro-2,6-dinitro-*N,N*-dipropyl-*p*-toluidine
k. Acetaldehyde polymer molluscicide	Metaldehyde	$(acetaldehyde)_n$
l. Pyrethroid insecticides (synthetics)	Cypermethrin	(RS)-α-Cyano-3-phenoxybenzyl (1RS-*cis,trans*-3-9-2,2-dichlorovinyl)-2,2-dimethyl-cyclopropanecarboxylate
	Deltamethrin	[1R-[1α(S*),3a]]-Cyano(3-phenoxyphenyl) methyl 3-(2,2-dibromoethenyl)-2,2-dimethylcyclopropanecarboxylate
	Permethrin	3-Phenoxybenzyl (1RS)-*cis,trans*-3-(2,2-dichlorovinyl)-2,2-dimethyl-cyclopropanecarboxylate
	Tetramethrin	3,4,5,6-Tetrahydrophthalimidomethyl (±)-*cis,trans*-chrysanthemate

[a]Modifed from Metcalf (1971), Entomological Society of America (ESA) (1980), and Briggs (1992).

prominent inorganic pesticides include the following:

1. *Bordeaux mixture* is a complex pesticide with several copper-based active ingredients, including tetracupric sulfate and pentacupric sulfate. Bordeaux mixture is used as a foliar fungicide for fruit and vegetable crops. It acts by inhibiting a variety of fungal enzymes.

2. *Various arsenicals*, including arsenic trioxide, sodium arsenite, and calcium arsenate, are used as nonselective herbicides and soil sterilants. Insecticides in this group include Paris green, lead arsenate, and calcium arsenate.

Organic Pesticides

Organic pesticides are a chemically diverse group of chemicals. Some are produced naturally by certain species of plants, but the great majority of organic pesticides has been synthesized by chemists.

Some prominent organic pesticides include the following:

1. *Natural organic pesticides* are chemicals that have been extracted from various species of plants. An important insecticide is the alkaloid nicotine and related nicotinoids, largely extracted from tobacco (*Nicotiana tobacum*) and often applied as the salt nicotine sulfate. Another insecticide is pyrethrum, a complex of six chemicals (pyrethrin I and II, cinerin I and II, and jasmolin I and II) extracted from the daisy-like insect or pyrethrum flowers, *Chrysanthemum cinerariaefolium* and *C. coccinium*. A third chemical complex that can be used as an insecticide, piscicide, or rodenticide is the rotenoids, especially rotenone, which is extracted from the tropical plants *Derris elliptica*, *D. malaccensis*, *Lonchocarpus utilis*, and *L. urucu*. Another rodenticide is red squill, composed of cardiac glycosides extracted from the sea onion *Scilla maritima*. A last example is the rodenticide strychnine, an alkaloid extracted from the tropical plant *Strychnos nux-vomica*.

2. *Synthetic organometallic pesticides* have been widely used, almost entirely as fungicides. Most important in this category are the organomercurials. Examples include phenylmercuric acetate, methylmercury, and methoxyethylmercuric chloride. Examples of environmental contaminations and human toxicity caused by the use of organomercurials are given in Chapter 3.

3. *Phenols* are fungicides used for the preservation of wood and other organic substrates. Prominent examples are trichlorophenols, tetrachlorophenol, and pentachlorophenol.

4. *Chlorinated hydrocarbons* are a diverse group of synthetic pesticides. Prominent subgroups are:

a. *DDT and its insecticidal relatives*, including DDD and methoxychlor. The related chemical DDE is noninsecticidal, but it is an important and persistent metabolic-breakdown product that is accumulated in organisms exposed to DDT and DDD. Residues of DDT and its relatives are persistent, typically having a half-life of about 10 years in soil environments, for example. A global contamination with these compounds has been caused by their persistence, their ability to weakly codistil with water, and a tendency to be dispersed with wind-blown dusts. In addition, their selective partitioning into lipids causes these chemicals to bioaccumulate. Persistence, coupled with bioaccumulation, results in the occurrence of the largest concentrations of these chemicals at the top of food webs, as described later in this chapter.

b. *Lindane*, the γ-isomer of 1,2,3,4,5,6-hexachlorocyclohexane, is the active insecticidal constituent of benzene hexachloride.

c. *Cyclodienes* are a group of highly chlorinated cyclic hydrocarbons that are used as insecticides. Prominent examples are the α-*cis* and β-*trans*-isomers of chlordane, heptachlor, aldrin, and dieldrin.

d. *Chlorophenoxy acid herbicides* have an auxin-like, growth-regulating property and are selective herbicides for broadleaved angiosperm plants. The parent compound is 2,4-D. Other important chemicals are 2,4,5-T (which is more effective than 2,4-D for the control of many angiosperm shrubs), MCPA, and silvex.

5. *Organic phosphorus pesticides* are a diverse group of chemicals, most of which are used as insecticides, acaricides, or nematicides. These generally have a high acute toxicity to arthropods, but a short persistence in the environment. Some of the insecticides are highly toxic to nontarget organisms such as fish, birds, and mammals. Some prominent examples are the insecticides parathion, methyl parathion, fenitrothion, malathion, and phosphamidon. An important herbicide is the phosphonoalkyl compound, glyphosate.

6. *Carbamate pesticides* generally have a high acute toxicity to arthropods, but a moderate environmental persistence. Important examples are aminocarb, carbaryl, and carbofuran.

7. *Triazine herbicides* are used in corn monoculture and for some other crops and as soil ster-

ilants. Prominent examples are simazine, atrazine, and hexazinone.

8. *Synthetic pyrethroids* are insecticides and acaricides used mostly in agriculture. In general, synthetic pyrethroids are highly toxic to fish and terrestrial and aquatic invertebrates, of variable toxicity to mammals, and of low toxicity to birds. Prominent examples are cypermethrin, deltamethrin, permethrin, synthetic pyrethrum and pyrethrins, and tetramethrin.

8.3
ENVIRONMENTAL EFFECTS OF PESTICIDE USE

Introduction

The intended ecological effect of any pesticide application is to manage a population of some pest species, usually by reducing its abundance to an economically acceptable level. In a few situations, this objective can be attained selectively and without important, nontarget damage. For example, the judicious use of a rodenticide in the local environment of human habitations can result in a selective kill of rats and mice. If care is taken in placement of the poison, direct exposures to nontarget mammals such as cats, dogs, and children can be minimized, although never eliminated. In agricultural situations, however, toxicity may be caused to predators or scavengers that ingest poisoned rodents. This sort of indirect exposure may, for example, have been important in the decline of barn owls (*Tyto alba*) over much of North America.

Most situations where pesticides are used are much more complex and less well controlled than the use of rodenticides in homes. Whenever a pesticide is broadcast sprayed over a field or forest, a wide variety of on-site, nontarget organisms is exposed to the chemical. In addition, some quantity of the sprayed pesticide invariably drifts away from the intended site of deposition, so that off-site, nontarget organisms and ecosystems are exposed.

The ecotoxicological risks of nontarget exposures to pesticides (and other chemicals) are influenced by a complex of variables, including the following:

1. *The biological sensitivity* of specific organisms to particular pesticides differs greatly. Some species are much more sensitive to particular chemicals than are other species. In addition, populations within species, and individuals within a population, may vary greatly in susceptibility. Such comparisons can be made using a variety of indicators of acute and chronic toxicity. For example, a commonly reported index of acute toxicity, known as LD_{50}, is based on the amount of chemical that is required to kill one-half of a population of organisms during a short-term exposure in a controlled, laboratory bioassay. Table 8.2 ranks the toxicity to rats of some pesticides and other chemicals using LD_{50}. Studies of chronic toxicity involve smaller doses, but longer exposure periods, and may examine teratogenesis, carcinogenesis, organ damages, and reproductive effects.

2. *The actual, environmental exposure* to any potentially toxic chemical can be influenced by many variables. These include the specific spray rates and equipment that are used during the application, weather, persistence of the chemical in the environment, and habitat choices and other behavioral attributes that affect exposure of nontarget species. Exposure has a fundamental influence on toxicity by influencing the dose of chemical that nontarget species receive. As noted in Chapter 2, this fundamental tenet of toxicology can be illustrated by paraphrasing Paracelsus (1493–1541): "Dosage alone determines poisoning." An exposure to a potentially toxic chemical must result in a dose that exceeds physiologically determined thresholds of tolerance if poisoning is to be demonstrated.

It is important to note that *if a sufficiently large dose is achieved, any chemical can poison any organism.* There are two corollaries of this principle:

1. All chemicals are potentially toxic. Even chemicals that are routinely encountered can cause a toxic effect, if the dose is large enough.

Table 8.2   Acute toxicity of selected chemicals to rats[a]

Chemical substance	Oral LD$_{50}$ for rats (mg/kg)
Sucrose (table sugar)	30,000
Fosamine (H)	24,000
Ethanol (drinking alcohol)	13,700
Benomyl (F)	10,000
Captan (F)	9,000
Picloram (H)	8,200
Maneb (F)	6,500
Simazine (H)	5,000
Glyphosate (H)	4,300
DDD (I)	4,000
Permethrin (I)	3,800
Sodium chloride (table salt)	3,750
Sodium bicarbonate (baking soda)	3,500
Sodium hypochlorite (household bleach)	2,000
Dicamba (H)	2,000
Malathion (I)	2,000
Atrazine (H)	1,750
Acethylsalicylic acid (aspirin)	1,700
Hexazinone (H)	1,690
Dalapon (H)	1,000
Tetramethrin (I)	1,000
DDE (I)	880
2,4-DP (H)	800
Mirex (I)	740
MSMA (H)	700
2,4,5-TP (H)	650
Trichlopyr (H)	650
Metaldehyde (M)	630
Carbaryl (I)	500
2,4,5-T (H)	500
Chlordane (I)	400
2,4-D (H)	370
Fenitrothion (I)	250
DDT (I)	200
Caffeine (alkaloid in coffee, tea, other foods)	200
Paraquat (H)	150
Diazinon (I)	108
Lindane (I)	88
Endosulfan (I)	75
Methiocarb (M)	65
Nicotine (alkaloid in tobacco)	50
Dinoseb (H, I)	40
Aminocarb (I)	39
Deltamethrin (I)	31
Strychnine (R)	30
Phosphamidon (I)	24
Methylparathion (I)	14
Parathion (I)	13
Carbofuran (I)	10
TEPP (I)	6.8

(continued)

Table 8.2   *(Continued)*

Chemical substance	Oral LD$_{50}$ for rats (mg/kg)
Aldicarb (I)	0.8
Saxitoxin (paralytic shellfish neuropoison)	0.26
Tetrodotoxin (Japanese globe fish toxin, *Spheroides rubripes*)	0.01
TCDD (dioxin isomer)	0.01

[a](H) Herbicide; (I) insecticide; (F) fungicide; (R) rodenticide; (M) molluscicide. Modified from S. D. Murphy (1980), Windholz (1983), and Walstad and Dost (1984).

For example, if enough is drunk in a short period of time, water can poison animals by overwhelming the capacity for osmoregulation of the blood plasma. Similarly, carbon dioxide, table sugar (sucrose), table salt (sodium chloride), and aspirin (acetylsalicylic acid) can all be poisonous, if the dose is large enough (Table 8.2).

2. Small doses of even highly toxic chemicals will not have a measurably poisonous effect unless the exposure exceeds some minimal threshold. Because tissues and organisms have mechanisms to repair damages and to sequester or detoxify many chemicals, the dose must be large enough to overwhelm these physiological mechanisms of tolerance. This notion of dosage thresholds was mentioned previously, in the context of the difference between contamination and pollution (see Chapter 1). In view of this interpretation, it is best to refer to "potentially toxic chemicals" in contexts in which the actual environmental exposures to chemicals are not known. [Note, however, that some toxicologists would not agree with this explanation of thresholds of toxicity. Some scientists believe that exposures to even single molecules of certain chemicals may be of toxicological importance, and that dose–response curves can therefore be linearly extrapolated to zero dosages (Lappe, 1991; Rodricks, 1992).]

In part, the ecological importance of toxicity caused to a nontarget, pesticide-sensitive species should be interpreted on the basis of the following considerations:

1. Are there measurable consequences of the toxic effect at the population level? The deaths of individuals and other toxic effects are regrettable and tragic in many respects (particularly for the individuals involved!), but there may not be measurable changes in the sizes of populations of that species. In this sense, populations of species and their communities may have capabilities of suffering a certain amount of pesticide-caused mortality, without exhibiting an overall change.

2. Is the species particularly important in maintaining the integrity of its community? All species have similar intrinsic values, but species can vary in terms of the contributions that they make to the structure and function of their ecological community. For example, animal ecologists refer to species that are critical in maintaining the structure of their community as "keystone species" (Paine, 1969; Krebs, 1985), while plant ecologists sometimes refer to "edificator species." Clearly, substantial changes in the abundance of these sorts of species may be judged as being relatively important, compared with changes in seemingly more minor species. (From other perspectives, however, the importance of non-target pesticide effects may also be judged on the basis of specific economic, aesthetic, and ethical considerations.)

It is also notable that many toxic chemicals are quite natural—not all poisons are synthesized by chemists! Among the substances listed in Table 8.2 are a number of naturally occurring biochemicals, including two of the three most toxic chemicals listed.

For example, saxitoxin is a potent neurotoxin produced by marine dinoflagellates. Saxitoxin and some other natural, marine poisons can be accumulated by filter feeders such as mollusks. As a so-called paralytic shellfish poison, saxitoxin can cause toxicity to mammals, including humans, who eat the shellfish (Anderson *et al.*, 1985; Tu, 1988; Hall, 1989; Okaichi *et al.*, 1989; Fritz *et al.*, 1992).

Toxic dinoflagellate biochemicals are also responsible for diarrhetic shellfish poison, while diatoms in the genus *Nitzchia* synthesize domoic acid and are responsible for amnesic shellfish poisoning.

The causes of the dinoflagellate blooms known as "red tides" are not well understood, but they are probably related to some combination of nutrient availability and water temperature. Red tides are ancient phenomena and were referred to in the Bible (Exodus 7: 19–21): " . . . and all the waters that were in the river were turned to blood." Because of the toxic risks to humans associated with eating marine shellfish, the toxicity of these foods are regularly monitored in many countries using mouse injection and/or biochemical bioassays.

Wildlife can be also affected by toxic marine algae. For example, in 1991 an outbreak of the diatom *Nitzchia occidentalis* in Monterey Bay, California, resulted in an accumulation of domoic acid in zooplankton. These were fed upon by small fishes such as anchovy (*Engraulis mordax*), which then poisoned large numbers of piscivorous birds, especially brown pelican (*Pelecanus occidentalis*) and Brandt's cormorant (*Phalacrocorax penicillatus*), along with some humans who ate contaminated shellfish (Fritz *et al.*, 1992). Similarly, blooms of the phytoplankton *Chrysochromulina polylepis* in the Baltic Sea in 1988 caused mass mortalities of various species of macroalgae, invertebrates, and fish (Rosenberg *et al.*, 1988), while *C. leadbeateri* caused an extensive mortality of caged salmon (*Salmo salar*) in Norway in 1991 (Aune *et al.*, 1991).

Even animals as large as whales can be killed by algal toxins. Over a 5-week period in 1985, 14 humpback whales (*Megaptera novaeangliae*) died at sea in Cape Cod Bay, Massachusetts, after eating mackerel (*Scomber scombrus*) polluted with algal-derived saxitoxin (Geraci *et al.*, 1989). The deaths of the whales were apparently quick and were symptomatic of neurotoxicity associated with saxitoxin. In one case, a whale was observed apparently behaving normally, and 90 min later it was dead. Concentrations of saxitoxin were 154 µg/100 g in a composite sample of livers of 17 mackerel caught in the area where whales were dying and 52 μg/100 g in the viscera of a sample of four fish (Geraci *et al.*, 1989).

There is such a great variety of pesticides available today, and such a diversity of uses and exposures, that it is impractical to specifically discuss the environmental effects for more than a few. Case studies can, however, be used to draw out some of the broader patterns and principles of pesticide-related ecotoxicology, and these will be used extensively in the remainder of this chapter.

The chemical complexity of environmental exposures can be illustrated by the observations of Frank *et al.* (1982), who studied pesticide use on 11 agricultural watersheds in Ontario. At least 81 different pesticides had been applied in agriculture or along right-of-ways, and other undocumented pesticides were used in and around homes (19% of the pesticide sold in North America in 1970 was for home use—many ordinary individuals use pesticides in large quantities; McEwen and Stephenson, 1979). On average, 39% of the land surface had been treated at a rate of 8.3 kg/ha-year. The rate of application in agriculture ranged from 0.005 kg/ha-year for hayfields and pasture to an average of 51 kg/ha-year for potato, tomato, and tobacco crops. The intensive use of pesticides contaminated surface waters in the study area. The herbicide atrazine accounted for 93% of the total pesticide flux in stream water of 2.2 g/ha-year. Although DDT had not been used since 1972, it occurred in 41% of the water samples (as DDT or DDE), while PCBs (polychlorinated biphenyls, a group of noninsecticidal chlorinated hydrocarbons that have largely been used as dielectric fluids in electrical transformers) were present in 78% of the samples.

In view of the extreme diversity of pesticides and their uses, a limited approach to the development of case studies is taken in this chapter. Initially, the ecological effects of pesticide use is characterized by reference to DDT and its related chlorinated hydrocarbons. Although the use of DDT was almost eliminated in most industrialized countries since the early 1970s, its use in large quantities continues elsewhere, especially in tropical countries. Moreover, DDT is one of the best-studied pesticides from an ecotoxicological perspective. Consideration of

its case gives insight into the movement, fate, and ecological damage that can be caused by contamination of natural environments with persistent, bioaccumulating, toxic chemicals. Therefore, DDT provides an excellent case study of the ecological effects that can accompany the use of certain types of pesticides.

Following the consideration of DDT, two case studies are developed for the use of pesticides in agriculture. The first case examines carbofuran, a more modern insecticide than DDT. Carbofuran is not as persistent in the environment as DDT, but it is extremely toxic to animals and causes substantial, nontarget mortality to exposed wildlife during the course of its normal, permitted, agricultural usage. The other agricultural case study briefly discusses decreasing populations of certain types of birds in Britain, attributed to the use of herbicides in conjunction with other practices in intensive agriculture.

Two other relatively detailed case studies of the ecological effects of several large-scale pesticide spray programs in forestry are then presented. The first describes the use of insecticides against eastern spruce budworm in mature conifer forests, and the second concerns the silvicultural use of herbicides. These cases were chosen because their spray programs affect complex, natural, or seminatural ecosystems and because there is a relatively broad base of information about their various ecological effects. As such, they allow an appreciation of the effects on complex ecosystems of large-scale pesticide spraying. It would be much more difficult to consider these effects in highly disturbed and relatively simple agroecosystems, and that is why nonagricultural pesticide programs were chosen as several of the case studies in this chapter.

Environmental Effects of the Use of DDT and Its Relatives

Background and History of Use

DDT was first synthesized in 1874. In 1939 its insecticidal qualities were recognized by P.H. Muller in Switzerland, who won a Nobel Prize in 1948 for that discovery and his subsequent research on the uses of DDT. The first important application of DDT was in human health programs during and after World War II, and at that time its use for agricultural and forestry purposes also began. The global production of DDT peaked in 1970, when 175 million kg was manufactured. The peak of DDT use in the United States was 35.8 million kg in 1959, while the peak of production was 90 million kg in 1964 (Edwards, 1975; Hayes, 1991). Because of the recognition of widespread and persistent environmental contaminations by DDT and its breakdown products, their ability to biomagnify, and the important nontarget damages that these caused, most industrialized countries banned the use of DDT after the early 1970s (with a minor exception; DDT can still be prescribed by North American physicians for control of the human body louse).

However, the use of DDT has continued elsewhere, especially in less developed countries of warmer latitudes, primarily for the control of the mosquito vectors of disease. In addition, although the uses of DDT, aldrin, dieldrin, and heptachor were greatly reduced after the early 1970s, these chemicals were not all banned in Britain until 1986 (Newton and Wyllie, 1992).

Although the use of DDT was banned in the United States in 1972, its manufacture continued for the export market, which was almost entirely in less developed countries. The export of DDT from the United States was 26 million kg in 1974, 21 million kg in 1975, and 12 million kg in 1976, but beginning in 1977 (0.23 million kg) exports were greatly reduced (ESA, 1980). Today there is no export of DDT from the United States, but its manufacture and use in large quantities continue in less developed countries. However, largely because of the widespread evolution of resistance to DDT by many insect pests, it is becoming a decreasingly effective pesticide. There has even been a resurgence of previously well-controlled diseases such as malaria (Chapin and Wasserström, 1981; NRC, 1986; Hayes, 1991). Ultimately, because of the emerging ineffectiveness of DDT, its uses will eventually be curtailed and it will be replaced by other insecticides.

DDT was the first pesticide to which a large number of insect pests developed resistance. This evolutionary process occurs because of the intense selection for resistant genotypes that occurs when populations of organisms are exposed to a toxic pesticide. Resistant genotypes may be present at a small frequency in unsprayed populations. In sprayed populations, however, they become numerically dominant after treatment because they are not killed by the pesticide and therefore survive to reproduce. According to NRC (1986), at least 447 species of insects and mites have populations that are known to be resistant to at least one insecticide, and there are more than 100 cases of resistant plant pathogens, 48 resistant weeds, and two resistant nematodes. Among arthropods, resistance is most frequent in the Diptera, with 156 resistant species. There are 51 resistant species of malaria-carrying *Anopheles* mosquito, including 47 that are resistant to dieldrin, 34 to DDT, 10 to organophosphates, and 4 to carbamates. The progressive evolution of resistance by *Anopheles* has been an important cause of the recent resurgence of malaria in warmer latitudes (NRC, 1986; Hayes, 1991).

Residues and Biological Uptake

DDT has several chemical and physical properties that profoundly influence the nature of its ecological effects. First, DDT is persistent in the environment. It is not easily degraded to other, less toxic chemicals by microorganisms or by physical agents such as sunlight and heat. The typical persistence of DDT and some other organochlorine insecticides in soil is given in Table 8.3. DDE is the primary breakdown product of DDT, and it is produced by dechlorination reactions that occur in alkaline environments or enzymatically in organisms. Unfortunately, the persistence of DDE is similar to that of DDT. Therefore, once it is released into the environment, DDT and its breakdown products persist for many years.

Another important characteristic of DDT is its small solubility in water (less than 0.1 ppm). Its sparse aqueous solubility means that DDT cannot be physically "diluted" into this ubiquitous solvent, which is so abundant on the surface of the Earth and

Table 8.3 Persistence of some organochlorine insecticides in soil[a]

Chemical	Typical annual dose (kg/ha)	Half-life (years)	Average time for 95% disappearance (years)
Aldrin	1.1–3.4	0.3	3
Isobenzan	0.3–1.1	0.4	4
Heptachlor	1.1–3.4	0.8	3.5
Chlordane	1.1–2.2	1.0	4
Lindane	1.1–2.8	1.2	6.5
Endrin	1.1–3.4	2.2	7
Dieldrin	1.1–3.4	2.5	8
DDT	1.1–2.8	2.8	10

[a]Modified from Edwards (1975).

in organisms. On the other hand, DDT is highly soluble in organic solvents such as xylene (60%) and kerosene (80%).

Also of great importance is the great solubility of DDT in fats or lipids, a characteristic that is shared with other chlorinated hydrocarbons. In the environment, most lipids are present in organisms. Therefore, because of its high lipid solubility, DDT has a strong affinity for organisms, and it tends to bioconcentrate by factors of several-or-more orders of magnitude. Furthermore, because organisms at the top of their food web are highly effective at assimilating DDT from their food, they tend to have especially large concentrations of DDT in their lipids.

The bioconcentration and food-web accumulation effects of DDT are illustrated in Fig. 8.1, which shows typical concentrations of DDT in a variety of atmospheric, terrestrial, aquatic, and biotic compartments of the environment. Note that the residues of DDT in air, water, and nonagricultural soil are relatively small, compared with the concentrations in organisms. Note also that residues in plants are smaller than in herbivores, and that residues are largest in animals that are at or near the top of their food web, such as humans and predatory birds.

A similar pattern is seen in Fig. 8.2, which summarizes the pattern of residues in an estuary on

Long Island, New York. The sources of the DDT were diffuse, but they included the direct spraying of salt marshes to control mosquitoes. The data indicate that there is a general concentrating of residues at the top of the food web. The largest concentration (75.5 ppm) was in an individual ring-billed gull (*Larus delawarensis*), an opportunistic bird that often feeds on small fish. Large residues were also present in other piscivorous birds, for example, 26 ppm in double-crested cormorant (*Phalacrocorax auritus*), 23 ppm in red-breasted merganser (*Mergus serrator*), and 19 ppm in herring gull (*Larus argentatus*) (Woodwell *et al.*, 1967).

A tropical example of food-web accumulation of DDT residues can be illustrated by the case of Lake Kariba, Zimbabwe (Berg *et al.*, 1992). DDT is still used for agricultural purposes and to control arthropod vectors of diseases throughout much of Africa and in other subtropical and tropical countries. In Zimbabwe, the use of DDT in agriculture was banned in 1982, but its use for control of mosquitoes and tsetse fly continues. The DDT in Lake Kariba is believed to have mostly resulted from nearby programs to control the tsetse fly, an important vector of disease of cattle and other large mammals. The concentrations of DDT in water of Lake Kariba were less than 0.002 ppb. Average concentrations in sediment were 0.4 ppm. Planktonic algae contained 2.5 ppm, while a filter-feeding mussel, *Corbicula africana*, contained 10.1 ppm (this and subsequent values are all for DDT in lipids). Two species of herbivorous fish, kapenta (*Limnothrissa miodon*) and Kariba tilapia (*Oreochromis mortimer*), contained 1.6 and 1.9 ppm, respectively, while a benthic-feeding fish, menyame labeo (*Labeo altivelis*), contained 5.7 ppm. The tigerfish (*Hydrocynus forskahlii*) and great cormorant (*Phalacrocorax carbo*) mostly prey on kapenta and tilapia, and contained 5.0 and 9.5 ppm, respectively. The top predator in the system (other than humans) is the Nile crocodile (*Crocodylus niloticus*), with 34.2 ppm. Therefore, the Lake Kariba example shows both a substantial bioconcentration from water and, to a lesser degree, from sediment, as well as a food-web concentration from herbivores to carnivores.

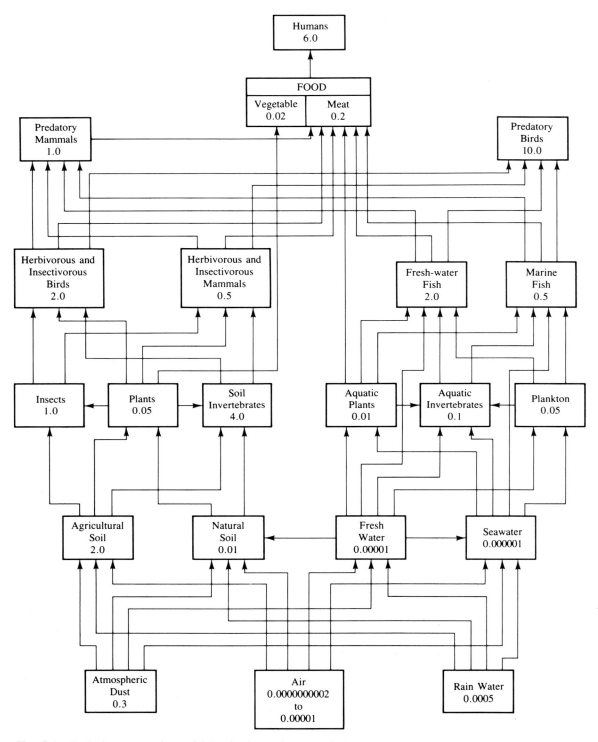

Fig. 8.1 Typical concentrations of DDT in the environment (ppm). Data were derived from a literature review of DDT residues, from Edwards (1975).

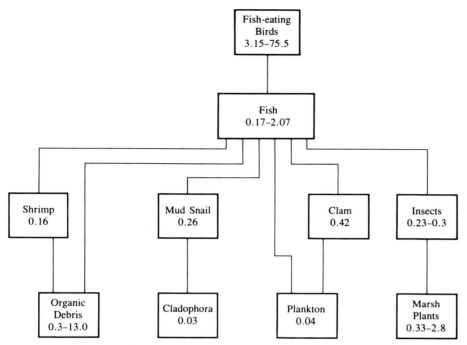

Fig. 8.2 Residues of total DDT (DDT + DDD + DDE: ppm) in various trophic levels of an estuary and salt marsh on Long Island, New York (Edwards, 1975, adapted from Woodwell *et al.*, 1967).

Another environmental feature of DDT is its ubiquity. DDT is now present throughout the biosphere in at least trace concentrations. The remarkably widespread environmental contamination with DDT and some related chlorinated hydrocarbons occurs because they can enter the global atmospheric cycle. This results from: (1) the ability of DDT to codistil with water in a small concentration; (2) the volatilization of DDT from sprayed surfaces (this is a relatively slow process, since its vapor pressure at 20°C is only 1.9×10^{-7} atm); (3) off-target, atmospheric drift of sprayed pesticide; and (4) the entrainment by wind of DDT-contaminated dusts into the atmosphere (Edwards, 1975; Taylor, 1978; McEwen and Stephenson, 1979).

Because of these processes, DDT and other pesticides can be present in trace atmospheric concentrations in remote locations. Atmospheric sampling for DDT between 1976 and 1978 found trace concentrations that ranged from 75 to 132×10^{-12} g

DDT/m³ at three locations in the Indian Ocean, while at three sites in the North Atlantic the contamination was $2.5–5.0 \times 10^{-12}$ g/m³ (Bidleman and Leonard, 1982). The difference reflected the contemporary patterns of DDT use, which had been banned in most countries of the temperate, northern hemisphere, but was still permitted in lower latitudes.

The ubiquity of the contamination with DDT can be illustrated by its concentration in antarctic wildlife, which live far remote from places where DDT has been used. Norheim *et al.* (1982) reported residues of total DDT (i.e., DDT + DDE, almost all of which is DDE) in some antarctic birds. The concentration in fat averaged largest in the southern polar skua (*Catharacta maccormicki*) at 4.6 μg/g w.w., compared with ≤0.43 μg/g in avian predators lower in the oceanic food web, such as the southern fulmar (*Fulmarus glacialoides*), cape pigeon (*Daption capense*), snow petrel (*Pagodroma nivea*), macaroni penguin (*Eudyptes chrysolophus*), and chinstrap penguin (*Pygoscelis antarctica*).

Much larger concentrations of DDT and its related compounds are present in wildlife that live closer to areas where the pesticides are manufactured and sprayed (Edwards, 1975). Marine mammals are at or near the top of the marine food web, and they frequently have large residues of chlorinated hydrocarbons. The concentration of DDT in the fat of seals from the coast of California was as large as 158 µg/g during the late 1960s (a period of great DDT use in that area), while in the Baltic Sea of northwestern Europe, residues were as great as 150 µg/g, and off eastern Canada they were up to 35 µg/g. Even larger residues were present in the fat of eastern Canadian porpoises (*Phocoena phocoena*), at up to 520 µg/g (Edwards, 1975). Concentrations of DDT in the lipids of seals increase markedly when the animals are not feeding well and are losing weight, including their fat volume. For example, gray seals (*Halichoerus grypus*) in the Baltic Sea had average concentrations of 84 µg/g of total DDT and 230 µg/g of PCBs (polychlorinated biphenyls) in lipids when they were feeding well and the fat concentration of their blubber was 90%, but 560 µg/g DDT and 2100 µg/g PCBs when they were starving and their fat concentration was 59% (Blomqvist *et al.*, 1992).

Large residues of DDT and other organochlorines have also been measured in birds, especially raptors. Residues as large as 356 µg/g (average of 12 µg/g) occurred in a sample of 69 U.S. bald eagles (*Haliaeetus leucocephalus*), up to 460 µg/g among 11 western grebes (*Aechmophorus occidentalis*), and up to 131 µg/g among 13 herring gulls (*Larus argentatus*) (Edwards, 1975). In some cases, remarkably large residues were measured. The white-tailed eagle (*Haliaeetus albicilla*) population of the Baltic Sea had individuals with as much as 36,000 µg/g of total DDT and 17,000 µg/g PCBs in their fat, and addled eggs had up to 1900 µg/g total DDT and 2600 µg/g PCBs (Koivusaari *et al.*, 1980).

Because of the cessation of most uses of DDT and other chlorinated hydrocarbons in the early 1970s in many countries, residues and toxic exposures have decreased significantly (Addison *et al.*, 1986; Noble and Elliott, 1990; Peakall *et al.*, 1990; Blomqvist *et al.*, 1992; Anonymous, 1993e). This trend is illustrated by changes in the concentrations of DDE and PCBs in the eggs of herring gulls nesting on various islands in the Great Lakes (Table 8.4a). More intense contamination with DDE occurred at Big Sister Island in Green Bay, Lake Michigan, than at Granite Island, Lake Superior or Muggs Island, Lake Ontario, while Big Sister and Muggs Islands were most contaminated with PCBs. All three locations, however, showed large decreases in DDE concentration during the monitoring period. During the period of 1971 to 1989, the concentrations of dieldrin in herring gull eggs were less than 1 µg/g at all three sites (Bishop and Weseloh, 1990).

Similar patterns of decline of residues are evident in eggs of double-crested cormorant from a larger area, ranging from the Strait of Georgia in southern British Columbia through the Great Lakes and to the estuary of the Saint Lawrence River (Table 8.4b). However, the concentrations in cormorant eggs are considerably smaller than in eggs of herring gulls. The cormorants are exclusively piscivorous, mostly eating relatively small, often juvenile fishes, while herring gulls are primarily scavengers and eat primarily large dead fish or fish offal. Because large, older fish have had more time than younger fish to accumulate residues in their tissues, the herring gulls have an increased exposure to chlorinated hydrocarbons than the cormorants.

The incidence of congenital deformities in chicks of double-crested cormorants is higher at breeding colonies on the Great Lakes than elsewhere in North America (Fox *et al.*, 1991). Between 1979 and 1987, a total of 31.2 thousand cormorant chicks was examined at 42 colonies on the Great Lakes, and 70 of those birds (0.22% of the number examined) had deformed or deflected bills, a debilitating affliction. Among the regions of the Great Lakes, the incidence of these deformities was highest in Green Bay on Lake Michigan, where 0.52% of 10 thousand cormorant chicks were afflicted. In comparison, only 2 out of 21.0 thousand chicks (i.e., 0.01%) examined in reference colonies

Table 8.4 Changes in concentration of some chlorinated hydrocarbons since the 1970s[a]

	Big Sister Island		Granite Island		Muggs Island	
	DDE	PCBs	DDE	PCBs	DDE	PCBs
(a) Herring Gull Eggs						
1971–1974	58	151	28	74	23	139
1975–1978	29	105	16	60	16	123
1979–1982	14	65	6	39	11	66
1983–1986	9	30	3	16	5	40
1987–1989	5	28	2	12	4	21
	Strait of Georgia		Great Lakes		St. Lawrence	
	DDE	PCBs	DDE	PCBs	DDE	PCBs
(b) Double-Crested Cormorant Eggs						
1971–1974	4	8	16	16	6	8
1975–1978	—	—	5	6	2	8
1979–1982	1	4	2	10	2	7
1983–1986	0.5	2	—	—	2	6
1987–1992	0.5	2	2	3	1	4

[a] Both herring gull (*Larus argentatus*) and double-crested cormorant (*Phalacrocorax auritus*) are top predators. See text for discussion. Data are μg/g, wet weight, in egg samples. Modified from Bishop and Weseloh (1990) and Anonymous (1993e).

in northwestern Ontario and the Canadian prairies had these deformities. The Great Lakes cormorants carry substantial body burdens of various chlorinated hydrocarbons, including DDT and its insecticidal relatives, TCDD and PCBs. However, Fox *et al.* attribute the cormorant deformities to exposure to PCBs, which induces a hepatic enzyme, aryl hydrocarbon hydroxylase, that may be important in increasing the frequency of the syndrome.

Effects on Birds

The large residues of DDT and related chlorinated hydrocarbons have had a number of important ecological effects. Among the most prominent of these has been wide-scale poisonings of birds.

Some of these poisonings were directly associated with sprays of these chemicals. For example,

there were many incidents in which dying or dead birds were found within a short time of the spraying of DDT, especially after urban sprays to kill the beetle vectors of Dutch elm disease. Spray rates for this purpose were intensive, ranging from 0.7 to 1.4 kg DDT per individual tree and resulting in foliar residues of DDT of 174–273 ppm, which persisted to 20–28 ppm in the autumn after leaf fall, while postspray concentrations in earthworms were 33–164 ppm (Cooper, 1991).

Not surprisingly, such intense exposures to DDT caused bird kills. Wurster *et al.* (1965) recovered 117 dead birds of various species in a 6-ha spray area in Hanover, New Hampshire. Undoubtedly many more individuals were killed by the spray, but were not found because they were hidden by vegetation or scavenged. It was not uncommon to have a greatly reduced abundance of songbirds in such a situation. For example, in the study of Wurster *et al.*, an estimated 70% of the breeding population of robins (*Turdus migratorius*) was lost. Hence, the title of Rachael Carson's (1962) seminal book, "Silent Spring."

However, the indirect effects of DDT and its related chlorinated hydrocarbons were more insidious than direct mortality. In some situations mortality was caused by chronic toxicity and it was therefore less immediate, and relatively detailed investigations were required to link the declines of bird populations to the use of a pesticide. A graphic example of this syndrome occurred at Clear Lake, California (Hunt and Bischoff, 1960). This is an important water body for recreation, but there were many complaints about the aesthetic nuisance associated with events of great abundance of a nonbiting midge (*Chaoborus asticopus*). This pest was dealt with in 1949 by application of DDD to the lake at about 1 kg/ha. Short-term bioassays were done prior to this treatment and had indicated that this dose of DDD would adequately control the midges, but would have no immediate, acute effects on fish. However, the unexpected happened. After a second application of DDD in 1954 to deal with a rebounding population of midges, about 100 western grebes (*Aechmophorus occidentalis*) were found dead on Clear Lake. The cause of mortality was not immediately identified, but an infectious disease was ruled out by autopsies and other forensic investigations. Sick and dead grebes were also found after later DDD treatments. Overall, the breeding population of western grebes on Clear Lake decreased from a prespray abundance of more than 1000 pairs to none in 1960 (although 30 nonbreeding adults were present in 1960). The catastrophic decline of grebes was linked to DDD when, in 1957, an analysis of fat taken from dead birds indicated residues as large as 1600 μg/g. Fish also proved to be heavily contaminated.

Therefore, the treatment of Clear Lake with DDD was an essentially unsuccessful attempt to combat a relatively trivial pest problem. The failure was caused by (1) unexpected nontarget damages and (2) the development of resistance to the insecticide by the pest midges. It must be appreciated that

in the early 1950s it was impossible to predict such an environmental "surprise," because of (1) the limited available information on the ecotoxicological effects of pesticides on birds and other wildlife and (2) society's limited experience with the insidious, chronic effects of persistent, bioaccumulating pesticides. An important lesson learned from cases such as Clear Lake is the need to understand the ecological implications of environmental management activities, such as pesticide use, before the practices become adopted for routine, extensive usage.

Chronic, ecotoxicological damage to birds also occurred in habitats that were remote from sprayed sites. This was especially true of raptorial birds of many species, because they are at the top of their food web and they can food-web accumulate chlorinated hydrocarbons to large concentrations. In some species, effects were sufficiently severe to cause declines in abundance beginning around the early 1950s and resulting in local or regional extirpations of breeding populations. Some prominent examples of birds that suffered population declines because of exposure to DDT, its metabolic breakdown product DDE, and other chlorinated hydrocarbons include the bald eagle, golden eagle (*Aquila chrysaetos*), peregrine falcon (*Falco peregrinus*), osprey (*Pandion haliaetus*), brown pelican (*Pelecanus occidentalis*), double-crested cormorant, and European sparrowhawk (*Accipiter nisus*), among numerous others (Ames, 1966; Hickey, 1969; Lockie *et al.*, 1969; Peterson, 1969; Cade and Fyfe, 1970; Ratcliffe, 1970; Cromartie *et al.*, 1975; Fyfe *et al.*, 1976).

Of course, in field situations birds and other wildlife were exposed not only to DDT. Depending on their habitat, nearby management practices, and other factors, there could also be exposures to other chlorinated hydrocarbons, including DDD, aldrin, dieldrin, heptachlor, and PCBs. The cyclodienes (e.g., aldrin, chlordane, dieldrin, and heptachlor) are more toxic than DDT to birds in laboratory exposures (Table 8.5; Hudson *et al.*, 1984; Noble and Elliott, 1990). There has been discussion about the relative importance of these various chemicals in causing the declines of populations of certain species of birds, especially raptors. In Britain, the

Table 8.5 The minimum critical concentrations of selected chlorinated hydrocarbons and mercury that are required to cause toxicity to raptors[a]

Contaminant	Minimum critical concentration		
	Brain	Liver	Eggs
Dieldrin	>5.0	3–10	>1.0
Oxychlordane	1.1–5.0	3–10	—
Heptachlor epoxide	3.4–8.3	3–10	>1.5
DDE	250	100	1.2–30
PCBs	500–3000	—	>50
HCB	—	—	>5.0
Mercury	>50	20–45	>0.5

[a]Brain and liver concentrations are considered diagnostic of acute toxicity, while egg concentrations indicate reproductive effects. Note that DDE is the main metabolite of DDT. Concentration data are in mg/kg, fresh weight. Modified from Noble and Elliott (1990).

crashes of raptors did not occur until the cyclodienes (especially dieldrin) came into common use, and the correlations of avian damage are stronger with cyclodiene residues than with DDT, DDD, or DDE (Ratcliffe, 1970; Moriarty, 1988; Newton and Wyllie, 1992). In North America, however, DDT use was more prevalent, and it was probably the most important cause of the avian declines related to chlorinated hydrocarbons (Moriarty, 1988; Cooper, 1991; Newton and Wyllie, 1992).

The greatly decreased exposures of wildlife to DDT and other chlorinated hydrocarbons since the early 1970s have allowed an encouraging recovery of abundance of various species, especially since the use of DDT was banned in most temperate countries around 1972 (Spitzer et al., 1978; Spitzer and Poole, 1980; Grier, 1982; Wallin, 1984; O'Connor and Shrubb, 1986). Unfortunately, there are indications that the continued use of DDT and related insecticides in the tropics is causing reproductive damage to raptors there (Tannock et al., 1983).

The damage to predatory birds was largely caused by the chronic effects of chlorinated hydrocarbons on reproduction, as opposed to direct toxicity to adults. The reproductive effects of chlorinated hydrocarbons include: (1) a decrease in clutch size; (2) the production of a thin eggshell (Table 8.6), which could break under the weight of an incubating parent; (3) a high death rate of embryos, unhatched and pipping chicks, and nestlings; and (4) aberrant adult behavior while incubating or raising hatchlings, causing a decrease in fledging success. Because of the reproductive pathology of DDT and its chemical relatives, many populations of raptors had a small frequency of juvenile individuals, and the abundance of those species decreased because of inadequate recruitment (Nelson, 1976; Peakall, 1976; Cooke, 1979; McEwen and Stephenson, 1979).

The syndrome of chronic toxicity to birds of DDT and other chlorinated hydrocarbons can be illustrated by describing the circumstances of the peregrine falcon, the decline of which attracted high-profile attention in North America and Europe. Decreased reproductive success and declining populations of this falcon were first noticed in the early 1950s in western Europe and soon after in North America.

In 1970, a North American census reported that there was virtually no successful reproduction by the eastern population of the *anatum* subspecies of the peregrine (*Falco peregrinum anatum*), while the arctic *tundrius* subspecies was declining in abundance. Only the *pealei* subspecies of the Queen Charlotte Islands of western Canada had a stable population and normal breeding success (Cade and Fyfe, 1970). The *pealei* race is nonmigratory, it lives in an area where pesticides are not used, and it feeds on an essentially nonmigratory food resource of seabirds. In contrast, the eastern *anatum* race bred in a region where chlorinated hydrocarbon pesticides were widely used, and its prey was generally contaminated. The northern *tundrius* race breeds in a region that is remote from situations where pesticides are used, but these falcons winter in sprayed areas in Central and South America where their avian foods can be contaminated (Fyfe et al., 1990; Peakall, 1990), and their prey of migratory waterfowl on the breeding grounds is also contaminated (Cade and Fyfe, 1970; Fyfe et al., 1976).

The implications of this latter characteristic of

Table 8.6   Changes in eggshell thickness of some populations of North American bird species: Change is percentage change comparing post-1945 with pre-1945 data, and shell thickness is indexed as shell weight (mg)/[length (mm) × width (mm)][a]

Species	Location	Change in thickness index (%)
Bald eagle (*Haliaeetus leucocephalus*)	Texas	−30
Double-crested cormorant (*Phalacrocorax auritus*)	Wisconsin	−30
Prairie falcon (*Falco mexicanus*)	New Mexico	−28
Peregrine falcon (*Falco peregrinus*)	California	−26
Brown pelican (*Pelecanus occidentalis*)	California	−25
Marsh hawk (*Circus cyaneus*)	Oregon, Alberta	−24
Osprey (*Pandion haliaetus*)	Northeastern U.S.	−21
Cooper's hawk (*Accipiter cooperi*)	Western Canada	−20
Black-crowned night heron (*Nycticorax nycticorax*)	New Jersey	−18
Great horned owl (*Bubo virginianus*)	Florida	−17
Common loon (*Gavia immer*)	Ontario	−15
White pelican (*Pelecanus erythrorhynchos*)	British Columbia	−14
Golden eagle (*Aquila chrysaetos*)	California	−11
Herring gull (*Larus argentatus*)	Great Lakes	−10

[a]Modified from Anderson and Hickey (1972).

the habitat of the *tundrius* peregrine can be illustrated by the study of Lincer *et al.* (1970), who compared pesticide residues in Alaskan peregrine falcons and rough-legged hawks (*Buteo lagopus*). On a dry-weight basis, three peregrines averaged 114 ppm DDE in muscle, 752 ppm in fat, and 131 ppm in eggs, while three rough-legged hawks averaged 1.2 ppm in muscle, 13.3 ppm in fat, and 7.1 ppm in eggs. These differences were related to residues in the prey of these raptors. The peregrines fed on migratory ducks, which had 10–20 ppm DDE in their fat, compared with less than 1 ppm in the nonmigratory small mammal prey of the rough-legged hawks.

Other studies conducted at about the same time showed that large residues of chlorinated hydrocarbons were widespread in North American peregrines (except for *pealei*). These residues were associated with eggshells that were thinner than the pre-DDT condition by 15–20% and with generally impaired reproductive capabilities of the adults (Hickey and Anderson, 1968; Ratcliffe, 1967, 1970; Berger *et al.*, 1970; Risebrough *et al.*, 1970).

In 1975, the North American peregrine survey was repeated. Again the *pealei* subspecies had a stable population. In contrast, the eastern *anatum* population was virtually, if not entirely, extirpated, and the *tundrius* subspecies had suffered a further decline in abundance and was clearly in trouble (Fyfe *et al.*, 1976). By 1985 there were only about 450 pairs of *tundrius* peregrines in all of the subarctic and arctic of North America, compared with a historical abundance of 5–8 thousand pairs (Peakall, 1990).

However, as with other raptors that have suffered from the chronic effects of chlorinated hydrocarbons, a recovery of peregrine populations has begun since the banning of DDT use in North America and most of Europe in the early 1970s. In 1985, northern North American populations of peregrines were stable or increasing compared with 1975, as were some southern populations although they remained small and endangered (Murphy, 1987, 1990; Cade, 1988).

This recovery has been enhanced by a captive-breeding and release program over much of the former range of the eastern *anatum* race (Fyfe, 1976; Anonymous, 1987; Cade, 1988; Halroyd and Banasch, 1980). In the United States, a falcon breeding facility at Cornell University began a program of experimental releases in 1974. This transformed to an operational program, and by the end of 1986 more than 850 peregrines had been released (Cade, 1988). In Canada, 563 young falcons had been released at 24 sites by 1986, and at least 35 (6.2%) had returned 1 or more years after their release. The Canadian releases included 264 birds that were released to suitable breeding sites in cities, where peregrines will nest on tall buildings as surrogate cliffs and feed on pigeons (rock dove, *Columba livia*) and other urban birds. By 1993, 1229 peregrines had been released in Canada (Holroyd, 1993). In 1988, 22 pairs of peregrines nested in North American cities and towns, and 21 of these fledged young (Cade and Bird, 1990). In 1986, at least 43 pairs of peregrine falcons occupied breeding territories in eastern North America, a region from which it had been virtually extirpated as a breeding species prior to the operational release of captive-reared individuals (Cade, 1988; White *et al.*, 1990).

Carbofuran and Birds

The organophosphate and carbamate insecticides (including carbofuran) poison arthropods by inhibiting the action of a specific enzyme, acetylcholine esterase (AChE). Birds and mammals are also sensitive to this so-called cholinesterase-inhibiting effect, if they ingest or otherwise absorb enough of these same chemicals (O'Brien, 1967; Corbett, 1974; Ludke *et al.*, 1975; McEwen and Stephenson, 1979; Baron, 1991; Gallo and Lawryk, 1991; Mineau, 1991).

Acetylcholine is a biochemical responsible for the transmission of neural impulses, by diffusing across synapses. After neurotransmission, acetylcholine reacts with AChE, and choline is generated as a reaction product. The choline diffuses back across the synapse, effectively regenerating the electrochemical, neurotransmission potential. Interference with this process impairs the function of the nervous system, and if the dose of poison is

large enough, it can cause tremors, convulsions, and ultimately death through impairment of the AChE function.

Exposure of birds to cholinesterase-inhibiting chemicals is assayed biochemically, using samples of brain tissue or blood plasma (e.g., Ludke *et al.*, 1975; Holmes and Sundaram, 1992; Forsyth and Martin, 1993). A reduction of brain-AChE activity of 20% is considered to be indicative of an exposure to a AChE-inhibiting pesticide, and a reduction of more than 50% (compared with a control value) in an avian carcass collected in the field is usually considered diagnostic of a pesticide-caused death (Ludke *et al.*, 1975; Robinson *et al.*, 1988).

Because most AChE-inhibiting pesticides are rapidly excreted and/or metabolized after ingestion, animals can generally recovery from a nonlethal exposure within 2–5 days, although there may be low-level, lingering effects for weeks (Hamilton *et al.*, 1981). In the interval, there may be important effects on behavior, and in the case of birds, nesting success may be impaired (Robinson *et al.*, 1988; Holmes and Boag, 1990; Hooper *et al.*, 1990; Busby *et al.*, 1983a, 1989; Kendall and Akerman, 1992; Forsyth and Martin, 1993).

Carbofuran is a carbamate pesticide, registered in North America and elsewhere for a variety of insecticidal uses in agriculture. Carbofuran is available as a flowable, liquid suspension, which can be broadcast-sprayed against epidemic pests, such as grasshoppers and leaf beetles. Carbofuran is also available in dry, granular formulations, in which the chemical is present on the surface of particles of grit. Granular carbofuran is used to protect tender, recently germinated, crop seedlings from insect damages, and for this purpose the insecticide is applied while the seeding is done.

The seed and carbofuran-containing granules are usually sown in one of two ways: (1) as a "banded application," which leaves a relatively large number of granules exposed on the surface, or (2) as an in-furrow application, so that most of the granules are covered by soil.

One type of banded application, favored by about 70% of corn farmers in Ontario, results in an estimated 15–31% of carbofuran granules remaining on the soil surface after application, equivalent to 515–1065 granules exposed/m of furrow [note that all data in this case study are from a hazard assessment of carbofuran by Mineau (1993)]. Another method of banded application, preferred by the other 30% of corn farmers, is also inefficient, leaving 7–16% of granules exposed. In-furrow applications, if they were required, would leave smaller numbers of granules exposed, about 0.5–0.8% of those applied, or 17–27/m of furrow. Similar observations have been made for other crops where carbofuran is used, for example, 4.7–5.3% of applied granules on the surface for rapeseed (*Brassica napus*) in western Canada.

From the avian perspective, there are important, ecotoxicological risks associated with having large numbers of carbofuran-containing granules on the soil surface. The granules are highly attractive to seed-eating birds, who actively seek out and ingest grit of that size and texture and retain them in their gizzard for use in grinding seeds. The carbofuran is, however, solubilized and absorbed and can poison the birds.

Carbofuran is highly toxic to birds, as is indicated by the following oral-LD_{50} values: (1) ducks, 0.2–0.7 mg/kg body weight, (2) seven small passerine species, 0.4–6 mg/kg; (3) three grouse and pheasant species, 1.7–5.0 mg/kg, (4) American kestrel (*Falco sparverius*), 0.6 mg/kg; and (5) screech owl (*Otus asio*), 1.9 mg/kg. Because a typical granule in one of its commonly used formulations can contain 0.032 mg of carbofuran, the consumption of only 1–5 granules can be fatal to small, seed-eating birds.

In addition, carbofuran applied to soil can contaminate invertebrates, which may later be eaten by birds. After an in-furrow application to a cornfield at 1.1 kg a.i./ha, earthworms averaged 85 ppm of carbofuran, and as much as 670 ppm. A single earthworm (*Lumbricus terrestris*) at 85 ppm carbofuran would supply a dose of 5.5 mg/kg to a robin (*Turdus migratorius*), while a worm at 670 ppm would yield 44 mg/kg. Both of these represent lethal exposures to a robin-sized bird. Of course, larger birds and mammals can then be secondarily poi-

soned, if they scavenge the carcasses of birds killed by a primary exposure to carbofuran.

Another type of lethal exposure can occur when fields that have had carbofuran applied become flooded. In such cases, the surface water can have large concentrations of the insecticide, while also providing superficially attractive habitat for waterfowl and some other birds. The risk of toxic exposure is especially great if the water is acidic, because under this condition carbofuran does not break down rapidly. At pH 9.5 it takes only 0.2 days to break down half of an initial quantity of carbofuran, but at pH 5.2 this takes 1700 days. Depending on soil type and land use, standing waters in agricultural fields can often be acidic.

Carbofuran is well known for the mortality it has caused to wildlife, especially birds, during its normal usages in agriculture. In North American agriculture, carbofuran probably causes more nontarget avian mortality during registered usages than any other pesticide.

Some information about ecotoxicity is available from experimental studies of the use of granular carbofuran. In two studies of cornfields in Iowa and Illinois, relatively great care was taken to minimize the surface occurrence of granules during the pesticide application, so that the risks of toxic exposure would have been relatively small, compared with operational, less careful treatments. One of the studies involved a 69-ha field and reported 103 carcasses of 17 species of bird (1.5 bodies/ha). The other study was of a 125-ha field and found 29 carcasses of 11 species (0.23 bodies/ha). These are all, of course, minimal estimates of mortality, because not all carcasses would have been found during the study. Another study of a cornfield in Texas reported a kill rate of 0.74 birds/ha.

There have also been many reports of bird kills following registered, operational uses of carbofuran products. Some of the North American incidents include the following [see Mineau (1993) for a comprehensive compendium]:

1. Incidents involving ingestion of granular carbofuran:

a. In May, 1984, more than 2000 Lapland longspurs (*Calcarius lapponicus*) were killed in a rapeseed field in Saskatchewan.

b. In September, 1986, an estimated 500–1200 seed-eating birds, mostly savannah sparrows (*Passerculus sandwichensis*), were killed in turnip (*Brassica rapa* var. *rapifera*) and radish (*Raphanus sativus*) fields in British Columbia.

c. In April, 1990, more than 200 passerine birds were killed in a cornfield in Virginia.

2. Incidents involving flooded fields polluted by granular carbofuran:

a. In December, 1973, 50–60 pintails (*Anas acuta*) and mallards (*A. platyrhynchos*) were killed in flooded turnip fields in British Columbia.

b. From November 1974 to January 1975, 80 ducks, mostly pintails, mallards, green-winged teal (*Anas carolinensis*), and widgeon (*Mareca americana*) were killed in flooded turnip fields in British Columbia.

c. From October to December 1975, more than 1000 green-winged teal were killed within a few hours of landing in a flooded turnip field in British Columbia.

d. During April and May 1990, 34 snow geese (*Chen hyperborea*), 7 ducks, a gull, and frogs were killed in a flooded cornfield in Delaware.

e. In January 1990, 155 ducks and grebes, 5 hawks, and uncounted songbirds were killed in a flooded cornfield in California.

f. Between 1984 and 1988, 22 incidents totalling 525 bird deaths were reported from carbofuran-treated ricefields in California.

3. Incidents involving flooded fields polluted by flowable carbofuran:

a. In March 1974, 2450 widgeon died the day after an alfalfa field was treated with carbofuran in California.

b. In April 1974, 79 coots (*Fulica americana*) were killed in an alfalfa field in Kansas.

c. In February 1976, 500 Canada geese (*Branta canadensis*) were killed in an alfalfa field in Oklahoma.

d. In May 1976, 750–1000 widgeon were killed in an alfalfa field in Kansas.

e. In March 1977, 1100 widgeon were killed in an alfalfa field in California.

f. In April 1985, 150 widgeon and 10 Canada geese were killed in an alfalfa field in Oklahoma.

g. In June 1986, 45 gulls died after eating carbofuran-contaminated grasshoppers in Saskatchewan.

The above represent only a fraction of the known incidents of events of bird kills associated with registered, operational uses of carbofuran in agriculture. There must also, of course, be much larger numbers of unknown bird kills, since there are no comprehensive, legal requirements to report such incidents of wildlife kills in agriculture, and many farmers would be reluctant to do so on a voluntary basis. Just as important, kills are difficult to detect without an intensive search for bodies, and the evidence is quick to disappear because of scavenging (Mineau and Collins, 1988).

Among the various pesticides that have been registered for regular agricultural usage in North America during the most recent decade, carbofuran has probably been the most problematic in terms of unintended damage to wildlife. In 1990, the American Ornithologists' Union passed a resolution demanding bans on the uses of carbofuran. Mostly because of the mortality caused to birds, the registration of carbofuran for agricultural usages is being reviewed by the federal governments of the United States and Canada. In 1993, the U.S. Environmental Protection Agency announced a negotiated settlement with the manufacturer of carbofuran to withdraw all but five minor uses of the granular formulation in the United States. As of late 1993, the use of carbofuran in liquid formulations continues, as does the use of granular formulations in Canada.

Agricultural Herbicides and Birds

A number of studies conducted in Britain have suggested that substantial decreases have occurred in the populations of some species of birds that breed on agricultural lands. It has been suggested that these changes in avian populations have been partly caused by the widespread use of herbicides since the late 1940s. This practice has resulted in substantial habitat changes, in terms of: (1) the abundance and species of weeds and other plants in the agroecosystem, a factor that affects habitat structure, nest-site availability, and amounts of food available for seed-eating birds, which largely rely on weed seeds and (2) changes in arthropod populations, which may feed on noncrop plants or rely on them for habitat.

Of course, extensive herbicide use is only one aspect of the increased intensity of agricultural practices in Britain and elsewhere. Occurring at the same time as the increased usage of herbicides in agriculture, and also influencing bird populations, have been the elimination or degradation of hedgerows from many landscapes, changes in crop species and cultivation techniques, insecticide and fungicide use, and improvements of seed cleaning, so that fewer weed seeds are sown with the crop seed (Potts, 1977, 1985; O'Connor and Shrubb, 1986; Sotherton and Rands, 1986; Freemark and Boutin, 1993).

Populations of gray partridge (*Perdix perdix*) have declined substantially in Britain during recent decades. In 1952, the average density of this species in Britain was about 25 pairs/km^2, but by the mid-1980s it was 5 pairs/km^2, an 80% decrease (Sotherton and Rands, 1986). This phenomenon has been widely attributed to the extensive use of herbicides in cereal agriculture in Britain. This weed-control practice results in smaller populations of noncrop plants and correspondingly small populations of arthropods. Insects and other arthropods are a critical food for partridge chicks until they are about 2–3 weeks old, after which they switch to a largely granivorous diet (Green, 1984; Sotherton and Rands, 1986).

Chick survival rate is usually the demographic parameter that has the strongest influence on population size of gray partridge in Britain, although nest predation can also be important in some circumstances (Southwood and Cross, 1969; Potts, 1980). This species is nidifugous, meaning that chicks leave the nest as soon as the clutch is fully hatched. When they are young, partridge chicks must catch their own food, and they forage mostly for arthropods. At an age of 10 days, chicks require at least 2 g d.w. of arthropods per day, but they grow and survive better if they are able to eat 3 g/day (Southwood and Cross, 1969). If there is an insufficient abundance of arthropods for young chicks to feed upon, they suffer an increased rate of mortality, and populations of partridge decrease. Several studies have demonstrated that the survival of chicks of gray partridge is positively related to the abundance of arthropods in their foraging habitat (Potts, 1980; Green, 1984; Rands, 1985, 1986a,b).

Several studies in Britain have demonstrated that survival rates of gray partridge and pheasant (*Phasianus colchicus*) are higher in fields in which small, nonherbicided refugia are left, in comparison with completely sprayed fields. These refugia can include unsprayed perimeters or other parts of grainfields, as well as perennial hedgerows, both of which have larger densities of arthropods than herbicided parts of fields (Hill, 1985; Rands, 1985, 1986a,b, 1987; Sotherton *et al.*, 1985).

Other species of birds whose population declines in Britain have been partly attributed to the agricultural use of herbicides are stock dove (*Columba oenas*), and the finches, linnet (*Carduelis cannabina*) and reed bunting (*Emberiza schoeniclus*). These largely granivorous species have suffered from a decreasing availability of weed seeds in herbicided fields, along with decreased nesting habitat in hedgerows and other factors related to intensive agricultural practices (O'Connor and Shrubb, 1986).

Spraying Forests Infested with Eastern Spruce Budworm

The Setting

Spruce budworms are important, lepidopteran defoliators of conifers in north-temperate and boreal forests. In North America, the economically most important species are the eastern spruce budworm (*Choristoneura fumiferana*) and the western spruce budworm (*C. occidentalis*), both of which feed on fir and spruce. Especially large amounts of forest damage are caused by the eastern spruce budworm. This is because of the tremendous areas of its recent infestations and because it causes extensive mortality in stands that have been continuously defoliated for a number of years. Defoliation by the western spruce budworm does not generally cause the death of trees, but it does cause important losses of forest productivity. In this case study, the focus will be on the eastern spruce budworm [see Brooks *et al.* (1985) and Sanders *et al.* (1985) for information on other budworm species].

Biology of Eastern Spruce Budworm

The eastern spruce budworm produces a single generation each year (MacLean, 1984; Blais, 1985; Sanders, 1991). Moths emerge from mid-July to early August, and each female lays an average of 10 egg masses, each containing about 20 eggs, on the foliage of host conifer species. The eggs hatch after about 10 days, and the first-instar larvae disperse locally on silken threads or sometimes beyond the stand by "ballooning" with the wind. Larvae that survive their high-risk dispersal manufacture a silken hibernaculum on a host tree and moult to the second instar. This stage overwinters, breaks its diapause in late April to early May and emerges and begins to feed by mining (i.e., feeding internally) current-year and 1-year-old foliage, and, if available, on seed and male flowers. Later instar larvae feed on the surface of foliage. Because of its relatively large size, the sixth instar does most of the damage, accounting for about 87% of the total defoliation. This instar prefers current-year foliage, but if this resource has been depleted these caterpillars will feed on older needles. Pupation takes place in early July, and the moths emerge after 8–12 days. Adult moths can undergo a long-range dispersal, typically travelling 50–100 km downwind, and as far as 600 km.

Population Dynamics

It appears that periodic irruptions of eastern spruce budworm have long been a feature of certain boreal landscapes. Blais (1965) established a budworm outbreak chronology that extended back to the 18th century in an area in Quebec. The evidence for historical outbreaks was based on ring-width measurements of living host trees, which recorded historical reductions of radial growth caused by defoliation. Blais concluded that outbreaks had taken place at rather even intervals and that infestations had begun in 1704, 1748, 1808, 1834, 1910, and 1947, for an average periodicity of about 35 years.

During its endemic phase of small population density, eastern spruce budworm is scarce. Only 10 larvae of eastern spruce budworm were found in more than 1 thousand insect collections made in New Brunswick during an endemic phase between 1939 and 1944 (Blais, 1985). This represents an estimated density of about 5 individuals per host tree (Miller, 1975). At the beginning of an outbreak, the density of larvae increases rapidly, to about 2 thousand larvae per tree within 4 years. The density then increases to more than 20 thousand per tree during the epidemic phase of the outbreak. This large density is typically sustained for 6–10 years, after which the outbreak collapses.

Outbreaks tend to be synchronous over large regions of susceptible forest (Royama, 1984). However, on more local scales the amplitudes of the oscillations vary, so that there are large differences among stands in the maximum density of budworm. According to Royama (1984), the occurrence and persistence of population outbreaks of budworm are governed by several intrinsic, density-dependent factors that affect mortality. Especially important are the abundance of insect parasitoids, disease, and an undefined "complex of unknown factors" (a recent outbreak collapsed for no discernible reason, hence the vague third group). Royama discounted the roles of predation, food shortage, weather, mortality during the dispersal phases, and the spraying of infested stands with insecticide. He also hypothesized that the invasion of stands by large numbers of egg-bearing moths was not important in upsetting the population equilibrium during the endemic phase. He felt that such invasions only played a role in accelerating already-occurring increases in abundance.

Forest Damage

The spatial extent of outbreaks of eastern spruce budworm appears to have increased in modern times. The outbreak that began in 1910 involved some 10 million ha while another that began in 1940 involved 25 × 10⁶ ha, and the one that began in 1970 involved 55–57 × 10⁶ ha (Fig. 8.3; Ketella, 1983; MacLean, 1990). Blais (1983, 1985) hypothesized that the trend to an increasingly widespread infestation was due to a combination of factors that may have been important in creating large areas of optimal budworm habitat. In no particular order, these are: (1) forest management practices such as clear-cutting; (2) fire protection; (3) regeneration of budworm-susceptible conifer stands on abandoned agricultural lands; and (4) the spraying of infested stands with insecticides, which keeps the habitat suitable for budworm and may thereby help to prolong the infestation [note that Royama (1984) has a contrasting view about factor 4]. These practices have had the effect of either producing or keeping

alive a mature fir-dominated forest, which is the preferred habitat of eastern spruce budworm.

Stands that are particularly vulnerable (i.e., apt to suffer damage) during an outbreak of eastern spruce budworm are mature and dominated by balsam fir (*Abies balsamea*) (MacLean, 1980, 1985, 1990). Immature stands of balsam fir are less vulnerable, followed by mature stands of spruce and then immature spruce stands. In terms of tree species, balsam fir is most vulnerable, followed by white spruce (*Picea glauca*), red spruce (*P. rubens*), and then black spruce (*P. mariana*).

Stand characteristics that are believed to increase susceptibility to eastern spruce budworm include the following: (1) species composition is mainly balsam fir, followed by white spruce; (2) stand age >60 years; (3) a large biomass of susceptible species; (4) an open stand with spike tops of host species extending above the canopy; (5) location within an extensive area of susceptible forest; (6) location downwind of an ongoing infestation; (7) elevation <700 m and south of 50°N latitude; and (8) a rela-

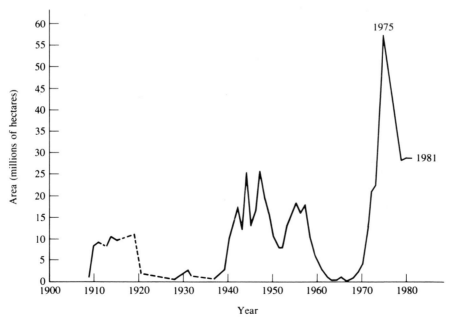

Fig. 8.3 The forest area moderately or severely infested by spruce budworm during the twentieth century, from Kettela (1983).

tively dry or wet site condition (Blum and Mac-Lean, 1985; MacLean and Ostaff, 1989).

Forests with a vulnerable character are widespread in eastern Canada and to a lesser extent in the northeastern United States, comprising about 60.7 $\times$ 10^6 ha. This is reflected in the distribution of budworm-damaged forest in 1981 (Fig. 8.4). Between 1977 and 1981, the average tree mortality caused by eastern spruce budworm in this region was equivalent to more than 38 $\times$ 10^6 m^3/year, and in 1983 substantial mortality occurred over an area of about 26.5 $\times$ 10^6 ha (Ostaff, 1985).

The progression of an infestation of eastern spruce budworm can be illustrated by the case of Cape Breton Island, Nova Scotia (Ostaff and Mac-

Lean, 1989). No defoliation was observed during surveys conducted in 1973. By 1974, however, moderate-to-severe defoliation of first-year foliage was found over 165 thousand ha, increasing to 486 thousand ha in 1975 and 1220 thousand ha in 1976, when virtually all vulnerable forest was infested.

In balsam fir trees, the first year of severe defoliation causes a decrease in volume growth of about 15–20%, followed by 25–50% in the second year of defoliation and 75–90% in subsequent years (MacLean, 1985, 1990). The growth loss of spruce trees is somewhat less, ranging up to 50–60%. Height growth stops entirely during a severe defoliation; in fact, tree height can decrease because of top-kill. Mortality caused by budworm begins in

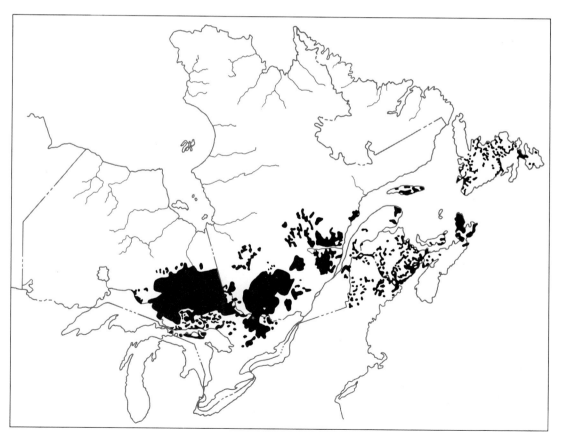

Fig. 8.4   Areas suffering mortality of balsam fir in 1981 as a result of spruce budworm infestation, from Kettela (1983).

Aerial view of an extensive area of conifer forest damaged by repeated defoliations by eastern spruce budworm on Cape Breton Island, Nova Scotia. The fringe of live trees around the bog on the left-hand side of the photo are black spruce, a relatively resistant species. The area in which most mature trees are dead was dominated by balsam fir, a relatively vulnerable species. This photograph was taken about 8 years after the collapse of budworm populations (B. Freedman photo).

A moribund forest of balsam fir, which has suffered at least 7 years of severe defoliation by spruce budworm. This Cape Breton forest was not sprayed with insecticide to reduce the abundance of spruce budworm. The gray cast of the forest indicates a great deal of mortality of canopy trees (photo: B. Freedman).

fir stands that have been continuously defoliated for 4–5 years, and after 7–8 years in spruce stands.

Mortality within the stand is progressive with time. Mortality of balsam fir was examined in unsprayed, mature stands on Cape Breton Island (Mac-Lean, 1984; Ostaff and MacLean, 1989). About 4% of balsam fir trees were dead after 2 years of heavy defoliation, 9% after 4 years, 22% after 5 years, 37% after 6 years, 48% after 8 years, 61% after 9 years, 75% after 10 years, and 95% after 12 years. Spruce budworm-caused mortality usually ends about twelve years after the beginning of the outbreak. The average final mortality was 85% (maximum of 100%) in mature balsam fir stands in eastern Canada,

A closer view of a moribund balsam fir stand on Cape Breton Island. Note the dense advance regeneration of balsam fir under the dead overstory. In about 40–50 years, this regeneration will provide the next harvestable balsam fir forest for spruce budworm or for humans (photo: B. Freedman).

42% in immature fir stands, 36% in mature spruce stands, and 13% in immature spruce stands (Mac-Lean, 1980, 1990).

Note that a great deal of mortality can occur after the budworm population has collapsed, as weakened trees succumb to environmental stresses that might otherwise be tolerated. Blais (1981) studied a situation in Quebec where a budworm outbreak had collapsed in 1975. At that time the mortality of balsam fir averaged 44%, but this increased to 91% after 4 postoutbreak years. Mortality of spruce increased from 17% in 1975 to 52% 4 years later.

In the understory of stands that have been damaged or killed by eastern spruce budworm, there is an advance regeneration of tree species that is generally dominated by the same species that comprised the original stand. For example, stands on Cape Breton Island that had been severely defoliated for 4–6 years had an average density of 45 thousand individuals of balsam fir/ha and 3.25 thousand spruce/ha (MacLean, 1984, 1988). Almost all of these small individuals of tree species survived the budworm infestation. After the death of the overstory, the advance regeneration is able to grow relatively freely, and it begins to establish the succeeding fir–spruce forest. Low-growing species of the ground vegetation also respond favorably to opening of the forest canopy by budworm-related mortality, especially shield fern (*Dryopteris austriaca*), goldenrod (*Solidago macrophylla*), sarsaparilla (*Aralia nudicaulis*), wood sorrel (*Oxalis montana*), and red raspberry (*Rubus strigosus*)

Therefore, the dynamic, budworm–fir system can be viewed as a cyclic succession with a longer-term ecological stability. This cycle has probably recurred on the landscape for thousands of years (Baskerville, 1975a; MacLean, 1984; Blais, 1985). Evidence that supports the hypothesis of longer-term stability includes: (1) the presence of large areas of relatively pure, even-aged, balsam fir forest; (2) the observed successional trajectory after stand mortality caused by eastern spruce budworm; and (3) dendrochronological data that indicate that budworm outbreaks are an ancient and periodic phenomenon.

In contrast to this longer-term ecological stabil-

Cordwood stacked along a logging road on the Cape Breton highlands. After the decision was made not to spray the spruce budworm-infested forest, a great deal of tree mortality occurred. Much of the dead forest was salvage harvested, using very large skyline-to-skyline clear-cuts. The wood in the photo had been stored at roadside for more than 5 years and, because of decomposition, it was then only marginally useful for the production of paper (photo: B. Freedman).

ity, eastern spruce budworm causes a severe economic instability, making it difficult for humans to plan for a longer-term exploitation and management of the forest. Industries that harvest the forest are in direct competition with budworm for the same natural resource. Because budworm causes an unstable wood supply, there is severe economic instability (Baskerville, 1975a,b; MacLean, 1984).

As a way of coping with this resource crisis, forests have been sprayed with insecticides in an attempt to limit defoliation and prevent some of the tree mortality. The objective of this management strategy is not to eradicate the eastern spruce budworm, but to lessen the degree of the damage to the forest resource and its dependent industries that it causes. In a study of fir-dominated stands in New Brunswick, Clowater and Andrews (1981) found that an average of 67% of the balsam fir volume was dead or severely damaged in unsprayed stands, compared with 38% in sprayed stands. In another study in New Brunswick, MacLean *et al.* (1984) found that balsam fir mortality in unsprayed spruce-dominated stands averaged 59%, while in sprayed stands it was 39%.

MacLean and Erdle (1984) modeled the effects of eastern spruce budworm on wood supply in New Brunswick. The model forecast that if forest protection were withdrawn as a management practice, the maximum, sustainable harvest of the spruce–fir forest would be reduced by 46–64% under a severe outbreak condition and by 23–36% under a moderate outbreak condition. Projections such as these have provided a powerful economic justification to mount insecticide spray programs in regions where the forest is infested with eastern spruce budworm. Deloitte and Touche Management Consultants (1992) estimated that the cost : benefit ratio of the use of insecticides against eastern spruce budworm was 1 : 5, i.e., for every dollar spent to kill budworm, there would be a net profit of $4.00. [This analysis only considers "conventional" economic costs and benefits and does not consider the nonvaluated, ecological costs of insecticide spraying against eastern spruce budworm (see Chapter 12).]

Insecticide Use and Residues

The first insecticide that was used against eastern spruce budworm was calcium arsenate. This was first used in 1927–1929 in Nova Scotia, where it was applied aerially at a rate of 10–40 kg/ha (Nigam, 1975; Armstrong, 1985b).

However, the real era of operational spraying of infested forests did not begin until after the Second World War, when DDT began to be used in large quantities. In 1949, about 108 thousand ha of forest was sprayed with DDT in Oregon and Washington; in 1952, 113 thousand ha was sprayed in Quebec and New Brunswick, and in 1953, 804 thousand ha was sprayed there (Armstrong, 1985b). Up to 1968, about 15-million ha of budworm-infested forest was sprayed at least once with DDT in eastern North America, and many stands more than once (Ennis and Caldwell, 1991). In New Brunswick alone between 1952 and 1968, a total of about 5.75 million kg of DDT was sprayed on infested stands (Pearce, 1975).

Other insecticides have also been used against eastern spruce budworm, especially after the use of DDT for this purpose was suspended in 1968. Up to 1985, the usage of insecticides against eastern spruce budworm was as follows (Ennis and Caldwell, 1991): DDT from 1945 to 1968, 15 million ha; phosphamidon between 1963 and 1978, 8.1 million ha; fenitrothion between 1969 and 1985, 64 million ha; aminocarb from 1972 to 1985, 19.4 million ha; mexacarbate between 1972 and 1975, <1 million ha; and *Bacillus thuringiensis* to 1985, 1.9 million ha.

The total area treated with insecticide up to 1985 was 118.5 million ha, of which 69.1 million ha were in New Brunswick, 37.0 million ha in Quebec, 10.0 million ha in Maine, and 1.7 million ha in

Newfoundland, with smaller areas elsewhere (Ennis and Caldwell, 1991). Note that these data are the sums of annual spray areas; many stands would have been treated a number of times during the budworm outbreaks.

New Brunswick has had the largest and most continuous program of forest spraying for "protection" against eastern spruce budworm, with large areas being treated with various insecticides in most years between the early 1950s and the early 1990s (Table 8.7). Much of this terrain was resprayed in consecutive years (Eidt and Weaver, 1986). For example, in 1982 about 1.7×10^6 ha were sprayed. Of this, 51% was resprayed in 1983; 17% was resprayed in 1984; 14% was resprayed in both 1983 and 1984; and 3.4% was resprayed in 1983, 1984, and 1985 (i.e., during 4 consecutive years).

Table 8.7   Insecticide spraying of forests in New Brunswick for the control of spruce budworm[a]

Year	Area sprayed (10^6 ha/year)	Year	Area sprayed (10^6 ha/year)
1952	0.08	1973	1.8
1953	0.73	1974	2.3
1954	0.45	1975	2.7
1955	0.45	1976	4.2
1956	0.81	1977	1.6
1957	2.1	1978	1.5
1958	1.1	1979	1.6
1959	0.00	1980	1.6
1960	1.1	1981	1.9
1961	0.89	1982	1.7
1962	0.57	1983	1.7
1963	0.28	1984	1.3
1964	0.81	1985	0.8
1965	0.85	1986	0.5
1966	0.81	1987	0.6
1967	0.40	1988	0.5
1968	0.20	1989	0.6
1969	1.3	1990	0.6
1970	1.7	1991	0.3
1971	2.4	1992	0.3
1972	1.9		

[a]DDT was the major insecticide used until 1968. After this time DDT was replaced by phosphamidon, fenitrothion, and aminocarb. Fenitrothion is currently the most important synthetic organic insecticide being used, but it is increasingly being supplanted by *B.t.* Modified from McEwen and Stephenson (1979), Mitchell and Roberts (1984), Eidt and Weaver (1986), and Carter (1992).

The typical spray rates and selected toxicity data for the most important insecticides that have been used against eastern spruce budworm are summarized in Table 8.8. Note that of these insecticides, DDT is least toxic to the budworm itself and is relatively toxic to salmonid fish. It was largely because of kills of commercially and recreationally important Atlantic salmon (*Salmo salar*) and brook trout (*Salvelinus fontinalis*) caused by DDT spraying that this chemical was banned for forestry purposes in New Brunswick (see below). The other insecticides listed in Table 8.8 are more toxic to budworm than is DDT, and therefore they can be sprayed at smaller operational rates. In addition, these insecticides are less toxic to fish than is DDT. Toxicity to other organisms varies; the relatively great toxicity of phosphamidon to birds resulted in much avian mortality during operational spraying of this chemical, and this was the major reason why it was no longer used after 1975.

Another important problem with DDT is its long persistence in the environment. As was discussed earlier, the persistence of DDT, coupled with its great solubility in lipids, meant that it could accumulate to large concentrations in organisms, especially in top predators.

Yule (1975) examined an area in New Brunswick that had been aerially treated with DDT between 1956 and 1967, for a cumulative spray equivalent to 4.9 kg DDT/ha. In 1968, the year after the last application of DDT, there was a residue of total DDT in soil and the forest floor that was equivalent to 0.77 kg/ha. Between 1968 and 1973 the half-life of DDT in this compartment was about 10 years. DDT was also persistent in conifer foliage, even if the tree had not been sprayed for several years (Table 8.9). The concentration of DDT was smallest in foliage produced after spraying stopped in 1967, but DDT was nevertheless present in fairly large concentrations in unsprayed tissues. Some of the DDT made its way into herbivores and predators. The residues in some mammal species in New Brunswick could be large, with a concentration of up to 156 ppm being measured in fatty tissue of a bobcat (Table 8.10).

The residues of insecticides that have been used more recently against eastern spruce budworm are much less persistent than DDT. Sundaram and Nott (1985) determined residues of aminocarb after two sprays, each of 0.07 kg a.i./ha (the usual operational procedure is to treat each stand twice, with about 5–7 days between sprays). Immediately after the first spray there was 2.2 ppm of aminocarb in foliage of balsam fir. The residues decreased to 0.73 ppm after 2 days, 0.40 ppm after 5 days, and 0.29 ppm after 7 days. A similar pattern was observed

Table 8.8 Data for acute toxicity of insecticides to fifth-stage spruce budworm larvae and selected vertebrate animals under controlled laboratory conditions[a]

Insecticide[a]	Typical Application rate (kg a.i./ha)	Contact toxicity to budworm (μg/cm²)		Rainbow Trout 96-hr LC$_{50}$ (ppm)	Pheasant oral LD$_{50}$ (mg/kg)	Rat oral LD$_{50}$ (mg/kg)
		72-hr LD$_{50}$	72-hr LD$_{95}$			
DDT (1)	0.3–2.2	1.3	6.6	0.0087	1334	87–500
Phosphamidon (2)	0.3	0.39	0.75	7.8	4.2	15–33
Fenitrothion (3)	0.21	0.31	0.67	2.4	56	250–600
Aminocarb (4)	0.07	0.04	0.11	13.5	42	30
Mexacarbate (5)	0.07	0.04	0.13	12.0	4.6	15–63

[a]Modified from Nigam (1975), Johnson and Finley (1980), and Hudson et al. (1984).
[b]Insecticides: (1) chlorinated hydrocarbon, 1,1,1-trichloro-2,2-*bis*(*p*-chlorophenyl) ethane; (2) organophosphate, 2-chloro-*N,N*-diethyl-3-hydroxycrotonamide, dimethyl phosphate; (3) organophosphate, *O,O*-dimethyl *O*-(4-nitro-*m*-tolyl) phosphorothionate; (4) carbamate, 4-dimethylamino-*m*-tolyl methylcarbamate; (5) carbamate, 4-dimethylamino-3,5-xylyl methylcarbamate.

Table 8.9 Total DDT residues in foliage of balsam fir and red spruce in an experimental spray plot in New Brunswick[a]

| Year foliage was produced | Concentration of DDT (ppm) | | | | | |
| | Balsam Fir | | | Red Spruce | | |
	1967	1969	1973	1967	1969	1973
1960	7.7					
1961	6.7					
1962	5.4			2.1		
1963	7.4			2.5		
1964	5.1			3.4		
1965	5.9	9.1		2.3	3.4	
1966	3.8	7.1		2.3	3.1	
1967	3.7	6.3		2.7	1.0	
1968		4.6			1.4	
1969		1.5	0.78		1.9	0.96
1970			0.76			0.54
1971			0.16			0.02
1972			0.13			0.01
1973			0.01			<0.01

[a]The plot was sprayed annually between 1956 and 1967 with a total of 4.9 kg DDT/ha. Foliage was collected in 1967, 1969, and 1973 and was sorted into year classes prior to its analysis for DDT. Modified from Yule (1975).

after the second spray. The initial concentrations in forest litter and soil averaged <0.11 and <0.06 ppm, respectively. In another field study in which soil was sprayed directly, A. Sundaram *et al.* (1985) found that the residues of aminocarb disappeared quickly. The initial concentration in the surface 1 cm of mineral soil was 0.050 ppm. The residue decreased to <0.007 ppm between 1 and 3 days and to <0.003 ppm after 4 days.

Residues of fenitrothion are similarly short lived. Immediately following two operational sprays of 0.21 kg a.i./ha, the concentration in foliage of balsam fir was as large as 2.9 ppm, while it was up to 1 ppm in the forest floor and 0.60 ppm in mineral soil (Sundaram and Nott, 1984). The residues in the forest floor and soil were not persistent and quickly decreased to <0.01 ppm (Ayer *et al.*, 1984; Sundaram and Nott, 1986). In contrast, the residues in needles of balsam fir appear to decrease to a threshold of about 0.5–1.5 ppm f.w., indicating that foliage serves as a weak sink for fenitrothion (Ayer *et al.*, 1984; Eidt and Mallet, 1986; Eidt and Pearce, 1986; Sundaram and Nott, 1986). The hydrolysis of fenitrothion in water is pH dependent,

Table 8.10 Average DDT residues in mammals in an area of New Brunswick that was sprayed annually between 1956 and 1967 with a total of 4.9 kg DDT/ha[a]

Species	Tissue	Sample size	Total DDT (DDD + DDE + DDT) (ppm wet weight)
Herbivores			
Red squirrel (*Tamiasciurus hudsonicus*)	Whole body	2	0.34 (0.20–0.48)
Beaver (*Castor canadensis*)	Fat	1	3.0
	Liver	1	0.08
Snowshoe hare (*Lepus americanus*)	Muscle	5	0.003 (tr–0.008)
White-tailed deer (*Odocoileus virginiana*)	Muscle	13	0.042 (0.014–0.11)
Carnivores			
Raccoon (*Procyon lotor*)	Fat	2	0.53 (0.31–0.75)
Short-tailed weasel (*Mustela erminea*)	Whole body	8	1.8 (0.65–3.0)
Mink (*Mustela vison*)	Whole body	2	3.9 (3.3–4.5)
Bobcat (*Lynx rufus*)	Fat	4	50 (9.3–156)
	Liver	10	6.1 (0.13–41)
	Muscle	2	1.8 (1.1–2.5)

[a]Modified from Buckner and McLeod (1975).

with a half-life of about 196 days at pH 5, 183 days at pH 7, and 100 days at pH 9 (Anonymous, 1993d).

Insecticides are often deposited to surface waters during operational spray programs. However, the concentrations in water disappear fairly quickly, as a result of dilution, downstream transport, and degradation. Kingsbury (1977) applied fenitrothion at 0.42 kg/ha to a small lake in Ontario. The initial postspray concentration was 21 ppb, but this declined to 8 ppb after 6 hr, 2–3 ppb after 12–48 hr, <0.6 ppb after 8 days, and <0.03 ppb after 1 year. Fish accumulated the fenitrothion rapidly, to a concentration of 1.0 ppm in whitefish (*Catastomus commersoni*), 0.76 ppm in fallfish (*Semotilus corporalis*), 0.44 ppm in brown bullhead (*Ictalurus nebulosus*), and 0.34 ppm in smallmouth bass (*Micropterus dolomieu*). However, no deaths or distress behavior were observed in these fish.

In recent years, there has been a substantial replacement of synthetic organic insecticides in budworm spray programs with a biological pesticide based on the bacterium *Bacillus thuringiensis* (B.t.), which can be applied as a high-potency, low-volume, aerial spray. Specific strains of B.t. can be effective against a wide variety of leaf-eating lepidopteran pests, including eastern spruce budworm, and some other susceptible insects such as blackflies and mosquitoes. B.t. can be applied using technology similar to that used for chemical insecticides, but the efficacy of B.t. is usually more variable and less effective overall, compared with chemical insecticides. However, the environmental effects of B.t. are considered to be more acceptable, because there is little nontarget toxicity.

Initially, B.t. was relatively expensive for budworm spraying, and its efficacy was substantially less than synthetic insecticides such as fenitrothion. These differentials have narrowed appreciably, however, so that in 1988 in Canada the costs of B.t. were (U.S.) $25.56/ha, compared with $20.38/ha for synthetics, and the effectiveness of formulations of B.t. and its application techniques were more predictable (van Frankenhuyzen, 1990; Cunningham and van Frankenhuyzen, 1991).

The use of B.t. against eastern spruce budworm has increased greatly in eastern North America (Armstrong, 1985a; Cunningham, 1985; Carter and Lavigne, 1986; van Frankenhuyzen, 1990; Cunningham and van Frankenhuyzen, 1991; Carter, 1992). B.t. is the only insecticide used against eastern spruce budworm in Nova Scotia in recent decades, with use between 1980 and 1984, when spray programs were relatively large, being 21–31 × 10^3 ha annually. B.t. has also been relatively important in Ontario (accounting for 17–91% of the 3.2–20.3 × 10^3 ha sprayed between 1979 and 1984), Newfoundland (100% of the 5.9–11.8 × 10^3 ha sprayed in 1979–1980, but <10% of the 48–240 × 10^3 ha treated in 1981 to 1983), and Maine (11–33% of the 332–491 × 10^3 ha treated in 1980 to 1984). In 1987, B.t. was the only insecticide used against eastern spruce budworm in Ontario, Quebec, Maine, and Nova Scotia. The use of B.t. has also increased in New Brunswick, the Canadian province with the largest budworm-spray programs, accounting for <1% of the 1.3–1.9 × 10^6 ha sprayed annually during 1979–1984, but 17–47% (average 28%) of the 0.3–0.6 × 10^6 ha treated between 1987 and 1992. Overall, B.t. was used in 63% of the spraying against eastern spruce budworm in eastern North America in 1988, compared with 2–4% between 1980 and 1984.

Effects on Nontarget Terrestrial Arthropods

An important characteristic of B.t. is its relatively small nontarget toxicity. Among nontarget terrestrial arthropods, the toxicity of B.t. is essentially limited to lepidopterans.

In contrast to *B.t.*, fenitrothion and aminocarb are both toxic to a wide spectrum of arthropods, and broadcast forest sprays with these chemicals cause large, nontarget kills of insects and spiders. According to Varty (1975), a typical spray of fenitrothion in early June can kill 2.4–7.5 million arthropods of several hundred species per hectare of fir–spruce forest (more than 90% of the biomass of the kill is typically eastern spruce budworm). Typically, a spray of fenitrothion causes a short-term decrease in arthropod biomass of 35% and a decrease in numbers of 50% (Anonymous, 1993d). Again, much of the decrease, especially of biomass, is composed of budworm larvae.

In spite of the large nontarget kill of arthropods, longer-term surveys in sprayed areas have indicated only temporarily reduced abundances of nontarget arthropods in sprayed forests in New Brunswick and elsewhere (Varty, 1975, 1977; Otvos and Raske, 1980; Millikin, 1990). Reductions of arthropod abundance may be large and significant at a local level, but not detectable at larger spatial scales. Because many of the forest arthropods sampled by conventional methods are univoltine, their apparent recoveries are probably due to movements of individuals into sprayed areas from unsprayed refugia in treated forests and not to population increases through the production of "new" individuals. In addition, there are large spatial and temporal variations that must be contended with during the quantitative sampling of forest arthropods, due partly to the widespread occurrence of unsprayed refugia. Relatively small ecological "signals" can be difficult to detect in such "noisy" sampling systems.

In the view of Varty (1975, 1977), the most important factors that affect the abundance of arthropods in balsam fir forests are: (1) the biomass of young foliage of fir, which is, of course, greatly diminished during and following severe infestations of eastern spruce budworm, so that heavily defoliated stands have a relatively small abundance of arthropods; (2) predator–prey relationships, which can cause cyclic or irruptive variations in abundance that overwhelm any spray effects; (3) density-independent influences on mortality, particularly inclement weather; and (4) insecticide intervention, which produces 3–5 days of increased mortality, usually followed by a rapid recovery. From his studies of nontarget arthropod effects in sprayed forests of New Brunswick, Varty concluded that "the arthropod community on balsam fir has not been drastically affected by fenitrothion larvicide treatments. Undoubtedly an unchecked budworm outbreak without insecticide usage would have had a far-more profound influence."

Effects on Pollination of Plants by Insects

Various species of bees have been negatively affected by the drift of insecticide to off-target, nonforest sites in New Brunswick (Kevan, 1975; Plowright et al., 1978; Wood, 1979). Bees are important commercially because they are the principal pollinators of lowbush blueberries (Vaccinium angustifolium and V. myrtilloides), a regionally important agricultural crop. Data from several studies have indicated a smaller abundance of bees and a decreased fruit set of blueberry in some sprayed areas. Subsequently, these damages were largely controlled by the prohibition of insecticide spraying within 3.2 km of any blueberry field.

A survey of plant fecundity in sprayed conifer forest demonstrated apparent effects on the reproductive success of insect-pollinated native plants, which were attributed to small populations of pollinators, especially bees (Thaler and Plowright, 1980). Fruit was produced by 71% of inflorescences of Aralia nudicaulis in unsprayed stands, but only 49% in sprayed stands; Clintonia borealis was 94% versus 78%; Cornus canadensis, 17% versus 9%; and Maianthemum canadense, 21% versus 15%. However, little is known about the effects of reduced fecundities of these magnitudes on the longer-term population dynamics of long-lived plants (Kevan and Plowright, 1989; Pauli et al., 1993). Vegetative growth and propagation are the most important mechanisms by which the above species maintain their populations in mature forest habitats, and in undisturbed situations their seedlings are rarely observed.

Effects on Nontarget Aquatic Arthropods

Effects of budworm spray programs on aquatic arthropods have also been studied, because sprays can reach surface waters by direct deposition and by drift from applications elsewhere. Either source of exposure results in a pulse of insecticide in water. In streams, the aquatic fauna is exposed to a pulse of insecticide, which rapidly diminishes in concentration because of dilution and downstream transport.

After applications of fenitrothion at 0.21–0.28 kg a.i./ha in New Brunswick, the observed concentrations in flowing water have been as large as 0.015–0.1 ppm, with a half-life of about 13 hr (Hall *et al.*, 1975; Fairchild *et al.*, 1989). In standing waters, concentrations as large as 1.1 ppm have been observed, with a half-life of 18 hr (Fairchild *et al.*, 1989).

Increased numbers of dead aquatic insects have often been observed after budworm sprays. This effect was especially severe with DDT, and in some streams in New Brunswick the density of grazing insects was so decreased that mats of filamentous algae developed over streambeds (Kingsbury, 1975). With fenitrothion the surface transport (i.e., "drift") of dead insects has sometimes been large, but there have not been measurable, longer-term reductions in the abundance of aquatic insects; the effects of aminocarb have been even smaller (Eidt, 1975, 1977; Hall *et al.*, 1975; Kingsbury, 1975; Holmes, 1979; Fairchild *et al.*, 1989). For example, the data of Table 8.11 show no apparent effects of spraying with fenitrothion or aminocarb on the drifts of aquatic and terrestrial insects in streams in Quebec. In another study, Eidt (1975) found that for a few days after spraying with fenitrothion there was a drift of aquatic insects of 95 thousand animals/24 hr, compared with a prespray drift of 13 thousand/24 hr. However, there were no measur-

able changes in total density of the benthos or of the benthic genera that were most vulnerable to suffering drift (i.e., the stoneflies *Leuctra* spp. and *Amphinemoura* spp. and the mayfly *Baetis* spp.).

Eidt (1985) studied the toxicity of B.t. to a variety of aquatic insects. He only found a toxic effect on larvae of the blackfly *Simulium vittatum*, but an unrealistically large dose was required to cause this effect. Mosquito larvae are also susceptible to toxicity caused by B.t.

Effects on Birds

Because larvae of eastern spruce budworm are a readily available and nutritious food for adults and nestlings, birds are relatively abundant in budworm-infested forests. Several bird species exhibit particularly strong numerical responses to the abundance of this insect (Kendeigh, 1948; Morris *et al.*, 1958; Crawford *et al.*, 1983). For example, the bay-breasted warbler increased in abundance from 2.5 pairs/10 ha in uninfested stands to 300 pairs/10 ha during an epidemic; blackburnian warbler increased from 25–50 pairs/10 ha to 100–125 pairs/10 ha; and Tennessee warbler increased from 0 pairs/10 ha to 125 pairs/10 ha (Morris *et al.*, 1958; see Table 8.13). A much larger number of bird species have functional responses to abundance of budworm, i.e., when this insect is readily available, birds feed

Table 8.11   Effect of insecticide spraying on the drift of aquatic and terrestrial arthropods in forest streams in Quebec[a]

Treatment	Drift of aquatic arthropods		Drift of terrestrial arthropods	
	4 Days prespray	4 Days postspray	4 Days prespray	4 Days postspray
Unsprayed	0.46	0.51	0.27	0.78
0.053 kg a.i./ha Aminocarb (a)	2.61	0.72	0.53	0.39
(b)	1.49	1.58	0.68	2.27
(c)	0.53	0.96	0.82	0.84
(d)	0.20	0.23	2.43	3.85
(e)	0.31	0.46	1.73	1.01
0.21 kg a.i./ha Fenitrothion	0.45	0.43	0.19	0.35

[a]Sampling was done for 4 days before and after the spray application. Calculated from Holmes (1979).

selectively and heavily on this tasty resource (Crawford and Jennings, 1989).

During outbreaks of eastern spruce budworm, these and other insectivorous species rely heavily on abundant budworm larvae for food for the rearing of nestlings, and they may have relatively successful rates of reproduction (actually, this latter demographic effect has not yet been demonstrated). One study in Maine estimated that in stands with epidemic populations of eastern spruce budworm, avian predation consumed about 89 thousand larvae and pupae/ha, compared with 54 thousand/ha in transitional stands and 5.6 thousand/ha in stands with endemic budworm populations (Crawford et al., 1983). However, it is believed that birds are only quantitatively important in causing reductions of budworm density in stands with small or transitional populations of budworm, in which they are estimated to remove 84–87% and 22% of the budworm population, respectively, compared with only 2.4% or less in epidemic stands (Crawford et al., 1983; Jennings and Crawford, 1985; Crawford and Jennings, 1989). The avifauna exerts no control over epidemic populations of budworm, because it is saturated by the abundant food resource.

Because budworm larvae present a superabundant resource during outbreaks, most species of forest birds focus their foraging efforts on this insect when it is abundant. Crawford and Jennings (1989) studied stands in New Hampshire and Maine with small or transitional populations of budworm (i.e., with $0.08–23 \times 10^6$ budworm larvae/ha). The most important avian predators were: blackburnian warbler (which consumed 22.5% of the total number of budworm eaten by birds), Cape May warbler (20.8%), bay-breasted warbler (10.7%), Nashville warbler (8.2%), black-throated green warbler, white-throated sparrow, magnolia warbler, solitary vireo, northern parula, and purple finch (1.8–6.1%), plus another 12 species that maintained populations smaller than the above. Only a few species did not eat budworm larvae, notably brown creeper, ruby-crowned kinglet, and black-throated blue warbler.

During the era of DDT spraying, measurable short-term changes in the abundance of nongame forest birds were not documented, even though some examples of avian mortality were identified (Pearce, 1975). The lack of measurable effects on avian abundance could have been influenced by such factors as: (1) movement of birds in and out of sprayed areas; (2) a floating surplus of nonbreeding individuals that can rapidly replace killed territorial birds; (3) errors inherent in forest bird censuses; and (4) an underestimation of direct mortality because of difficulties in finding sick or dead birds, due to their low density, scavenger activity, etc. (Stewart and Aldrich, 1951; Pearce et al., 1979; Richmond et al., 1979; Mineau and Peakall, 1987; Mineau and Collins, 1988; Busby et al., 1989). The uselessness of carcass counts as an indicator of avian mortality is underscored by the observation that, with all of the insecticide spraying conducted against budworm in New Brunswick between 1965 and 1987, only 125 dead birds are on record with the Canadian Wildlife Service (Busby et al., 1989).

One species that is believed to have suffered a reduced abundance as a result of DDT spraying is a gamebird, the woodcock (*Philohela minor*). This species had relatively large concentrations of DDT residues, with an average in breast muscle of 2.1 ppm f.w. in sprayed areas, compared with <0.1 ppm in unsprayed habitat (Dilworth et al., 1974).

The documented effects of spraying with phosphamidon, fenitrothion, and aminocarb on the territorial abundance of forest birds have generally been fairly small (Buckner and McLeod, 1977; Pearce et al., 1979; Kingsbury and McLeod, 1980, 1981; Pearce and Busby, 1980; Kingsbury et al., 1981; McLeod and Millikin, 1982; Spray et al., 1987; Millikin and Smith, 1990). As was noted previously, however, such population-level effects on forest birds are difficult to demonstrate.

In some cases, however, phosphamidon was an important exception to the above generalization about the field toxicity of spray programs against budworm. Because it is relatively toxic to birds (Table 8.8), phosphamidon caused large, estimated mortalities of some species after operational sprays (McLeod, 1967; Pearce and Peakall, 1977; Pearce et al., 1979). One calculation suggested that as many as 376 thousand individuals of a relatively susceptible species, the ruby-crowned kinglet, may have died in New Brunswick in 1975 as a result of exposure to phosphamidon and, to a lesser degree, fenitrothion (Pearce and Peakall, 1977). It is largely because of its effects on avifauna that phosphamidon is no longer used in budworm spray programs.

It should be noted, however, that the margin of toxicological safety for birds exposed to fenitrothion is small. Much greater mortality of adult and nestling white-throated sparrows was observed after an experimental double application of this chemical in New Brunswick, compared with the usual, operational spray rate, while surviving individuals exhibited impaired behavior, sometimes resulting in nest abandonment (Pearce and Busby, 1980; Busby et al., 1983a, 1990). Clutch size and hatchability were unaffected by spraying, but reproductive success was significantly less, as a result of the combined effects of direct pesticide toxicity and a presumed, reduced availability of food (Table 8.12).

Above-average exposures to pesticides are not an infrequent occurrence, because of: (1) the overlap of spray swaths caused by difficulties in the navigation of spray airplanes treating large areas of forest; (2) atmospheric conditions that affect spray deposition; and (3) calibration and mixing errors in the delivery systems. Some of these factors can result in relatively large rates of exposure even in places where spray swaths have not overlapped.

Toxic exposures of forest birds to insecticides

can also be influenced by: (1) spray droplet size and other effects related to the delivery system; (2) the particular pesticide and formulation that is being used; (3) differences in foraging height and other behavioral traits among species; and (4) variations of physiological sensitivity among species (Busby et al., 1981, 1982, 1983a,b, 1989; Grue et al., 1983).

The effects of a budworm spray treatment with fenitrothion on the censused abundance of breeding birds in an infested forest in Quebec are described in Table 8.13. When considering these data, it is necessary to compare the relative trends of the sprayed plot with those of the unsprayed, reference treatment. The reason for this is that during the prespray census in mid-May, not all of the migratory species had returned to this breeding habitat (in particular, Swainson's thrush and most of the warbler species). As a result, the total avian abundance at this time was smaller than the breeding density later in the season. If the data are considered with this relative comparison in mind, it is apparent that there were no clear, demonstrated effects of fenitrothion spraying on the abundance and species richness of the bird community of this fir–spruce forest.

In addition, because the various species of migratory birds return to their breeding habitats of conifer forest at different times, they may have different exposures to the insecticide sprays. For example, if eggs have not hatched at the time of a spray, only adult birds will be exposed to the insecticide. However, if hatchlings are present, both adults and relatively vulnerable, young birds will be exposed.

There is a large effect on the avian community when forests are badly damaged by uncontrolled defoliations by eastern spruce budworm. This is illustrated in Table 8.14, which compares the abundance of birds during and after a period of budworm infestation. Stand A suffered little longer-term damage to trees as a result of the defoliation. The decreased density of many bird species reflects their numerical responses to the reduction of insect prey after collapse of the outbreak (e.g., solitary vireo, many warbler species). Stand B suffered intense damage to trees because of the budworm infesta-

Table 8.12 Effects of aerial spraying with fenitrothion on breeding success of white-throated sparrows in New Brunswick[a]

	Reference	Sprayed
Clutch size	4.0 ($n = 7$)	3.7 (13)
Brood size	3.2 (6)	3.3 (8)
Number fledged per hatched egg	2.8 (6)	1.0 (8)
Number young raised per egg laid	2.4 (7)	0.6 (13)

[a]The application was intended to simulate a higher-than normal, approximately double-spray rate. Data are averages, with sample sizes indicated in parentheses. Modified from Busby et al. (1990).

Table 8.13 Effects of aerial spraying with fenitrothion on the populations of birds in a spruce budworm-infested fir–spruce forest in the Gaspé region of Quebec[a]

Species	Prespray Sprayed area	Prespray Reference area	Postspray 1 Sprayed area	Postspray 1 Reference area	Postspray 2 Sprayed area	Postspray 2 Reference area
Boreal chickadee (*Parus hudsonicus*)	2.5	4.3	1.5	1.5	1.4	1.1
Winter wren (*Troglodytes troglodytes*)	1.0	3.9	2.3	4.6	1.7	3.0
American robin (*Turdus migratorius*)	3.5	3.6	4.8	3.3	5.3	1.8
Swainson's thrush (*Catharus ustulatus*)	0.0	0.0	0.0	0.6	3.7	2.0
Ruby-crowned kinglet (*Regulus calendula*)	8.6	4.5	5.9	3.8	4.4	6.4
Tennessee warbler (*Vermivora peregrina*)	0.0	0.0	0.3	1.5	2.3	5.8
Nashville warbler (*Vermivora ruficapilla*)	0.0	0.0	0.3	0.4	0.3	0.6
Magnolia warbler (*Dendroica magnolia*)	0.0	0.0	0.1	1.5	0.8	3.1
Cape May warbler (*Dendroica tigrina*)	0.0	0.0	2.0	3.1	4.3	4.0
Yellow-rumped warbler (*Dendroica coronata*)	0.5	0.3	7.7	6.3	10.8	8.2
Black-throated green warbler (*Dendroica virens*)	0.0	0.5	0.5	2.3	2.6	3.9
Bay-breasted warbler (*Dendroica castanea*)	0.0	0.0	0.3	0.3	2.5	4.5
Blackpoll warbler (*Dendroica striata*)	0.0	0.0	0.0	0.0	1.1	2.8
Northern waterthrush (*Seiurus noveboracensis*)	0.0	0.0	0.6	0.2	1.3	1.6
Evening grosbeak (*Hesperiphona vespertina*)	1.3	0.4	1.8	2.4	0.8	1.4
Purple finch (*Carpodacus mexicanus*)	1.3	0.7	1.1	1.2	1.1	2.8
Pine grosbeak (*Pinicola enucleator*)	0.6	0.5	1.3	0.9	0.2	0.4
Pine siskin (*Carduelis pinus*)	0.2	0.0	0.9	1.8	1.6	1.5
Dark-eyed junco (*Junco hyemalis*)	9.6	4.1	4.2	1.8	4.5	0.5
White-throated sparrow (*Zonotrichia albicollis*)	8.9	9.6	9.7	10.5	10.4	8.4
Fox sparrow (*Paserella iliaca*)	2.1	3.8	3.3	4.2	2.5	3.1
Total bird abundance	46.1	39.1	52.1	57.7	68.8	75.8
Number of species	23	23	41	35	41	39

[a]Birds were censused for 5 days prior to spraying, for 7 days after the first spray on May 21, 1976, and for 5 days after the second spray on May 30. The total fenitrothion spray was 0.56 kg a.i./ha. Bird density is expressed as number/10 ha. Only prominent species are listed. Modified from Kingsbury and McLeod (1981).

tion, and death of the balsam fir overstory caused large changes in the plant-species composition and physical structure of the habitat. The most important habitat changes were the development of a large density of snags, a dense stratum of angiosperm shrubs, a vigorous regeneration of small fir trees in the understory, and a lush growth of monocotyledonous and dicotyledonous herbs. Correspondingly, in Stand B, some bird species naturally declined in abundance because of the decreased abundance of budworm prey and other habitat changes (e.g., solitary vireo, Tennessee warbler, black-throated green warbler, blackburnian warbler, and bay-breasted warbler). Other species responded positively to the habitat changes (e.g., least flycatcher, magnolia warbler, and white-throated sparrow).

Of course, any successful spray treatment reduces the abundance of the most important arthropod food of forest birds, i.e., larvae of eastern spruce budworm. This might also be expected to have indirect, nontoxic effects on birds, by affecting reproductive success (Millikin, 1990; Pascual and Peris, 1992). Such an indirect, ecotoxicological effect would be associated with any reductions of abundance of eastern spruce budworm, caused by: (1) a natural collapse of budworm populations at the end of an outbreak, (2) a spray of an insecticide that

Table 8.14 Abundance of selected bird species in spruce budworm-infested balsam fir stands in New Brunswick[a]

Species	Stand A		Stand B	
	Outbreak	Postoutbreak	Outbreak	Postoutbreak
Yellow-bellied flycatcher (*Empidonax flaviventris*)	4.6	p	5.1	0.0
Least flycatcher (*Empidonax minimus*)	3.9	2.8	2.1	15.9
Boreal chickadee (*Parus hudsonicus*)	2.0	4.7	2.0	p
Winter wren (*Troglodytes troglodytes*)	5.9	8.2	2.2	3.3
Swainson's thrush (*Catharus ustulatus*)	15.1	11.9	14.0	8.7
Golden-crowned kinglet (*Regulus satrapa*)	8.9	3.0	2.3	0.0
Ruby-crowned kinglet (*Regulus calendula*)	p	10.1	1.1	2.7
Solitary vireo (*Vireo solitarius*)	8.3	p	7.7	0.0
Tennessee warbler (*Vermivora peregrina*)	9.9	0.0	25.7	0.0
Nashville warbler (*Vermivora ruficapilla*)	p	3.0	0.0	p
Magnolia warbler (*Dendroica magnolia*)	22.0	7.9	1.3	14.9
Yellow-rumped warbler (*Dendroica coronata*)	7.1	1.6	5.1	2.1
Black-throated green warbler (*Dendroica virens*)	11.9	4.2	11.8	4.2
Blackburnian warbler (*Dendroica fusca*)	17.1	5.3	19.5	1.5
Bay-breasted warbler (*Dendroica castanea*)	55.8	16.4	78.2	2.0
Blackpoll warbler (*Dendroica striata*)	p	4.9	1.1	5.1
American redstart (*Setophaga ruticilla*)	3.9	2.3	2.1	p
Dark-eyed junco (*Junco hyemalis*)	5.9	3.5	8.4	4.1
White-throated sparrow (*Zonotrichia albicollis*)	9.3	5.0	11.4	19.5
Total density	191.6	94.8	201.1	84.0

[a]Stand A was censused for 8 years up to 1959 during an epidemic, and then for 6 postoutbreak years. No long-term damage was caused to trees, and the average age of balsam fir was >120 years during both time periods. Stand B was censused for 5 epidemic years up to 1959 and then for another 5 postoutbreak years. Intense damage was caused in Stand B: the average age of balsam fir declined from an initial >80 years, to <10 years after the collapse of the outbreak. Bird data are in number/40 ha; p, present. Modified from Gage and Miller (1978).

is not directly toxic to birds (e.g., B.t.), and (3) a spray with an insecticide that can cause toxicity to birds (e.g., phosphamidon and fenitrothion).

The influence of insecticide-caused reductions of food supply has been examined by studies in Spain of spraying oak (*Quercus pyrenaica*) forests infested with green moth (*Tortrix vividana*). The insecticide used was cypermethrin, a pyrethroid with a low toxicity to birds (LD_{50} of 0.75 to >10 g/kg body weight; Pascual and Peris, 1992). Blue tits (*Parus caerula*) breeding in nest boxes were monitored in sprayed and unsprayed stands, where they were able to fledge young from 35 and 100% of nests, respectively. There were no effects of spraying on clutch size or hatching success, but the weight gain of nestlings was less in sprayed forest, and this contributed to the smaller reproductive success.

It is difficult to come to definitive conclusions about the effects on birds of fenitrothion, the synthetic insecticide that is currently most used against eastern spruce budworm in New Brunswick. The laboratory and field studies relevant to cholinesterase inhibition indicate a likelihood of significant toxicity to many birds. Large numbers of birds probably die from their exposures to fenitrothion, while others suffer subacute effects. The field data for population-level effects are ambivalent, but because of the difficulties inherent in censusing forest birds, such effects would have to be rather large before they were measurable. The following conclusion from a report by a Canadian, multiagency risk-assessment team that reviewed the hazards of fenitrothion to wildlife is telling: "Because of the range of effects seen following fenitrothion applications, and because the AChE data indicate that these effects may occur frequently, concerns are raised for forest songbird populations in fenitrothion treatment areas" (Pauli *et al.*, 1993, p. XV).

Effects on Fish

During the era of DDT spraying in New Brunswick and elsewhere, fish kills were documented after spraying, and populations of sport fish were decreased in some areas (Kingsbury, 1975; Logie, 1975). Because of its commercial importance, Atlantic salmon (*Salmo salar*) was relatively intensively studied. Caged salmon in sprayed streams suffered a mortality rate of 63–91% within 3 weeks of spraying with DDT, compared with negligible mortality in unsprayed streams (Kingsbury, 1975). Young salmon were especially sensitive to DDT. After spraying, the abundance of stages younger than 1 year old was decreased by as much as 90%, while 2- and 3-year-old juvenile stages were reduced by 70 and 50%, respectively.

In the Miramichi River of New Brunswick, the abundance of juvenile salmon (i.e., ≤ 3 years old) decreased from $67/100$ m^2 of breeding habitat, to $15/100$ m^2 after a spray with 0.5 kg DDT/ha (Logie, 1975). It was predicted that two further years of spraying with DDT at this rate would have caused additional declines of abundance, to $3.1/100$ m^2 after 2 spray years, and to $1.0/100$ m^2 after 3 spray years. Eventually, DDT spraying in the watershed of this river caused a 50–60% decrease in the number of salmon migrating from the sea. After the use of DDT against eastern spruce budworm was suspended, the population of salmon in the Miramichi River recovered in about 3 years.

In contrast, the insecticides that replaced DDT in the budworm spray programs have caused no or few measurable effects on populations of freshwater fish (Kingsbury, 1975; Gillis, 1980).

Overview of Eastern Spruce Budworm

It appears that the demonstrated, ecological effects of the modern, post-DDT spray programs against eastern spruce budworm are relatively short term in duration and moderate in intensity. The persistence of insecticide residues is not long lived, and there is no food-web accumulation. There have been few documented, population-level effects on nontarget organisms (although such effects are difficult to demonstrate, especially at larger spatial scales). For example, although spraying causes large kills of nontarget insects and spiders, the abundance of these arthropods recovers quickly, and medium-term populations are not measurably affected. Similarly, although individual birds are affected by pesticide toxicity under certain conditions, there are not yet convincing data that demonstrate measurable effects on their breeding populations.

So far, the known damage of spray programs has been judged to be "acceptable" by resource-management decision makers and regulators, in view of the substantial economic benefits that are obtained by spraying forests infested by eastern spruce budworm. Of course, ecologists would come to different conclusions about the costs and benefits of insecticide spraying, because they would value certain ecological costs more highly than would resource managers and regulators (see Chapter 12 for a conceptual discussion of these issues). Ecologists are not, however, making the decisions about pesticide spray programs.

Beginning in the mid-1980s, there was a strong trend toward the replacement of synthetic insecticides with B.t., a bacterial insecticide with much smaller nontarget effects. Fenitrothion, the most commonly used synthetic insecticide in budworm spray programs, is still registered and used for that purpose. However, recent Canadian risk assessments of the ecotoxicity of fenitrothion in budworm spray programs have been negative in tone, as is evidenced by the following concluding statement from both Anonymous (1993d) and Pauli *et al.* (1993, p. XVI): "The weight of evidence accumulated with respect to the identified and potential negative impacts caused by the forestry use of fenitrothion on non-target fauna, . . . and their potential ecological implications, supports the conclusion that the large-scale spraying of fenitrothion for forest pest control, as currently practised operationally, is environmentally unacceptable." There will be a public review of these risk-assessment documents and their evidence and conclusions, and there is a strong possibility that the registration of fenitrothion for forest insect spraying in Canada will be withdrawn. Following that possible event, B.t. would be the only insecticide used for broadcast spraying against eastern spruce budworm, at least until viable alternatives are developed.

It must also be stressed that, apart from ecotoxicological effects, the budworm spray strategies have some important drawbacks from a forest management perspective. It is true that spraying infested stands with insecticide will keep the forest "alive and green," and therefore available to supply the needs of the economically important forest industry. However, spraying also maintains the prime habitat of budworm in a susceptible condition, and thus the infestation may be prolonged. Therefore, once spraying is begun, agencies become "locked" into a spray strategy and must continue the practice to maintain an economically viable forest resource. While this may be perceived to be the best available short-term strategy, it is clearly not preferable on the longer term to have to mount an annual spray program over a large area. [Note, however, that Royama (1984) has a fundamentally different view. Although spraying causes short-term decreases of abundance of budworm and therefore prevents some damage to the forest resource, according to his theory of budworm dynamics the longer-term infestation increases or decreases regardless of the nature of spray interventions.]

A possible alternative to spraying insecticides could involve some combination of forest management practices that reduce the susceptibility of stands to infestation. For example, less vulnerable species such as black spruce might be planted. In addition, the landscape could be structured so that large, continuous tracts of mature (>50 years old) stands do not occur. If a smaller-scale mosaic of less vulnerable stands were ideally structured, then the rate of harvesting by the forest industry could potentially equal the rate at which stands mature through succession. In this way, the area vulnerable at any time to infestation by budworm could be reduced. Any of these actions would, of course, cause substantial changes in the ecological character of the

forested landscape, and nothing is known about the longer-term stability, viability, or broader ecological consequences of these sorts of potential, forest management strategies (Baskerville, 1975b; Anonymous, 1976b; Batzer, 1976; Holling and Walters, 1977; Blum and MacLean, 1985).

An important perspective is gained from the observation that in the shorter term, some of the ecological effects of spraying appear to be less than those occurring when severely infested stands are not treated with insecticide. The consequences of not spraying are, of course, the natural ecological changes associated with budworm defoliation. These consequences can, nonetheless, be considered to be results of the do-nothing management option.

In unsprayed stands, a large proportion or all of the mature individuals of balsam fir may die, as may many of the spruce trees. For a number of years, the moribund forest of dead, dry, standing timber may present an explosive fire hazard, especially in early spring before the flush of new angiosperm foliage has developed. In a controlled-burn experiment in Ontario, it was found that in a forest in which most trees had been killed by budworm: (1) ignition took place readily; (2) the fire spread almost immediately to the tree crowns; (3) the rate of lateral spread was rapid, at up to 80 m/sec; (4) downwind "branding" was frequent as bark peeled, ignited, and was convectively transported downwind; and (5) a large fraction of the fuel was consumed by the intense combustion (Stocks, 1985, 1987). The fire hazard was severe for about 10 years after the trees died. Afterward, regeneration of the stand and decomposition of dead trees reduced the fire hazard.

In addition, budworm-killed stands can have a relatively small abundance of species of arthropods, birds, and mammals for as long as a decade or more after the death of the trees. For birds, this effect is related to decreased populations of the many species that feed on budworm, including the budworm specialists. After this time, a vigorous natural regeneration restores the quality of the habitat and wildlife populations recover. During the initial years of recovery the community includes species of wildlife that are early successional and are taking advantage of the disturbed nature of the habitat. It takes several decades for the forest to recover to the degree that it provides habitat for all of the late-successional species that were present before and during the budworm infestation (see Chapter 9 for a parallel discussion of the effects of forest harvesting by humans).

Clearly, many birds and other wildlife are poisoned and killed by spray programs using synthetic insecticides. However, the overall, population-level changes in the abundance and species composition of wildlife that occur in budworm-killed forests appear to be larger than those caused by the post-DDT spray programs (with the exception of the use of phosphamidon, which is a potent bird killer).

From a forest management, or industrial perspective, it is not desirable to allow large tracts of mature forest to be devastated by defoliation by eastern spruce budworm. Such events cause great difficulties of supply of raw materials for the forest industry and thereby cause great economic disruptions. In 1976, the government of Nova Scotia decided to not allow the spraying of insecticides to reduce the effects of an infestation of spruce budworm on the highlands of Cape Breton Island. As a result, trees died over a large area of infested forest, comprising most of the resource base of that province's largest pulp mill. The total loss of wood was about 21.5 million m³, equivalent to 10% of the total softwood resource of Nova Scotia, and half that of Cape Breton Island before the budworm infestation (Bailey, 1982). The longer-term, economic implications of this event are not yet clear. The pulp mill has managed to keep operating, in part by processing more costly wood obtained elsewhere.

In addition, the forest industry and provincial government were faced with the fact that there were enormous tracts of dead trees on Cape Breton Island. If harvested quickly, that wood was potentially useful for some purposes, and tremendous skyline-to-skyline clear-cuts were used to salvage some of the killed timber. Within the 5-year period of the salvage program, about 3.4 million m³ of wood was recovered from a clear-cut area of about 16 thousand ha. The largest, contiguous clear-cut is

6 thousand ha in area (E. Bailey, personal communication). There were important, but essentially undocumented, environmental consequences of the large-scale, intensive, salvage harvesting of the dead trees of Cape Breton Island. If the forest had been treated with insecticide during the budworm infestation, these extensive clear-cuts might not have been required, and large areas of mature forest would still occur in that area. However, areas of that forest might still be infested by eastern spruce budworm, and yearly spray programs might have been required to keep the trees alive. Alternatively, the epidemic might have collapsed, even if susceptible stands, and budworm habitat, were kept alive by spray interventions.

If insecticide spraying against eastern spruce budworm must be undertaken, it is desirable to pursue the least damaging but still effective option. At the present time, this seems to require the use of insecticides. Within this framework, the least undesirable option is the use of B.t., but in some circumstances synthetic insecticides such as fenitrothion may be required. Clearly, the forest management decision is more complicated than "To spray, or not to spray."

Silvicultural Use of Herbicides

Silvicultural Objectives

The most frequent objectives of the use of herbicides in forestry are: (1) to release young conifers from competition with economically undesirable (at least from the forestry perspective) angiosperm species, or (2) to prepare a site for the planting of young conifers. By these uses, herbicides can help to temporarily or permanently reduce the dominance of productive forest sites by unwanted, "weedy" vegetation, while allowing the desired conifer regeneration to more rapidly attain a greater site dominance (Newton, 1975; Daniel et al., 1979; Newton and Knight, 1981; Wenger, 1984; Malik and Vanden Born, 1986; Freedman, 1991). Other relatively minor uses of herbicides in forestry are for the spraying of vegetation beside forest roads and for manipulation of the habitat of wildlife that can injure conifer regeneration by browsing, for example, rabbits and hares.

Forestry usage comprises a relatively small proportion of the total use of herbicides in most areas, for example, 3.6% of the quantity of phenoxy herbicides used in Ontario in 1978, 5.1% in the United Kingdom in 1979, and less than 5% in the United States in 1980 (Kilpatrick, 1980; Mullinson, 1981; Stephenson, 1983). If herbicides other than the phenoxys are considered, the relative usage in forestry is even smaller. The silvicultural use of herbicides has, however, increased substantially since the mid-1970s. In 1988, about 2.2×10^5 ha of clearcuts were treated with herbicides for silvicultural purposes in Canada (Campbell, 1990), while about 2×10^6 ha were treated in forestry in the mid-1980s in the United States (Pimentel *et al.*, 1991).

In meeting the silvicultural objectives described above, herbicides have important advantages over alternative procedures such as mechanical scarification, the removal of vegetation by hand cutting, and prescribed burning. These advantages include: (1) economy and human safety, especially in comparison with manual treatments (e.g., Howard, 1992); (2) time savings, in that large areas can be treated relatively quickly; (3) fewer retreatments to control the regeneration of weeds; and (4) selectivity, so that damage to conifers can be minimized.

Important disadvantages of the use of herbicides for silvicultural purposes include: (1) success in meeting the silvicultural objectives depends on the appropriate selection of a herbicide, its formulation, and application rate—if these are not planned and executed properly, then conifers may be damaged, or weeds may not be adequately controlled; (2) the timing of the application is critical to maximizing conifer tolerance, while still achieving control of weeds; (3) there can be nontarget, off-site damage because of the drift of herbicide spray; (4) far fewer people are employed in vegetation management programs; (5) herbicide treatments may detract from the ability of weedy plants to serve as a biological "sponge" for mobile nutrients such as nitrate, which might otherwise leach from disturbed sites (see Chapter 9); and (6) there are often strongly

negative, social reactions to the broadcast spraying of herbicides (and insecticides) in forestry.

Deloitte and Touche Management Consultants (1992) estimated that the cost : benefit ratio of the use of herbicides in forestry in Canada was 1 : 2, compared with a net loss in the case of mechanical weeding (because of the relatively large costs of labor). [This analysis only considers "conventional" economic costs and benefits and does not consider the nonvaluated, ecological costs of herbicide use in forestry (see Chapter 12).]

Controversy

The reasons for the social controversy were discussed earlier in this chapter, and they include fears of: (1) potential, epidemiological consequences to humans, caused by exposure to herbicides; (2) losses of employment opportunities, which are much greater if alternative methods such as manual weed control are used; (3) landscapes dominated by monocultures of conifers; and (4) degradations of wildlife habitats and other aspects of ecological integrity [Daniel *et al.*, 1979; General Accounting Office (GAO), 1981; Newton and Knight, 1981; Freedman, 1982; Wenger, 1984; Freedman, 1991].

In some areas, opponents of herbicide spraying in forestry have caused interruptions, reductions, and even cessations of this silvicultural practice. There have been a few, isolated incidents of vandalism of spray equipment. In a few cases, groups of citizens have filed lawsuits and obtained injunctions against the silvicultural use of herbicides. In 1982–1983, a coalition of fifteen landowners took a forest company in Nova Scotia to trial to stop a proposed spray program using 2,4,5-T and 2,4-D. That legal action stimulated intense public interest and discussion, and there was a 1-month trial, with testimony from 49 witnesses, including international environmental and medical experts. In a 182-page decision, the judgement went strongly against the plaintiffs, and his legal conclusion was that 2,4,5-T and 2,4-D could be used safely for forestry purposes (Nunn, 1983). One could argue, however, whether a legal forum is appropriate for sorting out herbicide-related issues and whether the training and experience of a member of the judiciary provide adequate background for sensible interpretations of the voluminous and complex scientific data that are relevant to the ecological and other environmental effects of silvicultural herbicide use.

In any event, the silvicultural use of herbicides in Nova Scotia and elsewhere has remained a controversial and emotional issue, with a strong polarization of the attitudes of the opponents and proponents of the practice.

In the following, discussion will be made of some issues that are related to the silvicultural use of herbicides, with particular focus on the ecological aspects.

Weeds in Forestry

After the disturbance of a forested site by fire or harvesting, the regeneration is composed of many plant species that compete for space and other resources. During the first decade or so of secondary succession, the vigorous, young community is dominated by plants other than the conifers that are generally desired by foresters. Most important in this respect are various perennial herbs and woody angiosperm species.

This is illustrated for some 4- to 6-year old clearcuts of conifer forest in Nova Scotia (Table 8.15). Among the four sites, the economically desired regeneration of spruce and fir comprise only 3–8% of the total plant cover. Much more prominent in the young, seral community are ferns, monocots such as sedges and grasses, dicotyledonous herbs (various Asteraceae are especially prominent), raspberries and blackberries (*Rubus* spp.), and woody dicot species, especially birches (*Betula* spp.), red maple (*Acer rubrum*), and pin cherry (*Prunus pensylvanica*). In such a situation, the degree of site domination by the "undesirable" plants can be sufficient to inhibit the establishment and growth of the commercially desired conifers.

The ecological "strategy" of weedy plants is to take advantage of the relatively great abundance of environmental resources that are available for a

Table 8.15 Dominant vegetation of four 4- to 6-year-old conifer clear-cuts in Nova Scotia[a]

	Site 1	Site 2	Site 3	Site 4
Bryophytes	7	5	15	3
Lichens	1	<1	1	<1
Pteridophytes	22	29	15	1
Conifers	3	5	8	4
Monocots	9	5	5	26
Dicots, herbs	17	19	14	39
Dicots, woody shrubs	21	16	31	10
Dicots, *Rubus* spp.	20	21	11	17
Total plant cover (absolute)	137%	153%	110%	113%

[a]Data are relative cover (i.e., percent of total cover), average of 48 1-m² quadrats per site. Modified from Morash and Freedman (1987).

short time after the death of the forest overstory. After this initial stage of the postdisturbance secondary succession, competition from the regenerating forest canopy eliminates most of the relatively intolerant, ruderal species from the site.

Some of the species that dominate regenerating cutovers in Nova Scotia (Table 8.16) are also present in mature, uncut forest. They manage to survive the disturbance and, because they experience a release from the competitive stresses exerted by the previous, conifer-dominated overstory, they expand their relative degree of site dominance. Examples of such longer-term site occupants are hay-scented fern (*Dennstaedtia punctilobula*), and the tree *Acer rubrum*, which regenerates after cutting by a prolific stump sprouting. Other ruderals establish from a persistent seedbank (e.g., *Prunus pensylvanica* and *Rubus* spp.) or by an influx of airborne seed to the disturbed site (e.g., Asteraceae, *Betula* spp., *Populus* spp.).

The vigorous, regenerating vegetation of clearcuts can quickly reestablish substantial rates of net primary production and nutrient uptake. As such, the vegetation can act as a biological reservoir (or "sponge") for some of the site nutrient capital that might otherwise leach from the site during the reorganization phase of postcutting disturbance (Bor-

The vegetation of a 4-year-old clear-cut of a coniferous forest in Nova Scotia. The dominant plants in this vigorously regenerating ecosystem are red raspberry (*Rubus strigosus*), pin cherry (*Prunus pensylvanica*), red maple (*Acer rubrum*), hay-scented fern (*Dennstaedtia punctilobula*), and various herbs of the Asteraceae. The regeneration of commercially desired conifers can be inhibited or prevented by this intensely competitive "weedy" vegetation. An increasingly frequent silvicultural prescription for this sort of site is an aerial herbicide spray treatment (photo: B. Freedman).

mann and Likens, 1979). As normal development of the stand progresses, the early successional species are eliminated or reduced in abundance because of competitive stresses exerted by more tolerant species, or in some cases because of senescence. The nutrients that were immobilized in the biomass of early successional species are then made available for uptake by trees, after recycling of litterfall by decomposition (Marks and Bormann, 1972; Marks, 1974; Bormann and Likens, 1979; Crow *et*

al., 1991; Reiners, 1992; Crowell and Freedman, 1993). Note, however, that the ecological importance of nutrient conservation by the "weedy" vegetation of regenerating clear-cuts is a controversial notion. Although many ecologists perceive the noncommercial revegetation to be a desirable attribute of disturbed sites, most foresters perceive a vigorous regeneration of noncrop plants on clear-cuts to be an ecological value that is economically negative.

Many studies have shown that competing vegetation has the capacity to reduce the productivity of tree species and that the growth of conifers can be increased by a reduction of this stress [Stewart *et al.* (1984) assembled a lengthy bibliography of literature on the effects of competition on conifers]. This effect can be illustrated by the study of MacLean and Morgan (1983), who examined the vegetation of a site that had been treated with 2,4,5-T and 2,4-D 28 years previously (Table 8.16). Before the herbicide treatment, the study plots had a vigorous growth of competing angiosperm vegetation, which formed a closed canopy at a height of about 2 m that overtopped the then-shorter conifer species. The average density of raspberry (*Rubus* spp.) was 81 thousand stems/ha, mountain maple (*Acer spicatum*) was 7 thousand/ha, and other angiosperm shrubs totalled 2.5 thousand/ha. The silvicultural herbicide treatment released the suppressed conifers from some of the stresses of competition with intolerant hardwoods, and 28 years after spraying the volume of balsam fir on the two sprayed plots averaged about three times larger than on the unsprayed reference plot, while diameter averaged 21% greater. Overall, conifers comprised a much larger percentage of the stand basal area in the spray plots. Therefore, the herbicide treatment resulted in a more rapid establishment of a conifer-dominated forest, effectively shortening the length of the harvest rotation. Herbicide spraying did not eliminate the weedy angiosperms from the spray plots, but their relative abundance was reduced. After 28 postspray years, hardwoods still averaged 6–13% of the total basal area, compared with 56% on the reference plot. It is also important to recognize that because the most prominent of the competing hard-

Table 8.16   Effects of suppressing hardwoods on the plant community composition and growth of balsam fir in New Brunswick[a]

	Herbicide Plot A	Herbicide Plot B	Reference treatment
Total tree density (stems/ha)	4200	4460	3350
Total tree basal area (m²/ha)	44.7	36.6	23.3
Total volume, balsam fir (m³/ha)	191	135	52
Mean diameter, balsam fir (cm)	10.8	9.3	8.3
Species composition (% of basal area)			
Balsam fir (*Abies balsamea*)	85	83	42
Spruce (*Picea* spp.)	9	4	2
Pin cherry (*Prunus pensylvanica*)	2	2	19
White birch (*Betula papyrifera*)	2	3	30
Mountain ash (*Sorbus americana*)	2	8	7

[a]The study plots were three 0.4-ha, 6-year-old clear-cuts initially with 9500 stems/ha of hardwood trees and 81,000 stems/ha of *Rubus* spp. Two plots were treated with a 50:50 mixture of 2.4,5.-T and 2.4-D, and the other plot was a reference treatment. This table summarizes characteristics in 1981, 28 years after herbiciding. Modified from MacLean and Morgan (1983).

An aerial application of herbicide to a regenerating clear-cut, in a silvicultural conifer release program in Nova Scotia. The helicopter has to fly at a relatively high altitude of more than 20–30 m because of the uneven terrain, the surrounding forest, and the presence of residual uncut trees on the clear-cut. Partly as a result of this, there can be a considerable drift of herbicide, and nearby off-target vegetation can suffer some damage (photo: B. Freedman).

A 4-year-old clear-cut of a coniferous forest in Nova Scotia, 1 year after it was aerially treated with the herbicide glyphosate. The extensively defoliated stump-sprout shrubs are *Acer rubrum*, which was considered to be one of the most important "weeds" on the site. The abundance of all other plants (except conifers) was initially drastically reduced, but after only 1 postspray year, there was a substantial recovery of the weedy vegetation (photo: B. Freedman).

woods are intolerant species, they would eventually be eliminated from even the reference plot by the conifers, which are longer lived and more tolerant of the stressful conditions of a closed forest. In part, herbicide spraying changes the temporal dynamic of this successional change.

Of course, the broadcast spraying of herbicides on regenerating cutovers has important ecological consequences in addition to the release of economically desirable conifers from competition with weeds. These effects include direct toxicity to non-target vegetation and other biota and indirect influences on animals caused by changes to their habitat. These effects are discussed below, mostly in reference to the silvicultural use of the phenoxy herbicides 2,4-D and 2,4,5-T and the organic phosphorus herbicide, glyphosate.

Persistence of Herbicide Residues

Generally speaking, the herbicides used in forestry have a moderate persistence in the terrestrial environment, so that their residues decrease to small concentrations within a few months of application [Norris *et al.*, 1977, 1982; Norris, 1981; Smith, 1982; U.S. Department of Agriculture (USDA), 1984]. Norris *et al.* (1977) monitored the residues of 2,4,5-T after an aerial application in Oregon. The concentrations in graminoid plants, which are not killed by this herbicide, decreased from an initial postspray value of 120 μg/g to 3.4 μg/g after 1 month, 0.58 μg/g after 3 months, 0.14 μg/g after 6 months, 0.12 μg/g after 1 year, and to an undetectable concentration after 2 years. There was a similar pattern of disappearance in conifer foliage.

Morash *et al.* (1987) studied the persistence of glyphosate, another herbicide that is commonly used in forestry. The initial postspray residues were relatively small in conifer foliage (0.1 μg/g), but larger in the foliage of target species (5.2 μg/g in maple, 13 μg/g in graminoids and raspberry). The residues decreased fairly quickly in all types of foliage. In graminoid foliage, the concentration decreased to 1.4 µg/g after 1 week, to 0.76 µg/g after 2 weeks, and to 0.46 µg/g after 8 weeks.

Concerns are sometimes expressed about exposures to herbicides in wild fruits gathered for human consumption on sprayed clear-cuts. In a study in Ontario, initial glyphosate residues in fruits of raspberry (*Rubus strigosus*) were 7.9 µg/g, but these decreased to 3.7 µg/g after 13 days, 1.2 µg/g after 33 days, and 0.2 µg/g after 61 days (Roy *et al.*, 1989). The initial residues in blueberry (*Vaccinium myrtilloides*) were 20 µg/g, decreasing to 5.6, 1.2, and <0.1 µg/g after the same intervals. For comparison, the maximum permissible residues of glyphosate in food for human consumption in Ontario is 0.01 ppm (Roy *et al.*, 1989). These data suggest that it would not be prudent to pick wild fruits in clear-cuts in the same year that glyphosate had been sprayed.

Changes in Vegetation

The major objective of the silvicultural use of herbicides is to change the character of the vegetation on the site. An example of the longer-term changes in woody vegetation was previously described (Table 8.16). Examples of shorter-term effects on the plant community are summarized in Table 8.17, for two clear-cuts sprayed with glyphosate. On both clear-cuts, the vegetation of the reference treatment exhibited a vigorous and progressive development over the 3-year study period. In contrast, on the herbicided plots the abundance of the most important competing plants (woody shrubs and *Rubus* spp.) was initially decreased and then began to recover. In effect, the herbicide treatment set back the postcutting, secondary succession to an earlier stage of development, while releasing small conifer trees for that period of time. Morash and Freedman (1987) found that no plant species were eliminated within a grid of permanent quadrats located in these herbicided clear-cuts. However, there were, of course, large changes in relative abundance of the various plant species, because of differences in: (1) their susceptibility to glyphosate and (2) their rates of postspraying recovery by the growth of surviving plants and by the establishment of new individuals from seed.

It is usual for many individuals of some peren-

nial species of the ground vegetation to survive a herbicide treatment, in part because of reduced exposures associated with physical shielding by taller, overtopping foliage (Ingelog, 1978; Lund-Hoie, 1984, 1985; Freedman *et al.*, 1993). Examples of such low-growing species in Nova Scotia are *Cornus canadensis*, *Linnaea borealis*, *Maianthemum canadense*, *Viola* spp., and many forbs of the Asteraceae family (Freedman *et al.*, 1993b). These surviving plants experience relatively free growth for several seasons, because of a temporary decrease in the intensity of stresses associated with the overtopping canopy of shrub-sized plants.

Plants also survive herbicide applications because there typically are swaths of unsprayed vegetation that are missed during silvicultural herbicide applications (Morrison and Meslow, 1984a; Santillo *et al.*, 1989a,b). Such habitat constituted 1–10% of the sprayed area in a study in Maine (Santillo *et al.*, 1989a,b), and a similar range was estimated in a study in Nova Scotia (MacKinnon and Freedman, 1993).

In addition, after spraying with glyphosate and some other herbicides, regeneration from the seed bank can be prolific and diverse in species. This occurs because: (1) the seed bank in clear-cuts is large (Olmstead and Curtis, 1947; Marks, 1974; Graber and Thompson, 1978; Morash and Freedman, 1983) and (2) little toxicity is caused to the seed bank at doses of herbicide encountered during operational silvicultural sprays of glyphosate (Horsley, 1981; Ghassemi *et al.*, 1982; Morash and Freedman, 1989). Glyphosate does not affect the seedbank because it is immobile and not persistent in soil (Sprankle *et al.*, 1975; Torstensson and Stark, 1979; Caseley and Coupland, 1985), so that tender, recently germinated seedlings are not at much risk of suffering toxicity.

In a study in Nova Scotia, regeneration from a persistent seed bank was most important for *Rubus* spp., notably the red raspberry, *R. strigosus* (Freedman *et al.*, 1993b). Seeds of *Rubus* spp. are long-lived in the forest floor, and their populations can be large in clear-cuts that have had a few years of fruit production (Graber and Thompson, 1978; Morash and Freedman, 1983; Grignon, 1992). Most mature plants of *Rubus* spp. are killed by exposure to

Table 8.17   Changes in vegetation after aerial application of glyphosate at 1.4 kg a.i./ha for conifer release in Nova Scotia[a]

Vegetation category	Prespray cover	First-year postspray cover (% change)	Second-year postspray cover (% change)
		Site A	
1) Herbicided plot			
Monocots	8	8 (0)	14 (+75)
Dicot shrubs	29	12 (−59)	14 (−52)
Rubus spp.	28	9 (−68)	20 (−29)
Dicot herbs	21	24 (+14)	39 (+86)
Other plants[b]	10	12 (+20)	18 (+80)
Total	96	65 (−32)	105 (+9)
Species richness	58	56	60
Species diversity	2.7	3.1	2.9
(2) Reference plot			
Monocots	5	6 (+20)	8 (+60)
Dicot shrubs	13	28 (+115)	34 (+161)
Rubus spp.	17	20 (+18)	26 (+53)
Dicot herbs	12	21 (+75)	28 (+133)
Other plants[b]	32	41 (+28)	45 (+41)
Total	79	116 (+47)	141 (+78)
Species richness	59	56	57
Species diversity	2.9	3.0	3.2
		Site B	
(1) Herbicided plot			
Monocots	12	3 (−75)	16 (+33)
Dicot shrubs	38	11 (−71)	14 (−63)
Rubus spp.	12	1 (−92)	9 (−25)
Dicot herbs	11	8 (−27)	17 (+55)
Other plants[b]	26	15 (−42)	21 (−19)
Total	99	38 (−62)	77 (−22)
Species richness	57	47	54
Species diversity	2.8	2.9	3.2
(2) Reference plot			
Monocots	7	6 (−14)	8 (+14)
Dicot shrubs	19	30 (+58)	43 (+126)
Rubus spp.	2	4 (+100)	10 (+400)
Dicot herbs	8	16 (+100)	20 (+150)
Other plants[b]	37	46 (+24)	59 (+59)
Total	73	102 (+40)	140 (+92)
Species richness	35	37	40
Species diversity	2.4	2.6	2.7

[a]Plant-community data are: (a) percentage cover, grouped by major classes of vegetation, average of 40 permanent quadrats per treatment; (b) species richness, or the number of species encountered in the quadrats; and (c) species diversity, calculated as H′. Change in cover is expressed as postspray values relative to prespray. Site A was a 3-year-old clear-cut dominated by red maple (*Acer rubrum*) and herbaceous dicots. Site 3 was a 3-year-old clear-cut dominated by ericaceous shrubs. Modified from Morash and Freedman (1987) and Freedman *et al.* (1993b).

[b]Sum of bryophytes, lichens, pteridophytes, and conifers.

glyphosate, but the germination of seeds allows the establishment of large numbers of new individuals during postspraying succession (Sutton, 1978, 1985; Wendel and Kochenderfer, 1982; Lund-Hoie, 1984; Freedman *et al.*, 1993b).

The seed rain is another important source of postherbiciding regeneration for many plant species. At a site in Nova Scotia, birches (*Betula* spp.) had a prespray density of 3.5 thousand/ha, but only 30/ha after one postspray growing season (>99% mortality) (Freedman *et al.*, 1993b). Due to a prolific establishment of seedlings, however, birch density was back to 2.1 thousand/ha only two growing seasons after the herbicide treatment. The light, wind-blown seeds of birches confer a well-recognized ability to colonize disturbed sites (e.g., Marks 1974). A vigorous reinvasion by some species of angiosperm shrubs following herbicide and other weed-control treatments is a common observation in forestry (Cain and Yaussy, 1984; Newton *et al.*, 1989). Other species that mounted a vigorous, postherbiciding regeneration by wind-dispersed seeds in the Nova Scotia study included the graminoids *Agrostis scabra*, *Danthonia spicata*, and *Scirpus cyperinus* and the herbaceous dicots *Aster lateriflorus*, *A. umbellatus*, *Circium muticum*, *C. arvense*, *Epilobium* spp., *Erechtites hieracifolia*, *Solidago canadensis*, and *S. rugosa* (Freedman *et al.*, 1993b). These and other herbaceous species were more abundant for several years after spraying than initially, a factor that contributed to the species richness of the postspray plant communities.

It is typical that mixed communities of plants develop after a silvicultural herbicide treatment (MacLean and Morgan, 1983; Cain and Yaussy, 1984; Morrison and Meslow, 1984a,b; Freedman *et al.*, 1993b). In itself, the silvicultural use of herbicides is not intended to produce an absolute softwood monoculture. Rather, this forestry practice is used to enhance the rate of growth of conifers in the regenerating forest, so that the harvest rotation can be shortened.

Effects of Silvicultural Herbicide Use on Animal Wildlife

Toxicity to Animals of Silvicultural Herbicides. All pesticides are toxic substances, and they should

be handled with caution to avoid occupational exposures to applicators and to avert unnecessary environmental contaminations. However, compared with insecticides and other pesticides that are targeted to animals, most herbicides have relatively small toxicities to vertebrate wild life (Table 8.2). All of the herbicides that have been frequently used in forestry in North America (i.e., 2,4,5-T, 2,4-D, glyphosate, hexazinone, triclopyr, atrazine, and some other relatively minor chemicals in terms of quantity used) have toxicities to vertebrates that are less than those of caffeine and nicotine (Table 8.2; remember, however, that these data are only a comparative index of acute mammalian toxicity of these chemicals. The actual toxic effect is also related to the exposure, or dose.). This observation about the relative mammalian toxicities of chemicals does not, of course, mean that the use of herbicides is without ecotoxicological risks. Some of these chemicals may, for example, be toxic to animals other than mammals, for example, hexazinone and triclopyr in some aquatic environments.

Still, because of the relatively small toxicities of most herbicides to animals, it can generally be concluded that at the rates of application used for typical forestry applications, direct toxic effects are unlikely (Kenaga, 1975; Morrison and Meslow, 1983; Norris *et al.*, 1983; USDA, 1984; Freedman, 1991).

Consider the case of glyphosate, a commonly used silvicultural herbicide. Glyphosate is a potent plant poison, acting primarily by inhibition of a specific metabolic function, the Shikimic acid pathway, by which four essential, aromatic amino acids are synthesized (Grossbard and Atkinson, 1985). Only plants and a few microorganisms have this metabolic pathway—animals ingest these amino acids with their food. Consequently, glyphosate has a relatively small toxicity to animals, and there are large margins of toxicological safety, in comparison with environmental exposures that are realistically expected during operational silvicultural sprays with this chemical (Feng and Thompson, 1990; Feng *et al.*, 1990; Shipp *et al.*, 1986).

The acute toxicity of glyphosate to mammals is relatively small and is less than or similar to that of some chemicals that many humans voluntarily ingest, as is illustrated by the following series of rat-oral LD_{50}s (mg/kg; LD_{50} is the dose needed to kill one-half of a test population): nicotine 50; caffeine 366; acetylsalicylic acid 1700; sodium chloride 3750; glyphosate 5600; ethanol 13,000 (Table 8.2). The acute toxicity of the most common silvicultural formulation of glyphosate (i.e., 5400 mg/kg), which contains ingredients additional to glyphosate, is similar to that of the pure chemical.

The effects of longer-term, chronic exposures of mammals to glyphosate are also small, especially in comparison with exposures that humans or other animals might receive during an operational treatment in forestry. The following no-effect levels (NOELs) for exposure to glyphosate have been reported (i.e., chronic doses of glyphosate that have been tolerated in laboratory tests without observable biological damages; Shipp *et al.*, 1986; Sassman *et al.*, 1984): (1) rats and dogs fed for 90 days with food containing 2000 mg/kg; (2) mice fed food with 300 mg/kg for 18 months; (3) rats fed food with 100 mg/kg for 2 years; (4) dogs fed food with 300 mg/kg for 2 years; and (5) three generations of rats fed food with 300 mg/kg. These no-effect levels are large, in comparison with anticipated exposures from foods after silvicultural sprays of glyphosate (described earlier, in the section on herbicide residues).

Aqueous glyphosate is much less toxic when presented as pure technical material than in its commercial formulation, a difference that can be attributed to toxicity of the surfactant in the formulation. For example, the acute toxicities of pure glyphosate for rainbow trout (*Salmo gairdneri*) and bluegill sunfish (*Lepomis macrochirus*) (indexed by the LC_{50}, or the aqueous concentration required to kill 50% of a test population) are both 240 mg/liter, but the LC_{50}'s of the commercial formulation are 2.4 and 7.6 mg/liter, respectively (Mayer and Ellersieck, 1986). Typically, the concentrations of glyphosate formulation that are required to cause acute toxicity to aquatic animals is greater than 2 mg/liter (Sassman *et al.*, 1984; Grossbard and Atkinson, 1985; Mayer and Ellersieck, 1986; Feng *et al.*, 1990), a dose that is much larger than the aqueous residues observed even after deliberate over-

sprays of streams and ponds, which are generally less than 0.01 mg/liter (Newton *et al.*, 1984; Feng *et al.*, 1990).

Because herbicide use in forestry has been such a high-profile environmental issue, the practice has been scrutinized with respect to the occupational and more general epidemiological hazards to humans. The issue of human health hazards is controversial and has not been resolved to everyone's satisfaction. (Because of the nature of scientific investigations of epidemiological problems, few such controversial issues are ever resolved.) This issue was mentioned previously and is not dealt with here in detail. It is notable, however, that a number of critical reviews of the available scientific information by government and medical task forces in various parts of the world have concluded that when silvicultural herbicides are used in accordance with recommended operational guidelines, the practice is safe for sprayers and for people living in the vicinity of a spray area [E. G. McQueen *et al.*, 1977; Turner, 1977; Kilpatrick, 1979, 1980; Royal Society of New Zealand (RSNZ), 1980; Beljan *et al.*, 1981; Mullinson, 1981; Walsh *et al.*, 1983; Hatcher and White, 1984; Walstad and Dost, 1984; USDA, 1984; Shipp *et al.*, 1986; most of these studies examined the risks of 2,4,5-T]. This appears to be the "mainstream" scientific opinion on this matter. For alternative viewpoints, see Warnock and Lewis (1978), Whiteside (1979), or Hay (1982). Also, see Chapter 11 for a brief discussion of the toxicology of the poisonous trace contaminant of 2,4,5-T known as TCDD, a dioxin isomer.

Overall, considering the relatively small acute toxicity of glyphosate to animals, it is unlikely that animals inhabiting sprayed clear-cuts would be exposed to significant toxic risks as a result of a silvicultural application of this chemical. However, glyphosate causes great habitat changes through effects on plant productivity and by changing the distribution of biomass in three-dimensional space and among plant species. Therefore, wildlife such as birds and mammals could be secondarily affected through changes in vegetation and through tertiary effects on the abundance of arthropods or the availability of nesting sites. These indirect effects of the silvicultural spraying of glyphosate and other herbicides, which are legitimately considered within the purview of ecotoxicology, could affect the abundance and reproductive success of animal wildlife on the spray site, irrespective of a lack of direct effects due to glyphosate toxicity.

Effects on Birds and Their Habitat.   Several studies have investigated the effects of silvicultural herbicide use on the abundance of breeding birds. Morrison and Meslow (1984a) examined clear-cuts in Oregon that had been sprayed with 2,4-D or with a mixture of 2,4-D and 2,4,5-T, along with an unsprayed reference treatment (Table 8.18). One year after spraying with 2,4-D, and 4 years after spraying with 2,4,5-T plus 2,4-D, there were no large differences in the abundance, species richness, or diversity of birds between the herbicided and reference treatments. (Note, however, that because of the design of this study, it is not known how similar the avifauna of the plots was before spraying.) Certainly, any effects that might be attributable to herbiciding in this study are much smaller than the large changes in the avifauna that must have oc-

curred when the original, mature conifer forest was clear-cut (see Chapter 9).

A better design has been used in some more recent studies that examined the effects of glyphosate spraying, because there are prespray data that can be used to evaluate the initial similarities of the treatment and reference plots (Morrison and Meslow, 1984b; MacKinnon and Freedman, 1992, 1993). In a study in Nova Scotia, the aerial application of glyphosate caused relatively small changes in the abundance and species composition of birds (Table 8.19). Between the pre- and first postspray years, the abundance of all species decreased. Because this change also occurred on the reference plot, it may have been caused by some nonspray factor, perhaps weather, or some effect on the wintering grounds of these migratory species. In the second postspray year, bird abundance on the reference plot increased to about the prespray condition, while on the spray plots abundance stayed similar to the first postspray year and was smaller than the prespray population.

The two most abundant species, white-throated sparrow and common yellowthroat, decreased on

Table 8.18   Effects of herbicide treatment of clear-cuts in western Oregon on breeding birds[a]

Species	1 Year after 2,4-D treatment		4 Years after 2,4,5-T/2,4-D treatment	
	Spray	Reference	Spray	Reference
Willow flycatcher (*Empidonax traillii*)	37	34	26	29
American goldfinch (*Carduelis tristis*)	33	48	34	23
Rufous-sided towhee (*Piplio erythrophthalmus*)	46	23	28	45
Dark-eyed junco (*Junco hyemalis*)	21	13	6	19
White-crowned sparrow (*Zonotrichia leucophrys*)	39	72	76	23
Song sparrow (*Melospiza melodia*)	45	60	48	41
Swainson's thrush (*Catharus ustulatus*)	52	64	40	42
MacGillivray's warbler (*Oporornis tolmiei*)	20	29	21	22
Orange-crowned warbler (*Vermivora celata*)	32	30	27	34
Wilson's warbler (*Wilsonia pusilla*)	17	25	12	37
Rufous hummingbird (*Selasphorus rufus*)	48	43	54	72
Total density	416	460	396	410
Species richness	14	14	14	14
Species diversity (H')	2.5	2.4	2.4	2.5

[a]Data are in units of birds/40.5 ha. Modified from Morrison and Meslow (1984a).

Table 8.19 Abundance of breeding birds among four clear-cut blocks treated with glyphosate and one reference clear-cut in Nova Scotia[a]

Species	Year:	Reference plot				Spray plots			
		0	1	2	4	0	1	2	4
Alder flycatcher (*Empidonax alnorum*)		20	20	41	102	36	7	17	63
American robin (*Turdus migratorius*)		10	10	20	10	14	21	30	31
Red-eyed vireo (*Vireo olivaceous*)		0	10	31	41	0	0	0	4
Magnolia warbler (*Dendroica magnolia*)		0	20	20	143	5	5	5	102
Palm warbler (*Dendroica palmarum*)		0	10	51	31	0	4	18	53
Mourning warbler (*Oporornis philadelphia*)		71	41	20	31	50	13	12	19
Common yellowthroat (*Geothlypis trichas*)		122	112	122	163	151	140	90	136
Lincoln's sparrow (*Melospiza lincolnii*)		20	+	+	0	23	20	41	44
White-throated sparrow (*Zonotrichia albicollis*)		143	71	93	163	203	118	89	155
Dark-eyed junco (*Junco hyemalis*)		31	41	61	102	42	62	61	69
Song sparrow (*Melospiza melodia*)		41	20	10	10	43	28	60	86
American goldfinch (*Spinus tristis*)		61	41	20	20	52	24	15	13
All bird species		539	396	528	836	623	447	444	805

[a]Only abundant species are listed here. Data refer to abundance (pairs/km^2) determined by spot-map censusing, prior to spraying (0 years) and in the first, second, and fourth postspray years. Data for the spray plots are averages; + indicates less than 0.5 territory. Modified from MacKinnon and Freedman (1993).

all spray plots up to the second postspray year and then substantially recovered by the fourth postspray year (Table 8.19). Song sparrow and Lincoln's sparrow declined on the reference plot among the 4 study years, while on the spray plots these species were most abundant in the second and fourth postspray years. As the reference plot continued its postcutting successional development, it was colonized by some new species, including black-and-white warbler, red-eyed vireo, ruby-throated hummingbird, and palm warbler. This change in avifauna was inhibited on the spray plots, because the herbicide treatment set the vegetation back to an earlier successional state and onto a different, probably more conifer-dominated successional trajectory.

Overall, the data of Tables 8.18 and 8.19 indicate rather small effects on the avifauna of clearcuts, in spite of the large changes in vegetation caused by the herbicide treatment. Other field studies of the effects of silvicultural herbicide use have also reported fairly small effects on breeding birds (Beaver, 1976; Slagsvold, 1977; Freedman *et al.*,

1988; Santillo *et al.*, 1989a). An exception is the study of Savidge (1978) of the use of 2,4,5-T in eastern California. Savidge found that total bird abundance was only 30.9/10 ha on a 6-year-old herbicided plot, compared with 65.0/10 ha on a reference plot, while species richness was 8 versus 14.

It should be borne in mind, however, that information on the abundance of breeding pairs of birds, typically estimated by censusing unmarked, singing males, does not tell anything about the reproductive success of birds attempting to breed on herbicided clear-cuts, which might be influenced by habitat quality.

MacKinnon and Freedman surveyed the abundance of birds on herbicided clear-cuts late in the breeding season, after fledglings had left the nest and birds were foraging in family groups. In the first growing season after herbicide spraying, when vegetation damage is greatest, the abundance of birds on their reference plot averaged 27/ha, 7.1 times more than on the spray plots. Among the most abundant species, white-throated sparrow was 13.5

times more abundant on the reference plot, while common yellowthroat was 6.2 times more abundant. Note that these are the same plots for which no large differences in the density of avian territories were apparent between sprayed and reference plots in the first growing season after herbicide treatment (Table 8.19). MacKinnon and Freedman interpreted these apparently inconsistent results as follows:

1. the smaller abundance of birds utilizing the herbicided clear-cuts is indicative of a poorer quality habitat, associated with a smaller amount of foliage, with a consequently smaller abundance of arthropods and fruits, and perhaps an increased risk of nest predation.

2. The lack-of-effect on breeding density in the first growing season after herbiciding may be related to the phenomenon of site fidelity, in which many migratory passerines exhibit a strong tenacity to sites where they have previously bred successfully (e.g., Nolan, 1978; Greenwood, 1980; Wiens and Rotenberry, 1985; Wiens *et al.*, 1986; Morse, 1989). Site fidelity may reduce fitness, however, if unperceived habitat changes have reduced the quality of the breeding site (Wiens *et al.*, 1986; Petersen and Best, 1987). It is suggested that, when the migratory passerines arrived in the springtime to reclaim territories on previously high-quality habitat on clear-cuts, they may not be capable of distinguishing the relative qualities of herbicided and reference habitats. The amount or diversity of foliage would not be different, for example, because at the time of arrival of the birds in the spring the seasonally deciduous and herbaceous plants of the clear-cuts have not refoliated for the growing season. At the same time, the height, size, and density of leafless shrub stems are not distinguishable between herbicided and reference clear-cuts in the first spring after spraying. By the time it is apparent that the amounts of foliage and invertebrates on herbicided clear-cuts are of relatively poor quality, it may be too late to desert the territory in search of a better one, because of already substantial investments in nesting, and because higher-quality sites have already been claimed by other individuals.

In the absence of better studies, it is not possible to substantiate these speculations. However, it is reasonable to expect that the substantial changes in habitat on herbicided clear-cuts would affect the reproductive success of birds, and a study that examines that question would be useful.

Effects on Mammals and Their Habitat.
Studies of the effects of silvicultural herbicide use on deer have largely focused on the availability of browse on sprayed clear-cuts, rather than directly on the abundance of the animals (they are difficult to census).

Since angiosperm shrubs are important weeds in forestry, and also a preferred browse of *Odocoileus* deer, it should be anticipated that the quantity of at least some browse species would be reduced on sprayed clear-cuts. Lyon and Mueggler (1966) examined the effects of spraying with 2,4,5-T plus 2,4-D in Idaho and found a decreased abundance of the shrub *Ceanothus sanguineus*, the most desirable browse species. Similarly, Savidge (1978) found a decreased abundance of the important browse species *Ceanothus velutinus* and *Symphoricarpos* spp. on plots sprayed with 2,4,5-T in California. There was also a reduced abundance of fecal-pellet groups (an index of abundance) of mule deer (*Odocoileus hemionus*).

In another study of 2,4-D spraying, Krefting and Hansen (1969) found that the most affected shrubs (*Corylus* spp.) were not desirable as browse for deer. In their study area, the herbicide treatment stimulated the production of desired browse species and of grasses. As a result, white-tailed deer (*Odocoileus virgianus*) were attracted to the sprayed area for winter browsing and summer bedding, and pellet-group counts were more abundant in the herbicided area for 8 postspray years. Newton *et al.* (1989) also found substantial increases in the availability of browse on sprayed clear-cuts in Maine, partly because of a reduction of height of the canopy of angiosperm shrubs, which increased accessibility of the browse to deer.

Overall, the effects of herbicide treatment on deer browse appear to be rather site specific. To make predictions for management purposes, knowledge is required of the relative susceptibilities of particular browse species to the herbicide that is being used and of the height of the canopy, which influences the accessibility of browse to deer.

Studies have also been made of the effects of silvicultural herbicide use on small mammals. Most studies have found only small, sporadic effects on the abundance and species composition of small-mammal communities (Borrecco *et al.*, 1979; Sullivan and Sullivan, 1981, 1982, 1991; D'Anieri *et al.*, 1987; Freedman *et al.*, 1988; Santillo *et al.*, 1989b; T. P. Sullivan, 1990). However, Savidge (1978) found that small mammals were about twice as abundant on a sprayed plot, a phenomenon that was attributed to more favorable habitat because of an increased abundance of composites, graminoids, and the dicot shrub *Ribes*.

In an interesting but artificial study, Thalken and Young (1983) censused small mammals at a site in Florida that had been used to test aerial spray-delivery systems for the military herbicide program that was used during the Vietnam War (see Chapter 11). The 172-ha study area had received a massive dose of 2,4,5-T, equivalent to 426 kg a.i./ha, from numerous spray trials between 1962 and 1970 (this is more than 100 times the average rate of application during a single conifer release treatment in forestry). Because the spray area had been cleared of the regional pine–oak forest and since monocots are not sensitive to 2,4,5-T, the vegetation of the spray-test area developed into a distinct habitat island, dominated by the grasses *Andropogon virginicus*, *Panicum virgatum*, and *P. lanuginosum*. In studies conducted between 1973 and 1978, a total of 341 species of organism was identified on the intensively sprayed test area, including an abundant population of beach mouse (*Peromyscus polionotus*). Studies of the residues of TCDD and histopathological examination of 225 mice indicated minimal effects on the health and reproduction of this species. The small effects on the beach mouse, plus the general abundance and richness of biota on what must be the world's most intensively herbicided site, are notable.

Aquatic Effects. Finally, mention should be made of the apparent effects that occur in aquatic

systems as a result of the use of herbicides in forestry. The most frequently used herbicides in silviculture are not mobile in soil, and therefore most aquatic contamination occurs by the direct deposition of spray to the water body. (There are exceptions to this statement, however; hexazinone, for example, is mobile in soils.) Since direct spraying over water is usually avoided in spray programs in forestry (at least for larger water bodies), most investigations of operational spraying have not found detectable residues in water in the vicinity of sprayed areas (Matida *et al.*, 1975; Norris *et al.*, 1982, 1983; Morash *et al.*, 1987). Norris *et al.* (1982) studied an experimental situation in Oregon where direct deposition was not avoided and found that stream-flow discharge contained 0.35% of the aerially applied picloram and 0.014% of 2,4-D. This contamination took the form of a well-defined, short-lived pulse.

Reviews of aquatic toxicological data for the most frequently used herbicides in forestry indicate that the biological implications of aquatic contamination appear to be small (Norris *et al.*, 1983; USDA, 1984). In a field experiment, Matida *et al.* (1975) applied 2,4,5-T plus 2,4-D at 6.0 kg a.i./ha to a 9.5-ha forested watershed in Japan. The application did not result in detectable residues in streamwater (the detection limit was 0.06 ppm). No statistically significant changes were measured in the abundance of benthic invertebrates, and there was no mortality of fish.

Treatment of a harvested forest watershed in British Columbia with glyphosate found short-lived residues in directly sprayed sections of streams, but no detectible residues where the stream was protected by a 10-m buffer of unsprayed vegetation (Feng *et al.*, 1989). This caused no significant effects on abundance of stream invertebrates (Holtby and Baillie, 1989a; Kreutzweiser and Kingsbury, 1989; Scrivener and Carruthers, 1989). Some measurements suggested a possible (but statistically insignificant) effect on benthic invertebrates, but this was considered small compared with natural, spatial and temporal variations, and the much larger effects of logging. Caged fingerlings of coho salmon (*Oncorhynchus kisutch*) showed some behavioral stress immediately after exposure to the herbicide, and directly sprayed fish suffered a small amount of mortality (3%) (Holtby and Baillie, 1989a). However, fish protected by the spray buffer had no mortality, and no spray-associated deaths were observed in the natural, resident population of fingerling salmon.

Treatment of a harvested forest watershed in British Columbia with glyphosate found short-lived residues in directly sprayed sections of streams, but no detectable residues where the stream was protected by a 10-m buffer of unsprayed vegetation (Feng *et al.*, 1989). This had no significant effects on abundance of stream invertebrates (Holtby and Baillie, 1989a; Kreutzweiser and Kingsbury, 1989; Scrivener and Carruthers, 1989). Some measurements suggested a possible (but statistically insignificant) effect on benthic invertebrates, but this was considered small compared with natural, spatial, and temporal variations, and the much larger effects of logging. Caged fingerlings of coho salmon (*Oncorhynchus kisutch*) showed some behavioral stress immediately after exposure to the herbicide, and directly sprayed fish suffered a small amount of mortality (3%) (Holtby and Baillie, 1989a). However, fish protected by the spray buffer had no mortality, and no spray-associated deaths were observed in the natural, resident population of fingerling salmon.

Beyond the direct, toxic risks to aquatic biota associated with the use of herbicides in forestry, there may be some indirect effects, caused by changes in habitat. For example, use of herbicides in the riparian zone would reduce the amount of shading foliage, exposing streamwaters to the direct effects of sunlight and causing higher water temperatures, which may have detrimental effects on biota. Large reductions of vegetation may also predispose stream banks to erosion, causing siltation of aquatic habitats and biological damage. These topics are discussed in more detail in Chapter 9.

Integrated Pest Management

It could be concluded that the documented, ecological effects of the post-DDT spray programs against eastern spruce budworm and of the silvicultural use of herbicides are relatively small. Certainly, many

politicians, and public- and private-sector resource managers have decided that, at least in the shorter term, the environmental "costs" of these programs are "acceptable" in view of their substantial management and economic benefits. Of course, similar decisions have been made for the much larger-scale and more intensive spray programs that are annually mounted in most modern agroecosystems.

It is arguable, however, whether it is desirable on the longer term to rely on the broadcast spraying of broad-spectrum pesticides to cope with resource-management problems in agriculture, forestry, or for any other purpose. It would be much better if less reliance could be placed on such nonspecific practices.

A preferable approach is integrated pest management (IPM). Within the context of IPM, acceptable pest control is achieved, when possible, by employing an array of complementary approaches. These can include (CEQ, 1979; Bottrell and Smith, 1982; Flint and van den Bosch, 1983):

1. the use of natural predators, parasites, and other biological controls;

2. the use of pest-resistant varieties of crop species, which can sometimes be produced using standard breeding techniques, and modern, high-tech genetic engineering techniques (e.g., Fishhoff *et al.*, 1987);

3. the modification of ecological conditions so as to reduce the optimality of the pest habitat;

4. a careful monitoring of pest abundance; and

5. the use of pesticides only when they are required as a necessary component of the integrated, pest-management strategy.

If successfully implemented, an IPM program can greatly reduce, but not necessarily eliminate, the reliance on pesticides. For example, the widespread use of an IPM scheme for the control of boll weevil (*Anthonomus grandis*) in Texas cottonfields was largely responsible for a reduction of insecticide use for this purpose from 8.77 million kg in 1964 to 1.05 million kg in 1976 (Bottrell and Smith, 1982).

An important aspect of IPM is the use of procedures that are as pest specific as possible, so that nontarget damage can be avoided or reduced. There are some precedents for such pest-specific practices in resource management. Some examples are the biological control of certain introduced pests in agriculture:

1. The cottony-cushion scale (*Icyera purchasi*) was a serious threat to the citrus industry in the United States, where it had been introduced from its native Australia. However, virtually total control of this pest was achieved by the introduction in 1888 of the vedalia lady beetle (*Vedalia cardinalis*) and a parasitic fly (*Cryptochetum iceryae*) from Australia (Swan and Papp, 1972).

2. The klamath weed or common St. John's wort (*Hypericum perforatum*) was introduced from Europe to North America, where it became a serious weed of U.S. southwest pastures because of its toxicity to cattle. This pest was successfully controlled by the introduction in 1943 of the herbivorous leaf beetles *Chrysolina hyperici* and *C. gamelata* (Huffaker and Kennett, 1959; Swan and Papp, 1972).

3. The prickly pear cactus (*Opuntia* spp.) was imported to Australia from North America for use as an ornamental plant, but it became a serious weed of rangelands. This pest has been almost totally controlled by the introduction of the moth *Cactoblastis cactorum*, whose larvae feed on the cactus (Swan and Papp, 1972).

4. Ragwort (*Senecio jacobea*) is an important weed of pastures, because it displaces beneficial forage species and poisons livestock because of its content of hepatotoxic alkaloids. Ragwort is native to temperate Eurasia, but it has been introduced widely, and it is now considered a serious pest in the northwestern United States, southwestern Canada, eastern Canada, Argentina, New Zealand, Australia, and elsewhere. In Oregon, ragwort has been effectively controlled through the introduction of three herbivores: cinnabar moth (*Tyria jacobaeae*), ragwort flea beetle (*Longitarsus jacobaeae*), and ragwort seed fly (*Hylemya seneciella*). At one heavily infested pasture in Oregon, these control agents reduced the biomass of ragwort from an initial 700 g/m² in 1981, to 430 g/m² in 1982, 25 g/m² in 1983, and <1 g/m² during 1984–1988. The corresponding changes in perennial forage species were 80, 250, 1000, and 350–1020 g/m², respectively (McEvoy et al., 1991).

There are also cases of the successful, species-specific, biological controls of native pests of livestock and crops:

1. The common vampire bat (*Desmodus rotundus*) is a serious pest of livestock in Central and South America. It can be controlled by capturing individual bats and treating them with a topical application of petroleum jelly containing the anticoagulant, diphenadione. The pesticide is spread to other bats during social grooming in cave roosts. Other bat species are not affected by the treatment, including two other relatively uncommon and specialized species of vampire bats (*Diphylla acaudata* and *Diaemus youngii*) (Mitchell, 1986).

2. The screw-worm fly (*Callitroga hominivorax*) is a serious pest of cattle, because of damage done when its larvae feed on open wounds. In some areas this fly has been successfully managed by the release of large numbers of sterile individuals into the wild population. The sterile flies were produced by gamma irradiation of populations that were mass reared in the laboratory. Since the female fly only mates once, a copulation with a sterile male does not result in successful reproduction. The technique operates by the swamping of the wild population of the screw-worm fly with sterile males, so that there are relatively few successful matings and the abundance of the pest declines to an acceptable level (Baumhover et al., 1955).

Unfortunately, biological control has not been successful in the majority of cases in which it has been attempted, and the technique may not be suitable for all pest problems. This seems to be especially true of forest-pest problems, such as eastern spruce budworm and competing weeds. In the case of eastern spruce budworm, there have been active investigations of the potential roles of bacterial, viral, and other pest-specific disease agents, of the use of parasitoids, and of the use of species-specific sex pheromones to disrupt mating.

Recent research is examining the efficacy of an egg-parasitoid, *Trichogramma minuta*, for controlling eastern spruce budworm. In inundative releases, this wasp has achieved a rate of parasitism of 80% of budworm eggs, but the more usual rate is 5–25%, compared with about 1% in reference stands (Smith *et al.*, 1990; Ennis and Caldwell, 1991). This biological control method is promising, but because of its relatively high costs and variable efficacy, it is not yet ready for operational use.

With the exception of the use of pesticides formulated with the bacterium B.t. (discussed earlier), the biological methods have not yet proved to be sufficiently successful to serve as viable, operational alternatives to the broadcast spraying of synthetic pesticides to control eastern spruce budworm (Morris, 1982; Hulme *et al.*, 1983; L. K. Miller *et al.*, 1983; Sanders *et al.*, 1985; Hudak, 1991). However, this field of research is being actively pursued, and hopefully in the future biological controls will prove to be more useful. This will probably occur in conjunction with other methods in a program of integrated pest management including, for example, the management of habitat to make it less susceptible to budworm infestation (Schmidt *et al.*, 1983; Hudak, 1991). In the short term, however, IPM does not yet appear to be a viable alternative to the presently used control practices for eastern spruce budworm.

In general, unless there are substantial changes in the perceptions of decision makers about the importance of the known and hypothesized damage associated with the use of pesticides, the broadcast spraying of insecticides to control eastern spruce budworm will continue to be permitted, as will the use of herbicides in silviculture, and the especially intensive use of pesticides for many other pest-management problems, particularly in agriculture.

9

HARVESTING OF FORESTS

INTRODUCTION

Forests cover about one-third of Earth's land surface. A closed-canopy forest covers about 4–5 billion ha, while about 2 billion ha is relatively open woodland and savannah (Tables 9.1 and 9.2). Geographically, the most heavily forested regions are in North and South America, Europe, and Russia, each with more than 30% forest cover (Table 9.1). Temperate plus boreal forests cover an area comparable to that of tropical forests, but their production is only about 50% as large, and they have only 61% as much biomass (Table 9.2).

Humans clear forests for many reasons. Most important are the creation of new agricultural lands, the harvesting of biomass to manufacture lumber and paper, and burning to produce energy. Globally, huge areas of mature, forested ecosystems are affected each year by these various activities. In the late 1980s, about 25 million ha/year of forest was cleared, and in 1989 the global yield of wood was about 4.1 billion m^3/year, a 25% increase from 1975–1977. The 1989 global consumption of wood included 0.50 billion m^3 of sawn timber (a 13% increase from 1975–1977), 0.13 billion m^3 of wood panels such as plywood (27% increase), 1.7 billion m^3 of industrial roundwood (18% increase; used

mostly to manufacture 215 million tonnes of paper), and 1.8 billion m^3 of fuelwood (30% increase; about 85% of the fuelwood was consumed in underdeveloped countries) [Food and Agriculture Organization (FAO), 1982; WRI, 1986, 1990; Forestry Canada, 1992a]. The 1978 global harvest of 20 million ha represented an increase from 16×10^6 ha in 1950, 10×10^6 ha in 1900, and 6×10^6 ha in 1800 (FAO, 1982; Houghton *et al.*, 1983).

The vigorous, forest-based resource industries have a huge economic impact. In 1989, the value of forest products traded internationally was estimated at about $101 billion (U.S. dollars) (FAO, 1986; WRI, 1986; Forestry Canada, 1992a).

In many countries, forest resources are being severely depleted by excessively large rates of clearing. For example, between 1981 and 1985 the annual rate of forest clearing in Central America and northern South America ranged from 0.9%/year in Panama to 4.6%/year in Paraguay. In Nepal it was 3.9%/year, in Nigeria 4.0%/year, in the Ivory Coast 5.9%/year, and in Haiti 3.4%/year. Net global deforestation was about 33% from preindustrial times up to 1954, but between 1980 and 1985 it proceeded at about 1%/year, a rate that if extrapolated linearly would predict a forest-area half-life of only about 70 years (WRI, 1986). Note that the data in this paragraph refer to forest conversions. In typical forestry, another forest is allowed and usually encouraged to

Table 9.1   Global forest resources[a]

Region	Forested land (10⁶ ha)	Forests as % of total land area	Total average wood volume (m³/ha)	Percentage coniferous	Percentage broadleaf
North America	630	34	93	74	26
Central America	65	22	92	32	68
South America	730	30	172	1	99
Africa	800	6	133	1	99
Europe	170	30	94	67	33
U.S.S.R. (former)	915	35	106	83	17
Asia	530	15	96	16	84
Pacific	190	10	31	23	77
World total	4030	21	110		

[a]Modified from Persson (1974).

regenerate on harvested sites, although the plant community may be different in character from the original.

The global, net primary production of forests has been estimated as 48.7 billion tonnes/year, of which an extraordinary 28% is appropriated in vari-ous ways for human uses (Vitousek *et al.*, 1986). Of the appropriated production, 45% is accounted for by the clearing of forests for shifting cultivation in less developed countries, 18% by more permanent conversions of forests to agricultural lands, 16% by harvesting forest biomass, 12% by growth in plan-

Table 9.2   Forest vegetation by generalized ecosystem type[a]

	Area (10⁶ km²)	Total net primary production (10¹⁵ g C/year)[b]	Total mass of vegetation (10¹⁵ g C)
Tropical rain forest	17.0	16.8	344
Tropical seasonal forest	7.5	5.4	117
Temperate conifer forest	5.0	2.9	79
Temperate angiosperm forest	7.0	3.8	95
Boreal forest	12.0	4.3	108
Woodland and shrubland	8.5	2.7	22
Savannah	15.0	6.1	27
All other continental vegetation	77.0	10.8	35
Total continental	149	52.8	827
Total marine	361	24.8	1.74
Total global	510	77.6	828

[a]Modified from Woodwell *et al.* (1978).
[b]C, carbon.

tations of biomass destined for use by humans, while 10% was lost during harvests (Vitousek *et al.*, 1986).

To supply the global demand for tree fiber and to achieve the great economic benefits of forestry, large areas of mature forest must be harvested or otherwise disturbed each year. The scale and nature of this disturbance can be illustrated by the case of Canada, a nation whose economy is strongly dependent on industrial forestry. On average, 988 thousand ha was harvested each year between 1985 and 1990, 91% by clear-cutting (Forestry Canada, 1992a,b). This disturbance was followed by an average of 424 thousand ha/year of site preparation, 65% of which was mechanical scarification, 15% prescribed burning, and 20% other methods, including the use of herbicides. During the same time period, an average of 381 thousand ha/year were planted with conifer seedlings (674 million seedlings per year), while another 34 thousand ha was regenerated by direct seeding and the remaining areas (about 573 thousand ha/year, or 58% of the harvested area) by natural seeding and advance regeneration. In addition, an average of 298 thousand ha was subjected to stand tending, most of which involved the silvicultural use of herbicides. Finally, between 1975 and 1983, an average of 3.6 million ha/year of forest was treated to control pests, mostly in eastern Canada where insecticides are used in attempts to control an epidemic of spruce budworm in fir and spruce forests (Chapter 8).

Worldwide, an immense area of about 20 million ha of forest is cleared each year for various purposes, including forestry, agriculture, and rural and urban development. There are many longer- and shorter-term consequences of the environmental changes that are associated with this type of land use. In many cases the effects are relatively moderate, and they could be considered to be acceptable ecological "costs" that must be borne to harvest forests as a potentially renewable natural resource. Some of the longer-term effects include potential decreases in site fertility caused by nutrient losses, effects on biodiversity through conversion of the habitat of wildlife, permanent losses of old-growth forest, and effects on atmospheric CO_2, with associ-

ated climate effects (Chapter 2). Shorter-term effects can include increases in erosion, changes in watershed hydrology, inadequate forest regeneration, temporary changes in the habitat of wildlife, and depending on the perspective of the observer, degraded aesthetics of many harvested and managed sites. In some situations, these various environmental effects can be so severe that they should preclude the harvesting of particular stands or larger areas of forest.

In this chapter, some of the ecological effects of forest harvesting are described. The focus is on forestry in north-temperate latitudes, because the ecological effects are relatively well known in this region. However, it is important to realize that the less well documented effects of forest harvesting and conversions in other regions of the globe, particularly the humid tropics, can be severe and are very important in terms of resource degradation, climate change, and losses of biodiversity.

The topics discussed in this chapter are the implications of nutrient removals during harvesting for subsequent fertility of the site, effects of forest harvesting on erosion and hydrology, and implications for biodiversity at various levels (i.e., population, community, and landscape). Relevant topics that are dealt with in other chapters are: (1) the effects of insecticide and herbicide spray programs in forestry (Chapter 8) (2) the effects of deforestation on atmospheric CO_2 (Chapter 2), and (3) the effects of tropical deforestation on global species richness (Chapter 10).

When considering the material in this chapter, the following considerations should be borne in mind: (1) the management objectives in forestry are almost never to convert the site to a nonforested ecosystem, but to allow or encourage another forest to regenerate on harvested sites, and (2) forest harvesting and silviculture cause stand-level disturbances, but so do natural stressors such as wildfires, windstorms, and insect and disease infestations. It many respects it is prudent to judge the importance of the ecological effects of forestry in comparison with the analogous effects of these natural stressors. An obvious corollary is that ecologically appropriate forest management should strive to reflect the natural dynamics of the forested ecosystem and the structures and functions of natural communities and landscapes.

9.2
FOREST HARVESTING AND SITE FERTILITY

Background

The various methods of harvesting forests vary greatly in the intensity of biomass and nutrient removals. At the relatively "soft" end of this continuum are selection-tree cuts, in the middle are small-area harvests such as group shelterwood and small clear-cuts, and at the intensive end are extensive clear-cuts. The clear-cuts themselves vary in intensity, from stem-only harvests to whole-tree (all above-ground biomass) and complete-tree (above- and below-ground biomass) harvests.

The more intensive harvests increase the short-term yields of biomass from the forest, but these are accompanied by substantially larger removals of nutrients. Potentially, by decreasing site fertility, the nutrient removals can result in a deterioration of the inherent capability of the site to sustain the production of tree biomass (Kimmins, 1977, 1987; Norton and Young, 1976; Hornbeck, 1977; Gordon, 1981; Freedman, 1981a; Smith, 1985; Mann *et al.*, 1988; Brown and Brown, 1991). However, it should be realized that over the longer term, even the softer methods of harvesting result in substantial removals of site nutrients. In the softer systems, these removals occur at relatively small rates at each harvest. However, the harvests are more frequent than with the intensive systems, and to maintain the same flow of biomass they typically occur over larger areas of land and require a more extensive infrastructure of roads and other support systems.

In agriculture, the syndrome of site impoverishment caused by intensive cropping is a well-recognized problem. In severe cases, depleted land must be abandoned for some or all agricultural purposes. Usually the problem can be more-or-less managed by the application of fertilizers, but sometimes the structural degradation of the soil is too severe to allow this simple mitigation to be successful. In forestry, however, the harvest rotation is much longer than in agriculture (in which it is usually annual), and there are few data that allow comparisons of the productivity of subsequent forest rotations on the same site. Overall, the situation with respect to site-nutrient impoverishment by forest harvesting is less severe than it is in agriculture and should be viewed as a potential, longer-term problem.

Frequently cited examples of declines of site capability in forestry are for second-rotation plantations of introduced radiata pine (*Pinus radiata*) in certain places in Australia and New Zealand with relatively poor soils (Keeves, 1966; Whyte, 1973; Will, 1985). The cause of the reduced productivity of radiata pine is often attributed to nutrient impoverishment of the site caused by forest harvesting. However, the comparisons of subsequent rotations of radiata pine are complicated by such factors as: (1) soil changes other than those that directly involve nutrient supply, such as soil compaction, podsolization, changes in soil organic matter and acidity; (2) differences between rotations in the density and genetic quality of crop trees; and (3) the nature of weedy competition and any release treatments that may be used. It should be borne in mind that reduced productivity has not been observed in many second rotations of radiata pine in Australia, and in fact productivity is often improved on some sites (Squire *et al.*, 1991).

Because of the problems associated with the existing database, the syndrome of decreased site fertility caused by forest harvest is best viewed as a potential problem. Even though there is not yet hard evidence that this problem occurs on a large scale, from ecological principles it might be anticipated that decreased fertility could emerge after a number of forest-harvest rotations from the same site. Any wide-scale impoverishment of forested sites would, of course, have important implications for the management of forests as a renewable natural resource.

The question of nutrient impoverishment of forest sites will be examined by consideration of the magnitudes of nutrient removals during harvesting, the sizes of soil nutrient pools, the net accretions and depletions of nutrients in forests, and the ways in which these variables can interact to influence forest productivity.

Conceptual Models of Site Impoverishment

The conceptual problem of site impoverishment caused by intensive harvesting is illustrated by the models of Fig. 9.1. Longer-term decreases in the magnitude of nutrient capital would eventually result in a decreased capability of the site to sustain large rates of tree productivity. The effects of rotation length are conceptually illustrated in Fig. 9.1a. A relatively long rotation allows a sufficient passage of time for harvested nutrients to be replenished by inputs through rainfall, weathering, nitrogen fixation, etc. Such a post-harvest recovery period has been labeled an "ecological rotation" (Kimmins, 1977), because harvesting a forest on this basis is sustainable with respect to nutrient capital—the potentially renewable natural resource is not "mined" when an ecological rotation is practiced. Under the shorter rotation of Fig. 9.1a, the harvested nutrients are not replenished between successive harvests, and this causes a longer-term deterioration of site quality.

Such a deterioration can also occur if the intensity of the harvest is increased, even if a longer rotation is used (Fig. 9.1b). In this example, the whole-tree harvest removes twice as much nutrient as the stem-only harvest, and therefore the nutrients are not replenished before the next harvest takes place.

The influence of site quality is summarized in Fig. 1c, which indicates that inherently fertile sites can support relatively intensive harvests over shorter rotations, in comparison with less fertile sites. In general, fertile sites have relatively rapid rates of nutrient replenishment, because of some combination of: (1) rapid mineralization of nutrients from relatively immobile, organic, and mineral forms, to water-soluble ions that can be assimilated by plant roots; (2) large rates of input of nutrients from the atmosphere; (3) a large rate of nitrogen fixation; and (4) small rates of nutrient loss by leaching to below the tree-rooting depth.

Factors Affecting Nutrient Removals by Harvesting

When trees are harvested, there are substantial removals of nutrients that are organically fixed in the biomass. The quantities of nutrients that are removed are influenced by several factors, including tree species, age of the stand, fertility of the site, intensity of the harvest, and time of year, especially for seasonally deciduous species harvested during the dormant season. These effects are discussed below.

The influence of stand age is illustrated in Fig. 9.2 for a chronosequence of Scotch pine (*Pinus sylvestris*) stands in Britain (Ovington, 1959). As expected, the older the stand, the larger its above-ground contents of the macronutrients N, P, K, Ca, and Mg. Of course, this pattern reflects the net accumulation of above-ground biomass by the stand, since the nutrients are mostly organically bound. It can therefore be generalized that the biomass and nutrient removals by any given harvest method are influenced by age of the stand or by the length of the rotation. This is especially true of young and intermediate-aged stands younger than about 100 years. In older-growth stands, the net above-ground accumulations of biomass and nutrients may be in a state of dynamic balance between the death of old trees and the positive net production of living individuals (Likens *et al.*, 1977; Bormann and Likens, 1979).

The species of tree that is being harvested also influences the yields of biomass and nutrients by a forest harvest. Table 9.3 compares the above-ground quantities of biomass and nutrients in adjacent plantations of red pine (*Pinus resinosa*), jack pine (*Pinus banksiana*), and white spruce (*Picea glauca*) and a natural poplar forest (*Populus tremuloides* and *P. grandidentata*). The variations in the quantities of biomass and nutrients are large, amounting to a ratio of 1.4 for biomass (calculated as the largest/smallest value among the stands), 2.0 for nitrogen, 2.3 for phosphorus, 2.0 for potassium,

(a) Variation in Rotation Length: Fixed Utilization

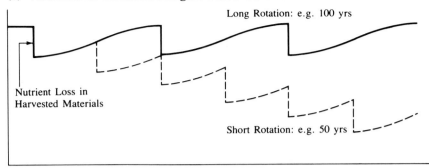

(b) Variation in Utilization: Fixed Rotation Length

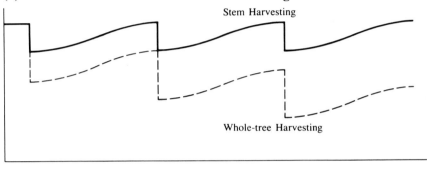

(c) Variation in Rates of Replacement of Nutrient Losses: Fixed Rotation and Utilization

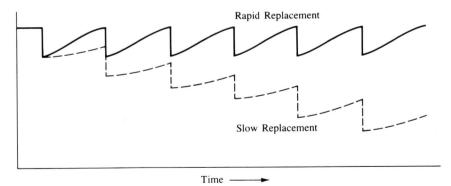

Time ⟶

Fig. 9.1 The influence of rotation length, harvest intensity, and rate of nutrient replenishment on site nutrient capital, and ultimately on the ability of the site to sustain a large rate of productivity. See text for discussion. Modified from Kimmins (1977).

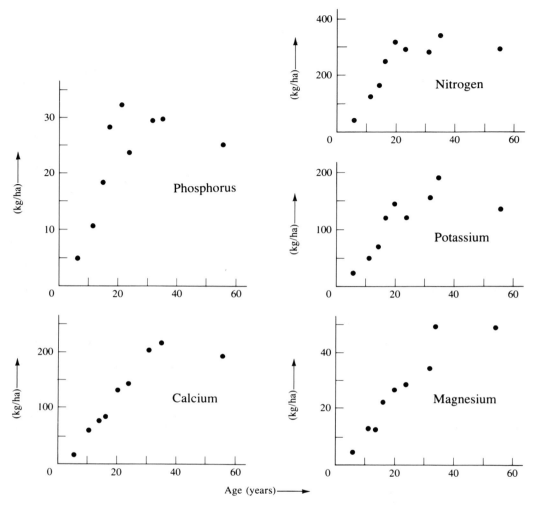

Fig. 9.2 The influence of stand age on the above-ground standing crops of nutrients in *Pinus sylvestris* plantations. Modified from Ovington (1959).

4.3 for calcium, and 1.5 for magnesium. Except for phosphorus, the most nutrient-demanding trees were the poplars. The poplar stand also had a relatively large net production of biomass. However, the nutrient-use efficiency (which is related to the ratio of biomass to nutrient quantities) of the poplar stand was less than that of the more productive red pine plantation.

Of course, in regions with longer growing sea-

sons, biomass can accumulate relatively quickly. This is especially true of tropical sites with a humid climate, where tree growth may be essentially continuous (Brown and Lugo, 1990, 1992; Lugo, 1992). Rates of above-ground accumulation of biomass and nutrients in trees in plantations in Puerto Rico are given in Table 9.4. With the exception of mahogany (*Swietenia macrophylla*), a slow-growing but valuable species, the growth rates of

Table 9.3 Effects of tree species on the above-ground quantities of biomass and nutrients in adjacent 40-year-old stands growing on a fine sandy-loam soil in Minnesota[a]

Species	Age	Biomass (tonnes/ha)	N (kg/ha)	P (kg/ha)	K (kg/ha)	Ca (kg/ha)	Mg (kg/ha)
Pinus resinosa	40	199	155	42	175	291	58
Pinus banksiana	40	141	118	25	97	199	38
Picea glauca	40	141	102	57	229	719	40
Populus tremuloides– P. grandidentata	40	167	199	47	287	848	58

[a] After Alban *et al.* (1978).

trees in tropical plantations are much faster than those given for temperate species in other tables in this section.

The influence of site fertility is illustrated by a comparison of the above-ground biomass and nutrients in plantations of sycamore (*Platanus occidentalis*) and Norway spruce, each growing on two different sites (Table 9.5). Net production and nutrient accumulation by the stands differ greatly between the sites. The inherent fertility of the sites is probably the most important determinant of these differences.

The influence of the intensity of harvesting is illustrated by the relative removals of biomass and nutrients by bole-only and whole-tree clear-cuts of a conifer stand in Nova Scotia (Table 9.6). The whole-tree clear-cut resulted in a 30% larger yield of biomass than the stem-only clear-cut. In terms of short-term economics, this can be an advantage of the whole-tree method, especially if the stand is being harvested to produce energy or pulp, for which quality of the biomass is not an important consideration. However, the increased yield of biomass is largely due to the removal of nutrient-rich tissues such as foliage and small branches. As a result, the removal of nutrients by the whole-tree clear-cut was increased by substantially larger degrees than for biomass: by a factor of 2.0 for nitrogen, 1.9 for phosphorus, 1.7 for potassium, 1.5 for calcium, and 1.8 for magnesium. To use an eco-

Table 9.4 Tree biomass and nutrient accumulations in plantations in Puerto Rico, established on former agricultural land[a]

Species	Stand age (years)	Biomass (a–g, tonnes/ha)	Nutrients (a–g, kg/ha)				
			N	P	K	Ca	Mg
Leucaena leucocephala	5.5	70	138	—	12	75	15
Albizia procera	5.5	102	159	—	11	77	23
Casuarina equisetifolia	5.5	162	256	—	18	150	29
Eucalyptus robusta	5.5	118	100	—	12	64	22
Pinus caribaea	6.0	62	44	2	8	—	—
	20	152	123	5	12	—	—
	27	272	187	9	46	102	37
Swietenia macrophylla	20	114	140	5	11	—	—

[a] Modified from Lugo (1992).

Table 9.5   Effect of site on the above-ground quantities of biomass and nutrients for particular tree species

Species	Site	Age	Biomass (tonnes/ha)	N (kg/ha)	P (kg/ha)	K (kg/ha)	Ca (kg/ha)	Mg (kg/ha)	Reference
Platanus	a	3	9.2	52	10	10	46	17	Wood *et al.* (1977)
occidentalis	b	3	13.7	90	21	53	53	24	
Picea abies	a	47	263	705	82	226	507	85	Ovington (1962)
	b	47	140	331	37	161	212	39	

nomic analogy: the relatively small increase in yield of biomass was "purchased" at the "expense" of much larger increases in the rates of nutrient removal.

Use of Simple Box Models to Evaluate the Consequences of Nutrient Removals

The implications of nutrient removal by forest harvests can be appreciated by comparing the quantities of harvested nutrients with the amounts in other important compartments and fluxes. Most important are: (1) the nutrient quantity in the forest floor and mineral soil within the tree-rooting depth; (2) nutrients in residual, nonharvested plants such as shrubs and ground vegetation; and (3) the fluxes of nutrients incoming from the atmosphere, moving about within the site as throughfall, stemflow, and

litterfall, and leaving the site with drainage, ultimately to groundwater or streamflow.

The conceptually simple, box-model approach of a comparison of compartments and fluxes is illustrated for a mature maple–birch stand in Nova Scotia (Table 9.7). The whole-tree biomass (154.6 tonnes/ha) is 1.4 times as large as the biomass of the merchantable stem (113.0 tonnes/ha). Therefore, for this stand a whole-tree clear-cut could potentially result in 40% more yield of biomass than a stem-only clear-cut. As in Table 9.6, the whole-tree / merchantable stem ratios for nutrients are considerably larger than those for biomass. The whole-tree quantity of nitrogen was 2.2 times as large as that in the stems, while phosphorus was 2.6 times as large, potassium 2.2 times, calcium 1.7 times, and magnesium 1.9 times. In summary, a moderate increase in yield of biomass (40%) would be achieved

Table 9.6   Effect of harvest intensity on the yield of biomass and nutrients from a *Picea rubens–Abies balsamea* stand in Nova Scotia[a]

Harvest	Component	Biomass (kg d.w./ha)	N (kg/ha)	P (kg/ha)	K (kg/ha)	Ca (kg/ha)	Mg (kg/ha)
Conventional	Merchantable stem	105,200	98.0	16.3	91.7	180.9	17.0
Whole tree	Merchantable stem	117,700	120.1	18.2	76.2	218.9	20.4
	Tops, branches, foliage	34,800	119.0	17.0	56.4	117.6	16.5
	Total harvest	152,500	239.1	35.2	132.6	336.5	36.9
	Percentage increase	29.6%	99.1%	93.4%	74.0%	53.7%	80.9%

[a] In the conventional clear-cut, only the stems of the trees were removed. The whole-tree clear-cut included the harvesting of tops, branches, and foliage. Modified from Freedman *et al.* (1981b).

Table 9.7 Standing crops and fluxes of biomass and nutrients in a mature stand of *Acer saccharum–A. rubrum–Betula alleghaniensis* in Nova Scotia[a]

Category	Biomass	N	P	K	Ca	Mg
Standing crops						
Trees, total above ground	154,600	355.4	38.4	183.1	448.1	45.4
	(1.0)	(1.0)	(1.0)	(1.0)	(1.0)	(1.0)
Trees, stem only	113,000	159.4	15.0	84.6	256.4	23.5
	(1.4)	(2.2)	(2.6)	(2.2)	(1.7)	(1.9)
Ground vegetation	218	3.2	0.3	2.8	0.5	0.3
	(719)	(111)	(128)	(65)	(896)	(151)
Forest floor, total	18,700	314	22.2	56.2	50.8	25.8
	(8.4)	(1.1)	(1.7)	(3.3)	(8.8)	(1.8)
Forest floor, available	—	1.0	0.3	4.3	20.8	3.5
	—	(355)	(128)	(43)	(22)	(13)
Mineral soil, total	324,600	4945	921	12,580	855	1139
	(0.48)	(0.072)	(0.042)	(0.0015)	(0.52)	(0.040)
Mineral soil, available	—	79.8	37.0	181	244	113
	—	(4.5)	(1.0)	(1.0)	(1.8)	(0.40)
Fluxes						
Litterfall	4,900	33.7	2.5	5.4	34.1	4.8
	(32)	(11)	(15)	(34)	(13)	(9.5)
Incident precipitation	—	3.7	<1.0	2.2	2.7	1.1
	—	(96)	—	(83)	(166)	(41)
Throughfall	—	0.9	0.3	9.9	3.6	1.2
	—	(395)	(128)	(18)	(124)	(38)
Stemflow	—	0.009	0.030	1.1	0.26	0.07
	—	(39,500)	(1280)	(166)	(1723)	(649)
Weathering[b]	—	0	0.6	5.0	18	5.0
	—	—	(64)	(37)	(25)	(9.1)
Nitrogen fixation[b]	—	10	—	—	—	—
	—	(36)	—	—	—	—
Streamflow[b]	—	0.32	<0.1	2.7	14.6	9.7
	—	(1110)	—	(68)	(31)	(4.7)
Net flux[b]	—	+7	+0.4	−0.6	−7	−2
	—	(51)	(128)	(305)	(64)	(23)

[a]Standing crops are in kg/ha; fluxes are in kg/ha-year; throughfall and stemflow are in kg/ha over the growing season. Data in parentheses are ratios of the quantity in the whole-tree compartment, relative to the quantity in other standing crops and fluxes. Modified from Freedman *et al.* (1986).
[b]These were not measured for this specific site; they were estimated by a review of the literature.

by a whole-tree harvest, but at the expense of much larger increases in the rates of nutrient removal.

Because the stand described in Table 9.7 was composed of seasonally deciduous angiosperm trees, the nutrient removals by harvesting would be smaller if the harvesting took place during the leafless dormant period. The midsummer quantities of foliar nutrients in this stand were 63.2 kg N/ha, 6.1 kg P/ha, 29.4 kg K/ha, 19.3 kg Ca/ha, and 5.8 kg Mg/ha (Freedman *et al.*, 1982). These are equivalent to 18% of the whole-tree quantity of N, 16% of

P, 16% of K, 4% of Ca, and 13% of Mg. However, some of these foliar nutrients would be resorbed back into perennial tissues or leached from the foliage before leaf drop in the autumn; the resorbed nutrients would be removed by a harvest of the stand during the dormant period. Ryan and Bormann (1982) examined a recently clear-cut and a 55-year-old hardwood stand in New Hampshire and found that about half of the N, P, and K was retained in foliage after senescence; the rest was resorbed or leached from foliage. Ca and Mg were largely retained in the senescing foliage. Therefore, only about half or less of the growing-season quantities of foliar N, P, and K would be removed by a harvest when the trees were leafless.

The largest pools of biomass and nutrients in a typical forest are in three compartments: (1) the trees, (2) the forest floor, and (3) the mineral soil. In aggregate, these essentially comprise the biomass and nutrient "capital" of the site, and therefore comparisons among them are important. In the maple–birch stand (Table 9.7), the whole-tree biomass is equivalent to 45% of the biomass content of the forest floor plus mineral soil, 6.8% of the total nitrogen, 4.1% of the total phosphorus, 1.4% of the total potassium, 50% of the total calcium, and 3.9% of the total magnesium. Only the calcium comparison indicates a shorter-term cause for concern with respect to an impoverishment of site nutrient capital by harvesting. This reflects the small calcium content of many soils that are derived from granite, gneiss, and other oligotrophic glacial tills and other parent materials (Cann et al., 1965; Boyle and Ek, 1972; Boyle et al., 1973; Weetman and Webber, 1972; Weetman and Algar, 1983; Dyck et al., 1986; Freedman et al., 1986). In some respects, the apparent "importance" of calcium seems anomalous, since available nitrogen (or much less often, phosphorus or potassium) is the nutrient to which temperate forests most frequently respond in fertilization trials (Weetman et al., 1974; Czapowskyj, 1977; Foster and Morrison, 1983).

Another important comparison to make is between the sizes of the potential, whole-tree nutrient removals and the "plant-available" nutrient pools in soil (Table 9.7). In this comparison, the whole-tree quantities are substantially larger, being equivalent to 4.4 times the size of the water-soluble ammonium–nitrogen plus nitrate–nitrogen content of the forest floor plus mineral soil, 1.0 times the extractable phosphate–phosphorus content, 1.0 times the exchangeable potassium, 1.7 times the exchangeable calcium, and 0.39 times the exchangeable magnesium. Clearly, quite different conclusions could be reached about the potential effects of a whole-tree harvest of this stand, depending on whether the nutrient removals are compared with the "available" or the "total" soil pools. The total pools, along with the biota, represent the gross nutrient "capital" of the site. However, most of the total nutrients occur in chemical forms that are not available to plants on the shorter term. Availability requires the mineralization of insoluble forms of the nutrients, by either microbial oxidations or inorganic weathering processes.

It is important to realize that although the available soil-nutrient pools are much smaller in magnitude than the total pools, they are relatively ephemeral because their turnover times are short. For example, in a northern-hardwoods forest in New Hampshire, the turnover time of available nitrogen was estimated as only 1.2 years and 7 years for calcium (Likens et al., 1977; Bormann and Likens, 1979). As such, it may not be important that apparent "depletions" of the available pools are calculated in simple box-model budgets. In fact, as described in Section 3, some studies have shown that there are short-term increases in the quantities of available nutrients (especially nitrate) following clear-cutting and other site disturbances (Likens et al., 1977; Vitousek et al., 1979; Krause, 1982; Mann et al., 1988).

Bulk precipitation is an important source of some nutrients to forests, and over the rotation period it helps to replace some of the nutrients that are removed by harvesting. For the maple–birch stand, 96 years of precipitation input are equivalent to the potential whole-tree removal of nitrogen, while the comparable data are 83 years for potassium, 166 years for calcium, and 41 years for magnesium (Table 9.7). Therefore, over a moderately long rotation (i.e., >50 years), substantial fractions of the nutrients removed during harvesting can be replaced by precipitation inputs. Other nutrient inputs take place by the weathering of minerals, by the fixation of atmospheric dinitrogen, and by the dry depositions of gases and particulates.

Of course, inputs of nutrients do not take place in the absence of outputs. The most important outputs are the leaching of soluble forms of nutrients to below the tree-rooting depth and ultimately to ground or surface water. In the case of the maple–birch site, the potential whole-tree harvest removals are equivalent to the potassium loss in 68 years of streamflow from an undisturbed forested watershed, 31 years for calcium, and 4.7 years for magnesium (Table 9.7). The streamflow losses of soluble nitrogen and phosphorus are of much smaller magnitude. (However, as discussed in Section 9.3, the flux of nitrate in streamflow can be greatly increased after disturbance.) The burning of residual tree biomass, called "slash," can also result in substantial losses of nutrients, especially nitrogen, which in some cases can be emitted to the atmosphere at rates of hundreds of kilograms per hectare (Kimmins and Feller, 1976; Feller, 1988, 1989).

To integrate the inputs and outputs of nutrients, net flux is calculated as the difference between the total inputs and the total outputs. A positive value of net flux (i.e., inputs > outputs) indicates that the nutrient in question is aggrading on the site over time, while a negative value (i.e., outputs > inputs) indicates that the quantity is decreasing. According to the estimates of net fluxes for undisturbed forested watersheds in Table 9.7, both nitrogen and phosphorus are aggrading; 51 years of net flux of nitrogen and 128 years of phosphorus would replace the removals by a whole-tree harvest. The other three nutrients are degrading, and in 305 years of net flux of potassium, 64 years of calcium, and 23 years of magnesium, a quantity of these nutrients equivalent to the potential whole-tree harvest removals would be lost from the site. Negative net fluxes of potassium, calcium, and magnesium also contribute to acidification of the site, a process that is exacerbated when additional quantities of these base cations are removed by harvesting (Oden, 1976; Rosenqvist et al., 1980; see also Chapter 4).

Overall, it appears that a single, whole-tree clear-cut of the maple–birch stand might not cause important depletions of site nutrient capital, since (1) the soil reserves are relatively large, and (2) over the rotation period large quantities of nutrients are regenerated. However, an important exception may be calcium. A large depletion of its site capital could occur, possibly resulting in decreased fertility and increased acidification of the site.

Use of Simulation Models to Evaluate the Consequences of Nutrient Removals

Several teams of forest researchers have developed computer simulation models of forest productivity, nutrient cycling, and forest-floor dynamics that can be parameterized to respond to variations in the intensity of stand harvest (e.g., Aber et al., 1978, 1979; Kimmins et al., 1981; Shugart, 1984; Binkley, 1986; Kimmins, 1986, 1987). These models predict the post-harvest successional dynamics of these variables, and they are based on the knowledge or estimation of the sizes of the biomass and/or nutrient compartments and the rates of transfer between compartments. Unlike the simplistic box-model approach described above, the simulation models have the valuable capability of a dynamic response to changes over time in the rates of forest processes.

The simulation models of Aber et al. (1978, 1979) incorporated the intensive database of the Hubbard Brook Ecosystem Study in New Hampshire (Likens et al., 1977; Bormann and Likens, 1979). Their simulations included six strategies of forest harvest: three cutting intensities and three rotation lengths. Figure 9.3a describes the effects on biomass of the forest floor of three harvest intensities (stem-only, whole-tree, and the rarely used complete-tree clear-cutting) over a 90-year rotation. The three types of harvest are predicted to have different effects on biomass of the forest floor, with the complete-tree treatment having as little as 33% the forest floor weight of the stem-only harvest during early to mid succession. However, by late succession (i.e., after 40–50 years) the forest floor biomasses of all three cutting treatments converge. These effects are largely due to differences in the fractions of the total site biomass that are removed by the three harvests: the whole-tree harvest removes foliage, twigs, and branches that are left behind as slash during the stem-only harvest; the complete-tree harvest additionally removes belowground biomass.

There are negative ecological implications of a decreased biomass of the forest floor during much of the post-harvest secondary succession. This

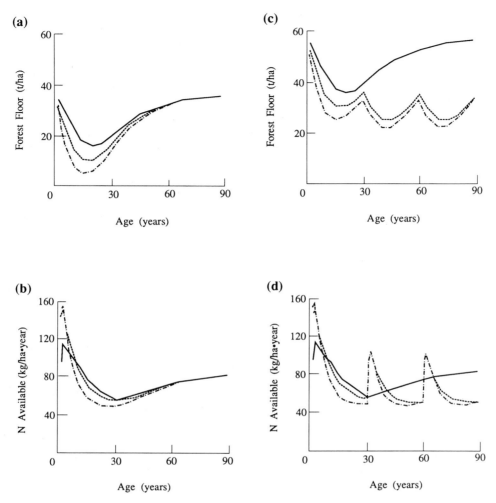

Fig. 9.3 Simulation models of forest floor dynamics, as affected by harvest intensity and rotation length. In all cases, —, is stem-only clear cut; ---, whole-tree clear cut; ·-·, complete-tree clear-cut (a) Effect of harvest intensity on forest floor biomass; (b) effect of harvest intensity on available nitrogen; (c) effect of rotation length on forest floor biomass; (d) effect of rotation length on available nitrogen. Modified from Aber *et al.* (1978).

compartment is important for cation exchange, moisture and nutrient retention, and other functions that may affect the overall capability of the site to sustain a large rate of ecosystem productivity.

The annual rate at which nitrogen is made available for plant uptake (Fig. 9.3b) was modeled on the basis of the net effects of: (1) mobilization of nitrogen from unavailable, organic N forms to available

forms such as nitrate and ammonium; plus (2) bulk inputs from the atmosphere; minus (3) the biological immobilization of soluble available N as unavailable organic N of decomposing litter and other organic fractions. The results of the nitrogen simulations are more complex than those for weight of the forest floor. For about the first 10 years, available nitrogen is present in larger quantities in the whole-tree and complete-tree treatments. This is probably due to some combination of: (1) larger rates of nitrogen immobilization in the relatively organic-rich forest floor of the stem-only treatment, because of decomposition of slash, and (2) a greater stimulation of nitrification by the more severe disturbances associated with the relatively intensive methods of harvest. Since there is relatively little aggrading plant biomass to take up the available nitrogen during the early postdisturbance years, this nutrient may leach from the site (as nitrate; see Section 9.3). Therefore, the residual organic matter from the stem-only harvest may play an important role in the conservation of site nitrogen capital, by immobilizing soluble, ionic forms of nitrogen as organic nitrogen, and thereby helping to prevent the leaching of inorganic nitrogen to below the rooting depth.

The effect on forest floor biomass of a decrease in rotation length by 30 years, while using a whole-tree or complete-tree clear-cut, is described in Fig. 9.3c. Compared with the 90-year stem-only rotation, the weight of the forest floor is greatly reduced by the shorter rotations, by a factor of about one-half. The relative effect of a shorter rotation is much larger than that of harvest intensity. This is likely caused by a generally enhanced condition for litter decomposition, because of the more frequent site disturbances associated with the use of a shorter rotation.

The effect of rotation length on the availability of nitrogen is shown in Fig. 9.3d. The 30-year rotation causes three times as many peaks of nitrogen availability as the 90-year cutting cycle. This results in more frequent pulses of nitrate loss to stream water and a larger loss overall. The net effect is that during the later stages of the postcutting succession, when net production is potentially large, nitrogen availability is smaller for the more frequent rotation by about 33% after 60 years and by 41% after 90 years. The model does not predict much of an effect of harvest intensity on this response. The overall implications of these effects on nitrogen availability are that the longer-term net production is likely to be smaller in frequently harvested stands.

In another simulation, Aber *et al.* (1979) varied harvest intensity and rotation length and examined the effects on total net ecosystem production and total harvest yield over a 90-year period (Table 9.8). The total net production was largest for the 90-year rotations, irrespective of harvest intensity. Of course, the yield of these three treatments was affected by harvest intensity. The next-largest net production was for a strategy of two 45-year rotations over the 90-year simulation period. These had a net production that was on average about 22% smaller than that of the 90-year rotation. The 30-year rotation ranked last in terms of net ecosystem production, averaging about 56% smaller than the 90-year rotation. Another important observation is that the

Table 9.8 Simulations of total net ecosystem production and total harvest yield over a 90-year period: There are three intensities of harvest and three rotation lengths[a]

Type of cutting	Length and number of rotations	Total net production (tonnes/ha per 90 years) and rank	Total harvest yield (tonnes/ha per 90 years) and rank	Percentage of total production harvested
Clear-cutting	90 (1)	1090 (2)	154 (4)	14
Whole tree	90 (1)	1120 (1)	197 (2)	18
Whole tree	45 (2)	853 (4)	108 (6)	13
Whole tree	30 (3)	478 (6)	93 (7)	19
Complete forest	90 (1)	1055 (3)	252 (1)	24
Complete forest	45 (2)	841 (5)	171 (3)	20
Complete forest	30 (3)	476 (7)	150 (5)	32

[a]From Aber *et al.* (1979).

relatively intensive harvests removed larger fractions of the total net production. Most efficient in this respect was the complete-tree, 30-year harvest, which removed 32% of the 90-year net production. Least efficient was the whole-tree 45-year treatment, which removed 13% of net production, and the stem-only 90-year treatment, which removed 14%.

If a best-management strategy was to be formulated solely on the basis of optimization of yield as predicted by the simulations summarized in Table 9.8, it would be to use the complete-tree treatment on a 90-year rotation. In fact, such decisions are often made in the real world of forest harvesting, and this is a reason why whole-tree and, occasionally, complete-tree harvesting are practiced. However, decisions to use such intensive methods of harvesting assume that the various potential, negative ecological effects, for example on tree regeneration, effects on the forest floor, effects on nutrients, etc., are either unimportant or less important than the short-term increases in yield.

9.3

LEACHING OF NUTRIENTS FROM DISTURBED WATERSHEDS

Various studies have demonstrated that there can be shorter-term increases in the rates of leaching of soluble nutrients from forested watersheds after extensive disturbance by cutting or wildfire (Tamm et al., 1974; Corbett et al., 1978; Hornbeck and Ursic, 1979; McColl and Grigall, 1979; Vitousek et al., 1979; Hornbeck et al., 1987a; Mann et al., 1988). This process causes reductions of site nutrient capital that are incremental to nutrients removed with harvested biomass. Leaching is especially important for nutrients that are relatively mobile in soil, particularly nitrate and potassium.

The most frequently cited example of this effect of forest disturbance is a study done at Hubbard Brook, New Hampshire. This large-scale experiment involved the clear-felling of all trees on a 15.6-ha watershed, but without the removal of biomass —the cut trees were left lying on the ground. Subsequent regeneration was prevented for the next three growing seasons by treatment of the watershed with the herbicides bromacil and 2,4,5-T. Clearly, the intent of this experiment was to examine the effects of a severe perturbation by devegetation, on biological control over nutrient retention and other watershed processes. This experiment was not designed to examine the effects of a typical, forest-management practice.

Over a 10-year post-disturbance period, the devegetated watershed had a stream-water loss of 499 kg/ha of NO_3-N, 450 kg/ha of Ca, and 166 kg/ha of K. These were much larger than the losses from an undisturbed, reference watershed of 43 kg/ha of NO_3-N, 131 kg/ha of Ca, and 22 kg/ha of K (Bormann et al., 1974; Likens et al., 1978) (see also Table 9.9). The increased nutrient fluxes in streamwater were partly due to an average 3-year increase of 31% in the yield of water from the cut watershed, which was caused by the disruption of transpiration. More important, however, were increases of nutrient concentrations in the streamwater, by factors of 40 for NO_3, 11 for K, 5.2 for Ca, 5.2 for Al, 3.9 for Mg, and 2.5 for H^+ (Bormann and Likens, 1979). In total, the net increases in the streamwater losses of N, Ca, and Mg from the disturbed watershed were similar in magnitude to the contents of these nutrients in the above-ground biomass of the hardwood forest (i.e., 371 kg/ha of N, 403 kg/ha of Ca, and 155 kg/ha of K; Whittaker et al., 1979).

As mentioned previously, this devegetation experiment did not involve a typical forest-management practice, and the measured effects were unrealistically large. However, some other watershed studies of the effects of more typical clear-cutting and other forestry practices have shown qualitatively similar effects on the leaching of nutrients. Streamflow losses in the first 2 years after clear-cutting a hardwood watershed in New Hampshire were 95 kg N/ha and 89 kg Ca/ha, compared with 144 kg N/ha and 221 kg Ca/ha that were removed with harvested biomass (Pierce et al., 1972; Likens et al., 1978).

In a more extensive study in New Hampshire, a comparison was made of nine clear-cut and five

Table 9.9 Annual net flux (atmospheric inputs minus streamwater exports) of dissolved substances for a deforested and a reference watershed at Hubbard Brook, New Hampshire[a]

Element	Net flux (kg/ha)	
	Deforested watershed	Reference watershed
Ca	−77.7	−9.0
Mg	−15.6	−2.6
K	−30.3	−1.5
Na	−15.4	−6.1
Al	−21.1	−3.0
NH_4-N	+1.6	+2.2
NO_3-N	−114.1	+2.3
SO_4-S	−2.8	−4.1
Cl	−1.7	+1.2
HCO_3-C	−0.1	−0.4
SiO_2-Si	−30.6	−15.9
Total	−307.8	−36.9

[a]Weighted average data over June 1966 to June 1969. Modified from Bormann and Likens (1979).

reference watersheds (Martin *et al.*, 1986). Averaged over these two treatments, 4 years of streamwater loss of NO_3–N from the clear-cuts were 71 kg/ha per 4 years, compared with 14 kg/ha per 4 years for the reference watersheds. Calcium export from these treatment watersheds averaged 111 kg/ha per 4 years, compared with 50 kg/ha per 4 years, while K exports were 23 kg/ha per 4 years compared with 8 kg/ha per 4 years. The 10-year streamwater losses from a strip-cut hardwood watershed in New Hampshire were increased by 48% for Ca, 135% for K, and 50% for N, while on a clear-cut watershed they were increased by 29, 218, and 128%, respectively (compared with values for an uncut reference watershed of 166 kg Ca/ha per 10 years, 22 kg K/ha per 10 years, and 45 kg N/ha per 10 years; Hornbeck *et al.*, 1975, 1987a). The 3-year loss of nitrate attributed to clear-cutting a 391-ha watershed in New Brunswick was shorter-lived and more moderate, at 19 kg NO_3–N/ha (Krause, 1982).

The above effects parallel, but are quantitatively much smaller than those observed for the devegetated watershed at Hubbard Brook that was de-

scribed first. In addition, it should be mentioned that some other studies of the effects of forest management on water quality have shown little or no effect, especially if only a portion of the watershed was cut (G.W. Brown *et al.*, 1973; Verry, 1972; Aubertin and Patric, 1974; Richardson and Lund, 1975; Hetherington, 1976; McColl, 1978; Burger and Pritchett, 1979; Stark, 1980; Mann *et al.*, 1988).

Effects on streamwater chemistry have also been observed after wildfire, but these have been relatively small in magnitude (Smith, 1970; Wagle and Kitchen, 1972; Grier, 1975; Wright, 1976; Tiedemann *et al.*, 1978; Schindler *et al.*, 1980; Neary and Currier, 1982). For example, the first-year export of NO_3–N from a burned 122-ha watershed in South Carolina was 0.67 kg/ha, compared with 0.05 kg/ha for an unburned reference watershed (Neary and Currier, 1982).

The effects of forest disturbance on nutrient losses via streamflow are influenced by such variables as soil and stand type, intensity of the disturbance, speed and vigor of the regeneration, watershed hydrology, and climate. Because operational forest harvests cause inconsistent losses of nutrients in streamflow, and these are generally fairly small in comparison with the nutrient capital of the site, some researchers have concluded that decreases in site fertility would not normally be anticipated through this mechanism (Sopper, 1975; McColl and Grigall, 1979). However, if most of the watershed is severely disturbed, if the harvest of biomass is intensive, and if the regeneration is not vigorous, then the leaching of soluble nutrients may be of greater importance.

The loss of nitrate is of most concern, because this is the nutrient that is most often lost in large quantities and because available nitrogen is the most frequent limiting nutrient for forest productivity. Nitrate is a relatively mobile ion in soils. For anions, the mobility series increases as $PO_4^{3-} < SO_4^{2-} < NO_3^- \simeq Cl^-$, and for cations the series is $Mg^{2+} \simeq Ca^{2+} < NH_4^+ < Na^+ \simeq K^+$ (Russell, 1973; McColl and Grigall, 1979).

There are several reasons why nitrate and other ions are leached from watersheds after disturbance.

1. After disturbance there is an increase in the rate of decomposition of organic matter, resulting in a release of soluble forms of nutrients. The rate of decomposition can be increased by a complex of environmental factors related to disturbance, including warmer surface soils, an influx of organic matter to the forest floor, and an increased availability of inorganic nutrients and moisture, due in part to decreased uptake by higher plants in the first years after disturbance (Cole and Gessel, 1965; Cole *et al.*, 1975; Likens *et al.*, 1970; Bormann *et al.*, 1974; Bormann and Likens, 1979; Piene, 1974; Harvey *et al.*, 1976, 1980; Jurgensen *et al.*, 1979; Wallace and Freedman, 1986).

2. Ammonification converts organic N to ammonium, which can be oxidized to nitrate by the bacterial process of nitrification. This process enhances the potential leachability of the fixed-nitrogen capital of the site, since nitrate is highly mobile in soil. In some situations, the rate of nitrification is greatly increased after disturbance; in other cases it is not, particularly if the soil is acidic (Likens *et al.*, 1970; Reinhart, 1973; Bormann and Likens, 1979; Wallace and Freedman, 1986).

From the perspective of many foresters, the vigorous plant regeneration that develops after clear-cutting is considered to be detrimental to silvicultural objectives. This is because the highly competitive situation may inhibit or preclude the successful establishment of a new forest dominated by economically desirable tree species (usually conifers in temperate latitudes). Indeed, most herbicide use and much mechanical site preparation in forestry is aimed toward reducing the degree of site dominance by the herb- and shrub-dominated stage of the secondary forest succession (see Chapter 8). However, the rapid revegetation of a disturbed watershed can reduce the loss of nutrients, because the growing plants take up mobile nutrients from soil and immobilize them in their aggrading biomass. Important actors in this initial process of ecological recovery are described in more detail later in this chapter, but they include both early-successional, ruderal species and some later-successional species that are

tolerant of the environmental conditions of the understory of the mature forest and that survive the disturbance by harvesting.

The various species of the plant community of revegetating clear-cuts can quickly reestablish substantial rates of net primary production and nutrient uptake. As such, the vegetation may act as a biological reservoir (or "sponge") for some of the nutrient capital that might otherwise leach from the site during the reorganization phase of postcutting succession (*sensu* Bormann and Likens, 1979). According to the nutrient-sponge hypothesis, as development of the stand proceeds after disturbance, the early successional species are progressively eliminated or reduced in dominance because of competitive stresses exerted by more tolerant species, or in some cases because of senescence. The nutrients that were immobilized in their biomass are then made available for uptake by trees, after recycling by litterfall and decomposition (Marks and Bormann, 1972; Marks, 1974; Bormann and Likens, 1979; Crow *et al.*, 1991; Reiners, 1992; Crowell and Freedman, 1993).

Foresters usually perceive a vigorous regeneration of plants other than the desired crop trees to be a negative economic value, but ecologists usually perceive the same vegetation to represent a desirable ecological value. Therefore, the ecological importance of nutrient conservation by the weedy vegetation of regenerating clear-cuts is controversial (see also the case study on herbicide use in forestry in Chapter 8).

9.4
SOIL EROSION RESULTING FROM DISTURBANCE

Forest harvesting has caused severe erosion in many watersheds, particularly those containing steep slopes. Sometimes, erosion has been triggered by improper forestry practices, including faulty planning and construction of logging roads, use of streams as skid trails, harvesting forest immediately adjacent to water bodies, running skid trails down slopes instead of along them, and removing forest from steep slopes that are hypersensitive to soil loss. In many countries, these sorts of irresponsible forestry activities are closely regulated, and they do not occur as frequently as in the past. In other places, however, these types of road building and logging activities are still recurrent.

Severe erosion of soil has many ecological effects, including: (1) the loss of mineral-soil substrate, which in severe cases can expose bedrock; (2) the loss of soil-nutrient capital; and (3) secondary effects on recipient aquatic systems, including siltation, flooding hazards, and the destruction of fish habitat. Because of the damage that can be caused by erosion after forest harvesting on some sites, the phenomenon has been the focus of much research, and several reviews have been published (Rice *et al.*, 1972; Fredriksen *et al.*, 1975; Anderson *et al.*, 1976; Corbett *et al.*, 1978; Hornbeck and Ursic, 1979; McColl and Grigall, 1979).

The most important features that characterize erosion from harvested lands are the following (after Rice *et al.*, 1972): (1) most logging activities, and disturbances in general, increase the rate of erosion from forested lands; (2) erosion is spatially variable on harvested sites; (3) initially there are large rates of sedimentation in streams after disturbance, followed by rapid reductions of erosion to the predisturbance condition, usually within 2–5 years; (4) landslides and creep are the quantitatively most important erosional processes in mountainous areas; (5) steep slopes are especially vulnerable; and (6) road building is an important factor in the initiation of erosion, especially if there is an inadequate number or size or improper installation and maintenance of culverts.

Of the above factors, road building is generally considered to have the largest influence on rates of erosion associated with forestry. However, the effects of road building, and of many of the other factors listed above, can be substantially prevented by careful planning and implementation to reduce the risks of erosion.

Studies of erosion from harvested watersheds have shown a wide range of effects, with soil losses measured as suspended sediment ranging from much less than 1, up to 5 tonnes/ha/year (McColl and Grigall, 1979). Megahan and Kidd (1972) studied a harvested *Pinus ponderosa* watershed in mountainous terrain in Idaho and found a 6-year average erosion loss of 4 tonnes/ha/year, compared with 90 kg/ha/year from an uncut, reference watershed. Haupt and Kidd (1965) reported a much smaller effect in another mountainous, *P. ponderosa* watershed in Idaho, where the 5-year postharvest sediment loss averaged 120 kg/ha/year, compared with essentially zero for an uncut watershed.

In Montana, a watershed with a mixed-conifer forest on an average slope of 24% was clear-cut, and then the logging slash was burned (DeByle and Packer, 1972; Packer and Williams, 1976). The treatment watershed had a sediment loss of 50 kg/ha in the first postcutting year, 150 kg/ha in the second year, 13–15 kg/ha in years 3 and 4, and about 0 kg/ha 7 years after harvesting (the loss was also essentially zero on a reference watershed).

E. L. Miller *et al.* (1985) studied a conifer-dominated forest in the Ouachita Mountains of Oklahoma and Arkansas and reported a 4-year sediment loss of 345 kg/ha per 4 years from a clear-cut watershed (250 kg/ha in the first postcutting year), compared with 73 kg/ha/4 years in a reference watershed. At a second location, they found a 3-year sediment loss of 450 kg/ha/3 years from a clear-cut watershed (215 kg/ha in the first postcutting year), compared with 150 kg/ha/3 years for a less intensively harvested selection-cut watershed and 82 kg/ha/3 years for a reference watershed.

In the relatively extreme devegetation experiment described earlier for a 16-ha hardwood watershed with 12–13% slopes at Hubbard Brook, New Hampshire, the 4-year post-treatment sediment yield averaged 190 kg/ha/year (maximum of 380 kg/ha/year), compared with 30 kg/ha/year for a reference watershed (Bormann *et al.*, 1974; Bormann and Likens, 1979). In this case, however, erosion from the disturbed watershed was somewhat minimized since there was no road building, and no damage was caused to the forest floor or stream banks by the skidding of logs, since wood was not removed from the cut watershed.

Many studies have shown that large erosional losses from watersheds managed for forestry purposes can be prevented by following certain operational guidelines (Packer, 1967a; Kochenderfer, 1970; Rothewell, 1971; Simmons, 1979; Miller and Sirois, 1986). These practices, most of which are now standard operating procedures in many areas, include the following: (1) careful planning of forest roads; (2) installation of a sufficient number of adequately sized culverts; (3) avoidance of direct disturbance to stream beds by heavy equipment; (4) leaving buffer strips of uncut forest along water courses; (5) use of skidding techniques that minimize the physical disturbance of the forest floor, for example, cable, skyline, or helicopter yarding; (6) allowing or encouraging a vigorous regrowth of vegetation so as to speed reestablishment of biological moderation of erosion; and ultimately (7) decisions to leave hypersensitive sites uncut.

Many researchers have identified logging roads that were poorly planned, constructed, or maintained as the primary factor causing erosion from harvested watersheds (Haupt and Kidd, 1965; Dyrness, 1967; Fredriksen et al., 1975; Corbett et al., 1978; McColl and Grigall, 1979). Dyrness (1967) found that following a severe rainstorm in a 6 thousand-ha watershed in mountainous terrain in Oregon, 72% of 47 massive-erosion events were associated with roads, even though roads accounted for only 2% of the terrain. Of 79 cases of erosion caused by forestry that were examined in a study in Maine, 37% were associated with roads, 55% with skidding in or across watercourses, and 8% with yarding [Land Use Regulatory Commission (LURC), 1979]. The highly compacted, mineral substrates of roads and skid trails are susceptible to erosion because they encourage the overland flow of water. Compacted soils of this sort were found to be present on 20% of a harvested area in Nova Scotia (Henderson, 1978) and 10% of another area in Newfoundland (Case and Donnelly, 1979).

To reduce erosion after harvesting of the forest, buffer strips of uncut forest can be left adjacent to streams, rivers, and lakes. Buffer strips can also reduce or eliminate temperature increases in water, preserve riparian habitats for wildlife, and reduce the aesthetic effects of harvesting (Van Groenwoud, 1977; McColl and Grigall, 1979). Ecologically, however, buffer strips are artificial, anthropogenic mitigations that are designed to deal with certain environmental problems associated with forestry. It should be recognized that many natural stand-level disturbances, such as wildfires, windstorms, and insect and disease infestations, do not usually create landscape patterns characterized by buffer strips along surface waters.

While it is generally accepted that buffer strips have useful mitigative effects, there is no consensus as to the width of uncut strips that should be used. This is an economically important consideration, since large areas of merchantable timber can be withdrawn from the potential harvest when uncut buffer strips are designated. Trimble and Sartz (1957) recommended a strip width of only 8 m between logging roads and streams on level sites in the northeastern United States and an increase in width of 0.6 m for each 1% increase in slope of the land. However, they suggested that the strip width should be twice as large on municipal watersheds, since the requirement for good water quality is greater in the case of drinking water. Wider strips have been recommended for eastern Canada: 15–20 m on each side of watercourses on level terrain and greater widths on slopes (Van Groenwoud, 1977). The requirements currently in place in New Brunswick are for a 15-m buffer on each side of watercourses, and there are also guidelines that suggest, but do not require, wider buffers beside access roads and recreationally important surface waters and the preservation of all deer yards.

9.5

HYDROLOGIC EFFECTS OF THE DISTURBANCE OF WATERSHEDS

Forest vegetation exerts a powerful influence on watershed hydrology, and this can be important in terms of preventing erosion and downstream flooding and in reservoir management. This hydrologic effect is caused because transpiration by vegetation evaporates large quantities of water into the atmosphere. In the absence of this process an equivalent quantity of water would exit the watershed as streamflow or seepage to deep groundwater (actually, this overstates the effect somewhat, because in the absence of a forest canopy the rate of non-biological evaporation is enhanced somewhat).

For example, Freedman *et al.* (1985) studied four gauged forested watersheds with shallow soil in Nova Scotia. On an annual basis, evapotranspiration was equivalent to 15–29% of the total precipitation inputs. Runoff via streams or rivers accounted for the remaining 71–85% of the atmospheric inputs of water (the flux to deep groundwater was assumed to be minimal, because of the relatively impervious nature of the bedrock).

There is a marked seasonality in the relative contributions of evapotranspiration and streamflow to the hydrologic budget of forested watersheds, especially in temperate and higher latitudes. This effect is illustrated in Fig. 9.4, which shows the average monthly hydrology for a 723-km² forested watershed in Nova Scotia. The total inputs of water as precipitation averaged 146 cm/year, of which 18%

arrived as snow. On an annual basis, 38% of the input was partitioned into evapotranspiration and 62% into riverflow. Although there was a slight tendency for November to January to have more precipitation, the seasonal variation in the quantities of atmospheric input was fairly small. In contrast, the seasonalities of evapotranspiration, runoff, and groundwater storage in the watershed were quite marked. Evapotranspiration was relatively large during the growing season of May to October, and as a consequence runoff was relatively small. Runoff was somewhat larger during autumn–early winter when transpiration was small; however, most of the precipitation input during this period served to recharge groundwater-storage capacity, which had been depleted by withdrawals by vegetation during the growing season. Runoff was largest during middle to late winter and especially in the spring, when the accumulated snowpack melted over a short period of time and caused a flush of stream and river flow.

This region of Nova Scotia experiences a relatively mild, maritime climate so that there are frequent episodes of snowmelt during the winter. In

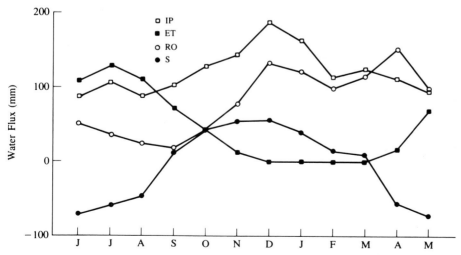

Fig. 9.4 Mean monthly hydrology for the 723-km² watershed of the Mersey River, Nova Scotia. Incident precipitation (IP) and runoff (RO) were measured between 1968 and 1982. Potential evapotranspiration (ET) was calculated using a climate-based model. Watershed storage (S) was calculated as IP − RO − ET. Modified from Ambler (1983) and Freedman *et al.* (1985).

more continental temperate climates, there are even more pronounced peaks of spring runoff, because snowpack accumulates to greater depths during the relatively cold winters, and it melts over a shorter period of time.

The influence of the forest on watershed hydrology is affected by disturbances caused by cutting or wildfire. Effects include changes in the timing and amounts of streamflow, with secondary downstream effects such as flooding and erosion. In addition, on imperfectly drained sites there can be changes in height of the water table (Hibbert, 1967; Douglass and Swank, 1972; Hornbeck and Ursic, 1979).

The increase in streamflow from a watershed is roughly proportional to the severity of the harvest, that is, in terms of the relative amount of foliar transpiration surface that is removed (Troendle, 1987). The increase in flow can be as large as 40% in the first year after the clear-cutting of an entire watershed. Rothacher (1970) found an increase of streamwater yield of 32% in a totally clear-cut Douglas fir (*Pseudotsuga menziesii*) watershed in Oregon, compared with a 12% increase in another watershed that was clear-cut over 30% of its area. Reinhart *et al.* (1963) compared four cutting methods that differed in the intensity of tree removal from mixed-hardwood watersheds in West Virginia. The most intensive method was clear-cutting, which caused an increase in water yield of 19% in the first postcutting year. In comparison, a less intensive, diameter-limit harvest caused a 10% increase, and a selection cut caused a 2% increase.

Usually, the largest increases in stream flow occur in the first postcutting year, with a rapid recovery to the precutting condition after 3–5 years because of revegetation and the consequent reattainment of the transpirational surface area of foliage. In many temperate regions, the largest increases in stream flow occur during late spring, summer, and early autumn, when transpiration normally exerts its strongest influence on watershed hydrology. Douglass and Swank (1975) monitored streamflows from a clear-cut, mixed hardwood watershed for 6 years. In the first postcutting year, there was an increase in water yield of 11.4 cm; this

decreased to 7.3, 5.0, 3.3, 2.0, and 0.9 cm in subsequent years.

Hydrologic effects were measured for the previously described, 16-ha hardwood watershed at Hubbard Brook that was clear-felled and then herbicided for 3 years. In the first 3 postdisturbance years, there were increases in streamwater yield of 40, 28, and 26%, respectively (Fig. 9.5). However, in the second full season after cessation of the herbicide treatment of the experimental watershed, the hydrologic effects largely disappeared because of the vigorous regeneration of vegetation (Likens, 1985; Reiners, 1992). A similar effect was observed in a clear-felled and herbicided, 60-ha watershed in West Virginia, where 5 years after the cessation of herbiciding the streamwater enhancement was reduced to about 20% of the initial impact, again because of a rapid revegetation of the site (Kochenderfer and Wendel, 1983).

Forest harvesting and management can also affect the size and timing of the peak streamwater flows from watersheds. These effects can occur both for storm flows and for spring meltwater flows (Hewlett and Helvey, 1970; Leaf and Brink, 1972; Verry, 1972; Hornbeck, 1973; Swanson *et al.*, 1986; Hornbeck *et al.*, 1987a). The stormflow effect occurs because the compacted soils that are often present in harvested watersheds have a relatively small ability to moderate the speed of the lateral flow of water and thereby enhance its rate of percolation into the ground and ultimately to streamwater. Therefore, harvesting can encourage overland water flows. The spring meltwater effect is caused when the rate of snow melt is increased by the relatively unshaded condition of clear-cuts. In addition, the accumulated snowpack is sometimes larger in forests than in clear-cuts, because the relatively exposed conditions of the clear-cuts can favor the evaporation of snow by sublimation (Golding and Swanson, 1986).

A change in the nature of the forest community can also affect hydrology. After the conversion of a mixed-hardwood forest to a white pine (*Pinus strobus*) forest in the southern Appalachians, there was a decrease in annual streamflow of about 20% (Douglass and Swank, 1975). The reason for this

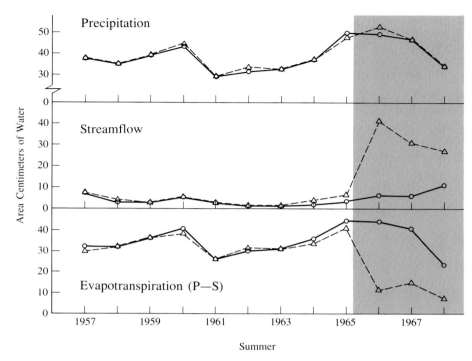

Fig. 9.5 Comparison of summer hydrology (June 1 to September 30) of a devegetated (dashed line) and reference (solid line) watershed at Hubbard Brook, New Hampshire. The shaded area represents the period when the devegetated watershed had all trees felled and herbicides applied for three growing seasons. The preceding period was used for calibration, and reflects annual variations in summer hydrology, from Bormann and Likens (1979).

effect was that the pines had a more extended transpiration season than did the seasonally deciduous angiosperms that they replaced. In the same region, the conversion of a mixed-hardwood forest to a grassland resulted in a longer-term increase in water yield (Douglass and Swank, 1975).

9.6
FORESTRY AND BIODIVERSITY

The disturbance of forested terrain by clear-cutting causes dramatic changes in the suitability of sites as habitat for some species of plants and animals and for their communities. At the same time, new hab-

itats and opportunities are created for early successional, ruderal species and their communities.

Habitat changes that are of particular importance to animal wildlife include: (1) modifications of the physical structure of the ecosystem by changes in the spatial distribution of plant biomass; (2) changes in the plant-species mixture; and (3) effects on the rates of processes, especially primary production. On the shorter term, the net effect of forest harvesting is that the habitats of some species of wildlife are improved, while at the same time the habitats of species that require mature, undisturbed forest are destroyed.

In this section, the effects of certain forest-harvesting and management practices on three elements of biodiversity are discussed: within species, within communities, and at the level of landscape.

The focus is on harvesting and nonpesticidal management (considerations relevant to silvicultural pesticide use are discussed in Chapter 8).

What is Biodiversity?

Biodiversity is often considered to be the number of species occurring in some area. Biodiversity is better defined, however, as being composed of the totality of the richness of biological variation, ranging from population-based genetic variation, through subspecies and species, to their communities, and also including the patterns and dynamics of these on the landscape. The geographic scales of biodiversity considerations can range from local, to regional, state or provincial, national, continental, and global [Office of Technology Assessment (OTA), 1987; Wilson, 1988a; Hunter, 1990; Organization of Economic Cooperation and Development (OECD), 1991a,b; CEQ, 1992b; Soule, 1991; Reid *et al.*, 1992]. For the purposes of the following discussion of the effects of forestry, biodiversity is considered within three levels of integration: (1) genetic variation within populations, (2) the richness of species within communities, and (3) the richness of community types on the landscape after Freedman *et al.*, 1994.

Genetic Variation within Populations of Trees

Individuals of almost all species differ genetically, and this variation contributes to genetic biodiversity at the level of populations and, ultimately, of the species. Most species contain a great deal of genetic variability, but some do not. For example, there is little within-population, genetic variance if the species has a strong reliance on nonsexual mechanisms of propagation, as is the case of some species of plants, for example poplars and aspens (*Populus* spp.).

For evolution to proceed through natural selection of better-fit phenotypes or through cultural selection of more desired phenotypes, there must be a genetic basis for the phenotypic variance. Genetic variation within populations is, therefore, the biological essence by which species adapt to environments that are changing naturally or because of the direct or indirect consequences of human activities.

Within the limits of their phenotypic plasticity, individuals may be able to cope with the changing stressor regimes of dynamic environments. However, for populations to respond to environmental changes through adaptive evolution, there must be a genetic variance for the ability to tolerate the challenges of environmental change.

The populations of almost all species contain genetic variation, making this topic universally relevant in biology. In this section, however, discussion of genetic variation within species will be limited to a few species of temperate trees and to some implications of the establishment of genetically narrow populations of trees in plantations (Freedman *et al.*, 1994).

Genetic variance within and/or among populations can be indicated by various measures. In studies of the genetic variance of trees, frequently reported indicators are: (1) the percentage of examined loci that are polymorphic, a trait usually assayed by allozyme heterozygosity, (2) the average number of alleles per locus, (3) the heterogeneity of phenotypic development under common-environment conditions, and (4) differences in any of these among populations or along geographical gradients.

Some studies have been made of genetic variance within and among populations of trees. In general, species that are outbreeding and dioecious and that reproduce primarily by the establishment of seedlings have substantial genetic variance at many gene loci. This phenomenon has been demonstrated by studies of phenotypic development in common-garden environments and by studies of isoenzyme systems. Most temperate species of tree exhibit substantial genetic variance. However, this is not the case for all species; some wind-pollinated conifer species appear to have little genetic variance, e.g., red pine (*Pinus resinosa*; Fowler and Morris, 1977) and eastern hemlock (*Tsuga canadensis*; Zabinski, 1992).

For most conifers, there is substantially more genetic variance within populations than among populations (Guries, 1989; Muona, 1990; Friedman, 1991). The relatively small genetic differentiation of populations of most species is presumably due to the efficiency of gene flow by pollen dispersal (Muona, 1990).

Many tree species, especially angiosperms, have a substantial capability for vegetative regeneration, in addition to an ability to establish new genets by dissemination of seed and establishment of seedlings. For example, mature individuals of red maple (*Acer rubrum*) produce large numbers of viable seeds each year, but if the tree is cut it is also capable of a vigorous regeneration by stump sprouting, producing hundreds of vegetative shoots, which then self-thin to only one to three stems after several decades (Prager and Goldsmith, 1977). Lees (1981) excavated entire root collars of red maple in New Brunswick and demonstrated at least three stump-sprout rotations of the sampled genets.

The capability of certain angiosperm species of tree for vegetative regeneration can also be illustrated by reference to poplars and aspens (*Populus* spp.), some species of which are capable of forming extensive, monogenetic or multiclonal stands by issuing vegetative shoots from underground rhizomes and roots after disturbance by fire or harvesting (Fowells, 1965). Some multistemmed genets of trembling aspen (*Populus tremuloides*) can cover several-to-many hectares (up to >40 ha) in area (Barnes, 1966, 1969; Perala, 1977; Kemperman and Barnes, 1976; DeByle, 1990), and they may be the world's largest "individual" organisms, in terms of biomass. Small stands or groves of aspen can be composed of a single clone, but extensive stands generally are multiclonal.

Genetic differentiation among populations of tree species can be related to geographic isolation, habitat differentiation, differences in rates of postglacial migration, and other factors (Cwynar and MacDonald, 1987; Cheliak *et al.*, 1988; Li and Adams, 1989; Pielou, 1991). Genetic differentiation of this sort can ultimately lead to the evolution of distinctly different populations. These may be recognized taxonomically, as is the case with Douglas fir, which has developed two varieties known as coastal and interior (*Pseudotsuga menziesii* var. *menziesii*, and var. *glauca*, respectively; Fowells, 1965; Li and Adams, 1989).

Intensive forestry can result in the establishment of genetically narrow, monocultural populations of trees in plantations. The potential for this type of management has been greatly increased by the de-velopment of techniques for the propagation of conifer trees by somatic embryogenesis, which results in the production of pseudo-seedlings known as "emblings" (Hakman and van Arnold, 1985; Cheliak and Rogers, 1990; Webster *et al.*, 1990). If this technology develops to the degree that it receives a widespread use in forestry, it could result in substantial decreases in genetic variance within tree populations in intensively managed plantations.

The cultivation of genetically narrow lineages of desired species of plants (and animals) is, of course, a common practice in intensive agricultural systems. Within limits, the practice has also been historically pursued in forestry, through the selection of phenotypically superior, so-called "plus trees" as sources of seed for the cultivation of seedlings in greenhouses. Such desired trees are often cultivated in seed orchards established by selectively thinning natural stands or by planting scions. However, because of the free pollen transfer among trees growing in seed orchards, substantial decreases in genetic variance do not necessarily occur (Kitzmiller, 1990).

It is well recognized that the intensive cultivation of populations of crop plants, including trees, that have selected, superior traits can lead to substantial, genetically based improvements in production and quality (e.g., Cheliak and Rogers, 1990; Kitzmiller, 1990; Muona, 1990). As noted above, these genetic gains are accompanied by substantial decreases in genetic variance in the case of clonally propagated trees, but this does not necessarily occur in the case of outbreeding seed orchards.

There are risks associated with the cultivation of genetically narrow populations of plants, because these may be relatively vulnerable to certain types of diseases and pest infestations (Ehrlich and Ehrlich, 1981; Heybroek, 1982; Ehrlich, 1988; Schoenwald-Cox *et al.*, 1983; Allendorf and Leary, 1988; Roberds *et al.*, 1990). In addition, genetically narrow populations may be more vulnerable to the effects of pervasive environmental changes, for example, in climate (e.g., Pollard, 1985; Harrington, 1987; Roberds *et al.*, 1990; Singh and Wheaton, 1991).

Consider, for example, the case of spruce bud moth (*Zeiraphera* spp., especially *Z. canadensis*), a forest pest in eastern Canada. It has been suggested that the recent (since ca. 1980) emergence of this pest is due to the widespread establishment of white spruce (*Picea glauca*) in plantations and abandoned pastures (Magasi, 1983, 1990; Neilson, 1985). Among the various spruce species of eastern Canada, white spruce is the primary, most favored host for spruce bud moth, because of the close phenological coupling of larval emergence and bud burst between these two species (Turgeon, 1986). It has been established that there is genetic heterogeneity within white spruce with respect to susceptibility to bud moth damage, a trait that could be taken advantage of to breed resistant genotypes (Quiring *et al.*, 1991). It is also possible that the seedling populations that have been planted widely and that resulted from plus-tree selection programs may be relatively susceptible to bud moth damage, but this hypothesis has not been tested through research.

To manage the risks associated with genetically narrow populations of plants, population geneticists have suggested the use of monocultural mosaics in plantations, established using emblings or vegetative scions, (e.g., Heybroek, 1982; Bentzer *et al.*, 1990). In the case of poplar plantations, the silvicultural establishment of monocultural mosaics would parallel the genetic structure of some wild populations, as was previously described for *Populus tremuloides*, which forms extensive, multiclonal stands.

The Richness of Species within Communities

Biodiversity within communities is related to the number of species present at a site or within a distinct ecological community. Often, community-level biodiversity is considered separately for (1) prominent groups of organisms, aggregated on a phylogenetic basis, e.g., plants, birds, mammals, arthropods, etc., or (2) on the basis of functional guilds, e.g., ground vegetation, woody plants, epiphytes, or foraging or nesting guilds of birds and small mammals.

The richness of species at a particular site, management area, or political entity (e.g., a park, country, or some other designated area) is usually expressed as the number of species encountered after substantial searches of all ecological communities. Often, longer-term, cumulative records are kept, with new species being added as they are discovered.

At the community or stand level, a similar indicator of species richness might be used, i.e., the number of species encountered in a particular community after a systematic search. Sometimes data for species richness within communities is expressed as the density of species per unit of area (e.g., species per square meter or species per hectare).

Another biodiversity-related indicator at the community level is species diversity, which accommodates both (1) the number of species present, i.e., richness, and (2) their relative abundance, usually estimated as relative biomass or as relative density. Compared with species richness, species diversity is usually considered to be the more useful indicator of biodiversity within communities, because it accounts for differences among species in rarity and commonness. Consider, for example, two theoretical communities, each composed of five species and 100 individuals:

Species	Abundance (Number of Individuals)	
	Community A	Community B
A	96	20
B	1	20
C	1	20
D	1	20
E	1	20
Species richness	5	5
Diversity (H')	0.2	1.3

In this example, both communities have the same species richness, but their species diversities differ because of the differing rarity of species, i.e., in the equitability of distribution of individuals among species.

A commonly used index of species diversity in ecological studies is the Shannon–Weiner Function (Shannon and Weaver, 1949):

$$H' = -\Sigma\, p_i \log p_i,$$

where H' is the index of species diversity in the community, and p_i estimates the probability that any particular individual encountered during a random search of the community will be of a designated, ith species.

In studies of community ecology, p_i is most often estimated as one of the following:

1. The relative abundance of the ith species, i.e., its abundance divided by the sum of abundances of all other species in the community. In the sense used here, abundance of animals is usually estimated on the basis of population. For plants, abundance is usually estimated as biomass, or as foliar cover, stem basal area, or some similar measure that is strongly correlated with biomass.

2. The relative density of the ith species, i.e., its density divided by the sum of densities of all other species in the community. Density is usually measured as the number of individuals per unit area of the community (or per unit volume for some aquatic communities). For some plant species that propagate by vegetative means, density is estimated on the basis of functional "individuals," for example, the number of stems issuing from the ground, regardless of their identity as genetic individuals.

The use of diversity indices in community ecology is somewhat controversial, and a variety of other indices has been proposed, and criticized (e.g., Hurlbert, 1971; Pielou 1975, 1977; Patil and Taillie, 1982; Krebs, 1985; Magurran, 1988; Begon et al., 1990). Consult these references for detailed comparisons of the various indices of species diversity, as the topic is not discussed in depth in this book.

Another element of within-community biodiversity is relevant to the heterogeneity of the vertical and/or horizontal distributions of distinctive patches. Within a forest community such patches might be associated with gaps in an otherwise-closed canopy or with food or habitat associated with a particular species of tree. Heterogeneity of within-community distributions of habitat are considered to be important for the interpretation of abundance and/or microdistribution of certain wildlife, such as birds (e.g., MacArthur and MacArthur, 1961; MacArthur et al., 1966; MacArthur, 1964; Karr and Roth, 1971).

In the following sections, effects of forestry on species richness and diversity are discussed for selected components of the larger ecological community. Emphasis is on vegetation, mammals, birds, and amphibians of the temperate zone, because these groups of organisms have been relatively well examined in the context of forestry (after Freedman et al., 1994).

Vegetation

Disturbance of the site and the harvesting of its dominant vegetation results in profound changes in the plant community, including changes in species composition and in physical structure. These changes are due to various influences associated with the disturbance, including:

1. removal of the previously dominant trees, and their substantial environmental influences;

2. the vigorous development of a species-rich assemblage of ruderal plants (sensu Grime, 1979), established from persistent seedbanks and/or the seed rain;

3. stimulation of the advance regeneration of tree species that may have survived the harvest;

4. vegetative regeneration by certain tree species, for example, by root or stump sprouting;

5. increased productivity and fecundity of some of the smaller species of the predisturbance ground vegetation;

6. decreased abundance of some of the stress-tolerant vegetation of the understory of the previous forest, especially bryophytes and lichens, which usually decline because of increased drought and exposure after harvesting. (In the context used here, stress tolerance specifically refers to tolerance of environmental constraints that the overstory vegetation exerts on understory vegetation, particularly in terms of the availabilities of light, nutrients, and moisture. Intolerant species require a relatively free availability of these site resources.)

Many studies have been made of the prominent plant species and communities of the secondary successions that follow forest harvesting and management. Of course, the particular species and their communities vary greatly among regions. To illustrate the general pattern, consider the following, functional groups of species, from a review of studies in temperate forests of eastern North America (from Shafi and Yarranton, 1973; Marks, 1974; Blair and Brunett, 1976; Zavitovski, 1976; MacLean and Wein, 1977; Bormann and Likens, 1979; Covington and Aber, 1980; Hibbs, 1983; Hornbeck *et al.*, 1987a; Martin and Hornbeck, 1989; Crow *et al.*, 1991; White, 1991; Reiners, 1992; Crowell and Freedman, 1993; Freedman *et al.*, 1994):

1. *Early successional, competition-intolerant species* can dominate the plant community during the early stages of post-harvesting succession, but they are absent or much reduced in abundance after the canopy closes and competitive stresses intensify. [Note, however, that some intolerant species may persist in older stands as individuals in a diapause state, occurring in a long-lived seedbank, e.g., *Rubus strigosus*, *Prunus pensylvanica*, and *Sambucus racemosa* (Marks, 1974; Marquis, 1975; Grignon, 1992)]. Prominent ruderal taxa in young, regenerating stands in eastern North America include the following:

(1.1) *Trees and woody shrubs*, such as alder (e.g., *Alnus rugosa*), birch (e.g., *Betula papyrifera*), cherry (e.g., *Prunus pensylvanica*), jack pine (*Pinus banksiana*), poplar and aspen (e.g., *Populus tremuloides*), dogwood (e.g., *Cornus stolonifera*), and shadbush (*Amelanchier* spp.). These are all intolerant taxa, which cannot reproduce after the canopy closes, although tree-sized individuals may persist for some time. [Designations of species tolerance follow Hall (1955), Reid (1964), Sparling (1967), Woods and Turner (1971), Bormann and Likens (1979), Spurr and Barnes (1980), and Crowell and Freedman (1993)];

(1.2) *Semi-woody shrubs*, such as blackberry and raspberry (e.g., *Rubus allegheniensis* and *R. strigosus*) and elderberry (e.g., *Sambucus racemosa*);

(1.3) *Herbaceous, dicotyledonous angiosperms*, composed of a great richness of annual and short-lived perennial species, dominated by species of the aster family (Asteraceae), the ginseng family (Araliaceae), the evening primrose family (Onagraceae), and the buckwheat family (Polygonaceae); and

(1.4) *Graminoids*, especially various species of grass (Poaceae), sedge and bulrush (Cyperaceae), and rush (Juncaceae).

2. *Species that have an intermediate tolerance of competitive stresses*, and that survive disturbance, and persist into later-successional stages. Such species are often present in a small abundance in the ground vegetation of mature stands. Individuals may survive the disturbance of forest harvesting, to be released from overstory-induced stresses and becoming relatively abundant and fecund until the next canopy is reestablished. These species can regenerate after harvesting by the establishment of seedlings and by vegetative regeneration and/or growth of surviving individuals. Some prominent species of this group include the trees, red maple (*Acer rubrum*), yellow birch (*Betula allegheniensis*), and white pine (*Pinus strobus*), along with many species of ground vegetation.

3. *Species that are tolerant of competitive stresses* may survive the disturbance and eventually dominate the later-successional stages. Some species in this group are present in all stands regardless of post-harvesting age. Under shaded conditions, these species have low light-compensation thresholds, a characteristic that allows survival under heavy shading, but restricts productivity under high-light conditions (Sparling, 1967; Spurr and Barnes, 1980). Prominent species in this group include the trees, balsam fir (*Abies balsamea*), sugar maple (*Acer saccharum*), beech (*Fagus grandifolia*), and hemlock (*Tsuga canadensis*). Small individuals of tolerant tree species may be present in mature stands as a so-called advance regeneration, and if these survive the disturbance they may be ecologically released from the stresses exerted by the previous canopy of trees, and be important in the natural regeneration of disturbed forests.

Many species of the ground vegetation are tolerant of the competitive stresses exerted by a closed canopy of trees, for example, white trillium (*Trillium grandiflora*), the fern *Dryopteris marginalis*, the feather-mosses *Pleurozium schreberi* and *Hylocomium splendens*, and certain epiphytic lichens, such as lungwort (*Lobaria pulmonaria*). Many of these tolerant species exhibit relatively little release after removal of the overstory. Such a response reflects a conservative strategy of slow growth rates and efficient use of resources to survive under competitively stressful conditions (Bormann and Likens, 1979; Grime, 1979). Because many of these species cannot tolerate full insolation, they may decrease greatly in abundance after disturbance and regenerate slowly or recolonize when the next forest canopy develops.

It is notable that, in general, the plant community of young, revegetating clear-cuts is usually more species rich and diverse than the community of the mature forest that was harvested (at least, once the clear-cuts have had an opportunity to begin to revegetate after the harvest, i.e., after 2–4 years or so). This phenomenon occurs because recent clear-cuts provide a habitat that is relatively rich in resources, notably light, water, and nutrients. Therefore, until a new, overtopping canopy develops, the intensity of ground-level competition is relatively small. Under such free-growing conditions many species of lower-growing plants can be temporarily supported on the site, including a diverse assemblage of ruderal species. In contrast, the stressful understory habitats of a mature, closed-canopied forest are capable of supporting relatively few species of plants.

The phenomenon of species-rich clear-cuts versus species-poorer forests can be illustrated using data from a post-clear-cutting chronosequence of hardwood forest in Nova Scotia. In that study, species richness and diversity of the ground vegetation of younger stands were generally greater than in mature stands (Fig. 9.6). The ground vegetation averaged 11 species/m^2 on two 1-year-old clear-cuts, increased to 14/m^2 at age 6, and was 3–6/m^2 by age 30 in maple-dominated stands, but

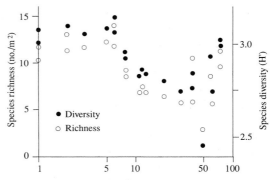

Fig. 9.6 Species richness and diversity of the ground vegetation in a chronosequence of 23 stands in a region of angiosperm forest in Nova Scotia. Stands older than 20 years were part of a fire-caused landscape mosaic, while younger stands originated with clear-cutting. Modified from Crowell and Freedman (1993).

9–11 in the relatively open-canopied birch-dominated stands. Shannon–Weiner diversity (H') of the ground vegetation was 3.0–3.2 among stands 1–6 years old and was generally smaller in older stands, reaching 2.5–2.8 in mature maple-dominated and 2.9–3.0 in birch-dominated stands. Other studies of temperate angiosperm forests have also reported greater species richness and diversity in young stands after harvesting or wildfire (e.g., Shafi and Yarranton, 1973; Ford and Newbould, 1977; Bormann and Likens, 1979; Hibbs, 1983; Albert and Barnes, 1987; Schoonmaker and McKee, 1988; Reiners, 1992).

Ungulates

Certain species of deer feed by browsing woody stems and therefore require brushy habitats for at least part of the year. These deer can benefit from human activities that create such habitats, for example, the harvesting of forests and the abandonment of agricultural lands, which can create habitats dominated by shrubs for 15–20 years, usually with a rich understory of forbs (i.e., herbaceous, dicotyledonous plants) and monocotyledonous species.

In North America, the best examples of such deer species are white-tailed deer (*Odocoileus virginianus*) and mule deer (*O. hemionus*). During most of the twentieth century, both of these species have increased greatly in abundance in many parts of their range because of an increased availability of suitable habitats, caused by regeneration following disturbance of forests by harvesting, abandonment of poor-quality farmlands, and in some cases by wildfire. In some areas, decreased populations of natural predators have also been important. In parts of their range, these species of deer are currently more abundant than they were prior to the European colonization, when the landscape was mostly covered with mature forest.

For example, white-tailed deer were uncommon in Nova Scotia at the time of its settlement by Europeans. The species was quickly extirpated by overhunting, but it was reestablished by introductions and natural immigration from adjacent New Brunswick (Banfield, 1974; Benson and Dodds, 1977). White-tailed deer are probably now more abundant in Nova Scotia than at any time since deglaciation, and it is the province's most important big-game species.

The modern abundance of white-tailed deer in Nova Scotia is due to a widespread availability of early successional, shrubby habitats associated with regeneration after clear-cutting and to a lesser extent after wildfire and the abandonment of depleted pastures. These habitats are generally distributed over the landscape as a mosaic of stands in various stages of secondary succession, within a dominant matrix of mature forest.

This spatial arrangement enhances the suitability of the landscape for white-tailed deer, because: (1) there are large amounts of ecotonal habitat, (2) there is a large production of nutritious and palatable browse species in the immature stands, (3) there is good "yarding" habitat of mature, coniferous forest, an important habitat component where the winter climate is severe, and (4) all of the above occur within a landscape mosaic of stands of different age. This sort of heterogenous landscape is more favorable to white-tailed deer than one that contains either small or large clear-cuts or an unbroken expanse of mature forest (Krefting, 1962; Telfer, 1967a,b, 1970, 1978b; Patton, 1974; Halls, 1978; Lyon and Basile, 1980; Lyon and Jensen, 1980; McNicol and Timmermann, 1981; Monthey, 1984; Hunter, 1990; Smith, 1991).

The central parts of large clear-cuts may not be well used by deer, as these animals prefer to not be excessively far from a protective forest cover. In a study of widely scattered stands in eastern Canada, it was found that the center of clear-cuts less than 80 ha in size averaged 3.8×10^3 browsed stems/ha (range $2.4-5.8 \times 10^3$ stems/ha), while larger clear-cuts of 80–410 ha were used much less intensively (average 0.54×10^3 stems/ha; $0.1-1.0 \times 10^3$ stems/ha) (Drolet, 1978).

Therefore, in larger clear-cuts, use by deer of the available habitat is most intensive near edges. For optimal white-tailed deer habitat in New Brunswick, Boer (1978) recommended that clear-cuts be no larger than 4 ha, while Euler (1978) recommended less than 2-ha clear-cuts in southern Ontario. Reynolds (1962, 1966, 1969) suggested 9–14 ha as the maximum size of clear-cut that deer would fully utilize in a variety of habitats in the southwestern United States. The operational sizes of clear-cuts in many parts of eastern North America are, in general, considerably larger than these suggested optimum sizes for deer. To improve habitat suitability for deer and moose in larger clear-cuts in Maine, Monthey (1984) recommended retention within the clear-cuts of uncut conifer "islands" at least 2 ha in area, for use as shelter in winter.

White-tailed deer eat a great variety of woody plants and herbs, along with some other foods such as fungi and lichens. The relative importance of particular foods varies between and within regions, but they are usually more abundant on cutover or burned sites than in mature forest (Martin et al., 1951; Telfer, 1972; Hoover, 1973; Speake et al., 1975; Drolet, 1978; Parker and Morton, 1978; Lyon and Jensen, 1980; Rogers et al., 1981; Monthey, 1984; Hodgman and Bowyer, 1985; Irwin, 1985; Crowell and Freedman, 1993).

Initially, the biomasses of browse and forbs increase progressively after cutting. Shrub biomass increases to a maximum after about 8–15 years, followed by a large decline as the maturing tree canopy begins to shade the understory. These temporal patterns are illustrated for 22 hardwood stands of different age in Nova Scotia (Table 9.10). Shrub biomass increased from less than 1 tonne/ha in 1-year-old stands, to a maximum of 17–19 tonne/ha

Table 9.10 Shrub and herb biomass in stands of different age in a region of maple–birch forest in Nova Scotia[a]

Stand age (year)	Number of stands sampled	Shrub biomass (tonnes d.w./ha)		Herb biomass (tonnes d.w./ha)	
		Average	Range	Average	Range
1	2	0.45	0.41–0.49	1.1	0.93–1.2
2	2	1.9	1.8–2.1	1.6	1.4–1.7
3–6	4	8.2	6.6–12.2	1.7	1.5–2.1
8–13	5	17.6	16.9–18.7	0.57	0.27–1.0
20	1	10.9	—	0.14	—
30–40	3	4.8	0.8–7.9	0.25	0.10–0.47
50–75	5	2.4	1.1–3.9	0.23	0.7–0.45

[a]Stands 20 years old and younger originated by clear-cutting of the mature forest. Older stands were part of a wildfire-created mosaic. Unpublished data of M. Crowell and B. Freedman.

at 8–13 years, and then decreased to less than 4 tonne/ha at 50–75 years. Herbaceous plants exhibited a similar pattern, but their biomass peaked earlier at 2–6 years after cutting.

In addition, the nutritional quality of browse on cutovers and burns is generally better than in mature forest. Recently sprouted, rapidly growing twigs have larger concentrations of protein, nitrogen, and phosphorus, and they are more succulent and more easily digested than is the older browse of mature forest (Halls and Epps, 1969; Short et al., 1975; Perins and Mautz, 1989; Thill et al., 1990). Therefore, food for deer is present in relatively large quantities and is of good nutritional quality on clear-cuts.

However, some other features of lands managed for forestry can restrict the use of otherwise suitable habitats by white-tailed deer. These can include the presence of logging roads with frequent traffic, harassment by hunters, large clear-cuts in which the central parts are far from protective forest cover, relatively deep snow on many clear-cuts because of a lack of canopy interception, especially by a conifer canopy, and physical obstructions on clear-cuts in the form of tangles of logging slash (Pengelly, 1972; Lyon, 1976; Drolet, 1978; Lyon and Basile, 1980).

It is generally felt that the more irregular shaped the clear-cut, the more favorable is the resulting

habitat for deer and some other species of wildlife (Halls, 1978; Lyon and Basile, 1980; Lyon and Jensen, 1980). Irregular shapes have a larger ratio of edge-to-surface area than do round, square, and rectangular shapes and therefore have relatively more ecotonal habitat. In addition, an irregular shape makes more of the central parts of clear-cuts accessible, if it is assumed that deer will not venture further than some maximum distance from forest cover into a clear-cut. Furthermore, irregular-shaped clear-cuts break up lines of vision and help to make deer and humans less visible to each other.

Sometimes, the snowpack effect mentioned above only occurs for a short time after snowfall events; under some climatic regimes the accumulated snowpack can be deeper in the forest, because of the greater amount of evaporation of snow from relatively open clear-cuts (Golding and Swanson, 1986). Snow depth is important, because the mobility of deer is restricted by snow depths greater than about 50–70 cm (Pengelly, 1972; Newman and Schmidt, 1980). It is important that forest management in areas with severe winters be conducted in ways that continue to provide suitable refuges or "yarding" habitats of mature-conifer forest in which microclimate is less severe, browse is available, and snow depths are not extreme. The availability of winter-yarding habitat may be more important than the quantity and quality of summer habitat in the annual use of a managed area by deer (Boer, 1978, 1992; Telfer, 1978b; Moore and Boer, 1979; Lyon and Jensen, 1980). Since particular winter yards are often used for many years by deer from a large surrounding area (Telfer, 1978b; Lyon and Jensen, 1980), these traditional and necessary habitat features should be identified and protected from any cutting that could detract from this use.

Finally, it should be mentioned that some types of ongoing disturbance in a forest management area can be detrimental to the use of otherwise suitable habitat by deer. These can include frequent vehicular traffic along roads, noise from harvesting operations, and excessive hunting pressure (Fletcher and Busnel, 1978; Halls, 1978; Lyon and Jensen, 1980).

Other important North American deer species that can benefit from certain forest-harvesting practices include moose (*Alces alces*) and elk (*Cervus elaphus*). Like the *Odocoileus* deer, moose are primarily browsers, although they also feed on aquatic and terrestrial herbs during the summer. Elk primarily graze on graminoids and herbaceous dicots during the growing season, but they browse in the winter when herbs are unavailable (Martin *et al.*, 1951; Peek, 1974; Crete and Bedard, 1975; Joyal, 1976; Telfer, 1978a; Irwin and Peek, 1983). Since the availability of browse and many herbs can be increased by forest harvesting, integrated programs of forestry and game management can potentially improve the habitat for these species (Dodds, 1974; Krefting, 1974; Parker and Morton, 1978; Telfer, 1978a; McNicol and Timmermann, 1981; Irwin and Peek, 1983). In general, however, moose and elk are somewhat less favored by extensive forest harvesting than are white-tailed deer and mule deer.

Moreover, in regions of eastern North America where white-tailed deer are abundant, moose may suffer a severe detriment from a debilitating, nematode parasite (brainworm, *Parelaphostrongylus tenuis*). This parasite is tolerated by the resistant deer population. However, when deer are abundant so is the brainworm, and it is effectively spread to the vulnerable moose population (Benson, 1958; Smith *et al.*, 1964; Robinson and Bolen, 1989; Smith, 1991).

Crouch (1985) examined the effects of clear-cutting a subalpine conifer forest in Colorado on the abundance of mule deer and elk (Table 9.11). He examined three habitat types: (1) four uncut, reference blocks of mature forest on mesic-xeric sites, (2) four mesic-xeric clear-cuts dominated by shrubs and herbs, and (3) one mesic-hydric clear-cut dominated by herbs. After clear-cutting of the mesic-xeric stands, the abundance of mule deer decreased for 1 year and then progressively increased until by the fifth post-cutting year it averaged 5.3 times larger than the precutting level. On the mesic-hydric site, abundance was also greater after 5 postcutting years, by a factor of about 3.7. Interpretation of these trends is complicated by the observation that mule deer were also more abundant in the reference habitat in the fourth and fifth postcutting years, by a factor of 4.3 in year 5. Although it cannot be unequivocally concluded that the abundance of mule deer was increased by clear-cutting, they certainly were not affected negatively within the period of this study. The effects on elk were more straightforward. After an initial decline in abundance for 1–2 years in the clear-cut habitats, their abundance increased by a factor greater than two times, while on the reference plots abundance was more stable.

In another study, Lyon and Jensen (1980) examined the abundance of deer in 87 clear-cuts of varying size and age and in adjacent, uncut coniferous forests in Montana. Abundance was indexed as the number of fecal groups per hectare. Averaged across all sites, they found a density of 77 elk fecal groups/ha on clear-cuts and 123 groups/ha in forests. The density of white-tailed deer plus mule deer was 81 fecal groups/ha on clear-cuts and 123 groups/ha in mature forest. Although deer use of clear-cuts averaged somewhat less in this study, certain types of clear-cuts were preferred over forest. These clear-cuts were generally characterized by good cover and growth of forage and ease of movement through logging slash, and they were not too close to an active logging road or other ongoing disturbances.

Another important big-game deer species is the woodland caribou (*Rangifer tarandus*; known as reindeer in Eurasia). For the most part, this species

Table 9.11 Deer density on uncut and clear-cut habitats of a subalpine forest in central Colorado[a]

	Index of abundance					
	Before logging	Years after logging (1978–1982)				
Species		1	2	3	4	5
Elk (*Cervus elaphus*)						
Uncut forest	57	82	44	49	69	44
Mesic-xeric clear-cuts	44	7	0	37	69	94
Mesic-hydric clear-cut	74	0	271	173	222	304
Mule Deer (*Odocoileus hemionus*)						
Uncut forest	32	44	37	44	86	136
Mesic-xeric clear-cuts	44	12	37	99	148	235
Mesic-hydric clear-cut	74	25	99	173	222	272

[a]The reference habitat ($n = 4$) was a mature forest of subalpine fir (*Abies lasiocarpa*), lodgepole pine (*Pinus contorta*), and Engelmann spruce (*Picea engelmannii*) growing on mesic-xeric sites. The mesic-xeric clearcuts ($n = 4$) were initially similar to the uncut habitat, but after 5 post-harvest years they were dominated by the shrubs *Vaccinium* spp., *Pachystima myrsinites*, and *Rosa* spp. and by a vigorous growth of graminoids and herbaceous dicots. The mesic-wet clearcut ($n = 1$) was initially a *Picea engelmannii* stand, but after 5 post-harvest years it was dominated by a lush growth of graminoids and herbaceous dicots. Deer and elk density were indexed as the mean number of fecal groups per hectare. Modified from Crouch (1985).

requires an extensive habitat of mature coniferous forest, particularly during winter when the so-called "reindeer moss" lichens (especially *Cladina alpestris*, *C. mitis*, *C. rangiferina*, and *C. uncialis*) can comprise the bulk of their diet (Scotter, 1967; Bergerud, 1972; Darby and Pruitt, 1984; Cumming and Beange, 1987). These lichens are most abundant in conifer stands that are 40–100 years old, but after canopy closure exceeds about 70%, lichens decline and less palatable feather mosses increase in abundance (Schaefer and Pruitt, 1991). Disturbance by wildfire and logging may be important in the successional regeneration of the lichen supply in closed-canopied stands.

In the short term, cutting the forest can increase the availability to caribou of highly palatable arboreal lichens, because these can be abundant in recently deposited logging slash (Klein, 1974; Eriksson, 1976). However, in the medium term, logging (and wildfire) greatly decreases the abundance of reindeer mosses and other lichens that are important as winter foods (Klein, 1974; Eriksson, 1976). In addition, by benefitting deer and moose,

logging may increase the population of timber wolves (*Canis lupus*), an important predator that may reduce populations of caribou (Bergerud and Elliot, 1986; Edmonds, 1988). Overall, it appears that woodland caribou would not benefit in the longer term from extensive forest harvesting and its associated activities, unlike white-tailed deer, mule deer, moose, and elk (McNicol and Timmermann, 1981; Eriksson, 1976).

Smaller Mammals

Hare and rabbits are important species of forested and shrubby habitats in many parts of the world, and they are economically important as pests, as small game, and aesthetically. In eastern North America, for example, cottontail rabbit (*Sylvilagus floridanus*) and snowshoe hare (*Lepus americanus*) are important small-game species, each sustaining a harvest of millions of individuals per year. Both of these species feed by browsing and grazing and can substantially benefit from increases in food availability resulting from disturbance of forests by harvesting, abandonment of agricultural lands, or wildfire. Both species can be sufficiently abundant in regenerating cutovers to cause economic damage by girdling and clipping small conifer regeneration (Monthey, 1986; Sullivan and Sullivan, 1988).

For most purposes, "small mammals" are considered to be mice, voles, shrews, and moles. These small mammals are of ecological and economic importance because they are components of terrestrial food webs, they can damage forest regeneration by girdling young trees and by consuming seed, and they can consume large quantities of potentially injurious arthropods (Silver, 1924; Tevis 1956; Wagg, 1963; Ahlgren, 1966; Ream and Gruel, 1980; Lloyd-Smith and Piene, 1981).

Studies of the effects of forest harvesting on small mammals have been made in many places. These studies have variously reported increases, decreases, or no measurable effects on population and community parameters, depending on the specifics of the forest ecosystem, harvest system, and small-mammal community (e.g., Tevis, 1956; Ahlgren, 1966; Kirkland, 1977; Goodwin and Hungerford, 1979; Ream and Gruel, 1980; Martell, 1984; Monthey and Soutiere, 1985; Clough, 1987; Probst and Rakstad, 1987; Medin and Booth, 1989; Parker, 1989; Corn and Bury, 1991a).

Most commonly, the effects on small mammals have been relatively small, indicating a substantial resilience to disturbance within this group. For example, a study done in an area of hardwood forest in Nova Scotia found no substantial differences in the abundance of small mammal or in the species richness or diversity of their communities among mature forest and habitats affected by harvesting, including 3- to 5-year-old clear-cuts, strip-cuts, and shelterwood cuts (Table 9.12).

Mustellids

Because they are believed to be at risk from habitat changes caused by forestry and because they are easily extirpated by trapping, several closely related species of the weasel family (Mustellidae) have received a great deal of attention from ecologists. The species at most risk from forestry-related activities

in North America are American pine marten (*Martes americana*) and fisher (*M. pennanti*). Sable (*M. zibellina*), Eurasian pine marten (*M. martes*), and Japanese marten (*M. melampus*) are similarly endangered in Eurasia by unsustainable trapping and disturbances of their habitat (Buskirk, 1992).

Sable, for example, is the source of a highly desirable fur. Although its initial range in northern Europe and Asia was expansive (more than 52 million km²), by the mid-1700s, the species was widely extirpated, surviving in only a few refugia (Buskirk, 1992). However, the sable has greatly increased its range and abundance in recent decades, because of protection and/or management of trapping pressure, coupled with release of more than 19 thousand captive-bred animals into suitable habitats in Russia (Buskirk, 1992).

Similarly, American marten and fisher have large ranges in North America. Both species have been intensively trapped over much of their range, resulting in many regional extirpations (Banfield, 1974; Buskirk, 1992). Marten and fisher are considered to be at risk from forest harvesting and management, because over much of their range they are dependent on older-growth, coniferous forests (Harris, 1984; Thompson, 1988; Buskirk, 1992). According to Buskirk (1992), both species have a "consistent, close association with mesic coniferous forests that have complex physical structure, most often in old, uneven-aged stands." This is particularly true in winter, when a complex habitat structure close to the forest floor is required for denning and hunting beneath the snow (Thompson, 1988; Buskirk, 1992). This sort of habitat structure is not generally maintained or provided when natural, mature or old-growth coniferous forests are converted to intensively managed plantations (this conversion is discussed in more detail later in this chapter).

Soutiere (1979) monitored the abundance of marten in three habitat types associated with different silvicultural systems in Maine: (a) selectively harvested forest, with about 40% basal-area removal; (b) a mostly clear-cut area with about 50% of the land harvested by clear-cutting, 25% by selective cutting, and the rest unharvested; and (c) a reference habitat of uncut spruce–fir–hardwood forest. The

Table 9.12 Comparison of small-mammal communities among habitat types affected by harvesting in a hardwood forest in Nova Scotia[a]

	Reference	Clear-cuts	Strip-cuts	Shelterwood
Short-tailed shrew, *Blarina brevicauda*	3.9	3.4	2.5	3.4
Masked shrew, *Sorex cinereus*	2.2	4.9	2.2	2.3
Gapper's red-backed vole, *Clethrionomys gapperi*	7.0	6.0	4.9	4.9
Deer mouse, *Peromyscus maniculatus*	1.0	0.7	0.4	3.0
Meadow vole, *Microtus pennsylvanicus*	0.2	2.8	2.6	0.6
Eastern chipmunk, *Tamias striatus*	0.2	0.2	0.7	0.4
Woodland jumping mouse, *Napaeozapus insignis*	2.9	0.3	2.9	4.7
Abundance	17.4	18.3	16.2	19.3
Species richness	7	7	7	7
Species diversity	1.54	1.56	1.75	1.73

[a]Abundance is indexed as No./100 trap-nights; species richness is the number of species observed; species diversity is H', calculated using relative abundance as p_i. Modified from Swan *et al.* (1984).

density of resident, adult marten was $1.2/km^2$ in both the uncut and selectively cut areas and $0.4/km^2$ in the mostly clear-cut area. These data suggest that in that region, marten may be tolerant of a substantial disturbance of their habitat, although not extensive clear-cutting. For that study area, Soutiere (1979) concluded the following: "harvesting methods that maintained a residual stand of $20–25$ m^2/ha of basal area as pole- and larger-sized trees provide adequate habitat for marten."

Birds

Many bird species use forests or plant communities that are part of postcutting forest successions as habitat for breeding, migrating, or wintering.

Some of these birds are hunted, or game species. Some North American examples include ruffed and spruce grouse (*Bonasa umbellus* and *Canachites canadensis*), wild turkey (*Meleagris gallopavo*), and common bobwhite (*Colinus virginianus*). These birds are all favored by habitat mosaics that include both mature forest and younger brushy stands, with extensive edges between these types. Ruffed grouse is the most important upland game bird in North America, with about 6 million harvested annually (the total harvest of all other grouse and ptarmigan species is about 2.4 million per year; Johnsgard, 1983).

Ruffed grouse live in a wide variety of habitats, but they prefer landscapes that are dominated by hardwood forest with some conifers mixed in, especially when there is a major component of poplars (particularly *Populus tremuloides*) and birches (especially *Betula papyrifera*) (Edminster, 1947; Gullion, 1967; Johnsgard, 1983). Ruffed grouse primarily feed on the foliage, young twigs, catkins, and buds of woody plants, but they also eat seasonally abundant fruits (Brown, 1946; Martin *et al.*, 1951). Clear-cut aspen stands in Minnesota become suitable for ruffed grouse after 4–12 years of regeneration and are then used as breeding habitat for 10–15 years, while older, mature stands are most important as wintering habitat (Gullion, 1969, 1986).

A 2-year-old clear-cut of a hardwood forest in Nova Scotia. Typical breeding birds in this habitat include white-throated sparrow (*Zonotrichia albicollis*), dark-eyed junco (*Junco hyemalis*), song sparrow (*Melospiza melodia*), and common snipe (*Capella gallinago*) (photo: B. Freedman).

Morgan and Freedman (1986) studied a postcutting chronosequence in Nova Scotia and found ruffed grouse in clear-cuts 5 years and older.

To optimize habitat for ruffed grouse in the northeastern United States, a mosaic of different-aged stands less than about 10 ha in area has been recommended, with adjacent blocks differing in age by 10–15 years (Gullion, 1977, 1986, 1988). Such an arrangement would provide cover at all seasons and abundant, accessible food in the form of browse and seasonal berries.

The much larger number of bird species of forested landscapes that are not hunted for sport are sometimes categorized as "nongame" species. These birds can, however, be economically important as predators of insects that are injurious to trees (Bruns, 1960; Martin, 1960; Crawford *et al.*, 1983), and in nonconsumptive recreation, such as bird watching. These benefits are considerable, but they are not well quantified in terms of dollars. It has been estimated that expenditures for the enjoyment of nongame birds (mainly bird watching and winter feeding) in North America could exceed several billions of dollars annually; sales of bird seed alone account for more than $200 million/year (Payne and DeGraaf, 1975; DeGraaf and Payne, 1975; George *et al.*, 1981). In 1980, an estimated 83 million people in the United States were engaged in some degree of "nonconsumptive" use of wildlife, compared with 47 million who were involved in hunting (WRI, 1987).

The effects of forestry on nongame birds can be considered from two directions: (1) effects on individual species and (2) effects on the avian community in terms of overall density, species richness, diversity, etc. Both bird species and the avian community are thought to be strongly influenced by the physical structure and plant species composition of their habitat.

Hypotheses relating vegetation structure to the size and species composition of avian communities have attracted much attention among ecologists. These theories have important implications for forestry, because of the great structural changes caused by harvesting and silvicultural practices. Many studies have shown strong relationships between avian community characteristics and vegetation structure, including studies of the relationships between: (1) spatial complexity and bird-species richness and diversity (MacArthur, 1964; MacArthur and MacArthur, 1961; MacArthur *et al.*, 1966; Karr and Roth, 1971; Roth, 1976; Morgan and Freedman, 1986), and (2) habitat structure and bird-community composition (Pitelka, 1941; Johnston and Odum, 1956; James, 1971; Shugart *et al.*, 1978; Collins *et al.*, 1982; Morgan and Freedman, 1986).

Important variables in terms of habitat structure are the vertical and horizontal distributions of plant

An 8-year-old clear-cut of a hardwood forest in Nova Scotia. Typical breeding birds in this habitat include alder flycatcher (*Empidonax alnorum*), chestnut-sided warbler (*Dendroica pensylvanica*), common yellowthroat (*Geothlypis trichas*), and white-throated sparrow (*Zonotrichia albicollis*). At this stage, an upper canopy is beginning to form and some species more typical of mature forest are beginning to breed, including red-eyed vireo (*Vireo olivaceus*), American redstart (*Setophaga ruticilla*), and veery (*Catharus fuscescens*) (photo: B. Freedman).

biomass, the amount of vegetation cover, and the distinctness of ecotonal gradients at habitat discontinuities. Vertical structure of temperate forests is often divided into three functional strata: the ground vegetation, shrubs/saplings, and trees (MacArthur and MacArthur, 1961; MacArthur, 1964; MacArthur *et al.*, 1966). Horizontal structure involves the spatial complexity of patches of distinct habitat within a larger stand or landscape mosaic. Considerations include the shape of patches, which influences the ratio of edge to area, and also patch size since small, isolated habitat islands will not sustain bird species that have large territories (Moore and Hooper, 1975; Forman *et al.*, 1976; Galli *et al.*, 1976; Crawford and Titterington, 1979; Noon *et al.*, 1979; Robbins, 1979; Temple *et al.*, 1979; Carey *et al.*, 1990).

The species composition of the vegetation is also an important feature of habitats, including the degree of mixture of coniferous and angiosperm tree species in temperate forests (Karr, 1968; Balda, 1975; Crawford and Titterington, 1979). The occurrence of snags (i.e., dead but erect trees) and logs on the forest floor are also critical habitat features for many bird species, as is discussed in more detail later in this chapter. Finally, the occurrence of an epidemic of a defoliating insect, such as spruce budworm, can dramatically increase the abundance of insectivorous birds (see Chapter 8).

These observations and hypotheses have suggested to ecologists that it should be possible to determine whether a given site is suitable for a particular bird species, or a community of species, by examining the structural and compositional character of the vegetation of the habitat. Birders have long used this principle in a qualitative sense,

A 3-year-old progressive strip cut of a hardwood forest in Nova Scotia. The clear-cut section is about 40 m wide and 300 m long, and there are intervening, similar-sized, uncut strips between the cut strips. After about 5–8 years, the uncut strips will also be harvested. An area composed of a mosaic of cut and uncut strips has a bird community that contains species typical of both clear-cut and mature forests, but these segregate among the strips (photo: B. Freedman).

through "bird watching by habitat"; that is, particular species and assemblages of birds are expected to be present in certain ecological situations.

It follows that it should be possible to manage a breeding-bird community or particular species by manipulation of their habitat (Verner, 1975; Noon *et al.*, 1979; Dennis *et al.*, 1991). The best forestry-related example of this principle is the use of prescribed fire in Michigan to create even-aged stands of jack pine (*Pinus banksiana*) to manage the habitat of the rare and endangered Kirtland's warbler (*Dendroica kirtlandii*). In 1950 the entire breeding population of this species comprised about 415 pairs, and it was 500 pairs in 1961, but there were only 165–240 pairs between 1971 and 1990, all of them nesting in central Michigan. The optimal habitat of Kirtland's warbler is even-aged, 1.5 to 6.0-m-tall, 7- to 23-year-old stands of jack pine. The availability of this habitat is maintained and en-

hanced by planting and by the deliberate burning of older stands (Mayfield, 1960; Buech, 1980). The manipulation of habitat structure and stabilization of its plant-species composition, coupled with intensive efforts to reduce the depredations of a nest parasite, the brown-headed cowbird (*Molothrus ater*), have allowed the maintenance of the small breeding population of Kirtland's warbler (Mayfield, 1961, 1977; Walkinshaw, 1983; Probst, 1988).

Therefore, following from the above discussion, the effects of forest harvesting on bird species follow indirectly from changes in the physical and botanical character of the habitat. Every bird species has particular habitat requirements, and on any particular site habitat suitable for species of mature forest is modified or destroyed by harvesting, while opportunities are created for early successional species (e.g., Webb *et al.*, 1977; Franzreb and Ohmart,

A 3-year-old shelterwood cut in a hardwood forest in Nova Scotia. During the cut, about 40% of the tree basal area was removed, leaving the "best" trees to promote their growth into high-value sawlogs and to encourage a good advance regeneration of desired tree species. This silvicultural treatment results in a habitat with a vertically complex structure. At this stage, there is a diverse and vigorous ground vegetation, a thick shrub layer largely composed of groups of stump sprouts of cut trees, and a sparse tree canopy. The bird community contains species typical of both clear-cut and mature forests (photo: B. Freedman).

Snags, or dead standing trees are an important habitat component for many species of animals. In this photo, four juvenile kestrels (*Falco sparverius*) have fledged from a cavity in a pine snag in a clear-cut in Nova Scotia. (Photo courtesy of I.A. McLaren.)

1978; Noon *et al.*, 1979; Szaro and Balda, 1979; Temple *et al.*, 1979; Titterington *et al.*, 1979; Niemi and Hanowski, 1984; Wetmore *et al.*, 1985; Morgan and Freedman, 1986; Thompson and Caper, 1988; Tobalske *et al.*, 1991; DeGraaf *et al.*, 1992).

These effects can be illustrated by comparison of the breeding birds of mature stands of hardwood forest and adjacent clear-cuts in Nova Scotia (Table 9.13). The uncut forest in this study was mixed maple–birch, and its bird community was dominated by ovenbird, least flycatcher, red-eyed vireo, black-throated green warbler, and hermit thrush. The total abundance of breeding birds in the forest stands averaged 663 pairs/km^2, while species richness averaged 12 (range 9–16), and species diversity 2.11 (1.84–2.48). Compared with the adjacent forest, the three 3- to 5-year-old clear-cuts had a different habitat structure and plant-species compo-

Table 9.13 Breeding birds of three mature hardwood forest plots and three adjacent clear-cuts 3–5 years old[a]

Species	Mature forest			Clear-cut plots		
	A	B	C	A	B	C
Common snipe (*Capella gallinago*)	0	0	0	10	15	0
Ruby-throated hummingbird (*Archilochus colubris*)	0	0	0	25	30	15
Least flycatcher (*Empidonax minimus*)	290	120	0	0	0	0
Hermit thrush (*Catharus guttatus*)	60	40	30	0	0	0
Veery (*Catharus fuscescens*)	50	10	0	25	0	0
Solitary vireo (*Vireo solitarius*)	60	30	0	0	0	0
Red-eyed vireo (*Vireo olivaceous*)	80	50	30	0	0	0
Black-and-white warbler (*Mniotilta varia*)	15	50	40	0	0	0
Northern parula warbler (*Parula americana*)	15	30	40	0	0	0
Black-throated green warbler (*Dendroica virens*)	50	30	30	0	0	0
Chestnut-sided warbler (*Dendroica pensylvanica*)	0	0	0	100	40	190
Ovenbird (*Seiurus aurocapillus*)	150	120	200	0	0	0
Mourning warbler (*Oporornis philadelphia*)	0	0	0	0	0	90
Common yellowthroat (*Geothylpis trichas*)	0	0	0	25	300	130
American redstart (*Setophaga ruticilla*)	15	80	100	0	0	0
Rose-breasted grosbeak (*Pheucticus ludovicianus*)	15	10	0	0	0	0
Dark-eyed junco (*Junco hyemalis*)	15	20	15	50	70	30
White-throated sparrow (*Zonotrichia albicollis*)	0	20	0	90	190	100
Song sparrow (*Melospiza melodia*)	0	0	0	90	70	0
Total density (pairs/km²)	815	660	515	435	745	585
Total number of species	12	16	9	10	8	7

[a]The mature forest had a closed canopy dominated by maple and birch (*Acer saccharum. A. rubrum. Betula papyrifera. B. alleghaniensis*). Regeneration on the clear-cuts was vigorous, with a shrub stratum dominated by maple and birch stump sprouts, raisinbush (*Viburnum cassinoides*), and pin cherry (*Prunus pensylvanica*), and a dense ground vegetation of raspberry (mainly *Rubus strigosus*), graminoids, and dicotyledonous herbs. Some uncommon bird species are not included. Modified from Freedman *et al.* (1981a).

sition (they were shrubby and had a dense ground vegetation, dominated by a diverse array of graminoids and forbs). In spite of the large differences in habitat, the avian density of the clear-cuts averaged only slightly less (average 588 pairs/km²; range 435–745 pr/km²) than forest stands, as did species richness (8; 7–10) and diversity (1.78; 1.65–1.99). However, the actual bird species were almost completely different on the clear-cuts, which were dominated by chestnut-sided warbler, common yellowthroat, white-throated sparrow, and dark-eyed junco. Overall, the differences in avian-community parameters were not substantial, and although species of mature forest were temporarily deprived of habitat by the clear-cutting, opportunities were created for a large number of early successional, native bird species.

Freedman *et al.* (1981a) also examined stands that were harvested in ways that created habitats that were intermediate in structure to those of young clear-cuts and mature forest. These harvest methods were (1) shelterwood cuts, in which about 50% of the tree basal area was removed, with the other trees left so as to encourage regeneration and to produce high-quality sawlogs, and (2) progressive strip-cuts, in which about one-half of the area was clear-cut in long (300 m) and narrow (30 m) strips, with intervening uncut strips left in order to encourage the regeneration of tolerant tree species. The shelter wood and strip-cut harvesting produces habitats intermediate to the uncut forest and clear-cuts, and these sites were correspondingly occupied by a mixture of bird species typical of either clear-cuts or mature stands, while total-bird density, species richness, and species diversity were not substantially different from the mature forest (see also Medin and Booth, 1989; Tobalske *et al.*, 1991).

In another study in the same area of Nova Scotia, bird populations and their habitat were examined in a chronosequence of 23 hardwood stands (Morgan and Freedman, 1986). All stands 20 years old and younger originated after clear-cutting, while older stands were part of a fire-caused landscape mosaic. Once the clear-cuts had regenerated for at least 3

years, there was little variation of overall bird community variables among stands of different age (Fig. 9.7a,b), in spite of great differences in habitat. The range of total avian density of clear-cuts 3 to 10 years old fell within the range of variation of density for the older, mature stands (Fig. 9.7a). Only the 1- and 2-year-old clear-cuts had bird populations that were smaller than those of the uncut stands. In addition, a distinct suite of bird species was present relatively early in the postclear-cutting succession;

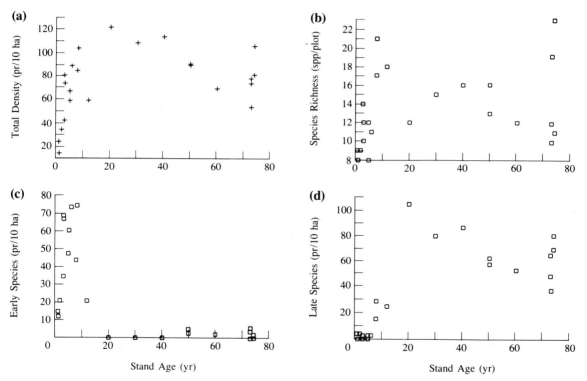

Fig. 9.7 Breeding birds of a chronosequence of 23 stands of different age in a region of hardwood forest in Nova Scotia. Note the replication of some stand ages. All stands younger than 20 years originated from clear-cutting of the mature forest. Stands 20 years and older were part of a fire-created landscape mosaic. (a) Total density, or the sum of density of all species; (b) species richness, or the number of species breeding on each plot; (c) sum of densities of species that were prominent in young stands (i.e., alder flycatcher, *Empidonax alnorum*; chestnut-sided warbler, *Dendroica pensylvanica*; common yellowthroat, *Geothlypis trichas*; dark-eyed junco, *Junco hyemalis*; white-throated sparrow, *Zonotrichia albicollis*; and song sparrow, *Melospiza melodia*); (d) sum of densities of species that were prominent in older stands (i.e., least flycatcher, *Empidonax minimus*; hermit thrush, *Catharus guttatus*; veery, *Catharus fuscescens*; red-eyed vireo, *Vireo olivaceous*; black-throated green warbler, *Dendroica virens*; American redstart, *Setophaga ruticilla*; and ovenbird, *Seiurus aurocapillus*). Modified from Morgan and Freedman (1986).

these were later replaced by another group as the forest matured (Fig. 9.7c,d).

The transition between these two bird communities took place in stands 12–20 years old. During this period, hardwood stump sprouts and saplings had thinned to relatively few stems with a canopy at 8–12 m. At the same time, the shrub and ground-vegetation strata were declining in prominence because of shading and other stresses exerted by the developing tree canopy. These habitat changes allowed the progressive invasion of the stand by bird species characteristic of mature forest. The distinctness of the "early" and "late" bird communities was shown by a cluster analysis of the data matrix of bird species abundance versus stand age. This multivariate procedure strongly separated the bird communities of stands 1–12 years old from those of stands 20–74 years old. Clusters of stands within these two age groups were much weaker, indicating that their bird communities were similar.

A similar pattern of change in the breeding-bird community has been reported in postlogging chronosequences of conifer forest. Welsh and Fillman (1980) described the changes in birds among nine clear-cuts that ranged in age from 3 to 24 years, plus a reference black spruce (*Picea mariana*) forest in northern Ontario (Table 9.14). The greatest abundance of breeding birds occurred in moderate-aged cutovers. Stands aged 11–24 years had a range of 1020–1970 pairs/km^2, considerably more than were present in the mature, uncut forest (561 pairs/km^2). The smallest avian abundance was in the 3-year-old cutover with about 200 pairs/km^2, but by 5 postcutting years, the abundance of birds had recovered to 690 pairs/km^2.

The stands studied by Welsh and Fillman had not been completely clear-cut—they all had occasional uncut islands of trees or scattered uncut stems. As a result, some of the birds that were dominant in the mature forest were also present on relatively young clear-cuts. Obviously, these species did not require a continuous, closed forest as their habitat. A good example of such a species is the Tennessee warbler, which was present in all stands except the 3-year-old cutover. It was the most prominent species in the uncut stand, where it comprised 18% of all breeding

birds. However, the Tennessee warbler was most abundant in cutovers 11–24 years old, where its density ranged from 217 to 678 pairs/km^2. The most prominent species in the 3-year-old cutover were Lincoln's sparrow, savannah sparrow, and song sparrow. Only the latter two species were restricted to this young habitat—Lincoln's sparrow was present on almost all stands in the chronosequence. Overall, the effects on breeding birds of the clear-cutting of this lowland black spruce forest were not catastrophic. If anything, the overall avian abundance was greater in the vigorously regenerating, moderately aged cutovers.

Few studies have examined the effects of forestry on birds in tropical habitats. One study of selective logging of Amazonian forest in French Guiana found relatively moderate changes in community-level, avian parameters, but a large degree of species turnover (Table 9.15; Thiollay, 1992). Overall, there was a 45% difference in avian-species composition between primary forest and selectively logged stands, and 42% of the primary-forest species decreased or disappeared after logging, especially birds associated with the understory. In this study, only three trees per hectare were removed. However, the intensity of disturbance was much greater than this harvest rate might superficially suggest, since 38% of the understory vegetation was damaged, and cover of the forest canopy was decreased by more than 40%. Of course, clear-cutting of tropical forest would lead to much more drastic changes, and these would be essentially permanent if the site was then converted to agriculture, as often occurs. Beyond the direct effects of habitat changes, the construction of forestry roads greatly improves access to the interior of the tropical forest, so local hunters can deplete populations of game animals by market hunting (Wilkie *et al.*, 1992), and slash-and-burn farmers can convert the disturbed forest to agriculture. These topics are discussed in more detail in Chapter 10.

As mentioned previously, the density of dead trees in the forest, occurring either as standing snags or as logs lying on the forest floor, are important habitat features that influence the presence and success of certain species of wildlife (Conner, 1978;

Table 9.14 Breeding birds in a post-clear-cutting chronoseqence in a lowland black spruce (*Picea mariana*) forest in northern Ontario[a]

Species	Uncut forest	24-year cutover	19-year cutover	17-year cutover	13-year cutover	11-year cutover	9-year cutover	6-year cutover	5-year cutover	3-year cutover
Tennessee warbler (*Vermivora peregrina*)	100	489	678	283	622	217	78	106	94	
Yellow-rumped warbler (*Dendroica coronata*)	67		28	11	28	11		p		
Least flycatcher (*Empidonax minimus*)	50	61	39	44	22					
Nashville warbler (*Vermivora ruficapilla*)	50	78	50	67	67	50	67	6	11	
Cape May warbler (*Dendroica tigrina*)	50	117	22	67						
Ruby-crowned kinglet (*Regulus calendula*)	34		22	11	11					
Alder flycatcher (*Empidonax alnorum*)	25	6	200	39	206	133	83	56	67	
White-throated sparrow (*Zonotrichia albicollis*)	25	128	256	267	211	183	228	78	111	14
Lincoln's sparrow (*Melospiza lincolnii*)	25		133	28	133	50	100	67	78	61
Magnolia warbler (*Dendroica magnolia*)	17	167	139	83	67	106	39	56	22	
Common yellowthroat (*Geothlypis trichas*)	17		244	6	222	133	106	111	150	11
Swainson's thrush (*Catharus ustulatus*)		67	67	17	50	22				
Hermit thrush (*Catharus guttatus*)		61	22	28						
Red-eyed vireo (*Vireo olivaceus*)		44					11	p		
Northern waterthrush (*Seiurus noveboracensis*)		39	44		50					
Yellow-bellied sapsucker (*Sphyrapicus varius*)		28					11	11		
Purple finch (*Carpodacus purpureus*)		11	17	33	17	17				
Wilson's warbler (*Wilsonia pusilla*)			11	22	44	83	39	11	50	11
Swamp sparrow (*Melospiza georgiana*)				6			11	11	56	6
Mourning warbler (*Oporornis philadelphia*)							33	11		6
Chestnut-sided warbler (*Dendroica pensylvanica*)							17	72	44	
Song sparrow (*Melospiza melodia*)								6		25
Savannah sparrow (*Passerculus sandwichensis*)										33
Total number of species	15	21	16	18	17	13	18	17	12	14
Total density	561	1518	1972	1034	1778	1022	851	619	694	196

[a]Some less common species are not listed here. Modified from Welsh and Fillman (1980). Data are in pairs/km^2; p, present.

Table 9.15 Avian community parameters in primary and selectively logged stands in French Guiana[a]

Parameter	Primary forest	10-year-old logged stand	1-year-old logged stand
Total species recorded	239	163	166
Species richness/count	4.6	3.1	3.4
Density/count	7.7	5.1	5.7
Species diversity (H')	4.9	4.5	4.5

[a]Birds were censused using 20-min point counts, with sample sizes of 328 in primary forest, 273 in 10-year-old logged areas, and 336 in 1-year-old logged areas. Modified from Thiollay (1992).

Maser *et al.*, 1979, 1988; Evans and Conner, 1979; Miller and Miller, 1980; Scott *et al.*, 1980; Dickson *et al.*, 1983; Raphael and White, 1984; DeGraaf and Shigo, 1985; Bull *et al.*, 1986; Hunter, 1990). These habitat features are especially critical to certain species of birds, as they are used for nesting in excavated and/or natural cavities, as foraging substrates, and as perches for hunting, resting, feeding, and singing. For example, Scott *et al.* (1980) estimated that 30–45% of the breeding-bird species of conifer and aspen forests of the northwestern United States are cavity nesters, requiring snags as a necessary habitat feature. These species include primary excavators such as woodpeckers, secondary cavity users, and species that use natural cavities. Clearly, snag management is an important variable that influences the effects of forest harvesting on birds.

Maintenance of the habitat of cavity-dependent birds has become an important consideration in the multiple-use, integrated management of many forests in North America. This issue has been particularly prominent in the Pacific northwest of the United States, where as many as six species of woodpeckers can cooccur in relatively large populations along with other cavity-dependent species (Davis *et al.*, 1983; Bull *et al.*, 1986; Lundquist and Mariani, 1991). Cavity management is also an important concern in southeastern U.S. pine forests, especially loblolly pine (*Pinus taeda*) forests, which are habitat of the endangered red-cockaded woodpecker (*Picoides borealis*) (Hooper *et al.*, 1980; Conner and Rudolph, 1989; Walters, 1991).

In old-growth, Douglas-fir-dominated forests of the Blue Mountains of eastern Oregon, woodpeckers require the following snag characteristics as a component of their habitat: pileated woodpecker (*Dryocopus pileatus*), 0.32 snags/ha; common flicker (*Colaptes auratus*), 0.93 snags/ha; black-backed three-toed woodpecker (*Picoides arcticus*), 1.5 snags/ha; and hairy woodpecker (*Picoides villosus*), 4.5 snags/ha (Thomas *et al.*, 1979; Maser *et al.*, 1988). For the same region, Bull *et al.* (1980) recommended that at least 4 snags/ha be left on harvested sites to ensure maintenance of at least 70% of the potential woodpecker population.

If a forested landscape is to be managed for both fiber and birds, a number of practices should be followed (after Evans, 1978; Freedman, 1982): (1) a proportion of the management area should be reserved, or at least maintained as mature, natural forest, in blocks that are as large as possible; (2) cut portions should generally be arranged to produce a mosaic of habitats of different age abutting each other; (3) dead, dying, or girdled cull trees should be left standing whenever possible, or plantations could be otherwise suitably managed to provide habitat for birds that require snags and dead logs; (4) less intensive harvest techniques such as thins, shelterwood cuts, strip cuts, and especially selection cuts will produce less severely modified habitats, and in some cases these methods could be used with relatively little detriment to avian species composition or abundance; and (5) if rare or endangered bird species are present in the management area, specific management plans should be developed to preserve or enhance their habitat.

Except for item (5), the rationale for the use of these various practices in a multiple-use management system that involves fiber, birds, and other wildlife has been discussed previously. Point (5) is illustrated later in this chapter, with reference to the cases of the spotted owl (*Strix occidentalis*) and red-cockaded woodpecker (*Picoides borealis*).

The aesthetic and economic benefits of multiple-use management practices can be further enhanced by providing bird-watching opportunities. This could be accomplished by identifying and publicizing unique or diverse birding areas, by providing facilities where birds are easily visible to people, or by developing interpretation programs to promote an understanding and appreciation of birds. Such management practices would, of course, also enhance the value of wildlife other than birds, both animals and plants. If implemented, such a scheme would have sociological, educational, and economic benefits that would increase the value of the overall management area and of the wildlife resource itself.

Amphibians and Reptiles

Studies in the northeastern United States have indicated that amphibians can contribute substantially to the total productivity of vertebrates in hardwood and mixedwood forests, in some cases exceeding that of birds and mammals (Burton and Likens, 1975a,b; Hairston, 1987). There have been a number of studies of the effects of forest harvesting and management on populations of amphibians, and to a lesser extent reptiles, mostly in the northwestern and southern United States (Bury, 1983; Enge and Marion, 1986; Pough *et al.*, 1987; Ash, 1988; Bury and Corn, 1988a,b; Corn and Bury, 1989, 1991b; Petranka *et al.*, 1993). Most of these studies have found substantial effects in the abundance of these animals after disturbance of mature stands by forestry, with the nature of the changes being related to the intensity of the changes in habitat.

The effects of forest harvesting on amphibians and reptiles can be illustrated using data for 30 stands of various age in coastal Oregon and Washington (Table 9.16). This study indicates that most of the common species of salamander, frog, lizard, and snake in this region can tolerate a wide range of habitats, ranging from old-growth forest to clear-cuts. In general, however, amphibians tended to be more abundant in forested habitats, while reptiles were more abundant in clear-cuts.

Amphibians breeding in streams are also affected by forest harvesting, depending on the intensity of the disturbance of shading, riparian vegetation, and on physical disturbance of the streambed during the skidding of logs (if this latter, destructive practice occurs). For example, Corn and Bury (1989) studied 23 streams in stands of intact forest and 20 in logged stands and found significant decreases in the abundance of amphibians. On average, the Pacific giant salamander (*Dicamptodon ensatus*) had a density of $2.28/m^2$ in streams in intact forest but only $0.50/m^2$ in logged stands, while tailed frog (*Ascaphus truei*) averaged $0.76/m^2$ and $0.37/m^2$, respectively; Dunn's salamander (*Plethodon dunni*) $0.41/m^2$ and $0.17/m^2$; and Olympic salamander (*Rhyacotriton olympicus*) $0.29/m^2$ and $0.0.04/m^2$. In this study, detrimental effects of sedimentation were relatively small, because none of the harvested sites had upstream road crossings.

Arthropods

Arthropods such as insects and spiders are the most species-rich phylogenetic groups in temperate forests. Forest arthropods are of great ecological and economic importance, because of their critical trophic roles as herbivores, predators, and detritivores and because a few species are important pests in forestry. However, except for a few species of insects that are of economic importance, the terrestrial arthropod biota of temperate forests has been little studied, and virtually nothing is known about the effects of forest harvesting and management on this group (information relevant to insecticide spraying against spruce budworm is discussed in Chapter 8). The paucity of information about arthropods is a major deficiency in terms of understanding the effects of forestry on biodiversity.

Table 9.16   Populations of the most abundant species of amphibians and reptiles in various stands dominated by Douglas fir (*Pseudotsuga menziesii*) in Oregon and Washington[a]

Species	Old growth	Mature	Young	Clear-cut
Northwestern salamander (*Ambystoma gracile*)	4.2	1.3	3.5	2.2
Pacific giant salamander (*Dicamptodon ensatus*)	0.7	0.0	1.0	0.7
Oregon ensatina (*Ensatina eschscholtzi*)	19.1	10.7	21.5	4.8
Rough-skinned newt (*Taricha granulosa*)	22.8	6.0	21.5	8.6
Tailed frog (*Ascaphus truei*)	34.8	11.2	10.2	1.4
Red-legged frog (*Rana aurora*)	3.8	27.8	7.8	6.6
Pacific tree frog (*Hyla regilla*)	0.2	1.0	1.2	3.2
Western skink (*Eumeces skiltonianus*)	0.2	0.0	0.0	4.0
Northern alligator lizard (*Gerrhonotus coeruleus*)	0.7	0.2	2.0	5.0
Western fence lizard (*Sceloporus occidentalis*)	0.1	0.0	0.0	1.6
Northwestern gartersnake (*Thamnophis elegans*)	0.5	0.1	0.7	2.4
Common gartersnake (*Thamnophis sirtalis*)	0.8	0.1	0.5	0.4
Total salamanders	49.8	18.3	46.5	17.2
Total frogs	38.8	40.0	19.0	10.8
Total reptiles	2.9	0.5	3.5	11.8
Total abundance, all species	91.5	58.8	69.0	39.8
Average species richness	6.1	4.8	5.8	7.6

[a]Old-growth stands were 195–450 years old ($n = 13$), mature stands were 105–150 years old ($n = 6$ stands), young forest was 30–76 years old ($n = 6$), and clear-cuts were <10 years old ($n = 5$). Amphibian abundance is indexed as average numbers captured per stand by standard sampling efforts, using pitfall traps. Modified from Bury and Corn (1988a).

A study in western Oregon compared the spider community of old-growth stands and clear-cuts aged 4–31 years old (McIver *et al.*, 1992). This study found substantial changes in the arachnid community, with visual predators tending to dominate young clear-cuts, and sit-and-wait, microweb, and trap-door predators tending to dominate mature and older stands. Most of the common species of the oldest stands reinvaded the clear-cuts when they reached a post-harvest age of about 30 years.

Freshwater Biota

Forestry practices, including road building, can detrimentally affect populations of freshwater biota in three major ways: (1) by causing siltation of habitats; (2) by causing increases in water temperature by the removal of streamside, shading vegetation; and (3) by blockage of stream channels with logging slash and other debris. In some cases, damage can also be caused by chemical and fuel spills and as a result of pesticide spraying.

Compared with reference watersheds, clear-cut watersheds in New Brunswick had 17% fewer brook trout (*Salvelinus fontinalis*) and 26% fewer benthic invertebrates, but 200% more sculpin (*Cottus cognatus*), a nongame species of fish (Welch *et al.*, 1977). In Oregon, there were no consistent differences in the abundance of salmonid fish and insects between streams running through uncut conifer forest or through logged areas (Murphy and Hall, 1983). Overall, however, stream sections in clear-cuts 5–17 years old tended to have a larger biomass, density, and species richness of benthic insects, compared with reaches in old-growth forest. In another case, a doubling of the abundance of aquatic insects was reported in an Appalachian stream running through habitat that 10 years previ-

ously had been converted from a mature hardwood forest to a grassland, by clear-cutting, herbiciding, and fertilization (Haefner and Wallace, 1981).

A study of a region of conifer-dominated forest in southeastern Alaska examined streams affected by clear-cutting with or without riparian buffer strips, plus reference areas of old-growth forest (Murphy *et al.*, 1986). Streams running through clear-cuts supported larger populations of young salmon fry, although parr up to 2 years of age were not enhanced (Table 9.17). The stimulation of fry was attributed to moderately warmer waters and an increased abundance of arthropod foods. The amount of debris left in streams was important to parr, with densities of 10–22/100 m^2 occurring when logging debris was left in clear-cut reaches of streams, but only <2/100 m^2 if the debris was removed (Murphy *et al.*, 1986).

Erosion resulting from poorly controlled forestry operations can cause serious problems of turbidity and siltation of freshwater habitats (Packer, 1967b; Burns, 1972; Corbett *et al.*, 1978; Murphy *et al.*,

Table 9.17 Habitat and density of salmonid fish[a] among 18 streams in southeast Alaska that were affected by clear-cuts with or without riparian buffer strips, plus a reference habitat of old-growth, conifer-dominated forest[b]

	Old-growth forest	Clear-cut with buffer	Clear-cut without buffer
Canopy density over stream (%)	74 (61–86)	65 (52–87)	27 (14–39)
Periphyton biomass (dry mg/m²)	2.0 (1.2–3.2)	2.9 (1.8–4.7)	4.7 (2.9–7.4)
Benthos density (10³/m²)	5.0 (2.6–9.6)	7.8 (4.1–15)	7.9 (4.1–16)
Summer fish density (No./100 m²)			
Coho fry (Oncorhynchus kisutch)	66 (28–120)	168 (102–249)	164 (99–245)
Coho parr	8.6 (4.3–15)	10 (5.2–16)	5.6 (2.3–11)
Dolly varden parr (*Salvelinus malma*)	15 (2.0–38)	39 (15–75)	16 (2.8–41)
Trout (Salmo gairdneri & S. clarki)	6.0 (1.1–15)	1.9 (0.0–7.1)	4.4 (0.5–12)

[a]Six treams per treatment, overall average, with range of stand averages in parentheses
[b]Modified from Murphy *et al.* (1986).

1981). Streamwater turbidity caused by fine inorganic particulates is rarely lethal to fish, since fish can tolerate large concentrations of suspended solids, especially if the particulates are not abrasive. Cordone and Kelley (1961) found no detectable behavioral responses by salmonids to short-term bioassay concentrations of turbidity of less than 20 thousand ppm, and no lethal effects in short exposures unless the concentration exceeded 175 thousand ppm. In contrast, Phillips (1970) found that longer-term bioassay exposures of only 200–300 ppm were lethal to salmonids. Even these latter concentrations of turbidity are considerably larger than are likely to occur for a protracted period in streams affected by erosion caused by forestry. However, indirect effects of turbidity could be important to fish. By reducing light transmission, turbidity could decrease primary productivity and make feeding difficult for sport fish, most of which are visual predators (Hesser *et al.*, 1975; Corbett *et al.*, 1978).

Coarser particles that have been mobilized by erosion can be of greater importance, since they can settle out of the water and in-fill the critical breeding habitats of fish (Beschta and Jackson, 1979). Salmonid and many other types of fish require a clean gravel streambed in which to spawn in the autumn. Salmonid eggs hatch in early spring in the gravel interstices, where the alevins remain until middle to late spring, when they emerge to the streamwater as fry. The gravel bed must be well flushed to maintain large oxygen concentrations and to remove toxic metabolites. If the interstices are filled with eroded sediment, the circulation of water is impeded, with deleterious effects on these early life-history stages. Even if the eggs and alevins do manage to survive, their emergence from the gravel can be prevented by silt deposits (Hall and Lantz, 1969; Corbett *et al.*, 1978). Clean gravel is also required to provide young fish with cover from predators and from high-water velocities and to provide overwintering sites.

Clearly, erosion caused by forestry operations, especially road construction, can have severe effects on the habitat of freshwater biota, particularly for gravel-spawning fish such as salmonids. However, erosion rates can be greatly reduced if operational care is taken during the construction of roads and culverts, in the skidding of logs, and by leaving buffer strips of uncut forest beside water courses (Packer, 1967a; Kochenderfer, 1970; Rothewell, 1971; Simmons, 1979; Miller and Sirois, 1986).

Excessive accumulations of slash and other debris can degrade freshwater habitats by creating obstructions to the movements of fish. However, in smaller quantities, this debris can be beneficial by providing habitat structure and cover (Narver, 1970; Slaney *et al.*, 1977a,b). The use of uncut buffer strips beside water courses and the directional felling of trees away from water can almost eliminate debris inputs into streams (Slaney *et al.*, 1977a).

Many studies have shown that the temperature of water courses can increase substantially after the removal of streamside vegetation by forest harvesting. This effect, which can be virtually eliminated by leaving buffer strips of uncut forest beside streams, is caused by direct heating of the water after the removal of shading vegetation (Brown, 1969, 1970). Studies in West Virginia showed that streamwater temperatures averaged 4.4°C higher in the first year after clear-cutting of the adjacent forest, compared with uncut, reference parts of the watercourse. There was a maximum-summer temperature of 26°C on the cut portion. Water temperatures returned to the reference condition after 4 years of regeneration (Kochenderfer and Aubertin, 1975). Another study in West Virginia found that the maximum-daily streamwater temperature increased to 26°C after clear-cutting and herbiciding, compared with a reference value of 18°C (Corbett *et al.*, 1978). In North Carolina, the maximum streamwater temperature reached 29°C after clear-cutting, which was as much as 7°C higher than the reference value. The temperature returned to the reference level in the fourth postcutting year (Swift and Messer, 1971). After a streamside clear-cut in Oregon, there was a water temperature increase of as much as 16°C. The annual-maximum temperature increased to 29°C in the first postcutting year, compared with a reference value of 14°C (Brown and Krygier, 1970).

Holtby (1988) measured changes in water temperature in a stream draining a watershed in British

Columbia that had 41% of its area clear-cut. The increase in average water temperature ranged from 0.7°C in December to 3.2°C in August. In this case, the warmed water resulted in an earlier postwinter emergence of fry of coho salmon (*Oncorhynchus kisutch*), allowing an extended growing season of about 6 weeks, a consequently larger size, and increased overwinter survival of fingerlings and then a 47% increase in the abundance of smolt-sized fish. However, the warmed water also caused an earlier seaward migration of smolt, which may have caused a decreased rate of survival to adulthood. Overall, it was estimated that the net effect of logging in this case was to cause a 9% increase in the abundance of adult salmon.

In general, large increases or fluctuations of water temperature in forest streams are not desirable, since salmonid fishes such as trout and salmon are sensitive to high temperatures. This is particularly true during late summer, when the water of even undisturbed rivers and streams can attain quite high temperatures. For example, Huntsman (1942) reported lethal high temperatures in several rivers in Nova Scotia during the droughty summer of 1939. He attributed the deaths of both adult and immature (parr) Atlantic salmon (*Salmo salar*) and brook trout to the high temperatures that occurred on sunny days during a period of low flow.

The optimum range of water temperature for adult Atlantic salmon and brook trout is about 15–20°C. These salmonids begin to show signs of stress at warmer temperatures (Fisher and Elson, 1950; Stroud, 1967; Swift and Messer, 1971; Hokenson *et al.*, 1973). For adult and juvenile life stages of trout and salmon the upper lethal limit is about 25°C (Fry *et al.*, 1946; Alabaster, 1967; Hynes, 1971), although Huntsman (1942) reported that Atlantic salmon parr could survive up to 33°C and grilse up to 29°C. However, salmonid eggs have an upper lethal temperature of only about 14°C (Hynes, 1971). Other fish species can tolerate much higher temperatures: for example, up to 36°C by white suckers (*Catostomus commersoni*) and 38°C by carp (*Cyprinus carpio*) (Huntsman, 1942; Hynes, 1971). Water temperatures as high as these have been reported after the removal of vegetation from

the immediate vicinity of forest streams in many areas. However, this effect can easily be avoided by leaving buffers of uncut, shading forest beside watercourses.

Old-Growth Forest and Its Dependent Species

In the sense used here, an old-growth forest is considered to be late-successional and characterized by trees of great age, an uneven-aged population structure, and a complex physical structure, including multiple layers in the canopy, large trees (usually), and a substantial component of large-dimension snags and dead logs lying on the forest floor (Harris, 1984; Maser *et al.*, 1988; Franklin and Spies, 1991a,b; Buskirk, 1992; Kaufmann *et al.*, 1992; Moir, 1992). In some ecological contexts, the term old-growth could also refer to senescent stands of shorter-lived species, such as stands of intolerant species of trees (e.g., cherries, birches, and poplars), and even stands of perennial herbs such as dandelions or goldenrods. This is not, however, the usual meaning of old-growth forests, which are a terminal-successional, or climax-ecosystem type, with the general characteristics just described.

The U.S. Forest Service has adopted the following definition of old-growth forest (after Kaufmann *et al.*, 1992, p. 4): "Old-growth forests are ecosystems distinguished by old trees and related structural features. Old-growth encompasses the later stages of stand development that typically differ from earlier stages in several ways, including tree size; accumulations of large, dead, woody material; number of canopy layers; species composition; and ecosystem function."

Some characteristics of old-growth forests, including elements of biodiversity, can be accommodated by so-called "new-forestry" harvesting systems that are relatively "soft" in terms of the intensity of disturbance (for example, selection-cutting systems with snag retention), so that the integrity of the forest is always substantially intact (Gillis, 1990; Franklin and Spies, 1991a,b; Hopwood, 1991; Swanson and Franklin, 1992). There are limits, however, to what can be achieved through new-forestry practices. The special values of old-growth forests are best accommodated by preserving large, landscape-scale, protected areas in which the ecological dynamics that permit the development of old-growth forest are allowed to occur over the longer-term (Harris, 1984; Gillis, 1990; Franklin and Spies, 1991a,b; Hopwood, 1991). This landscape perspective is important because, in general, particular stands of old-growth forest cannot be preserved over the longer-term, because of the inevitable effects of stand-level disturbances and/or environmental changes. Therefore, according to this model, old-growth forests can only be preserved through the designation of large ecological reserves that preserve the necessary, ecological dynamics.

Old-growth forests are an extremely valuable natural resource, because of their considerable standing crops of large-dimension timber of desirable tree species. Old-growth stands are rarely, however, managed by foresters as a renewable, natural ecosystem. Rather, old-growth forests are invariably "mined" by harvesting, which along with subsequent silvicultural management converts the ecosystem to a forest of younger, second-growth character, which is only allowed to successionally develop into middle-aged forest before it is again harvested.

The rationale for this management strategy is that old-growth forests have little or no, positive net production, because at the stand level, the production by living trees is approximately balanced by the deaths of other individuals by disease, senescence, or accident. If the primary management objective is to optimize the productivity of tree biomass in the stand, then it is much better to harvest the middle-aged stand soon after its net productivity suffers a large decrease, i.e., before it becomes an old-growth stand.

Because of the particular, structural characteristics of old-growth habitats, some species of wildlife have an obligate requirement for extensive tracts of old-growth forest to provide all or a major portion of their range. Some well-known examples of species that are often considered to be substantially depen-

dent on old-growth forests in North America are: northern spotted owl (*Strix occidentalis caurina*) and marbled murrelet (*Brachyramphus marmoratus*) of the northwestern United States and southwestern Canada, red-cockaded woodpecker (*Picoides borealis*) of the southeastern United States, and American marten and fisher (*Martes americana* and *M. pennanti*). Some species of plants may also be more abundant in old growth than in younger, mature forests, for example, Pacific yew (*Taxus brevifolia*) and lungwort lichen (*Lobaria pulmonaria*) in Douglas fir old growth of western North America (Spies, 1991). Large trees with dead tops and large-dimension snags and logs on the forest floor are important habitat components for many species of old-growth forests, and these characteristics are absent or substantially deficient in intensively managed, second-growth forests (Spies and Cline, 1988; Spies *et al.*, 1988; Hansen *et al.*, 1991).

The northern spotted owl is a nonmigratory raptor with an apparently obligate requirement of large tracts of old-growth, mesic-to-wet, conifer forest, with each breeding pair requiring more than about 600 ha of forest older than 140–170 years and each viable population requiring a minimum of about 20 breeding pairs (Lee, 1985; Dawson *et al.*, 1987; Carey *et al.*, 1990, 1992; Murphy and Noon, 1992). Because old-growth forests of this character are valuable as an exploitable natural resource, their extent has been greatly reduced and fragmented by logging, and the abundance of spotted owls has been critically reduced in many areas (Harris, 1984; Gutierrez and Carey, 1985; Carey *et al.*, 1992; Lamberson *et al.*, 1992).

The northern spotted owl has been officially recognized as a "threatened" species in the United States. Because the habitat of this species is threatened by the logging of old-growth forests of Washington, Oregon, and California, specific management plans have been formulated for its longer-term protection (as is required under the U.S. Endangered Species Act). These plans have resulted in the withdrawal of large tracts of old-growth, Douglas-fir (*Pseudotsuga menziesii*)-dominated forests from the economically exploitable forest resource to ensure that sufficiently large tracts of suitable habitat will always be available to support a viable population of northern spotted owls. If this strategy is successfully implemented over the longer term, it will preserve ecological reserves of old-growth forest to provide sufficient habitat for spotted owls and thereby preserve a high-profile component of the ecological heritage of the United States (Gutierez and Carey, 1985; Murphy and Noon, 1992). Of course, at the same time important, largely shorter-term, economic detriments are suffered by the forest industry because of the withdrawal of high-value timber from the "working forest."

The red-cockaded woodpecker has an obligate requirement for old-growth pine forests in the southeastern United States, in which it excavates nest cavities in large, living, but heart-rotted trees. The red-cockaded woodpecker breeds colonially and has a relatively complex social system, involving clan helpers that aid in the rearing of broods (Hooper et al., 1980; Conner and Rudolph, 1989, 1991; Walters, 1991). Forests that satisfy the rather specific habitat requirements of the red-cockaded woodpecker have been greatly diminished in extent because of their conversion to agricultural lands, plantation forests, and residential developments. The resulting, diminished populations of red-cockaded woodpeckers have exacerbated the threats to survival of this species caused by natural disturbances, such as hurricanes and wildfires intense enough to kill mature stands of pine. As a result of the large decreases in its habitat and populations, the red-cockaded woodpecker is considered to be endangered.

Unlike the spotted owl, the red-cockaded woodpecker is tolerant of limited disturbances within its broader habitat. Provided that substantial buffers (i.e., 800 m) are provided around its nesting colonies and that sufficient foraging habitat is available, it may be possible to harvest trees in stands in which this species breeds (e.g., Walters et al., 1988; Kulhavy et al., 1990; Conner and Rudolph, 1991). Scientific information in support of such an integrated-management strategy is still deficient, however, and at the present time the ecologically

most prudent strategy toward preservation of the red-cockaded woodpecker is through the implementation of large ecological reserves of its natural habitat.

Further Effects of Plantations

When clear-cutting is combined with intensive silviculture, the net ecological effect is usually the conversion of natural, mature or old-growth, mixed-species forests, to an anthropogenic forest of a more simple character. Ecological conversions of this sort have longer-term implications for biodiversity and for site quality, including effects on hydrology and microclimate, degradation of site capability through nutrient losses and acidification, and changes in the communities of wildlife. Furthermore, if conversions are practiced over large areas, there are landscape-scale implications for these same variables (Hansen et al., 1991; Freedman et al., 1994).

Establishment of a plantation on land previously occupied by a natural forest will cause substantial changes in the plant community. The nature of the changes in vegetation is influenced by many factors, but the most important of these are the degrees of difference between the tree-species composition and physical structure of the plantation forest and those of the natural forest that is being replaced.

In temperate regions, changes in the plant community are: (1) greatest when a hardwood-dominated or mixed-hardwood–conifer forest is replaced by a conifer plantation and (2) somewhat less when a natural, mixed-conifer forest is replaced by a conifer plantation. Changes in vegetation are most substantial in the mature phases of succession, when the physical influences of the conifer canopy and the chemical influences of the conifer litter are most intense. The changes in plant community are smaller in the initial stages of succession, because the abundance of site resources allows the development of a lush and species-rich plant community, even in intensively managed plantations. The silvicultural use of herbicides for plantation management has an important influence on the nature of the vegetation

in plantations (see Chapter 8), although the influence of longer-term changes in tree-species composition and dominance are larger.

Some effects of plantations on animal wildlife can be illustrated by comparing mature, natural forests with intensively managed conifer plantations in New Brunswick. In that region, conifer plantations can quickly develop an abundant population of breeding birds (Table 9.18). It takes somewhat longer for species richness and diversity to recover, but a 15-year-old spruce plantation supported a larger population of birds than nearby, natural-forest stands, while species richness and diversity were similar. In another study in New Brunswick, there were no substantial differences among natural forests and plantations in avian abundance, species

richness, or diversity, once the plantations had developed for about a decade (Table 9.19). In addition, there were no substantial differences among 4- to 8-year-old, naturally regenerated cutovers and 2- to 8-year-old plantations in species composition of the early successional avifauna (Table 9.19). In both of these studies, many bird species of the mature, natural, conifer forest began to invade the developing plantation forests when they were about 10 years or older. Examples of these birds are yellow-bellied flycatcher, hermit thrush, olive-backed thrush, black-and-white warbler, Tennessee warbler, Nashville warbler, magnolia warbler, yellow-rumped warbler, and ovenbird.

Intensively managed plantations are, however, deficient in certain elements of habitat. In particu-

Selective logging in an area of old-growth, tropical rain forest in the province of West Kalimantan, Indonesian Borneo. **Top left:** A view of intact rain forest. The species composition of trees and other biota is highly diverse, and the stand structure is complex, with all size and age classes being represented and large numbers of snags and dead logs lying on the forest floor. **Center left:** A recently constructed logging road into this tropical forest. The forest has been selectively logged, with about 30 trees with a diameter at breast height (DBH) larger than 50 cm harvested per hectare. Regulations require that a minimum of 25 trees/ha with DBH >20 cm must be left undamaged by the logging. **Bottom left:** The observer is standing on a recently felled tree. Note the large gap that has been created overhead, which is substantially larger than the canopy space filled by the felled tree. This is due to secondary damage that has been caused to other trees, partly because much of the canopy is laced together with woody lianas. **Top right:** A nursery in which seedlings, mostly of valued Dipterocarpaceae dug from intact forest, are cultivated for 3–6 months until they are large enough to transplant into logged sites. This nursery has a capacity of about 200 thousand seedlings/year. One year after a stand is logged, foresters working for the company that holds the logging concession survey the regeneration of desirable species in the harvested stands. If the regeneration is inadequate, then seedlings are fill planted into the gaps. On average, about 30% of the cut is planted to supplement the natural regeneration of desirable species of trees. Survival of the seedlings is monitored, and they are replaced if necessary. The abundance of competing plants is managed by hand weeding. **Center right:** Logs harvested from the forest are shipped by river to a coastal mill that processes them into veneer for export and lumber for local use. No unprocessed logs are exported. About 15 species of tree comprise most of the selective harvest, and of this about 80% is of *Shorea* spp., especially *S. stenoptera* and *S. leprosula*. **Bottom right:** This forestry system is designed to sustain a harvest of desirable species of trees over the longer term, and it could probably accomplish this. However, the system fails, in large part for the following reasons: (1) after the managed, commercial logging, local people take advantage of the road access into the forest to engage in unregulated, secondary harvesting of smaller trees for their own use and to earn a cash income, and (2) once most of the valuable sawtimber has been harvested, poor, landless people move into the area, slash and burn any forest growing on fertile sites, and convert the land into a mixed-cropping agricultural system. Most of these are local, indigenous people, but some are recent migrants to this "frontier" area from other parts of Kalimantan, and even from other Indonesian islands where land is extremely scarce. These people engage in agriculture for subsistence purposes and to raise a small income of cash by exporting their limited, surplus production to nearby towns. Of course, this agricultural conversion destroys the remaining integrity of the forest, and makes sustainable forestry impossible (photos: B. Freedman).

Table 9.18   Breeding birds in natural conifer forest and in spruce and pine plantations in southern New Brunswick[a]

| Species | Age: | Plantations | | | | Forest | |
		3	6	7	15	60	60
Yellow-bellied flycatcher, *Empidonax flaviventris*		0.0	0.0	0.0	6.9	4.1	1.3
Alder flycatcher, *Empidonax alnorum*		0.0	5.3	4.3	13.4	0.0	0.0
Ruby-crowned kinglet, *Regulus calendula*		0.0	V	0.0	5.0	1.1	0.0
Olive-backed thrush, *Hylocichla ustulata*		0.0	V	0.0	V	4.1	1.3
Hermit thrush, *Hylocichla guttata*		V	0.0	0.0	2.0	2.2	2.6
Red-eyed vireo, *Vireo olivaceus*		0.0	0.0	0.0	0.0	V	6.0
Nashville warbler, *Vermivora ruficapilla*		V	0.0	0.0	3.5	0.8	0.0
Magnolia warbler, *Dendroica magnolia*		0.0	1.9	V	13.4	9.3	2.6
Black-throated blue warbler *Dendroica caerulescens*		0.0	0.0	0.0	0.0	3.4	1.7
Yellow-rumped warbler, *Dendroica coronata*		0.0	0.0	0.0	5.5	2.2	1.3
Black-throated green warbler *Dendroica virens*		0.0	0.0	0.0	V	2.2	5.5
Blackburnian warbler, *Dendroica fusca*		0.0	0.0	0.0	0.0	4.5	6.0
Palm warbler, *Dendroica palmarum*		0.0	V	V	9.5	0.0	0.0
American redstart, *Setophaga ruticilla*		0.0	V	+	1.5	V	4.3
Ovenbird, *Seiurus aurocapillus*		0.0	0.0	0.0	0.0	1.5	4.3
Mourning warbler, *Oporornis philadelphia*		+	1.5	2.8	0.0	0.8	0.0
Common yellowthroat, *Geothlypis trichas*		0.9	15.5	9.2	15.3	0.0	V
Song sparrow, *Melospiza melodia*		4.8	4.4	7.9	0.0	0.0	0.0
Lincoln's sparrow, *Melospiza lincolnii*		4.4	1.2	17.2	6.0	0.0	0.0
White-throated sparrow, *Zonotrichia albicollis*		+	12.6	5.3	8.4	1.5	0.4
Northern junco, *Junco hyemalis*		3.5	1.0	0.8	2.5	2.2	0.4
Total bird density		15.7	53.9	47.2	102	57.8	50.6
Species richness		16	20	22	38	42	32
Species diversity		1.6	1.9	1.7	2.6	3.0	2.9

[a]Only relatively abundant species are listed here. Data are in pairs/10 ha; stand age is in years; +, <0.5 territory/plot; V, visitor; richness is the number of breeding species; diversity is H'. Unpublished data of G. Johnson and B. Freedman.

Table 9.19   Breeding birds in natural conifer forest, 4- to 8-year-old naturally regenerated cutovers, and spruce and pine plantations in New Brunswick[a]

Clear-cut age (year): No. stands:	Mature forest (3)	Natural regen. (2)	Spruce plantation			Pine plantation	
			2–7 (2)	11–12 (2)	15–17 (3)	4–8 (4)	10–11 (2)
Total density	48.8	33.3	34.1	49.3	40.8	33.7	38.0
Species richness	26	15	7	24	22	19	24
Species diversity	2.7	2.3	1.4	2.6	2.7	2.1	2.7

[a]Data are in pairs/10 ha, average for the number of stands indicated; richness is the number of breeding species; diversity is H'. Modified from Parker *et al.* (1993).

lar, plantations typically have few snags or dead logs on the forest floor, and they may not support species of wildlife that require these as components of their habitat. For example, in natural forests dominated by Douglas fir in the northwestern United States, old-growth stands (i.e., aged 200–900 years) typically have a substantially larger biomass of large-dimension woody debris than younger stands, largely because of the gap-disturbance dynamics of old forests (Table 9.20). The larger quantity of woody debris in younger stands compared with mature stands is due to the substantial inputs of large-dimension, dead biomass after natural, stand-level disturbances, such as wildfire.

In contrast to natural-forest conditions such as these, silvicultural plantations support much smaller quantities of woody debris, because their sites are typically site prepared prior to planting, and the second-growth stands are harvested before the trees get old (Maser *et al.*, 1988; Spies and Cline, 1988; Spies *et al.*, 1988; Freedman *et al.*, 1994). For example, intensively managed conifer plantations in New Brunswick have virtually no snags (Table 9.21). Although the plantations initially have large quantities of large-dimension debris that is deposited to the ground as slash during the harvest, this dead-biomass habitat element is depleted by decomposition. There are few prospects for substantial inputs in the future, unless changes are made in the management system to specifically accommodate this habitat value.

Although it is likely that snag-dependent species

A 28-year-old plantation of white spruce trees (*Picea glauca*) in New Brunswick. The plantation is a conifer forest, but it has a relatively simple physical and biological structure. The original, natural forest was dominated by several species of conifer and angiosperm trees, and it was more complex in structure and species composition than the plantation (photo: B. Freedman).

Clear-cuts and plantations are deficient in the habitat requirements of some species of animal wildlife. For example, these habitats are lacking in nesting cavities in snags or dead logs lying on the ground. This photograph is from a study in New Brunswick that is using artificial cavities as a method to assess the deficiency of this habitat feature in conifer plantations of different age (photo: B. Freedman).

of birds, such as woodpeckers, might find adequate foraging habitat in older plantations, they would not be likely to breed there because of the lack of cavity trees. Because most woodpeckers have large territories, they may be able to utilize nesting opportunities off the plantations (for example, in uncut forest of riparian-buffer strips), while utilizing the plantation as a component of their habitat, perhaps for foraging. Alternatively, the snag deficiency of plantations might be mitigated by providing small,

Table 9.20 Quantities of large-dimension woody debris in natural forests dominated by Douglas fir (*Pseudotsuga menziesii*) in coastal Oregon and Washington[a]

	Old growth (*n* = 85 stands)	Mature (*n* = 51)	Young (*n* = 30)
Average stand age (years)	65	121	404
Biomass of dead logs	66	20	43
Biomass of snags	57	23	35
Total large woody debris	123	53	78
Density of dead logs	415	447	600
(with diameter >60 cm)	59	28	53
Density of snags	60	121	171
(with diameter >50 cm)	27	16	27

[a]Old-growth stands are aged 200–900 years, mature stands 80–190 years, and young stands <80 years. Biomass is in tonnes/ha; density is in logs/ha or snags/ha; data are averages. Modified from Spies *et al.* (1988).

Table 9.21   Snags and logs on the forest floor in natural forest and plantations in southern New Brunswick[a]

	Snags		Logs	
	Basal area (m²/ha)	Density (no./ha)	Surface area (m²/ha)	Density (no./ha)
Stands of mature, natural forest				
Mixed wood–A	3.5	90	77	108
Mixed wood–B	10.0	415	289	415
Hardwood	3.3	70	108	130
Conifer	17.6	635	341	460
Stands of plantation forest				
3-year Plantation–A	0.0	0	510	1933
3-year Plantation–B	0.0	0	272	256
6-year Plantation	0.0	0	119	724
7-year Plantation	0.0	0	198	867
15-year Plantation–A	0.0	0	51	122
15-year Plantation–B	0.0	0	2	12

[a]Data are for snags and logs with diameter greater than 10 cm. Unpublished data of T. Fleming and B. Freedman.

uncut forest islands to provide snags for these wildlife or less desirably, by provision of artificial cavities or nest boxes. Regrettably, little field research has investigated these important questions, and it is possible that woodpeckers and other snag- and log-dependent wildlife may be at substantial risk because of these habitat deficiencies of plantations.

If there is good winter cover in conjunction with an adequate supply of browse, snowshoe hare (*Lepus americanus*) can maintain large populations in conifer plantations in New Brunswick and elsewhere, sometimes becoming a silvicultural pest because of damages caused by feeding on the bark and shoots of small conifers (Sullivan and Sullivan, 1981; Bergeron and Tardif, 1982; Parker, 1984, 1986; Monthey, 1986). Red squirrel (*Tamiasciurus hudsonicus*) also find these conifer plantations to be acceptable habitat as soon as the trees mature and start to produce sizable cone crops (Sullivan and Sullivan, 1981; G. Johnson and B. Freedman, unpublished).

Parker (1989) compared small-mammal populations in natural, mixed-conifer forest and in conifer plantations in New Brunswick, and found a similar abundance and community composition (Table 9.22), indicating great resilience to these particular habitat changes within this group of wild animals.

In South Carolina, pine plantations are being established on many sites, and they provide a different habitat than the natural, mixed-angiosperm forests. Comparison of amphibian populations indicates that although total abundance was smaller in plantations, there is a substantial tolerance to these varying habitat conditions by the most common species (Table 9.23).

Sometimes, plantations are established on previously agricultural or industrial lands. Depending on the sort of habitat mosaic that results, the plantations may enhance the diversity of wildlife populations by providing opportunities for forest species. For example, since 1958, afforestation in the United Kingdom has proceeded at about 30 thousand ha/year, resulting in an extensive conversion of upland pastures into conifer forests. Some of the medium-term implications of this process for wildlife were investigated in a 140-ha study area in northern England, where beginning in 1972, peaty pastures and blanket heaths were progressively replaced by plantations of sitka spruce (*Picea sitchensis*) (Sykes *et al.*, 1989). Many species of plants

Table 9.22   Small mammals in natural conifer forest and in spruce and pine plantations in New Brunswick[a]

| Species | Mature forest | Spruce plantation | | | Pine plantation | |
No. of stands:	(2)	24 years (1)	7–12 years (3)	15–17 years (3)	4–6 years (3)	8–11 years (3)
Masked shrew, *Sorex cinereus*	6.3	7.0	3.5	5.3	5.5	3.3
Pygmy shrew, *Sorex hoyi*	1.1	1.2	0.4	0.1	1.0	0.4
Short-tailed shrew, *Blarina brevicauda*	0.6	0.9	0.5	0.1	0.9	0.4
Deer mouse, *Peromyscus maniculatus*	0.2	2.4	0.7	0.3	0.2	0.2
Red-backed vole *Clethrionomys gapperi*	2.3	2.9	1.8	1.7	1.9	2.0
Meadow vole, *Microtus pennsylvanicus*	0.6	0.0	3.0	0.0	0.7	0.1
Total abundance	11.7	15.6	11.8	4.6	29.0	7.4
Species richness	6.5	9.0	9.7	6.3	7.7	7.0
Species diversity (H')	1.4	1.6	1.7	1.4	1.3	1.3

[a]Data are average No./100 trap-nights; only abundant species are represented here. Modified from Parker (1989).

increased in abundance because of the habitat changes, but other species decreased because of the conversion of their habitat to forest. Eventually, once the tree-canopy closes, the plantations will not support many species of ground vegetation, because of the densely shaded conditions that develop in spruce stands under these conditions (e.g., Wallace *et al.*, 1992).

Sykes *et al.* also studied changes in animal populations. In 1972 there were nine species of mammals in their 140-ha study area, and by 1984 these were joined by another nine species in the developing plantations. There were some local reductions of breeding-bird species, most notably of meadow pipit (*Anthus pratensis*), which decreased from 200–300 pairs between 1972 and 1978, to 19–27 pairs over 1982–1984. Local extirpations included skylark (*Alauda arvensis*), from 27–38 pairs in 1972–1973 to 1 in 1979–1980 and none in 1981, and curlew (*Numenius arquata*), from 8 pairs in 1972 to none by 1978. During the same period, some birds that did not initially breed in the study area colonized the developing forests and increased in abundance. The species that colonized in largest numbers were willow warbler (*Phylloscopus trochillus*), 156 pairs in 1984; goldcrest (*Regulus regulus*), 103 pairs; chaffinch (*Fringilla coelebs*), 86 pairs; robin (*Erithacus rubecula*), 83 pairs; coal tit (*Parus ater*), 73 pairs; wren (*Troglodytes troglodytes*), 68 pairs; wood pigeon (*Columba polumbus*), 31 pairs; and song thrush (*Turdus philomelos*), 26 pairs.

The Richness of Community Types on the Landscape

Biodiversity at the level of landscape is related to the heterogeneity of the spatial distribution of a richness of distinct community types, including their patch dynamics over time. In this sense, a landscape that is uniformly covered with one or a few types of community has little biodiversity at this level, while a landscape with a complex and dynamic mosaic of distinctive communities has more landscape-level biodiversity.

As with species diversity within communities, landscape-level biodiversity can be indicated as: (1)

Table 9.23 Populations of the most abundant species of amphibian in natural mixed-hardwood forest dominated by white oak (*Quercus alba*) and mockernut hickory (*Carya tomentosa*), a 24-year-old plantation of slash pine (*Pinus elliottii*), and a 15-year-old plantation of loblolly pine (*Pinus taeda*)[a]

Species	Slash pine	Loblolly pine	Oak-Hickory
Red-spotted newt	655	977	1445
Notophthalmus viridescens			
Mole salamander	61	189	144
Ambystoma talpoideum			
Slimy salamander	16	10	122
Plethodon glutinosus			
Dwarf salamander	2	0	8
Eurycea quadridigitata			
Marbled salamander	2	0	2
Ambystoma opacum			
Total salamanders	736	1196	1709
Narrow-mouthed frog	1352	1357	1486
Gastrophryne carolinensis			
Southern toad	1126	697	1489
Bufo terrestris			
Southern leopard frog	16	24	40
Rana utriculata			
Eastern spadefoot toad	10	17	50
Scaphiopus holbrooki			
Barking treefrog	13	32	13
Hyla gratiosa			
Total frogs	2519	2138	3083
Number of individuals	3255	3334	4792
Species richness	12	13	14
Species diversity	1.79	1.98	2.05

[a]Amphibians abundance is indexed as numbers captured by standard sampling efforts, using a combination of drift fences and pitfall traps. Modified from Bennett *et al.* (1980).

the number of distinctive community types occurring within some management or political area (i.e., community richness); (2) on the basis of community richness and the relative abundance of community types (i.e., community diversity); and (3) according to the shape, size, edge:area, connectedness, and age–class adjacency of patches on the landscape (Romme and Knight, 1982; Noss, 1983; Risser *et al.*, 1984; Forman and Godron, 1986; Turner, 1989; Shafer, 1990).

These attributes are all influenced by the nature of the stand-level disturbance regime. Natural agents of catastrophic, stand-level disturbance in-

clude wildfire, windstorms, volcanic events, and insect defoliation, while anthropogenic agents include agriculture, defoliation by introduced insects and pathogens, and forestry. The study of landscape-scale patterns and dynamics of communities has emerged as an important sub-discipline within ecology, called landscape ecology (Noss, 1983; Risser *et al.*, 1984; Forman and Godron, 1986; Turner, 1989; Shafer, 1990).

Forest harvesting can be important in patterning the landscape, while subsequent forest-management activities influence the community-level characteristics of the patches that result. Sometimes, for-

Forestry can create mosaics of stand and habitat types on the landscape. **Top:** A patch-work of clear-cuts and unharvested forest in New Brunswick. **Middle:** A riparian buffer strip of uncut forest beside a stream, also in New Brunswick. **Bottom:** Vertical stratification in an area of temperate conifer rainforest on central Vancouver Island in coastal British Columbia. The tops of these mountains have not yet been harvested, and are still covered by old-growth forest. The valley floor was clear-cut about 15 years previously and was then site prepared by burning. Conifer seedlings were planted, and these in combination with abundant, natural seeding by conifers are rapidly

est harvesting is designed to mimic, to some degree, the natural patch-disturbance regime. For example, the natural-disturbance dynamics of many pine-dominated forests is controlled by wildfire, and to some degree this stressor can be mimicked during the preparation of forest-harvesting and management plans for pine forests.

Usually, however, forestry imposes an anthropogenic patch dynamic onto the landscape. This can be done, for example, by creating an unnatural, checkerboard mosaic of clear-cuts and plantations of various age, interspersed within a mosaic of residual, natural forest and nonforest communities. Such a landscape mosaic may sometimes be recommended by wildlife managers, because it can favor certain game species, such as ruffed grouse and white-tailed deer, as has been previously described.

There are important implications for elements of community-level biodiversity of the patch dynamics of forest harvesting and management. For example, if residual patches of unharvested natural forest are too small or isolated, they may not be capable of sustaining all native species and communities over the longer term. If these same patches are intended to function as reserves for the preservation of natural-ecological values, then there would be important implications for the ecological sustainability of the forestry system (Chapter 12 contains a discussion of the notion of ecologically sustainable systems).

Forest harvesting and management creates fragmented landscapes, composed of successionally dynamic patches of silvicultural- and natural-forest habitats. Many species of native wildlife will find the silvicultural habitats to be adequate for their purposes. However, such habitats and their dynamics will be deficient for some other elements of bio-

restoring another mixed-species forest. The stands at midslope were clear-cut about 3 years previous to the photo and were also managed by burning and planting. Note the tongues of burned forest on the right-hand side of the photograph. This damage was caused when ground fires used to burn residual slash spread into intact forest. Note also the landslides associated with the road at midslope. Erosion is often associated with logging roads built in mountainous terrain (photos: B. Freedman).

diversity, and if these are to be protected they have to be accommodated within natural-forest patches, which may have to be preserved as ecological reserves. The size, shape, spatial arrangement, and dynamics of these various patches, but particularly the ecological reserves, are all important considerations with respect to the preservation of all elements of biodiversity. For example, if the patches are too small, isolated, or young to accommodate all biodiversity objectives, then it will be necessary to design a landscape that is more ecologically appropriate. Generic design options that have been suggested for the fulfilment of biodiversity objectives of ecological reserves are discussed in more detail in Chapter 10.

It has recently been suggested that there have been substantial declines in population of several species of birds that winter in tropical habitats but migrate to temperate habitats to breed (Robbins *et al.*, 1989; Terborgh, 1989, 1992; Finch, 1991; see also Chapter 10). Species that breed in the interior of mature forests are considered to be especially at risk. The reasons for the declines of forest-interior birds are hypothesized to include: (1) a substantial net loss of mature-forest habitat in the northern breeding ranges as well as in the southern wintering ranges, along with (2) fragmentation of the breeding habitats into "islands" that are too small to sustain populations over the longer term and that are easily penetrated by forest-edge predators and parasites. If these landscape-ecology scenarios are correct, it is clear that further losses or fragmentation of natural, mature-forest habitats through forestry could have serious consequences for these declining species of birds.

A few field studies in the United States have provided corroborating evidence in support of the hypothesis that declines in population of neotropical forest migrants are partly related to landscape characteristics and dynamics in their breeding ranges. These studies have found that avian communities in small forest islands are relatively depauperate and that birds attempting to nest in small woodlots or near forest ecotones are at greater risk of predation and/or parasitism (Wilcove, 1985,

1988; Freemark and Merriam, 1986; Blake and Karr, 1987; Freemark, 1988; Yahner and Scott, 1988; Robinson, 1992).

It has also been suggested that many of the national parks in North America, which are among the largest ecological reserves in the world, are not sufficiently large to sustain: (1) viable populations of some species of native wildlife or (2) certain ecological dynamics that are required for the occurrence of certain communities, for example old-growth forest (see Chapter 10). Species that are especially at risk require large areas of suitable habitat to sustain unmanaged, viable populations. A few North American examples of species that will be difficult to sustain within the boundaries of smaller ecological reserves include grizzly bear (*Ursus arctos*), wolf (*Canis lupus*), spotted owl, and marbled murrelet. These and similar "flagship" species must be preserved in the context of large ecological reserves, and the implications of anthropogenic activities (such as forestry) in the peripheral areas of the reserves must also be considered. The ecological reserves and their surrounding areas must be managed as "greater reserves," i.e., as a single, integrated ecosystem, with a view to the longer-term viability of the populations of species- and communities-at-risk.

The notion of an integrated, ecological approach to the management of "greater" national parks and other ecological reserves, which incorporates concerns about activities occurring in the peripheral areas, is becoming increasingly influential in North America and elsewhere (e.g., Romme and Knight, 1982; Berger, 1991; Marston and Anderson, 1991; Woodley, 1991). This important concept is also being examined within the forestry community, along with the notions of ecological integrity and the design of ecologically sustainable systems (Chapter 12). It remains to be seen, however, whether these considerations will result in substantive changes in the ways that the forestry community accommodates some of the biodiversity-related considerations that require the establishment of large ecological reserves, for example, spotted owls and old-growth forests.

10

BIODIVERSITY AND EXTINCTIONS

10.1

INTRODUCTION

Biodiversity was defined in the previous chapter as the total richness of biological variation. Biodiversity considerations range from population-based genetic variation, through subspecies and species, to their communities, but also the patterns and dynamics of these on the landscape, on geographic scales ranging from local, to regional, state or provincial, national, continental, and global (OTA, 1987; Wilson, 1988a; Hunter, 1990; OECD, 1991a,b; CEQ, 1992b; Soule, 1991; Groombridge, 1992; Reid *et al.*, 1992; Freedman *et al.*, 1994). In this chapter, the topic of biodiversity is considered again, but with an emphasis on permanent losses of this value, occurring through the extinction of species and the loss of distinctive, ecological communities.

Extinction refers to the loss of some taxon of organism, over all of its range on Earth (extirpation refers to a more local disappearance, with the taxon still surviving elsewhere). The extinction of any taxon represents an irrevocable and regrettable loss of a portion of the biological richness of Earth, the only place in the universe known to support such life. Extinction can be a natural process, being

caused by random catastrophic events, by biological interactions such as competition, disease, and predation, and by chronic physical stresses. However, with the recent ascendance of humans as the dominant large animal on Earth and the perpetrator of global environmental changes, there has been a dramatic increase in the rate of extinction.

The recent spasm of anthropogenic extinctions has included well-known cases of species loss as the dodo, passenger pigeon, and great auk. There are many other high-profile species that humans have brought to the brink of extinction, including the plains buffalo, whooping crane, ivory-billed woodpecker, and various species of marine mammals, such as the right whale. Most of these instances were caused by an unregulated and insatiable over-exploitation of species that were incapable of sustaining such high rates of mortality, sometimes coupled with an intense disturbance or conversion of their habitat. In the future, however, anthropogenic habitat destructions will be increasingly dominant causes of extinction.

Beyond the tragic and well-known cases of extinction or endangerment of large species of vertebrate animals, Earth's biota is facing, and is already experiencing, an even more tragic loss of species richness. In large part, this ruin is being caused by the conversion of large areas of tropical ecosystems, particularly moist forests, to agricultural and other ecologically degraded habitats. A large fraction of the species richness of tropical biomes is composed of endemic taxa with limited distributions. Therefore, the conversion of natural tropical forest to habitats unsuitable for the continued presence of these specialized taxa inevitably causes the extinction of most of the locally endemic biota.

Remarkably, the species richness of tropical forests is so enormous, particularly in insects, that most of it has not yet been described taxonomically. Earth is therefore faced with the prospect of a mass extinction of perhaps millions of species before they have even been recognized by science.

All species have genuine intrinsic values, which do not have to be defended or rationalized. This is a central tenet of an ecological or biocentric world view. Therefore, from an ethical perspective, the individual or mass extinction of species are reprehensible consequences of the ways that humans are using their powers to exploit and manage ecosystems. This is a also foolish way for humans to administer their global dominance, because unique organisms are disappearing before their importance as components of ecosystems has been discovered, and before they have been examined for their potential utility in medicine or agriculture.

In this chapter, the problem of the anthropogenic impoverishment of Earth's biota is examined.

10.2
SPECIES RICHNESS OF THE BIOSPHERE

Globally, about 1.7 million organisms have been identified and designated with a binomial name. About 6% of the identified species live in boreal or polar latitudes, 59% in the temperate zones, and the remaining 35% in the tropics (Table 10.1).

However, knowledge of species richness is incomplete, especially in tropical latitudes. According to some estimates of the number of undescribed

Table 10.1   The estimated number of species in three major climatic zones[a]

Zone	Number of identified species ($\times 10^6$)	Estimated total number of species	
		Assume 5×10^6	Assume 10×10^6
Boreal	0.1	0.1	0.1
Temperate	1.0	1.2	1.3
Tropical	0.6	3.7	8.6
Total	1.7	5.0	10.0

[a]Modified from World Resources Institute (WRI) (1986).

tropical taxa, global species richness could range as large as 30–50 million species, and the fraction of global species richness that lives in the tropics increases to more than 90% (WRI, 1986; Wilson, 1988b; Erwin, 1991). Clearly, tropical and subtropical biotas are much more species rich than those of temperate or higher-latitude regions.

Invertebrates comprise the largest proportion of described species, with insects making up the bulk of that total, and beetles (Coleoptera) comprising the greater fraction of the insects (Table 10.2; see

Table 10.2   Estimated number of species in various classes of organisms[a]

	Number of identified species	Estimated total number of species
Nonvascular plants	150,000	200,000
Vascular plants	250,000	280,000
Invertebrates	1,300,000	4,400,000[b]
Fishes	21,000	23,000
Amphibians	3125	3500
Reptiles	5115	6000
Birds	8715	9000
Mammals	4170	4300
Total	1,742,000	4,926,000

[a]Modifed from World Resources Institute (WRI) (1986).
[b]This figure is conservative. Some recent estimates (see text) suggest that there may be more than 30 million species of insects in tropical forests alone.

also Wilson, 1988b). After a public lecture, the famous geneticist and evolutionist, J. B. S. Haldane, was asked by a theologian to succinctly tell what he could discern of God's purpose, from his knowledge and understanding of biology. Haldane reputedly said that God has "an inordinate fondness of beetles," reflecting the fact that in any random sampling of all known species, there is a strong likelihood that a beetle would be sampled (Hutchinson, 1959; this story may be apocryphal—see Gould, 1993). Furthermore, it is believed that there is a tremendous richness of undescribed insects in the tropics, possibly as many as another 30 million species, again mostly beetles (Erwin, 1982, 1983a, 1988, 1991).

Erwin's remarkable conclusions about species richness have emerged from experiments in which tropical-forest canopies were fogged with an insecticidal mist and the "rain" of dead arthropods collected using ground-level sampling trays. This innovative sampling procedure indicated that: (1) a large fraction of the insect species richness of moist tropical forests is undescribed; (2) most insect taxa are confined to a single type of forest or even particular plant species that are themselves restricted in distribution; and (3) most tropical-forest insect species have a limited dispersal ability (Erwin, 1982, 1983a,b, 1988).

For example, Erwin (1983a,b) fogged four types of tropical rain forests in Amazonian Brazil. Beetles comprised the largest fraction of the total number of previously known or newly discovered species, and 58–78% of these taxa were restricted to only one forest type and were probably narrowly endemic (Table 10.3). The tree *Luehea seemanii* had more than 1100 species of beetle in its canopy, of which 14.5% were specific to that species. The field work associated with these canopy fogging experiments is expensive and logistically difficult, and it takes years to sort the samples and name the many new species. As a result, few of these measurements have been made. The emerging conclusion from this descriptive work, however, is that there is a remarkable number of undescribed species of insects and other invertebrates in moist tropical forests.

Table 10.3   Distribution of species of adult Coleoptera among four forest types at Manaus, Brazil[a]

Forest type	Number of restricted species	Expressed as % of the total number of species in each forest type
Black-water forest	179	64%
White-water forest	129	58%
Mixed-water forest	325	71%
Terra firma forest	266	78%
Total	899	

[a]The samples were obtained by intensively sampling the "rain" of dead insects after fogging the forest canopy with insecticide. Out of the total of 1080 species that were observed among 24,350 individuals in the four forests, 83% were restricted to only one forest type. Modified from Erwin (1983a).

These estimates of enormous numbers of undescribed species of tropical insects have been challenged as being unrealistically large (e.g., Gaston, 1991). In some respects, however, the actual numbers of insect species are less important than the fact that so many of them are becoming extinct through loss of their tropical-forest habitats.

Although it is often downplayed in conservation programs that tend to focus on large, charismatic vertebrates or on particularly noteworthy plants, the species richness of insects and other invertebrates is ecologically important in its own right. According to Janzen (1987), these animals "are more than just decorations on the plants; rather they are the building blocks and glue for much of the habitat." These conclusions were based on a number of observations, including the following: (1) insects are the primary foods of most of the small, vertebrate carnivores of tropical forest; (2) insects are important predators of seeds and they thereby influence the plant-species composition of the forest; and (3) insects are important pollinators, often in obligate, species-specific relationships with particular plant species. These and other observations indicate that insects have a strong influence on the structure and functioning of tropical ecosystems.

Compared with arthropods, the species richness of other tropical-forest organisms are better known. For example, a plot of only 0.1 ha in a Ecuadorian moist-tropical forest contained 365 species of vascular plants (Gentry, 1986). Consider also some examples of the species richness of woody plants in tropical rainforests: (1) 98 woody species with diameter at breast height (DBH) larger than 20 cm, in 1.5 ha of moist forest in Sarawak, Malaysia (Richards, 1952); (2) 90 species >20 cm DBH in 0.8 ha of forest in Papua New Guinea (Paijmans, 1970); (3) 44–61 species >20 cm DBH (total of 112 species) among five 1-ha plots of forest on Barro Colorado Island, Panama (Thorington *et al.*, 1982); (4) 80 species of tree with DBH >15 cm in 0.5-ha plots of lowland forest in Sumatra, Indonesia (Whitten *et al.*, 1987); (5) 742 species with DBH >10 cm in 3 ha of moist forest in Sarawak, with 50% of the species being recorded as single individuals (Primack and Hall, 1992); (6) more than 300 species of woody plants on a 50-ha forest plot on Barro Colorado Island (Hubbell and Foster, 1983); and (7) 283 tree species in a 1-ha plot in Amazonian Peru, with 63% of species represented by only one individual and another 15% by only two (Gentry, 1988).

These tropical forests are much more species rich than temperate forests, which typically have fewer than 12–15 tree species. The Great Smokies of the United States harbor some of the richest temperate forests in the world, and they typically contain 30–35 species (Leigh, 1982).

Note, however, not all tropical forests are rich in species. In Sumatra, for example, mangrove forests only have about six species of tree, while some lowland stands dominated by ironwood (*Eusideroxylon zwageri*) are virtually monospecific, with as many as 96% of trees being this valuable species, present in all size and age categories (Whitten *et al.*, 1987).

A few systematic studies have been made of the richness of avian species in plots of moist tropical forest. Terborgh *et al.* (1990) found 245 resident and another 74 transient species in a 97-ha plot of Amazonian floodplain forest in Peru. Thiollay (1992) recorded 239 species of birds in a primary Amazonian rainforest in French Guiana. Whitten *et*

al. (1987) reported 151 species in a 15-ha plot of lowland forest in Sumatra. These species richnesses are substantially greater than what is found in typical, temperate forests in North America. For example, during a 15-year monitoring period, the number of bird species breeding in a 10-ha plot of hardwood forest at Hubbard Brook, New Hampshire, ranged from 17 to 28 (Holmes *et al.*, 1986).

There have been few systematic studies of all biota of particular tropical ecosystems. In one case, a savannah-like, dry-tropical forest in Costa Rica was studied for several years (Janzen, 1987). It was estimated that a particular 108-km^2 reserve had about 700 plant species, 400 vertebrate species, and a remarkable 13 thousand species of insects, including 3140 species of moths and butterflies.

10.3
RATIONALIZATION OF THE PRESERVATION OF SPECIES

Biodiversity is valuable and important for many reasons (Ehrlich and Ehrlich, 1981; Myers, 1983; Bolandrin *et al.*, 1985; OTA, 1987; Farnsworth, 1988; Plotkin, 1988; Raven, 1990; OECD, 1991a; CEQ, 1992a; Reid *et al.*, 1992), including the following:

1. *Intrinsic value.* Biodiversity has its own, intrinsic values, regardless of any direct or indirect values in terms of human welfare. Central questions concerning the ethics of impoverishment of biodiversity are: (a) whether humans have the "right" to significantly impoverish or exterminate unique and irretrievable elements of biodiversity, even if our species may be empowered to do so, and (b) whether the human existence is somehow impoverished by extinctions caused by our activities. These are deeply philosophical issues, and they are not scientifically resolvable. Enlightened humans do not, however, applaud the extinction of species or their distinctive communities.

2. *Direct utilitarian functions.* Humans are not isolated from the biosphere, and they have an ab-

solute requirement for the products of certain components of biodiversity. This need results in the use of species and communities in myriad ways, but especially as sources of sustenance, biomaterials, and energy. The exploitation of wild biodiversity for these purposes can be conducted in ways that allow renewal. However, potentially renewable, biodiversity resources are often managed as nonrenewable resources (i.e., they are "mined"), through excessive harvesting and/or inadequate fostering of the post-harvest regeneration. In such cases, the resource is typically degraded in quantity and quality, often beyond a threshold of economic extinction (see also Chapter 12). Sometimes, overexploited species are locally extirpated or even rendered extinct globally, in which case their unique biochemical, ecological, and other values are no longer available for actual or potential utilization by humans.

3. *Provision of ecological services.* Species and their communities provide essential ecological services of many types. Although these may not be valuated in the conventional, economic sense, they can nevertheless be important to human welfare. Examples of such ecological services include nutrient cycling, biological productivity, trophic functions, cleansing of water and air, control of erosion, provision of atmospheric oxygen and removal of carbon dioxide, and other functions related to maintenance of the stability and integrity of ecosystems. In only a few cases is there sufficient knowledge to evaluate the ecological "importance" of particular species or communities in roles such as these, but this does not diminish the importance of the services that are provided. According to Raven (1990, p. 770): "In the aggregate, biodiversity keeps the planet habitable and ecosystems functional."

There are many cases where research on previously unexploited species of plants and animals has revealed the existence of bioproducts of utility to humans as food, medicinals, or for other purposes. This topic is described comprehensively in recent books and monographs (NAS, 1975b, 1979b; Ehrlich and Ehrlich, 1981; Myers, 1983; Bolandrin *et al.*, 1985; Farnsworth, 1988; Plotkin, 1988), and is

not dealt with here in detail. It is worth noting, however, that between 1965 and 1990, about one-quarter of the prescription drugs dispensed in the United States contained active ingredients from angiosperm plants, and these contributed about $14 billion/year to the U.S. economy and $40 billion/year worldwide (Miller and Tangley, 1991).

To illustrate the importance of medicinal plants, only one example is described: the rosy periwinkle (*Catharantus roseus*), a herbaceous angiosperm that is native to the island of Madagascar (Myers, 1983; OECD, 1985; Miller and Tangley, 1991).

During the course of an extensive screening of a diverse array of wild plants for the occurrence of possible anti-cancer properties, an extract of the rosy periwinkle was found to counteract the reproduction of cancer cells. Subsequent research identified the active ingredients as several alkaloids that are present in foliage of the rosy periwinkle, probably to deter herbivores. These natural biochemicals are now used to prepare the important drugs known as vincristine and vinblastine, which counteract the reproduction of cancerous cells by interfering with mitotic division. Chemotherapy based on these chemicals has been especially successful in the treatment of childhood leukemia and a cancer of the lymph system known as Hodgkin's disease. Leukemia victims now have a 94% chance of remission and the Hodgkin's victims a 70% chance, compared with 5% and <1%, respectively, prior to discovery of the therapeutic properties of the periwinkle extracts. About 530 tonnes of plant material are required to produce 1 kg of vincristine, which in 1985 had a value of about $200 thousand/kg, and annual sales of $100 million. So far, the anti-cancer alkaloids of the rosy periwinkle cannot be economically synthesized by chemists, and all of the chemotherapeutic material must be extracted from plants grown for that purpose. Therefore, a species of plant whose existence was until recently only known to a few botanists, has proven to be of great benefit to humans. This once-obscure plant prevents many deaths from previously incurable diseases, and is the basis of a multi-million-dollar economy.

Undoubtedly, there is a tremendous, undiscovered wealth of other biological products that are

A powerful justification for the preservation of species richness is that there are a great many uses of plants and animals that have not yet been discovered by scientists. If species become extinct before their utility is discovered, then humans are irrevocably deprived of their potentially beneficial function. Illustrated here is the rosy periwinkle (*Catharantus roseus*), an angiosperm plant with a very restricted natural distribution on the island of Madagascar. An extensive screening of plants for potential anti-cancer properties identified the possible utility of this rare species, which subsequently proved to be highly efficacious for the treatment of several types of human cancers. This valuable plant is now cultivated in abundance for the production of its chemotherapeutic chemicals. However, because it is endangered in its natural habitat, the rosy periwinkle could easily have become extinct prior to the discovery of its utility to humans. Most of the natural habitat of Madagascar has been converted to agricultural uses or disturbed in other ways, and this degradation is continuing apace (photo courtesy of Steele, World Wildlife Fund).

of potential utility to humans. Many of these natural bioproducts occur in tropical species that have not yet been "discovered" by taxonomists.

10.4
EXTINCTION AS A NATURAL PROCESS

Almost all of the species that have ever lived on Earth are now extinct, having disappeared "naturally" for some reason or other. Most likely, they could not successfully cope with inorganic or biotic changes occurring in their environment (e.g., a change in climate or in the nature and intensity of predation, competition, or disease). Alternatively, many of the extinctions could have occurred synchronously, as mass events caused by unpredictable, catastrophic disturbances (Fisher *et al.*, 1969; Raup, 1984, 1986a,b; Vermeij, 1986).

From the geological record it is well established that species and larger phylogenetic assemblages have periodically appeared and disappeared over time (Clark and Stern, 1968; Dobzhansky *et al.*, 1977). For example, many phyla of invertebrates proliferated during an evolutionary explosion at the beginning of the Cambrian era about 570 million

Virtually all species and most populations of large terrestrial carnivores are endangered in the wild. Illustrated here is a jaguar (*Panthera onca*), whose formerly extensive range included the southwestern United States and most of the northern half of South America. Jaguars have suffered precipitous decreases in abundance, largely because of a loss of habitat, persecution as predators of livestock, and hunting for their valuable pelage. This individual was photographed in a reserve that was specifically created to preserve jaguar habitat in Belize (photo courtesy of World Wildlife Fund).

years ago, but most of these are now extinct. The 15–20 extinct metazoan phyla from that period of early invertebrate evolution, known from the Burgess Shale of British Columbia, represent novel and fantastic experiments in invertebrate form and function. These extinct creatures were only discovered because of extraordinary circumstances that allowed their fossil preservation, including soft-bodied species and structures (Conway Morris and Whittington, 1979; Gould, 1989). Similarly, entire divisions of plants have appeared, radiated, and disappeared, e.g., the seed ferns Pteridospermales, the cycad-like Cycadeoidea, and the woody Cordaites. Of the 12 classically recognized orders within the class Reptilia, only three are extant. Clearly, the fossil record is replete with evidence of extinctions, including several within the hominid lineage (Walker, 1984).

The rates of extinctions, and subsequent radiations, have not been uniform over geological time. The geological record indicates that long periods characterized by fairly uniform rates of loss of taxa have apparently been punctuated by about nine catastrophic episodes of mass extinction (Eldredge and Gould, 1972; Gould and Eldredge, 1980; Knoll, 1984; Rampino and Stothers, 1984; Stanley, 1984; Raup, 1988; Jablonski, 1991).

The most intense event of mass extinction appears to have occurred at the end of the Permian period some 245 million years ago, when a remarkable 54% of marine families, 84% of genera, and 96% of species are estimated to have become extinct (Erwin, 1990; Jablonski, 1991).

Another well-known example of mass extinction is the apparently synchronous extinctions of many vertebrate animals that occurred over a geologically

The giant panda (*Ailuropoda melanoleuca*) is perhaps the most popularly recognized example of a rare and endangered species. Its native range is restricted to a region of mountainous forest in the interior of China, but the extent of its habitat has decreased because of agricultural conversion, logging, and other disturbances, and the species has also been hunted. The government of China is now committed to the preservation of the giant panda through the establishment of protected reserves and, with international cooperation, it is engaging in an intensive program of research on the ecology of this animal, which will hopefully allow for effective management in the future (photo courtesy of MacKinnon, World Wildlife Fund).

short period of time about 65 million years ago, at the end of the Cretaceous period. The most notable extinctions were of the reptilian orders Dinosauria and Pterosauria, but many plants and invertebrates also became extinct at about the same time. In total, perhaps 76% of species and 47% of genera became extinct during the end-of-Cretaceous spasm (Raup, 1986a,b, 1988; Jablonski, 1991). A popular hypothesis with which to explain the cause of this catastrophic event of mass extinction involves a ca. 10-km-wide meteorite impacting the Earth, causing great quantities of fine dust to be spewed into the atmosphere, secondarily causing a climatic deterioration that most large animals could not tolerate (Alvarez *et al.*, 1980, 1984). This theory is controversial, however, and some scientists believe that the extinctions of the last of the dinosaurs were more gradual and/or not caused by a rogue meteorite (Paul, 1989; Briggs, 1991).

There is a relatively high rate of both natural and anthropogenic extinctions in small and isolated populations, such as those that exist on oceanic islands (Diamond, 1984; Richman *et al.*, 1988). In general, the smaller and more isolated the island, the higher is the rate of extinction (Fig. 10.1). This is a primary reason for the area-dependent impoverishment of the biota of oceanic islands (Fig. 10.2), and of habitat islands within large land masses.

There have been no observations of naturally occurring extinctions, but a few researchers have recorded local, often shorter-term extirpations of populations of wide-ranging species on oceanic islands or on continental "habitat islands" (Diamond, 1984). These events were usually caused by some catastrophic disturbance, such as a brief period of inclement weather. For example, Ehrlich *et al.* (1972; see also Ehrlich, 1983) had been studying an isolated, subalpine population of the butterfly *Glaucopsyche lygdamus* in Colorado for several years. Early in the summer of a particular year, a late snowstorm killed the flower primordia of the lupine *Lupinus amplus*, the only food plant of the larvae of this butterfly. This event caused the local extirpation of the population of the butterfly, although because of its great vagility it later recolonized the site.

Extirpations after the physical isolation of populations of species in relatively small, habitat islands have also been documented. Local extirpations of birds were documented after Barro Colorado Island was isolated from continuous forest by the creation of Lake Gatun in 1914 during construction of the Panama Canal (Terborgh and Winter, 1980; Karr, 1982). Initially, there were at least 218 resident bird species on Barro Colorado Island. By 1981, at least

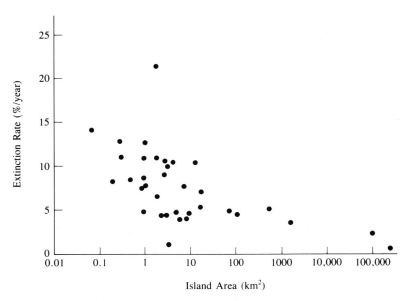

Fig. 10.1   The risk of extirpation of species of breeding landbirds of northern European islands, as a function of island area. The vertical scale is the average percentage of the bird species richness that becomes extirpated from one year to the next [from Diamond (1984)].

56 species or 26% of the resident avifauna were known to have become extirpated. Of the extirpated species, 26 are typical of young, disturbed habitats, which disappeared over time as the forest on the island developed successionally. In addition, eight aquatic birds disappeared because of a reduction in the abundance of their prey after the introduction of a piscivorous fish. The other 22 known extirpations were of bird species that occur in mature forest (the actual number may have been as large as 50–60 species, since the birds of mature forest were incompletely known in 1921). The reasons for the extirpations of the forest species are unclear, especially in view of the fact that the area of mature forest on Barro Colorado Island actually increased between 1921 and 1981, because of succession. The forest species presumably became extirpated because of such factors as: (1) changes in habitat quality; (2) inadequate areas of suitable habitat to support breeding populations over the longer term; and (3) sporadic extirpation events, possibly caused by unpredictable episodes of inclement weather;

coupled with (4) limited vagility, i.e., an inability to disperse across water to recolonize the island after a local extirpation.

10.5
ANTHROPOGENIC LOSSES OF SPECIES RICHNESS

Many species have been brought to, or beyond, the brink of extinction by the direct and indirect consequences of human activities. The most important anthropogenic influences that have caused the extinction or endangerment of species are: (1) excessive exploitation; (2) the effects of introduced predators, competitors, or diseases; and (3) habitat destruction and conversion. These stressors can result in small and fragmented populations, which are subject to the deleterious effects of inbreeding and demographic instability. The populations then decline further, ultimately to extirpation or extinction (Fig. 10.3).

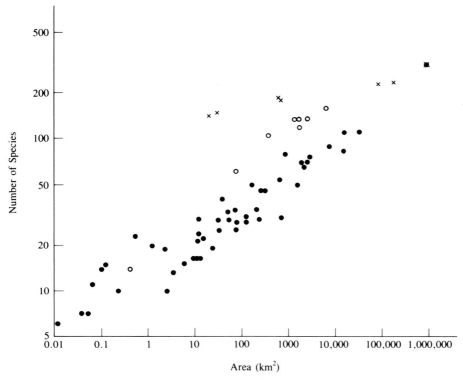

Fig. 10.2 Plot of bird species richness versus island area in the New Guinea region. The oceanic islands occur variously around New Guinea. The land-bridge islands were isolated from the New Guinea land mass by a rise in sea level at the end of the Pleistocene. The mainland census areas were habitat islands. x, New Guinea mainland; ○, land-bridge islands; ●, oceanic islands [from Diamond (1984)].

Because of these anthropogenic influences, the rate of extinction and the number of species threatened with extinction have increased greatly in the last few centuries. This phenomenon is best documented for vertebrates (as noted previously, most invertebrate species, particularly insects, have not yet been described). Over the most recent 4 centuries, there has been a global total of more than 700 known extinctions, including about 100 species of mammals and 160 birds, all because of anthropogenic influences (Fitter, 1968; Wood, 1972; Soule, 1983; Reid *et al.*, 1992).

In addition, a distressingly large number of plant and animal species are facing imminent extinction.

Consider, for example, the conservation status of the major orders of resident terrestrial vertebrates of Latin America, as summarized in Table 10.4. Note that the percentage of the total fauna that is threatened or endangered appears to be larger for mammals and birds than for reptiles and amphibians. However, this apparent trend does not represent the reality of the situation—it only reflects the relatively impoverished state of knowledge of the conservation status of the reptiles and amphibians of Latin America. In fact, because of incomplete information for all of the classes of animals in Table 10.4, the data for percentage threatened or endangered should be considered as minimal indications

The American ginseng (*Panax quinquifolium*) is a formerly abundant species of the understory of rich angiosperm forests of eastern North America. Similarly, the true ginseng (*P. ginseng*) is now very rare in its native forest habitat of eastern Asia. Both species have been badly overharvested throughout their natural range, because of a lucrative and virtually inexhaustible demand in China, Korea, Japan and elsewhere for the dried roots of this plant, which are believed to have many medicinal properties, including those of an aphrodisiac. Both ginseng species are now abundant in cultivation, but they remain very rare in the wild (photo courtesy of Charron, World Wildlife Fund).

of risk. The actual state of affairs is cause for greater pessimism, and it is rapidly deteriorating as increasingly more of the native habitat is disturbed or converted to agricultural lands.

The most comprehensive global catalogues of threatened organisms are the Red Data Books of the World Conservation Union, along with specialized compendia such as those prepared by the International Council for Bird Protection (e.g., King, 1981; Groombridge, 1982; Thornback and Jenkins, 1982; Wells *et al.*, 1983; Collar and Andrew, 1988). These catalogues are most accurate and complete for relatively large and conspicuous species, particularly those that live in relatively well researched temperate and higher-latitude ecosystems. In contrast, the information base is woefully incomplete for the multitude of relatively small and poorly known species that live in tropical ecosystems.

Many of these endangered tropical species, particularly insects, will disappear before they have been described by taxonomists. This will occur as their geographically restricted and unexplored tropical habitats are converted to agriculture or other uses of the land.

In the following, a few cases of extinctions of species caused by anthropogenic influences are examined. For more comprehensive treatments of cases of anthropogenic extinction and endangerment, see Ziswiler (1967), McClung (1969), Myers (1979), Ehrlich and Ehrlich (1981), or Wilson (1988b).

Unsustainable Exploitation

The most prominent and most easily appreciated cause of anthropogenic extinction, which largely

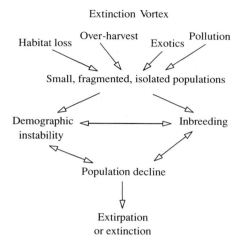

Extinction Vortex

Fig. 10.3 Extinction can be caused by various anthropogenic stressors, including habitat loss, overharvesting, exotic predators and diseases, and pollution. These can cause larger, metapopulations to fragment into small isolated populations, which are subject to deleterious effects of inbreeding and demographic instability. The populations then decline further, ultimately to extirpation or extinction. This syndrome has been labeled the "extinction vortex." Modified with permission from lecture notes of R. Lacy, Chicago Zoological Society, Brookfield, Illinois.

applies to relatively large and conspicuous species, is exploitation by hunting or harvesting at a grossly unsustainable rate. The insatiable and unsustainable "mining" of potentially renewable biological resources has caused the extinction of many species, as described below.

Pleistocene Extinctions

At about the end of the Pleistocene Epoch, there were great waves of apparently synchronous extinctions of large-animal species in various places. These mass-extinction events occurred at different times and places, but many of them apparently coincided with the colonization of land masses that were previously uninhabited by humans. The compelling implication is that overhunting caused or contributed to the extinctions of many species of large animals, because they could not adapt to a sudden onslaught of anthropogenic predation and habitat changes (Martin, 1967, 1984; Diamond, 1982).

In North America, the wave of extinctions began about 11 thousand years ago, roughly coincident with the postglacial colonization of the continent by people migrating across a Siberian land bridge.

Table 10.4   Status of resident terrestrial vertebrates in 1985 in selected countries in Latin America[a]

Region or country	Mammals Total number	%	Birds Total number	%	Reptiles Total number	%	Amphibians Total number	%	Total Total number	%
Latin America + Caribbean	1234	11	3812	11	2420	5	1833	1	9299	7
Mexico + Central America	559	8	1318	12	930	4	505	1	3312	7
Tropical Andean region	602	14	2518	10	823	6	818	0	4761	8
Guiana region	330	12	1354	11	286	10	229	0	2199	10
Tropical Atlantic region	419	19	1598	13	475	8	491	0	2983	11
Southern South America	284	17	1007	12	266	8	156	0	1713	11
Caribbean islands	141	13	639	16	434	12	138	1	1352	13
Mexico	439	7	958	13	702	5	273	1	2372	8
Brazil	394	20	1561	13	462	8	479	0	2896	11
Peru	359	17	1640	10	291	8	233	0	2523	10
Ecuador	280	18	1449	11	344	8	349	0	2422	10

[a]The designation of the total number of species as threatened or endangered (i.e., % in table) was based on their listing as such by one or more of the U.S. Department of Interior, the Committee on Trade of Endangered Species (CITES), or the International Union for Conservation of Nature and Natural Resources (IUCN). According to the IUCN, a species is endangered if it is in imminent danger of extinction; vulnerable implies that the species will soon be endangered if nothing is done to protect it. Modified from World Resources Institute (WRI) (1986).

The peregrine falcon (*Falco peregrinus*) became extirpated or endangered over much of its natural range in North America and Europe because of various toxic effects caused by chlorinated hydrocarbons, especially the insecticides DDT and dieldrin and the industrial chemicals, polychlorinated biphenyls (PCBs). Peregrine falcons are beginning to experience a recovery of abundance, because of a great decrease in the use of these chemicals, coupled with a vigorous program of captive breeding and release of young birds (see Chapter 8) (photo courtesy of Oliphant, World Wildlife Fund).

Within a relatively short time, extinctions of at least 27 genera comprising 56 species of large mammals (i.e., larger than 44 kg), two genera and 21 species of smaller mammals, and several large birds had occurred. Some of the notable losses include: 10 species of the horse *Equus*, the giant ground sloth *Gryptotherium listai*, four species within the camel family Camelidae, two species of *Bison*, a native species of cow in the genus *Bos*, the saiga antelope

Saiga tatarica, four species within the elephant family Proboscidea, including the relatively widespread mastodon *Mammut americanum* and mammoth *Mammuthus primigenius*, the sabertooth tiger *Smilodon fatalis*, the American lion *Panthera leo atrox*, and the 25-kg scavenger raptor *Terratornis merriami*. Many of these extinct, large animals were discovered as part of a fossil fauna in the Rancho la Brea tar pits in southern California. Today, the only extant representative of the large-animal fauna of those tar pits is the critically endangered California condor (*Gymnogyps californianus*) (Martin, 1967, 1984; Diamond, 1982).

Mass extinctions elsewhere also coincided with the first arrivals of aboriginal hunters. In Australia and New Guinea there were waves of extinctions about 50 thousand years ago that involved many species of large marsupials, large flightless birds, and tortoises (Martin, 1984; Diamond, 1982).

In New Zealand an extinction wave took place less than 1 thousand years ago (McCulloch, 1982; Diamond, 1984; Holdaway, 1989; McGlone, 1989). This swept away 30 large bird species, including the 3-m-tall, 250-kg giant moa *Dinornis maximus*, 26 other species of moa, and other large birds, including a flightless goose (*Cnemiornis calcitrans*), a swan (*Cygnus sumnerensis*), a flightless giant coot (*Fulica chathamensis*), a pelican (*Pelecanus novaezealandiae*), and an eagle (*Harpagornis moorei*), along with large-sized species of lizards and frogs, and fur seals (*Arctocephalus forsteri*) from North Island. The moas and their eggs were probably killed and eaten as quickly as they could be found by humans and their dogs; their extinctions progressed as a wave from North Island to South Island over about a 200-year period following the colonization of New Zealand by Polynesians. The prehistoric moa extinctions are evidenced by large bone deposits at butchering sites, which were later mined by European colonists and processed into a phosphate fertilizer. Today, moas are remembered in the following New Zealand verse:

"No moa, in old Aotearoa.
Can't get 'em, they've et 'em.
They-re gone and there ain't no moa."
[W. Chamberlain, cited in McCulloch (1982); Aotearoa is the Maori name for New Zealand].

Conservation issues are not usually straightforward; they can be greatly complicated by political, economic, and cultural considerations that are related to the historical patterns of use of natural resources. The bowhead whale (*Balaena mysticetus*) of the Western Arctic is one of the world's most endangered species of cetacean, numbering perhaps 5000 individuals, or 5–10% of its preexploitation abundance. In spite of its precarious status, the bowhead whale still sustains an aboriginal hunt that results in the death of about 20–46 individuals per year. Its habitat may also be undergoing degradation by chemical and acoustic pollution associated with exploration for hydrocarbons in the offshore of the North Slope of Alaska and in the Beaufort Sea. Although Alaskan Inuit have been engaged in bowhead hunting since prehistoric times and the practice is an important component of their cultural identity, there has been controversy over the longer-term sustainability of the continued exploitation of this rare species. This 1980 photo depicts a 9-m-long bowhead whale being hauled onto shorefast ice near Barrow, Alaska, prior to its butchering (photo courtesy of T. F. Albert).

The colonization of Madagascar by humans about 1.5 thousand years ago coincided with a wave of extinction on that island, involving 14 species of large and giant lemurs (another 10 species of lemur are still extant on Madagascar), about 6–12 species of giant, flightless elephant birds (Aepyornithidae), two giant tortoises, and various other large animals.

Other well-known events of prehistoric, mass extinctions occurred on Hawaii, New Caledonia, Fiji, the West Indies, and South America. These are also believed to have been caused by overhunting by newly colonizing, aboriginal humans (Martin, 1984; Diamond, 1982; Steadman, 1991).

The biotas of smaller islands of the Pacific Ocean, especially birds, were especially hard hit by prehistoric, Polynesian colonizations (Steadman, 1991). Each of the approximately 800 Polynesian islands may initially have had two to three endemic species of rails (family Rallidae), plus other unique species. Most of these species, probably totalling in the thousands, became extinct as a result of prehistoric overhunting, depredations by introduced rats, and habitat conversions. An excavation of bird bones from an archaeological site on the island of Ua Huka in the Marquesas of the central, equatorial Pacific, revealed that 14 of the 16 original species of

The natural range of the orangutan (*Pongo pygmaeus*) is forests of the islands of Sumatra and Borneo in Indonesia and Malaysia. The abundance of this species has decreased to a level of endangerment because of hunting, the capture of individuals for the pet trade and for export to zoos and circuses, and the conversion of its habitat to agricultural purposes. The photo illustrates an orangutan rehabilitation center on Sumatra, which was established to facilitate the return of individuals to the wild after they had been kept in captivity and had lost many of their essential survival skills (photo courtesy of the World Wildlife Fund).

birds are now extinct or extirpated from the island, including 10 endemic species. The surviving species of birds on the Polynesian islands are now threatened by modern stressors, including further habitat conversions, introduced diseases, competitors, and predators, and sometimes, military activities, including warfare.

The above examples clearly suggest that the unsustainable exploitation of potentially renewable, biological resources, often culminating in unnecessary extirpations of economically valued species, is not just a modern phenomenon. Prehistoric, aborig-

inal peoples could be as rapacious as modern peoples, given circumstances of suitable ecological opportunities of naive, but edible species.

More Recent Anthropogenic Extinctions on Islands

In general, plants and animals that live on islands are relatively vulnerable to anthropogenic extinction (Fisher *et al.*, 1969; Ehrlich and Ehrlich, 1981; Soule, 1983; Gentry, 1986; Richman *et al.*, 1988; Norton, 1992). In many cases, these species had not

The whooping crane (*Grus americana*) is the tallest and one of the rarest birds in North America. This species breeds in boreal muskeg habitat in Wood Buffalo National Park in northern Canada, migrates through the Great Plains, and winters in coastal salt-marsh habitat of the Gulf of Mexico near Aransas, Texas. Largely because of habitat loss and overhunting during its migration and in its wintering habitat, the whooping crane decreased in abundance to only 15 individuals in 1941. However, vigorous conservation efforts by agencies of the United States and Canadian governments have encouraged an increase in abundance to more than 160 individuals in 1988. This photograph taken in the Northwest Territories shows an immature bird flying between its parents (photo courtesy of the Canadian Wildlife Service/World Wildlife Fund).

been subject to intense predation during their recent evolutionary history and hence they were poorly adapted to coping with sudden onslaughts of mortality caused by human predation.

For example, the avifaunas of undiscovered oceanic islands often had many species that were flightless, relatively large, and fearless of novel predators. Furthermore, species on islands often did not cooccur with closely competing organisms, so that they were easily displaced by more capable, introduced species. In addition, the colonization of oceanic islands by humans, but particularly Europeans, usually resulted in great degradations of the natural habitats. These changes have been caused by the clearing of endemic vegetation for agricultural, ur-

ban, and tourism developments and because of habitat degradations caused by introduced plants, mammalian herbivores and predators, and diseases (this is discussed later in more detail).

Because of these influences, the biotas of many remote islands have been subject to especially intense rates of anthropogenic extinctions. This can best be illustrated by extinctions of birds. Of 167 bird taxa (including 95 full species) considered to have become extinct worldwide since 1600, only 9 were continental in their distribution (Fisher *et al.*, 1969; King, 1981). Four of the continental species disappeared from Asia: (1) pink-headed duck, *Rhodonessa caryophyllacea*, in 1944; (2) Himalayan mountain quail, *Ophrysia superciliosa*, in 1868; (3)

An aerial view of a pristine rainforest covering an expansive lowland near Tortugeuro, Costa Rica. Mature tropical forest of this type contains a tremendous richness of plant and animal species, most of which are of relatively local distribution. By far, the greatest fraction of the species richness of the biosphere occurs in tropical forests. Moreover, a very large amount of carbon is stored as organic carbon of plant biomass. If a mature forest of this or any other type is converted to an agricultural ecosystem with a much smaller standing crop of organic carbon, then the difference in carbon content is made up by a large emission of carbon dioxide to the atmosphere, with possible secondary climatic changes taking place through the so-called "greenhouse effect" (see Chapter 2, Section 2.5) (photo courtesy of E. Greene).

Jerdon's courser, *Cursorius bitorquatus*, in 1900; and (4) forest spotted owlet, *Athene blewitti*, about 1872. The other 5 species disappeared from North America: (5) Labrador duck, *Camptorhynchus labradorium*, in 1875; (6) Cooper's sandpiper, *Pisobia cooperi*, in 1833; (7) passenger pigeon, *Ectopistes migratorius*, in 1914; (8) Carolina parakeet, *Conuropsis carolinensis*, in 1914; and (9) Townsend's finch, *Spiza townsendii*, in 1833. The other 158 bird taxa that have become extinct since 1600 only lived on islands.

Similarly, of the 1029 species of birds considered to be "threatened" by the International Council for Bird Protection, 46% are island species, and 54% are continental (Collar and Andrew, 1988). Birds of tropical forest comprise 43% of the threatened avian species, while wetland species account for 21%, grassland and savannah species 19%, and other habitats 17%. Only 1.5% of the threatened species of bird are North American, while 4.2% are European and Russian, 33.1% Central and South American, 17.8% African, 29.6% Asian, and 13.8% Australasian and Pacific.

The syndrome of extinction-prone island biotas can be illustrated by the case of the Hawaiian Islands, an ancient archipelago of volcanic origin in the Pacific Ocean that is 1.6 thousand km from the nearest island group and 4.0 thousand km from the

A very large area of previously forested landscape in the tropics has been converted to various types of agricultural agroecosystems. Some notable effects are a great loss of species richness and a large flux of carbon dioxide to the atmosphere. This grazing ecosystem in Costa Rica was created after the clearing of a continuous expanse of dry tropical forest (photo courtesy of E. Greene).

nearest major land mass (Simon, 1987). At the time of the discovery of the Hawaiian Islands by Polynesian explorers, there were at least 68 endemic species of birds, out of a total richness of land birds of 86 species. Of the initial 68 endemics, 24 are now extinct and 29 are barely surviving (Ehrlich and Ehrlich, 1981; King, 1981; Vitousek, 1988). Especially hard hit has been an endemic family of birds, the Drepanididae or Hawaiian honeycreepers. Thirteen species of this endemic family are believed to be extinct, and 12 are endangered (King, 1981; Freed *et al.*, 1987). Since 1780, more than 50 alien species of birds have been introduced by humans to the Hawaiian Islands (Vitousek, 1988), but this gain can hardly be considered to compensate for the loss and endangerment of native, endemic species.

Similarly, the native flora of the Hawaiian archipelago is estimated to have comprised 1765–2000 species of angiosperm plants, of which a remark-

able 94–98% were endemic. (In addition, it is estimated that humans have introduced another 4.6 thousand species of vascular plants to the Hawaiian islands, of which about 700 can survive in the wild.) In the last several hundred years, more than 100 native plant species have become extinct, and at least an additional 500 species are threatened or endangered (Ayensu, 1978; Fay, 1978; Vitousek, 1988). The extreme rarity and endemism of the extant Hawaiian flora can be illustrated by the 6 surviving species of the endemic tree genus *Hibiscadelphus*, which have an aggregate total of only 14 known individuals in the wild, and one species that is represented by only a single wild individual (Gentry, 1986). The most important causes of extinction of Hawaiian biota have been habitat conversions to agricultural and urban landscapes and the introduction of alien biological agents, such as predators, competitors, destructive herbivores such as feral

An expanse of cleared, slashed, and burned tropical forest in Indonesia, in the process of conversion to an agricultural use. Virtually none of the species that comprised the previously tremendous richness of species of the primary forest will be able to survive in the new agricultural landscape (photo courtesy of Lloyd, World Wildlife Fund).

goats (*Capra hircus*), and diseases to which native species were not tolerant (Simon, 1987; Vitousek, 1988).

There are many other examples that describe how the unique and fragile biotas of remote islands have become impoverished because of the direct and indirect depredations of humans. Consider the island of Madagascar, which has been continuously isolated from the African continent for at least 20 million years and has developed a unique biota. Madagascar has a native flora of about 10 thousand species, of which 90% are endemic, including 95 endemic genera [World Wildlife Fund (WWF), 1984]. Within the fauna, perhaps the most interesting group is the lemurs (Lemuridae), an endemic family of primitive primates. It was previously noted that the 14 largest species of lemur disappeared during a wave of extinctions about 1500 years ago, after aboriginal colonists arrived on Mad-

agascar. However, the 15 extant species of lemur are also decreasing in abundance, and most are in danger of extinction, as are the members of two primitive families of primates (the Indriidae with eight species and the Daubentoniidae with only one species—the aye-aye, *Daubentonia madagascariensis*, one of the world's rarest mammals) (Fitter, 1968; Fisher *et al.*, 1969; Mittermeier, 1988). The extreme endangerment of the native biota of Madagascar is further evident from the fact that, in the early 1980s, only about 5% of that island's primary vegetation was still intact and essentially undisturbed (Myers, 1988).

Species Made Extinct or Endangered by Overharvesting

Hunting at an unsustainable rate has been the most prominent cause of the anthropogenic extinction or

In 1979, the World Wildlife Fund began an innovative research program called the Biological Dynamics of Forest Fragments Project, in which forest "islands" of various size were created during the operational conversion of continuous Amazonian rain forest to pastureland. The purpose of the research is to study the longer-term ecological dynamics and stability of isolated forest fragments, with a view toward determining the optimal size and shape that reserves of tropical forest should be to provide suitable habitat for various species. This airphoto shows newly created forest islands of 1 ha and 10 ha (photo courtesy of R. Bierregaard).

endangerment of large species that have value as a market commodity. If the rate of exploitation is greater than the rate of recruitment, then a precipitous decline in abundance of a species can occur in a remarkably short time. This cause-and-effect relationship can hold for even extremely abundant species. This phenomenon will be illustrated by reference to the dodo, passenger pigeon, great auk, and a few other notable cases.

The Dodo.   The first well-documented extinction of an animal caused by overhunting by humans occurred in 1681. This involved the dodo (*Raphus cucullatus*), a turkey-sized, flightless member of the pigeon order, Columbiformes (McClung, 1969; Campbell and Lack, 1985). The extinction of the dodo has been immortalized in our everyday language by the phrase "dead as a dodo," which con-

notes an irrevocable loss, and by the general usage of the term "dodo" to describe an old-fashioned or stupid person. This etymology recognizes the unfortunate dodo's apparent inability to adapt to or even recognize the threats posed by the human intruders on their remote-and-only island.

The dodo lived on Mauritius, a small island east of Madagascar in the Indian Ocean. Mauritius was discovered by the Portuguese in 1507 and was colonized by the Dutch in 1598. The colonists hunted the dodo as game, gathered its eggs, cleared its habitat for agriculture, started devastating ground-fires, and released feral cats, pigs, and monkeys that predated dodos and their eggs in ground-level nests. As a result, dodo populations rapidly declined to extinction. Not even a complete skeleton of the species exists. There had been a stuffed individual at Oxford University, but it was damaged by a fire in

A 1-year-old, 100-ha reserve of tropical forest created as part of the Biological Dynamics of Forest Fragments Project of the World Wildlife Fund. This fragment is connected to a continuous rainforest by a corridor, and it has a fairly heterogenous shape in order to increase its ratio of edge:area over that of a square or a circle (photo courtesy of R. Bierregaard).

1755, and only the head and foot were retrieved from the ashes. The only other member of the family Raphidae was the dodo-like Rodrigues solitaire (*Pezophaps solitaria*), but between 1715 and 1720 it too disappeared from its habitat on several islands near Mauritius, for reasons similar to those of the dodo.

Interestingly, it appears that the dodo may have coevolved with the endemic tree, tambalacoque (*Calvaria major*). The dodo benefitted in this mutualism from a food of abundant and nutritious seeds, while the plant benefited from passage through the crop and gut of dodos, which scarified its seeds and allowed them to germinate (Temple, 1977, 1979). The tambalacoque is now perilously rare in the wild, with only about 10 surviving mature individuals, each older than 300 years. No young individuals are known in the wild, even though the mature trees produce fertile fruit. It has been suggested that the need for dodoaceous scarification was obligate,

so that natural recruitment of the plant became impossible after the demise of the dodo. Temple (1977) found that seeds of tambalacoque germinated well after they had passed through the gizzard and gut of a turkey, and he speculated that he may have grown the first seedlings of this species since the disappearance of the dodo. Human intervention may save the day for this tree by the use of artificial scarification followed by the outplanting of seedlings (Owadally, 1979).

The Great Auk.   The first well-described anthropogenic extinction of an animal whose range was at least partly North American was that of a seabird, the great auk (*Pinguinus impennis*) (McClung, 1969; Nettleship and Evans, 1985). The great auk was commonly known to mariners as the original "pennegoin," although it is a member of the family Alcidae and not related to the superficially similar southern hemisphere penguin family, Spheniscidae.

The edge of a forest fragment, just after clearing of the adjoining forest to create pasture for cattle. At first, the ecotone is very sharp, but with time, a dense growth of secondary vegetation develops into a 10- to 25-m-wide windbreak that softens the transition between the forest and the pasture (photo courtesy of R. Bierregaard).

For many centuries, the great auk had been exploited by aboriginal Newfoundlanders and European fishermen as a source of fresh meat, eggs, and oil. However, when its feathers became a valuable commodity for the stuffing of mattresses in the mid-1700s, a systematic and relentless overexploitation began that quickly caused the extinction of the species.

A harvesting and processing operation was described in 1785, at one of the largest breeding colonies of the great auk, along with multitudes of other alcids, on Funk Island off eastern Newfoundland (from Nettleship and Evans, 1985, p. 68):

> It has been customary of late years, for several crews of men to live all summer on that island, for the sole purpose of killing birds for the sake of their feathers, the destruction of which they have made is incredible. If a stop is not soon put to that practice, the whole breed will be diminished to almost nothing, particularly the penguins.

The slaughter of great auks and other seabirds on Funk Island was so great during the late 1700s that much of the soil that presently occurs on the island has been formed from their composted carcasses (Olson *et al.*, 1979; Kirkham and Montevecchi, 1982). About 85 years after the extirpation of the great auk from Funk Island, the common puffin (*Fratercula arctica*) began to breed there. This seabird requires soil that is sufficiently deep for the excavation of a nesting burrow. On Funk Island, the auk-derived soil is apparently suitable for burrowing, and today puffins can sometimes be observed carrying bones of the extinct great auk out of their excavations.

The great auk became extinct on Funk Island in the early 1800s. The last known individuals of this species were killed on June 4, 1844, on the Island of Eldey Rock, by three Icelanders who were searching for specimens for a bird "collector" in Reykjavik. They killed the only two adult birds that they saw and smashed the only egg they found, because it had been cracked and therefore was not a good specimen. They are reputed to have said: "What do you mean the great auk's extinct? We just killed two of them!"

In memorial to this unfortunate species, the journal of the American Ornithological Union has been named "The Auk."

Originally, the great auk had a pan-Atlantic distribution, breeding on a few north-Atlantic islands off eastern Newfoundland, in the Gulf of St. Lawrence, around Iceland, and north of Scotland. Because the great auk bred on only a few islands, it was vulnerable to extirpation by uncontrolled hunting. Although the great auk was initially abundant on its breeding islands, it was easy to catch or club because it was flightless, and therefore it could be killed in large numbers.

The conversion of moist tropical forests to agriculture is a great risk to biodiversity, and one that is intensifying as time passes. These are views of tropical rain forests that have been converted to agriculture in West Sumatra, Indonesia. (**Top**) A system in which paddy rice (*Oryza sativa*, planted as young shoots and cultivated in a wetland system) is grown in small areas of lowland valley and on terraces, while cassava (*Manihot esculenta*) is continuously cropped on much of the deforested slopes. An attempt was made to reforest these hilltops with pine (*Pinus merkusii*), a few trees of which can be seen on the skyline. However, this initiative failed, largely because the local people did not see it to be in their interests, because land would be taken out of annual, agricultural production. Agronomists from Andalas University are attempting to develop perennial and mixed-cropping agricultural systems, which would supply a more-or-less continuous income to farmers, while being substantially less degrading of soil and site capability than the continuous cropping of cassava. However, these systems have not yet become popular. (**Middle**) Cultivation of paddy rice in the foreground and a degraded hillside in the background. The slopes are little used for agriculture in this area, because they have been ecologically degraded through the development of highly competitive grasslands dominated by *Imperata cylindrica*, one of the most important agricultural weeds in tropical Asia. (**Bottom**) A tea (*Thea sinensis*) plantation at higher altitude. Tea is a perennial crop, so once the shrubs are well established, erosion is not much of a problem, and soil capability does not degrade excessively. However, because very few native species of wildlife find this simple habitat to be acceptable, there are substantial ecological consequences of the conversion of natural forests to this type of perennial agroecosystem (photos: B. Freedman).

The Passenger Pigeon.   One of the few birds to have become extinct from a continental range is the passenger pigeon, *Ectopistes migratorius* (Schorger, 1955; Ziswiler, 1967; McClung, 1969; Blockstein and Tordoff, 1985). Prior to the beginning of its demise some 300 years ago, the passenger pigeon may have been the world's most abundant landbird, possibly numbering some 3–5 billion individuals and comprising one-fourth of the total population of birds of North America.

The passenger pigeon migrated in tremendous flocks that reputedly were sufficiently dense to obscure the sun and could take many hours to pass. In 1810, the American naturalist Alexander Wilson guesstimated 2 billion birds in a single flock that was 0.6 km wide and 144 km long.

The breeding range of the passenger pigeon was the northeastern United States and southeastern Canada, in a habitat of mature, mast-producing forests containing oak, beech, and chestnut (*Quercus* spp., *Fagus grandifolia*, and *Castanea dentata*). The passenger pigeon wintered in mast-rich forests of the southeastern United States. When they occurred in large numbers, roosting pigeons would break the limbs of trees under their weight, and trees would be killed by the excessive deposition of guano.

Early naturalists were most impressed with the extraordinary abundance and colonial habits of passenger pigeons. The Englishman Mark Catesby published many natural history observations of eastern America in the early 1700s, including the following paragraph about wintering passenger pigeons in the southeastern United States (cited in Feduccia, 1985, p. 61):

> Of these there came to winter in Virginia and Carolina, from the North, incredible numbers; insomuch that in some places where they roost, they often break down the limbs of oaks with their weight, and leave their dung some inches thick under the trees they roost on . . . they so effectively clear the woods of acorns and other mast, that the hogs that come after them . . . fare very poorly. In Virginia I have seen them fly in such continued trains three days successively, that there was not the least interval in losing sight of them, but that somewhere or other in the air they were to be seen continuing their flight south.

At about the same time, in 1709, John Lawson described an impressive passage of pigeons in the Carolinas (Feduccia, 1985, p. 63):

> I saw such prodigious flocks of these pigeons . . . in 1701–2 . . . that they had broke down the limbs of a great many trees all over these woods, whereupon they chanced to sit and roost . . . These pigeons, about sun-rise . . . would fly by us in such vast flocks, that they would be near a quarter of an hour, before they were all passed by; and as soon as that flock was gone, another would come; and so successively one after another, for the rest of the morning.

The seemingly overwhelming abundance of passenger pigeons, coupled with their tastiness and certain biological characteristics that made it easy to kill them in large numbers (i.e., their propensity to migrate, breed, and winter in large and dense groups), made this species an attractive target for commercial hunters who sold the carcasses in urban markets. The passenger pigeon was killed in incredible numbers by clubbing, shooting, netting, and incapacitation with smoke. Of the early 17th century hunt, Catesby wrote (Feduccia, 1985, p. 61):

> In their passage the people of New York and Philadelphia shoot many of them as they fly, from their balconies and tops of houses; and in New England there are such numbers, that with long poles they knock them down from their roosts at night in great numbers.

Somewhat later, in the early 1800s, there was a well organized market hunt of passenger pigeons to supply urban markets with cheap, if somewhat seasonal meat. According to A. Wilson in 1829 (Feduccia, 1985, p. 62): "Wagon loads of them are poured into the market . . . and pigeons became the order of the day at dinner, breakfast, and supper, until the very name became sickening."

The reported sizes of some of the annual kills is staggering to the imagination—for example, an estimated 1 billion birds were taken in 1869 in Michigan alone. In 1874, at a particular nesting colony in Michigan, professional netters killed 25 thousand birds per day for 28 days, for a total of 700 thousand pigeons.

Because of this intensity of exploitation, coupled with destruction of its breeding habitats, the passenger pigeon suffered a precipitous decline in abundance. The last observed nesting attempt was in 1894. Martha, the last known individual, died a lonely death in the Cincinnati Zoo in 1914. In only a few decades, the world's most abundant species of bird had been rendered extinct.

It has recently been suggested that in spite of the terrific slaughter of passenger pigeons, it is possible that the enormous death rate caused by overhunting was not the primary cause of its extinction (Blockstein and Tordoff, 1985). Instead, it is possible that the annual interference with reproduction at its colonial breeding sites caused a succession of nesting failures, so that the rate of recruitment was much too small to make up for the intense anthropogenic predation. In addition, the passenger pigeon was deprived of much of the original area of its breeding and wintering habitats by the conversion of mature, mast-producing forests to agricultural purposes.

The Eskimo Curlew.   The eskimo curlew (*Numenius borealis*) is a large North American sandpiper. As recently as 1 century ago, the eskimo curlew was abundant. This fact, coupled with its relatively large size, fine taste, tameness, and gregarious nature, encouraged its relentless exploitation by market hunters during its migrations through the prairies of Canada and the United States and on the pampas and coasts of South America, where the eskimo curlew wintered. Because of the uncontrolled hunting, this species became rare by the end of the nineteenth century. The last observation of a nest of the eskimo curlew was in 1866, and the last "collection" of a specimen was in Labrador in 1922 (McClung, 1969). It seems, however, that the eskimo curlew may not be extinct, although it certainly perches on the precipice of existence. There have been a number of reliable sightings of individuals and small flocks of this species, mostly in migratory habitat in Texas and elsewhere, but also in breeding habitats in the Canadian arctic (McClung, 1969; Blankenship and King, 1984; Gollop *et al.*, 1986).

Some Overharvested Marine Mammals.   Marine mammals have suffered the consequences of overharvesting in many parts of the world (see also Chap-

ter 12). As is described later in this chapter, some heavily exploited and badly depleted species have made encouraging recoveries. However, some species of marine mammals have become extinct because of overexploitation, including the following:

1. Steller's sea cow (*Hydrodamalis stelleri*) occurred in the Bering Sea and was discovered by explorers employed by the Russian Czar in 1741. This species was apparently extinct by 1768, after only 26 years of hunting (McClung, 1969; Fitter, 1968).

2. The Caribbean monk seal (*Monachus tropicalis*) was formerly abundant in the Caribbean Sea and Gulf of Mexico, where it was "discovered," and eaten, on Christopher Columbus' second voyage to the New World in 1494. This species was decimated by an eighteenth century seal "fishery" and then exterminated by subsequent hunting by fishermen and loss of habitat (Thornback and Jenkins, 1982).

Many other species of marine mammals have been endangered by human activities, especially hunting. Some of the more notable of these are described below (after Thornback and Jenkins, 1982):

1. The Guadalupe fur seal (*Arctocephalus townsendi*) was historically abundant on the west coast of Mexico, possibly numbering as many as 200 thousand individuals. This seal was hunted relentlessly for its valuable pelage, and in the 1920s it was believed to be extinct. However, a breeding colony was discovered in 1965 off Baja California, and the species may currently number 1.3–1.5 thousand individuals.

2. The Juan Fernandez fur seal (*Arctocephalus philippii*) was historically abundant off southern Chile. It was intensively harvested for its pelt; between 1797 and 1804 more than 3 million individuals were killed. This seal was believed to have been extinct until it was rediscovered in 1965. It presently numbers at least 2.4 thousand individuals.

3. The Mediterranean monk seal (*Monachus monachus*) ranges over most of the Mediterranean Sea and the Black and Adriatic Seas, while the Hawaiian monk seal (*M. schauinslandi*) lives off

the Hawaiian Islands. Neither species was ever abundant, and both suffered large decreases in population because of hunting, deterioration of their habitat, and pollution in the case of the Mediterranean monk seal. Both species currently number in the low thousands.

4. All four extant species of the order Sirenia are considered to be endangered or vulnerable (Steller's sea cow was also in this order). The dugong (*Dugong dugon*) has a paleotropical distribution, and it has suffered large population decreases over its entire range. The manatees (genus *Trichechus*, including *T. manatus* of the Caribbean Sea and coastal Florida) are also considered to be vulnerable to extinction, because of declines in their populations caused by hunting, habitat loss, and pollution.

Some of the great whales have also been greatly reduced from their preexploitation abundances. These include the following (after Northridge, 1984; Best, 1993):

1. The right whale (*Eubalaena glacialis*), which ranges over all temperate waters of the northern and southern hemispheres. The current, estimated world population is about 2 thousand individuals. The only estimate of the preexploitation abundance is for the north Pacific, where the right whale is thought to have numbered 10 thousand individuals, compared with 200–500 at present. There may be as many as 500 right whales in the northwest Atlantic Ocean.

2. The bowhead whale (*Balaena mysticetus*) of Arctic waters, currently numbering about 8 thousand individuals, or 8–16% of its estimated original abundance.

3. The blue whale (*Balaenoptera musculus*), of virtual worldwide distribution, and currently estimated to number 10–20 thousand individuals, compared with an estimated initial abundance of 250 thousand.

Extinctions of Competitors of Humans

Some cases of anthropogenic extinctions have involved species that were relentlessly persecuted be-

cause humans perceived them to be economically important competitors for some common resource, usually an agricultural product or big game. These can be considered to be special cases of extinction caused by overexploitation.

Many species of large predator have been persecuted because they were considered to be important competitors (in a few cases, they were also predators of humans). Examples of such animals that are now extirpated or endangered over much of their range include the wolf (*Canis lupus*) and other members of the genus *Canis*, brown and grizzly bears (*Ursus arctos*), and most of the large cats within the Felidae, such as the cougar (*Felis concolor*) in North America and the tiger (*Panthera tigris*) in Asia.

Often, pest-control measures for predators have involved the use of poisoned baits and carcasses, a practice that can cause large, incidental kills of nontarget scavengers. Indiscriminate poisoning (along with hunting, lead poisoning from ingested bullets and shot from hunter-killed carcasses, and excessive collecting of eggs and adults by ornithologists) has been an important cause of the decline and possibly imminent extinction of the California condor (King, 1981; Wiemeyer *et al.*, 1988).

In at least two cases, species that were viewed as important competitors of humans were made extinct largely through pest control efforts:

1. The original range of the Carolina parakeet (*Conuropsis carolinensis*) was the southeastern United States (Ziswiler, 1967; McClung, 1969). This brightly colored frugivore was fairly common, living mostly in mature bottomland and swamp forests, where it foraged and roosted communally. Although they were occasionally killed for their colorful feathers, Carolina parakeets were not a valuable natural commodity. This bird became extinct because it was considered to be an agricultural pest, as a result of the damage that flocks would cause while feeding in fruit orchards and grain fields. In the early 1700s, Mark Catesby wrote of this species in the Carolinas (cited in Feduccia, 1985, p. 64): "They feed on seeds and kernels of fruit; particularly those of . . . apples.

The orchards in autumn are visited by numerous flocks of them; where they make great destruction for the kernels only." For this reason, the Carolina parakeet incurred the enmity of humans, and it was relentlessly persecuted. Unfortunately, this parakeet was an easy mark for extirpation because of its conspicuousness, communal nesting and feeding, and the propensity of birds to aggregate around wounded colleagues, so that an entire flock could be wiped out by hunters. The last definite record of a flock of Carolina parakeets was in Florida in 1904, and the last-known individual of the species died in the Cincinnati Zoo in 1914, at the same place and in the same year that Martha, the last passenger pigeon expired.

2. The Tasmanian wolf or thylacine (*Thylacinus cynocephalus*) became extinct in Australia about 4–5 thousand years ago, probably because of its inability to compete with anthropogenically introduced dingoes (*Canis dingo*). However, dingoes were absent on the island of Tasmania, so the thylacine survived there and was the largest extant carnivore when that island was colonized by Europeans (Fisher *et al.*, 1969; Thornback and Jenkins, 1982). Unfortunately, the thylacine developed a taste for the sheep and chickens that were raised on Tasmanian ranches, and it was therefore perceived to be an important agricultural pest. A bounty was established for dead thylacines, and it was relentlessly persecuted by stockholders. The thylacine was extirpated over most of its range by the early 1900s. The last definite record of a wild animal was made in 1933, and this individual was captured and kept in a zoo, where it died in 1936.

Extinctions Caused by Introduced Species

Vulnerable biotas in many places have been decimated by introduced predators such as mongooses (Viverridae), domestic cats (*Felis catus*), and domestic dogs (*Canis familiaris*), by omnivores such as humans and pigs (*Sus scrofa*), and by herbivores such as goats (*Capra hircus*) and sheep (*Ovis aries*).

A recent, tragic event of mass extinction oc-

curred in Lake Victoria, Africa's largest and the world's second-largest lake (Baskin, 1992; Kaufman, 1992). Although Lake Victoria has been affected by eutrophication and other stressors associated with a regional population of 30 million people, the mass-extinction event appears to have been precipitated by the introduction of Nile perch (*Lates niloticus*). This fish can grow to a length of 2 m and a weight of 60 kg, and was introduced to establish a large-fish resource to supply an export market. The Nile perch was first introduced to Lake Victoria in 1954, but it was not until the early 1980s that its population exploded (Fig. 10.4a). This species now sustains a harvest of 2–3×10^5 tonnes/year. Revenues from this new fishery are much larger than from the previous, regional subsistence fishery. However, relatively few of the local populace are involved in the new fishery, and because of the substantial cash value of commodities with an export potential, such as Nile perch, locals cannot afford to purchase much of the harvest from its fishery.

Unfortunately, the initial, spectacular increase of productivity of Nile perch was substantially based on predation upon the extremely diverse assemblage of native fishes in Lake Victoria. This fish community comprised more than 400 species, mostly of the haplochromine group of the family Cichlidae, with 90% of the species being endemic to Lake Victoria. In only a few years during the explosion of abundance of Nile Perch, the community of native fishes in Lake Victoria collapsed in abundance and richness. Prior to 1978, the diverse native cichlids comprised about 80% of the fish yield of Lake Victoria and Nile perch only 2% (Fig. 10.4b). Between 1983 and 1986, however, Nile perch comprised 80% of the catch, and another introduced species (Nile tilapia, *Orechromis niloticus*) and a native minnow (omena, *Rastrineobola argentea*) accounted for the remaining 20%. By this time, the previously diverse cichlid species contributed <1% of the catch and were commercially extinct.

Because of the precipitous nature of the collapse of the native fish community of Lake Victoria, the risks to the endemic cichlids were not immediately recognized, and substantive rescue efforts were not

(a)

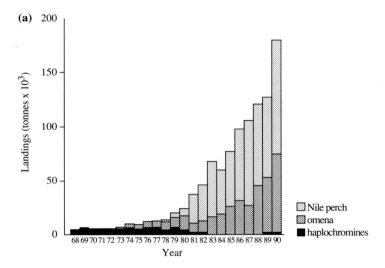

(b)

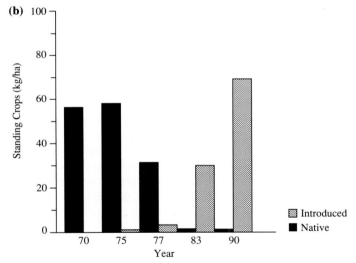

Fig. 10.4 Changes in the fish community of Lake Victoria. Tanzania, Uganda, and Kenya all have shorelines on Lake Victoria, but these data are for Kenyan waters only. (a) Landings of haplochromine cichlids, omena (a native minnow), and introduced species (mostly Nile perch). (b) Standing crops of introduced and native fishes [from Kaufman (1992)].

undertaken until most of the damage had been done. Consequently about one-half of the endemic cichlid species became extinct in Lake Victoria at this time, although other species survive in aquaria and a few in the lake.

As was noted previously, many species living on islands are particularly vulnerable to mortality caused by predators. This can be illustrated by the case of the accidental introduction of the boid snake *Boiga irregularis* in the late 1940s to the island of Guam in the Pacific Ocean (Pimm, 1987; Savidge, 1987; Fritts, 1988; Haig *et al.*, 1990). The biota and natural habitats of Guam have been affected by many stressors, including warfare, residential and agricultural development, agricultural pesticides, and introduced predators other than *Boiga*. It appears, however, that predation by the large populations of this snake may have been the key factor in the recent decline of Guam's avifauna.

Prior to the introduction of *Boiga*, the avifauna of Guam comprised 18 species, of which 11 were native, and most were abundant. However, by the mid-1980s, 7 of the native species were extinct on Guam and another 4 were critically endangered, including two endemics of the island. The Guam rail (*Rallus owstoni*), for example, is extinct in the wild, although it survives in captivity and will hopefully be captive bred and released to the nearby, and *Boiga*-less, island of Rota. The pattern of avian decline on Guam closely correlated with the pattern and timing of the range expansion of *Boiga*, and careful observations of the snake's natural history showed it to be an effective predator of birds and their nests. It appears that the severe decline of the extinction-prone avifauna of Guam has been caused by the effects of many anthropogenic stresses, but the introduction of *Boiga* was the most important, precipitating factor.

A botanical case of extinction caused by an introduced species involves the recent extinction of an endemic mistletoe, *Trilipidea adamsii*, in New Zealand (Norton, 1991). In this case, depredations by the brushtailed possum (*Trichosurus vulpecula*), a generalist herbivore introduced from Australia as a fur bearer, may have finished off this rare species, which was already threatened by habitat loss.

Habitat Destruction as a Cause of Extinctions

Many species have become extinct or are endangered because of conversion of their natural habitats to agricultural purposes or because of other anthropogenic disturbances that rendered the habitat unsuitable for their survival. This phenomenon is illustrated first with some notable examples of species that have suffered from this type of stressor. The broader implications for global biodiversity of

anthropogenic destructions of tropical forests are then considered.

American Ivory-Billed Woodpecker. The ivory-billed woodpecker (*Campephilus principalis principalis*) was probably never abundant in its North American range, which comprised most of the southeastern United States (McClung, 1969; King, 1981). The habitat of this large woodpecker was mature, bottomland angiosperm forests and cypress (*Taxodium distichum*) swamps, most of which was heavily logged and/or converted to agriculture by the early 1900s. There have been no published reports of this species since the early 1960s, and it may now be extinct in North America. The closely related Cuban ivory-billed woodpecker (*Campephilus principalis bairdii*) is also critically endangered, as is the congeneric Imperial woodpecker (*C. imperialis*) of Mexico.

Dusky and Cape Sable Seaside Sparrows. The dusky seaside sparrow (*Ammodramus maritimus nigrescens*) was a locally distributed subspecies of the seaside sparrow that lived in *Spartina bakerii*-dominated salt marshes on the east coast of Florida (King, 1981). In early systematic treatments, this bird was considered to be a distinct species, *Ammospiza nigrescens*, but recent treatments lumped it and other related taxa within a seaside sparrow "superspecies" [American Ornithologists' Union (AOU), 1983].

The dusky seaside sparrow became functionally extinct in the early 1980s, when several male birds still lived in captivity, but there were no known female individuals. The last known individual died in 1987. The causes of its extinction were habitat losses caused by drainage and construction activities and possibly toxic effects of the spraying of insecticides to control salt-marsh mosquitoes in its habitat, much of which was close to areas used for tourism or residential land.

The closely related Cape Sable seaside sparrow (*Ammodramus maritimus mirabilis*, formerly considered to be a separate species, *Ammospiza mirabilis*) of south Florida has also suffered from extensive habitat losses (King, 1981). These have been caused by invasions of introduced shrubs and trees into its habitats of salt-marsh and freshwater wet-prairies, by habitat deteriorations caused by saltwater intrusion and fire and by drainage of habitat for construction and other purposes. The Cape Sable subspecies as a whole is endangered, and several of its former populations are extirpated.

Black-Footed Ferret. The black-footed ferret (*Mustela nigripes*) was first "discovered" in the prairie of North America in 1851 (McClung, 1969; Thornback and Jenkins, 1982). This mustelid predator may never have been abundant during historical times. It has become endangered because of habitat changes caused by widespread conversions of the short-grass and mixed-grass prairies to agriculture and because of the great reductions in abundance of its principal food, the prairie dog (*Cynomys ludovicianus*). The prairie dog has declined because of losses of habitat and because it was relentlessly poisoned as a perceived competitor of cattle on rangelands.

Furbish's Lousewort. Furbish's lousewort (*Pedicularis furbishiae*) is endemic to a 230-km reach of the valley of the St. John River in Maine and New Brunswick (Lucas and Synge, 1978; Day, 1983). The last known observations of this rare plant had been made in the early 1940s, and it was considered to be extinct until a botanist "rediscovered" it in Maine in 1976. At that time, the only known habitat of the lousewort was threatened with obliteration by inundation from a proposed hydroelectric development on the St. John River. This dam was controversial for a number of reasons in addition to the lousewort (it would have damned one of the largest remaining free-flowing rivers in the northeastern United States), and its development has been put on long-term hold.

Tropical Deforestation and Global Species Richness. As was described previously, most of the global richness of species is composed of millions of as-yet undescribed taxa of tropical insects. Because of the extreme endemism of many of these arthropods and of other tropical biota, there is a likelihood of the extinction of innumerable species as a result of the clearing of tropical forests and their conversion to other types of habitat, especially agricultural.

The extent of global deforestation in the mid-1980s was about 20%, including a 30% loss of closed forest in temperate and boreal latitudes (WRI, 1986). These were considerably larger than the 15–20% loss of closed tropical plus subtropical forests up to that time. In the late-1980s, the total area of tropical forests was about 11.6×10^6 km², or 89% of their preagricultural area of 13.0×10^6 km²) (WRI, 1986; Newman, 1990). However, much of the intact tropical forest has been significantly disturbed by selective logging and other extractions, and only about 6.7×10^6 km² is considered to be primary, undisturbed forest, or 58% of the total remaining (Newman, 1990).

However, it is well known that rates of deforestation in the moist tropics are increasing alarmingly, in contrast to the situation at higher latitudes where the total forest cover is relatively stable. This situation is illustrated for recent decades in the Americas in Table 10.5. Between the mid-1960s and the mid-1980s there was little net change (-2%) in the total forest cover of North America. However, in Central America forest cover decreased by 17%, and in South America it decreased by 7% (but by larger percentages in equatorial countries of South America). The global rate of clearing of closed tropical rainforests in the mid-1980s was about $6-8 \times 10^6$ ha/year, equivalent to 6–8% of that biome per year (Grainger, 1983; Melillo *et al.*, 1985; Richards, 1986).

Of course, some of the cleared tropical forests would regenerate by a secondary succession, which would ultimately produce another mature, forested ecosystem. However, little is known about the rates and biological characters of successions in the tropical rainforest biome or how long it takes to restore fully diverse ecosystems after disturbance (Lugo, 1988). Nevertheless, if a 6–8% rate of clearance is accepted and simplistically projected in a linear fashion into the future, a biome half-life of only 9–12 years is calculated.

The most important causes of tropical deforesta-

Table 10.5 The change in forested area between 1964 and 1985 in selected countries in North and South America[a]

	Total land area (10^3 km²)	Forested area in 1964–1966 (10^3 km²)	Forested area in 1981–1983 (10^3 km²)	Percentage change (%)
North America	18,348	6056	5965	−2
Canada	9221	3135	3227	+3
United States	9127	2921	2738	−6
Central America	2617	831	691	−17
Mexico	1923	558	481	−14
Nicaragua	119	62	42	−31
Cuba	115	16	20	+25
Honduras	112	53	39	−26
Guatemala	108	51	38	−26
Panama	76	46	41	−11
Costa Rica	51	30	17	−43
Others	114	15	14	−7
South America	17,437	9880	9231	−7
Brazil	8457	6004	5666	−6
Argentina	2737	602	602	0
Peru	1280	742	704	−5
Bolivia	1084	596	564	−5
Colombia	1039	655	520	−21
Venezuela	882	397	344	−13
Chile	749	157	157	0
Paraguay	397	214	206	−4
Ecuador	277	172	144	−16
Guyana	197	181	164	−10
Uruguay	174	5	5	0
Suriname	161	155	155	0

[a]Modified from World Resources Institute (WRI) (1986).

tion are their conversion to subsistence agriculture, conversion to pasture and other types of market agriculture, and logging and fuelwood harvesting. Globally less important causes of tropical deforestation include the development of hydroelectric reservoirs, the manufacture of charcoal, and mining.

In Central America, for example, the number of cattle doubled to more than 9 million between 1950 and 1975, and two-thirds of farmland is composed of cattle ranches (Miller and Tangley, 1991). Much of this beef is raised for export. In the United States in 1982, about 24 million tonnes of beef were consumed, but less than 3% of this was imported, and only <0.5% from Latin American ranches established on converted forest. Still, this represented

25% of Latin American beef production, and 90% of their beef exports in that year. There is a clear connection between the conversion of tropical forests to pasture and a desire to earn profits by selling beef in foreign markets.

However, it is important to point out that most conversions of tropical forests to agriculture are carried out by poor farmers trying to establish subsistence farms and to supply local markets. This type of conversion is estimated to account for the clearing of about 10^5 km² of tropical forest per year (Miller and Tangley, 1991). The agricultural system is frequently based on shifting cultivation, in which the land is cultivated for 2–3 years until fertility declines and weeds become an overwhelming problem, followed by a fallow period of 15–30 years. Increasingly, however, a more permanent agricultural conversion is used, called slash and burn, in which the forest is cleared and used to grow crops on a longer-term, sometimes continuous basis. Conversions of tropical forests by subsistence farmers are greatly enhanced whenever access to the forest interior is improved. This occurs whenever highways are built and when logging roads are constructed for the extraction of timber. The primary driving force behind this relentless process of forest conversion is, of course, the increasingly larger populations of poor people that require access to land on which food can be grown for subsistence and some cash.

Fuelwood harvesting is also an important stressor of tropical forests. Wood is the predominant source of energy for cooking in tropical countries, with an estimated two-thirds of tropical peoples using this fuel, particularly poorer peoples (Miller and Tangley, 1991). Fuelwood contributes an estimated 26% of the total energy usage in developing countries, with a maximum of 52% in Africa (Miller and Tangley, 1991). For most of the world's humans, energy crises involve fuelwood, not fossil fuels.

It is readily apparent that modern rates of disturbance and conversion of tropical forests are having important consequences for global biodiversity, and this effect will be increasingly more important in the future. Because of widespread awareness and concern about this important ecological problem, there has recently been much research and other activity in the conservation and protection of tropical forests. As of 1985, several thousand sites, comprising more than 160 million ha, have received some sort of "protection" in low-latitude countries (Table 10.6).

In general, the effectiveness of the protection of ecological reserves varies greatly, depending on factors that influence the real commitments of governments to conservation. These factors include: (1) political stability, (2) political priorities, and (3) the finances available to mount an effective warden system to control poaching of animals and lumber and to prevent other disturbances.

China, for example, had only 45 ecological re-

Table 10.6 Protected areas in 1985 in selected low-latitude biomes[a]

Biome	Region	Number of protected areas	Total area protected (ha $\times$ 10^6)
Tropical humid forest	Afrotropical	44	8.91
	Indomalayan	122	5.09
	Australian	53	7.78
	Neotropical	61	17.28
	Total	280	39.06
Tropical dry forest and woodland	Afrotropical	240	48.67
	Indomalayan	238	10.42
	Australian	10	0.93
	Neotropical	93	5.50
	Total	581	65.52
Evergreen sclerophyllous forest	Nearctic	6	0.05
	Palaearctic	122	3.37
	Afrotropical	41	1.62
	Australian	301	6.92
	Neotropical	5	0.04
	Total	475	12.00
Subtropical and temperate rainforest and woodland	Nearctic	18	8.13
	Palaearctic	48	1.74
	Australian	26	0.90
	Antarctic	145	2.78
	Neotropical	38	8.85
	Total	275	22.40
Tropical grassland and savannah	Australian	18	2.04
	Neotropical	12	7.01
	Total	30	9.05
Mixed island systems	Palaearctic	9	0.05
	Afrotropical	4	0.02
	Indomalayan	177	10.43
	Oceanian	51	4.11
	Neotropical	26	1.19

[a]Modified from World Resources Institute (WRI) (1986).

serves in 1965, comprising 1.5 million ha, but by 1987 these had increased to 481 reserves and 23 million ha (Zhu, 1989). However, these reserves are not necessarily secure, as can be illustrated by the case of the Dinghushan Reserve of subtropical forest, established in 1956 and incorporated into the International Biosphere Reserve network in 1980 (Guo-hui et al., 1991). The forest at this site had been protected from earlier logging and agricultural conversion by Buddhist monks, who have long maintained a temple in the reserve. The site is also a famous scenic destination for tourists in south China, and it receives as many as 700 thousand visitors each year, with diverse, attendant effects on the integrity of the reserve. The reserve supports at least 1740 species of vascular plants, but 15 species have become locally extirpated in the last 35 years. All of these species are used in traditional medicine and were presumably poached to extirpation. Other Chinese reserves have problems with the poaching of timber, medicinal plants, and animals, including the giant panda (*Ailuropoda melanoleuca*), a distinguished icon of conservation. Therefore, although progress toward conservation and protection is being made, the prospects of native wildlife are precarious in China and in most other places.

The importance of poaching in hindering the protection of endangered wildlife can be further illustrated by the cases of elephant (*Loxodonta africana*) and black rhino (*Diceros bicornis*) in national parks in the Luangwa Valley of Zambia (Leader-Williams *et al.*, 1990). In the early 1970s, this area contained an estimated 100 thousand elephants and 4–12 thousand black rhinos. However, these populations have subsequently collapsed because of poaching, motivated by high prices for horns and tusks in foreign, wealthy markets. In spite of highly motivated conservation efforts under difficult political circumstances, it has so far proved impossible to effectively control the poaching, because the high prices of the commodities have spawned a well-organized chain of poaching, smuggling, and end-use, and hunters have not been punished with realistic deterrents.

Nevertheless, in spite of the problems, an encouraging level of conservation is beginning in some areas. There appears to be a real, emerging commitment to the protection of threatened biodiversity in the tropics.

For example, within Central America, Panama, Costa Rica, and the Dominican Republic are the nations with the most progressive policies with respect to the conservation and protection of natural ecosystems. As of 1989, 17.3% of the area of Panama had been given park or reserve status, as had 12.0% of Costa Rica and 11.4% of the Dominican Republic (WRI, 1990). This sort of vigorous conservation activity is badly needed in Costa Rica, since the current rate of deforestation of closed forests is about 7%/year (it is about 0.9%/year in Panama and 0.6%/year in the Dominican Republic; WRI, 1990). For perspective, note that the relative areas of protected land in these Central American countries compare favorably with those of Canada (3.7%) and the United States (8.6%), in spite of great differences in wealth and industrialization among these countries.

In other countries in Central America, conservation efforts have been badly disrupted by civil wars, other political instabilities, and government and social priorities (Hartshorn, 1983). For example, the percentage of territory that had received protected status in 1989 was only <0.1% in Barbados and Jamaica, 0.3% in Haiti, 0.4% in Nicaragua, 0.9% in Guatemala, 1.1% in El Salvador, 2.9% in Mexico, 3.1% in Trinidad and Tobago, and 5.2% in Honduras (WRI, 1990).

The Amazonian basin comprises the world's largest expanse of tropical rain forests. It has been broadly estimated that Amazonia could contain as much as 10% of the biotic diversity of the Earth (Lovejoy, 1985). At present, this rich tropical region still largely comprises mature rain forests that have been relatively little influenced by modern agriculture, lumbering, and other anthropogenic stressors. There is, however, an accelerating exploitation of the Amazonian forests, including: (1) agroindustrial conversions of great expanses of rain forest to establish cattle ranches and (2) expansive clearings of smaller patches of forest by poor farmers who have migrated from other, more heavily populated regions of Amazonian countries in search of "new" agricultural lands. Amazonian forests have also been degraded by hydroelectric developments and by extensive harvesting to manufacture charcoal for the production of iron.

In Amazonian Brazil, for example, the population increased 10-fold between 1975 and 1986, to more than 1 million persons, and in 1985 the rate of forest clearing was about 17 thousand km^2/year (Myers, 1988). It was officially estimated that up to 1979, only about 1% of the closed Amazonian forest had been "deeply altered" by modern anthropogenic disturbance (de Gusmao-Camara, 1983). However, the actual extent of disturbance is not well quantified. Lovejoy (1985) estimated that deforestation in Amazonian Brazil was as great as 10% in the early 1980s. In any event, while the government of Brazil is clearly committed to a vigorous program of economic "development" of its Amazonian frontier, there also appears to be an emerging commitment to conservation of its biodiversity. There are official plans to keep 1.8×10^6 km² of the Amazonian rain forest of Brazil in a natural state, and up to 1989 the government had established about 180 thousand km² of parks and reserves, mostly in Amazonia (de Gusmao-Camara, 1983; WRI, 1990).

An important aspect of any conservation strategy that may be used in the Amazon basin is the selection of reserves using criteria that depend strongly on the existence of centers of endemism and great species richness. It has been theorized that areas with a high frequency of endemism represent the likely locations of "forest-island" refugia during the Pleistocene and earlier glacial eras (Haffer, 1969, 1982; Prance, 1982; but see Endler, 1982, for a contrasting view). At those times, the regional climate may have been relatively dry, so that the tropical rainforests were contracted into relatively small nuclei surrounded by savannah. Because the forest-island refugia were periodically isolated during glacial epochs, this circumstance may have favored enhanced rates of speciation of tropical-forest biota. On this basis, the putative refugia might be reasonable locations for the siting of large ecological reserves. If the reserves are sufficiently large in area, they could preserve much of the species richness of Amazonia. Unfortunately, the "boundaries" of these forest refugia are not accurately known, and their determination (if they exist at all) could be a major stumbling block in the utility of such a system for the designation of ecological reserves (Endler, 1982).

As discussed in more detail later in this chapter, the size of ecological reserves is an important consideration with respect to their longer-term stability and capability to protect species richness. As might be expected from predictions of the theory of island biogeography (MacArthur and Wilson, 1967; Diamond, 1972; Diamond and May, 1976), newly created, but ecologically isolated reserves of tropical or other forests that are "islands" could lose some of their species richness (a process called "relaxation") if they are too small to accommodate the following (Terborgh, 1974; Soule, 1986):

1. Viable breeding populations of species that have large home ranges. Note that within this factor, consideration must be made not only of space, but also of time, since habitats change successionally.

2. The relatively large rates of extinction that occur on small islands. This is due to the deleterious effects of genetic drift and to the tendency of small, isolated populations to be extirpated by random or sporadic catastrophic events.

The relaxation of bird species richness on Barro Colorado Island after its isolation by construction of the Panama Canal was described earlier (Terborgh and Winter, 1980; Karr, 1982). In another longer-term study that began in 1979, a series of tropical-forest fragments of various size has been created experimentally to investigate the importance of size and amount of habitat edge to the stability of Amazonian forest reserves (Lovejoy et al., 1984, 1986; Bierregaard and Lovejoy, 1989; Bierregaard et al., 1992). Isolated forest remnants have been created from continuous rainforest in the vicinity of Manaus, Brazil, by controlling the operational conversion of forest to pasture for the grazing of cattle. When the experiment is fully established, there will be eight reserves of 1 ha, nine of 10 ha, five of 100 ha, two of 1000 ha, and one of 10,000 ha. Some of the shorter-term observations from this experiment are as follows:

1. After isolation, there are large microclimatic changes, caused by hot, dry winds that blow into the forest reserves from the surrounding clearings. These effects are especially pronounced close to the edges of the reserves, and they are important in causing increases in the rate of tree mortality after the stand is isolated. The tree mortality is greatest along the windward edges of

fragments and extends to about 60 m into the isolated stand. With time, a dense growth of regenerating, secondary vegetation produces a 10- to 25-m-wide windbreak, which effectively shades and protects the interior of the reserve.

2. Initially, isolation of the reserve causes a large but short-term increase in the abundance and species richness of forest birds, due to an influx of individuals that have been displaced from the surrounding, clear-cut forest. This effect is only measurable in 1-ha and 10-ha reserves, but it is followed within about 200 days by a relaxation of the bird species richness, and a decreased abundance of birds in these small stands. There is a "negative edge effect" on bird species richness, with only about 28 species present within 10 m of the edge of a reserve, 47 species at 50 m, and 50 species further into the stand. The negative edge effect translates into a smaller species richness for relatively small reserves, since these have relatively large ratios of edge:area compared with larger reserves.

3. Primates are strongly affected by the area of the reserve. Most species decline in reserves that are ≤10 ha in size. Only one species, the red howler (*Alouatta seniculus*), was able to persist for the first 4 years after isolation of a 10-ha reserve. Within a few years of isolation, a 1-ha reserve could only support three species of nonflying mammal, while a 10-ha stand had five species, and intact forest had 20 species.

4. There are large, positive edge effects for some insects, notably for large species of butterfly. This is due to the use of the vigorous, secondary vegetation by many light-loving butterfly species, some of which penetrate 200–300 m into the interior of the reserve. However, other insects decline in reserves that are ≤10 ha in area. These include Euglossine bees, an important group of pollinators, and army ants (*Echiton burchelli*) and their associated avifauna, which follow army ants to feed on displaced arthropods.

5. Forest islands that are surrounded by pasture become much more impoverished in species than isolates surrounded by regenerating, second-growth forest. In part, this happens because many

bird and mammal species of the forest are reluctant to cross forest–pasture gaps of 100 m and even less, a factor which impedes recolonization of small, isolated fragments.

Songbird Declines

It is well known that the populations of some groups of birds have declined greatly over large areas. In North America, for example, most species of waterfowl are much less populous today than they used to be. The obvious causes of the waterfowl declines have been excessive hunting, coupled with extensive losses of their wetland habitats (Bellrose, 1976; Terborgh, 1989).

It has also recently been suggested that substantial declines are occurring in the populations of neotropical, migratory birds, i.e., species that spend most of the year in tropical habitats, but migrate to temperate habitats to breed (Robbins *et al.*, 1989; Terborgh, 1989, 1992; Finch, 1991). Species that breed in the interior of stands of mature, temperate forests are considered to be especially at risk of declining. The reasons for the declines of neotropical migrants are hypothesized to include: (1) substantial net losses of mature-forest habitats in the northern breeding ranges; (2) changes occurring in their tropical, wintering habitats, especially deforestation; (3) fragmentation of the breeding habitats into "islands" that are too small to sustain populations over the longer term, and that are easily penetrated by forest-edge predators and nest parasites; (4) losses of key staging and migratory habitats; and (5) exposures to pesticides and other toxic chemicals.

A species of neotropical migrant that has probably become extinct because of the loss of its tropical habitat is Bachman's warbler (*Vermivora bachmanii*) (Hamel, 1986; Terborgh, 1989). This species bred in mature, bottomland, hardwood forests in the southeastern United States. Although it could still be found at a few sites in the mid-1950s, this warbler has not been seen for years and is probably extinct. There still appears to be adequate breeding habitat for Bachman's warbler, including forests where it used to breed successfully. Its apparent extinction has most likely been caused by the disappearance of its tropical-forest wintering habitats, which are believed to have been on the island of Cuba, and were lost through conversion to sugarcane fields. Terborgh (1989) has suggested that once the wintering populations of Bachman's warbler were decreased to below some critical threshold, potential breeding pairs were not able to find each other in their relatively large breeding range, and the residual population collapsed to extinction.

Much of the evidence for declining populations of neotropical migrants is anecdotal—birders feel that they are not seeing as many individuals of some species as they used to. More quantitative evidence is difficult to come by, because there have been few longer-term censuses of bird populations in mature-forest habitats. Terborgh (1989) compiled and analyzed the best of the existing information and presented a strong case to substantiate the putative declines of neotropical migrants. One of the longer-term data sets is that of Hall (1984) for a spruce–mixed-hardwood stand in appalachian West Virginia. These data show substantial decreases in the breeding populations of birds over a 37-year period, particularly of the neotropical migrants (Table 10.7). These are, however, longer-term data from only one site.

Table 10.7 Changes in bird-community parameters during a longer-term census of a red spruce–mixed-hardwood stand in West Virginia[a]

| | Number of species | | Density (pairs/100 ha) | |
Date	All species	Neotropical species	All species	Neotropical species
1947	25	14	325	238
1948	27	14	376	243
1953	28	16	369	238
1958	18	9	233	119
1964	13	8	288	175
1968	17	9	373	181
1973	15	8	273	105
1978	15	8	253	164
1983	15	8	273	150

[a]Modified from Hall (1984) and Tarborgh (1989).

Important, longer-term census data for forest birds in North America are for a 10-ha census plot at Hubbard Brook, New Hampshire (Holmes *et al.*, 1986). Between 1969 and 1984, 70% of the breeding species had an overall population decline at this site. However, successional dynamics may have influenced the results in this study because although mature, the forest at Hubbard Brook is in an aggrading stage. And again, these are data for only one site.

Data that integrate relatively larger geographical areas are also available. Robbins *et al.* (1989) analyzed data from across North America, gathered from 1966 to 1987 during the North American Breeding Bird Survey, which monitors roadside point counts of singing birds during the breeding season (Table 10.8). This study found that between 1966 and 1978, the populations of many neotropical migrants tended to increase, while between 1978 and 1987 their populations tended to decrease. During the same periods, there were no obvious trends in the abundance of resident, nonmigratory species. Interestingly, many of the neotropical migrants that increased during 1966–1978, breed in large numbers in forests infested with spruce budworm, and they could have been taking advantage of the irruption of that insect during that period in northeastern North America. The declining numbers of those same species during 1978–1987 could be partly related to the decreased intensity of the budworm outbreak during the middle to late 1980s (see Chapter 8).

Another useful data set comes from a bird observatory located at Long Point, on the north shore of Lake Erie. Long Point is a 32-km-long peninsula that juts into Lake Erie, and it effectively focuses migrating birds that fly over the lake during migrations. During the southerly autumn migration, the tip of Long Point is a good place to begin the translake crossing, because the traverse is relatively short, while during the northerly spring migration the tip is the first land that many birds see after their open-water crossing, and they tend to landfall there in large numbers. Consequently, the bird observatory at Long Point samples a large population of birds that breed in vast habitats to the north.

Table 10.8 Trends in the abundance of neotropical migrants in North America, as indexed by percentage changes in roadside point counts over 1966 to 1987 during the North American Breeding Bird Survey[a]

Species	Trend (% change/year) 1966–1978	1978–1987
Black-billed cuckoo *Coccyzus erythropthalmus*	13.4	−5.9
Yellow-bellied flycatcher *Empidonax flaviventris*	14.9	3.6
Tennessee warbler *Vermivora peregrina*	18.6	−11.6
Cape May warbler *Dendroica tigrina*	19.3	−2.3
Bay-breasted warbler *Dendroica castanea*	10.2	−15.8
Blackpoll warbler *Dendroica striata*	18.8	−6.3
Wilson's warbler *Wilsonia pusilla*	9.8	−6.5
No. neotropical migrants with increasing population	47	18
No. neotropical migrants with decreasing population	15	44
No. non-migrants with increasing population	6	8
No. non-migrants with decreasing population	7	5

[a]Only species with changes of >10%/year are shown here. Modified from Robbins *et al.* (1989).

Changes in the abundance of selected, neotropical migrants at Long Point are summarized in Table 10.9. Overall, between 1961 and 1988, 29 neotropical species declined in abundance, while 4 species increased, compared with 12 decreases and 11 increases for resident, nonmigratory species. Again, there appears to be a spruce budworm-related signal in the data, as indicated by large increases in abundance during the 1970s of Tennessee (*Vermivora peregrina*), Cape May (*Dendroica tigrina*), Blackburnian (*Dendroica fusca*), and bay-breasted warblers (*Dendroica fusca*), and some other species, followed by decreases during 1979–1988.

A few field studies in the United States have provided evidence in support of the hypothesis that declines in populations of neotropical-forest migrants are partly related to changes in their breeding landscapes. These studies have found that avian communities in small forest islands are relatively depauperate and that birds attempting to nest in small woodlots or near forest ecotones are at greater risk of predation and/or parasitism (Wilcove, 1985, 1988; Freemark and Merriam, 1986; Blake and Karr, 1987; Freemark, 1988; Yahner and Scott, 1988; Robinson, 1992).

The breeding habitats of many neotropical migrants have become much reduced by conversion to other land uses, and much of the remainder exists as relatively small and fragmented islands of natural habitat. It has been suggested that the breeding success of birds in small forest fragments is less, because of a greater vulnerability to predation of nests. Wilcove (1985) found that there was a greater likelihood that nests located in small woodlots would be predated, compared with larger woodlots (Fig. 10.5). This effect was ascribed to a greater abundance of nest predators near forest edges, particularly blue jay (*Cyanocitta cristata*), crow (*Corvus brachyrhynchos*), and common grackle (*Quiscalus quiscula*).

Another negative influence on the breeding success of many neotropical migrants is the brown-headed cowbird (*Molothrus ater*), a nest parasite that has greatly expanded its range and increased its abundance in North America (Fig. 10.6). Many species of birds in the northern and eastern parts of the range of the cowbird are especially vulnerable, because they have only recently come in contact with this debilitating parasite, and these species often suffer reproductive failure as a result (Brittingham and Temple, 1983; Terborgh, 1989). The brown-headed cowbird can cause large reductions of population of vulnerable species, as was mentioned in Chapter 9 for Kirtland's warbler (*Dendroica kirtlandii*), which can suffer a parasitism rate of 70% (Walkinshaw, 1983). In a study in Illinois, 10 nests of wood thrush (*Hylocichla mustelina*) had an average of only 1.2 host eggs, but 4.6 cowbird eggs and only a single fledgling thrush was pro-

Table 10.9   Trends in the abundance of neotropical migrants in northeastern North America, as indexed by percentage changes in counts made at Long Point, a 32-km-long peninsula that juts into the northern part of Lake Erie and focuses migrating birds[a]

| Species | Trend (% change/year) | | | |
	1961–1970	1970–1979	1979–1988	1961–1988
Black-billed cuckoo *Coccyzus erythropthalmus*	−9.1%	—	10.2%	−3.9%
Yellow-bellied flycatcher *Empidonax flaviventris*	−15.1	13.2	−6.4	−5.4
Least flycatcher *Empidonax minimus*	−6.7	8.4	−0.9	−2.3
Wood thrush *Hylocichla mustelina*	−9.4	−0.6	−15.0	−6.0
Tennessee warbler *Vermivora peregrina*	−13.3	16.5	−4.5	−2.3
Cape May warbler *Dendroica tigrina*	−10.3	13.3	4.6	0.8
Nashville warbler *Vermivora ruficapilla*	−5.0	2.0	−6.3	−2.9
Yellow warbler *Dendroica petechia*	−3.6	9.1	4.2	1.2
Magnolia warbler *Dendroica magnolia*	−8.9	9.8	−2.0	−1.4
Black-throated green warbler *Dendroica virens*	−1.3	4.2	−7.2	−1.5
Blackburnian warbler *Dendroica fusca*	−8.1	14.7	−0.8	−0.6
Bay-breasted warbler *Dendroica castanea*	−12.2	19.8	−5.1	0.8
Blackpoll warbler *Dendroica striata*	−1.7	4.8	6.0	0.7
Wilson's warbler *Wilsonia pusilla*	−5.9	9.1	−5.3	−3.2
Rose-breasted grosbeak *Pheucticus ludovicianus*	−2.0	−4.9	−4.6	−2.4

[a]Only species with relatively large changes are shown here. Modified from Hussell *et al.* (1992).

duced by this population (Robinson, 1992). Overall, two-thirds of 75 nests of all species in that study were parasitized by cowbirds, including 76% of 49 nests of neotropical migrants. Another study of cowbird parasitism involved Nuttall's white-crowned sparrow (*Zonotrichia leucophrys nuttali*) near San Francisco (Trail and Baptista, 1993). This study found a parasitism rate of 40–50% in 1990–1991, a substantial increase from 5% 15 years previously and considerably larger than the estimated 20% rate of parasitism that the population of sparrows could potentially sustain without decline.

From the above discussion, it is apparent that important declines in abundance may be occurring for a large number of species of migratory songbirds in North America. Unfortunately, it is not yet possible to make conclusive statements about the reality of the declines, because the population-level monitoring data are mostly shorter term, spatially restricted, or confounded by environmental influences that are unrelated to permanent population declines, such as the dynamics of succession and/or insect abundance. Some reasonable suggestions have been made about the causes of the putative

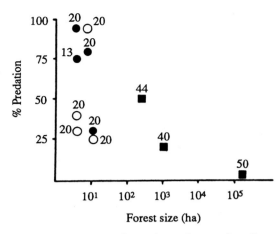

Fig. 10.5 Predation of experimental nests of quail eggs as a function of forest size. Square symbols denote large tracts; open circles are rural fragments; solid circles are suburban fragments; the numbers above points are the number of experimental nests [from Wilcove (1985)].

songbird declines, involving various aspects of habitat changes in the southern wintering grounds and in the northern breeding ranges. None of these suggested causes of the avian declines would be easy to manage because, for example, they involve exten-

sive habitat conversions by humans or complex relationships of predation and/or parasitism. Further monitoring of avian populations, and more research into potential management options, will probably be required before political decision makers can be convinced of the importance of this probable ecological change and the need to take actions to deal with it.

10.6

BACK FROM THE BRINK: A FEW CONSERVATION SUCCESS STORIES

Some species have been close to extinction because of anthropogenic stressors, but have recovered in abundance after protection was implemented. In some cases, the recoveries have been sufficiently vigorous that the species are no longer in imminent danger of extinction. A few examples of these successes of conservation are described below.

The Northern Fur Seal and Some Other Pinnipeds

The range of the northern fur seal (*Callorhinus ursinus*) is the northern Pacific Ocean (Walker *et al.*, 1968; Northridge, 1984; Anonymous, 1992e). This seal was relentlessly exploited for its pelage, to the degree that it was reduced in abundance from several million individuals, to about 130 thousand in 1920. At that time, it was believed that further exploitation would place the northern fur seal in danger of extinction. As a result, an international treaty that strictly regulated the harvest of these animals, by implementing large reductions in the total kill, by a ban of the pelagic catch, and by only allowing fur seals to be taken by land-based hunters was signed in 1911. The population of northern fur seals responded vigorously to these conservation measures, so that there are now about 875 thousand individuals. This presently abundant species is again being exploited for pelts, hides, and oil, and the harvest is closely regulated so that it will hope-

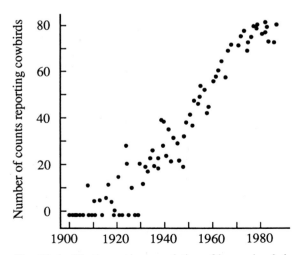

Fig. 10.6 The increasing population of brown-headed cowbirds (*Molothrus ater*) in North America, as indicated by the frequency of reports of this species in Christmas Bird Counts [from Brittingham and Temple (1983)].

fully be sustainable on the long term. However, the northern fur seal also suffers considerable mortality from entanglement with pelagic drift-fishing nets, which will affect the size of the commercial harvests.

The northern fur seal is the best documented example of a seal that was excessively exploited, but then rebounded in abundance after the cessation or strict regulation of hunting. Other notable examples of seals that had been badly depleted by unregulated exploitation up to the first decade or so of the present century and that then recovered after conservation measures were initiated, include (after Walker *et al.*, 1968; Northridge, 1984): (1) the southern fur seal (*Arctocephalus australis*) of sub-Antarctic waters off South America; (2) the harp seal (*Phoca groenlandica*) of the North Atlantic and Arctic; (3) the gray seal (*Halichoerus grypus*) of Atlantic temperate waters; and (4) the northern elephant seal (*Mirounga angustirostris*) of the Pacific coast of Mexico and southern California.

The Whaling Industry

The best-documented recovery of a whale species after the cessation of unregulated exploitation is that of the gray whale (*Eschrichtius robustus*) of the Pacific coast of North America (Northridge, 1984; Ellis, 1991; NMFS, 1991a). The population of gray whales is now about 21 thousand individuals, which is about the same as its estimated abundance prior to exploitation (<20 thousand), but substantially recovered from the 1880s to 1930s when there were only about 1–2 thousand individuals. However, another small population of gray whales in the western Pacific is still threatened, and an Atlantic gray whale was extirpated several centuries ago.

Other large baleen whales remain greatly depleted in abundance as a result of commercial exploitation. Some prominent examples include (after Northridge, 1984; Anonymous, 1992e; see also Chapter 12): (1) the blue whale (*Balaenoptera musculus*), with a prewhaling abundance of about 250 thousand, but currently numbering only 10–20 thousand; (2) the fin whale (*B. physalus*), initially 700 thousand, but now 163 thousand; (3) the hump-back whale (*Megaptera novaeangliae*), initially >100 thousand, but now 10 thousand; and (4) the sperm whale (*Physeter macrocephalus*), initially 2 million, but now fewer than 1 million.

Although the abundances of these species of large whales (and of other marine mammals) are notoriously difficult to measure, it appears that they may be starting to recover in response to several decades of complete or partial protection, which has resulted from economic and political pressures in favor of more "scientific" management and conservation of these animals. This rebound of abundance will hopefully intensify in the near future due to the almost complete cessation during the late 1980s of the legal hunting of large whales.

It is possible, however, that the attainable, equilibrium abundance of some of the recovered stocks of large baleen whales, particularly in the Antarctic Ocean, will be smaller than their population prior to commercial exploitation (Laws, 1977; Bengston and Laws, 1985). The reason for this is that there appear to have been substantial increases in the abundances of some other, much smaller marine vertebrates with which the baleen whales may compete for an important pelagic food, the crustacean macrozooplankter known as krill (*Euphausia superba*). The most notable krill-eating species whose abundance appears to have increased in response to decreased competition from large whales are the crabeater seal (*Lobodon carcinophagus*), which may now number about 15 million individuals, several other less numerous species of seal, and several species of seabirds, especially penguins (Spheniscidae) and possibly petrels and shearwaters (Procellariidae).

Unfortunately, some badly depleted species of large whales have shown little or no signs of recoveries of abundance, despite the almost complete elimination of hunting for at least 50 years (Northridge, 1984; Anonymous, 1992e). These include: (1) the Atlantic right whale (*Eubalena glacialis*), with an estimated population of only 200–400 individuals; (2) the Pacific right whale with only a few hundred individuals; and (3) the bowhead whale (*Balaena mysticetus*), now numbering about 5000 individuals. The bowhead whale is still exploited by a legal aboriginal hunt in northern coastal Alaska, while the most important cause of mortality to right whales appears to be collisions with ships. Both of these species presently number less than 5–10% of their abundance prior to commercial exploitation.

The American Bison

Prior to its intensive commercial exploitation, the American bison or buffalo (*Bison bison*) was North America's most populous large mammal, with an estimated abundance of 60 million individuals (after McClung, 1969; Grzimek, 1972; Wood, 1972). An appreciation of the remarkable abundance of the plains bison can be gained from descriptions of the sizes of some of its migratory herds, one of which was characterized as being 40 × 80 km in size, another as 320 km deep, and another as advancing over a 160-km-wide front! The density of animals in these migrating groups was variously described as being "as thick as they could graze," but it probably ranged from about 25 to 38 animals/ha.

At the time of settlement of North America by Europeans, the bison ranged over most of the continent. However, the bison was exploited relentlessly for its meat and hide, and it suffered a precipitous decline in abundance. The eastern subspecies (*B. b. pennsylvanicus*) originally ranged over much of the eastern United States (witness the name of Buffalo, the largest city in western New York State). However, because of overhunting and the conversion of its habitats to agricultural purposes, the eastern bison declined and by 1832 it had probably been extirpated east of the Mississippi River.

The plains bison (*B. b. bison*) was by far the most populous subspecies, ranging throughout the prairies of North America. Its near extermination by commercial hunting in the nineteenth century may actually have been strategically encouraged by the United States and Canadian governments of the time, because: (1) the plains bison disrupted the emerging, prairie-agricultural systems by its mass migrations, and (2) extermination of the bison was viewed as a means of disrupting the ecosystem of the Plains Indians in order to ease their displacement in favor of European agricultural colonists.

There are many descriptions of the wanton destruction of the plains bison. During construction of the transcontinental Union Pacific Railroad (beginning in 1869), bison were killed in large numbers to feed railroad workers. The most famous hunter was William F. Cody or "Buffalo Bill," who reportedly killed 4280 bison in 18 months and 250 in a single day. After the railroads were built, sporting excursions were organized in which a train would be stopped near a buffalo herd and passengers would leisurely shoot animals through the open windows of the coaches. Later, the tongues (a delicacy) would be cut from some of the dead animals, but in the main the carcasses were left where they had fallen. Much more importantly, the railroad allowed for well-organized commercial hunts of the plains bison, since it made possible the rapid shipment of carcasses to urban markets. It was estimated that between 1871 and 1875, market hunters killed about 2.5 million bison per year. Furthermore, by this time the Plains Indians had acquired rifles, and they also initiated more effective hunts of bison.

Obviously, this sort of unregulated exploitation of a wild population of large animals proved to be unsustainable, and the plains bison underwent a precipitous decline in abundance. In 1889 it was estimated that there were fewer than 1 thousand bison left in the United States, including a protected herd of 200 individuals in newly proclaimed Yellowstone National Park. The numbers of plains bison surviving in Canada were not determined, but they were believed to be small. Because of concerns over the imminent extinction of the plains bison, a number of closely guarded preserves were established, and captive-breeding programs were initiated. As a result, the numbers of plains bison have increased, to the point where they now number more than 50 thousand individuals. Because their original habitat is largely gone, the plains bison will never again achieve an abundance close to its preexploitation population. However, its future as a viable species appears to be secure, because (1) it breeds easily in captivity, and (2) it appears that viable breeding populations can be maintained in relatively small, but managed reserves.

The wood bison (*B. b. athabascae*) is a third subspecies of American bison. This animal was also hunted intensely, and it declined to a small abundance. When the demise of the wood bison seemed imminent, its only known wild population was preserved in and around a large reserve in Wood Buffalo National Park in northwestern Canada [Reynolds and Hawley, 1987; Federal Environmental Assessment and Review Office (FEARO), 1990]. However, the genetic integrity of this population has suffered because of intergradation with plains bison, introduced to the region in the late 1920s. Fortunately, in 1960 in a remote area of Wood Buffalo National Park, a previously unknown population of wood bison that had not yet suffered from interbreeding with the plains subspecies was discovered. Some of the pure wood bison have been used to establish another, isolated population northwest of Great Slave Lake.

Unfortunately, in recent years introduced diseases such as tuberculosis and brucellosis, along with predation by wolves (*Canis lupus*) and humans, have taken a toll on the wild-ranging bison populations of northwestern Canada, and there are concerns over the longer-term viability of their remote, protected populations.

There is also an incredible, but quite serious proposal by agricultural interests within the Government of Canada to exterminate virtually all of the bison in the vicinity of Wood Buffalo National Park, except for pure wood bison occurring in isolated populations that are known to be free of bovine diseases. There are two fundamental rationales for this slaughter: (1) to prevent the spread of several debilitating diseases, brucellosis and tuberculosis, from bison to the cattle herds that are spreading northward in the region, and (2) to protect the genetic integrity of the wood bison, since the hybrid wood-plains population would be exterminated, leaving the more northerly, isolated, nondiseased, genetically pure wood-bison population to repopulate the bison habitats (FEARO, 1990). It seems unlikely that such an extermination could ever be successful, because of the enormous efforts that would be required to find each and every bison in the target area, with its enormously expansive boreal forests and muskeg.

Nevertheless, this highly controversial proposal is being seriously considered and may proceed.

Overall, however, it appears that vigorous conservation efforts have preserved the American bison. The species will survive, but in relatively small populations and in captivity.

The Ginkgo

The ginkgo or maidenhair tree (*Ginkgo biloba*) is an unusual gymnosperm, with broad leaves, seasonally deciduous foliage, and a dioecious habit, in which male and female functions are performed by separate trees. The ginkgo is also a so-called "living fossil," because it is the only surviving species of the class Gingkoales, a group of gymnosperms with a fossil lineage extending to the lower Jurassic, some 190 million years ago, and probably once having a worldwide distribution (Tredici *et al.*, 1992).

The modern, natural distribution of the ginkgo was restricted to a small area of southern China. However, it is likely that no truly natural populations of ginkgo still exist. The only reason why this species, and class, is still extant is because it was preserved and/or cultivated in small stands around a few Buddhist temples, by prescient monks who recognized the ginkgo as a special and different species, and who valued its edible and putatively medicinally active fruits (Tredici *et al.*, 1992). Today, the ginkgo is no longer a rare species, and it again has a virtually worldwide distribution, because it is often cultivated in cities and elsewhere for its attractive foliage in autumn, its resistance to diseases and air pollution, and its special allure to botanists and others. However, the ginkgo no longer occurs in truly natural habitats.

Some Other Recoveries from Depleted Abundance

It is useful to briefly consider a few other examples, all taken from North America, of species that have exhibited substantial recoveries of abundance after the cessation of overexploitation (after McClung, 1969; Anonymous, 1992e).

1. The sea otter (*Enhydra lutris*) of the west coast of North America was the object of intense exploitation for its dense and lustrous fur, and as a result it was greatly reduced in abundance. In fact, the species had been considered extinct until the 1930s, when small populations were discovered along the west coast of California and the Aleutian Islands. The sea otter is now exhibiting a substantial rebound of abundance along the west coast of North America, and it is being reintroduced to some areas from which it had been extirpated. The total population of sea otters is now about 100 thousand individuals.

2. The pronghorn antelope (*Antilocapra americana*) of the western plains was badly overhunted by the late-nineteenth century, and had been diminished to an estimated 20 thousand individuals in its range north of Mexico. However, because of strong conservation efforts it now numbers more than 0.5 million individuals, and it sustains a sport hunt over most of its range.

3. The trumpeter swan (*Olor buccinator*) is the world's largest member of the Anatidae. It was greatly diminished by harvesting for its meat and skin. However, because of the cessation of a legal hunt and efforts at habitat conservation, it has recovered to an abundance of more than 5 thousand individuals.

4. The wild turkey (*Meleagris gallopavo*) was greatly diminished in abundance by hunting and habitat loss (a domestic variety is, of course, abundant in agriculture). However, in many places the wild turkey has recovered substantially, because of conservation measures and its introduction to areas from which it had been extirpated. Many populations of this large game bird now sustain a sport hunt.

5. The wood duck (*Aix sponsa*) was greatly diminished in abundance by hunting for its beautiful feathers and for meat, and by deteriorations of its habitat caused by lumbering and the drainage of wetlands. However, the wood duck has recovered substantially in abundance, in part because of widespread programs of provision of nest boxes for this cavity-nesting species. The same nest-box program benefits another relatively uncommon duck, the hooded merganser (*Lophodytes cucullatus*). An unrelated, terrestrial nest-box program has been important in recent increases of abundance of eastern and western bluebirds (*Siala sialis* and *S. mexicana*).

6. The fur trade that stimulated much of the exploration of northern and western North America was largely based on harvesting of the valuable pelage of the American beaver (*Castor canadensis*). Because of its overexploitation, this species declined to a small abundance, and it was extirpated from much of its range. However, because of conservation measures and decreased demands for its fur, the beaver has recovered to its original abundance over most of its range where the habitat remains suitable.

7. Finally, the whooping crane (*Grus americana*) may be a case of an incipient conservation success story. The whooping crane was probably uncommon prior to the time of the maximum anthropogenic effects on its abundance in the 1890s, with a likely original population of only about 1350 individuals (Allen, 1952; Johns, 1993). However, because of hunting for food and feathers, conversion of prairie breeding habitats to agriculture, egg and specimen collecting, and deteriorations of its wintering habitat, the whooping crane declined greatly in abundance, to as few as 15 individuals in 1941. Since then, strong preservation measures for the wild population and for their breeding and wintering habitats, coupled with a captive-breeding program, have increased the number of whooping cranes to more than 150 individuals in the mid-1980s and 250 birds in 1993 (145 of which were in captivity). As a result, there is guarded optimism for the survival of this species.

10.7
ECOLOGICAL RESERVES

In the generic sense, ecological reserves are protected areas that are established for the preservation of natural values, usually to protect the known habitat of threatened species, threatened ecological communities, or representative examples of widespread communities.

Globally, in the early 1990s there were about 6.9 thousand protected areas with a total area of 651 million ha (WRI, 1992). Of this total area, about 2.4 thousand sites comprising 379 million ha were fully protected and could be considered to be ecological reserves.

Ideally, systems of ecological reserves within any country should be designed with the aim of providing for the longer-term protection of all native species and natural communities, including terrestrial, freshwater, and marine systems. To ensure adequate representation within a system of ecological reserves, knowledge is required of all species and community types in the larger geographic or political area, and these types should be accommodated within a comprehensive system plan for a network of reserves.

These are, of course, ideal criteria. No countries have yet designed and implemented a comprehensive system of ecological reserves, in terms of adequacy of the representation and protection of natural communities and species. Moreover, in many cases: (1) the ecological reserves that have been established are relatively small and directly focused on rare communities or the local habitat of endangered species, (2) insufficient attention has been paid to challenges to the integrity of reserves associated with internal stressors and with activities occurring in the area peripheral to the protected area, and (3) there has not been enough ecological research and monitoring designed toward understanding the environmental factors that influence the valued species and communities. Consequently, without intensive management, it is doubtful whether many of the smaller ecological reserves will be able to sustain their present ecological values over the longer term, in the face of disturbance and inexorable environmental changes (Peters, 1991; Freedman, 1992).

In many countries, parks and other protected areas serve an important ecological-reserve function. Sometimes, however, the ecological-reserve function of parks can be partially compromised by land-use and management activities associated with other objectives, especially tourism. For example, it could be argued that the ecological-reserve function of parks is incompatible with certain consumptive uses of natural resources (e.g., sport fishing) or with the development of tourism-related infrastructure (e.g., campgrounds, hotels, golf courses, interpretation facilities, etc.).

In addition, the ecological-reserve function of protected areas can be significantly challenged by intensive harvesting and management of natural resources in their surrounding, peripheral area, for example, through activities associated with forestry, agriculture, or hunting. Sometimes, harvesting activities may even occur within the protected area, as in the case of regulated logging and fishing in some parks, and illegal poaching in others.

Some considerations relevant to these issues were previously discussed, in the context of old-growth forests (Chapter 9). It was, for example, suggested that particular stands of old-growth forest cannot be preserved over the longer term, because they inevitably become degraded by stand-level disturbances and/or environmental changes. Therefore, the special values of old-growth forest must be accommodated by preserving large, landscape-scale, ecological reserves. The large scale is required to accommodate, over the longer term, the dynamics that allow the development of old-growth ecosystems.

It has been suggested that many of the National Parks in the United States and Canada, which are among the largest ecological reserves in those countries and in the world, are not sufficiently large to sustain: (1) viable populations of some species of wildlife or (2) ecological dynamics that are required for the longer-term occurrence of certain communities, especially old-growth forests (Glenn, 1990; Berger, 1991; Woodley, 1991; Freedman *et al.*, 1994).

For example, although Yellowstone National Park is large at about 9 thousand km², it does not appear to be large enough to sustain, over the longer term and without intensive management, viable populations of some of its large-mammal species, such as grizzly bear and some ungulates (Berger, 1991; Mattson and Reid, 1991). Yellowstone may also not be large enough to accommodate some natural, catastrophic disturbances without undergoing profound changes in character, as is suggested by the fact that 45% of the park burned during the great fire of 1988 (Christensen *et al.*, 1989).

Similarly, Everglades National Park may not be sufficiently large to sustain its populations of panthers (*Felis concolor*) or possibly even its dominant community type, wet prairies dominated by sawgrass (*Cladium jamaicense*), because of the deleterious effects of water management outside of its boundaries. In the Maritime Provinces of eastern Canada, the largest ecological reserves are their five national parks, but these are all considered too small to sustain viable populations of some species of wildlife within their own boundaries (Woodley, 1991; Freedman *et al.*, 1994).

Considering the above, it is most appropriate to manage the species-at-risk in the context of the national parks plus their surrounding, peripheral areas as single, integrated ecosystems, with a view to the longer-term viability of populations of these and other species. A conceptually similar framework can also be developed toward sustaining certain types of natural communities-at-risk, for example, old-growth forests. The notion of an integrated, ecological approach to the management of ecological reserves and their peripheral areas is becoming increasingly influential in conservation science.

For example, the ecological values of Yellowstone National Park are increasingly being managed in the context of the park and its surrounding area. This region is called the Greater Yellowstone Area, and it comprises two national parks, six national forests, two national wildlife refuges, and state and private lands, collectively amounting to 48 thousand km^2, more than five times the area of Yellowstone National Park itself (Marston and Anderson, 1991). Similarly, preservation of the important, wetland values of Everglades National Park requires intimate comanagement of water flows in a large region north of the park boundaries, which supplies most of the water that flows southward through the national park itself.

Aspects of the spatial design of ecological reserves have generated a great deal of debate among ecologists in recent years, and a number of generic suggestions have been made, and criticized (Diamond, 1975; Shaffer, 1985; Forman and Godron, 1986; Noss, 1987; Simberloff, 1988; Simberloff and Cox, 1987; Simberloff *et al.*, 1992; Shafer, 1990; Saunders *et al.*, 1991; Hobbs, 1992). The least-controversial recommendations are that ecological reserves be as large as possible, and as numerous as possible. However, other aspects of the design of ecological reserves are debatable and have not yet been subject to much critical research. These include the following:

1. Is it better to have a single large reserve, or a number of smaller ones of the same aggregate area (Fig. 10.7a) ? This question has become identified with the acronym, *SLOSS—single large, or several small*? According to the theory of island biogeography, it is expected that populations in larger reserves would have a smaller risk of extinction. It also seems reasonable that eco-

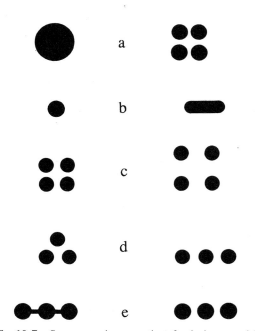

Fig. 10.7 Some generic suggestions for the improved design of ecological reserves. In each comparison, it has been suggested that the designs on the left are superior to those on the right [Adapted with permission from Simberloff (1988)].

logical dynamics associated with stand-level disturbances may be better accommodated in larger reserves. However, if there are many reserves, redundancy may protect against catastrophic losses of endangered populations or of the ecological values of particular sites.

2. Reserve size is also important in terms of the relative amount of "interior" habitat, i.e., habitat that is relatively unaffected by the influences of reserve-edge ecotones. For any particular two-dimensional shape, larger reserves will have proportionately more habitat that is uninfluenced by activities and land uses in the peripheral area, and therefore interior species and habitats would be better protected over the longer term.

3. Of all two-dimensional shapes, a circle has the smallest ratio of edge to area. If relatively extensive edges are a generally negative feature of ecological reserves, because they decrease the amount of mature-ecosystem interior and allow penetrance of invasive species and predators, would it be better to have circular reserves, in preference to any other shape (Fig. 10.7b) ?

4. If a population of some species becomes extirpated in a reserve, would the chances of its natural recolonization be better if there was another reserve with a surviving population nearby? If so, would it be better to have disjunct reserves arranged relatively close to each other rather than farther away (Fig. 10.7c) ?

5. Following from (4) above, the average distance among ecological reserves would be minimized if they are aggregated, rather than arranged in a linear fashion (Fig. 10.7d).

6. If reserves are connected by corridors, are there better opportunities for gene flow and for recolonization after local extirpation (Figure 10.7e)? Corridors may also, however, provide better opportunities for the spread of diseases and invasive weeds and predators.

Population ecology is also contributing substantially to conservation science, in part through determination of the minimal population sizes that are required to preserve species (Shaffer, 1985; Simberloff, 1988). Franklin (1980) suggested that to avoid the deleterious effects of inbreeding depression over the shorter term, a population should not be composed of fewer than 50 individuals, and to avoid the effects of genetic drift over the longer term, not less than 500 individuals. This too-easy conclusion became known as the "50/500 rule," although it is not supported by empirical research and would certainly not apply in most particular, real-world cases because of confounding influences of unpredictable demographic variations, other species-specific biological traits, and environmental disturbances and longer-term changes in habitat.

Of course, it is not sufficient to simplistically declare tracts of natural ecosystem to be ecological reserves. The integrity of reserves must also be monitored, and research toward proper management, if required, must be conducted. For example, the populations of any endangered species and the composition of endangered communities must be monitored. If deleterious changes are identified though monitoring, the environmental causes must be discovered through research and then managed, if possible.

This need for integrated programs of monitoring, research, and active management can be illustrated by the previously described case of Kirtland's warbler (Chapter 9). In this case, monitoring has revealed that this endangered species has declined precipitously in abundance. Research has revealed that the breeding habitat of this species, stands of jack pine (*Pinus banksiana*) of a particular age and structure in Michigan, can be successfully managed, and this is actively being done. Research has also indicated that the brown-headed cowbird, a debilitating nest parasite, must be controlled, and within limits this is also being done. Further research into the factors that affect the integrity of this species in its largely unknown wintering range and into other potential environmental influences on Kirtland's warbler is also required. And of course, monitoring will have to continue as long as Kirtland's warbler remains threatened, so that information is always available concerning changes in the abundance of the endangered species and in its habitat.

There are often substantial conflicts between the

types of land and resource uses that are compatible with ecological reserves and those that are desired by many people. In particular, there may be powerful financial incentives for local people to selectively exploit valuable trees or game from nearby ecological reserves. In the absence of such sources of cash flow, local peoples may not perceive ecological reserves to be in their direct, economic interest, but rather to be a mechanism by which they are deprived of opportunities.

The overexploitation of valuable ecological resources is a well-known problem in many protected areas. Even apparently intact tropical forests can be quite empty of large mammals and other game species, if they are exploited by skilled, unregulated hunters. These people may overharvest game for subsistence purposes, or if there is good access to and from the forest interior, to sell to urban populations to whom meat can be supplied by a market hunt (Redford, 1992; Wilkie *et al.*, 1992).

The numbers of animals killed by aboriginal hunters can be substantial (Redford, 1992; Redford and Stearman, 1993). For example, to supply 1-year's wild meat to three Amazonian villages in Ecuador with a population of 230 persons, subsistence hunters killed 3165 mammals, birds, and reptiles, including 562 wooly monkeys (*Lagothrix lagothrica*), 313 Cuvier's toucans (*Ramphastos cuvieri*), and 152 white-lipped peccaries (*Tayassu pecari*) (Redford, 1992). Overall, Redford estimated that for all of the Brazilian Amazon, about 14 million mammals and 5 million birds and reptiles are harvested each year for subsistence purposes, another 4 million are killed for trade, and another 38 million suffer fatal wounds but are not recovered by hunters. Because of the substantial rates of mortality in hunted tracts of Amazonian forest, the abundance of large animals is much less than in unhunted areas, sometimes to the degree that superficially intact forests are wildlife "deserts."

Excessive hunting is also, of course, an important problem in wealthier countries of temperate regions, where poaching of game species in protected areas is a frequent problem.

Climate change is another stressor that challenges the integrity of ecological reserves (Peters and Darling, 1985; Peters, 1991). The structure and function of natural communities is disrupted by climate changes, and the survival of species and unique communities are therefore at risk. Changes in species and communities must be monitored, and if risks are identified, mitigative actions will have to be designed through research and then implemented. The risks of climate change for a hypothetical nature reserve in a mountainous area are illustrated conceptually in Fig. 10.8. In response to climatic warming, the altitudinal distributions of species and communities would change, with some becoming extirpated from the reserve and new ones invading in response to the emerging ecological opportunities. Conceptually similar scenarios could

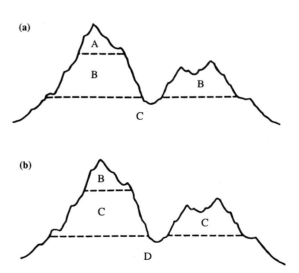

Fig. 10.8 Conceptualization of the changes in plant species and communities to climate changes. A, B, C, and D represent species or communities, each having a distinct tolerance of climatic conditions and distributed altitudinally in accordance with those tolerances. (a) The present distribution of these types; (b) their distribution after climate warming. The climatic tolerance of species/community A has been exceeded by the warming, and it has become locally extirpated. Type B occupies a much smaller range after warming and has been extirpated from one site. The previously continuous range of type C has become fragmented, while a new type, D, has invaded and is now dominant. Modified from Peters and Darling (1985).

be developed for latitudinal and other gradients of climate.

Within the contexts of economic, social, political, and ecological constraints, the biodiversity-related objectives of most ecological-reserve systems would be favored by the following, generic design and management options (Freedman, 1992): (1) establish as large a number of ecological reserves as possible, (2) each as large as possible, and (3) in aggregate protecting as large a fraction as possible of biodiversity at the genetic, species, community, and landscape levels, with (4) as much redundancy as possible to guard against catastrophic losses caused by disturbance. Furthermore, the reserve system would be enhanced if there are (5) peripheral, buffer zones of restricted resource exploitation that are as large as possible, and perhaps if (6) reserves are interconnected, by corridors. There must also be (7) research and monitoring, directed toward both basic ecological principles and to preservation through management, and (8) ecological intervention, whenever necessary, to cope as much as possible with demonstrated threats to the integrity of the reserves.

10.8
CONSERVATION ACTIVITIES MORE BROADLY

The conservation of wild species of plants and animals is regarded by almost all human societies as a laudable and important objective. As a result, there is: (1) a great deal of government activity in this field, (2) much research and training of ecologists and other scientists at universities and other educational institutions, and (3) many nongovernment organizations that are active at the local, regional, national, and/or international levels. All of these contribute to important conservation activities, such as: (1) identifying, acquiring, and managing sites where rare and endangered organisms live or where endangered communities occur; (2) increasing the awareness of the broader citizenry about important conservation issues; (3) monitoring and

research into the biology and ecology of rare and endangered species and communities; and (4) raising or providing funds for all of the above.

The intensity of these conservation activities is greatest in relatively wealthy, developed countries, but awareness and activity are also beginning to emerge in less-developed countries. Of course, a respect for nature and a conservation ethic has always been an integral component of many religions that developed in tropical countries, for example, Buddhism, Hinduism, and Jainism. In spite of this attitude of many peoples of tropical countries, wildlife and natural habitats have suffered badly, mostly because of the conversion of natural ecosystems to agricultural ones, but also for other reasons. Hopefully, the progressive, worldwide development of conservation activities will itself come to be regarded as an emerging, "success story."

It is beyond the scope of this chapter to describe the many agencies and activities that are important in the protection of Earth's biodiversity. It will be sufficient to briefly describe a joint program of three of the most important, international agencies whose mandates center on the conservation and protection of the world's biota and ecosystems, namely, the World Conservation Union, the World Resources Institute, and the United Nations Environment Program. Jointly, these organizations have developed a framework for a global program called the "Global Biodiversity Strategy" (Reid *et al.*, 1992). The broad objectives of this program are to: (1) maintain essential ecological processes and life-support systems on Earth; (2) preserve biotic diversity; and (3) ensure the sustainable development of Earth's natural resources. These are general but important goals, partly because of the direct linkage of biological conservation and the sustainable development of global resources and human societies (see also Chapter 12).

The *Global Biodiversity Strategy* is a potential mechanism by which all countries and peoples can initiate meaningful actions to conserve and protect Earth's biodiversity for the benefits of present and future generations of humans, and for intrinsic reasons as well. To achieve this end, 85 specific actions have been recommended, of which the following

Table 10.10 Key elements of the *Global Biodiversity Strategy*[a]

1. Catalyze action through international cooperation and national planning
2. Establish a national, policy framework for biodiversity conservation
   - Reform existing public policies that invite the waste or misuse of biodiversity
   - Adopt new public policies and accounting methods that promote conservation and the equitable use of biodiversity
   - Reduce demand for biological resources
3. Create an international policy environment that supports national biodiversity conservation
   - Integrate biodiversity conservation into international economic policy
   - Strengthen the international legal framework for conservation to complement the *Convention on Biological Diversity*
   - Make the development-assistance process a force for biodiversity conservation
   - Increase funding for biodiversity conservation, and develop innovative, decentralized, and accountable ways to raise funds and spend them effectively
4. Create conditions and incentives for local biodiversity conservation
   - Correct imbalances in the control of land and resources that cause biodiversity loss, and develop new resource-management partnerships between government and local communities
   - Expand and encourage the sustainable use of products and services from the wild for local benefits
   - Ensure that those who possess local knowledge of genetic resources benefit appropriately when it is used
5. Manage biodiversity through the human environment
   - Create the institutional conditions for bioregional conservation and development
   - Support biodiversity conservation initiatives in the private sector
   - Incorporate biodiversity conservation into the management of biological resources
6. Strengthen protected areas
   - Identify national and international priorities for strengthening protected areas and enhancing their role in biodiversity conservation
   - Ensure the sustainability of protected areas and their contribution to biodiversity conservation
7. Conserve species, populations, and genetic diversity
   - Strengthen capacity to conserve species, populations, and genetic diversity in natural habitats
   - Strengthen the capacity of off-site conservation facilities to conserve biodiversity, educate the public, and contribute to sustainable development

(continued)

Table 10.10   *(Continued)*

8. Expand human capacity to conserve biodiversity
   - Increase appreciation and awareness of biodiversity's values and importance
   - Help institutions disseminate the information needed to conserve biodiversity and mobilize its benefits
   - Promote basic and applied research on biodiversity conservation
   - Develop human capacity for biodiversity conservation

[a]Modified from Reid *et al.* (1992).

Table 10.11 Essential elements of the *Convention on Biological Diversity*[a]

- A commitment by governments to survey their natural living resources, both domesticated and wild, and to conserve sites noted for their rich biological diversity, as well as threatened species and domesticated varieties
- Recognition that both *in situ* conservation and *ex situ* preservation of biodiversity are key tools in any effective biodiversity conservation strategy
- A commitment by governments to ensure that any use of biodiversity is sustainable and equitable
- Recognition that conservation of biodiversity is a common concern of all humankind and that states have the sovereign right to use their biological resources
- Recognition that access to biodiversity is contingent upon prior informed consent of the country concerned, and that those who possess traditional knowledge about genetic resources and farmers who have contributed to and maintained diversity in crops and livestock deserve just compensation for the use of their knowledge or their varieties
- The establishment of a financial mechanism that would provide both technical and financial assistance to developing countries in need of support for surveying, characterizing, and conserving their biodiversity
- The establishment of an administrative structure giving equal control to developed and developing countries that are Parties to the Convention in the distribution of funds under the Convention, and ensuring participation of scientists, governments, and nongovernmental organizations to advise on funding priorities
- Arrangements by which the commercial exploiters of biodiversity help finance much of its conservation in the countries that give it refuge
- Mechanisms to ensure access for developing countries to technologies for conserving and using biodiversity
- The establishment of a monitoring and early-warning system to alert governments and the public to potential threats to biodiversity

[a]Modified from Reid *et al.* (1992).

five are considered to be catalytic (after Reid *et al.*, 1992): (1) ratify and implement the recommendations of the international *Convention on Biological Diversity*, as presented in 1992 by the United Nations Environment Program at the *United Nations Conference on Environment and Development* held at Rio de Janeiro, Brazil; (2) implement the actions detailed in the *Global Biodiversity Strategy*, beginning immediately with a decade of concentrated efforts toward conservation and protection of biological diversity; (3) create an international administrative mechanism to ensure broad participation in international decisions concerning biodiversity, with representation from governments, the scientific community, citizens, industry, the United Nations, and nongovernmental organizations; (4) establishment of a network-based capability, linked to the *Convention on Biological Diversity*, to monitor acute threats to biological diversity, so that individuals and organizations can be alerted and take appropriate actions; and (5) achieve an integration of biodiversity considerations into national-planning processes. Other key elements of the *Global Biodiversity Strategy* are summarized in Table 10.10,

while essential elements of the *Convention on Biological Diversity* are in Table 10.11.

It is too early to appraise the success of the *Global Biodiversity Strategy*, since the program only began in the late 1970s (through an initial, progenitor program, called the *World Conservation Strategy*). However, the mere existence of this comprehensive international effort is encouraging, as is the participation in it of most of Earth's countries, representing nations in all stages of socioeconomic development.

If the use of Earth's resources for the benefit of humans is to occur on an ecologically sustainable basis, there must be adequate protection of rare-and-endangered species and communities. Clearly, a program like the *Global Biodiversity Strategy* will have to be a key element of the system that guides the process of development.

> The one process ongoing in the 1990s that will take millions of years to correct is the loss of genetic and species diversity by the destruction of natural habitats. This is the folly that our descendants are least likely to forgive us.
> E.O. Wilson (cited in Reid *et al.*, 1992, p. 19).

11

ECOLOGICAL
EFFECTS OF
WARFARE

11.1

INTRODUCTION

War has long been a fashionable human enterprise. The earliest written records of our cultures often refer to war as a pervasive activity of humans and gods, for example: "The Lord is a man of war" (*Exodus* **15**, 3). The first, prehistoric battles probably involved conflicts between extended families or clans. However, in modern times warfare has progressed to contests with an intensity just short of what might be required to literally achieve Armageddon.

Like modern wars, early conflicts involved much carnage, but also mayhem, through attempts to destroy the culture and agricultural and industrial capacities of vanquished enemies. An early example of ecocide as a tactic of warfare occurred when the Romans, led by *Scipio Africanus Minor*, defeated the Carthaginians in the Third (and final) Punic War of 146 *BC* and then razed the city of Carthage in what is now Tunisia in North Africa and salted the surrounding, agricultural fields.

The destructiveness of warfare to humans and

their civilizations is well appreciated. Warfare can, of course, cause tremendous losses of human life. For example, the First World War of 1914–1918 caused an estimated 20 million fatalities, the Second World War of 1939–1945 about 38 million, the Korean War of 1950–1953 about 3.0 million, and the Second Indochina War of 1961–1975 about 2.4 million (Sivard, 1986, 1987, 1989). In terms of killing humans, these have been the most prominent of the many 20th century wars, which up to the mid-1980s had resulted in the deaths of about 84 million people (Sivard, 1986).

Military activities appropriate remarkably large amounts of economic activity (Sivard, 1982, 1986, 1987, 1989, 1993). The all-time peak of global military spending occurred during 1987, when about $715 billion was spent (83% by developed countries and 17% by developing countries). In 1991, global military expenditures had decreased somewhat to about $655 billion (in 1987 U.S. dollars), or about 3.6% of all economic activity on Earth. Total arms sales in 1985 were $250 billion, including $55 billion in the international arms trade. Other military-related economic activities involved expenditures for salaries, training, infrastructure, logistic support, research, and other ventures. The global military spending of about $655 billion in 1991 represented a tremendous 103% increase from the 1960 expenditure of $322 billion and a 44% increase over the 1970 expenditure of $454 billion (all in 1987 U.S. dollars).

Between 1960 and 1987, global expenditures for military purposes were $17 trillion (i.e., 17×10^{12}), compared with 15×10^{12} for education and 10×10^{12} for health (all in 1988 U.S. dollars; Sivard, 1989). In 1987 alone, global spending for these purposes was about 850×10^9, 750×10^9, and 640×10^9, respectively (Sivard, 1989).

Military activities also appropriate a large human resource (Sivard, 1982, 1987, 1989, 1993). In 1989 there were about 27 million people in various armed forces worldwide. Another 77 million people were in military reserves or in paramilitary forces, or were civilians producing weapons or providing other direct services to the military.

In some parts of the world, the pace of warfare did not appreciably slow during the 1980s and early 1990s. However, beginning in the late 1980s, rates of arms-reductions negotiations and treaty-signing between the United States and the U.S.S.R. and, more recently, Russia and other former Soviet nations, accelerated. As a result, there are good reasons to be optimistic about decreased likelihoods of global catastrophe through nuclear warfare. Such risks have not disappeared, however, because the global nuclear arsenals and armed forces remain huge in spite of large, recent reductions. Moreover, with the destabilization of governments in the former Soviet Union, the possibilities of nuclear proliferation and nuclear terrorism may be greater than they were prior to the break-up of that former superpower.

In spite of the tremendous destructions that have been caused by warfare there is remarkably little hard information on the ecological damage that has been caused. Nevertheless, in this chapter some of the ecological effects of warfare are investigated. Topics that are covered include effects of the explosion of conventional munitions, the use of poisonous gases, the military use of herbicides, and the direct and indirect consequences of nuclear warfare. As will be seen, the potential ecological implications of extensive explosions of nuclear weapons are particularly horrific. In fact, some predictions have been made of a collapse of biosphere processes, caused by the potential climatic consequences of a nuclear war.

11.2. DESTRUCTION BY CONVENTIONAL WARFARE

Tremendous quantities of explosive munitions are used in warfare. During the Second World War, an estimated 21.1×10^9 kg of explosives was expended by the combatants, equivalent to about 25.3×10^{15} J of explosive energy (36% of this was used by the United States, 42% by Germany, and the rest by other nations' combatants) (Westing, 1985b). During the Korean War, the total expenditure of munitions was about 2.9×10^9 kg, of which 90% was used by U.S. forces (Westing, 1985b). During the Second Indochina War, the total expenditure was about 14.3×10^9 kg, more than 95% by U.S. and South Vietnamese forces (Westing, 1985b). In the relatively brief, but intense, Gulf War of 1990–1991, an estimated 0.12×10^9 kg of explosives was expended, almost all by the coalition forces (Barnaby, 1991). Explosion of these great quantities of munitions during these conflicts caused direct, though poorly documented, damage to ecosystems.

The World Wars

In terms of the loss of human lives, the most prominent wars waged in human history were the First and Second World Wars. Both of these widespread conflicts caused a terrible mortality, destroyed cities, and disrupted civilization in many other ways. Also among the important effects of these wars was damage caused to ecosystems. These ecological effects have never been subject to detailed scientific documentation or scrutiny, and therefore they are poorly quantified. Moreover, although they were substantial, the ecological effects have not been

"The way up to Passchendaele Ridge, and some of the German defences," in France in 1917. Long-term static confrontations between large armies typically caused this sort of churned-up morass of mud over most of the Western Front on the poorly drained coastal lowlands of western Belgium and France. The water-filled depressions are old explosion craters of artillery shells. In this case, the terrain was initially agricultural fields (photo courtesy of *The Times*, 1919).

popularly perceived as notable impacts of these wars—the human tragedy has been emphasized most strongly.

To a certain degree, an appreciation of the ecological effects of these human conflicts can be gained from literary images of the time, which allow an appreciation of the intensity of the devastation caused to forests and other ecosystems. This appears to be especially true of literature that emerged from the Western Front of the First World War. That particular conflict involved virtually static confrontations between huge, well-equipped armies. The most massive battles took place in the coastal-plain lowlands of Belgium and France. For as long as 4 years, armies fought back and forth over a well-defended terrain that was webbed with elaborate trenchworks. In these circumstances, territorial gains were difficult, in spite of the wastage of many men and much materiel during offensives. The intense and long-lasting confrontations devastated the agricultural terrain and woodlands of the battlefields.

These effects were especially great in the flat and poorly drained lowlands of the Flanders region of Belgium. The city and vicinity of Ypres was devastated:

In this landscape nothing existed but a measureless bog of military rubble, shattered houses, and tree stumps. It was pitted with shell craters containing fetid water. Overhead hung low clouds of smoke and fog. The very ground was soured by poison gas (Wolff, 1958).

The soils of Flanders were largely developed from deep deposits of clay. Such soils were readily churned into a sticky, gelatinous morass by intense bombardments and the movements of men and machinery:

Because of the impervious clay, the rain cannot escape and tends to stagnate over large areas . . . the low-

"German trenches in a wood wrecked by artillery," probably in Belgium in 1918. The lowlands of the Western Front were largely agricultural, but the small remnant woodlands were strategically important because they provided cover for troops and equipment, fuelwood, and lumber for the construction of fortifications and other structures. In the foreground is a latticework made of local forest materials, intended to strengthen the walls of the trenches to keep them from collapsing inward. In the background is a woodland devastated by repeated artillery barrages (photo courtesy of *The Times*, 1918).

lying, clayey soil, torn by shells and sodden with rain, turned to a succession of vast, muddy pools . . . the ground remains perpetually saturated . . . gluey, intolerable mud . . . liquid mud . . . molasses-like topsoil (excerpts from military dispatches and other literature of the time; from Wolff, 1958).

The churned-up lands were largely devegetated by warfare. However, some ruderal plants managed to flourish briefly during summer, most notably the red-flowered poppy (*Papaver rhoeas*):

I've seen them, too, those blossoms red,
   Show 'gainst the trench lines' screen,
A crimson stream that waved and spread
   Thro' all the brown and green:
I've seen them dyed a deeper hue
   Than ever nature gave,
Shell-torn from slopes on which they grew,
   To cover many a grave (W. C. Galbraith).

Other prominent ruderals of the battlefields of Flanders were coltsfoot (*Tussilago farfara*), lesser celandine (*Ranunculus fricaria*), cornflower (*Centauria cyanus*), and various graminoids (Gladstone, 1919).

Literature from the battlefields of the Western Front also speaks of the devastation of forests. For example, from military dispatches and other descriptions (Wolff, 1958):

The scene in No Man's Land . . . was indeed a chilling one . . . . Woods were empty fields masked by what seemed to be a few short poles stuck in the ground. . . . Gaunt, blackened remnants of trees drip in the one-time forests. . . . Houthulst Forest, shelled day and night throughout six hundred acres of broken tree stumps, wreckage, and swamps—the acme of hideousness, a Calvary of misery.

Similar images can be excerpted from poetry of the Western Front:

"Gun crew . . . firing 37 mm gun during an advance," somewhere in France in 1918. Another view of a woodland devastated by artillery barrages (photo courtesy of *The Times*, 1919).

Seared earth, stark, splintered trunks, proclaim the Bois-Etoile once was fair (Bois-Etoile, by E.M. Hewitt);

Ruins of trees whose woeful arms vainly invoke the sombre sky—stripped, twisted boughs and tortured boles (Ruins, by G. H. Clarke).

A final passage describes the dismemberment of a forest during a bombardment:

When a copse was caught in a fury of shells the trees flew uprooted through the air like a handful of feathers; in a flash the area became, as in a magician's trick, as barren as the expanse around it (Wolff, 1958).

Forests in the battle zones of World War I were pretty much devastated by the fighting and by the harvesting of wood for fuel and lumber. The deforestation of the Western Front and elsewhere was described in a series of essays in the journal *American Forestry* that were well illustrated with graphic photographs of the damage and the stark landscapes (Dana, 1914; Risdale, 1916, 1919a–d; Buttrick, 1917; Graves, 1918).

In addition, beyond the battle zones, forests were harvested at a frantic rate to supply the war effort. Belgium lost almost of her forested area, while France lost about 10%. In the British Isles, where there was no fighting, the rate of forest harvest was increased by a factor of at least 20, and about one-half of the forest area was harvested during the period of the war. The most important demand on the British forests was to supply timber for use as pit props in underground coal mines, because there was a tremendous demand for this strategic fuel from industry and the navy (Risdale, 1919a–d). Similarly large increases in the rates of harvesting of the forests of Europe occurred during the Second World War.

A few botanists described the flora and ecological successions that occurred in and around the city of London after bombing during the Second World War (Salisbury, 1943; Castell, 1944, 1954; Wrighton, 1947, 1848, 1949, 1950, 1952; Jones, 1957). Salisbury (1943) noted that sites that had been bombed and also burned were initially dominated by the fireweed *Epilobium angustifolium* and the bryophytes *Ceratodon purpureus, Funaria hygrometrica*, and *Marchantia polymorpha*. Unburnt bombed areas were more species rich and were dominated by various urban ruderals, including members of the aster family (especially *Erigeron canadensis, Galinsoga parviflora, Senecio vulgaris, Sonchus oleraceus, Taraxacum officinale*, and *Tussilago farfara*), grasses (especially *Agrostis alba, Lolium perenne*, and *Poa annua*), clovers (*Trifolium pratense* and *T. repens*), chickweed (*Stellaria media*), and plantain (*Plantago major*). Eventually, the London bomb sites supported a flora of more than 350 species (Jones, 1957). Depending on particular site characteristics (especially the availability of a soil substrate, as opposed to concrete or brick rubble), the vegetation progressively developed into a shrub-dominated community or into an open grassland–herb community (Jones, 1957).

Animal wildlife has also been affected by warfare, but again there is little in the way of documentation. Gladstone (1919) discussed some of the direct effects on birds during the First World War. He described mortality of seabirds (alcids, gulls, and sea ducks) caused by oil pollution from torpedoed wrecks, a common but essentially undocumented cause of avian deaths during both of the world wars and other modern conflicts. Gladstone also described concussion injuries to birds during gunnery practice, because gulls would be attracted to the target sites by the large numbers of dead fish that could be scavenged from the surface of the sea.

Remarkably, Gladstone (1919) described birdlife on land as "almost normal" within a short distance of the front line trenches of the Western Front and elsewhere. Only in the active bombardment zones were there large decreases in species richness. Some bird species were remarkably resilient to the disturbance caused by warfare on the Western Front, and they persisted in the battle zones, but undoubtedly at relatively small densities. A notable example was the house sparrow (*Passer domesticus*), which was "plentiful and unconcerned" in "half-felled orchards and ruined houses." Other prominent landbirds of the front, including "No-Man's Land," were the kestrel (*Falco tinnunculus*),

woodlark (*Lullula arborea*), skylark (*Alauda arvensis*), reed warbler (*Acrocephalus scirpaceus*), blackcap (*Sylvia atricapilla*), willow warbler (*Phylloscopus trochilus*), robin (*Erithacus rubecula*), nightingale (*Luscinia megarhynchos*), blackbird (*Turdus merula*), song thrush (*Turdus philomelos*), great tit (*Parus major*), whitethroat (*Sylvia communis*), and others. Many anecdotes described how individuals of these species nested in the midst of seeming devastation, and how they sang and otherwise went about their business in the midst of bombardments and assaults. Some of these observations found their way into the poetry of the time:

> In Flanders fields the poppies blow
> Between the crosses, row on row,
>   That mark our place; and in the sky
>   The larks, still bravely singing, fly
> Scarce heard amid the guns below.
> (In Flanders Fields; J. McCrae).

Some species of wildlife have been brought close to or beyond the brink of extinction as direct or indirect consequences of warfare. The last semiwild Pere David's deer (*Elaphurus diavidianus*) were killed by foreign troops during the Boxer Rebellion in China in 1898–1900 (the species still survives in captivity) (Westing, 1980). The European bison or wisent (*Bison bonasus bonasus*) was taken close to extinction during the First World War by hunting to provide food for troops. Its population then recovered somewhat in response to protection and management, but during the Second World War it was again decimated. There are now, however, several thousand animals in managed reserves in Poland, Belarus, and elsewhere in eastern Europe (Grzimek, 1972; Westing, 1980).

In addition, a number of commercially valuable large-mammal species are presently endangered in Africa as a direct consequence of: (1) warfare and rebellion; (2) the associated political instability and inadequate enforcement of game laws; and (3) the widespread availability of automatic weapons to poachers. The financial incentives for poaching are especially great in the case of endangered rhinos (*Ceratotherium simum* and *Diceros bicornis*) because of the great value of their horns as dagger sheaths and as an aphrodisiac, and for African elephants (*Loxodontia africana*) because of their valuable ivory tusks (Curry-Lindahl, 1971, 1972; Fitter, 1974; Westing, 1980).

As a result of the conflict in the Pacific Ocean during the Second World War, the avifauna of a number of remote islands was badly damaged. On the small island of Iwo Jima, only three species of landbird and one shorebird were observed 1 month after a devastating bombardment and invasion by United States forces, and most of the individuals that were "collected" for examination had healing or recently healed injuries (R. H. Baker, 1946).

On the Pacific island of Guadalcanal "the destruction of the jungle by the fighting drove out the forest-loving species, and those frequenting the open country made their appearance" (Donaghho, 1950). In the initial stage of postwar succession on Guadalcanal, the forest was open and "completely devastated," and there was essentially no bird life. As plants began to develop on the disturbed site, the first bird to occur was the little starling (*Aplonis cantaroides*), followed by the common myna (*Acridotheres tristis*), the whiskered tree swift (*Hemiprocne mystacea*), and the dollar bird (*Eurystomus orientalis*). When a dense underbrush developed, the avian community was dominated by the chestnut-bellied monarch (*Monarcha castaneiventris*) and the broad-billed flycatcher (*Myiagra ferrocyanea*). As the site developed into a open forest, and then a dense rainforest, a characteristic and diverse avifauna reappeared (Donaghho, 1950).

On some of the smaller and relatively isolated Pacific islands, warfare and its associated disturbances caused a loss or extreme endangerment of endemic bird species (see also Chapter 10). On the twin islands of Midway, there were extinctions of the Laysan rail (*Porzanula palmeri*) and the Laysan finch (*Telespyza cantans*) (Fisher and Baldwin, 1946). On the island of Guam, the Guam rail (*Rallus ownstoni*), the Guam broadbill (*Myiagra freycineti*), and the Marianas mallard (*Anas oustaleti*) were greatly diminished in abundance (Baker, 1946; Savidge, 1984). Wake Island lost its endemic species of rail (*Rallus wakensis*) (Westing, 1980).

In a few cases, the abundance of certain wild animals increased as an indirect consequence of warfare. Usually this happened because of decreased exploitation. During the First and Second World Wars, the abundance of gamebirds in Britain increased substantially because of decreased hunting pressures (Gooders, 1983). The populations of some raptorial birds also increased, because most British gamekeepers were recruited for military service, so there was no control of these supposedly injurious predators of game species (Gooders, 1983). During the Second World War, fish stocks in the North Atlantic increased, because of smaller fishing efforts (R. S. Clark, 1947; Westing, 1980). During that same conflict, populations of fur-bearing mammals increased in northern and eastern Europe because of decreased hunting and trapping (e.g., polar bear *Ursus maritimus*, red fox *Vulpes vulpes*, wolf *Canis lupus*, and wolverine *Gulo gulo*; Westing, 1980). Lastly, it was reported that during the Second Indochina War, tigers (*Panthera tigris*) increased in abundance in certain areas because of the ready availability of human corpses as food (Westing, 1980).

The Second Indochina War

Although there are few data on the magnitude of the ecological disturbances that were caused by this conflict, some insight can be gained from knowledge of the tremendous amounts of munitions that were expended. During the Second Indochina War of 1961–1975, the total quantity of munitions used by the United States forces alone was more than 14.3 million tonnes, equivalent to about twice the amount used by the United States during World War II (Martin and Hiebert, 1985; Westing, 1985b). About half of the explosive tonnage was delivered from the air, half by artillery, and less than 1% from offshore ships (Martin and Hiebert, 1985). In terms of the numbers of explosive devices, the United States dropped about 20 million aerial bombs of various sizes, fired 230 million artillery shells, and used more than 100 million grenades, plus additional millions of rockets and mortar shells (Martin and Hiebert, 1985; Westing, 1976, 1984a, 1985a,b).

This vast outpouring of munitions caused great damage to the landscape of Indochina. One effect was the creation of explosion craters (Orians and Pfeiffer, 1970; Westing, 1976, 1982). In 1967 and 1968, an estimated 2.5 million craters were formed by the explosion of 225- and 340-kg bombs dropped in saturation patterns by B-52 bombers flying higher than 10 thousand m. Each sortie produced a bombed-out area of about 65 ha/plane. The total area of terrain affected in this way was about 8.1 million ha or 11% of the landscape of Indochina, including 4.5 million ha or 26% of the area of South Viet Nam. Westing (1976) quoted the following impression of a military observer:

> The landscape was torn as if by an angry giant. The bombs uprooted trees and scattered them in crazy angles over the ground. The tangled jungle undergrowth was swept aside around the bomb craters.

Each crater typically had a diameter of about 15 m and a depth of 12 m. They usually filled with freshwater and produced habitat for aquatic biota, including mosquitoes and other insects. In agricultural areas, some of the water-filled bomb craters were subsequently developed for aquaculture (Westing, 1982).

In addition, bomb explosions often started forest and grassland fires. It was estimated that more than 40% of the area of South Viet Nam's pine plantations was burned during the war, with most fires being ignited by exploding bombs (Orians and Pfeiffer, 1970).

The Gulf War

The Gulf War of 1990–1991 was a brief but intense conflict between Iraq and a coalition of nations led by the United States. The cause of the war was the Iraqi invasion and subjugation of Kuwait on August 2, 1990. Among the nations of the anti-Iraq coalition, the public justification for this war was largely based on the sanctity of sovereign states, i.e., that of Kuwait prior to the Iraqi invasion.

However, the real, underlying reason for the conflict was the need to safeguard the reliability of supply of Middle Eastern petroleum to Europe, North America, and Japan. The world's known, recoverable reserves of petroleum are about 124.6 billion tonnes, of which about 63% is in the Middle East (WRI, 1990). Prior to the invasion of Kuwait, Iraq had petroleum reserves of 13.6 billion tonnes, or 11% of the global reserve. After the invasion, Iraq added the Kuwaiti reserve of 12.7 billion tonnes, so it then controlled 21% of the world's petroleum, while also threatening another 29% in Saudi Arabia and the United Arab Emirates. Clearly, invasion or threat of imminent invasion by Iraq was unacceptable to the governments of other Middle Eastern states, and it was equally unacceptable to oil-dependent industrial nations to have such a large fraction of the global petroleum reserve under the control of a capricious Iraqi dictatorship.

An estimated 100–120 thousand Iraqi soldiers were killed during the Gulf War, about one-half during the 4-day ground war and one-half during the preceding allied bombardment (Barnaby, 1991). In comparison, 343 coalition soldiers died, including 143 deaths through noncombat accidents (Barnaby, 1991). An additional 49–76 thousand Iraqi civilians died, about 15% of these during the coalition bombings and the ground war, and the rest during the war's aftermath of an Iraqi civil war (Barnaby, 1991). A further 2 thousand Kuwaitis were killed or are missing and presumed dead (Barnaby, 1991).

Few data about the munitions used during the Gulf War are available. One estimate is that about 120 thousand tonnes of explosives were expended, virtually all by the coalition forces (Barnaby, 1991). On average, the coalition bombardment resulted in about one enemy-military death per tonne of explosive, compared with about one-half/tonne in the Viet Nam War and one-quarter/tonne in Korea (Barnaby, 1991). The increased "efficiency" of the bombardment in the Gulf War is presumably due to the use of relatively "smart" weapons that could be accurately guided to their targets, coupled with the relatively open exposure of most of the Iraqi military units in the desert.

The Legacy of Unexploded Munitions

An important secondary consequence of the delivery of such great quantities of explosives is the presence of large numbers of unexploded devices, which cause lingering hazards on the landscape. The legacy of explosive remnants following World War II is still considerable (Westing, 1985a). For example, in Poland more than 88 million explosive items have been discovered and removed since 1945, and this cleansing still proceeds at a rate of about 200 thousand items per year. A similar problem exists in all of the theaters of battle during that war. In the Second Indochina War an estimated 10% of U.S. munitions did not explode, causing an explosive legacy of about 2 million bombs, 23 million artillery shells, and tens of millions of other high-explosive items (Martin and Hiebert, 1985; Westing, 1984a, 1985a).

The use of buried, explosive mines in warfare adds to the lingering hazards, because many mines are not recovered after the hostilities cease. For example, even in the brief Gulf War an estimated 6 million mines were laid, and these will pose an explosive hazard for many years (McKinnon and Vine, 1991). Modern anti-personnel mines can be virtually impossible to detect magnetically, because these explosive devices can contain as little as 1 g of metal (Stover and Charles, 1991).

Cambodia has sustained a long-running civil war through the 1970s to the early 1990s, and has a population of about 35 thousand amputees (out of a national population of 8.5 million), most of whom were injured by mines (Stover and Charles, 1991). The amputation rate due to mine injuries was 6 thousand/year in Cambodia in 1990 (Stover and Charles, 1991), and will remain similarly large for years after the actual hostilities cease.

11.3
CHEMICAL WEAPONS IN WARFARE

Anti-personnel Agents

Chemical warfare began on a large scale during World War I, when more than 100 million kg of

"A gas and flame attack seen from the air," somewhere in the coastal lowlands of Belgium or France in 1918. An estimated 100 million kg of lethal anti-personnel chemicals were used during the First World War, and these caused at least 1.3 million casualties. These blanket, ground-level fumigations with toxic chemicals also caused severe but undocumented ecological damage (photo courtesy of Freidel, 1964).

lethal, anti-personnel agents were used. These were mainly the devastating lung agents chlorine, phosgene, trichloromethyl chloroformate, and chloropicrin, and the dermal agent bis(2-chloroethyl) sulfide, or mustard gas, most of these being typically used at a dose of about 100 kg/ha. These agents caused an estimated 1.3 million casualties, including 100 thousand deaths (Westing, 1977; Dyer, 1991).

Lethal gases were also used during the later stages of the Iran–Iraq War of 1981–1987, especially by Iraqi forces. The best-known incident concerned the aerial gassing of the northern Iraqi town of Halabja, which had been overrun by Kurds backed by Iran. In this incident, the Iraqi military used a mixture of the nerve gases sabin and tabun and caused about 5 thousand mostly civilian deaths (McKinnon and Vine, 1991).

The nonlethal, "harassing agent" CS or *o*-chlorobenzolmalononitrile was used by the United States military during the Second Indochina War. A total of 9 million kg of CS was sprayed aerially at 1 to 10 kg/ha over more than 1 million ha of South Viet Nam. A sprayed site was rendered uninhabitable by humans for 15 to 45 days (Westing, 1977).

There has been no documentation of ecological damages caused by these various chemical agents of warfare. However, effects on sprayed areas must have been severe, and they would have included the deaths of large numbers of animal wildlife (Westing, 1977).

Herbicides in Viet Nam

Somewhat more complete information is available about the use by the U.S. military of broadcast-sprayed herbicides to deprive their enemy of food production and forest cover in Viet Nam. This was a large-scale program, with more than 1.4 million ha being sprayed at least once (equivalent to about one-seventh the land area of South Viet Nam), including more than 100 thousand ha of cropland (Boffey, 1971; Westing, 1971, 1976, 1984b,c; Turner, 1977). The program began in 1961, peaked in 1967, and was stopped in 1971 (Table 11.1).

The most frequently used herbicide formulation was a 50:50 mixture of 2,4,5-T plus 2,4-D that was

Table 11.1 The use of herbicides by the U.S. military during the Second Indochina War[a]

Year	Area sprayed (ha × 10³)	Quantity of 2,4,5-T + 2,4-D (10⁴ kg acid equivalent)
1961	<10	<10
1962	<10	<10
1963	10	10
1964	30	40
1965	31	45
1966	167	595
1967	416	1214
1968	267	845
1969	409	1245
1970	75	214
1971	0	0
Total	1425	4228

[a]After Westing (1971) and Turner (1977).

known as Agent Orange, but picloram and cacodylic acid were also sprayed (Orians and Pfeiffer, 1970; Boffey, 1971). In total, more than 25 million kg of 2,4-D, 21 million kg of 2,4,5-T, and 1.5 million kg of picloram were sprayed in this military program (Westing, 1971, 1984b,c). The rates of application were relatively large, averaging about 12 kg 2,4,5-T plus 13 kg 2,4-D per hectare, or about 10 times the application rate for conventional silvicultural purposes (NAS, 1974; Turner, 1977). About 86% of the spray missions were targeted against forests and the remainder against croplands (Westing, 1984b,c).

As was essentially the military intention, great ecological damage was caused by this defoliation program. The effects could be so severe that opponents of the practice labeled it "ecocide," that is, the intentional use of anti-environmental actions, carried out for years over a large area, as an important, tactical component of a military strategy (Westing, 1976). Although the ecological effects of herbicide spraying in Viet Nam were not documented in detail, impressions can be gained from several cursory surveys that were made by ecologists. These are described below.

The most extensively sprayed type of vegetation was forest, an ecosystem that covered more than 10 million ha in South Viet Nam, comprising about

60% of the land surface (Westing, 1971, 1976). Mangrove forest is especially sensitive to herbicide application. About 110 thousand ha of coastal mangrove was sprayed at least once, or about 36% of the area of that type of forest (NAS, 1974). Virtually all of the diverse assemblage of halophytic plant species in mangrove forests (including the dominant tree, *Rhizophora apiculata*) is extremely sensitive to phenoxy herbicides. As a result, spraying devastated the mangrove ecosystem and created large areas of poorly vegetated or unvegetated coastal barrens (Orians and Pfeiffer, 1970; Westing, 1971, 1976, 1984b,c; NAS, 1974; Hiep, 1984; Snedaker, 1984). The regeneration that followed was typically slow. By the early to middle 1980s, revegetation of the sprayed mangrove areas was essentially complete, but it usually did not involve the commercially preferred mangrove tree *Rhizophora apiculata*. More frequently the forest was dominated by less desirable mangrove trees such as *Avicennia alba* or commercially undesirable species such as the palm *Phoenix paludosa*, the mangrove *Ceriops decandra*, the shrub *Acanthus ebracteatus*, the fern *Acrosticum aureum*, and grasses (Westing, 1982; Hong, 1984a,b, 1987). There has been some reforestation, in which seedlings of the favored mangrove tree *Rhizophora apiculata* were planted over about 100 thousand ha. However, the success rate of these plantations has only been about 50% in some places (Hong, 1984b, 1987). Overall, the damage caused to coastal mangrove forests by herbicide spraying rendered it the ecosystem type that was most severely affected by the Second Indochina War (Snedaker, 1984; Westing, 1984b,c).

Effects were also severe in the much more species-rich uplands, including rain forest with a total area of 10.5 million ha. Mature, tropical forest of this type has many angiosperm species, especially of the families Dipterocarpaceae (tree species of *Dipterocarpus*, *Hopea*, *Anisoptera*, and *Shorea*) and Leguminoseae (*Dalbergia*, *Pterocarpus*, *Sindora*, and *Pahudia*). Tree height is up to 40 m or taller, and diameter at breast height is up to 2 m (NAS, 1974; Galston and Richards, 1984). A single aerial application of herbicide initially killed about 10% of the trees of the upper canopy layer. The total

defoliation was much more intensive than this, but defoliated trees were not necessarily killed. Subsequent resprays (about 34% of the sprayed land was treated more than once) killed additional canopy trees, plus understory trees, shrubs, and ground vegetation (Westing, 1971, 1984b,c; Galston and Richards, 1984). The total loss of potentially merchantable timber caused by the military use of herbicides in South Viet Nam was about 47 million m^3 (Westing, 1971).

Some plant species are relatively tolerant of phenoxy herbicides, and of course these were released from competitive stresses when the upper tree canopy was killed. Examples of relatively herbicide-tolerant tree species are *Irvingia malayana* and *Parinari annamense*. In addition, some important understory plants survived, and these plus invading ruderal species dominated many stands in the post-herbiciding secondary successions. Particularly important in this respect were several species of bamboo, especially *Bambusa* spp., *Thrysostachys* spp., and *Oxytenanthera* spp., and the aggressive grasses *Imperata cylindrica* and *Pennisetum polystachyon*. Pioneer woody dicots included *Adina sessilifolia*, *Randia tomentosa*, *Colona auriculata*, and *Sindera cochinchinensis*. The dense stands of these and many other early successional plants made forest regeneration a slow process (Westing, 1971; Ashton, 1984; Galston and Richards, 1984). Some reforestation to high-yielding species such as *Acacia auriculaeformis* is being used in Viet Nam to supplement the natural regeneration of herbicide-treated forests (Ashton, 1984).

Secondary effects on wildlife of the large-scale herbicide treatments of forest habitats in South Viet Nam have not been well documented. In herbicide-treated mangrove forests, there were anecdotal accounts of great decreases in the abundances of most species of birds, mammals, reptiles, and other wildlife, and reduced catches in the nearshore fishery, of which the mangrove ecosystem is an integral component (Hong, 1984a,b, 1987; Quy, 1984).

A study of an inland valley that had been converted by herbicide treatment from continuous upland tropical forest to 80% cover of grassland found only 24 bird and 5 mammal species, compared with 145–

170 bird and 30–55 mammal species in two un-sprayed, reference valleys (Quy *et al.*, 1984). Another study of the conversion of a diverse and luxuriant tropical forest to a degraded *Imperata cylindrica*–low shrub savannah found decreased abundances of some mammal species, but increases in others (Huynh *et al.*, 1984). Species described as "common" prior to spraying but rare subsequently were wild boar (*Sus scrofa*), wild water buffalo (*Bubalus bubalis*), Sumatran goat (*Capricornis sumatraensis*), tiger (*Panthera tigris*), and various species of deer (family Cervidae). Many other, initially less common mammals also decreased in abundance. The relatively few species that increased markedly in abundance after the herbicide-caused habitat changes included opportunistic species such as the fawn-colored mouse (*Mus cervicolor*), Sladen's rat (*Rattus sladeni*), and rats in general (*Rattus* spp.).

In addition to these effects on wildlife, which may have been largely caused by herbicide-caused habitat changes, there were many anecdotal reports of illnesses in domestic animals after military herbicide spraying (Westing, 1980). Large domestic mammals that apparently became ill included water buffalo (*Bubalus bubalus*), zebu (*Bos indicus*), and adult pig (*Sus scrofa*). More intense illnesses and occasional mortality were reported in smaller domestic animals such as young pig, chicken (*Gallus gallus*), and domestic duck (*Anas* sp.). The specific causes of these reported illnesses were not documented, but they were commonly attributed to exposure to herbicides, directly and/or in foods.

Another important aspect of the use of phenoxy herbicides in Viet Nam was the contamination of the 2,4,5-T formulation by the very toxic dioxin isomer known as TCDD (2,3,7,8-tetrachlorodibenzo-*p*-dioxin). TCDD is an incidental by-product of the manufacturing process of 2,4,5-T. Using post-Viet Nam manufacturing technology, the contamination by TCDD in 2,4,5-T solutions can be kept to concentrations that are well below the maximum level of <0.1 ppm set by the U.S. Environmental Protection Agency (Norris, 1981). However, the 2,4,5-T used in Viet Nam was considerably more grossly contaminated with TCDD. A concentration as large

as 45 ppm was measured, and the weighted-average concentration was about 2.0 ppm (Kilpatrick, 1980; Beljan *et al.*, 1981). A total of perhaps 110–170 kg of TCDD was sprayed with herbicides onto Viet Nam (Westing, 1982).

TCDD is well known as being extremely toxic, and it can cause birth defects and miscarriages in laboratory mammals at small doses (Kilpatrick, 1980; Norris, 1981; Walsh *et al.*, 1983). However, as is often the case, the direct effects of TCDD on humans are less well understood. The most useful data on humans have come from observations of persons who were: (1) occupationally exposed to TCDD during the manufacture of 2,4,5-T or (2) exposed to TCDD after a chemical explosion in a trichlorophenol factory in Seveso, Italy, in 1976. The latter accident caused the deaths of livestock and other animals within 2–3 days of the accident. Remarkably, it was not until 2.5 weeks had passed that about 700 people were evacuated from a heavily contaminated residential area close to the factory. The exposure to TCDD at Seveso caused 187 cases of a skin condition known as chloracne, which resembles adolescent acne but is more persistent and severe. However, there were apparently no statistically significant increases in the rates of embryotoxic or teratogenic effects in children born of women from the contaminated area (that is, compared to the average worldwide rate; unfortunately, there were no suitable control data for the study area in Italy) (Kilpatrick, 1980).

Because the 2,4,5-T formulation that was sprayed in Viet Nam was significantly contaminated with TCDD, there has been much concern and debate over the possible effects on people exposed to these chemicals. Although there have been claims of epidemiological effects in exposed populations, not all of the studies have been scientifically rigorous, and thus there is still controversy (Kilpatrick, 1980; Westing, 1984c; White *et al.*, 1984). The most statistically valid epidemiological studies were done on U.S. military veterans who had served in Viet Nam. These studies did not find clear evidence of an increased incidence of health problems in Air Force personnel involved with the spray program (Lathrop *et al.*, 1984) or an increased probability that veterans would father children with a birth defect (Walsh *et al.*, 1983; Erickson *et al.*, 1984). However, observations such as these do not rule out the possibility of small rates of human-health responses caused by exposure to 2,4,5-T, 2,4-D, or formulations of these contaminated by TCDD. More importantly, no rigorous studies were made of women exposed to TCDD in Viet Nam. The likelihood of teratogenic or embryotoxic effects are greater in the case of maternal exposures to TCDD than in that of paternal exposures (White *et al.*, 1984). Nevertheless, the apparent, mainstream opinion from the scientifically most rigorous epidemiological studies that do exist suggests that large effects have not occurred in relatively heavily exposed human populations, and this is encouraging (White *et al.*, 1984).

It also seems likely that the specific effects of TCDD added little to the direct and indirect ecological effects of the herbicides that were sprayed in Viet Nam.

Petroleum as a Weapon in the Gulf War

As was described in more detail in Chapter 6, large quantities of petroleum have been spilled during warfare. Most commonly this has been caused by the deliberate targeting of tankers or offshore production platforms. However, in the Gulf War, oil spills were used to give some tactical advantage and as a way to inflict economic disruptions on the postwar Kuwaiti economy.

The world's largest-ever, oceanic oil spill occurred during the brief Gulf War of 1991, when Iraqi forces deliberately released about 0.8×10^6 tonnes of crude oil into the Persian Gulf from several tankers and an offshore, Kuwaiti loading facility (Holloway and Horgan, 1991; Horgan, 1991; Jones, 1991; Popkin, 1991; Hopner *et al.*, 1992; Sorkhoh *et al.*, 1992).

Much effort was expended to ensure that the seawater intakes of Saudi Arabian desalinization plants were not affected by oil, because these supply about 80% of that country's freshwater and have great strategic value (Holloway and Horgan, 1991). In contrast, relatively little effort was expended in preventing or treating ecological damage caused by the Gulf oil spills (Holloway and Horgan, 1991).

About 770 km of coastline was eventually affected by residues of this spill, mostly in Saudi Arabia (Hopner *et al.*, 1992; Sorkhoh *et al.*, 1992). In places, a stable, asphaltic, pliable, oil-in-sand matrix developed that was up to 35 cm thick and could support the weight of a vehicle without collapsing (Hopner *et al.*, 1992). This substrate contained as much as 20 kg/m^2 of asphaltic residues (Sorkhoh *et al.*, 1992). One year after the spill, the surface of this substrate supported a flourishing mat of blue-green bacteria, dominated by *Microcoleus* sp. and *Spirulina* sp. (Hopner *et al.*, 1992; Sorkhoh *et al.*, 1992).

The Gulf spill killed at least 20–30 thousand seabirds, mostly cormorants and grebes, and including a large toll of the rare and endangered Socotra cormorant (*Phalacrocorax nigrogularis*) (Holloway and Horgan, 1991; Horgan, 1991; Pellew, 1991; Anonymous, 1992d). In addition to the seabirds, another 260 thousand shorebirds may have been killed by this oil spill (Anonymous, 1992d). About two thousand oiled seabirds were rescued from beaches in Saudi Arabia and cleaned, but only 20% survived their ordeal (Holloway and Horgan, 1991; Horgan, 1991). Marine mammals, sea turtles, and sea snakes were also killed in substantial numbers by the Gulf oil spill (Anonymous, 1991; McKinnon and Vine, 1991; Pellew, 1991).

Another, even larger oil spillage began in January 1991, when about 788 Kuwaiti producing wells (out of a total of 1116 wells) on land were sabotaged and ignited by Iraqi forces, causing tremendous releases of petroleum and combustion residues to the land and atmosphere (Pellew, 1991; Earle, 1992). Emissions of petroleum from the oil wells were estimated as $2–6 \times 10^6$ tonnes/day, with about half of the wells being capped within 6 months and the last one in early November 1991 (Bakan *et al.*, 1991; Johnson *et al.*, 1991; Popkin, 1991; Warner, 1991; Earle, 1992).

Much of the oil and virtually all of the associated methane burned in the atmosphere, but huge amounts of petroleum accumulated on land in oil lakes up to 7 m deep, which in November 1991 contained an estimated $5-21 \times 10^6$ tonnes of crude oil (Hoffman, 1991; Horgan, 1991; Earle, 1992).

The burning wells produced an oily smoke, equivalent to about 10% of the mass of the burning oil (Bakan *et al.*, 1991). These plumes typically ascended to 3–5 thousand m in the atmosphere and maximally to 6.7 thousand m, and they could sometimes be detected more than 1 thousand km from Kuwait (Hoffman, 1991; Earle, 1992). There was initial speculation about whether the oil-well plumes could potentially cause global changes through effects on albedo and/or the greenhouse effect. Consider, for example, the following statement made by the well-known physicist Carl Sagan, during a televised interview on January 22, 1991, shortly after the oil wells were ignited:

> We think the net effects will be very similar to the explosion of the Indonesian volcano Tambora in 1815, which resulted in the year 1816 being known as the year without a summer. There were massive agricultural failures in western Europe, and serious human suffering, and in some cases starvation. Especially for south Asia, that seems to be in the cards, and perhaps for a significant fraction of the northern hemisphere as well (cited in Zimmer, 1992).

Fortunately, this and other doomsday scenarios proved to be overly pessimistic, because the plumes did not reach high enough into the atmosphere to cause those global changes (Barnaby, 1991; Browning *et al.*, 1991; Small, 1991; Warner, 1991). Locally, however, the smoke and other fumes caused intense pollution and were dense enough to blot out the noonday sun, resulting in relatively cool weather. Regionally, the smoke caused black rains and snows as far as several thousands of kilometers away (Barnaby, 1991; Hoffman, 1991; Small, 1991; Warner, 1991). One set of airborne measurements found the following densities of pollutants in the plumes: smoke particles, $0.5-1.0$ mg/m^3; hydrocarbons, up to 0.46 ppm (as carbon); SO_2, $0.5-1.0$ ppm; NO_x, $0.03-0.06$ ppm; and O_3, 0.13 ppm (compared with 0.08 ppm outside of the plume) (Johnson *et al.*, 1991).

At ground level, there were awesome vistas of the burning oil fields and their surrounding pools of petroleum. According to William Reilly, then Administrator of the U.S. Environmental Protection Agency:

> If hell had a national park, it would be those burning oil fires. . . . I have never seen any one place before where there was so much compressed environmental degradation (cited in Popkin, 1991).

11.4
EFFECTS OF NUCLEAR WARFARE

Most people have at least an inkling of the enormously destructive capabilities of nuclear weapons. An all-out or even a limited exchange of strategic nuclear weapons would cause tremendous damages to humans and their civilization, as was suggested by John F. Kennedy during a speech to the General Assembly of the United Nations in 1961: "Mankind must put an end to war, or war will put an end to mankind."

Nuclear warfare also poses a tremendous threat to natural ecosystems. Westing (1987) described nuclear warfare as "the ultimate insult to nature."

Nuclear Arsenals

The world's nuclear arsenal is enormous. In the early 1980s, the explosive yield of all nuclear weapons was about $11-20 \times 10^3$ megatons (MT) of TNT equivalent (TNT is 2,4,6-trinitrotoluene, the explosive ingredient in dynamite) (Table 11.2). This was more than 1000 times larger than the total 11-MT yield of all conventional explosives used in the Second World War (6.0 MT), the Korean War (0.8 MT), and the Second Indochina War (4.1 MT) (Westing, 1985b; Sivard, 1989). If the late 1980s nuclear weapons arsenal is standardized on a per-capita basis, there was the equivalent of 3–4 tonnes of TNT per person on Earth (Grover and White, 1985; Sivard, 1989). Note that the average explosive yield of a strategic weapon (about 0.60 MT of TNT equivalent; these are delivered over thousands of kilometers) is considerably larger than that of the more numerous tactical warheads (0.20 MT; used locally in a battlefield) (calculated from Table

Table 11.2   The estimated numbers and explosive yield of the world's nuclear arsenal in the mid-1980s[a]

Country	Strategic weapons		Tactical weapons		Totals	
	Warheads (10^3)	Yield (10^3 MT)	Warheads (10^3)	Yield (10^3 MT)	Warheads (10^3)	Yield (10^3 MT)
United States	9–11	3–4	16–22	1–4	25–33	4–8
Soviet Union	6–7.5	5–8	5–8	2–3	11–15.5	7–11
United Kingdom					0.2–1	0.2–1
China					0.3	0.2–0.4
France					0.2	0.1
Total					37–50	11–20

[a]Strategic weapons are delivered over thousands of kilometers; Tactical weapons are designed for battlefield or theater use. One megaton (MT) of explosive yield is equivalent to the energy contained in about 1 million tonnes of TNT, and it equals 4.2×10^{15} J of energy. Modified from Grover and White (1985).

11.2). In the mid-1980s there were more than 4300 strategic-delivery vehicles (mainly missiles, but also long-range aircraft), carrying an average of about six warheads each and with an aggregate yield of more than 11 thousand MT (Barnaby, 1982; Sivard, 1986).

The global total of nuclear weapons has been substantially reduced from the 1983 total of 64.4 thousand (strategic plus tactical), to 26.7 thousand in mid-1993 (Sivard, 1993). Further global reductions are anticipated in accordance with the most recent arms reductions treaties between the United States and the states of the former USSR, to 10.6–19.4 thousand weapons in 2003.

Hiroshima and Nagasaki

There is little direct information that can be used to describe or predict the possible environmental consequences of the explosions of large numbers of nuclear weapons during a human conflict. There have, however, been two cases of the use of nuclear weapons in warfare. Although there were no studies of their purely ecological effects, some insight can be gained by considering other environmental damage caused by these explosions.

Both of these devices were used by the United States against Japan, and their use probably resulted in an earlier end of the Second World War than

would otherwise have occurred. The bomb dropped on the city of Hiroshima on August 6, 1945 had an explosive yield of about 0.015 MT of TNT, and the one dropped on Nagasaki a few days later had a yield of 0.021 MT (Barnaby and Rotblat, 1982; Pittock et al., 1985). Although enormous in comparison with any conventional explosive devices, these nuclear bombs were relatively "small" in comparison to the typical yield of today's strategic warheads, which average about 0.6 MT, and range up to 6 MT (Grover and White, 1985).

Each explosion occurred as an air burst at about 500–580 m. The Hiroshima bomb killed an estimated 140 thousand people or about 40% of that city's population, while the Nagasaki device killed 74 thousand or 26% of the population (Barnaby and Rotblat, 1982). The leading causes of death and physical destruction were the combined effects of blast and thermal energy. An estimated 50% of the explosive energy was expended as physical blast, which induced a shock wave that traveled at the speed of sound (about 11 km in 30 sec). Blast pressure caused severe structural damage to buildings located as far as 2–3 km from the epicenters of the explosions. Thermal energy accounted for about one-third of the energy of these explosions, and each created a fireball that was sufficiently intense to vaporize people near the blast epicenter, to ignite wood at about 2 km, and to cause human-skin burns as far as 4 km away. The ensuing fires created an areas of "burnout" of 13 km² at Hiroshima and 6.7 km² at Nagasaki. The combined effects of blast and thermal energy were sufficient to destroy about two-thirds of the 76 thousand buildings in Hiroshima, and one-fourth of the 51 thousand buildings in Nagasaki.

Ionizing radiation comprised about 15% of the explosive yield of these bombs. About one-third of the ionizing radiation was emitted within 1 min of the explosion and the rest more gradually by the decay of radioactive fallout material. After both explosions, there were "black rains" that were induced by upward-moving convective air masses caused by the tremendous fires. The rains had been blackened by soot and were highly radioactive. Among other effects, the ionizing energy caused radiation sickness in many survivors of the initial explosions. Symptoms included nausea, vomiting, diarrhea, hair loss, fever, weakness, septicemia, and bleeding from the bowels, gums, nose, or genitals. The presence of latent epidemiological damages in the exposed populations that were not affected by short-term symptoms is controversial. However, there have been claims of increased incidences of eye diseases, blood disorders, malignant tumors, psychoneurological disturbances, and leukemia, compared with the rates of these afflictions in nonexposed populations of Japanese (Barnaby and Rotblat, 1982; Pittock et al., 1985).

Nuclear Test Explosions

There are limited amounts of ecological information from observations made after above-ground nuclear test explosions conducted by the United States during the 1950s. These detonations caused severe damage to ecosystems, which decreased more-or-less geometrically with increasing distance from the blast epicenter. After the explosions, succession restored ecosystems in the damaged areas. For example, a test area in the Mohave Desert of Nevada was subjected to at least 89 relatively small above-ground detonations (most were equivalent to about 10 kT of TNT; the largest was 67 kT) (Shields and Wells, 1962; Sheilds et al., 1963). Initially, the

explosions cleared a central area of 73–204 ha of all life. There was severe vegetation damage over an additional 400–1375 ha, but no damage beyond an area of 3255 ha. Beginning in the first postblast year, the denuded and disturbed areas were progressively invaded by pioneer species of plants. Vagile, annual species such as the tumbleweed *Salsola kali*, along with *Mentzelia albicaulis*, *Erodium cicutarium*, and other ruderals, invaded rapidly. With time, these were replaced by longer-lived species. Overall, the patterns and rates of succession were similar to those expected in that desert region following severe disturbances (Shields and Wells, 1962; Sheilds et al., 1963).

The United States also conducted above-ground nuclear tests on islands in the south Pacific. Palumbo (1962) studied tagged plants of seven species on Belle Island in the Eniwetok archipelago, located 4.3 km from a 1952 blast site on Elugelab Island. He observed initial plant damage after the detonation, but recovery was rapid and essentially complete within 6 months. Elugelab Island was essentially obliterated and was transformed into a water-filled crater (Westing, 1980).

In another study, Fosberg (1959) reported on the effects of radioactive fallout from a large, above-ground test on Bikini Island in 1954 and on other islands in the Marshall group. On the most heavily exposed island, 13 species out of the total monitored flora of 15 species showed conspicuous pathological or other abnormal symptoms that were attributed to radiation damage. Three species were especially sensitive: *Suriana maritima*, *Cordia subcordata*, and *Pisonia grandis*. Two widespread species were tolerant: *Scaevola sericea* and *Tournefortia argentea*. As expected, the damage was less severe on islands located relatively far from the blast epicenter.

Experimental Exposure to Ionizing Radiation

Several experimental field studies provide another important source of information that can be used to predict the ecological effects of ionizing radiation from nuclear explosions or other sources. One of

Aerial view of damage to an oak–pine forest on Long Island, New York, caused by chronic exposure to ionizing gamma radiation for 20 hr/day from a 9500-Ci source of cesium-137 (located in a metal pipe visible in the center of the damaged area: The source can be lowered by remote control into a shielded container below the ground, so that the irradiated area can be entered to take measurements). The intensity of exposure decreases more or less geometrically with distance from the point source of irradiation, with dosage ranging to as much as several thousand R/day within a few meters of the source. The photo was taken in 1962, about 6 months after initiation of the experiment. A clear concentric pattern of death and damage to trees is visible in the photograph. This experiment is designed to run for many years, with an aim at elucidating the longer-term patterns of ecological damage caused by chronic exposure to ionizing radiation, as well as the more specific genetic and physiological effects of this stress on organisms (photo courtesy of Brookhaven National Laboratory).

the best-known projects involved the experimental treatment of an oak–pine forest on Long Island with gamma radiation from an elevated, 9.5 thousand-Ci, cesium-137 point source (Woodwell and Whittaker, 1968; Woodwell, 1970, 1982).

Because the dose of ionizing radiation decreased geometrically with increasing distance from the source (with the complication of physical shielding by tree trunks, at a rate of about 5%/cm of wood), there was a distinct zonation of vegetation damage. After 6 months of chronic irradiation, an essentially sterilized zone was created at a radiation dose of >345 R (Roentgen)/day. In addition, five well-defined vegetation zones were created:

Aerial view of the irradiated oak–pine forest on Long Island, photographed in 1976 after 15 years of experimental exposure to gamma radiation. Over this time period, there was a progressive ecological deterioration of sites close to the point source of radiation stress. The photograph shows a clear concentric pattern of a decreasing intensity of damage with increasing distance from the source. At this time, the rate of ecological change had slowed markedly, as the ecosystems had more or less adjusted to the intensities of radiation stress that occur at various distances from the source (photo courtesy of Brookhaven National Laboratory).

1. There was a central devastated zone at >200 R/day. Vascular plants were not present, but at least some bryophytes and lichens survived up to 1000 R/day. A notable resistant species was the British soldier lichen, *Cladonia cristatella*.

2. Between <200 and >150 R/day, there was a sedge-dominated zone with almost continuous cover of *Carex pensylvanica*.

3. Between <150 and >40 R/day, there was a shrub zone dominated by heaths (*Vaccinium angustifolium*, *V. heterophyllum*, and *Gaylusaccia baccata*) and a short-statured species of oak (*Quercus ilicifolia*).

4. Between <40 and >16 R/day, there was a zone dominated by tree-sized oaks (*Quercus alba* and *Q. coccinea*).

5. At <2 R/day, there was an oak–pine (*Pinus rigida*) zone with some acute damage, but little mortality.

The overall effect of ionizing radiation on ecosystem structure was to cause a systematic dissection of the vertical layers of the original oak–pine forest, similar to the broad pattern described for forests affected by gaseous air pollutants in Chapter 2. The initial, relatively heterogenous vertical structure of the mature forest was reduced to a simple, almost two-dimensional bryophyte–lichen community in the heavily irradiated inner zone.

In part, this spatial pattern reflects differences in species sensitivity, as there were progressive replacements of species along the irradiation gradient, based on their relative susceptibility to the ionizing stress. However, the physical layering of vegetation targets in the highly structured forest also determined which species were initially affected by the gamma radiation and which were shielded. This also influenced the temporal patterns of layer-by-layer "peeling" of the forest. Of course, the irradiated zones also had a smaller net productivity, live biomass, and species richness and diversity, compared with reference vegetation. Field studies of the ecological effects of ionizing radiation on other types of vegetation have come to qualitatively similar conclusions about differential species sensitivity and effects on ecosystem structure and function (McCormick and Platt, 1962; Platt, 1965; Monk, 1966; Sparrow *et al.*, 1970; Dugle and Mayoh, 1984).

The irradiation of various ecosystem types has revealed differences in their sensitivity to gamma radiation. On average, coniferous forest is relatively sensitive, followed by temperate mixed forest, tropical rain forest, shrub vegetation, grassland, moss and lichen communities, and pure lichen communities (Table 11.3). The ranges of sensitivity of various systematic groups of organisms are summarized in Fig. 11.1. Vertebrates and vascular plants are generally more sensitive than less complex organisms, and unicellular organisms and mollusks are the least sensitive. Therefore, any chronically irradiated situation would be expected to

Table 11.3   The sensitivity of various types of vegetation to ionizing radiation[a]

	Dose (10^3 rad[b])		
Vegetation type	**Low damage**	**Moderate damage**	**High damage**
Coniferous forest	0.1–1	1–2	>2
Mixed forest	1–5	5–10	10–60
Tropical rain forest	4–10	10–40	>40
Shrub community	1–5	5–20	>20
Grassland community	8–10	10–100	>100
Moss and lichen community	10–50	50–500	>500
Lichen community	60–100	100–200	>200

[a]From Harwell and Hutchinson (1985), using data of Whicker and Schultz (1982).
[b]Dose: rad = Roengten absorbed dose, a measurement of the dose of ionizing radiation, in terms of energy absorbed. Rad varies with the type of radiation and the physical medium through which it passes.

develop a structurally simple ecosystem, generally dominated by simpler life forms.

Possible Consequences of a Large-Scale Nuclear Exchange

Several approaches have been used to predict the probable human and environmental consequences of a large-scale nuclear exchange. Relatively direct approaches have attempted to budget the likely magnitudes of the loss of human life, the destruction of property, and the devastation of ecosystems that would result from the blast, thermal, ionizing, and other relatively short-term effects of the actual detonations. Another approach, first attempted in the early 1980s, uses complex simulation models to predict the global climatic consequences of large-scale detonations of nuclear weapons. Following from the climatic predictions of the simulations, the likely consequences for the biosphere were predicted by expert panels of scientists.

In general, the short-term predictions of the direct effects of a nuclear holocaust are, of course, for terrible consequences. However, it seems likely that life (including a substantially degraded human

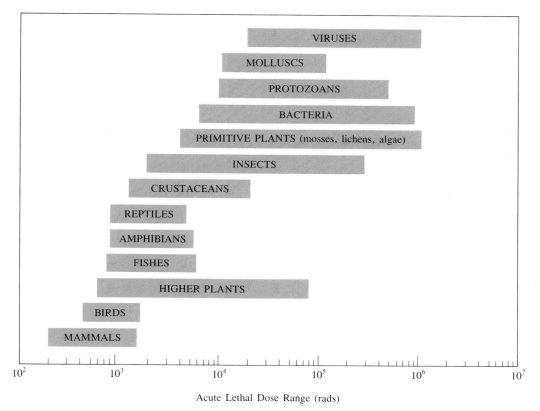

Fig. 11.1 Lethal dose ranges of ionizing radiation for various groups of organisms. From Harwell and Hutchinson (1985), using data of Whicker and Schultz (1982).

civilization) would "go on" in some form and that ecological recovery would eventually occur. In contrast, some of the computer-based simulations of climatic changes and their ecological consequences have predicted a more substantial, possibly longer-term devastation of the biosphere. In the following, the direct, shorter-term effects of a nuclear exchange will be dealt with first. After this, some of the secondary climatic effects of a wide-scale nuclear war and their tertiary ecological consequences are discussed.

Shorter Term Effects

Estimates of the size of the global nuclear arsenal are given in Table 11.2. In a large-scale nuclear

exchange, not all of this capacity would actually be delivered, since (1) many nuclear devices would be destroyed by preemptive strikes (probably causing much less damage than if the weapons had exploded), and (2) some capacity would be held back for strategic reasons. Scenarios of the likely scale of the realized yield of the nuclear explosions vary greatly, but for representative purposes an estimate of 5–6 thousand megatons can be cited, almost all of which would be targeted in the Northern Hemisphere (Grover and White, 1985; Harwell and Grover, 1985).

The predicted short-term effects on people are for the deaths of 20% of the world's population. In the United States, as much as 75% of the population could die from the immediate effects of the nuclear explosions, and there would be many additional injuries (Harwell and Grover, 1985). The principal cause of death would be thermal radiation, but effects of blast, fire, and ionizing radiation would also be important (Harwell and Grover, 1985). This overwhelming loss of human life, in conjunction with the physical devastation and incapacitation of social systems, would transform civilization.

The estimated distributions of energy release from air bursts of strategic nuclear devices of various size are summarized in Table 11.4. In all cases, about 50–54% of the energy is dissipated as a shock wave, 35–38% as thermal radiation, and 6–7% as ionizing radiation. Table 11.5 summarizes estimates of the likely causes of damage to forested terrain as a result of atmospheric or surface detonations of nuclear devices of the size considered in Table 11.4. Close to the blast epicenter, initial damage to trees would be caused by the power of the blast wave, and to a lesser extent by ionizing radiation. However, much larger areas of forest would be consumed by fires ignited by the intense thermal radiation, and those conflagrations would account for most of the shorter-term damage caused to vegetation. The most important effects on vertebrates would be caused by thermal radiation, and the area affected in this way would extend 30–34% beyond the area of ignition of vegetation by thermal radiation.

Therefore, a single detonation of any of these variously sized nuclear devices would result in an extensive devastation of forests and other ecosystems. The simultaneous detonations of nuclear weapons at other sites would cause similar effects. The aggregate ecological damage would be enormous. However, these would largely be limited to the northern hemisphere, because most strategic targets are located there.

Climatic Effects

There have been attempts to model the global climatic consequences of a large-scale nuclear exchange. The complex computer models of atmospheric physical properties and circulation processes that have been used in these studies were developed to simulate the potential climatic consequences of increasing concentrations of atmospheric CO_2 (Chapter 2). In the secondary application of these models to predict the climatic effects of nuclear warfare, key parameters were modified to reflect the expected atmospheric changes that would be caused by a nuclear holocaust. Especially important in this respect are injections of small carbonaceous smoke particulates and inorganic dusts into the upper atmosphere (Crutzen and Birks, 1982; Covey, 1985; NRC, 1985a; Pittock et al., 1985; Stephens and Birks, 1985; Turco et al., 1990).

Table 11.4 Distribution of energy released by a nuclear bomb exploded in the troposphere[a]

Energy	Yield: Bomb size:	Energy released for bomb size		
		(10^{12} J) 18 kT	(10^{15} J) 0.91 MT	(10^{15} J) 9.1 MT
Blast (shock)		41.9	2.28	22.8
Thermal radiation		29.3	1.59	15.9
Nuclear radiation (first minute)		4.2	0.10	1.0
Nuclear radiation (residual)		8.4	0.21	2.1
Total		83.7	4.19	41.9

[a]From Westing (1977). Note that the data refer to typical air bursts and that the kT bomb size is a fission bomb, while the other two are half fission and half fusion.

Table 11.5   Damage to biota caused by a nuclear bomb exploded in the troposphere or on the surface[a]

| | Area suffering the given type of damage (ha) | | | | | |
| | Tropospheric air burst | | | Surface detonation | | |
Type of damage	18-kT bomb	0.91–MT bomb	9.1-MT bomb	18-kT bomb	0.91–MT bomb	9.1-MT bomb
Craterization by blast wave	0	0	0	1	12	57
90% of trees blown down by blast wave	565	14,100	82,000	362	9040	52,500
Trees killed by nuclear radiation	129	648	1250	148	12,800	63,800
All vegetation killed by nuclear radiation	18	312	759	43	2830	12,100
Vegetation ignited by thermal radiation	1170	33,300	183,000	749	21,300	117,000
Vertebrates killed by blast wave	43	591	2740	24	332	1540
Vertebrates killed by nuclear radiation	318	1080	1840	674	36,400	177,000
Vertebrates killed by thermal radiation	1570	42,000	235,000	1000	26,900	150,000

[a]Modified from Westing (1977).

One commonly used (but of course, not necessarily correct) scenario is for a 6.5 thousand-MT exchange (after NRC, 1985a; Stephens and Birks, 1985), which would cause the explosive injection of 330–825 million tonnes (Tg) of particles into the atmosphere, of which 35% would enter the stratosphere and therefore have a long residence time. These largely inorganic particulates would be accompanied by some 180–300 Tg of sooty smoke from fires, much of which would also end up in the stratosphere. The inorganic dust would increase the albedo of the atmosphere and thereby decrease the amount of sunlight that penetrates to ground level. However, probably more important would be effects of the sooty smoke, which would absorb incoming solar radiation at a high altitude in the atmosphere and then reradiate much of the absorbed energy back to space as long-wave infrared radiation. In addition, a high-altitude layer of warm air would create a strong atmospheric stability and thus retard the rates of processes by which particles are removed from the atmosphere. The two mechanisms of (1) reflection by dust and (2) absorption and reradiation by sooty smoke could reduce the amounts of energy received at the Earth's surface by more than 90%.

It is impossible to predict accurately the climatic consequences of effects such as these. This is largely because of uncertainty over the persistence of the particulates at various altitudes in the atmosphere, and because there are various scenarios concerning the scales of the potential exchange of nuclear weapons, the resulting quantities and particle-size distributions of dust and smoke, and other important complications. However, a number of studies using different atmospheric-process models and nuclear-exchange scenarios have concluded that the climatic effects could be severe and that persistent subfreezing temperatures could occur over much of the Northern Hemisphere and, to a lesser extent, in the Southern Hemisphere (Thompson et al., 1984; Covey, 1985; Pittock et al., 1985; Turco et al., 1990). The phenomenon of potentially widespread climatic deteriorations has been labeled "nuclear winter." Some other studies have predicted less consequential cooling effects, by assuming smaller emissions of sooty dusts and more moist smoke, which would be relatively dense and would therefore not penetrate so effectively to the stratosphere (Cess, 1985; Small and Bush, 1985; Penner et al., 1986; Schneider and Thompson, 1988). This smaller climatic effect might be considered a "nuclear autumn."

Atmospheric effects of a nuclear holocaust would

include the generation of large quantities of gaseous pollutants. An estimate for the 6.5 thousand-MT nuclear exchange scenario is for the production of 32.5 Tg of gaseous NO from the oxidation of atmospheric N_2 by the heat and pressure of explosions, plus 4.4 Tg of NO produced by fires (NRC, 1985a; Stephens and Birks, 1985). In addition, 225 Tg of CO would be produced by the combustion of some 4500 Tg of organic fuel, and there would be large emissions of SO_2 and other oxides of sulfur, hydrocarbons, asbestos fibers, metals, chlorinated hydrocarbons, and other toxic substances. Some of these pollutants (especially NO_x) would consume stratospheric ozone, resulting in an increased penetration of ionizing, solar ultraviolet radiation to the Earth's surface, with secondary biological effects (see Chapter 2). For the simulation of a 6.5-thousand-MT exchange, it was estimated that there would be a 17% decrease in the concentration of stratospheric O_3 (NRC, 1985a; Stephens and Birks, 1985). On the other hand, the concentration of O_3 in the troposphere could increase due to photochemical reactions involving the large concentrations of NO_x and hydrocarbons (Crutzen and Birks, 1982; see Chapter 2).

Several studies have attempted to assess the likely ecological consequences of a nuclear winter (Harwell and Hutchinson, 1985; Grover and Harwell, 1985; Grime, 1986; Westing, 1987). Important damage to natural and agricultural vegetation would be caused by chilling and freezing injuries. That damage would be especially severe in vegetation at low latitudes, since tropical and subtropical plants are not adapted to cold temperatures. For example, during the growing season a typical, woody, tropical plant cannot tolerate a temperature of 0 to 4°C for 24 hr without suffering lethal damage, while cultivated tropical plants cannot tolerate 3 to 7°C (Larcher and Bauer, 1981). Somewhat lower temperatures can be tolerated by tropical plants for shorter periods of time, especially if they are exposed during a dormant season (e.g., as low as −3 to −5°C for tropical trees, −4 to −6°C for tropical grasses). Plants of higher latitudes can easily tolerate much colder temperatures than the above when they are in a winter-hardened condi-

tion. However, damage can even be caused to arctic plants if they are exposed to prolonged cold temperatures during their growing season (Larcher and Bauer, 1981).

Therefore, if the predictions of a prolonged period of low, possibly subfreezing temperatures during a nuclear winter are correct, there would be devastating effects on terrestrial vegetation. These effects would be especially great if the nuclear winter occurred during the growing season, when plants of higher latitudes are not hardened to tolerate cold temperatures. The results could include a catastrophic collapse of agricultural production, and the devastation of most natural ecosystems. The temperature-related effects on aquatic plants would likely be smaller, since large water bodies provide thermal buffering.

Vegetation would also be affected by the small rates of insolation that are predicted to occur in the aftermath of a nuclear holocaust. The immediate effect would be decreased rates of photosynthesis and a loss of net production. However, it is notable that the ability of some plants to survive under a condition of darkness is enhanced by low temperatures (Hutchinson, 1967).

Damage to vegetation would also be caused by (1) an increased surface flux of solar ultraviolet radiation, because of the deterioration of stratospheric ozone; (2) large doses of ionizing radiation from longer-lived thermonuclear fission products; and (3) effects of phytotoxic air pollutants, such as O_3, SO_2, and NO_x.

These various effects of the nuclear aftermath would, of course, add to the great ecological destruction caused by blast, thermal radiation, fire, and ionizing radiation within a short time of the nuclear exchange.

Plants that survive the initial and aftermath conditions of a nuclear holocaust would most probably have some combination of the following characteristics (after Grime, 1986): (1) a persistent seed-bank that would allow regeneration even if all adult plants were killed; (2) a ruderal strategy, characterized by a large reproductive output, great dispersal ability, and a general ability to do well under disturbed, but not excessively stressful, conditions;

(3) adaptation to stressful habitats of small potential net productivity, since plants of such sites are often relatively tolerant of low temperature, low insolation, and other stresses; and (4) be components of an ecosystem that is naturally resilient to the effects of severe, periodic or cyclic disturbance, since such plants are generally capable of effective regeneration after an event of mass mortality. Plants with these various characteristics would be relatively well adapted to survive or regenerate after a nuclear winter, and they would be prominent in the ensuing ecological succession.

Animal life would also be devastatingly affected by the environmental conditions occurring during a nuclear winter. This would largely be due to an intolerance of prolonged low temperatures, the effects of ultraviolet and ionizing radiation, and shortages of food and other critical habitat features as a result of damage to vegetation. Terrestrial animals (including humans) would probably be more greatly affected than aquatic animals.

In summary, it is clear that the predicted environmental consequences of all-out nuclear warfare would be horrific. Because of the many possible war scenarios and the incomplete knowledge of atmospheric and ecological responses, it is not possible to predict the effects with accuracy. However, the general consensus of several independent modeling experiments is for a bleak post-holocaust future for both humans and the biosphere. Hopefully, these predictions will never be tested in a real-life experiment.

Warfare has always been a part of the human enterprise, but this has invariably been deeply regretted by at least some of the more enlightened participants and observers. Awareness of the catastrophic effects of warfare, including the terrible potential of any nuclear war, is essential to deter use of this strategy to resolve conflicts. Recent signings of treaties on nonproliferation of nuclear weapons and nuclear arms reductions provide real hope in this regard.

Optimistic, war-less hopes for the future spring eternal, and are exemplified by this passage:

> They shall beat their swords into plowshares, and their spears into pruninghooks: nation shall not lift up sword against nation, neither shall they learn war any more. (Isaiah **2**: 4).

12

BIOLOGICAL RESOURCES

12.1
INTRODUCTION

The Earth is finite, and so therefore are the planetary resources available to sustain the economic endeavors of humans. This fact is especially obvious for nonrenewable natural resources, such as coal, petroleum, and metal ores, which can only be mined. However, the stocks of potentially renewable natural resources, such as forests, hunted animals, and flowing water, can also be rapidly depleted by harvesting that exceeds the rate of replenishment.

The overall, ecological reality is that humans have an absolute dependence on sustained flows of natural resources to support their livelihoods and economic systems. It is self-evident that over the longer term, sustainable economic systems can only be based on the sensible use and management of renewable natural resources, that is, in ways that do not diminish their quality and availability to future generations.

However, the predominant pattern of use by humans of potentially renewable, natural resources has been anything but sustainable. Almost always, these resources are harvested excessively, or managed inappropriately, so that they are quickly exhausted by "mining." The essential reasons for this behavior, which is clearly self-destructive over the longer term, are:

1. An ethic in most cultures that presumes that humans have an unalienable right to take whatever is desired from nature. This world view is symptomized by the Judeo-Christian ethic (White, 1967), based on the biblical view of creation, in which God directs humans to "be fruitful, and multiply, and replenish the earth, and subdue it," and to "have dominion over the fish of the sea, and over the fowl of the air, and over the cattle, and over all the earth and over every creeping thing that creepeth upon the earth" (Genesis 1:28). The self-aggrandizing and ecologically arrogant attitudes of the Judeo-Christian culture are not unique among the world's major societies and religions. Judeo-Christian culture has, however, been especially influential in the development of the dominant, technological ethic that has legitimized overexploitations of potentially renewable natural resources.

2. Self-interests of individuals and their societies, which are manifest in desires and actions to optimize profits over the shorter term and to deemphasize the longer-term implications of actions and policies. The maximization of profits is favored by the fact that often

3. Investments of money in some sectors of the larger economy may grow faster than many natural resources. As a result, it may appear to be economically prudent, as a short-term strategy, to

liquidate natural resources through overexploitation and then invest the "profits" in some other, more profitable part of the economy. This strategy only works if

4. Not all of the costs are accounted for. In particular, costs that are indirect and ecological, involving resource and environmental degradation, are not taken into account in the calculation of apparent profit.

These factors, when operating within a context of free access to common-property resources, inevitably lead to overuse and resource collapse. This commonly observed phenomenon was called "the tragedy of the commons" by Hardin (1968). In a highly influential essay, Hardin used the analogy of a publicly owned pasture to which sheep farmers have free, unrestricted access to graze their animals. This system causes a degradation of the pasturage resource by overgrazing, because individual farmers perceive that there are shorter-term, economic benefits from having as many of their own livestock as possible grazing the pasture, causing an excessive, aggregate use. One of Hardin's conclusions was that "freedom in a commons brings ruin to all."

The tragedy of the commons is a well-known phenomenon. Societies everywhere bemoan the diminished supplies and other degradations of the natural resources that sustain them. Yet, in so many cases, little of substance is done to effectively manage the problems of resource degradation. So far, with remarkably few exceptions, the actual design and implementation of sensible resource-management strategies have proved to be beyond the capability of our societies.

Potentially, humans have the capability to develop longer-term world views and to implement resource systems that could protect the health of the ecosystems that sustain them. Resource-related predicaments clearly require solutions that integrate scientific knowledge, social changes, political will, and the development and adoption of new, ecologically based, economic approaches that pursue sustainability, rather than growth and unabated overexploitation of resources.

In this chapter, a few case studies of potentially renewable, biological resources that have been depleted by excessive harvesting are examined. The major reasons for overexploitation as a seemingly universal but clearly inappropriate style of resource use are then explored, and better pathways are identified through brief examination of the emerging discipline of ecological economics.

12.2
EXAMPLES OF DEGRADATION OF BIOLOGICAL RESOURCES

Easter Island

A prehistoric example of resource degradation occurred on Easter Island, a small (389 km^2) island in a temperate-climate region of the Pacific Ocean (Jennings, 1979; Ponting, 1991). Easter Island is remote and isolated, being 3.2 thousand km west of South America and 2 thousand km from Pitcairn Island, the nearest inhabited place. When Europeans first observed Easter Island in 1772, there were about 3 thousand indigenous people, apparently living in squalid conditions in caves and reed huts, engaged in perpetual warfare among clans, and engaged in cannibalism, possibly to supplement the meager supplies of exploitable food on the then-treeless island.

Prior to its colonization by humans, most of Easter Island was probably covered by a dense forest. It is estimated that Polynesians first discovered Easter Island in the fifth century, bringing with them chicken (*Gallus gallus*), Polynesian rat (*Rattus exulans*), and sweet potato (*Ipomoea batatas*) as their only viable crops, the climate being too temperate for their other, tropical-crop species, such as breadfruit (*Artocarpus incisus*), coconut (*Cocos nucifera*), taro (*Colocasia antiquorum*), and yam (*Dioscorea batatas*).

There is archaeological evidence of an initially flourishing, prehistoric Polynesian society. The population may have been as large as 7 thousand in 1550. At that time, an important cultural activity, probably central to the identity and cohesion of the social system, was the laborious erection of about 600 large, stone, human-visaged monoliths at coastal locations. The stones were quarried at an inland place, but they were heavy (more than 20 tonnes each) and could only be moved with great human effort (there were no draught animals) to their coastal repositories by rolling them upon logs cut from the island's forests.

The cutting of trees for use as stone rollers, and as timber for the construction of buildings and fishing boats, eventually led to a prehistoric deforestation of Easter Island. Loss of the essential forest resource meant that monoliths could no longer be moved, sturdy homes could no longer be constructed of wood, and fishing was no longer possible, because there was no way to build boats that were strong enough to use in the surrounding ocean. Consequently the deforestation of Easter Island led to a virtual collapse of this Polynesian society. Warfare among rival, resource-limited clans became endemic. The cultural collapse was so substantial that at the time of the European discovery of the island, none of the indigenous residents could remember who had erected the stone monoliths, or why.

The lesson of Easter Island is that human societies, even prehistoric aboriginal ones, are quite capable of causing collapses of the vital, ecological-resource system that sustains them. The inhabitants of Easter Island must have been acutely aware of their isolated and precarious circumstances and of the limited resources that were available to sustain themselves and their society on a small island. As these natural resources began to become obviously diminished in quantity, the Easter Islanders may have engaged in heated discussions about conservation. Clearly, however, any such deliberations ultimately came to naught, and their resource system, economy, and culture suffered irretrievable collapses.

There are obvious parallels between the resource limitations, and the resource failure of Easter Island and that of Earth as a planetary island. Globally, there are limited quantities of energy, materials, and biological resources available to sustain human societies. All of these natural resources can be rapidly depleted by excessive usage. The Easter Islanders

had no alternative resource-rich refuge to which they could escape from their self-inflicted ecological catastrophe. There is also no alternative refuge to Earth.

The Mediterranean Region

There are many references in the biblical, historical, and anthropological literatures to ecologically rich places that became degraded through excessive harvesting or conversion of their ecological resources. This phenomenon is especially true of places that were degraded by deforestation. For example, as early as 8 thousand years ago, villages in what is now central Jordan may have been abandoned because of soil erosion associated with deforestation, and there are many historical references to flourishing cities and rich agriculture in presently near-barren landscapes of Mesopotamia and the valleys of the Tigris and Euphrates Rivers (Thirgood, 1981; Ponting, 1991).

The biblical regions of the Middle East used to support extensive, primeval forests, replete with now extirpated animals, but overharvesting and ecological conversions have caused these communities and their species to vanish. There are many references in the Bible that evoke these ecological changes, including the following examples (Tristram, 1873):

1. There are mentions in the Bible of places that are now semidesert or desert, but had biblical names that suggest forest cover, for example The Forest of Hamath, The Wood of Ziph, and The Forest of Bethel. A few of these places retained their forest for relatively long times. For example, during the Crusades of the 11th to 13th centuries, there were still extensive pine forests in the now deforested area between Jerusalem and Bethlehem, and some of the interior valleys of Lebanon still had cedar-dominated (*Cedrus libani*) forests into the 19th century, although these have since been harvested and converted to other ecosystems. There are a few biblical references to the harvesting of cedars in this region, including the following declaration of the Assyrian king, Sen-

nacherib (705–681 *BC*): "I am come up to the height of the mountains, to the sides of Lebanon, and will cut down the tall cedar trees thereof, and the choice fir trees thereof" (2 Kings, XIX 23). The cedars of Lebanon were famous for their abundance, height, girth, and quality as timber for the construction of buildings and ships, but today they only survive in four, scattered, endangered groves of small trees (Ponting, 1991).

2. There are mentions in the Bible of plants and animals that are now extirpated from the biblical region. Exodus refers to extensive marshes of the reed-like papyrus (*Cyperus papyrus*) in the lower reaches of the Nile River. The infant Moses, for example, was discovered by an Egyptian princess in what was probably a papyrus-dominated marsh, where he had grounded after being set afloat by his mother in a small boat, probably constructed of the same reed. Papyrus marshes no longer occur in lower Egypt. The aurochs (*Bos primigenius*) is an extinct species of cow that is mentioned in the Bible. There are other citations of animals that are now extirpated from the biblical region, including lion (*Panthera leo*), bear (*Ursus arctos*), hippopotamus (*Hippopotamus amphibius*), and large antelopes such as addax (*Addax nasomaculatus*), bubale (*Alcephalus bucelaphus*), and oryx (*Antilope leucoryx*).

Most other regions around the Mediterranean Sea have become extensively degraded by unsustainable harvesting of forests for fuel and timber and by unsuitable agricultural practices on the deforested lands (Thirgood, 1981; Maser, 1990; Ponting, 1991). These ecological degradations were well advanced in the earliest, historical times. Ecological changes in Greece were remarked upon by Plato (ca. 429–347 *BC*) in *Critias* (after Rubner, 1985; Ponting, 1991):

> This present land (is) justly called a remnant of what once existed. What now remains . . . is like the skeleton of a sick man, with all the fat and soft earth having wasted away, and only the bare framework of the land being left . . . [In early times] the mountains were high and covered with soil, and likewise the plains, which now have come to be called stony fields, were full of rich earth; furthermore the mountains bore

thick forests, of which visible evidence survives to the present day. There are some mountains which now have nothing but food for bees [i.e., herbaceous wild-flowers], but they had trees not very long ago that were felled for rafters in the largest buildings, the roofs of which still stand intact . . . . There were many lofty trees of cultivated species and . . . boundless pasturage for flocks. Moreover, [the land] was enriched by the yearly rains from Zeus, which were not lost to it, as now, by flowing from the bare land into the sea; but the soil it had was deep, and therein it received the water, storing it up in the retentive loamy soil, and . . . providing all the various districts with abundant supplies of springwaters and streams, whereof the shrines still remain even now, at the spots where the fountains formerly existed.

Plato's descriptions are rather intuitive in terms of their linkages of human activities and their direct and indirect ecological consequences. The descriptions are also prescient of further, more extensive ecological damage in the same and other regions around the Mediterranean Sea, where deforested and otherwise degraded landscapes are now common and extensive.

Excessive harvesting of forests, coupled with overgrazing and agricultural conversions, have caused deforestations and associated ecological degradations in many other regions. Areas have been affected in this way throughout the world, including most of the Mediterranean region, the Middle East, much of Europe, south Asia, much of temperate North and South America, and increasingly, many parts of the subtropical and tropical world.

Recent Depletions of Some Particular, Biological Resources

In previous chapters, a few cases of biological resources that were exhausted by unsustainable harvesting were described. The Pleistocene megafaunal extinctions and those of the dodo (*Raphus cucullatus*), passenger pigeon (*Ectopistes migratorius*), and great auk (*Pinguinus impennis*), among other species, are examples of biological extinctions caused at least in part by excessive harvesting. The American bison (*Bison bison*) and various species of marine mammals are examples of exploitation to the level of unprofitability (i.e., economic extinction of biological resources) (see Chapter 10). The fishery changes in Lake Erie (Chapter 7) involved a multispecies resource that was depleted by excessive harvesting of a series of progressively less desirable species, a process that was exacerbated by degradations of habitat quality. In the following sections, some additional marine examples are given to illustrate the exhaustion of bioresources, caused by excessive, uncontrolled harvesting.

Whales

Virtually all populations of marine mammals that have been hunted on an industrial scale became severely overexploited. A few species were overharvested to such small abundances that the species became extinct, as described in Chapter 10 for Steller's sea cow (*Hydrodamalis stelleri*) and the Caribbean monk seal (*Monachus tropicalis*). In other cases, subspecies or particular populations were extirpated, such as the Atlantic population of the gray whale (*Eschrichtius robustus*) (Ellis, 1991).

The first species of whale to be hunted were the slowest-moving ones, such as the northern right whale (*Balaena glacialis*; this was considered the "right" whale to kill, because it was slow moving and floated when dead), bowhead whale (*Balaena mysticetus*), and gray whale. In the earliest European hunts, men would row or sail beside a whale, harpoon it, and allow it to tow their boat until the animal exhausted itself (New England sperm whalers called this a "Nantucket sleigh ride"). The whale would then be repeatedly lanced until it bled to death. This was a rather primitive hunting technique, but it was sufficiently effective to kill these sorts of whales in large numbers and to eliminate them from European waters.

Much to their probable frustration, the earliest, sailing whalers could only gaze longingly at the much larger, and much faster species of whale, especially the rorquals (*Balaenoptera* spp., such as blue, fin, sei, and minke whales). It was not until the development of steam-assisted sailing ships in the 1850s, and the later full-steam ships, that it became possible to catch these swifter whales. Just as important, the invention in 1873 of the harpoon gun, and later the exploding-head harpoon by the Norwegian, Sven Foyn, made it possible to rapidly dispatch even the largest rorquals. Once killed, these could be quickly inflated with compressed air to keep the carcass from sinking, and then processed at the side of the vessel or on shore. The acme of sophistication of whaling technology was reached in 1925 with the introduction of factory ships, which could spend long periods of time in remote, pelagic waters, processing whales brought to them by small fleets of fast, catcher-killer boats, often guided by spotter aircraft. These modern, industrial methods of whaling allowed whalers to go wherever their prey might happen to be and to locate, kill, and process animals of all species and sizes, in numbers large enough to deplete the stocks rapidly, and profitably.

Generally, populations of whales and seals were not overharvested to biological extirpation. Rather, they were hunted to levels of abundance below which it was no longer profitable to kill and process them. Where the hunting grounds supported a number of species of marine mammals, the hunting focused initially on the largest, most desirable species (i.e., from the perspective of short-term profits). As the stock of these was exhausted, there would be a switch to smaller, less profitable species, which would also be depleted. This process of sequential exploitation of a mixed-species biological resource is called "working up" and is described in more detail later in this chapter.

The sequential exploitation of a marine-mammal community is illustrated in Fig. 12.1 for five species of whales in antarctic waters (Ellis, 1991). Initially, these whales were subjected to an unregulated, open-access hunt by whalers from various countries, in which the harvest was limited only by the attainable speed of finding, killing, and processing the animals. After the International Whaling Commission (IWC) was established in 1949, attempts were made to exert conservation-related controls over this profitable, fiercely competitive, highly capitalized, multinational industry. These efforts were not successful, because of difficulties in esti-

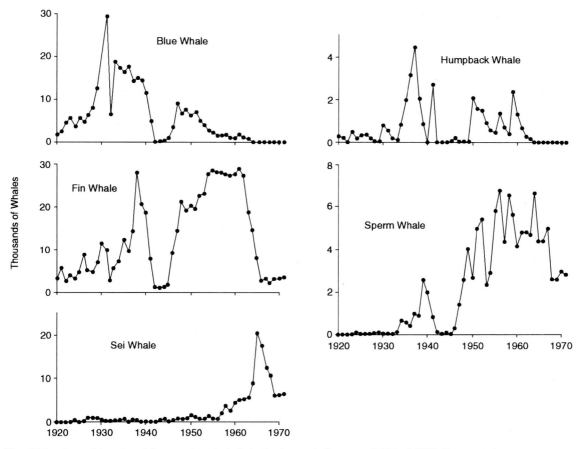

Fig. 12.1   Annual harvest of five species of whale in the Antarctic between 1920 and 1971. Data are
from Ellis (1991).

mating stock sizes and rates of recruitment, in se-
curing cooperation among the whaling nations, and
a general lack of will on the part of the IWC in the
setting and enforcing of realistically small quotas.

These sorts of problems are, of course, anticipated when a profit-driven industry is expected to
regulate and police itself. According to J. L. Mc-
Hugh, a former U.S. commissioner and chairperson
of the IWC: "From the time of the first meeting of
the Commission . . . almost all major actions or
failures to act were governed by short-range economic considerations rather than by the requirements of
conservation" (cited in Ellis, 1991, p. 390).

Initially, the blue whale (*Balaenoptera mus-*

culus) was the most profitable species to hunt in the
antarctic seas, because of its enormous size, reaching up to 32 m and 136 tonnes for the largest male
animals (Burt, 1976). The initial population of blue
whales in antarctic waters is estimated to have been
about 180 thousand animals (NMFS, 1991a). About
2 thousand individuals were killed each year in
1920–1921, and a peak harvest of 29 thousand/year
was reached in 1931 (Fig. 12.1; note the brief re-
spite during 1941–1945, when as few as 59 ani-
mals/year were taken because of the Second World
War). Between 1955 and 1962, the harvest was 1.1–
2.2 thousand/year, but the harvest collapsed thereafter, to 112 animals in 1964, 20 in 1965, 1–4 in

1966–1967, and zero after 1968 (the killing of blue whales was prohibited after 1965). The total kill of blue whales in antarctic waters between 1909 and 1967 was about 331 thousand animals (NMFS, 1991a). The present population of antarctic blue whales has been estimated with much uncertainty as 450–1900 animals, less than 1% of the initial abundance (NMFS, 1991a; Best, 1993). The global population of blue whales is less than 10 thousand individuals (NMFS, 1991a).

The fin whale (*B. physalus*; up to 21 m long) was the next most favored species. The initial population of this species in antarctic waters was about 400 thousand animals (NMFS, 1991a). The fin whale sustained annual harvests as large as 29 thousand (Fig. 12.1.), and it also declined greatly in abundance (as indexed by the annual catch), although not to the point of commercial extinction. The total kill of fin whales in antarctic waters between 1904 and 1975 was more than 704 thousand animals (NMFS, 1991a). The present population of antarctic fin whales is 85 thousand animals, about 21% of the initial abundance (NMFS, 1991a). The global population of fin whales is more than 110 thousand animals (NMFS, 1991a).

As these largest species became more difficult or impossible to catch because of their increasing rarity in Antarctica, they were replaced as primary objects of the hunt by initially less desirable species, such as sei whale (*B. borealis*), humpback whale (*Megaptera novaeangliae*), sperm whale (*Physeter macrocephalus*), and minke whale (*B. acutorostrata*). (Note, however, that if encountered during hunts for smaller species, the larger whales were taken opportunistically.) The smaller species were also overharvested, as is suggested by their patterns of decreasing harvests over time, and the ban on humpback killing in the early 1960s. The total kill of humpback whales in these waters between 1904 and 1978 was about 148 thousand animals, while 704 thousand sei whales and more than 928 thousand sperm whales were killed, worldwide (NMFS, 1991a). The initial population of humpback whales of 100 thousand animals was reduced by hunting to a present antarctic population of about 20 thousand, while sei whales were reduced from >65 thousand to >12 thousand animals and sperm whales from 1.18 million to 780 thousand (NMFS, 1991a; Best, 1993).

By the early 1980s, the remaining antarctic whalers were mostly killing the relatively small (to 9.1 m) but still numerous minke whale, of which about 760 thousand were taken in Antarctica. In 1982, the International Whaling Commission announced a moratorium on antarctic whaling, to begin in the summer of 1985–1986. However, Japan and the then-USSR objected to the moratorium and continued their commercial hunt until 1986–1987. Thereafter, the only antarctic whaling was by Japan, which killed 2–3 hundred minke whales each year for "scientific" purposes during 1987–1989 (Horwood, 1990; WRI, 1990). In 1994 the IWC declared the Antarctic Ocean to be a whale sanctuary (as is the Indian Ocean), although it remains to be seen whether any research whaling will yet be conducted in those waters.

Overall, the intense overexploitation of antarctic whales is estimated to have reduced populations by factors of >99% for blue whale, 97% for humpback whale, 82% for sei whale, and 79% for fin whale (Allen, 1990; WRI, 1990; NMFS, 1991a).

A northern hemisphere example of the industrial depletion of a whale stock is illustrated in Fig. 12.2 for the bowhead whale of the Arctic Ocean of northwestern North America and far-eastern Siberia. As was noted previously, this slow-moving whale could be easily overtaken and killed, even by sailing

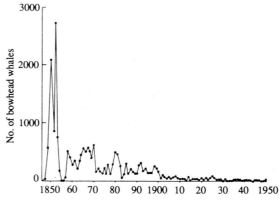

Fig. 12.2 Annual catches of bowhead whales from the beginning of the commercial hunt in 1848 to the mid-1950s. Modified from Cushing (1988).

ships. The initial population of bowhead whales in the western Arctic was about 18 thousand animals (there were another 37 thousand animals in the north Atlantic and eastern Arctic; NMFS, 1991a). After a bonanza of killing in the 1850s of the newly discovered Alaskan and western Canadian population of this highly vulnerable species, which winters in the Bering Sea and summers in the Beaufort Sea, its abundance declined precipitously and remains small, estimated at perhaps 5–8 thousand individuals (Ellis, 1991; NMFS, 1991a). The commercial hunt for bowhead whales in the western arctic ended early in the twentieth century, after about 19 thousand animals had been killed (NMFS, 1991a). However, in recent decades this population of bowhead whales has still been subject to an aboriginal hunt that kills about 20–46 animals/year (see also Chapter 10). Each year most of the population of bowhead whales migrates past Point Barrow in northern Alaska to their summering waters in the Beaufort Sea, and in 1978 about 2.3 thousand animals are estimated to have passed that place, giving a minimal estimate of the population size (Cushing, 1988). There are now about 8 thousand bowhead whales, including about 7.5 thousand in the Bering, Chukchi, and Beaufort Seas (Best, 1993).

In spite of the relatively small, modern kill of bowhead whales, the population of this species appears to be recovering very slowly. Nor has the depleted population of a similar species, the northern right whale, substantially recovered since its commercial exploitation was stopped. The right whale hunt during 1600 to 1700 was the first commercial whaling enterprise. This species was extirpated from the European Atlantic Ocean by Basque whalers during that period, as was the northeastern population of gray whale at about the same time.

When populations of northern right whales were discovered in the northwest Atlantic of North America, they were also severely depleted. The initial abundance of this species in the western north Atlantic was about 10 thousand animals (NMFS, 1991a). About 25–40 thousand animals are estimated to have been killed during 1530 to 1610 in the Basque hunt and only occasional animals afterward because of the severely depleted stock. The endangered population of northern right whales now to-

tals only about 350 animals (NMFS, 1991a). Despite protection from whaling for more than 50 years, the abundance of this species does not appear to have increased, probably because of mortality associated with collisions with ships and entanglement with fishing gear on the busy, eastern North American coast (Krause, 1990; Gaskin, 1991).

A final example of stock depletion of a species of whale is that of the population of the gray whale of the west coast of North America. This species winters and breeds in the warm waters of Baja California and migrates up the coast of the United States and Canada to spend the summer in the western Arctic Ocean. The commercial killing of gray whales began in 1845, with most hunting occurring on its wintering range and during its twice-yearly migrations. By 1900 the hunt had largely ended, with the abundance of gray whales being reduced from an initial <20 thousand animals to an endangered 1–2 thousand animals during the 1880s to 1930s (Northridge, 1984; Ellis, 1991; NMFS, 1991a). However, more recently the population of gray whales has increased, and its present abundance is more than 21 thousand animals, about the same as its preexploitation population (NMFS, 1991a; Best, 1993).

It has been estimated that, in total, more than 2.5 million whales of all species were killed during the commercial hunts of the last 400 years (NMFS, 1991a).

Seals

Some species of seal were much more numerous than any whales, and because they must breed and raise their young on land, their populations are seasonally aggregated. Moreover, seals are awkward and ungainly on land and sea ice, where they can be killed easily with clubs, lances, or guns. As a result, seals have been slaughtered in tremendous numbers for their skins which can be manufactured into fur or leather, for their blubber which can be rendered into oil, and for their meat, teeth, bacula, and other useful products.

Unregulated, open-access hunting caused great depletions of seal abundance. There were some regional or local extirpations and in a few cases extinctions of species that had initially small populations (see Chapter 10). Some of the severely depleted populations, for example, northern elephant seal (*Mirounga angustirostris*) and northern fur seal (*Callorhinus ursinus*), have substantially recovered their abundance following the adoption of strict conservation measures. A few species, most notably the harp seal (*Phoca groenlandica*) in the northwest Atlantic part of its range, were able to maintain relatively large populations throughout the most intense period of onslaught by hunters and were never endangered. Some of these examples are discussed in more detail below.

The Caribbean monk seal (*Monachus tropicalis*) is an example of a species of seal, initially occurring in relatively small and local populations, that was rendered extinct by overharvesting. The last observation of this species was in 1952. It has been suggested, however, that the critical blow was delivered to this species when its last, large breeding population was discovered by a scientific expedition off Yucatan in 1886, and 40 of the animals were taken as "specimens" (Busch, 1985). Another, non-seal example of an extinction of a marine mammal is that of Steller's sea cow (*Hydrodamalis stelleri*) of the Bering Sea, discovered by Europeans in 1741, subsequently hunted, and apparently extinct by 1768 (Chapter 10).

Other species of initially relatively uncommon seals that have been rendered endangered by overharvesting include the Guadalupe fur seal (*Arctocephalus townsendi*) of the west coast of Mexico, the Juan Fernandez fur seal (*Arctocephalus philippii*) of southern Chile, and the Hawaiian monk seal (*Monachus schauinslandi*). The endangered Mediterranean monk seal (*M. monachus*) has been endangered by a combination of overharvesting and habitat loss (see also Chapter 10).

The above species never occurred in large populations. Some other species of seal, such as the northern fur seal of the Bering Sea, sustained much more substantial hunts. This species has an especially desirable pelage, with a hair density of about 45 thousand/cm², about one-half that of the sea otter (*Enhydra lutris*) (Busch, 1985). About 80% of the population of the northern fur seal breeds on the

Pribolof Islands of Alaska, while the rest procreate on a few other islands off eastern Siberia and northern Japan. Because of the seasonal concentrations of this species and its awkwardness on land, it can easily be herded into large groups to be clubbed or shot.

From the beginning of the commercial hunt to 1911, an estimated 6 million northern fur seals were killed for their pelts (Busch, 1985). By 1920 this species had been depleted to only 120 thousand animals, but strict regulation of the hunt since about 1911 has allowed a substantial recovery of the abundance of these seals, to about 1.8 million individuals (Northridge, 1984). Until recently this species was subject to a regulated hunt, with about 25 thousand animals killed each year (Busch, 1985). Except for an aboriginal subsistence hunt, the exploitation of northern fur seals has since been stopped because of population declines, possibly related to overexploitation of its food resource by human fishers.

The southern fur seal (*Arctocephalus fosteri*) was also subjected to a massive hunt that eliminated the species from many places and decimated its populations everywhere. In total, an estimated 5.2 million southern fur seals were killed for their pelage and blubber, but the populations of this species are now increasing rapidly in antarctic and subantarctic waters (Busch, 1985).

Other substantial hunts were for southern and northern elephant seals (*Mirounga leonina* and *M. rostris*, respectively). These are large species of seal, with the biggest males being about 5.5 m long and weighing 3 tonnes (Busch, 1985). These animals were killed almost exclusively for their blubber, which was rendered into oil for use in lamps, lubrication, and the manufacture of soap and paint. In total, about 1 million elephant seals were killed, and both species were taken to a small abundance.

The hunt for northern elephant seals began in 1846 and was essentially finished only 30 years later because of the depleted resource (Busch, 1985). By that time, only a few stragglers could be harvested in places where the species had formerly been abundant. As an example of the absence of a conservation ethic in the marine-mammal hunts of those times, it is worth noting that in 1880, after several years during which no elephant seals could be found to kill, a small population of these animals was discovered at San Cristobal Bay in Baja California. All 300 animals on that beach were dispatched and rendered into oil. Between 1884 and 1892 no elephant seals were observed, in spite of several searches by museum-based scientists. In 1892, however, a team from the Smithsonian Museum found 8 animals on a beach on Guadalupe Island off Baja California, and according to the chief scientist, Charles H. Townsend: "some of the elephant seals were secured." As it turned out, the team killed seven of the eight animals as specimens. At the time, there may have only been about 20 individuals of this species alive. In another scientific triumph, a small herd was discovered in 1907 by a team from the Rothschild Museum, and 10 animals were killed as specimens.

In spite of the best efforts of the scientists of the time, protection from commercial hunting allowed the northern elephant seal to achieve a remarkable recovery of abundance. In 1922, a herd of 264 animals was discovered on Guadalupe Island. The government of Mexico immediately gave complete protection to the species and posted a garrison on the island to enforce that regulation. By 1957 there were an estimated 13 thousand northern elephant seals. The population increased to 48 thousand in 1976 and to more than 70 thousand in the mid-1980s.

In terms of the raw numbers of animals killed, the largest hunts of any marine mammal have been mounted for the harp seal. This is an abundant, northern species that migrates in the wintertime to subarctic, north-Atlantic waters, where it pups on pack ice. Tremendous hunts for this species have been conducted in the northwest Atlantic off Newfoundland and Labrador, in the Gulf of the Saint Lawrence River, off southeastern Greenland, and in the White Sea off western Russia. During March through April, the harp seal is highly vulnerable to human hunters. The recently born seals, called "white coats" because of the color of their pelage, are not yet aquatic and can be easily found and clubbed to death, while adult animals can be caught in nets, shot, and sometimes clubbed if they try to defend their young.

The hunt for harp seals by Newfoundlanders was especially intense, and during its heyday more than 600 thousand animals/year were killed (this occurred in 1831, 1840, 1843, and 1844; the maximum was 686 thousand animals; Busch, 1985). Between 1800 and 1914, about 21 million harp seals were killed. The magnitude of this slaughter is only comparable with that inflicted upon American bison (*Bison bison*) at about the same time (see Chapter 10). The hunts of these two species are among the most intense slaughters of large, wild animals ever undertaken by humans. (Additional examples include the modern killing of about 5 million kangaroos of several species each year in Australia, and perhaps the overhunting of the megafauna of North and South America when these landmasses were discovered by humans about 11–12 thousand years ago, although this latter killing was spread out over a much longer period of time than the hunts of buffalo, harp seals, and kangaroos; see Chapter 10.)

The hunt for harp seals also extended into more modern times. Between 1915 and 1982 about 12 million harp seals were killed, at a maximum rate of 380 thousand animals in 1956 (Busch, 1985). Since 1983, the hunts for this species have been much smaller, because of controversies about conservation, the ethics of such a large, commercial slaughter of the babies of wild animals, and other issues. In 1984 the most profitable, European markets for the fur of this species collapsed because of a legislated ban enacted on the import of whitecoat pelts into the European Economic Community (EEC). The loss of the most important market for seal pelage greatly reduced the harvests of harp seals, from 170–202 thousand animals per year during 1981–1982 to 19–80 thousand per year during 1983–1990 (Hart, 1991). Note that because young harp seals are not considered to be whitecoats after they are about 9–10 days old (they begin shedding their white fur at that time), older young can be killed and legally exported to the EEC.

In total, between 1817 and the mid-1980s, about 33 million harp seals were killed in the western North Atlantic (Busch, 1985).

The Newfoundland seal hunt was also dangerous for the hunters, because of the unpredictable perils of the weather of the wintertime North Atlantic, especially wind shifts that could trap and crush wooden vessels, and even steel-hulled ones. Between 1800 and 1865, about 400 vessels were lost, mostly to the ice, and 1 thousand men died in the Newfoundland seal hunt.

There is a pervasive imagery in the media and much of the public consciousness of a cruel seal hunt, occurring in Newfoundland and elsewhere. For example, during the frenzied rush of the hunt for harp seals, some animals might only be stunned by clubbing, rather than killed, and then perhaps be skinned while still alive. Young harp seals have rather thin skulls, which can easily be crushed by a well-aimed, heavy blow from a club. In almost all cases this should kill the animal, and it is quite possible to conduct a seal hunt without resorting to the skinning of live animals.

There are contrary views to the widespread portrayal of a cruel seal hunt. Such an opinion is evidenced by the following quotation from W. T. Grenfell, a missionary–doctor from Newfoundland and Labrador, who observed the hunt of harp seals on the pack ice in 1896 (from Busch, 1985, p. 61):

> Now the killing of young seals has been frequently described as brutal and brutalizing, and the seal hunters depicted as inconceivable savages, and this not only by shrieking faddists or afternoon tea drinkers. But, to my mind, the work is not nearly so brutalizing as the ordinary killing of sheep, pigs, or oxen, (brought) terror-stricken to the shambles (i.e., slaughter house), . . . already reeking with their fellows' blood. Here (i.e., the seal hunt) the animal is too young to feel fear, and evinces no sign of it; no animal is wounded and left to die in vain.

The above view suggests that the killing of wild seals is no more brutal that the killing of domestic livestock, of which hundreds of millions are routinely slaughtered each year, often cruelly, to provide meat and other products for use by humans. In terms of the ethics of killing animals, this logic is essentially correct. However, seals are wild creatures, while livestock are specifically bred, raised, and killed for human consumption. It is the purview of philosophers to discuss which of these, if either, is the greater moral outrage.

Remarkably, the intense hunting of harp seals in eastern Canada does not seem to have excessively depleted the abundance of this species, at least not to the point of endangerment. This was not a result of any conservation ethic on the part of the sealers. In general, the sealers killed as many as they could of any harp seals that were encountered, especially prior to 1970, when regulation through a quota system was instituted. Rather, the logistical difficulties of operating in the treacherous pack ice of the North Atlantic limited the numbers of seals that could be located and killed. When the commercial hunt was substantially discontinued in the late 1980s by a legislated end to the European markets for seal furs, the global abundance of harp seals was about 3 million animals, of which about 2 million were in Canadian waters (Busch, 1985). In 1993, the estimated abundance of this species in Canadian waters was about 3.5–4 million individuals (P. Brodie, Fisheries Canada, personal communication).

To bring this section on the industrial seal hunt to a close, it is worth noting that, in total between 1790 and 1980, at least 50 million seals of all species were killed during the various commercial hunts (Busch, 1985).

Fisheries

There are numerous examples of the discovery of new fishery resources, followed by their overexploitation to commercial extinction. These instances have been especially frequent during recent times, because of the great technological prowess of modern, industrial fishers. Historically, when fishing technologies were less effective, the stocks of fish were not so substantially depleted by harvesting. Figure 12.3 illustrates the most important methods of fishing, ranging in intensity from angling and the use of hand lines to industrial drift netting, seining, and trawling.

Between 1985 and 1987, the global harvest of marine fish averaged 79 million tonnes/year, while that of freshwater fish was 11 million tonnes/year (WRI, 1990). These harvests were substantially larger than those that occurred during 1975–1977, when the harvest of marine fish was about 23% smaller and that of freshwater fish 49% smaller

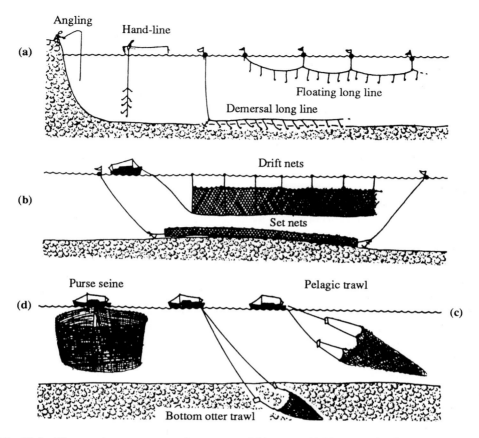

Fig. 12.3 The most important types of open-water fishing gear. (a) Lines are of various sorts and can range from hand-held, single, or multiple lines, to floating, pelagic lines that can be many kilometers in length. (b) Gill nets are long net curtains into which fish swim and, depending on their size, get caught by their gills. These nets can be set along the bottom or as drift nets that can extend for many kilometers, floating at some desired depth. (c) Trawls are conical nets with a broad mouth that are dragged along the seabed or through the water column. The mouth of the net is kept open using various devices. (d) Seines are nets that are used to encircle a school of fish and are then closed from below by ropes laced through rings at the bottom of the net. Modified from Cushing (1988).

(WRI, 1990). The most important reasons for these changes are increased effort and, especially in freshwaters, the extensive development of new aquaculture facilities. The major fishing nations in 1985–1987 were Japan (14.6% of global landings), the United States (6.4%), Chile (6.4%), Peru (6.0%), China (5.9%), South Korea (3.6%), and Canada, Denmark, Iceland, India, Indonesia, Mexico, North Korea, Norway, Philippines, Spain, and Thailand (all less than 3%; WRI, 1990).

It is likely that the global harvests of wild fish are about as large as, and perhaps larger than can be sustained over the longer term. Potentially, however, enormous future harvests might be made of some marine invertebrates, such as the seasonally abundant, pelagic, antarctic shrimp known as krill (*Euphasia superba*). Antarctic waters sustain a stock of krill that has been estimated to be in the range of 250–600 million tonnes, with an annual production of 100–500 million tonnes (WRI, 1990).

At the present time, the two major krill-fishing nations are Japan and Russia, and together they harvest about 0.5 million tonnes/year of these crustaceans. This is a small fraction of the estimated, presumably sustainable consumption of more than 100 million tonnes/year by the preexploitation populations of antarctic whales (WRI, 1990).

In the following sections, a few cases of extremely abundant stocks of fish that became severely depleted by harvesting are described. Usually, these resource depletions occurred as a result of a rapacious, uncontrolled, open-access fishery. Sometimes, however, regulated fisheries have also collapsed, usually because of quotas that were too large to be sustainable, but sometimes because of insufficient understanding of the habitat requirements of the hunted species, and of habitat degradations caused by natural and anthropogenic stressors.

The Cod Fishery in the Northwest Atlantic Ocean.   Other than the Norse in the 11th century, the earliest, known European explorers of the waters around Newfoundland were the Cabot brothers, Genoans sailing during 1497–1499 on behalf of England. Soon after, Portuguese fishers began to sail to North America in 1505, as did Basques of the Bay of Biscay of southwestern France and northern Spain, who began whaling and fishing in the area around 1525.

Among the most important natural resources discovered by these adventurers was an extraordinary abundance of cod (*Gadus morhua*) in the shallow, off-shore banks and near-shore coastal waters of Newfoundland. The most important of the cod stocks occurred on the Grand Banks, which comprise about 2.5×10^5 km^2 of water shallower than 100 m (Mowat, 1984; Cushing, 1988). Smaller, but important populations of cod also occurred off Labrador, Nova Scotia, the Gulf of Saint Lawrence, and New England.

The abundance of cod at that time is evidenced by enthusiastic entries in the journals of explorers and sea captains (Mowat, 1984). In 1497, John Cabot wrote that the Grand Banks were so "swarming with fish (that they) could be taken not only with a net but in baskets let down (and weighted) with a

stone." The island of Newfoundland was initially named Baccalaos, or "land of cod," by Basque and Portuguese sailors. Around 1516, Peter Martyr wrote that "in the sea adjacent (to the island of Newfoundland, John Cabot) found so great a quantity . . . of great fish . . . called baccalaos . . . that at times they even stayed the passage of his ships." These statements suggest an abundant natural resource of cod, even after the prose is discounted for the often exaggerated reports of explorers, who propagandized because they had economic stakes in colonizing and developing their discovered lands.

By 1550, hundreds of ships were sailing each year from western European ports for these waters, mostly in search of cod. During those times cod were readily caught in great abundance, cured by drying or salting, and taken to the hungry markets of Europe. At the beginning of the 1600s, as many as 650 ships were fishing cod in the waters off Newfoundland, by 1620 there were more than 1 thousand vessels, by 1783 at least 1.5 thousand, and in 1812 about 1.6 thousand (Mowat, 1984).

During this earliest era of the European exploitation of northwest Atlantic cod stocks, individual fish were typically large (Mowat, 1984; Cushing, 1988). The highest quality fish were called "gaffe" and "officers" cod and were 1–2 m long and weighed more than 100 kg. So-called "market" cod for the commoner Europeans were 36–65 cm long—anything else was considered unmarketable and was discarded. The largest sizes of cod are called "mother cod" by Newfoundlanders, because they represent important spawning capability (May, 1967). These large cod are today rare.

During the 18th and 19th centuries the cod fishery on the Newfoundland banks was an intensive, open-access, unregulated enterprise, involving hundreds of ships sailing from European ports of various nations, many other ships from Newfoundland, Canadian, and New England ports, as well as flotillas of smaller boats working the near-shore populations of cod. It has been estimated that between 1750 and 1800, the average landings were 1.9×10^5 wet tonnes/year, compared with 4.6×10^5 tonnes/year during 1800–1850 and 4.0×10^5

tonnes/year over 1850–1900 (Cushing, 1988). Between 1899 and 1904, about 1.6 thousand vessels were catching close to 1 million tonnes/year (Mowat, 1984). During these times, most of the cod were caught using hand lines and long lines. Although this fishing technology was not efficient, the substantial effort during this period probably led to the depletion of some local cod stocks, especially in near-shore waters. Overall, however, up until the intensified efforts that began at the end of the 19th century, it appeared that the large harvests of cod could be sustained by most of the exploited populations.

In the present century, the industrial cod fishery on the banks off Newfoundland was intensified through a number of technological improvements, including: (1) the development of netting technologies that are highly efficient at capturing cod, especially the use of trawls and gill nets, coupled with (2) sonar-based equipment used to locate schools of the fish, and (3) a large increase in ship-borne storage and processing capacity, which permitted vessels to stay at sea for extended periods (Hutchings and Myers, 1993).

The largest harvests of cod in the northwest Atlantic occurred during the 1960s. At that time, the fishery was essentially unregulated and open access. Because of the technological efficiencies of the fleet, enormous catches could be made, during what amounted to a frenzied, multinational, liquidation of the cod stocks. In 1968, the total catch was more than 2 million tonnes (Mowat, 1984). This intensity of fishing mortality could not be sustained by the cod populations, and the stocks quickly began to collapse.

The recent patterns of catch and population biomass are summarized in Fig. 12.4 for the so-called "northern cod" stock, occurring over most of the northern Grand Banks and halfway up the coast of Labrador. This was the largest and most important of the cod stocks of the northwest Atlantic, and because relatively good data are available, it serves to represent the pattern of change in the larger cod resource in this region (Hutchings and Myers, 1993).

In response to the collapse of the cod stocks and

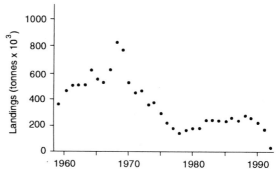

Fig. 12.4 Total landings of cod in the so-called "northern cod" stock, occurring over most of the northern Grand Banks and half-way up the coast of Labrador. Until the resource collapsed, this was the largest and most important of the cod stocks of the northwest Atlantic. Beginning in 1993 there has been a moratorium on cod fishing within Canada's 320-km-wide fisheries-management zone, but some fishing has continued beyond this limit. Data are courtesy of R. Myers, Canada Department of Fisheries and Oceans.

other crises in its economically important, offshore fisheries, in 1977 the government of Canada unilaterally declared a 320-km-wide fisheries-management zone and began to more effectively control access and allocate quotas of fish, including cod. Initially, these conservation actions allowed small, but short-term increases in cod stocks and landings (Fig. 12.4). However, the cod fishery has suffered a subsequent, more serious collapse. Because of the small populations of adult cod that are now available for spawning, it is likely that recovery of the stocks will be quite slow, even considering the large reductions in fishing quotas that were instituted in the controlled-management zone in 1992, and an outright moratorium on cod fishing in the management zone that began in June 1992. This latter step recognized the fact that the stocks of northern cod, once one of the world's great, potentially renewable natural resources, had been fished to commercial extinction.

A number of reasons, based on more-or-less convincing logic and information, have been advanced to explain the reason(s) for the collapse of cod stocks in the northwestern Atlantic Ocean. The most important of the putative causes of the collapse of cod stocks are as follows:

1. There has been a history of overexploitation of cod, leading to a decline of cod stocks that began to be especially acute in the 1970s (Hutchings and Myers, 1993). As is discussed below, the excessive harvesting was caused by a number of factors.

2. Although substantial efforts have been made toward science-based estimations of the sizes of the stocks of cod on the Grand Banks and elsewhere in the northwest Atlantic and of the yields that could be sustainably taken from those stocks, the scientific information has been imperfect. It is, quite simply, extraordinarily difficult to count fish in the open ocean. In addition, in 1992 a systematic error was discovered in a Canadian cod-population model that had been used briefly to determine stocks sizes and to set quotas. This modeling problem led to an overestimation of the available biomass, an underestimation of fishing mortality, and recommendations by scientists of fishing quotas that were too large to be sustainable (Steele et al., 1992).

3. In Canada and elsewhere, there are considerable economic and social influences on decision makers and politicians to permit the largest possible intensities of fishing effort. These pressures originate with individual fishers and their associations, because there are large, fishery-related debts to pay and because there are usually few alternative employment possibilities. Of course, private companies also lobby for higher quotas, for similar reasons—they need to have a minimal flow of revenues to service their debts, and they can only generate those revenues by fishing. Because of these powerful, socioeconomic influences, decisions are often made to set quotas that are substantially larger than are recommended by fishery scientists.

4. Although most of the Grand Banks falls within Canada's 320-km management zone, some parts of the banks extend beyond, into international waters. In these areas there was a multinational fishery for cod and other species, which

until recently was essentially unregulated. Because cod is a mobile species, which does not recognize the boundaries of management zones, the overfishing in the international waters compromises the efforts, albeit imperfect, toward conservation-based management of the cod stocks. However, to put this effect into perspective it should be pointed out that between 1977 and 1991, Canadians landed 85% of the cod caught in both the 320-km management zone and the nearby international waters, and the activities of the Canadian fishers were regulated.

5. Important physical and biological damage to the habitat of cod and other species may be caused by the use of certain types of fishing technologies, especially the use of bottom trawls. Not much is known about the nature and consequences of these types of damage, because little research has been carried out. Bottom-trawling equipment is, however, known to affect ocean bottoms by scraping and plowing the substrate, resuspending sediment, and destroying benthic life, sometimes leading to longer-term ecological changes in disturbed areas (Messieh et al., 1991; Jones, 1992). It is not known whether such damage to their physical habitat and prey have contributed significantly to the collapse of cod stocks in the northwest Atlantic.

6. Substantial by-catches of cod, including both adult and immature animals, occur during seining and trawling for this and other species, but it is not known whether this has contributed significantly to the collapse of the cod stocks. The topic of fishery by-catches is discussed in more detail later in this chapter.

7. Nonhuman predators also have the potential to deplete fish stocks. The most abundant species of seal in the northwest Atlantic is the harp seal, which has recovered a population estimated at about 3.5–4 million animals, similar to its preexploitation abundance (P. Brodie, personal communication). Based on this population size, the present-day population of this seal may consume an estimated 800–900 thousand tonnes/year of food, but most of this is composed of a wide variety of species other than cod, especially the small

fish known as capelin (*Mallotus villosus*) and polar cod (*Boreogadus saida*), and the pelagic crustaceans, *Pandalus borealis* and *Thysanoessa* spp. (Sergeant, 1991). Therefore, although harp seals eat cod, it appears to be a relatively minor species in their diet. The most recent collapse of cod stocks occurred at a time when populations of harp seals were increasing in abundance, suggesting a correlation between the two phenomena. Nevertheless, because of the unimportance of cod in the diet of harp seals, it seems unlikely that seals have been important in the collapse of the cod stocks of the northwestern Atlantic Ocean.

8. Prior to the recent collapse of cod stocks, research had suggested that the annual abundance of cod was positively correlated with the salinity of surface waters (Sutcliffe *et al.*, 1983; Myers *et al.*, 1993). The suggested, ecological mechanism for this putative relationship is that: (a) larger concentrations of nutrients are associated with deeper, higher-salinity waters, and (b) the mixing of these fertile waters into the surface, euphotic zone is inhibited by large inputs of freshwater from northern rivers and from icebergs calved from glaciers in the Arctic. It has been suggested that changes in these nutrient inputs, and possibly also changes in surface-water temperatures, may have contributed to the catastrophic decreases of the stocks of cod in the northwest Atlantic Ocean. This effect would have been caused by reproductive failures, possibly associated with a lack of food for the vulnerable, pelagic, larval stage of the life history of cod. So far, however, there is little supporting evidence for this environment-related hypothesis.

9. A few other theories have been advanced to explain the disappearance of the cod stocks of the northwestern Atlantic Ocean. These explanations commonly serve to externalize blame away from human influences and include themes such as: "the cod have gone somewhere else."

Overall, of all of the suggestions that have been offered to explain the collapse of cod stocks on the Grand Banks, the most parsimonious and compelling hypothesis appears to be that the populations of this species were exploited at intensities that exceeded their capability for renewal, i.e., the cod were mined.

Sardines, Anchovies, and Other Small Fish.
At various places around the world, initially superabundant stocks of some smaller species of fish have undergone spectacular collapses during periods of time that they were supporting large, industrial fisheries. In some cases, the collapses were clearly linked to overexploitation. However, some population changes may have occurred in response to natural episodes of water-temperature change. Some of these cases are described below.

One of the most famous collapses of a fishery in North America was that of the Pacific sardine (*Sardinops sadax*) of California. For a time, this population supported enormous catches, peaking at 719 thousand tonnes in the 1936–1937 fishing year (Fig. 12.5; McEvoy, 1986). The fishery then underwent a series of progressively more severe declines. For example, 133 thousand tonnes of sardine were landed in 1951, but only 13.5 thousand in 1952. More recently, the stocks of this species have substantially recovered, and they again supported a large fishery in the 1980s (Fig. 12.5).

The most frequently cited reason for the collapse of this fishery is overexploitation of the resource. Some researchers, however, have suggested that changes in water temperature may also have been important (Rothschild, 1986; Kawasaki, 1989). The role of overfishing is suggested by the following quotation from the novelist, John Steinbeck (1954): "The canneries themselves fought the [Second World War] by getting the limit taken off the fish and catching them all. It was done for patriotic reasons, but that didn't bring the fish back . . . Cannery Row was sad when all the pilchards [i.e., sardines] were caught and canned and eaten."

There have been equally spectacular changes in the small-fish landings of the Pacific coast of South America. Around 1970, the Peruvian anchovy (*Engraulis ringens*) was the basis of the world's largest, single-species fishery. The stocks of this species subsequently collapsed, although not to below the

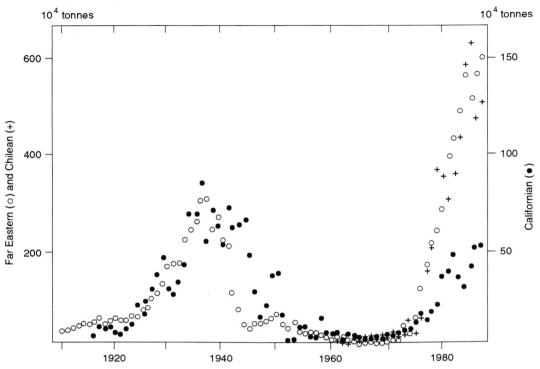

Fig. 12.5 Changes in the landings of Pacific sardine (*Sardinops sadax*) in three Pacific fisheries—the Asian Far Eastern, The Chilean, and the Californian. According to Kawasaki, the strong synchrony among these patterns of change for widely dispersed stocks suggests a climatic influence on their populations. Modified from Kawasaki (1989).

threshold of commercial extinction, and there has been a substantial recovery during the 1980s (Figs. 12.5 and 12.6). The collapse of the stocks of this species has been variously attributed to the combined effects of environmental changes associated with warm-water, El Niño events of 1972, 1976, and 1982–1983, along with excessive fishing pressures (Gulland and Garcia, 1984; Kawasaki, 1989; Serra, 1989).

During the collapse of the Peruvian anchovy populations, there were substantial increases in the size of the stocks of sardine (*Sardinops sadax*). Both of these fish are pelagic species that feed on plankton, and it possible that the sardine was ecologically released by the demise of the anchovy populations (Gulland and Garcia, 1984). Similar patterns of replacement of ecologically similar spe-

cies of fish have been suggested elsewhere [but see Daan (1980) for a dissenting view]. For example, large decreases in the abundance of Atlantic herring (*Clupea harengus*) in the northeast Atlantic were accompanied by increases in abundance of European pilchard (*Sardina pilchardus*) (Kawasaki, 1989). The reasons for the collapse of herring are not yet resolved, but the prominent hypotheses include overfishing and water-temperature changes that affected the survival of larvae of this species (Bailey, 1989).

The Bluefin Tuna Fishery in the Western Atlantic Ocean. Bluefin tuna (*Thunnus thynnus*) are large, pelagic fish that can achieve weights of 700 kg and are capable of attaining speeds of 90 km/hr. The species is wide ranging, undertaking transo-

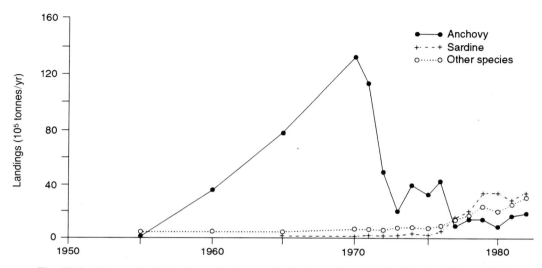

Fig. 12.6   Changes in the landings of major species in the southeast Pacific fisheries. Illustrated here are anchovy (*Engraulis* spp.), sardine (*Sardinops sadax*), and other species. Data are from Gulland and Garcia (1984).

ceanic migrations, and is somewhat endothermic, maintaining body temperatures of 24–35°C, even in water as cold as 6°C (Safina, 1993).

The flesh of bluefin tuna is valuable, because of strong markets to supply Japanese sushi restaurants. In the early 1990s, a single, prime fish could sell for U.S. $30 thousand in North America, for more than $60 thousand at auction in Tokyo, and ultimately selling for $770/kg as sushi, equivalent to about $231–385 thousand/fish, if a dressed weight of 300–500 kg is assumed for a large individual (Safina, 1993).

Because of its enormous per-fish value, bluefin tuna has been exploited heavily, leading to a collapse of its stocks. Between 1975 and the early 1990s, the population of this species in the western Atlantic decreased by 85%, from an initial abundance of 150 thousand animals, to 22 thousand (Fig. 12.7). Stocks in the eastern Atlantic have also declined, by about 50% compared with their abundance in 1970 (Safina, 1993).

The collapse of the stocks of bluefin tuna is an archetypal example of a potentially renewable, biological resource that was greedily overexploited and mismanaged instead of being sustainably har-

vested. Although an international commission was established to manage the fishery for this species (the Commission for the Conservation of Atlantic Tunas), the managers of this administrative body consistently ignored the advice of their scientists regarding sustainable fishing efforts (Safina, 1993). The regulated overfishing, coupled with unregulated, pirate fishing by ships flying flags-of-convenience of nations that were not members of the commission, resulted in the rapid demise of the stocks of bluefin tuna.

An Invertebrate Example.   In 1987, an international conservation organization, the World Wildlife Fund (WWF), reported that a coral reef in a bay on the coast of Irian Jaya (Indonesian New Guinea) in the south Pacific Ocean had a large population of an endangered species, the giant clam (*Tridacna gigas*) (Anonymous, 1989). In cooperation with government, the site was proposed as a national, marine, ecological reserve. However, when the news about the clam population was discovered by a shop owner in a nearby town, he immediately hired local fishers to harvest the resource. As quickly as possible, they gathered about 7 tonnes of clams

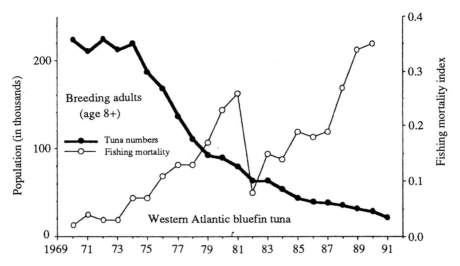

Fig. 12.7 Changes in the abundance of mature bluefin tuna (*Thunnus thynnus*) in the western Atlantic Ocean, compared with an index of mortality associated with commercial fishing. Modified from Safina (1993).

ranging in size up to 1 m. Most of the meat was flown to Japan, where it is regarded as a delicacy, and the rest was used to feed crocodiles that were being cultivated for their hides. The residual, non-harvested clams typically ranged in size from 10 to 20 cm and would take about 100 years to mature. In only a few weeks, this local resource was exhausted.

By-catches.   The use of certain types of fishing technologies can result in large by-catches of unwanted fish, which are usually jettisoned, dead, back into the ocean. This nonselectivity of harvesting is especially problematic with the use of trawls, seines, and drift nets. The substantial by-catch consists of unwanted species of fish and other animals, but there can also be a by-catch of undersized, immature individuals of commercial species of fish. For example, (1) in the North Sea fishery for cod and related fish, the percentage of discarded, non-market fish during 1974–1982 averaged 42% of the catch (range 20–56%; (2) the discard rate is 44–72% in the Mediterranean fishery; and (3) the discard rate is 75% for a trawl fishery for shrimp, and 60–70% for other species-targeted fisheries in

Senegal, west Africa (Gulland and Garcia, 1984; Cushing, 1988).

Some fishing practices also result in substantial by-catches of marine mammals, seabirds, and sea turtles (NMFS, 1991b). Worldwide, pelagic drift nets, each as long as 90 km, may have killed 0.3–1.0 million dolphins, porpoises, and other cetaceans during the fishing year of 1988–1989 (Anonymous, 1990). To cite a specific example, during a 24-day monitoring period, a typical set of drift net of 19 km/day in the Caroline Islands of the south Pacific entangled 97 dolphins, 11 larger cetaceans, and 10 sea turtles (LaBudde, 1989). Few data are available, but hundreds of thousands of seabirds may have recently drowned each year in pelagic drift nets (LaBudde, 1989).

Purse seining for tuna has also caused tremendous mortalities of some species of dolphin and porpoise, killing an average of more than 200 thousand animals/year since the 1960s, but perhaps about one-half that number more recently, and depleting some populations of these animals (Humane Society of the United States, 1989).

In addition, large quantities of drift and other nets are lost each year. Because the synthetic materials of which the nets are manufactured are resistant to degradation, these go on fishing for many years, as so-called "ghost nets" (LaBudde, 1989). Clearly, fishery by-catches can cause substantial, nontarget mortalities, which can threaten numerous marine species.

12.3
RESOURCE MANAGEMENT

Sustainable harvests can potentially be made from populations of animals and plants and their communities, without causing a substantial degradation of the capability for renewal of the resource. In all species, potential fecundity and productivity are larger than the actual recruitment, growth, and maturation of new individuals and biomass. Some part of the excess of productive capability over actual survival and growth can be utilized to sustain humans and their activities. Moreover, the potential harvest can often be substantially enhanced through management practices that increase the renewable yield of biological resources.

Obviously, ecology has much to offer to the design of sustainable systems of resource harvesting and management. Applied ecological disciplines, such as agricultural, forestry, and fisheries sciences have developed a rich literature on this topic, including many textbooks. In this section the subject of resource management is dealt with in a conceptual manner, but not in depth.

The productivities of populations of plants and animals are governed by intrinsic, biological parameters, especially natality, mortality, and growth rates of individuals. Potential harvests are related to productivity and to the proportion of individual biomass that is useful to humans. These biological factors can be substantially influenced by many environmental variables (Fig. 12.8).

If the influences of environmental factors on the productivity of biological resources are understood, and accommodated within a harvesting and/or management system, then the yield of harvested products can be increased. In a sustainable system, these increases must be accomplished without degrading the inherent capability for renewal of the resource through the production of new individuals and/or biomass. Some management activities that can be used to manage the rates of productivity, recruitment, and mortality of biological resources are summarized below.

Growth Rate

In both natural and human-influenced ecosystems, the productivities of virtually all plants and animals are constrained by environmental stressors, both biological and inorganic. If the most important constraints are ameliorated through management, then the growth rates of individual animals or plants and their populations can be increased (see also Chapters 8 and 9). For example:

1. In agriculture, the productivity of desired species of plants is often increased by a variety of management practices. High-yield varieties of plant species may be sown at some optimal density and sustained by some combination of fertilization, irrigation, tillage or herbicide use to reduce the intensity of competition from weeds, use of fungicides or other practices to reduce the intensity of diseases, and use of insecticides or other actions to reduce the amount of defoliation caused by insects and other animal pests.

2. In forestry the intensity of management varies greatly, but the productivity of crop trees may be increased by using a silvicultural system that involves some combination of site preparation, planting high-yield varieties or species, use of herbicides to reduce the intensity of interspecific competition, thinning of immature stands to reduce the intensity of intraspecific competition, and use of insecticides to cope with pests.

3. In highly managed aquaculture, there are also opportunities to increase the productivity of desired species. High-yield varieties may be grown in ponds or pens, fed to satiation, and protected from diseases and parasites by the use of antibiotics and other chemicals. Ponds may be

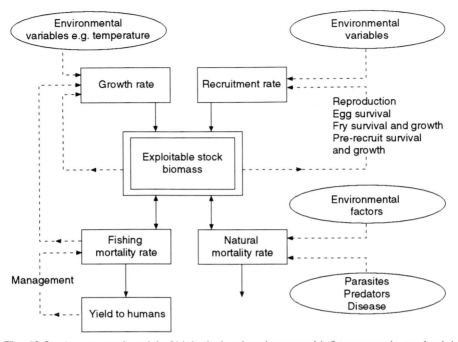

Fig. 12.8 A conceptual model of biological and environmental influences on the productivity and potential harvest from a population of fish. In this model, the major biological variables are the rates of: (a) productivity of individuals, (b) recruitment of new individuals, and (c) mortality. Deaths can be caused by disease, accidents, and natural predators, or mortality can be associated with harvesting by humans, another type of predation. Note that all of the biological variables can be enhanced or constrained by diverse environmental influences, including harvesting and management by humans. In this model, anthropogenic management is only shown as an influence on the rate of harvesting. However, for some fish and other biological resources, management can also influence the rates of growth, recruitment, and mortality from natural causes. Similar models can be developed for other plant and animal resources that can be harvested by humans. Modified from Pitcher and Hart (1982) and Begon *et al.* (1990).

fertilized to enhance the trophic base of the productivity of fish or shrimp.

4. In fisheries involving wild stocks of fish or invertebrates, the usual focus of productivity-related management is to ensure that harvesting is not so excessive as to deplete the regenerative capacity of the stock. There are relatively few opportunities to enhance rates of productivity. Sometimes, care is also taken to avoid physical degradations of the habitat caused by the method of harvesting, but again this type of management focuses on regenerative capability, rather than on enhancement of productivity.

Recruitment Rate

In many cases, the rate of recruitment of new individuals into exploited populations is considerably smaller than can potentially be realized through active management. Often, recruitment can be increased or maintained by planting, stocking, site preparation, and other practices. For example:

1. In intensively managed systems in agriculture, forestry, and aquaculture, the species composition of the crop is usually controlled by enhancing the recruitment of one or several species. In the most intensively managed systems the

intention may be to achieve a virtual monoculture of some desired species, so that its productivity is not constrained by competition from other species.

In agriculture, recruitment of the crop is most often accomplished by sowing seeds under relatively controlled, cultivated conditions that favor germination and establishment of the desired species. In other cases, young individuals may be grown under controlled conditions and later outplanted, a practice widely used, for example, to cultivate paddy rice, to establish fruit orchards, and to establish plantations in forestry. Sometimes, a particular harvest system will facilitate the vegetative regeneration of perennial-plant crops, which may be able to resprout from roots, rhizomes, or stumps after the mature plant is cut and harvested. This regeneration system can be used effectively to manage sugar cane and some species of trees that are intensively managed in plantations, such as high-yield, hybrid varieties of aspens and poplars.

.   2. To increase the recruitment of wild populations, other strategies may be used:

a. In forestry, certain types of harvesting and management systems favor the recruitment of particular, desired species of trees. For example, as long as there is an adequate supply of seeds, some species of pine will recruit effectively onto clear-cuts that have been site prepared by burning. Other species of tree recruit effectively onto exposed mineral soil, a condition that can be enhanced by mechanical site preparation. In addition, the post-harvest regeneration of some species of trees can be accomplished through vegetative mechanisms. For example, many species of poplars, aspens, maples, ashes, and other species can regenerate profusely by stump, root, and/or rhizome sprouting after the mature stem is cut, a regeneration practice known as coppicing.

b. Recruitment of some species of hunted animals may be maintained or favored by only killing male animals, or individuals within a particular size range, or by limiting the hunting season to particular times of the year. Deer, for example, are typically polygynous. If a deer herd is relatively small, the hunt may be restricted to male individuals, on the assumption that the surviving bucks will be capable of impregnating all of the females in the population.

c. Recruitment in many fishes, for example, salmon and trout, is often actively enhanced by stripping wild individuals of their eggs and milt, which are then fertilized under controlled conditions and incubated to hatching. The larvae and subsequent juvenile stages are then cultivated until they reach some minimal size, after which they are released to supplement the natural recruitment of wild individuals.

Mortality Rate

Mortality directly affects the size of stocks, and can indirectly influence the growth rates of unharvested individuals by influencing the intensity of intraspecific competition. Both of these influences can affect the productivity of populations of exploited species.

Mortality can result from natural influences and from exploitation by humans. The total rate of mortality cannot exceed the regenerative capability of wild or managed stocks or the resource will be depleted. As is described below, a number of practices can be used to manage the rates of mortality associated with both natural and exploitation-related factors.

Natural Mortality Rate.   The potential harvests of stocks of plants or animals can often be increased, if the rates of mortality associated with natural predators, parasites, and diseases are actively decreased. The influences of natural sources of mortality are sometimes managed in the following ways:

1. Especially in intensively managed systems, there are strict controls over losses of crops due to

natural predators. For example, coyote, wolf, cougar, and bear are often considered to be important predators of agricultural livestock and also of some hunted mammals. As a result, their abundance has been relentlessly reduced by hunting, trapping, and poisoning. In addition, access of predators to the desired animals may be restricted by building fences or by using guard animals such as dogs or donkeys.

2. Depredations due to parasites may be reduced through the use of pesticides or by changing the cultivation system to decrease their abundance. For example, domestic sheep infested with ticks may be dipped in toxic chemicals to kill those potentially debilitating parasites.

3. Mortality associated with disease may be reduced through the use of antibiotics or prophylactic chemicals, by using palliatives to manage the symptoms of disease, or by changing the cultivation system to decrease vulnerability.

Harvesting Mortality Rate. If any resource of plants or animals is to be harvested in a renewable fashion, the total rate of mortality, including that associated with exploitation, must be equal to or smaller than the regenerative capability. Both rates may vary over a wide range, some parts of which could be considered to represent underutilization and some overexploitation.

For an ideal population the largest amount of harvest (i.e., mortality associated with harvesting) can be set as the maximum sustainable yield (MSY) that can be taken from the stock without causing a diminution of its productivity (Watt, 1968; Clark, 1981, 1989; Begon *et al.*, 1990). An ideal case of MSY under conditions of a constant rate of harvesting, occurring regardless of population size, is illustrated in Fig. 12.9. For any population size, a rate of harvesting that is greater than the upper bounds of the curve is unsustainable and would drive the population to extinction. A harvest yield equal to MSY is sustainable, but any larger harvest rate is not. Any harvest rate smaller than MSY will force the population toward one of two equilibrium points. One of these is unstable because it forces the population to extinction, while the other is stable and sustainable,

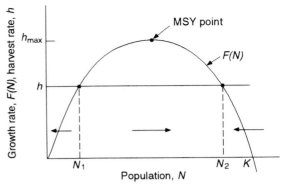

Fig. 12.9 A simple, diagrammatic model of maximum sustainable yield (MSY). The model is based on the growth and harvesting of an ideal population of a single species, whose pattern of growth under resource-limited conditions is described by the logistic equation. The logistic equation is a continuous-change, equilibrium model, relevant to populations in which births and deaths occur continuously, but under conditions in which there is a limited capacity of the environment to sustain the population (Begon *et al.*, 1990). The logistic equation is given as:

$$\frac{dN}{dt} = rN \frac{(K - N)}{K},$$

where N is the population size at any given time, t is time, dN/dt is the rate of change of the size of the population, i.e., the difference in population size standardized to a given interval of time, r is the intrinsic rate of natural increase, i.e., the population growth rate that would occur as the difference of intrinsic reproductive capability minus intrinsic mortality (these intrinsic rates occur in the absence of environmental constraints associated with factors such as the availability of resources, predation, disease, etc.), and K is the carrying capacity of the environment, i.e., the maximum number of individuals that can be sustained without causing a degradation of the ability of the environment to support the species. The MSY model describes the relationship between growth rate of the population and harvest rate (both vertical axis) with population size (horizontal axis). When the population is at carrying capacity (i.e., K), there is no net change in abundance, and no excess production is available for harvesting. For obvious reasons, there is also no growth available for harvesting when the population size is zero. When the population is at the apex of the curve, growth rates are maximal, and MSY (h_{max}, occurring when $N = \frac{1}{2} K$) can be achieved. Note that a particular, fixed intensity of harvest rate (e.g., h) can intersect the curve at two places, and there are two population equilibria in the curve, N_1

because it is approached from two directions. Therefore, under conditions of a constant rate of harvesting, there are several, sustainable intensities of yield, the maximum of which is MSY.

From a resource-management perspective, it is more realistic to consider the effects of harvesting at a constant effort, rather than at constant yield (except, perhaps, in the cases of by-catch and opportunistic harvesting). Over short periods of time, resource-extraction industries generally consist of fixed amounts of personnel and harvesting infrastructure (e.g., fish traps and fishing boats, or chainsaws, feller-bunchers, and skidders in forestry, etc.). Because of the value of the investments in personnel and harvesting machinery, there is a tendency for harvesting efforts to be relatively stable from year to year. Because the amount of harvest depends on the interaction of effort and stock sizes, a constant effort of harvesting results in a variable yield (Fig. 12.10). The constant-effort model predicts that there is a harvesting effort that will result in a maximum sustainable yield and a stable equilibrium size of the exploitable stock. Within bounds set by the growth rates of the stock at various population sizes, there are also smaller, sustainable levels of harvest at other intensities of effort. However, greater intensities of harvest effort are unsustainable and would drive the stock to extinction.

There are other, more sophisticated models of population growth that consider multispecies stocks,

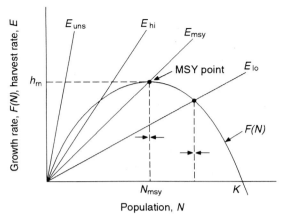

Fig. 12.10 Maximum sustainable yield under a condition of constant harvesting effort, in which the yield varies depending on size of the exploited population. The population growth model, $F(N)$, is based on the logistic curve (see Fig. 12.9). In the model depicted here, a single, stable equilibrium point is associated with all efforts of harvesting that intersect the curve. At an effort of E_{msy} the maximum sustainable yield (i.e., h_m) is harvested, and this yield is achieved at a population of N_{msy}. Harvesting at E_{msy} when the population is larger or smaller than N_{msy} results in movement of the population toward N_{msy}, indicating that this is a stable, equilibrium point. Harvesting at efforts of E_{hi} or E_{lo} also leads to stable equilibrium points, but these have smaller sustainable yields than are achieved by harvesting at E_{msy}. Harvesting at E_{uns} does not intersect the curve and represents an unsustainable effort that would drive the stock to extinction. Adapted from Clark (1981).

and N_2. Harvesting (at a constant rate, h) from the population when it is smaller than N_1 will drive the resource toward extinction (as is indicated by the arrow) and represents overharvesting. (Note also that at any particular population size, any effort of harvesting that is larger than the upper bound of the curve is also unsustainable.) Harvesting (at h) when the population is larger than N_2 will drive the resource toward N_2 and is a sustainable practice, although the yields are smaller than can potentially be achieved. Harvesting (at effort h) from a population that is larger than N_1 and smaller than N_2 will allow the population to grow toward N_2 and is also sustainable. The maximum, sustainable harvest occurs at the apex of the curve. Note that if the harvesting effort (h) changes, so do the two population equilibria. Modified from Clark (1981).

changes in harvesting efforts, environmental dynamics, regulatory strategies, and other variables associated with demographics, environment, technology, and regulation. These have mostly been developed in fisheries science, but to some extent they have also been used to guide the management of other resources, such as forests and terrestrial game animals. However, it is beyond the scope of this section to describe these more advanced models. Specialized texts can be consulted for details.

For obvious reasons, the notion of MSY is intuitively attractive, and it has been important in the development of the theory and practice of forest, fish, and wildlife management. MSYs are relatively easy to model in the ideal world of the logistic equation, where the environment and its carrying capacity are known and sometimes constant, and one or a few species are growing without complex behavior, such as social systems. However, MSY is difficult to determine under operational conditions, where the quality of the environment changes constantly, and sometimes dramatically, and where there are corresponding adjustments by an interacting community of species.

An especially serious failure of the MSY model based on the logistic is associated with the fact that the logistic is an "instantaneous" model in which the balance between mortality and density-dependent recruitment is continuously maintained. The introduction of time lags, especially reproductive time lags, greatly destabilizes such models. This problem is likely to be especially severe in late-maturing species.

Moreover, in the real world, changes in population size can be influenced by social systems, interactions with other species, and other uncertain, often unpredictable variables. MSY is only a concept, from which useful insights have been gathered about factors, including harvesting, that influence intrinsic rates of growth and mortality of simple populations and communities. Often, however, MSY models cannot be successfully applied to the management of real, biological resources.

In any event, managers often need to regulate or adjust the rates of mortality associated with harvesting. They typically do this by controlling the harvesting effort, which is an integrated function of the methods (e.g., types of fishing boats and their gear) and intensity (e.g., number of boats) of harvesting. Some ways by which this can be accomplished are summarized below:

1. Changes in harvesting technology can have a substantial influence on the mortality rates associated with harvesting. Consider, for example, the various methods of catching fish, which are summarized in Fig. 12.3. The efficiencies of these technologies vary greatly, in terms of the quantity of fish caught per person fishing, per unit of energy expended, and per unit of capitalization. Therefore, the fishing mortality associated with

the most intensive methods, such as drift nets, trawls, and seines, can be much more intense than that associated with, for example, hand lines or long lines. The by-catch associated with these methods also varies greatly, with the more efficient methods also being much less selective of species and sizes, and therefore catching and killing substantial quantities of species that are not the target of the harvest. A corollary related to the efficiency of the various technologies is that to achieve some constant catch, many more people could be employed in a fishery using simpler harvesting methods, although if labor is expensive the fish would be more expensive to consumers.

2. Some other changes in harvesting methods can also result in greater selectivity of the species or sizes of exploited resources, which can be an important consideration in terms of stock management. In a fishery, for example, a change in the mesh size of netting will have a substantial influence on the sizes of animals that are caught. For some exploited species, it may be advantageous to not harvest smaller individuals, which may not yet have bred, and may also have a relatively small per-unit-weight value compared with larger animals. In some cases, larger individuals may have a much larger reproductive capability, and to sustain the productivity of the stock it might be sensible not to harvest the bigger animals. Similarly, in forestry, size-selective harvesting practices such as selection cutting might be used, in preference to clear-cutting. Selective harvesting practices may encourage the regeneration of more desirable tree species, while avoiding or reducing many environmental damages by keeping the integrity of the forest relatively intact.

3. Another, obvious way to adjust the mortality rates associated with harvesting is to limit the numbers of persons or other harvesting units participating in the harvest. In a fishery, for example, the numbers of fishers could be limited, as could the number of boats using particular harvesting technologies.

To achieve control over harvesting effort, and the mortality associated with exploitation, regulators must use socioeconomic tools. In a marine fishery, for example, there may be a need to manage the harvesting effort by influencing: (1) the methods that are used to catch fish (e.g., the size of fishing boats and the amounts and types of gear that are used), (2) the intensity of fishing (e.g., number of fishers and the number of boats using a particular technology), and (3) the times of year and places that fishing is allowed, by regulating seasons and areas.

If the social climate and political system allow direct regulation by a governing body (e.g., in some sort of a centrally planned system), there can be strict restrictions on the numbers of participants, the technology they can use, the resource quotas they have access to, and the times and places they can harvest, with noncompliance severely penalized.

In more open, market driven, socioeconomic systems, the potential profits of different harvesting strategies could be influenced. Ways by which this could be accomplished include: (a) fines for noncompliance, which would decrease profit by raising costs, (b) taxes on more harmful harvesting methods, or subsidies on less harmful ones, which influence profit by increasing or reducing costs, and (c) buyouts of inappropriate or excess harvesting capacity (either equipment or licenses), which would influence profit for the remaining fishers by increasing their relative allocation.

Although these strategies would result in smaller harvests on the shorter term, the economic benefits might be enhanced by regulation of the nature of the outputs of a regionally integrated, resource-dependent industry, with a focus on value-added products. In forestry, for example, there would be no exports of logs, but valuable products such as furniture or violins would be manufactured and exported. Similarly, a fishery might focus on highly valued products such as prepared foods, rather than whole unprocessed fish. Such a strategy would optimize the regional economic benefits of resource-based industries, while allowing smaller, sustainable harvests of the resource.

There is, of course, an imperfect understanding of the various intrinsic and environmental factors that influence the sizes and productivities of poten-

tially renewable, biological resources. As a result, the harvesting and management prescriptions suggested by applied ecologists, such as foresters and fishery and agricultural scientists, are imperfect. Nevertheless, enough is often known about ecological relationships to design harvesting and management systems that would allow sustainable harvesting of populations of animals and plants, and their communities, without causing a substantial degradation of the capability for renewal.

At the least, conservative predictions could be made of harvesting and management intensities that would clearly be small enough to not result in overexploitation of the resource, even though the harvest might also be smaller than the potential, maximum sustainable yield. There are no intrinsic reasons why the harvests of natural resources must be as large as are potentially attainable. If accurate MSYs cannot be reliably determined by scientists, then it will always be ecologically prudent to harvest at relatively smaller, but surely sustainable rates.

Regrettably, in the real world the predominant result of bioresource exploitation, even with management, has been overexploitation. This is clear from the various examples of resource degradation presented earlier in this chapter, the several others presented in previous chapters, and the many other examples that are common knowledge.

In fact, one is hard pressed to find examples of valuable, potentially renewable, biological resources that have been exploited by humans in a fashion that is sustainable over the longer term. This statement assumes that there was a technology that would allow the overexploitation to be accomplished, and that unlimited markets existed for the commodity, so that anything harvested could be sold at a profit.

Often, the pattern of depletion has been one of sequential overexploitation of mixed-species resources, a process sometimes called "working-up." Previously unexploited, or "virgin" biological resources are often characterized by: (a) a mixed-species composition, of which only a few are initially desirable economically, and (b) individuals of large size, especially in the case of old-growth eco-systems. For example, the old-growth forests of western North America are dominated by large individuals of some valuable species of trees, as well as other species (Chapter 9). Preexploitation communities of fish, whales, and other species are also typically dominated by large individuals of desirable species.

During the sequential exploitation of biological resources, some variation of the following sequence of events usually takes place. Note that the process usually results in the progressive harvest of commodities with smaller unit-individual and unit-weight values and less potential for the gaining of value added during manufacturing.

1. First, the largest individuals of the most desirable species are selectively harvested, and almost always depleted. In an old-growth forest, the economic products have great value per unit of biomass and might include large-dimension lumber of the most precious species, or perhaps valuable veneer products.

2. Next, smaller individuals of the most desired species are selectively harvested, along with the largest individuals of secondarily desired species. In a forest, the economic products might involve smaller-dimension lumber.

3. Area-harvesting methods are then used to harvest virtually all individuals of all species, for the production of bulk commodities. In the case of forestry, clear-cutting might be used to harvest biomass for the production of pulp, industrial fuel, charcoal, or domestic fuelwood.

4. The intensively harvested system may then be managed to establish a productive biological resource, but of a different ecological character from the original, natural community. In forestry, the natural forests might be converted into plantation forests, or sometimes to agriculture.

5. Intensive harvesting can also lead to a virtual collapse of biological productivity that is useful for human purposes and can result in degraded ecosystems. Sometimes, for example, harvested forests are converted into shrub-dominated barrens that may resist the invasion of tree seedlings, unless intensive efforts are made toward regeneration of a forest.

The predominance of shorter-term considerations and values, over longer-term ones, is the most important reason for the seemingly universal overexploitation of potentially renewable biological resources, sequentially or otherwise. This topic is discussed in more detail in the next section of this chapter.

12.4
ECOLOGICAL ECONOMICS

Concepts and Criteria

Conventionally, economics is a social science that examines the allocation of scarce resources among competing uses. Economics is concerned with understanding and predicting the consumption of goods and services, as well as the broader, commercial activities of societies.

Within the broader scope of economics, microeconomics is concerned with particular commodities, companies, or individuals and the relationships among these. Macroeconomics is concerned with larger levels of integration such as regional, national, or global economies.

Within the economic system, a broader goal of individuals and corporations is to maximize the profits that they gain within the marketplace.

A broader goal of economics is to maximize the utility of goods and services to society. In almost all cases, however, these products are valuated and measured in units of currency, such as dollars. This is the case for manufactured products such as automobiles and buildings, harvested and processed natural resources, such as wood and fish, and the services provided by farmers, doctors, and professors.

Potentially, economics can also take into account goods and services that are not often assigned value in the marketplace, for example, the aesthetics of natural landscapes, ecological services, such as nutrient and water cycling, and the integrity of rare species, such as spotted owls. However, it has proven difficult for the conventional economic marketplace to assign values to these sorts of goods and services, so that their merits can be compared with those of other, conventionally valuated goods and services. Some contingency valuations have been based on notions such as: (a) the amount of money that people would be willing to spend to visit some ecological resource for purposes of recreation, (b) the potential values of plants and animals as undiscovered sources of medicines or foods, or (c) the amount of money that people would be willing to pay for the happiness of knowing that rare species such as spotted owls and blue whales exist safely somewhere. These valuations are, of course, strongly influenced by the perspectives of individuals and societies. As a result, the assigned worths of these nonconventional resources are highly contentious, and they are not often considered seriously during the accounting of costs and benefits of alternative actions in the marketplace.

In comparison with conventional economics, the principal distinction of ecological economics is that it attempts to find a nonanthropocentric system of valuation. In ecological economics, accountings are made of important social and environmental costs that are associated with resource depletion and ecological degradation, but that are not considered during conventional economic accountings. An important failure of conventional economics is that the marketplace has not recognized the values of important, but nontraded services, especially when these are common-property ecological services. As discussed in more detail in the rest of this section, these costs are accounted for in ecological economics.

In economic terms, capital can be regarded as actual or potential wealth that is capable of being applied toward the production of further wealth (Costanza, 1991; Costanza *et al.*, 1991; Costanza and Daly, 1992). Capital can be divided up into three broad types:

1. Manufactured capital (MC) is the industrial means of production. Examples of manufactured capital are factories, mines, harvesting machinery, buildings, tools, and other types of physical infrastructure.

2. Human capital (HC) is the cultural means of production and is composed of knowledge, skills, and a workforce with certain capabilities.

3. Natural capital (NC) refers to stock or quantities of raw, natural resources that can be harvested, processed, and utilized to yield flows of goods or services for some time into the future. In the economic context, "goods" refers to a quantity of valuable material, such as metal-containing ore, coal and petroleum, water, or biomass of trees, fish, or deer. Although not usually valuated by economists (i.e., they are not priced), resource "services" can refer to functional capabilities, such as biological productivity, nutrient cycling, hydrologic cycling, erosion control, and cleansing and detoxifying of air, water, and soil. These services are important and therefore have value, but the market-based mechanisms of conventional economics have generally failed to reflect these values as costs or prices.

There are two broad types of natural capital:

3.1 Nonrenewable natural capital (NNC) is finite in its quantity, so the stock must inexorably diminish as a result of its extraction, a process that is generically referred to as "mining." NNC is always subject to liquidation and can never be used in a sustainable fashion. The rate of exhaustion of NNC depends on the initial size of the recoverable stock and the rate of mining, although for some types of NNC, recycling and reuse can extend the lifetime of the resource. Note that NNCs do not yield economic value in the form of goods or services unless they are actually mined and utilized. Examples of NNC are metal ores, coal, and petroleum.

3.2 Renewable natural capital (RNC) can, at least potentially, sustain harvesting indefinitely. This requires that the rate of extraction is smaller than the natural rate of renewal of the resource. An example of nonbiological RNC is flowing surface water, which can be utilized for the production of hydroelectricity, for irrigation, and as a medium for transportation. Examples of biological RNCs that are harvested to produce valuable products are more numerous, and include trees and hunted animals, such as fish, waterfowl, and deer. These are obvious, direct examples of RNC, because their products and their many uses are so familiar. It is important, however, to note the following, additional characteristics of RNC:

a. Other aspects of the structure and function of ecosystems should also be regarded as legitimate types of RNC, even if these do not trade in markets and are therefore conventionally unpriced. This notion is discussed in more detail below.

b. Overharvesting, that is, exploitation that exceeds the regenerative capacity, will degrade the stock of any RNC. In such a case, the potentially renewable resource is effectively being mined or managed as if it were a nonrenewable resource.

c. When a previously unexploited or virgin stock of RNC is initially harvested, there may be a degradation of resource "quality," even if productivity is not degraded by overharvesting. The change in quality may be manifest in terms of the average size of organisms, in total biomass, in soil capability, or in other ways. For example, when an old-growth forest is harvested and converted to a second-growth forest, the net ecosystem productivity usually increases substantially. However, the average size of trees and the total ecosystem biomass will be smaller, and the community changes in important ways. As discussed below, and previously in Chapter 9, these changes have implications for ecological sustainability, which is not the same as resource sustainability.

d. Many types of RNC provide important services in their natural state, even when they are not harvested. For example, biological productivity, nutrient cycling, hydrologic cycling, erosion control, aesthetics, cleansing and detoxifying of the environment, and other ecological services all occur in unharvested and unmanaged watersheds. These services are important to ecological and environmental health, but their values are not usually recognized by the conventional marketplace.

The ecological reality is that humans have an absolute and undeniable dependence on a continuous flow of NC to sustain their economic systems. Economic and ecological systems are inextricably interrelated. This is particularly true of RNC, because sustainable economic systems cannot, by definition, be based on NNC, which is inexorably depleted by usage.

The size and vigor of an economic system are substantially determined by the wealth that is generated by the flows and interactions of all of the various types of capital. Also important are demands by consumers for goods and services and some other factors, such as entrepreneurship and labor relationships. In economic systems, all of the types of capital (i.e., MC, HC, RNC, and NRC) are interconnected, and within limits they can be substituted or converted from one type to another. The latter point is important, because as particular resources become scarce, their role in the economic system must be realized by substituting alternative resources or by developing and implementing technological changes that obviate their requirement.

In conventional economics, "income" is simplistically considered to be the flow of money over time, and "profit" is calculated as the difference between income and conventionally considered costs. In this context, "money" can be regarded as any medium that can be exchanged for goods and services. Usually, money is valuated in units of currency, but in barter systems, goods or services may be exchanged for other goods or services.

In conventional economic accountings, income can be generated and profits made, even though these may be achieved by an unsustainable mining of RNC. The basis for this ecologically false accounting is that the degradations of ecological resources and values are considered to be externalities, i.e., exoteric, nonvalued costs that do not have to be directly borne by the individuals or corporations who are using the resources. Therefore, these resources and values are free for the taking, and the ruining.

There are, of course, substantial social and environmental costs associated with resource and ecological degradation. However, these costs are borne by society as a whole. An important failure of conventional economics is that the marketplace fails to recognize the value of important, but nontraded services, especially when these are common-property or public-good ecological services.

In summary, in the context of the competitive markets of conventional economics, aggressive depletions of renewable natural resources by individuals or corporations can be enormously profitable over the shorter term. This is especially true of common property resources. In spite of the longer-term societal foolishness that is obviously inherent in the mining of RNC, this has been the most common way in which these resources have been exploited and managed. Case studies were presented earlier in this chapter to illustrate this remarkable fact.

The three most important reasons for the depletion of RNC by individuals and corporations are the following:

1. *Natural, free goods.* No investments of manufactured or human capital were made toward the building of most stocks of RNC or of NNC. Therefore, these resources are considered to be naturally occurring, free goods. In such cases, the costs of exploitation are only considered to be those associated with the direct expenses of mining or harvesting, and not with subsequent investments in the renewal that is required for sustainable use and management of RNC. This means that on the shorter term, the costs of development of virgin natural resources are zero.

For example, humans have undertaken none of the activities, and have borne none of the expenses involved in the geological development of stocks of NNCs, such as coal, petroleum, and metal ores. These resources of materials and potential energy are therefore available for only the costs of mining and processing, plus whatever levies, tariffs, or other fees governments or owners might choose to impose for the right of exploitation, or to pay for environmental damage associated with exploitation.

Similarly, no human investments have been made toward the development of virgin stocks of RNC such as natural forests, fish stocks, populations of hunted animals, groundwater aquifers, or fertile soils. The costs of developing these resources are considered to be externalities that are subsidized by ecological processes. (There are, however, substantial costs associated with subsequent renewal of the secondary natural resources, after the virgin stock of RNC is exploited and the resource is integrated into the "working," economic-production system. These costs may be borne by the individuals or corporations that harvested the original resource, or they may be externally borne by society.)

For example, if harvested, an old-growth rainforest has great conventional economic value because of its content of large individuals of desirable species of trees. Because no money was invested in the development of that natural forest, the only costs of utilization of the resource are those associated with harvesting the trees and converting their biomass into saleable products. Therefore, the shorter-term profits appear to be enormous, because of the great differential between the cash values of the sales and the expenses. Those profits are, of course, ecologically fraudulent because they are subsidized by the "free" ecological services that allowed the original forest to develop.

2. *Opportunity costs.* Whenever money is invested in some enterprise there are costs of foregone opportunities, because that money is not available to invest in other ventures. Opportunity costs are associated with any investments that are made toward the development of some future, profitable resource or activity.

The present use of natural resources has opportunity costs, because both NNC and RNC may have future values to humans, even if they are temporarily left unexploited. Any resources of coal, oil, or metal ores that are left in the ground by the present generation will be available for exploitation by future generations. These nonrenewable resources will, in fact, be much more valuable in the future than they are today, because their quantities will be smaller, and demands will be larger.

Similar arguments can be developed about the future values of RNC, such as undiminished stocks of biological resources. These do not, however, only have future values as exploitable resources. There are also costs associated with lost opportunities of ecological services, such as nutrient and hydrologic cycling, erosion control, cleansing of air, water, and soil, etc.

Because of the future worth of unexploited resources, the notion of zero-opportunity costs of resource exploitation is a shorter term, individualistic view. If the longer time frame and the proper social and ecological contexts are considered, the opportunity costs are actually considerable.

3. *Growth rates.* Often, biological RNCs are growing more slowly than the rate of growth of money invested elsewhere in the larger economy. In such a case, shorter-term economic considerations might suggest that an individual person or corporation could maximize profits by liquidating the natural resource and investing the money gained elsewhere in the economy, where the returns or profits are greater than the growth rate of the natural resource (Clark, 1989).

Consider, for example, a simple case in which marketable biomass in a forest is accreting at a rate of 5%/year, but real interest rates (i.e., adjusted for inflation) on secure investments are 10%/year. Because the forest resource is growing at 5%/year, its value would double every 14 years. If, however, the forest was harvested and the products sold, and the resulting money invested at 10%/year interest, the quantity of money would double in only 7 years, and profits would be made twice as quickly.

Obviously, this way of thinking only works if (a) the temporal perspective is short term; (b) the social perspective is that of individual persons or corporations, and not of society at large; (c) it is assumed that the natural resource only has value if it is harvested; and (d) only the costs of extraction are considered in the calculation of profit, while other ecological costs are treated as externalities.

Clearly, over the longer term the liquidation of potentially renewable natural resources is a losing strategy for society at large, and for future generations (Burton, 1993). For individuals, firms, and local economies, however, this can clearly be a profitable strategy because wealth can be accumulated during a short period of time. Therefore, this tactic is often pursued and advocated by powerful people. Consider the following statement in 1986 by Bill Vander Zalm, a former Premier of British Columbia, one of the world's greatest exporters of forest products: "Let's cut down the trees and create jobs" (Luinenberg and Osborne, 1990, p. 14). This is what has been happening, more or less, to many potentially renewable natural resources.

If certain types of RNC are not valuated, changes in their integrity will not have any influence in the conventional economic equations by which costs and profits are calculated. Ecologically, this is clearly incongruous, because the degradation of RNC has profound consequences for the longer-term sustainability of the flows of income and profit and of the health of the economic system. Some examples of ecological resources that markets fail to value (and that consequently are unpriced) and are therefore considered to provide "free" ecological goods and services include the following:

1. *Nonexploited wildlife,* such as plants, animals, and the "invisible" biodiversity of microbes, that are not often or ever used to sustain cash flows. Consequently, changes in the integrity of their populations and communities do not influence conventional economic cost accountings. These components of biodiversity are, however, valuable because they: (a) carry out important ecological services, (b) may have as-yet undiscovered uses to humans, (c) are important in the aesthetic environment, and (d) have intrinsic worth or existence value, irrespective of their direct utility to humans (see Chapter 10).

2. *Biological productivity* in natural communities that are not part of the "working" ecosystem, i.e., productivity that is not associated with directly valuated natural resources, such as that of agroecosystems and exploited forests and fish-

eries. This nonvalued productivity may nevertheless be important in maintaining natural biodiversity and in fostering ecological services such as biological control over erosion, hydrologic cycling, and chemical composition of the atmosphere.

3. *Nutrient cycling*, an important ecological service that is performed by diverse species of microbes, animals, and plants. Nutrient cycling helps to sustain the productivity of ecosystems, but in spite of its importance this service is usually taken for granted by markets, and its value is not reflected in the prices of commodities based on harvests from natural ecosystems. (In agriculture, however, the costs of degraded nutrient cycling may be taken account in the pricing of commodities, through the costs of fertilizer use.)

4. *Clean air, water, and soil.* These can be maintained, within limits, by ecological processes that reduce the concentrations of pollutants. For example, (a) aggrading vegetation removes substantial quantities of carbon dioxide from the atmosphere, and (b) microbes can metabolize many toxic chemicals and transform them to less toxic substances.

5. *Losses of soil through erosion*, able to be substantially prevented by maintaining an intact cover of vegetation. Losses of soil can have severe consequences for site fertility, and the resulting sedimentation can degrade aquatic habitats.

In conventional economics, the costs of degradation of these ecological resources are rarely valuated or considered to be important, for the following reasons:

1. *The nonvaluated resource is considered to be enormous*, so that any anthropogenic diminutions in its quantity or quality are unimportant, and need not be accounted for in the economic system. This world view is the essence of so-called "frontier economics," which assumes that natural capital will always be available in quantities that are sufficiently large to not limit the growth of economic systems. According to this attitude, there will always be fish to harvest from the oceans, trees to harvest in the hinterlands, air

and water to dilute gaseous emissions, and soil fertility to support agricultural and forestry systems. This view assumes that if resources are degraded in certain places, then new resources can always be exploited elsewhere, or the damage can be managed through mitigations. Of course, the logic behind this conventional economic view has always been fundamentally flawed in a finite world where there are limits to the quantities of materials and where the laws of thermodynamics govern the efficiencies of conversions of energy. This fact is becoming ever-more apparent, as anthropogenic activities significantly diminish the capacities of NNC and RNC to sustain the flows of goods and services that are required to support increasing populations of human and their societies, as well as other species and their communities.

2. *The nonvaluated resource is not considered to be important to human welfare*, and therefore its degradation is inconsequential and need not be considered as a cost. According to this view, (a) the extinctions of multitudinous species of tropical-forest beetles are of no demonstrated importance to human welfare, (b) the extinct dodo and great auk are not missed because they have been replaced by other sources of food and feathers, and (c) degradations of wilderness and other natural-ecological values are inconsequential because people, given time, can learn to love other types of landscapes, sceneries, and values. This opinion of the unimportance of ecological and resource degradation reflects an anthropocentric world view and rejects eco- and biocentric ethics and world views.

3. *The costs of degradation are considered to be externalities*, as was previously discussed.

In contrast, ecological economics assesses these services from a social and ecological perspective and assigns positive values. Therefore, degradations of these services are considered to be negative factors in the determination of profit. The assigned values are not necessarily measured in terms of dollars (although, in principle, they could be), largely because of difficulties in measuring disparate ecological resources in a common currency (Usher, 1986). The important point is that in ecological economics, consumption that runs down the net amount or quality of RNC does not contribute to income, because the value of any degradations must be subtracted from the revenues generated by the mining of RNC.

By definition, the consumption of NNC can never occur in a sustainable fashion, because the stock of the resource is inexorably depleted by its use. Therefore, sustainable economies can never be based on the forced throughput of NNC. The depletion of NNC can only occur within the context of a sustainable economic system, if it is somehow accompanied by an equivalent, integrated increase of RNC. For example, the use of coal to generate electricity might occur within the context of a sustainable system, if forests were planted or managed to increase the net production of biomass at a rate that was sufficient to (1) absorb the emissions of carbon dioxide from the fossil-fueled power plants and (2) restore the natural capital of potential energy for the production of electricity, in this case as RNC of tree biomass. Of course, other effects of coal burning, associated perhaps with emissions of toxic gases and particulates, reclamation of mined areas, etc., would also have to be mitigated or otherwise dealt with if the coal mining and burning were to be considered legitimate components of a sustainable economy.

Sustainable Economic Systems

As suggested above, sustainable economic systems must ultimately be based on the longer-term use of RNC, but in a fashion that does not diminish the stock of RNC or compromise its capability for renewal and its availability for use by future generations. A sustainable economic system must, therefore, be designed as an integral component of the ecological system of which it is only a part. This philosophy is central to ecological economics (Pearce *et al.*, 1989; Pearce and Turner, 1990; Rees, 1990; Costanza, 1991; Goodland *et al.*, 1991; Costanza and Daly, 1992; Costanza *et al.*, 1992; Daly and Townsend, 1993).

In ecological economics, there is an important

distinction between economies that are growing and those that are developing, as follows:

1. Economic growth is achieved by the consumption of natural capital, usually accompanied by its destruction or impoverishment. Growth involves a forced throughput of NC, including the mining of both NNC and RNC. There are physical and thermodynamic limits to growth. Therefore growth can never be sustained over the longer term, because it can only be achieved by the depletion of resources. Nevertheless, economic growth is regarded favorably by most economists, politicians, and citizens, because over the shorter term it can result in substantial, if unsustainable, accumulations of wealth. However, in the perspective of ecologists and ecological economists, growth is not desirable: "Economic growth as it now goes on is more a disease of civilization than a cure for its woes" (Ehrlich, 1989, p. 481).

2. Economic development, in contrast, does not occur at the expense of a degradation of NC. In the context of ecological economics, development implies a quantitative or qualitative improvement of NC, and it is achieved through efficiency of use, renewal, and fostering of RNC. If the natural-resource base does not change over time, then the economy is in a steady-state condition. Ecological degradation must also be managed within acceptable limits.

In the context of ecological economics, the notion of sustainable development refers to an economic system based on the longer-term exploitation of natural capital, occurring at some continuous rate, but in such a way that does not compromise the availability of those resources for use by future generations. In this sense, the existing human economy is clearly unsustainable, because it involves economic growth achieved through the vigorous mining of both NNC and RNC.

The notion of sustainable development is often advocated by politicians, economists, and resource managers, but many of these persons confuse sustainable development with "sustainable economic growth," which by definition is not possible. Even the widely acclaimed Brundtland report [World Commission of Environment and Development

(WCED), 1987], which first popularized the term sustainable development, obfuscates some of the important differences between economic growth and development.

The Brundtland report calls for a substantial expansion of the global economy: "It is . . . essential that the stagnant or declining growth trends of this decade (i.e., the 1980s) be reversed." It was suggested that economic growth, coupled with redistributions of wealth, are required to improve the living standards of the poorer peoples of the world. This would achieve social and economic conditions that are suitable for stopping population growth and the overexploitation of NC, so that a no-growth, equilibrium condition can be reached.

The Brundtland report recommends a growth of average, global, per-capita income of 3%/year, which would result in a doubling of per-capita income about every 23 years. However, because the global human population is growing at about 2%/year, overall economic growth would have to compensate and would have to increase at about 5%/year, doubling every 14 years. In regions where population growth is most rapid, for example much of Africa, west Asia, and Latin America, economic growth would have to be as large as 6%/year (doubling time of 12 years) or more, to achieve a 3%/year increase in real per-capita income. Ultimately, it was predicted that a global, economic expansion of 5–10 times would be required to set the stage for the achievement of a condition of sustainable development.

The authors of the Brundtland report suggested that this growth should be achieved by "policies that sustain and expand the environmental resource base." Those policies would include the development and implementation of technologies that could allow economic growth to be achieved, while consuming fewer material and energy resources. As noted previously, redistributions of wealth from richer peoples and regions to poorer ones would also be central to achieving the growth of per-capita income that is championed by the World Commission of Environment and Development.

Because the Brundtland report is essentially a consensus document, prepared by representatives of many countries and cultures and involving wide-ranging consultations, it is not surprising that the growth-related aspects of the "development" strategy that it advocates are relatively "easy" for politicians to support. Ecologically, however, there are profound reservations about whether a 5–10 times increase in the scale of the human economy could be sustained. It would be much more sensible to aggressively pursue more difficult and unpopular solutions. These would have to include population control, less intensive resource usage by richer peoples of the world, and redistributions of wealth.

Some ecological economists have suggested more appropriate, less growth-focused pathways toward the design and implementation of sustainable economic systems (Clark, 1986; Rees, 1990; Costanza, 1991; Costanza et al., 1991; Costanza and Daly, 1992). In such systems, it is recognized that the human scale (which is a combined function of population and per-capita environmental effects) must be limited within the carrying capacity of Earth's remaining natural capital. In fact, that scale may already be too large, and further economic growth is probably not desirable.

Scientific and technological progress is required toward the design of systems of resource use and management that can be maintained over the longer term. This is required to foster economic development, as opposed to maximizing the throughput of resources, which only achieves nonsustainable growth. Elements of the sustainable system should include the following:

1. Technological progress must increase the efficiency of use and recycling of NNC to extend the lifetime of nonrenewable resources.

2. RNC must be exploited at or below its sustainable capacity. As stocks of NNC become exhausted, RNC will become much more important in the economic system. RNC can be directly degraded if its rate of harvesting exceeds its rate of regeneration or indirectly if emissions of pollutants and other wastes exceed the renewable, assimilative capacity of ecosystems.

3. In a sustainable system, the rate of exploitation of NNC must be equal to or less than the rate of creation of new, renewable substitutes, i.e., of RNC.

4. Conventionally nonvaluated ecological resources must also be sustained. Some species and types of natural communities will decline because of the use and management of natural resources. However, an ecologically sustainable economic system must accommodate viable populations of all native species and other elements of biodiversity. Because some of these ecological values cannot be accommodated within landscapes that are managed for economic resource use, they will have to be obliged by designating nonextractive ecological reserves of sufficient numbers and sizes to be self-maintaining over the longer term. These accommodations must be made if any economic system is to be considered truly sustainable in the ecological sense.

It must be recognized that sustainable systems would not be very popular among the public, government administrators, politicians, and industry, because these would all experience short-term pains to achieve a longer-term, societal gain. Humans and their societies are self-interested, and they think on the shorter term. The longer-term benefit is achievement of an economic system that could sustain human society for a long time. The short-term pains would be associated with substantially less use of natural resources, abandonment of the economic-growth paradigm, and a rapid stabilization and perhaps downsizing of the human population.

Costanza and Daly (1992) suggested that an important element of the economic transition could be a natural-capital depletion tax, by which there would be large financial disincentives to net consumptions of NC. Any such system of depletion taxes would require a comprehensive, international agreement to prevent a free-market access to goods from countries or regions without a natural-capital depletion tax. In an open, global market, this restriction would recognize that the economic "playing field" must be level in terms of environmental protection and resource depletion. In such a system, individual lifestyles would adjust to a more ecologically appropriate condition, as consumers purchase cheaper, "green" goods that are manufactured without the depletion of natural capital and avoid more expensive goods that are produced through highly taxed depletions of NC.

As has been noted repeatedly in this chapter, humans have an undeniable reliance on natural resources to sustain their endeavors. Throughout human history, whenever there was a technological capability to do so, essential resources have been consumed to exhaustion. There are clearly better, sustainable systems that use renewable natural resources without depletion and without degrading ecological integrity in the broader sense.

Human societies desperately require these sustainable systems, but it remains to be seen if these systems will ever be developed and implemented. René Descartes noted in 1637: "It is not enough to have a good mind. The main thing is to use it well." To finish this chapter, Descartes' statement could be paraphrased as: 'It is not enough to have a good plan. The main thing is to use it well.'

13

SOME APPLICATIONS OF
ENVIRONMENTAL ECOLOGY

INTRODUCTION

Changes in the intensity of environmental stressors cause many responses in terms of the structure and function of ecosystems. This fact has been illustrated and discussed repeatedly in previous chapters of this book. As discussed in Section 13.2 of this chapter, some of these ecological effects are generic in their wider patterns and can therefore be broadly predicted. This observation has utility in the prediction of ecological effects of proposed human activities, an important component of the process of environmental impact assessment (Section 13.3). Knowledge of the causes and consequences of ecological responses to changes in stressors is also critical for ecological monitoring programs, as discussed in Section 13.4.

Human activities are causing enormous ecological changes on Earth. These are resulting in great challenges to present societies, and will do so more in the future. Still, knowledge of the ecological effects of stressors associated with human activities is incomplete, and ecologically sustainable systems of resource use and management have not yet been designed. These important deficiencies of understanding will have to be addressed, in part, by a more vigorous participation by ecologists in research, monitoring, and education in environmental ecology. This topic is discussed in Section 13.5 of this chapter.

13.2
EFFECTS OF STRESSORS
ON ECOSYSTEM STRUCTURE
AND FUNCTION

The intensities of ecological stressors vary in space and time. Whenever a variable concatenation of stressors affects particular species or communities there are, of course, unique ecological responses. However, in spite of particular circumstances, there are commonly observed patterns of ecological responses to changes in stressor intensities. These generic responses allow a certain degree of predictability that transcends the specifics of unique situations. Some of the frequently occurring patterns of ecological response to stressors are examined below.

Ecological Effects of Longer-term, Chronic Stress

Particular environments can be inimical to most organisms if intense stress occurs for a long period of time. In extreme cases, little or no ecological development may be sustained. Examples of such situations include the following:

1. Sites where severe, climatic stresses constrain ecological development, as in the Arctic, the Antarctic, at high altitude, and in deserts.
2. Sites where natural or anthropogenic pollution has caused intense toxic stress, such as the most polluted sites at the Smoking Hills (Chapter 2), surface exposures of minerals with large concentrations of toxic elements (Chapter 3), the immediate vicinity of smelters where depositions of toxic elements and gases have been substantial (Chapters 2 and 3), and some hazardous waste disposal sites.
3. Sites that are disturbed too frequently to allow organisms to establish and persist, such as the ice scour zone of northern oceans and lakes, and severely trampled paths.

More often, ecosystems are subject to less-intense exposures to longer-term stresses, which impoverish but do not preclude ecological development. Situations of this sort can be characterized by convergent, i.e., qualitatively similar patterns of ecological structure and function.

For example, the high-arctic tundra occurs in a natural environment that is chronically stressed. Ecological development is typically constrained by: (1) a short and cool growing season and (2) limited availability of moisture and nutrients, the latter because of oligotrophic soils and climatic suppression of nutrient cycling and litter decomposition (Bliss *et al.*, 1973; Chapin and Shaver, 1985; Henry *et al.*, 1986a). Because of these chronic environmental limitations, there are expansive landscapes of sparsely vegetated terrain in the high Arctic. Advanced ecological development is limited to only a few, relatively moist and warm situations called "oases," which comprise only a few percent of the land area (Bliss, 1977; Freedman *et al.*, 1983, 1993d; Henry *et al.*, 1986b).

The plants that are most common in the high Arctic are archetypal "stress tolerators" (*sensu* Grime, 1979), and the structure of the tundra ecosystem reflects their morphological and ecophysiological characteristics. Typically, these stress-tolerant plants are long lived, and they have a decumbent stature, small annual productivity, little allocation of biomass below the ground, long-lived green foliage and persistent dead foliage, a small requirement for and efficient recycling of nutrients, and reliance on vegetative propagation to persist in the community (Bliss, 1971; Savile, 1972; Maessen *et al.*, 1983; Nams and Freedman, 1987a,b).

Field experiments have shown that these characteristics of stress-tolerant, tundra plants are conservative. When field greenhouses or fertilization are used to ameliorate environmental stresses, the phenotype of these plants does not change much (Henry *et al.*, 1986a). In contrast, the relatively productive species that grow in less stressful arctic habitats, such as wet meadows, produce more biomass and flower profusely under experimentally ameliorated conditions (Henry *et al.*, 1986a).

Because of the severe climate and small primary production of the high-arctic tundra, there is a small species richness and a minuscule biomass of resident animals. In contrast, the avian biota is largely composed of about 10–20 migratory species that take advantage of the brief, seasonal abundance of arthropods and plant production (Freedman and Svoboda, 1982; Svoboda and Freedman, 1994).

In environments where the climatic limitations to ecological development are less severe than in the Arctic, stresses associated with pollution can cause the affected communities to exhibit a structure and function that superficially resemble those of the tundra. For example, the soil of sites rich in serpentine minerals is stressful to plants because of the toxic effects of nickel and chromium, coupled with nutrient deficiencies (Chapter 3). In an otherwise forested landscape, serpentine sites can be distinct, tundra-like, ecological islands, with plant communities dominated by low-growing shrubs and herbaceous plants that are tolerant of the stressful edaphic conditions. Serpentine vegetation is not only physiognomically distinct—it also has a species composition that differs from that of nearby, nonserpentine terrain. In many cases the serpentine taxa are endemics that are physiologically tolerant of the toxicity of these sites, and that cannot survive elsewhere because they are poor competitors in less stressful environments.

Under the longer-term influences of anthropogenic pollution and other stresses, ecosystem structure is generally simplified. This phenomenon can result in striking gradients of ecosystem structure, distributed concentrically around point sources of toxic stress. Consider, for example, the structure of terrestrial vegetation along radial transects from large smelters that emit toxic elements and/or gases, such as those at Sudbury and Gusum (Chapters 2 and 3). In unpolluted reference terrain near both of those smelters the landscape is dominated by mature forest, with a species composition and physical structure typical of the geographical region. In contrast, the immediate vicinity of the smelters is often characterized by virtually complete ecological degradation, because so few organisms can tolerate the inimical toxic stresses. As distance from the point sources increases, the intensity of toxic stresses decreases geometrically. The result is a pattern of progressive invasion and/or survival of species, depending on their physiological tolerance of the toxic conditions at various distances from the smelters.

Often, species that are tolerant of the toxic stresses are absent or uncommon in the non-polluted, reference habitats. Frequently, the most tolerant plants are a few species of lichens and bryophytes, plus a few tolerant species or ecotypes of monocots and herbaceous dicots. Together, these comprise a relatively sparse, low-growing community close to the point sources. At greater distances, shrubs may dominate, and still further away particular species of tree may form an open forest. Finally, at distances beyond which the toxic stresses are too small to exert important ecological effects, the reference forest type is present. It is important to note that the clinal variations of these vegetation types is continuous, although it may be possible to mathematically aggregate stands into identifiable "communities," occurring along gradients of continuous environmental and ecological change.

The syndrome of spatial declines of the vertical strata of terrestrial vegetation along transects from smelters has been described as a "peeling" or "layered vegetation effect" (Gorham and Gordon, 1963). Qualitatively similar patterns of community decline are observed along transects from point sources of stress associated with ionizing radiation, as was described for an oak–pine forest affected by gamma radiation by Woodwell and co-workers (Chapter 11).

In addition to these structural characteristics, severely stressed ecosystems sustain relatively small rates of certain functions, such as productivity, nutrient cycling, and litter decomposition. As was the case for species composition and physical structure of the ecosystem, the rates of these functional properties can vary predictably along gradients of severe toxic stress around large point sources, such as the smelters at Sudbury and Gusum.

Ecological Effects of an Intensification of Stress

When an ecosystem is subjected to a substantial intensification of pollution or another stress, it may be capable of serving as a "sink," by absorbing the increased stress without undergoing measurable changes. However, if some threshold of tolerance or resistance is exceeded, the ecosystem may display a syndrome of disruptions of its structure and function (Bormann, 1982; see also Fig. 13.3). In a qualitative sense, these ecological responses are observed generically, regardless of the type of ecosystem or the nature of the stress. Depending on the intensity of the stress, the affected ecosystem may become unbalanced energetically, nutrient cycles may become less conservative and "leaky" of nutrient capital, there may be changes in the species composition and dominant ecological "strategy" of the biota, and the system generally becomes less stable and less complex (Auerbach, 1981; Odum, 1981, 1985; Bormann, 1982; Smith, 1981, 1984; Rapport *et al.*, 1985; Schindler, 1987; Kelly and Harwell, 1989).

Many of the ecological consequences that might be expected to result from a substantial intensification of stress are summarized in Table 13.1. Only some of these effects are likely to be manifest in any particular situation of increasing environmental

Table 13.1.   The trends that may be expected
in ecosystems upon an intensification of stress[a]

Energetics
1. Community respiration increases.
2. Production/respiration becomes unbalanced (i.e., P/R becomes greater than or less than 1).
3. Production/biomass and respiration/biomass (i.e., maintenance to biomass structure) ratios increase.
4. Importance of auxiliary energy increases.
5. Exported or unused primary production increases.

Nutrient cycling
6. Nutrient turnover increases.
7. Horizontal transport increases and vertical cycling of nutrients decreases.
8. Nutrient loss increases (system becomes more "leaky").

Community structure
9. Proportion of r-strategists increases.
10. Size of organisms decreases.
11. Life spans of organisms or parts (leaves, for example) decrease.
12. Food chains shorten because of reduced energy flow at higher trophic levels and/or greater sensitivity of predators to stress.
13. Species diversity decreases and dominance increases; if original diversity is low, the reverse may occur, and at the ecosystem level, redundancy of parallel processes declines.
14. General biotic impoverishment by extirpation of sensitive species and increased dominance by a few tolerant species.

General system-level trends
15. Ecosystem becomes more open (i.e., inputs and outputs become more important as internal cycling is reduced).
16. Successional trends reverse (succession reverts to earlier stages).
17. Efficiency of resource use decreases.
18. Parasitism and other negative interactions increase, and mutualism and other positive interactions decrease.
19. Functional properties (such as community metabolism) may be more robust (homeostatic resistant to stressors) than are species composition and other structural properties. In systems dominated by long-lived perennial plants (e.g., forests), the reverse may be true.

[a]Modified from Odum (1985) and Schindler (1990).

stress. Important factors include the intensity of the stress, thresholds of tolerance of particular ecological responses, and the period of time during which the ecosystem is exposed to the environmental change.

In many freshwater systems, an intensification of environmental stress will cause large changes in the species composition of communities (e.g., of phytoplankton) before changes take place in eco-

logical functions (e.g., in primary production or respiration) (Schindler, 1988). In such a case, the toxic thresholds of sensitive species occur at a smaller intensity of stress than the thresholds for the ecological functions with which they are involved. Other, more stress-tolerant species may be able to take over the functional role of the sensitive species. Therefore, within limits, that function is carried out at a community level, irrespective of species composition. As a result, changes in species composition and demographics may be relatively sensitive indicators of initial, ecological responses to small or moderate intensifications of stress. Functional responses may be less sensitive.

In many terrestrial ecosystems, however, the reverse may be true (Schindler, 1988). In forests, which are dominated by long-lived plants, a moderate intensification of stress will typically affect functional attributes such as photosynthesis, respiration, and nutrient cycling well before there is an extensive mortality of individual plants or elimination of species from the community.

Many of the generic, ecological effects of intensifications of environmental stresses are revealed by consideration of the various stressors that have been examined in this book. For example:

1. Toxic stress associated with ozone is eliminating the relatively sensitive genotypes of ponderosa pine from montane forests in southern California. Previously dominant but pollution-sensitive individuals are being replaced by more tolerant genotypes and by other species of tree, resulting in a longer-term change in community composition (Chapter 2);

2. Clear-cutting of forests allows a species-rich community of ruderal plants and animals to temporarily dominate the site and utilize its resources (Chapter 8). This change in community composition results from destruction of the initial ecosystem dominants, which creates ephemeral opportunities for less competitive species. Similar effects are caused by wildfire. However, in the case of forestry, the change in community composition may only be a shorter-term one, or it may be an initial stage in a longer-term conversion of a

natural forest to an anthropogenic forest. For example, intensive management involving site preparation, planting of desirable species, and herbicide spraying might be used to establish plantations of only one or a few species of trees (Chapters 8 and 9).

3. The felling of all trees and subsequent herbiciding of a forested watershed at Hubbard Brook caused large losses of dissolved nutrients, especially nitrate, in streamflow (Chapter 9). This functional response reflects a decreased nutrient-retention capacity and results in a loss of nutrient capital from the system.

4. The exposure of certain species of trees to undetermined complexes of predisposing stressors may cause stand-level declines, often secondarily accompanied by infestations of pathogenic fungi and destructive insects that are normally resisted by healthier trees (Chapter 5). This syndrome represents indirect, functional responses to an intensification of the stressor regime.

5. The acidification of water bodies as a result of atmospheric depositions causes large changes in species composition and trophic relationships of biotic communities, and sometimes an oligotrophication of the ecosystem (Chapter 4). These represent a complex of structural and functional responses to an intensification of the toxic stresses associated with acidification.

Ecological Effects of Reductions in the Intensity of Stress

After the intensity of stress is alleviated, ecosystems that have been affected by pollution, disturbance, and other stressors will recover, although not necessarily to their original condition. Successions that follow deglaciation (Chapter 4), wildfire, clearcutting (Chapter 9), herbicide spraying (Chapters 8 and 11), and other intense events of stress are well known and readily appreciated phenomena.

Broadly speaking, during the relatively early and dynamic stages of successional recovery there are: (1) progressive accumulations of biomass and nutrients in the aggrading vegetation; (2) substantial changes in species composition of communities and

in the prominence of particular species; (3) a tightening of nutrient cycling, so that in older stages of succession there are smaller losses of nutrient capital and more internal storage and recycling; and (4) many other changes in community structure and function (Likens *et al.*, 1977; Bormann and Likens, 1979; Odum, 1983; West *et al.*, 1981; Kelly and Harwell, 1989; Begon *et al.*, 1990; Schindler, 1990).

Human intervention can ameliorate the conditions of stressed environments in order to encourage recovery, or to otherwise manage the situation so that more "acceptable" ecosystems develop. An obvious way to avoid stresses associated with pollution is to reduce or eliminate the emissions of toxic substances. Strategies of pollution abatement have, for example: (1) allowed a vigorous recovery of damaged ecosystems close to the Sudbury smelters (Chapter 2); (2) greatly reduced the ecotoxicological effects of the insecticide DDT on birds and other wildlife (Chapter 8); (3) reduced the amounts of petroleum and refined hydrocarbon products entering the oceans (Chapter 6); and (4) alleviated or reversed the eutrophication of many surface waters (Chapter 7).

In many other situations of severe, anthropogenic stressing of ecosystems, societies could potentially adopt laudable, but economically and politically difficult, management strategies to avoid environmental threats. For example: (1) more vigorous abatements of emissions of SO_2 and NO_x are obvious ways to reduce the ecological effects of toxic gases and acidification (Chapter 4); (2) banning of the use of chlorofluorocarbons would eliminate their effects on stratospheric ozone (Chapter 2); (3) vigorous protection of tropical rainforests and what remains of other threatened ecosystems would prevent the extinction of much of Earth's biodiversity (Chapter 10); (4) decreased atmospheric emissions of CO_2 from the combustion of fossil fuels and deforestation would help prevent predicted global climate change through an intensification of the greenhouse effect (Chapter 2); and (5) the conversion of all nuclear weapons into plowshares would avoid the direct and indirect effects of a thermonuclear holocaust (Chapter 11).

In many cases, decision makers believe that the abatement of certain anthropogenic stressors is not feasible, because of political, economic, social, or other constraints. The resulting (or potential) ecological degradations are considered (by the decision makers) to be "acceptable" costs of some desirable human endeavors that unfortunately carry substantial environmental risks.

In other cases, secondary management strategies may be used to mitigate the ecological responses to certain anthropogenic stresses, often while still allowing their "causes" to continue. Usually mitigative strategies are adopted on the shorter term in order to cope with an unacceptable state of environmental degradation, while continued research and political lobbying might hopefully lead to a more sustainable, longer-term solution to the environmental problem. Examples of this sort of "coping" strategy include: (1) the periodic liming of acidified surface waters in order to allow the survival of fish until an effective program of emissions abatement is implemented (Chapter 4); (2) the spraying of spruce budworm-infested forests with insecticides to keep the economic resource alive, while working toward the design of a more acceptable system of pest management for use in the future (Chapter 8); and (3) the use of prolonged and often tedious international negotiations to control or reduce inventories of conventional, chemical, and nuclear weapons, until a more comprehensive state of disarmament might be implemented (Chapter 11)

Mitigations can be used to successfully cope with many actual or potential environmental problems, including many of those that are ecological. However, in terms of avoiding deteriorations of ecological values, mitigations are always less desirable than not undertaking the degradation-causing activities in the first place.

13.3

ENVIRONMENTAL IMPACT ASSESSMENT

Environmental impact assessment is an interdisciplinary process that is used to identify and eval-

uate the environmental consequences of proposed actions and various alternatives [Canter, 1977; Environmental and Social Systems Analysis Ltd. (ESSA), 1982; Canadian Environmental Assessment Research Council/National Research Council (CEARC/NRC), 1986). Environmental impact assessment may involve considerations of ecological, physical/chemical, sociological, economic, and other environmental effects. Such assessments may be conducted in anticipation of proposed: (1) individual projects, such as the construction of a particular dam, smelter, power plant, incinerator, or airport; (2) integrated schemes, such as a proposal to initiate a complex development involving many projects within a watershed, an industrial park, or a pulp or lumber mill with its attendant wood-supply and forest-management plans; or (3) government policies that carry a risk of having substantial environmental effects.

An enormous number of ecological effects can potentially be caused by any proposed project, program, or policy. As a result, it is impractical to consider all of the potential effects when engaged in an assessment exercise. Usually, valued ecosystem components (VECs) are selected for study, on the basis of their perceived importance to society (Beanlands and Duinker, 1983; Conover *et al.*, 1985). Criteria for the selection of VECs for consideration vary greatly among assessments. Usually, the valued ecosystem component is one or more of the following:

1. An economically important resource or process, such as the biomass or productivity of agricultural crops, forest, or hunted fish, mammals, or birds.

2. A rare or endangered species or community, whose local, regional, or global survival may be at risk.

3. Of cultural or aesthetic importance, which can apply to a particular species or community, but also to landscape and wilderness values.

4. An indicator that integrates the potential effects on a complex of ecological variables. Spotted owls, for example, are considered to be an indicator of the integrity of certain types of old-growth forest in western North America. If a pro-

posed forest-harvesting plan is judged to pose a threat to the viability of a population of these owls, this could be considered as indicating a challenge to the larger, old-growth forest community.

5. Other characteristics identified as important during consultations with the public, the scientific community, concerned nongovernmental organizations, or representatives of regulatory agencies.

During the initial, screening phase of an environmental impact assessment, attempts are made to judge whether there will be any important interactions between project-related activities and the valued ecosystem components. This is done by comparing (1) the known spatial and temporal boundaries of the VECs, with (2) the predicted spatial and temporal dimensions of environmental stressors associated with the proposed development.

During a scoping exercise of this sort, ecologists often have to make professional judgements about the intensity and importance of potential interactions between predicted stressors and VECs. These opinions should be based on the best-available information, while recognizing that it is usually incomplete. If time and funding are available, it may be possible to undertake field, laboratory, or computer-based simulation research to more deeply investigate risks associated with some of the potential interactions identified during the screening process.

Once potentially important risks to VECs are identified, certain planning options will have to be considered. During this component of the impact assessment, ecologists and other environmental specialists provide information and professional judgements to decision makers to influence the choices that must be made to deal with the conflicts. The three, broad types of choices are the following:

1. *Avoid the predicted damages.* It is always, of course, possible to avoid the exposure of the VEC to the project-related stressor. This can be accomplished by choosing to not proceed with the development, or by modifying its characteristics. Avoidance is not always a desirable option from the perspective of the proponents of a develop-

ment, because: (a) substantial capital and operating costs can be involved, and (b) projects can sometimes be cancelled because of irreconcilable conflicts with ecological values. Regulators and politicians may also dislike this option, again because substantial socio-economic opportunities can be foregone, and because there is often intense controversy.

2. *Mitigate the predicted damages.* In many cases, mitigations can be designed and implemented, so that there would be no or few risks to the VEC. For example: (a) if the habitat of an endangered species is threatened, it may be possible to move the population-at-risk to acceptable habitat at another site or to create or enhance habitat elsewhere, so there would be no net detriment; (b) if a proposed industrial activity is expected to cause acidification of a lake, appropriate mitigations could include liming to reduce the acidification or fertilization to enhance certain ecological processes, again with an intention of causing no net detriment; or (c) emissions of carbon dioxide from a coal-fired generating station could be offset by reforesting land elsewhere, so there would be no net change in atmospheric CO_2 concentrations and no enhancement of Earth's greenhouse effect. Mitigations are the most common mechanisms of resolution of potential conflicts between project-related stressors and VECs. It must be understood, however, that mitigative actions are never foolproof, because of incomplete understandings of ecological processes and responses to environmental stressors.

3. *Accept the predicted damages.* Another option sometimes selected by decision makers is to allow or "eat" the damage to the VEC, because of the perceived socioeconomic benefits that would result from proceeding with the threatening development. In this case, regulators and/or politicians are of the opinion that the costs to society of not proceeding with a proposed development are larger than the costs associated with the predicted ecological and other environmental degradations. (Of course, the decision-making process is influenced by public opinion and the popular media, and the context is dominated by conventional economic costs and profits; see Chapter 12.) The

choice to accept the ecological damage is a common one, because: (a) not all ecological damage can be mitigated or avoided, and (b) some activities that cause ecological damage are perceived to be profitable on the shorter term.

A famous case of an almost cancelled project involved the snail darter (*Percina tanasi*), a 7- to 8-cm-long endangered minnow that was identified as threatened by extirpation through construction of the proposed Tellico Dam on the Little Tennessee River in Tennessee (Ehrlich and Ehrlich, 1981). An injunction was obtained in the courts on behalf of this fish, and for several years this action prevented completion of the 80% built dam. However, because of the perceived economic and political losses, the U.S. Endangered Species Act was amended by Congress in 1979 to allow completion of the dam. To mitigate the effects of completion of the dam on the snail darter, the threatened population was transplanted to suitable habitat elsewhere. In addition, subsequent field surveys found other, natural populations of snail darters in nearby streams, suggesting that the ecological effects predicted during the assessment were not as severe as originally thought.

Sometimes, developments are not permitted because irreconcilable conflicts with important VECs are discovered during impact assessments. A few examples of resource-related developments that have been halted include (Ehrlich and Ehrlich, 1981; Freedman *et al.*, 1993a): (1) foregoing of the Dickey–Lincoln Dam in Maine, partly because of concerns for an endangered plant, Furbish's lousewort (*Pedicularis furbishae*); (2) not allowing a proposal to mine a peat bog in Nova Scotia, because that site is one of the only sites in Canada for a rare and endangered plant, the thread-leaved sundew (*Drosera filiformis*); and (3) the withdrawal of large amounts of old-growth forest from logging allocations in the Pacific Northwest of the United States, on behalf of the spotted owl (*Strix occidentalis*).

It must be recognized that these proposed industrial developments are not necessarily foregone into perpetuity—the option always exists to proceed with the threatening activity at a later time. In this sense the endangered spotted owl, lousewort, and

sundew are always at risk, because of the short-term profits that can potentially be made by exploiting their habitats.

It is useful for ecologists involved in impact assessments to understand that stressed ecosystems exhibit a broad syndrome of qualitatively similar ecological changes. Such knowledge allows a degree of predictability of the broader ecological consequences of the advent or intensification of environmental stressors, such as pollution, in previously little-affected areas.

Of course, the structural and functional attributes of particular ecosystems respond in specific ways to changes in environmental stressors. The ability of ecologists to accurately predict specific ecological responses to changes in the stressor regime is immeasurably improved if there is knowledge of: (1) the tolerances of particular species and communities to the new stressor regime, (2) the predicted intensity and temporal and spatial variations of the stressors, and of mitigating environmental circumstances, and (3) the role and importance of any sensitive biota or processes in the functioning of the affected ecosystem.

Some understanding of these variables can be gained from knowledge of the prior effects of the same or similar stressors in other field situations. These field exposures can result from naturally occurring stressors, or because of an anthropogenic influence. Insights can also be gained from laboratory experiments that examine the ecophysiological tolerances and responses of key species of potentially affected communities to exposure to project-related stressors.

Consider, for example, a situation where ecologists are asked to participate in an interdisciplinary study of the potential environmental effects of a new smelter, whose proposed location is in forested terrain. Engineers and chemists can use mass-balance models to predict the anticipated emissions of toxic metals and gases from the smokestacks of the proposed facility. Subsequent modeling by atmospheric and other scientists can predict the dispersion of those emissions into the environment. These predictions would include the likely frequency of ground-level fumigation events at various places, the concentrations and deposition rates of toxic substances at various distances and directions from the point source, and their progressive accumulations (if any) in soil, the forest floor, vegetation, and surface waters.

Ecologists can then reasonably predict the likely ecological effects of the proposed smelter. This can be done by using the physical/chemical scenarios to predict exposure to project-related stressors, and then interpreting the likely ecological effects through knowledge of: (1) the toxicity thresholds of key, indicator biota, gained from laboratory studies; (2) the effects of similar toxic stresses on forest vegetation and surface waters near previously studied, operating smelters; and (3) the effects of relevant, natural pollutions by, for example, sulfur dioxide, metals, and acidification.

In large part, the process just described is the basis of most ecological studies within environmental impact assessments. Once the project is underway and the environmental stresses are initiated (e.g., after the smelter is built), there may be compliance monitoring of effects for which regulatory criteria have been established (e.g., emissions of chemicals and/or their accumulations in soil, water, or biota). Sometime, nonchemical, ecological effects predicted during an impact assessment are also monitored after a project is undertaken. Unfortunately, ecological effects monitoring is not always required by regulating agencies and therefore may not be carried out (Beanlands and Duinker, 1983).

If monitoring of ecological effects is required, the nature of the studies is largely guided by predictions of the impact assessment. It is important to recognize, however, that the structure of monitoring studies should not be too rigid, because "surprises" often occur. In this context, surprises are unpredicted ecological responses that are unprecedented, because they may be particular to the interactions of ecosystems and project-related stressors that are being investigated (Loucks, 1985). Any surprises that emerge must be investigated through adaptive changes in the monitoring program and perhaps by the initiation of further research.

Clearly, it is important that ecologists accumulate knowledge of the effects of stressors in both field and laboratory situations. This understanding allows ecologists to better predict the likely ecological effects of proposed anthropogenic activities.

13.4
ECOLOGICAL MONITORING AND RELATED ACTIVITIES

Because of widespread concerns about ecological changes, many countries are designing and implementing longer-term programs of ecological monitoring and research. These programs deal with changes over large areas and are different from the compliance and ecological effects monitoring described in the previous section (which are associated with particular projects or facilities).

In the sense used here, *environmental monitoring* is a multi- and interdisciplinary activity, involving repeated measurements of inorganic, ecological, social, or economic variables, with a view to documenting or predicting important changes over time. The interpretation of environmental-monitoring data requires an understanding of linkages among the components and processes of complex systems. Within this broader context, *ecological monitoring* focuses on temporal changes in the structure and functioning of ecosystems.

As considered in this section, ecological-monitoring programs should consist of two, integrated activities—monitoring and research (Freedman *et al.*, 1993c; Staicer *et al.*, 1993). Monitoring investigates questions that are conceptually simple, involving changes over time. However, monitoring is not necessarily a simple activity, because: (1) the initial choices of a few, astute indicators from multitudinous possibilities can determine the ultimate success of monitoring programs, and (2) the actual monitoring process, i.e., data collection, can be expensive and difficult to accomplish, and longer-term commitments may be required because important changes may be subtle and not detectable by studies of short duration.

In comparison with monitoring, the hypotheses investigated in ecological research are relatively diverse and complex. Often, however, those questions will deal with the causes and consequences of important changes that are detected during ecological monitoring.

The ultimate goals of ecological-monitoring programs are: (1) to detect or anticipate ecological changes, by measuring appropriate indicators, and (2) to understand the causes and consequences of those changes. Changes in indicators are determined by comparison with their known historical condition, or with a reference or control situation.

Environmental indicators are surrogate measurements that are considered to be related to important aspects of environmental quality. It is important to note, however, that the cause-and-effect relationships among stressors and indicators may not be well understood.

For example, monitoring of forest health in some region may indicate an obvious and widespread decline of some species of tree. The causes of the decline may not be known, but it might be hypothesized that they are related to some intrinsic biological factor, such as senescence of a cohort of trees, and/or to some combination of environmental stressors, such as air pollution, insect damage, climate change, or harvesting (Chapter 5). The resulting degradation of ecological integrity (this term is discussed later in this section) of the forest is associated with a complex of changes, including productivity, biomass, nutrient cycling, soil erosion, age–class structure, and species composition and richness of all levels of biodiversity. However, for the purposes of ecological monitoring, only one or a few well-chosen indicators would actually be measured over time. For example, the net-annual production of trees might be chosen as an indicator of changes in the economic-forest resource, while a species of bird with a specific habitat requirement might be chosen as an indicator of the integrity of mature or old-growth forests.

For the purposes of ecological monitoring and research, indicators can be generically classified according to a simple, widely used, SER model of *stressor–exposure–response* (Kelly and Harwell, 1989; Hunsaker and Carpenter, 1990):

1. In the context of ecological monitoring, stressors are agents of change and are associated with physical, chemical, and biological constraints on ecological integrity. Stressor indicators are mostly associated with human activities, such as atmospheric emissions of sulfur dioxide and other primary pollutants, emissions of precursors of secondary pollutants such as ground-level ozone, the use of pesticides and other potentially toxic substances, and rates of habitat change by disturbance, for example, by forest harvesting or conversion. Natural processes such as wildfire, hurricanes, volcanic eruption, and climate change are also environmental stressors.

2. Exposure indicators reflect changes in (a) the intensity of stressors at a place or in a region or (b) the accumulated dose over time. Qualitative exposure indicators measure only the occurrence of a stressor. Quantitative exposure indicators reflect the actual intensity or accumulation of stressors and include the concentrations of potentially toxic substances (e.g., milligrams per kilogram of lead in the forest floor; Chapter 3). Exposure indicators of disturbance and conversion could include the extent of specific habitat changes associated with forest fires, clear-cutting, urbanization, etc.

3. Response indicators reflect changes in organisms, communities, processes, or landscapes that are caused by exposure to stressors. Response indicators can include changes in the physiology, productivity, or mortality of organisms, in the diversity of species within communities, or in the rates of nutrient or soil loss from entire watersheds.

The SER model is conceptually useful because it implies intuitive cause-and-effect linkages, with ecological changes occurring in response to exposure to environmental stressors. However, the SER model is a simplistic model with important deficiencies. Most importantly, the cause-and-effect linkages of stressor–exposure–response are often not understood and/or not quantified, and the seemingly linear SER model does not deal effectively with more complex interactions of stressors and effects.

Alternative, conceptual variants of stressor–response models include the following:

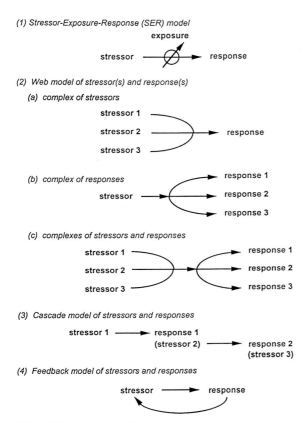

(1) Stressor-Exposure-Response (SER) model

exposure

stressor ⟶⊘⟶ response

(2) Web model of stressor(s) and response(s)

(a) complex of stressors

stressor 1

stressor 2 ⟶ response

stressor 3

(b) complex of responses

response 1

stressor ⟶ response 2

response 3

(c) complexes of stressors and responses

stressor 1 ⟶ response 1

stressor 2 ⟶ response 2

stressor 3 ⟶ response 3

(3) Cascade model of stressors and responses

stressor 1 ⟶ response 1
(stressor 2) ⟶ response 2
(stressor 3)

(4) Feedback model of stressors and responses

stressor ⟶ response

Fig. 13.1 Conceptual variants of stressor–exposure–response models. See text for explanation. Modified from Freedman *et al.* (1993c).

1. As was described above, the linear SER model (Fig. 13.1.1) suggests that exposure to a stressor may cause an ecological response. The intensity of exposure is important in determining ecological effects, but there may be nonlinear thresholds of tolerance or resistance (e.g., Fig. 13.3). For example, exposure of plants to ozone will only cause acute toxicity if physiologically determined thresholds of dose are exceeded (Chapter 2).

2. Web-models (Fig. 13.1.2) incorporate complexes of stressor(s) and/or ecological response(s) (Fig. 13.1.2). For example, some forest declines are thought to be caused by as-yet undescribed complexes of environmental stressors (Chapter 5; Fig. 13.1.2a). Analogously, a complex of ecologi-

cal responses can be caused by exposure to a relatively simple stressor (Fig. 13.1.2b). Acidification of a lake, for example, results in a large number of ecological responses, including direct toxicity to phytoplankton, invertebrates, fish, and other biota (Chapter 4). Of course, there can also be simultaneous webs of stressors and responses (Fig. 13.1.2c).

3. A cascade model (Fig. 13.1.3) acknowledges that ecological responses can in themselves become secondary (and higher-order) stressors, causing subsequent ecological changes. Consider, for example, an extension of the acidification example described above: directly toxic effects of acidification to certain biota can result in secondary and tertiary effects on other biota, as a result of changes in the trophic structure and dynamics of the ecosystem. In this case, changes in phytoplankton occurring in direct response to acidification can be a secondary stressor affecting herbivorous zooplankton, resulting in tertiary influences on planktivorous fish and quaternary effects on piscivorous fish and birds.

4. A feedback model (Fig. 13.1.4) suggests that some ecological responses can result in modifications of the intensity of the stressor. Wetlands, for example, might become drier through environmental responses to climate changes that may be forced by increased concentrations of carbon dioxide in the atmosphere (Chapter 2). Much of the organic carbon accumulated over the longer term in the wetlands might be oxidized under the drier conditions, resulting in a large flux of carbon dioxide to the atmosphere and representing a positive feedback loop of stressor–response.

Of course, these alternative, conceptual models of stressor–exposure–response are also simplistic in view of the complexity of the real, ecological world. When designing indicators for ecological-monitoring programs, simplicity can be an important, operational asset. However, during the critical interpretation of monitored changes in simple indicators, it must always be borne in mind that they represent complex ecological changes that are occurring in the real world.

Indicators can also take the form of composite

indices, which aggregate information of related or disparate types. Composite indices are commonly used to monitor complex trends in finance and economics, e.g., the Consumer Price Index and the Dow-Jones and other stockmarket indices. Composite indices of environmental quality and ecological integrity are especially desirable for reporting to the broader public, because they present complex changes in a simple manner. However, any proposed, composite indices of ecological integrity engender scientific controversy, mostly because of difficulties in selecting the component variables and in weighting their relative "values." Nevertheless, progress is being made in the design of composite indices of ecological integrity, especially for freshwater ecosystems (Karr, 1981, 1991; Steedman, 1988; Steedman and Regier, 1990).

In ecological-monitoring programs, interpretation of the causes and consequences of measured changes in indicators, or of predicted future changes, are evaluated using: (1) the accumulated understanding of ecological interactions, including the effects of stressors, coupled with (2) research that is specifically designed to address emergent questions. Therefore, according to this conceptual framework, important questions to investigate through ecological research can emerge from changes that are detected or predicted from ecological monitoring. The strategic linkage of monitoring and research can be illustrated by the following two examples:

1. Atmospheric monitoring in some region might detect an extensive acidification of precipitation. Interpretation of the cause(s) of this change might involve an examination of contemporaneous changes in (a) the concentrations in precipitation of chemicals in addition to hydrogen ion, and (b) anthropogenic emissions of gases and particulates to the atmosphere. Interpretation of the consequences of increased loadings of acidifying substances to ecosystems would be based on (a) existing knowledge of how acidification can be caused by atmospheric depositions and of the ecological effects of acidification, coupled with (b) new research targeting risks that are not yet understood. Specific research questions could relate to changes in the chemistry, biota, and eco-

logical processes in aquatic and terrestrial ecosystems, occurring in response to actual or predicted intensities of acidification (Chapters 2–5).

2. Monitoring might indicate an extensive conversion of natural forests into silvicultural plantations, occurring through clear-cutting and intensive management. Interpretation of the consequences of those changes would be based on existing knowledge of the ecological effects of forestry, plus novel research investigating poorly understood issues. Specific questions might address the effects of ecological conversion on biodiversity, productivity, soil quality, watershed hydrology and chemistry, disease and insect infestations, and global environmental changes (Chapters 8 and 9).

The information and understanding that can result from ecological-monitoring programs are important, because they can assist in the development and implementation of strategies to prevent further deteriorations of environmental quality and of ecological integrity. This objective is based on the unarguable premise that healthy ecosystems are necessary to sustain a healthy socioeconomic system.

Ecological monitoring and research can influence environment-related decisions by administrators, politicians, corporations, and individuals and therefore affect environmental quality (Fig. 13.2). For decision makers in government and industry, environmental reporting and interpretation (which includes information from ecological monitoring and research) should focus on the causes and consequences of actual or predicted environmental changes and the balance of the costs and benefits of activities required to deal with those changes. Decision makers have the responsibility to make decisions to avoid, mitigate, or accept environmental damages. Their decisions are based on the balance of perceived environmental costs associated with the damage and the shorter-term benefits of the activity that is causing the environmental degradation (almost always, the perspective has been of conventional, rather than ecological economics; Chapter 12).

Information from environmental monitoring and

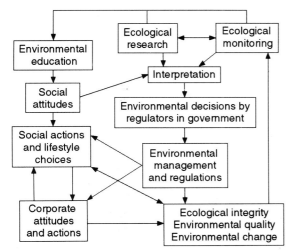

Fig. 13.2 A conceptual model of the relationship between an ecological monitoring program and environmental quality and change. The knowledge from ecological monitoring and research is interpreted and reported to decision makers in government, who then implement regulations and directly undertake management activities that affect environmental quality and ecological integrity (the changes in which are monitored to close the loop). The knowledge about environmental and ecological change is also used to inform the public, through the educational system, state-of-the-environment reporting by government, environmental nongovernment organizations, and the media. Social attitudes regarding the environment are influenced by this knowledge, resulting in more appropriate choices of lifestyle and resource use, and a public influence on corporations and government. Ecologists have much to contribute to this complex but necessary process, by being involved in monitoring and research that is relevant to environmental issues, and by having a more pervasive influence across the educational system.

research should also be objectively interpreted and reported for the broader public. Vehicles of communication to the public include the media, educational institutions, government state-of-the-environment reporting, and nongovernmental organizations. Environmental literacy could eventually have a pervasive influence on public attitudes, so that informed opinions will (1) influence choices of lifestyle, which can have important effects on environmental quality, and (2) influence decision makers to implement appropriate programs of environmental management, which will also affect en-

vironmental quality (Fig. 13.2). Poorly informed public opinions can encourage less appropriate environmental actions, including (1) rampant consumerism and wasteful usage of energy and material resources and (2) the occurrence of counterproductive "red herrings" of environmental management, which can displace attention from more important problems. Examples of environmental red herrings are the general misunderstanding of the difference between contamination and pollution and many instances of the closely related syndrome known as NIMBY (i.e., *not-in-my-backyard*).

Usually, ecological-monitoring programs address issues and concerns related to changes in: (1) the abundance and productivity of economically important, ecological resources such as agricultural products, forests, and hunted fish, mammals, and birds; and (2) aspects of the environment that relate to the health and integrity of ecosystems and their components, for example water, air, soil quality, and biota. These aspects may not be conventionally perceived as having direct economic value, but they are nevertheless of great intrinsic worth and ecological importance.

The context of ecological monitoring is, of course, change. As discussed repeatedly in this book, ecological dynamics are driven by both acute and chronic changes in the environment, caused naturally or through the activities of humans. For example:

1. Relatively rapid changes are associated with catastrophic disturbance and ensuing succession at the stand or landscape levels, e.g., episodes of fire, windstorm, clear-cutting, ecosystem conversion, etc. Sometimes extensive, rapid changes can be initiated by introductions of disease or exotic species. Microsuccession can also occur, at the more local level of gaps, as when individual trees or groups of trees die because of senescence, disease, insect attack, or a lightning strike. Succession is initiated by an intense event of stress, but the ensuing community dynamics, especially in middle and later stages of the sere, are largely driven by stressors associated with biological interactions, such as competition, predation, and herbivory.

2. Slower changes are associated with chronic stressors, both inorganic and biological. Some examples include climate change, acidification, stratospheric ozone and UV-b flux, nutrient loading, and accumulations of chemicals beyond toxic thresholds.

The dynamic character of ecosystems is an important and pervasive consideration in the shorter- and longer-term interpretations and management of ecological changes. Moreover, increasing intensities of environmental stress are the context within which environmental quality and ecological integrity must be managed and/or preserved into the future. This is occurring, of course, because anthropogenic stressors are becoming increasingly more important in forcing ecological changes and in degrading ecological integrity.

Ecological integrity is a notion, and there is no consensus on its definition. Any changes caused by human activities will enhance some species, communities, and processes, while simultaneously causing detriment to others. However, from consideration of the generic changes in disturbed and stressed ecosystems (e.g., Table 13.1), it might be suggested that greater ecological integrity is displayed by systems that, in a relative sense: (1) are resilient and resistant (see glossary) to changes in the intensity of environmental stress; (2) are biodiverse; (3) are structurally and functionally complex; (4) have large species present; (5) have higher-order predators present; (6) have controlled nutrient cycling, i.e., are not "leaky" of nutrient capital; (7) are efficient in the use and transfer of energy, e.g., are not degrading in terms of fixed-carbon capital; (8) have an intrinsic capability of maintaining their natural ecological values, without anthropogenic interventions through management; and (9) are components of a "natural" sere, as opposed to one strongly influenced by human activities (Odum, 1985; Schindler, 1987, 1990; Hunsaker *et al.*, 1990; Woodley, 1990, 1993; Anderson, 1991; Cairns *et al.*, 1991; Karr, 1991, 1993; Freedman *et al.*, 1993c; Regier, 1993). Note that the last indicator, in particular, involves judgements about worth and suggests a nonscientific bias. Nevertheless, most ecologists would consider that the attributes of native species and their communities represent greater ecological integrity, compared with systems that are strongly influenced by introduced species and anthropogenic activities.

In many respects, the notion of ecological integrity is analogous to that of health. A healthy body is physically and mentally vigorous and free from disease. A healthy organism, and an ecosystem with integrity, is indicated by diagnostic symptoms bounded by ranges of values considered to be normal, and by attributes that are regarded as desirable, while pathological conditions are indicated by the opposite (Schaeffer *et al.*, 1988; Rapport, 1990, 1992; Ryder, 1990).

Indicators of ecological integrity can include measures at the metabolic, organismic, population, community, and landscape levels, and holistic composites of these. All of these variables can respond to changes in environmental conditions. Ecological responses to environmental change can vary from simple monotonic responses to complex curvilinear responses (Fig. 13.3). The nature of the ecological responses is partly dependent on the resistance to environmental change and on thresholds of tolerance to stress. Ideally, to understand what an indicator is measuring, it is necessary to determine the shape of its response curve when exposed to variations of a particular stressor (achieved by experiment or by examination of existing environmental gradients). In practice, however, accurate response curves are rarely known.

Sometimes, important ecological changes are detected during monitoring. However, the causes of those changes may be uncertain because: (1) they are part of an undiscovered complex of environmental stressors, or (2) they are extrinsic to the monitored ecosystem. Once important changes are documented, however, hypotheses can be identified and research directed to determine the causes and consequences. At any stage during the process, monitoring and related research can be used to identify important risks to ecological integrity and to prevent or mitigate the hypothesized or measured damage.

As noted previously, ecological-monitoring programs must have the capability of (1) detecting recent changes and (2) predicting future changes.

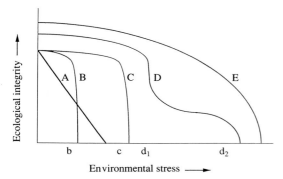

Fig. 13.3 A conceptual model of responses of ecological integrity to changes in environmental stress. Ecological integrity can be indicated by measures at the physiological, organismic, population, community, and landscape levels, all of which can respond to changes in environmental conditions. The curves depict variations in the resistance of ecological indicators to environmental change. Curve height represents ecological integrity, while slope represents the rate of change of integrity in response to an increasing intensity of environmental stress. Ecological integrity is initially lower in examples A–C and higher in D and E. Example A responds to increasing environmental stress with a steady decline in integrity, but B and C only begin to show a notable response after a threshold of tolerance (b and c, respectively) is exceeded, with C having a higher threshold. Example D shows a complex, curvilinear response to environmental change, with rapid loss of integrity after one threshold is exceeded (d_1), followed by relative stability (resistance) and then another rapid change at a second threshold (d_2). Example E has the highest integrity initially, as well as over the longer term. For the purposes of ecological monitoring, components A and B would be most useful as early indicators of a loss of ecological integrity. Note that the shape of the response curves for decreasing stress would probably be different than for increasing stress; i.e., there is hysteresis in the stressor–response function. Modified from Freedman *et al.* (1993c).

In their design, ecological-monitoring programs should: (1) deal with the most important stressors that are known or potential threats to ecological integrity; (2) have a program in place to measure exposure to those stressors over time and space; and (3) monitor or predict the responses of populations, communities, and ecological processes to changes in exposure to the stressors.

In terms of this conceptually simple framework, there are some major deficiencies in existing ecological-monitoring programs, and in ecological knowledge more broadly. As a result, many important environmental issues that involve ecological changes cannot be effectively addressed, because of a lack of appropriate monitoring, research, and/or understanding. A few examples selected from preceding chapters include the following:

1. Is a widespread decline of migratory, neotropical, forest-interior songbirds taking place in North America? If so, what are the causes, and how can society manage those stressors?

2. What is the magnitude of the mass extinction that is now presumably occurring? What species are affected, where, and why? How, specifically, are these species important to the health of the biosphere, and to human welfare? Much of the mass extinction is occurring because of the conversion and disturbance of tropical forests, but how will people of higher latitudes be affected? How are the people of richer countries connected to the biodiversity-depleting stressors in poorer countries?

3. What are the importance, extent, and causes of forest declines and diebacks? Which of the declines and diebacks are caused by anthropogenic stressors, and how can these be managed?

4. What are the biological and ecological risks associated with increased exposures to ultraviolet radiation, resulting from depletions of stratospheric ozone caused by anthropogenic emissions of chlorofluorocarbons?

5. What constitutes an acceptable exposure to toxic chemicals? For naturally occurring toxins, such as metals, what are the exposure thresholds beyond which industrial emissions should not increase concentrations? Is any increase in exposure acceptable for nonnatural toxins, such as TCDD, PCBs, synthetic pesticides, and radionuclides? Or are there acceptable thresholds of exposure to these chemicals as well?

6. What are the ecological and other environmental consequences and costs of warfare at local, regional, and global levels?

7. Is it possible to objectively valuate species, communities, and ecological services, so that these considerations can be integrated into economic cost-benefit models?

8. How intensively can potentially renewable, ecological resources be harvested and managed, without unacceptable risks to their sustainability over the shorter and, especially, longer term?

These are just a few of the important ecological problems that will have to be addressed. Ecologists, of course, have a great deal to contribute to their resolution.

There have already been some notable cases where ecological monitoring and research have contributed to the resolution of important environmental problems. Three cases taken from preceding chapters are:

1. The identification of eutrophication as an important problem in freshwaters, followed by research that uncovered phosphate as the primary, causal agent, and then the development of sewage-treatment facilities and low-phosphorus detergents to manage the problem.

2. The recognition of acidification as a widespread problem with diverse ecological consequences. Research demonstrated a sufficiently clear linkage with atmospheric depositions of sulfur and nitrogen compounds to convince decision makers to implement transnational strategies of emissions reduction to attempt to manage the problem.

3. The identification of contamination with persistent, chlorinated hydrocarbons such as DDT, dieldrin, and PCBs as a widespread phenomenon. Research and monitoring provided a linkage with toxic effects in some wildlife and possible effects on humans. The toxicological evidence has been sufficiently convincing to result in bans and other actions to reduce environmental exposures to these substances.

To provide the ecological information and knowledge that is required to deal with environmental problems, many countries are now designing programs of longer-term ecological monitoring and research.

In the United States, a network of Long-Term Ecological Research (LTER) sites has been established by the National Science Foundation (Franklin *et al.*, 1990; Van Cleve and Martin, 1991). This is

an important initiative, because longer-term research is essential for the formulation and testing of ecological theories and for achieving an understanding of the population biology of long-lived species and of gradual and transient ecological responses to environmental changes (Franklin, 1989; Likens, 1989; Risser, 1991). The LTER program does not, however, have a strategic focus on monitoring and research directed toward a better understanding of anthropogenic stressors (although particular projects within the network do have such a focus). Each site in the LTER network is examining unique hypotheses and therefore has its own design of ecological research and monitoring, although to some degree networking relationships are fostered among the LTER sites, and comparative studies and analyses have some priority. In addition, the research and monitoring effort in the LTER network is not integrated with the more extensive EMAP program (see below).

The Environmental Monitoring and Assessment Program (EMAP) of the U.S. Environmental Protection Agency is intended to provide information on ecological changes across large areas of the United States (Hunsaker and Carpenter, 1990). EMAP does this by monitoring ecological indicators at a large number of sites arranged in a statistical-sampling grid over the entire country. This is a rigorous sampling design, but it would be difficult for many countries to emulate such a system, because of the great expense and effort that is required.

In Canada, for example, the population and economy are both about one-tenth that of the United States, and most people and economic activity are focused along the southern margin of that large country, as is the greatest intensity of anthropogenic stressors. Although sampling locations for monitoring programs should be chosen from a probability sample to statistically represent a region, existing monitoring programs in Canada have often been designed to investigate local problems, to study terrain or particular ecosystems that are sensitive to particular stressors, or to take advantage of local administrative resources. In part, this happens because logistical support is lacking over vast expanses of Canada. The logistical and financial constraints of Canada make an EMAP-style, grid-based sampling design impractical. In addition, Canada has not yet committed to supporting a large program of longer-term ecological research, such as LTER.

To address its deficiencies in ecological monitoring and research, governments in Canada are beginning to design a national, ecological-monitoring program. One conceptual framework that has been proposed involves: (1) an integrated network of two classes of monitoring sites, intensive and extensive, and (2) ecological monitoring and research that focus on environmental problems (Freedman *et al.*, 1993c; Staicer *et al.*, 1993). The sites in the network would be spatially arranged within each of 15 terrestrial ecozones, the largest units within the ecological land classification of Canada.

In this framework, each ecozone would contain at least one intensive-monitoring site, which would be used for relatively detailed monitoring of indicators relevant to structural and functional ecology, including reference monitoring in relatively mature and undisturbed areas. Reference monitoring is important to understanding the effects of regional or global stressors, such as climate change and acidic deposition. Intensive-monitoring sites would also be used to determine ecological responses to environmental changes. This role would be accomplished, in part, by conducting experimental research into questions examining the causes and consequences of ecological responses to important stressors. Extensive monitoring sites would be more numerous and located throughout the ecozones. Data collected from these sites would provide overviews of larger-scale changes in the ecological character of landscapes. Relatively localized stressors occurring at widely spaced locations are particularly relevant to extensive monitoring, for example, agriculture, forestry, mining, industrialization, and urbanization.

Important aspects of this ecological-monitoring framework are the integration of extensive and intensive monitoring and research and the focus on environmental problems (Freedman *et al.*, 1993c; Staicer *et al.*, 1993). Integrated monitoring can refer to: (1) the integration of activities that monitor different indicators at a particular site and (2) the integration of monitoring activities among different sites. Both of these concepts are fundamental to national ecological-monitoring programs, because activities, indicators, and methodologies should be harmonized within and among sites, to the extent possible. Furthermore, research must be integrated into the monitoring program, particularly through the pursuit of questions regarding the causes and consequences of monitored changes. Achievement of these integrations requires a great deal of cooperation among ecologists involved in monitoring and research and other environmental scientists in government, universities, and nongovernment agencies.

13.5
ECOLOGISTS AND ENVIRONMENTAL PROBLEMS

Environmental Policy

The environmental crisis is not an imagined phenomenon, and its ultimate consequences are potentially cataclysmic for the biodiversity of Earth and for human society. The causes of the environmental crisis are fundamentally social and economic, but the critical responses are biological and ecological. As a result, ecologists should be intimately involved in defining the dimensions of the environmental crisis and in designing pathways to avert or manage its consequences to whatever degree is still possible.

It is widely acknowledged that ecological considerations are central to environmental science and to the development of sensible environmental policies. In view of this fact, it is quite remarkable that many professional ecologists, especially academics, have been somewhat unenthusiastic in their pursuit of science relevant to the environmental effects of human activities and in contributing to the formulation of environmental policies (Ehrlich, 1989; Schlesinger, 1990; Ehrlich and Wilson, 1991). This counterintuitive phenomenon was recently remarked upon by Ehrlich (1989, p. 481):

> If most biologists, including those trained in ecology . . . cannot bestir themselves to put a major effort into changing the course of society, what hope is there for [government] or the general public?

It must be recognized, however, that the proper role of most ecologists in the formulation of environmental policies is a professional one. In this sense, ecologists should contribute their scientific advice and judgements toward the development of environment-related policies. The usual role of ecologists would not be to actually develop the policies, but to contribute expert counsel toward that goal (Cowling, 1992). There are exceptions to this generalization, and they apply to ecologists who specifically move from science and into the role of policy and decision makers.

The professional counsel of ecologists should consist of objective, expert opinions and should not be expressions of personal ethics or world views. This is not to say that ecologists should not state personal opinions about environmental issues. However, when they do so it should be clearly acknowledged that the opinions are not professional advice, but individual expressions and interpretations within the context of a free-thinking society. The personal opinions of ecologists are, of course, strongly influenced by ecological world views and would be recognized as such, but they are still personal opinions.

This is an important context for the professional involvement of ecologists in the formulation of environmental policies, and it has previously been remarked upon: " . . . ecologists must clearly separate their views based on personal feelings from those that can be substantiated empirically" (Schlesinger, 1989, p. 1).

Environmental Research

It is abundantly clear that humans are causing tremendous changes to the ecosystems of Earth and that in large part those changes involve degradations of ecological integrity. It is also clear that scientific understanding of ecological responses to anthropogenic stressors is incomplete—there are some very consequential unknowns about the ecological effects of human activities.

Considering the above, it is remarkable that so many ecologists have a sparse involvement in research that examines the degradative effects of hu-

man activities. The truth of this statement is suggested by the following observations:

1. A recent survey of the membership of the Ecological Society of America found that only 4% of 3695 respondents (56% of the ESA membership) identified their primary research field as involving global change or biodiversity, and only 2% identified ecologically sustainable systems (Morrin *et al.*, 1993). These are the primary research areas that are highlighted by the Sustainable Biological Initiative of the ESA (see below).

2. Examination of recent contents of some of the most prestigious journals in ecology, such as *Ecology*, *Ecological Monographs*, *Journal of Ecology*, *Journal of Animal Ecology*, and *Oecologia*, shows that few of the papers deal directly with issues related to anthropogenic stressors and ecological degradation.

A tendency to pursue interesting research questions, regardless of their direct relevance to the needs and problems of society, is especially characteristic of academic ecologists and of the research opportunities that tend to be presented to their graduate students. The freedom to pursue curiosity-driven research is a cherished institution in academia, and this should certainly not end. However, considering the present and future dimensions of the environmental crisis, and the central role of ecological issues, there is a need for a more appropriate balance between applied and fundamental research by academic ecologists, and for environment-related work to be better supported by funding agencies (Bliss, 1984; Mooney, 1991; Schlesinger, 1990; Schindler, 1991; Ehrlich, 1992; Levin, 1992; Solbrig, 1992):

> The need for accelerating environmental research is patent . . . The training of graduate students to do nothing but basic research should end . . . all scientists, and especially all biologists, need to spend at least a portion of their careers working at the science-society interface . . . It is the environmental crisis that most demands the attention of biologists today and is most likely to shape their careers tomorrow (Ehrlich, 1992, p. 704).

Some of the professional activities that are required of ecologists are relatively descriptive, relating, for example, to a more complete description and cataloguing of biodiversity, so that anthropogenic threats can be better understood. Ecologists must also be intimately involved in the design and execution of monitoring programs, so that important ecological changes will be detected and their causes and consequences understood. All of these needs are important. However, some of the activities are relatively mundane, in terms of depth of the scientific questions that are asked.

This is not always the case, however, because many research challenges in environmental ecology can involve insightful questions about environmental influences on the structure and function of ecosystems. Exciting and important ecological research can, indeed, be designed around anthropogenic stressors. There are many intellectually stimulating research questions and opportunities to pursue in environmental ecology. And society requires the answers.

Ecologists have a critical and central role to play in the design of ecologically sustainable systems of resource use and management. Hopefully, this profession will soon come to grips with the importance and opportunities of this challenge. It is time for ecologists to more effectively apply their collective expertise, experience, and research skills toward the solution of environmental problems.

There are some indications that ecologists as a group are beginning to deal with these issues and are developing mechanisms for a more substantial involvement in environmental science. This trend can be illustrated with two recent examples from North America.

The Sustainable Biological Initiative (SBI) is an emerging program of the Ecological Society of America (Lubchenko *et al.*, 1991; Levin, 1992). Within the broader context of ecological sustainability, SBI will focus on research and education, by fostering individual studies, rather than large and integrated projects. SBI is intended to serve as a "wellspring of creativity and scientific innovation" directed toward problems of global change, biodiversity, and the maintenance of ecologically sustainable systems (Levin, 1992).

A much bigger science program in North America was the National Acid Precipitation Assessment Program (NAPAP). NAPAP was strategically tar-geted to a particular suite of environmental and ecological changes, and was "designed to provide an improved understanding of the phenomenon of acid precipitation and its effects on lakes, streams, fish, crops, forests, visibility, human health, engineering materials, and cultural resources," with a focus on the United States (Cowling, 1992). NAPAP involved a decade of activities between 1980 and 1991, involving thousands of scientists, many of them ecologists, and costing more than \$530 million in U.S. federal funds. Also integrated with the NAPAP program were State and Canadian research and private-sector activities, especially by the Electric Power Research Institute, an industrial research consortium.

A major goal of NAPAP was to conduct an assessment of the state of scientific understanding of the causes and consequences of acidification caused by atmospheric depositions. Science assessment is a process by which the available scientific evidence is critically reviewed and interpreted, in order to predict the outcomes of alternative strategies of environmental action, for example, various emissions-control scenarios for SO_2 and NO_x. During assessments of the costs and benefits of alternative actions, the quality of scientific understanding is analyzed, uncertainties are bounded, and critical research needs are identified, and sometimes pursued.

NAPAP was a science assessment, but it was also a major research program intended to fill in many of the gaps of understanding of the causes and consequences of acidification, including ecological responses. Science assessments can influence the nature of some expensive environmental-management actions, which may have to be advocated and pursued in the face of substantial scientific uncertainties. The science assessments and research of the NAPAP program were highly influential in the design and passing of the Clean Air Act amendments of 1990 by Congress and their signing by President George Bush, resulting in commitments to achieve a 50% decrease of U.S. SO_2 emissions beginning in 1995 (equivalent to a 10 million tonne/year reduction) and a 20% decrease in NO_x emissions (2 million tonne/year), as well as other clean-air actions (Cowling, 1992).

The initiatives described above are two important examples of a more substantial involvement of ecologists in research and assessments associated with environmental issues. There is, however, a critical need and obligation for deeper commitments by ecologists to these sorts of activities.

Environmental Literacy

Environmental literacy is an important social goal. As was noted previously, a well-informed public will formulate opinions and make choices that will improve the quality of the environment and reduce anthropogenic stresses on ecosystems. Environmental literacy could be systematically achieved by an institutionalized incorporation of objective information on changes in environmental quality into the educational curriculum, at all levels. Ecologists have much to contribute toward achieving an environmentally literate citizenry.

In all countries there are substantial deficiencies in the awareness and education of citizens about environmental issues. In part, these shortcomings exist because the public largely relies on the mass media for environmental education. The media communicates a great deal of useful environmental information, but its presentation of some issues may be biased and sometimes inaccurate. The media is a for-profit business and is often driven to report what sells and to interpret accordingly. As such, red herrings can become high-profile issues that dominate the environmental agenda and thereby detract from efforts to solve more important problems. To some degree, this dilemma can be managed if formulations of individual and public opinions are based on better information and understanding of facts and issues.

For these and other reasons, there is a serious need in all countries for an institutionalized exposure to objective and critical environmental information. This exposure should occur throughout the educational system, from primary and high schools, through universities, to continuing education for the out-of-school public.

The lack of access to environmental education can be illustrated by the observation that in North American universities, most undergraduates do not presently have, or do not take advantage of, opportunities to attend interdisciplinary, environmental studies classes: "At major universities . . . a large proportion of the students graduate still ignorant of (environmental) basics" and their pervasive influence on their daily lives and society (Ehrlich, 1989).

All ecologists have much to offer in terms of environmental education. In particular, ecologists who are faculty in universities can contribute in many ways, including the following:

1. By participating in environmental-studies classes. Environmental studies is an interdisciplinary field, but its foundation is ecological, and this core material should be interpreted and taught by ecologists.

2. By helping other faculty to integrate environmental case materials across the curriculum. There are opportunities to include environment-related information into the teaching materials of virtually all university-level subjects, from the physical sciences, through the natural sciences and medicine, to the social sciences. Ecologists have much to contribute to achieving this pervasive integration. They can do this by encouraging the process, by participating in multidisciplinary workshops, and by helping academics in other disciplines to discover and interpret appropriate case materials.

3. By integrating case material from environmental ecology into their own specialized classes in ecology. Many ecological principles can be effectively illustrated using case materials from environmental ecology. A pedagogical integration of fundamental and environmental ecology would allow ecologists to be trained in the basics of their discipline, while still achieving a core literacy about the ecological effects of human activities.

4. By facilitating the development of environmental curricula in public and high schools. Usually, students at preuniversity levels of the educational system have little exposure to ecology or to environmental issues. There are many opportunities to integrate ecological and especially environmental issues and information into the cur-

riculum at all levels of the educational system. The involvement of ecologists would facilitate this important process.

Achievement of a public that is literate about the causes and consequences of the environmental crisis is a necessary component of any strategy to resolve these predicaments. Ecologists must be centrally involved in this educational process.

13.6
CONCLUSION

Over the longer term, human societies can only prosper on Earth if they institute limitations on their population and ensure that the harvesting and management of ecological resources are undertaken in ways that are truly sustainable.

It is critical to achieve an understanding of the ways that ecological degradations are caused by the direct and indirect consequences of human activities. Means must also be designed to prevent or effectively mitigate those effects.

The coupling of population control with sensible strategies of environmental and ecological management will be crucial to achieving a sustainable prosperity for humans, while still accommodating other species and their natural communities, on the only planet in the universe that is known to sustain life.

REFERENCES

Aber, J. D., Botkin, D. B., and Melillo, J. M. (1978). Predicting the effects of different harvesting regimes on forest floor dynamics in northern hardwoods. *Can. J. For. Res.* **8**, 306–315.

Aber, J. D., Botkin, D. B., and Melillo, J. M. (1979). Predicting the effects of different harvesting regimes on productivity and yield in northern hardwoods. *Can. J. For. Res.* **9**, 10–14.

Aber, J. D., Nadelhoffer, K. J., Steudler, P., and Melillo, J. M. (1989). Nitrogen saturation in northern forest ecosystems. *BioScience* **39**, 378–386.

Aber, J. D., Melillo, J. M., Nadelhoffer, K. J., Pastor, J., and Boone, R. D. (1991). Factors controlling nitrogen saturation in northern temperate forest ecosystems. *Ecol. Appl.* **1**, 303–315.

Aber, J. D., Magill, A., Boone, R., Melillo, J. M., and Steudler, P. (1993). Plant and soil responses to nitrogen additions at the Harvard Forest, Massachusetts. *Ecol. Appl.* **3**, 156–166.

Abercrombie, M., Hickman, C. J., and Johnson, M. L. (1978). "The Penguin Dictionary of Biology." Penguin Books, New York.

Abernethy, S., Bobra, A. M., Shiu, W. Y., Wells, P. G., and Mackay, D. (1986). Acute lethal toxicity of hydrocarbons and chlorinated hydrocarbons to two planktonic crustaceans: The key role of organism-water partitioning. *Aquat. Toxicol.* **8**, 163–174.

Abrahamsen, G. (1980). Acid precipitation, plant nutrients, and forest growth. *In* "Ecological Impact of Acid Precipitation" (D. Drablos and A. Tollan, eds.), pp. 58–63. SNSF Project, Oslo, Norway.

Abrahamsen, G., and Stuanes, A.O. (1980). Effects of simulated rain on the effluent from lysimeters with acid, shallow soil, rich in organic matter. *In* "Ecological Impact of Acid Precipitation" (D. Drablos and A. Tollan, eds.), pp. 152–153. SNSF Project, Oslo, Norway.

Abrahamsen, G., Bjor, K., Horntvedt, R., and Tveite, B. (1976). "Impact of Acid Precipitation on Forest and Freshwater Ecosystems in Norway," Res. Rep. FR6/76. SNSF Project, Oslo, Norway.

Abrahamsen, G., Horntvedt, R., and Tveite, B. (1977). Impacts of acid precipitation on coniferous forest ecosystems. *Water, Air, Soil Pollut.* **8**, 57–73.

Adams, H. S., Stephenson, S. L., Blasing, T. J., and Divick, D. N. (1985). Growth-trend declines of spruce and fir in mid-Appalachian subalpine forests. *Environ. Exp. Bot.* **25**, 315–325.

Addison, R. F., Zinck, M. E., and Smith, T. G. (1986). PCBs have declined more than DDT-group residues in arctic ringed seals (*Phoca hispida*) between 1972 and 1981. *Environ. Sci. Technol.* **20**, 253-256.

Advisory Committee on Oil Pollution of the Sea (ACOPS) (1980). "Annual Report 1979." ACOPS, London.

Aerts, R., Wallen, B., and Malmer, N. (1992). Growth-limiting nutrients in *Sphagnum*-dominated bogs subject to low and high atmospheric deposition supply. *J. Ecol.* **80**, 131–140.

Agosti, J. M., and Agosti, T. E. (1973). The oxidation of certain Prudhoe Bay hydrocarbons by microorganisms indigenous to a natural oil seep at Umiat, Alaska. *In* "Proceedings of the Symposium on the Impact of Oil Resource Development on Northern Plant Communities," pp. 80–85. Institute of Arctic Biology, Fairbanks, AK.

Agrios, G. N. (1969). "Plant Pathology." Academic Press, New York.

Ahlgren, C. E. (1966). Small mammals and reforestation following prescribed burning. *J. For.* **64**, 614–618.

Ahlgren, I. (1978). Response of Lake Norrviken to reduced nutrient loading. *Verh.—Int. Ver. Theor. Angew. Limnol.* **20**, 846–850.

Alabaster, J. S. (1967). Survival of salmon (*Salmo salar*) and sea trout (*S. trutta*) in fresh and saline waters at high temperatures. *Nat. Resour.* **1**, 717–730.

Alban, D. H. (1982). Effects of nutrient accumulation by aspen, spruce, and pine on soil properties. *Soil Sci. Soc. Am. J.* **46**, 853–861.

Alban, D. H., Perala, D. A., and Schlaegel, B. E. (1978). Biomass and nutrient distribution in aspen, pine, and spruce stands on the same soil type in Minnesota. *Can. J. For. Res.* **8**, 290–299.

Albert, D. A., and Barnes, B. V. (1987). Effects of clearcutting on the vegetation and soil of a sugar maple—dominated ecosystem, western upper Michigan. *For. Ecol. Manage.* **18**, 283–298.

Alcock, B., and Allen, L. H. (1985). Crop responses to elevated carbon dioxide concentrations. *In* "Direct Effects of Increasing Carbon Dioxide on Vegetation," DOE/ER-0238, pp. 53–97. U.S. Department of Energy, Washington, DC.

Alexander, V., and Van Cleve, K. (1983). The Alaska pipeline: A success story. *Annu. Rev. Ecol. Syst.* **14,** 443–463.

Allan, A. A., Schlueter, R. S., and Mikolaj, P. J. (1970). Natural oil seepage at Coal Oil Point, Santa Barbara, California. *Science* **170,** 974–977.

Allan, R. J. (1971). Lake sediments: A medium for regional geochemical exploration of the Canadian Shield. *CIM Bull.* **64**(715), 43–59.

Allen, D. C., Barnett, C. J., Millers, I., and Lachance, D. (1992). Temporal changes (1988–1990) in sugar maple health, and factors associated with crown condition. *Can. J. For. Res.* **22,** 1776–1784.

Allen, K. R. (1990). "Conservation and Management of Whales." Univ. of Washington Press, Seattle.

Allen, R. P. (1952). "The Whooping Crane," Res. Rep. No. 3. National Audubon Society, New York.

Allen, S. E. (1989). "Chemical Analysis of Ecological Materials." Blackwell, London.

Allen, S. E., Grimshaw, H. M., Parkinson, J. A., and Quarmby, C. (1974). "Chemical Analysis of Ecological Materials." Blackwell, London.

Allendorf, F. W., and Leary, R. F. (1986). Heterozygosity and fitness of natural populations of animals. *In* "Conservation Biology. The Science of Scarcity and Diversity" (M. E. Soule, ed.), pp. 57–76. Sinauer Assoc., Sunderland, MA.

Almer, B., Dickson, W., Ekström, C., Hornström, E., and Miller, U. (1974). Effects of acidification of Swedish lakes. *Ambio* **3,** 330–336.

Almer, B., Dickson, W., Ekström, C., and Hornström, E. (1978). Sulfur pollution and the aquatic ecosystem. *In* "Sulfur in the Environment" (J. O. Nriagu, ed.), Part II, pp. 271–311. Wiley, New York.

Altshuller, A. P. (1983). Review: Natural volatile organic substances and their effect on air quality in the United States. *Atmos. Environ.* **17,** 2131–2165.

Alvarez, L. W., Alvarez, W., Asaro, F., and Michel, H. V. (1980). Extraterrestrial causes for the Cretaceous-Tertiary extinction. *Science* **208,** 1095–1108.

Alvarez, W., Kauffman, E. G., Surlyk, F., Alvarez, L. W., Asaro, F., and Michel, H. V. (1984). Impact theory of mass extinctions and the invertebrate fossil record. *Science* **223,** 1135–1141.

Alvo, R., Hussell, D. J. J., and Berrill, M. (1988). The breeding success of common loons (*Gavia immer*) in relation to alkalinity and other lake characteristics in Ontario. *Can. J. Zool.* **66,** 746–752.

Ambler, D. C. (1983). Surface water hydrology. *In* "Kejimkujik Calibrated Catchments Program" (J. Kerekes, ed.), pp. 15–23. Environmental Protection Service, Dartmouth, Nova Scotia.

Amdur, M. O. (1980). Air pollutants. *In* "Toxicology. The Basic Science of Poisons" (J. Doull, C. D. Klaassen, and M. O. Amdur, eds.), pp. 608–631. Macmillan, New York.

American Conference of Governmental and Industrial Hygienists (ACGIH) (1982). "Threshold Limit Values for Chemical Substances in Work Air Adopted by ACGIH for 1982." ACGIH, Washington, DC.

American National Standards Institute (ANSI) (1976). Evaluation of total sulfation in atmosphere by the lead peroxide candle. *In* "Annual Book of ASTM Standards," pp. 533–535. ANSI, Washington, DC.

American Ornithologists' Union (AOU) (1983). "Checklist of North American Birds." AOU, Washington, DC.

Ames, P. L. (1966). DDT residues in the eggs of osprey in the north-eastern United States and their relation to nesting success. *J. Appl. Ecol.* **3,** Suppl., 87–97.

Amiro, B. D., and Courtin, G. M. (1981). Patterns of vegetation in the vicinity of an industrially disturbed ecosystem, Sudbury, Ontario. *Can. J. Bot.* **59,** 1623–1639.

Amthor, J. S. (1984). Does acid rain directly influence plant growth? Some comments and observations. *Environ. Pollut., Ser. A* **36,** 1–6.

Amundson, R. G., and MacLean, D. C. (1982). Influence of oxides of nitrogen on crop growth and yield: An overview. *In* "Air Pollution by Nitrogen Oxides" (T. Schneider and L. Grant, eds.), pp. 501–510. Elsevier, Amsterdam.

Amundson, R. G., Hadley, J. L., Fincher, J. F., Fellows, S., and Alscher, R. G. (1992). Comparisons of seasonal changes in photosynthetic capacity, pigments, and carbohydrates of health sapling and mature red spruce and of declining and healthy red spruce. *Can. J. For. Res.* **22,** 1605–1616.

Anderson, D. W., and Hickey, J. J. (1972). Eggshell changes in certain North American birds. *Proc. Int. Ornithol. Congr., 15th, 1970,* pp. 514–540.

Anderson, D. M., White, A. W., and Bader, D. G., eds. (1985). "Toxic Dinoflagellates." Elsevier, New York.

Anderson, H. W., Hoover, M. D., and Reinhart, K. G. (1976). "Forests and Water: Effects of Forest Management on Floods, Sedimentation, and Water Supply," USDA For. Serv. Gen. Tech. Rep. PSW-18. Pacific Southwest Forest and Range Experiment Station, Berkeley, CA.

Anderson, J. E. (1991). A conceptual framework for evaluating and quantifying naturalness. *Conserv. Biol.* **5,** 347–352.

Andersson, F. (1986). Acidic deposition and its effects on the forests of Nordic Europe. *Water, Air, Soil Pollut.* **30,** 17–29.

Andresen, A. M., Johnson, A. H., and Siccama, T. G. (1980). Levels of lead, copper, and zinc in the forest floor in the northeastern United States. *J. Environ. Qual.* **9,** 293–296.

Aneja, V. P., Adams, D. F., and Pratt, C. D. (1984). Environmental impact of natural emissions. *J. Air Pollut. Control Assoc.* **34,** 799–803.

Anlauf, K. G., Fellin, P., Wiebe, H. A., and Melo, O. T. (1982). The Nanticoke shoreline diffusion experiments, June 1978. IV. A. Oxidation of sulfur dioxide in a power plant plume. B. Ambient concentrations and transport of sulfur dioxide, particulate sulfate and nitrate, and ozone. *Atmos. Environ.* **16,** 455–466.

Anonymous (1973). "National Inventory of Sources and Emissions of Lead (1970)," Publ. No. APCD 73–7. Air Pollution Control Directorate, Environment Canada, Ottawa, Ont.

Anonymous (1976a). "A Nationwide Inventory of Emissions of Air Contaminants (1974)," Rep. EPS 3-AP-76–1. Air Pollution Control Directorate, Environment Canada, Ottawa, Ont.

Anonymous (1976b). "Report to the Task-force for Evaluation of Budworm Control Alternatives." Prepared for the Cabinet Committee on Economic Development, Province of New Brunswick, Fredericton.

Anonymous (1981). "National Inventory of Natural Sources and Emissions of Primary Pollutants." Air Pollution Control Directorate, Environment Canada, Ottawa, Ont.

Anonymous (1982). "Ecological Study of the *Amoco Cadiz* Oil Spill." U.S. Department of Commerce, National Oceanic and Atmospheric Administration, Washington, DC.

Anonymous (1986). "Human Actions and the Environment: A Statistical Compendium." Statistics Canada, Ottawa, Ont.

Anonymous (1987). "National *anatum* Peregrine Falcon Recovery Plan." Federal-Provincial Wildlife Conference of Government Wildlife Directors, Ottawa, Ont.

Anonymous (1988). Untitled cartoons. *Bull. Br. Ecol. Soc.* **2**, 78.

Anonymous (1989). Giant clams nabbed. *Environesia* 3(4), 15.

Anonymous (1990). "Large-scale Pelagic Driftnet Fishing and its Impact on the Living Marine Resources of the World's Oceans and Seas," Rep. A/45/663. United Nations General Assembly, New York.

Anonymous (1991). Kuwait oil fires, Phillipine volcanoes, and NAPAP. *Natl. Acid Precip. Assess. Newsl.* **2**(1), 2–4.

Anonymous (1992a). "The World Almanac and Book of Facts, 1993." Pharos Books, New York.

Anonymous (1992b). "Two Years After: Conditions in Prince William Sound and the Gulf of Alaska, October 1991." Exxon Co., Houston, TX.

Anonymous (1992c). "Three Years After. Conditions in Prince William Sound and the Gulf of Alaska." Exxon Co., Houston, TX.

Anonymous (1992d). Gulf War impacts. *Conserv. Biol.* **6**, 168.

Anonymous (1992e). "Proposed Regime to Govern Interactions Between Marine Mammals and Commercial Fishing Operations." National Oceanic and Atmospheric Administration, National Marine Fisheries Service, Silver Spring, MD.

Anonymous (1993a). Our numbers. *Populi* **20**(8), 16.

Anonymous (1993b). "1992 Report to Congress." National Acid Deposition Assessment Program, Washington, DC.

Anonymous (1993c). "National Atmospheric Deposition Program (IR-7)/National Trends Network." NADP/NTN Coordination Office, Colorado State University, Fort Collins.

Anonymous (1993d). "Registration Status of Fenitrothion Insecticide," Discuss. Rep. 093–01. Food Production and Inspection Branch, Plant Industry Directorate, Agriculture Canada, Ottawa, Ont.

Anonymous (1993e). "Toxic Contaminants in the Environment. Persistent Organochlorines." State of the Environment Reporting, Environment Canada, Ottawa, Ont.

Antonovics, J., and Bradshaw, A. D. (1970). Evolution in closely adjacent plant populations. VIII. Clinal patterns in *Anthoxanthum odoratum* across a boundary. *Heredity* **25**, 349–362.

Antonovics, J., Bradshaw, A. D., and Turner, R. G. (1971). Heavy metal tolerance in plants. *Adv. Ecol. Res.* **7**, 1–85.

Armstrong, F. A. J. (1979). Mercury in the aquatic environment. *In* "Effects of Mercury in the Canadian Environment," NRCC No. 16739, pp. 84–100. Associate Committee on Scientific Criteria for Environmental Quality, National Research Council of Canada, Ottawa, Ont.

Armstrong, J. A. (1985a). Tactics and strategies for larval suppression and prevention of damage using chemical insecticides. *In* "Recent Advances in Spruce Budworms Research," pp. 301–319. Canadian Forestry Service, Ottawa, Ont.

Armstrong, J. A. (1985b). Spruce budworm control program in eastern Canada. *In* "Recent Advances in Spruce Budworms Research," pp. 384–385. Canadian Forestry Service, Ottawa, Ont.

Ash, A. N. (1988). Disappearance of salamanders from clearcut plots. *J. Elisha Mitchell Sci. Soc.* **104**, 116–122.

Ashton, P. S. (1984). Long-term changes in dense and open inland forests following herbicide attack. *In* "Herbicides in War: The Long-term Ecological and Human Consequences" (A. H. Westing, ed.), pp. 33–37. Taylor & Francis, London.

Atkins, R. P. (1969). Lead in a surburban environment. *J. Air Pollut. Control Assoc.* **19**, 591–595.

Atlas, R. M. (1982). Microbial hydrocarbon degradation within sediment impacted by the Amoco Cadiz oil spill. *In* "Ecological Study of the Amoco Cadiz Oil Spill," pp. 1–25. U.S. Department of Commerce, National Oceanic and Atmospheric Administration, Washington, DC.

Atlas, R. N. (1985). Effects of hydrocarbons on microorganisms and petroleum biodegradation in Arctic ecosystems. *In* "Petroleum Effects in the Arctic Environment" (F. R. Engelhardt, ed.), pp. 63–99. Am. Elsevier, New York.

Aubertin, G. M., and Patric, J. H. (1974). Water quality after clear-cutting a small watershed in West Virginia. *J. Environ. Qual.* **3**, 243–249.

Auclair, A. N. D. (1987). The distribution of forest declines in eastern Canada. *In* "Proceedings of the Workshop on Forest Decline and Reproduction: Regional and Global Consequences" (L. Kairiukstis, S. Nilsson, and A. Stroszak, eds.), pp. 307–319. IIASA, Laxenburg, Austria.

Auclair, A. N. D., Warrest, R. C., Lachance, D., and Martin, H. C. (1992). Climatic perturbation as a general mechanism of forest dieback. *In* "Forest Decline Concepts" (P. D. Manion and D. Lachance, eds.), pp. 35–58. American Phytopathological Society, St. Paul, MN.

Auerbach, S. I. (1981). Ecosystem response to stress: A review of concepts and approaches. *In* "Stress Effects on Natural Ecosystems" (G. W. Barrett and R. Rosenberg, eds.), pp. 29–41. Wiley, New York.

Aumento, F. (1970). "Serpentine Mineralogy of Ultrabasic Intrusions in Canada and on the Mid-Atlantic Ridge," Pap. 69–53. Geological Survey of Canada, Department of Energy, Mines and Resources, Ottawa, Ont.

Aune, T., Skulberg, O. M., and Underdal, B. (1991). A toxic phytoflagellate bloom of *Chrysochromulina* cf *leadbeateri* in coastal waters in the north of Norway, May–June, 1991. *Ambio* **21**, 471–474.

Ayensu, E. S. (1978). Calling the roll of the world's vanishing plants. *Smithson. Mag.* **9**(8), 122–129.

Ayer, W. C., Brun, G. L., Eidt, D. C., Ernst, W., Mallet, V. N., Matheson, R. A., and Silk, P. J. (1984). "Persistence of Aerially Applied Fenitrothion in Water, Soil, Sediment, and Balsam Fir Foliage," Inf. Rep. M-X-153. Maritimes Forest Research Centre, Fredericton, New Brunswick.

Bache, B. W. (1980). The acidification of soil. *In* "Effects of Acid Precipitation on Terrestrial Ecosystems" (T. C. Hutchinson and M. Havas, eds.), pp. 183–20. Plenum, New York.

Baes, C. F., Bjorkström, A., and Mulholland, P. J. (1985). Uptake of carbon dioxide by the oceans. *In* "Atmospheric Carbon Dioxide and the Global Carbon Cycle," DOE/ER-0239, pp. 81–111. U.S. Department of Energy, Washington, DC.

Bailey, R. E. (1982). "The Current Status of the Softwood Resource on Cape Breton Island," For. Res. Note 3. Nova Scotia Department of Lands and Forests, Truro.

Bailey, R. S. (1989). Changes in the North Sea herring population over a cycle of collapse and recovery. *In* "Long-term Variability of Pelagic Fish Populations and Their Environment" (T. Kawasaki, S. Tanaka, Y. Toba, and A. Taniguchi, eds.), pp. 191–198. Pergamon, New York.

Bakan, S., Chloud, A., Cubasch, U., Feichter, J., Graf, H., Grassl, H., Hasselmann, K., Kirchner, I., Latif, M., Roeckner, E., Sausen, R., Schlese, U., Schriever, D., Schult, I., Schumann, U., Sielmann, F., and Welke, W. (1991). Climate response to smoke from the burning oil wells in Kuwait. *Nature (London)* **351**, 367–371.

Baker, B., Campbell, B., Gist, R., Lowry, L., Nickerson, S., Schwartz, C., and Stratton, L. (1989). Exxon Valdez oil spill: The first eight weeks. *Alaska Fish & Game* **21**(4), 2–37.

Baker, J., and Schofield, C. L. (1982). Aluminum toxicity to fish in acidic waters. *Water, Air, Soil Pollut.* **18**, 289–301.

Baker, J. M. (1971a). The effects of a single oil spillage. *In* "The Ecological Effects of Oil Pollution in Littoral Communities," pp. 16–20. Institute of Petroleum, London.

Baker, J. M. (1971b). Successive spillages. *In* "The Ecological Effects of Oil Pollution in Littoral Communities," pp. 21–32. Institute of Petroleum, London.

Baker, J. M. (1971c). Seasonal effects. *In* "The Ecological Effects of Oil Pollution in Littoral Communities," pp. 44–51. Institute of Petroleum, London.

Baker, J. M. (1971d). Effects of cleaning. *In* "The Ecological Effects of Oil Pollution in Littoral Communities," pp. 52–57. Institute of Petroleum, London.

Baker, J. M. (1971e). Growth stimulation following oil pollu-tion. *In* "The Ecological Effects of Oil Pollution in Littoral Communities," pp. 71–77. Institute of Petroleum, London.

Baker, J. M. (1971f). The effects of oils on plant physiology. *In* "The Ecological Effects of Oil Pollution in Littoral Communities," pp. 88–98. Institute of Petroleum, London.

Baker, J. M. (1971g). Seasonal effects of oil pollution on salt marsh vegetation. *Oikos* **22**, 106–110.

Baker, J. M. (1973). Recovery of salt marsh vegetation from successive oil spillages. *Environ. Pollut.* **4**, 223–230.

Baker, J. M. (1976). Ecological changes in Milford Haven during its history as an oil port. *In* "Marine Ecology and Oil Pollution" (J. M. Baker, ed.), pp. 55–66. Wiley, New York.

Baker, J. M. (1983). "Impact of Oil Pollution on Living Resources," Comm. Ecol. Pap. No. 4. International Union for Conservation of Nature and Natural Resources, Gland, Switzerland.

Baker, J. M., Clark, R. B., and Kingston, P. F. (1991). "Two Years After the Spill: Environmental Recovery in Prince William Sound and the Gulf of Alaska." Institute of Offshore Engineering, Heriot-Watt University, Edinburgh, Scotland.

Baker, J. P. (principal author) (1990). Biological effects of changes in surface water acid-base chemistry. *In* "Acidic Deposition: State of Science and Technology. Vol. II. Aquatic Processes and Effects," pp. 13–1 to 13–381. Superintendent of Documents, U.S. Govt. Printing Office, Washington, DC.

Baker, L. A. (principal author) (1990). Current status of surface water acid-base chemistry. *In* "Acidic Deposition: State of Science and Technology. Vol. II. Aquatic Processes and Effects," pp. 9–1 to 9–367. Superintendent of Documents, U.S. Govt. Printing Office, Washington, DC.

Baker, L. A., Herlihy, A. T., Kaufmann, P. R., and Eilers, J. M. (1991). Acidic lakes and streams in the United States: The role of acidic deposition. *Science* **252**, 1151–1154.

Baker, R. H. (1946). Some effects of the war on the wildlife of Micronesia. *Trans. North Am. Wildl. Conf.* **11**, 205–213.

Balda, R. P. (1975). Vegetation structure and breeding bird diversity. *In* "Management of Forest and Range Habitats for Non-game Birds," USDA For. Serv. Gen. Tech. Rep. WO-1, pp.59–80. U.S. Department of Agriculture, Washington, DC.

Banfield, A. W. F. (1974). "The Mammals of Canada." National Museum of Natural Sciences, National Museums of Canada, Univ. of Toronto Press, Toronto.

Barica, J. (1980). Why hypertrophic ecosystems? *In* "Hypertrophic Ecosystems" (J. Barica and L.R. Mur, eds.), pp. IX–XI. Junk, The Hague, The Netherlands.

Barnaby, F. (1982). The effects of a global nuclear war: The arsenals. *Ambio* **11**, 76–83.

Barnaby, F. (1988). Refrigerants and the ozone layer. *Ambio* **17**, 354.

Barnaby, F. (1991). The environmental impact of the Gulf War. *Ecologist* **21**, 166–172.

Barnaby, F., and Rotblat, J. (1982). The effects of nuclear weapons. *Ambio* **11**, 84–93.

Barnard, J. E. (principal author) (1990). Changes in forest health and productivity in the United States and Canada. *In* "Acidic Deposition: State of Science and Technology. Vol. III. Terrestrial, Materials, Health, and Visibility Effects," pp. 16–1 to 16–186. Superintendent of Documents, U.S. Govt. Printing Office, Washington, DC.

Barnard, W. R. (1986). Alkaline materials flux from unpaved roads: Source strength, chemistry, and potential for acid rain neutralization. *Water, Air, Soil Pollut.* **30,** 285–293.

Barnes, B. V. (1966). The clonal growth habit of American aspen. *Ecology* **47,** 439–449.

Barnes, B. V. (1969). Natural variation and delineation of clones of *Populus tremuloides* and *P. grandidentata* in northern lower Michigan. *Silvae Genet.* **18,** 130–142.

Barney, G. O., dir. (1980). "The Global 2000 Report to the President." U.S. Govt. Printing Office, Washington, DC.

Barnola, J. M., Raynaud, D., Korotkevich, Y. K., and Lorius, C. (1987). Vostok ice core provides 160,000-year record of atmospheric CO_2. *Nature (London)* **329,** 408–414.

Baron, R. L. (1991). Carbamate insecticides. *In* "Handbook of Pesticide Toxicology. Vol. 3. Classes of Pesticides" (W.C. Hayes and E.R. Laws, eds.), pp. 1125–1190. Academic Press, San Diego.

Barr, J. F. (1986). "Population Dynamics of the Common Loon (*Gavia immer*) Associated with Mercury-contaminated Waters in Northwestern Ontario," Occas. Pap. No. 56. Canadian Wildlife Service, Ottawa, Ont.

Barsdate, R. J., Alexander, V., and Benoit, R. E. (1973). Natural oil seeps at Cape Simpson, Alaska: Aquatic effects. *In* "Proceedings of the Symposium on the Impact of Oil Resource Development on Northern Plant Communities," pp. 91–95. Institute of Arctic Biology, Fairbanks, AK.

Baskerville, G. L. (1975a). Spruce budworm: Super silviculturalist. *For. Chron.* **51,** 138–140.

Baskerville, G. L. (1975b). Spruce budworm: The answer is forest management: Or is it? *For. Chron.* **51,** 157–160.

Baskin, Y. (1992). Africa's troubled waters. *BioScience* **42,** 476–481.

Batcheler, C. L. (1983). The possum and rata-kamaki dieback in New Zealand: A review. *Pac. Sci.* **37,** 415–426.

Bates, T. S., Cline, J. D., Gammon, R. H., and Kelly-Hansen, S. R. (1987). Regional and seasonal variations in the flux of oceanic dimethylsulfide to the atmosphere. *J. Geophys. Res.* **92,** 2930–2938.

Batzer, H. O. (1976). Silvicultural control techniques for the spruce budworm. *Misc. Publ.—U.S., Dep. Agric.* **1327,** 110–116.

Bauce, E. and Allen, D. C. (1991). Etiology of a sugar maple decline. *Can. J. For. Res.* **21,** 686–693.

Bauce, E., and Allen, D. C. (1992). Role of *Armillaria calvescens* and *Glycobius speciosus* in a sugar maple decline. *Can. J. For. Res.* **22,** 549–552.

Baumhover, A. H., Graham, A. J., Bitter, B. A., Hopkins, D. E., New, W. D., Dudley, F. H., and Bushland, R. C. (1955). Screw-worm control through release of sterilized flies. *J. Econ. Entomol.* **48,** 462–466.

Bazzaz, F. A. (1990). The response of natural ecosystems to the rising global CO_2 levels. *Annu. Rev. Ecol. Syst.* **21,** 167–196.

Bazzaz, F. A., Garbutt, K., and Williams, W. E. (1985). Effect of increasing carbon dioxide concentration on plant communities. *In* "Direct Effects of Increasing Carbon Dioxide on Vegetation," DOE/ER-0238, pp. 155–170. U.S. Department of Energy, Washington, DC.

Beamish, R. J. (1974). Growth and survival of white suckers (*Catostomus commersoni*) in an acidified lake. *J. Fish. Res. Board Can.* **31,** 49–54.

Beamish, R. J., and Harvey, H. H. (1972). Acidification of the La Cloche Mountain Lakes, Ontario, and resulting fish mortalities. *J. Fish. Res. Board Can.* **29,** 1131–1143.

Beamish, R. J., Lockhart, W. L., Van Loon, J. C., and Harvey, H. H. (1975). Long-term acidification of a lake and resulting effects on fishes. *Ambio* **4,** 98–102.

Beanlands, G. E., and Duinker, P. N. (1983). "An Ecological Framework for Environmental Impact Assessment in Canada." Institute for Resource and Environmental Studies, Dalhousie University, Halifax, Nova Scotia.

Beaton, J. D., Tisdale, S. L., and Platov, J. (1971). "Crop Response to Sulfur in North America," Tech. Bull. No. 18. Sulfur Institute, Washington, DC.

Beaton, J. D., Bixby, D. W., Tisdale, S. L., and Matov, J. S. (1974). "Fertilizer Sulfur. Status and Potential in the U.S.," Tech. Bull. No. 21. Sulfur Institute, Washington, DC.

Beaver, D. L. (1976). Avian populations in herbicide treated brush fields. *Auk* **93,** 543–553.

Becker, M. (1989). The role of climate and present and past vitality of silver fir forests in the Vosges Mountains of northeastern France. *Can. J. For. Res.* **19,** 1110–1117.

Beeton, A. M., and Edmondson, W. T. (1972). The eutrophication problem. *J. Fish. Res. Board Can.* **29,** 673–682.

Beggs, G. L., and Gunn, J. M. (1986). Response of lake trout (*Salvelinus namaycush*) and brook trout (*S. fontinalis*) to surface water acidification in Ontario. *Water, Air, Soil Pollut.* **30,** 711–717.

Begon, M., Harper, J. L., and Townsend, C. R. (1990). "Ecology. Individuals, Populations, and Communities." Blackwell, London.

Beilke, S., and Gravenhorst, G. (1978). Heterogenous SO_2-oxidation in the droplet phase. *Atmos. Environ.* **12,** 231–239.

Beljan, J. R., Irey, N. S., Kilgore, W. W., Kimura, K., Suskind, R. R., Vostal, J. J., and Wheater, R. H. (1981). "The Health Effects of "Agent Orange" and Polychlorinated Dioxin Contaminants." Council on Scientific Affairs, American Medical Association, Chicago.

Bellrose, F. C. (1959). Lead poisoning as a mortality factor in waterfowl populations. *Bull.—Ill. Nat. Hist. Surv.* **27,** 235–288.

Bellrose, F. C. (1976). "Ducks, Geese, and Swans of North America." Stackpole Books, Harrisburg, PA.

Bengston, G., Nordström, S., and Rundgren, S. (1983). Population density and tissue metal concentration of lumbricids in forest soils near a brass mill. *Environ. Pollut., Ser. A* **30**, 87–108.

Bengston, J. L., and Laws, R. M. (1985). Trend in crabeater seal age at maturity: An insight into Antarctic marine invertebrates? *In* "Antarctic Nutrient Cycles and Food Webs" (W. R. Siegfield, P. R. Candy, and R. M. Laws, eds.), pp. 661–665. Springer-Verlag, Berlin and New York.

Bengtsson, B., Dickson, D., and Nyberg, P. (1980). Liming lakes in Sweden. *Ambio* **9**, 34–36.

Benkovitz, C. M. (1982). Compilation of an inventory of anthropogenic emissions in the United States and Canada. *Atmos. Environ.* **16**, 1551–1563.

Benndorf, J. (1987). Food web manipulation without nutrient control: A useful strategy in lake restoration? *Schweiz. Z. Hydrol.* **49**, 237–248.

Bennett, S. H., Gibbons, J. W., and Glanville, J. (1980). Terrestrial activity, abundance, and diversity of amphibians in differently managed forest types. *Am. Midl. Nat.* **103**, 412–416.

Benoit, L. F., Skelly, J. M., Moore, L. D., and Dochinger, L. S. (1982). Radial growth reductions of *Pinus strobus* L. correlated with foliar ozone sensitivity as an indicator of ozone-induced losses in eastern forests. *Can. J. For. Res.* **12**, 673–678.

Benoit, P., and Lachance, D. (1990). "Gypsy Moth in Canada: Behaviour and Control," Inf. Rep. DPC-X-32. Forestry Canada, Ottawa, Ont.

Benson, D. A. (1958). Moose "sickness" in Nova Scotia. *Can. J. Comp. Med. Vet. Sci.* **5**, 287–296.

Benson, D. W., and Dodds, G. D. (1977). "Deer of Nova Scotia." Nova Scotia Department of Lands and Forests, Halifax.

Bentzer, B. G., Foster, G. S., and Helberg, A. R. (1990). Impact of clone mixture composition on stability of 7th-year mean height in a series of Norway spruce clone tests. *Can. J. For. Res.* **20**, 757–763.

Berg, H., Kilbus, M., and Kautsky, N. (1992). DDT and other insecticides in the Lake Kariba ecosystem, Zimbabwe. *Ambio* **21**, 444–450.

Berger, D. D., Anderson, R. W., Weaver, J. D., and Risebrough, R. W. (1970). Shell thinning in eggs of Ungava peregrines. *Can. Field Nat.* **84**, 265–267.

Berger, J. (1991). Greater Yellowstone's native ungulates: Myths and realities. *Conserv. Biol.* **5**, 353–363.

Bergeron, J. M., and Tardif, J. (1982). Winter browsing preferences of snowshoe hares for coniferous seedlings and its implication for large-scale reforestation programs. *Can. J. For. Res.* **18**, 280–282.

Bergerud, A. T. (1972). Food habits of Newfoundland caribou. *J. Wildl. Manage.* **36**, 913–923.

Bergerud, A. T. (1988). Caribou, wolves, and man. *TREE* **3**, 68–72.

Bergerud, A. T., and Elliott, J. P. (1986). Dynamics of caribou and wolves in northern British Columbia. *Can. J. Zool.* **64**, 1515–1529.

Berggren, B., and Oden, S. (1972). "1. Analysresultat Rorande Fungmetaller Och Klorerade Kolvaten I. Rotslam Fran Svenska Reningsverk 1968–1971." Institutionen fur Markvetenskap Lantbrukshogskolan, Uppsala, Sweden.

Berne, S., and Bodennec, G. (1984). Evolution of hydrocarbons after the Tanio oil spill—a comparison with the Amoco Cadiz accident. *Ambio* **13**, 109–114.

Berne, S., Marchand, M., and D'Ozouville, L. (1980). Pollution of sea water and marine sediments in control areas. *Ambio* **9**, 287–293.

Bernier, B., and Brazeau, M. (1988a). Foliar nutrient status in relation to sugar maple dieback and decline in the Quebec Appalachians. *Can. J. For. Res.* **18**, 754–761.

Bernier, B., and Brazeau, M. (1988b). Magnesium deficiency symptoms associated with sugar maple dieback in a Lower Laurentian site in southeastern Quebec. *Can. J. For. Res.* **18**, 1265–1269.

Berresheim, H., and Jaeschke, W. (1983). The contribution of volcanoes to the global atmospheric sulfur budget. *J. Geophys. Res.* **88**, 3732–3740.

Berrow, M. L., and Webber, J. (1972). Trace elements in sewage sludges. *J. Sci. Food Agric.* **23**, 93–100.

Berry, C. R. (1973). The differential susceptibility of eastern white pine to three types of air pollution. *Can. J. For. Res.* **3**, 543–547.

Beschta, R. L., and Jackson, W. L. (1979). The intrusion of fine sediments into a stable gravel bed. *J. Fish. Res. Board Can.* **36**, 204–210.

Best, P. B. (1993). Incease rates in severely depleted stocks of baleen whales. *ICES J. Mar. Sci.* **50**, 169–186.

Beyer, W. N., Pattee, O. H., Sileo, L., Hoffman, D. J., and Mulhern, B. M. (1985). Metal contamination in wildlife living near two zinc smelters. *Environ. Pollut., Ser. A* **38**, 63–86.

Bidleman, T. F., and Leonard, R. (1982). Aerial transport of pesticides over the northern Indian Ocean and adjacent seas. *Atmos. Environ.* **16**, 1099–1107.

Bierhuizen, J. F. H., and Prepas, E. E. (1985). Relationship between nutrients, dominant ions, and phytoplankton standing crop in prairie saline lakes. *Can. J. Fish. Aquat. Sci.* **42**, 1588–1594.

Bierregaard, R. O., and Lovejoy, T. E. (1989). Effects of forest fragmentation on Amazonian understory bird species. *Acta Amazonica* **19**, 215–241.

Bierregaard, R. O., Lovejoy, T. E., Kapos, V., dos Santos, A. A., and Hutchings, R. W. (1992). The biological dynamics of tropical rainforest fragments. *BioScience* **42**, 859–866.

Bij der Vaate, A. (1991). Distribution and aspects of population biology of the zebra mussel, *Dreissena polymorpha* in the Lake Ijsselmeer area (The Netherlands). *Oecologia* **86**, 40–50.

Billings, W. W., Luken, J. O., Mortensen, D. A., and Peterson, K. A. (1982). Arctic tundra: A source or sink for atmospheric carbon dioxide in a changing environment. *Oecologia* **53**, 7–11.

Binkley, D. (1986). The value of simplicity and complexity in nutrient cycling models: A comparison of FORNUTS and FORCYTE. In "Predicting the Consequences of Intensive Forest Harvesting on Long-term Productivity" (G. I. Agren, ed.), Rep. No. 26, pp. 85–93. Swedish Univ. Agric. Sci., Uppsala.

Birdsey, R. A. (1990). Potential changes in carbon storage through conversion of lands to plantation forests. *North Am. Conf. For. Responses Clim. Change,* Washington, DC, *1990.*

Birge, W. J. (1983). Structure—activity relationships in environmental toxicology. *Fundam. Appl. Toxicol.* **3,** 341–342.

Birkhead, M. (1982). Causes of mortality in the mute swan (*Cygnus olor*) on the River Thames. *J. Anim. Ecol.* **52,** 727–741.

Birkhead, T. R., and Harris, M. P. (1985). Ecological adaptations to breeding in the Atlantic Alcidae. In "The Atlantic Alcidae" (D. N. Nettleship and T. R. Birkhead, eds.), pp. 205–232. Academic Press, New York.

Bishop, C., and Weseloh, D. V. (1990). "Contaminants in Herring Gull Eggs from the Great Lakes," SOE Fact Sheet 90–2. State of the Environment Reporting, Environment Canada, Ottawa, Ont.

Bjor, K., and Tiegen, O. (1980). Effects of acid precipitation on soil and forest. 6. Lysimeter experiment in greenhouse. In "Ecological Impact of Acid Precipitation" (D. Drablos and A. Tollan, eds.), pp. 200–201. SNSF Project, Oslo, Norway.

Bjork, S. (1977). Recovery and restoration of damaged lakes in Sweden. In "Recovery and Restoration of Damaged Ecosystems" (J. Cairns, K.L. Dickson, and E.E. Herricks, eds.), pp. 110–133. Univ. Press of Virginia, Charlottesville.

Black, C. A., Evans, D. D., White, J. L., Ensminger, L. E., and Clark, F. E., eds. (1965). "Methods of Soil Analysis," Monogr. No. 9. Am. Soc. Agron., Madison, WI.

Blair, R. M., and Brunett, L. E. (1976). Phytosociological changes after timber harvest in a northern pine ecosystem. *Ecology* **57,** 18–32.

Blais, J. R. (1965). Spruce budworm outbreaks in the past three centuries in the Laurentide Park, Quebec. *For. Sci.* **11,** 130–138.

Blais, J. R. (1981). Mortality of balsam fir and white spruce following a spruce budworm outbreak in the Ottawa River watershed in Quebec. *Can. J. For. Res.* **11,** 620–629.

Blais, J. R. (1983). Trends in the frequency, extent, and severity of spruce budworm outbreaks in eastern Canada. *Can. J. For. Res.* **13,** 539–547.

Blais, J. R. (1985). The ecology of the eastern spruce budworm: A review and discussion. In "Recent Advances in Spruce Budworms Research," pp. 49–59. Canadian Forestry Service, Ottawa, Ontario.

Blake, D. R., and Rowland, F. S. (1988). Continuing worldwide increase in tropospheric methane, 1978 to 1987. *Science* **239,** 1129–1131.

Blake, J. G., and Karr, J. R. (1987). Breeding birds of isolated woodlots: Area and habitat relationships. *Ecology* **68,** 1724–1734.

Blakeslee, P. A. (1973). "Monitoring Considerations for Municipal Wastewater Effluent and Sludge Application to the Land," Universities Workshop, July 9–13, 1973, Urbana, IL. U.S. Environmental Protection Agency, U.S. Department of Agriculture, Washington, DC.

Blank, L. W. (1985). A new type of forest decline in Germany. *Nature (London)* **314,** 311–314.

Blank, L. W., Roberts, T. M., and Skeffington, R. A. (1988). New perspectives on forest decline. *Nature (London)* **336,** 27–30.

Blankenship, D. R., and King, K. A. (1984). A probable sighting of 23 eskimo curlews in Texas. *Am. Birds* **38,** 1066–1067.

Blasing, T. J. (1985). Background: Carbon cycle, climate, and vegetation responses. In "Characterization of Information Requirements for Studies of CO_2 Effects: Water Resources, Agriculture, Fisheries, Forests, and Human Health," DOE/ER-0236, pp. 9–22. U.S. Department of Energy, Washington, DC.

Bliss, L. C. (1971). Arctic and alpine plant life cycles. *Annu. Rev. Ecol. Syst.* **2,** 405–438.

Bliss, L. C., ed. (1977). "Truelove Lowland, Devon Island, Canada. A High Arctic Ecosystem." Univ. of Alberta Press, Edmonton.

Bliss, L. C. (1984). Ecologists need to increase their involvement in society. *Bull. Ecol. Soc. Am.* **65,** 439–444.

Bliss, L. C., Courtin, G. M., Pattie, D. L., Riewe, R. R., Whitfield, D. W. A., and Whidden, P. (1973). Arctic tundra ecosystems. *Annu. Rev. Ecol. Syst.* **4,** 359–399.

Blockstein, D. E., and Tordoff, H. B. (1985). Gone forever. A contemporary look at the extinction of the passenger pigeon. *Am. Birds* **39,** 845–851.

Blomqvist, G., Roos, A., Jensen, S., Bignert, A., and Olsson, M. (1992). Concentrations of sDDT and PCB in seals from Swedish and Scottish waters. *Ambio* **21,** 539–545.

Blouin, A. C. (1985). Comparative patterns of plankton communities under different regimes of pH in Nova Scotia, Canada. Unpublished Ph.D. Thesis, Department of Biology, Dalhousie University, Halifax, Nova Scotia.

Blouin, A. C., Collins, T. M., and Kerekes, J. (1984). Plankton of an acid-stressed lake (Kejimkujik National Park, Nova Scotia, Canada). Part 2. Population dynamics of an exclosure experiment. *Verh. Int. Ver. Theor. Angew. Limnol.* **22,** 22–27.

Blum, B. M., and MacLean, D. A. (1985). Potential silviculture, harvesting, and salvage practices in eastern North America. In "Recent Advances in Spruce Budworms Research," pp. 264–280. Canadian Forestry Service, Ottawa, Ont.

Blus, L. J., Stroud, R. K., Reiswig, B., and McEneaney, T. (1989). Lead poisoning and other mortality in trumpeter swans. *Environ. Toxicol. Chem.* **8,** 263–271.

Bobra, A., Shiu, W. Y., and Mackay, D. (1985). Quantitative structure-activity relationships for the acute toxicity of chlorobenzenes to *Daphnia magna. Environ. Toxicol. Chem.* **4,** 297-305.

Bodaly, R. A., Hecky, R. E., and Fudge, R. J. P. (1984). Increases in fish mercury levels in lakes flooded by the Churchill River diversion, northern Manitoba. *Can. J. Fish. Aquat. Sci.* **41,** 682–691.

Boehm, P. (1982). The Amoco Cadiz analytical chemistry program. *In* "Ecological Study of the Amoco Cadiz Oil Spill," pp. 35–99. U.S. Department of Commerce, National Oceanic and Atmospheric Administration, Washington, DC.

Boer, A. H. (1978). Management of deer wintering areas in New Brunswick. *Wildl. Soc. Bull.* **6,** 200–205.

Boer, A. H. (1992). Transience of deer wintering areas. *Can. J. For. Res.* **22,** 1421–1423.

Boffey, P. M. (1971). Herbicides in Vietnam: AAAS study finds widespread devastation. *Science* **171,** 43–47.

Bolandrin, M. F., Klocke, J. A., Wurtele, E. S., and Bollinger, W. H. (1985). Natural plant chemicals: Sources of industrial and medicinal materials. *Science* **228,** 1154–1160.

Bolin, B., Doos, B. R., Jager, J., and Warrick, R. A. (eds.). (1986). "The Greenhouse Effect, Climatic Change, and Ecosystems," SCOPE Rep. No. 29. Wiley, Chichester, UK.

Bonnor, G. M. (1978). "A Guide to Canadian Forest Inventory Terminology and Usage." Forest Management Institute, Canadian Forestry Service, Ottawa, Ont.

Bordeleau, C. (1986). Evaluation of sugar maple dieback in Quebec. Presented at the meeting of Northeastern Forest Pest Control Council, Albany, NY., 1986.

Bormann, F. H. (1982). The effects of air pollution on the New England landscape. *Ambio* **11,** 338–346.

Bormann, F. H., and Likens, G. E. (1979). "Pattern and Process in a Forested Ecosystem." Springer-Verlag, New York.

Bormann, F. H., Likens, G. E., Siccama, T. G., Pierce, R. S., and Eaton, J. S. (1974). The export of nutrients and recovery of stable conditions following deforestation at Hubbard Brook. *Ecol. Monogr.* **44,** 255–277.

Borrecco, J. E., Black, H. C., and Hooven, E. F. (1979). Response of small mammals to herbicide-induced habitat changes. *Northwest Sci.* **53,** 97–106.

Borror, D. J., DeLong, D. M., and Triplehorn, C. A. (1976). "An Introduction to the Study of Insects." Holt, New York.

Borucki, W. J., and Chameides, W. L. (1984). Lightning: Estimates of the rates of energy dissipation and nitrogen fixation. *Rev. Geophys. Space Phys.* **22,** 363–372.

Bottrell, D. G., and Smith, R. F. (1982). Integrated pest management. *Environ. Sci. Technol.* **16,** 282A–288A.

Bourne, W. R. P. (1976). Seabirds and pollution. *In* "Marine Pollution" (R. Johnston, ed.), pp. 403–502. Academic Press, London.

Bowen, H. J. M. (1979). "Environmental Chemistry of the Elements." Academic Press, New York.

Bowman, K. P. (1988). Global trends in total ozone. *Science* **239,** 48–50.

Box, E. O. (1982). Biosphere processes and natural emissions to the atmosphere: A quantitative geographic modelling basis. *J. Air Pollut. Control Assoc.* **32,** 954–957.

Boyce, R. L. (1988). Wind direction and fir wave travel. *Can. J. For. Res.* **18,** 461–466.

Boyle, J. R., and Ek, A. R. (1972). An evaluation of some effects of bole and branch pulpwood harvesting on site macronutrients. *Can. J. For. Res.* **2,** 407–412.

Boyle, J. R., Phillips, J. J., and Ek, A. R. (1973). Whole tree harvesting: Nutrient budget evaluation. *J. For.* **71,** 760–762.

Boyle, R. W. (1971). Environmental control. The land environment. *CIM Bull.* **64**(712), 46–50.

Boyles, D. T. (1980). Toxicity of hydrocarbons and their halogenated derivatives in an aqueous environment. **In** "Hydrocarbons and Halogenated Hydrocarbons in the Aquatic Environment" (B. K. Afghan and D. Mackay, eds.), pp. 545–557. Plenum, New York.

Bradshaw, A. D. (1977). The evolution of metal tolerance and its significance for vegetation establishment on metal contaminated sites. *Int. Conf. Heavy Met. Environ. [Symp. Proc.], 1st, 1975,* Vol. 2, pp. 599–622.

Bradshaw, A. D., and Chadwick, M. J. (1981). "The Restoration of Land." Blackwell, Oxford.

Bradshaw, A. D., McNeilly, T. S., and Gregory, R. P. G. (1965). Industrialization, evolution, and the development of heavy metal tolerance in plants. *Br. Ecol. Soc. Symp.* **5,** 327–343.

Brakke, F. H., ed. (1976). "Impact of Acid Precipitation on Forest and Freshwater Ecosystems in Norway," Res. Rep. FR6/76. SNSF Project, Oslo, Norway.

Brand, D. G., Kehoe, P., and Connors, M. (1986). Coniferous afforestation leads to soil acidification in central Ontario. *Can. J. For. Res.* **16,** 1289–1391.

Breen, P. A., and Mann, K. H. (1976). Destructive grazing of kelp beds by sea urchins off eastern Canada. *J. Fish. Res. Board Can.* **33,** 1278–1283.

Briggs, J. C. (1991). A Cretaceous–Tertiary mass extinction? *BioScience* **41,** 619–624.

Briggs, S. A. (1992). "Basic Guide to Pesticides: Their Characteristics and Hazards." Taylor & Francis, Washington, DC.

Brittingham, M. C., and Temple, S. A. (1983). Have cowbirds caused forest songbirds to decline? *BioScience* **33,** 31–35.

Brodin, Y. W., and Kuylenstierna, J. C. I. (1992). Acidification and critical loads in Nordic countries: A background. *Ambio* **21,** 332–338.

Brooks, J. L. (1969). Eutrophication and changes in the composition of the zooplankton. *In* "Eutrophication: Causes, Consequences, Correctives," pp. 236–255. National Academy of Sciences, Washington, DC.

Brooks, M. N., Colbert, J. J., Mitchell, R. G., and Stark, R. W., Tech. Co-ord. (1985). "Managing Trees and Stands Susceptible to Western Spruce Budworm," Tech. Bull. 1695. U.S. Department of Agriculture Forest Service, Washington, DC.

Brooks, R. R. (1983). "Biological Methods of Prospecting for Minerals." Wiley, Toronto, Ont.

Brooks, R. R. (1987). "Serpentine and Its Vegetation." Dioscorides Press, Portland, OR.

Brooks, R. R., Wither, E. D., and Zepernick, B. (1977). Cobalt and nickel in *Rinorea* species. *Plant Soil* **47**, 707–712.

Brown, A. E., and Brown, S. (1991). Comparing tropical and temperate forests. *In* "Comparative Analysis of Ecosystems" (J. Cole, G. Lovett, and S. Findlay, eds.), pp. 319–330. Springer-Verlag, New York.

Brown, C. P. (1946). Food of Maine ruffed grouse by seasons and cover types. *J. Wildl. Manage.* **10**, 17–28.

Brown, E. R., Sinclair, T., Keith, L., Beamer, P., Hazdra, J. J., Nair, V., and Callaghan, O. (1977). Chemical pollutants in relation to diseases in fish. *Ann. N.Y. Acad. Sci.* **298**, 535–546.

Brown, G. W. (1969). Predicting temperatures of small streams. *Water Resour. Res.* **5**, 68–75.

Brown, G. W. (1970). Predicting the effect of clear-cutting on stream temperature. *J. Soil Water Conserv.* **25**, 11–13.

Brown, G. W., and Krygier, J. T. (1970). Effects of clear-cutting on stream temperature. *Water Resour. Res.* **6**, 1133–1140.

Brown, G. W., Gahler, A. R., and Marston, R. B. (1973). Nutrient losses after clear-cut logging and slash burning in the Oregon Coast Range. *Water Resour. Res.* **9**, 1450–1453.

Brown, L. R., and Jacobson, J. (1987). Assessing the future of urbanization. *In* "State of the World 1987," pp. 38–56. Worldwatch Institute, Norton, New York.

Brown, L. R., and Postel, S. (1987). Thresholds of change. *In* "State of the World 1987," pp. 3–19. Worldwatch Institute, Norton, New York.

Brown, R. G. B., Gillespie, D. I., Lock, A. R., Pearce, P. A., and Watson, G. H. (1973). Bird mortality from oil slicks off eastern Canada, February–April 1970. *Can. Field Nat.* **87**, 225–234.

Brown, R. S., Wolke, R. E., Saila, S. B., and Brown, C. W. (1977). Prevalence of neoplasia in 10 New England populations of the soft-shell clam (*Mya arenaria*). *Ann. N.Y. Acad. Sci.* **298**, 522–534.

Brown, S., and Lugo, A. E. (1990). Tropical secondary forests. *J. Trop. Ecol.* **6**, 1–32.

Brown, S., and Lugo, A. E. (1992). Aboveground biomass estimates for tropical moist forests of the Brazilian Amazon. *Interciencia* **17**, 8–18.

Browning, K. A., Allam, R. J., Ballard, S. P., Barnes, R. T. H., Bennetts, D. A., Maryaon, R. H., Mason, P. J., McKenna, D., Mitchell, J. F. B., Senior, C. A., Slingo, A., and Smith, F. B. (1991). Environmental effects from burning oil wells in Kuwait. *Nature (London)* **351**, 363–367.

Bruck, R. I. (1985). Boreal montane ecosystem decline in the Southern Appalachian Mountains: Potential role of anthropogenic pollution. *In* "Air Pollutant Effects on Forest Ecosystems," pp. 137–155. Acid Rain Foundation, St. Paul, MN.

Bruck, R. I. (1989). Forest decline syndromes in the southeastern United States. *In* "Air Pollution's Toll on Forests and Crops" (J. J. MacKenzie and M. T. El-Ashry, eds.), pp. 113–190. Yale Univ. Press, New Haven, CT.

Bruns, H. (1960). The economic importance of birds in forests. *Bird Study* **7**, 193–208.

Bryan, G. W. (1976). Some aspects of heavy metal tolerance in aquatic organisms. *In* "Effects of Pollutants on Aquatic Organisms" (A. P. M. Lockwood, ed.), pp. 7–34. Cambridge Univ. Press, London and New York.

Buchauer, M. J. (1973). Contamination of soil and vegetation near a zinc smelter by zinc, cadmium, and lead. *Environ. Sci. Technol.* **7**, 131–135.

Buckner, C. H., and McLeod, B. B. (1975). The impact of insecticides on small forest mammals. *In* "Aerial Control of Forest Insects in Canada" (M. L. Prebble, ed.), pp. 314–318. Department of the Environment, Ottawa, Ont.

Buckner, C. H., and McLeod, B. B. (1977). "Ecological Impact Studies of Experimental and Operational Spruce Budworm (*Choristoneura fumiferana* Clemens) Control Programs on Selected Non-target Organisms in Quebec, 1976," Rep. CC-X-137. Forest Pest Management Institute, Environment Canada, Sault Ste. Marie, Ont.

Buech, R. R. (1980). Vegetation of a Kirtland's warbler breeding area and 10 nest sites. *Jack Pine Warbler* **58**, 59–72.

Buhler, D. R., Claeys, R. R., and Mate, B. R. (1975). Heavy metal and chlorinated hydrocarbon residues in Califirnia sea-lions *Zalophus californianus*. *J. Fish. Res. Board Can.* **32**, 2391–2397.

Bull, E. L., Twombly, A. D., and Quigley, T. M. (1980). Perpetuating snags in managed mixed conifer forests of the Blue Mountains, Oregon. *In* "Management of Western Forests and Grasslands for Non-game Birds," USDA For. Serv. Gen. Tech. Rep. INT-86, pp. 325–336. Intermountain Forest and Range Experiment Station, Ogden, UT.

Bull, E. L., Peterson, S. R., and Thomas, J. W. (1986). "Resource Partitioning Among Woodpeckers in Northeastern Oregon." USDA For. Serv. Res. Note PNW-444. Pacific Northwest Research Station, Portland, OR.

Bunce, H. F. (1979). Fluoride emissions and forest growth. *J. Air Pollut. Control Assoc.* **29**, 642–643.

Bunce, H. F. (1984). Fluoride emissions and forest survival, growth, and regeneration. *Environ. Pollut., Ser. A* **35**, 169–188.

Burger, J. A., and Pritchett, W. L. (1979). Clearcut harvesting and site preparation can dramatically decrease nutrient reserves of a forest site. *In* "Impact of Intensive Harvesting on Forest Nutrient Cycling," p. 393. College of Environmental Science and Forestry, State University of New York, Syracuse.

Burns, J. W. (1972). Some effects of logging and associated road construction on northern California streams. *Trans. Am. Fish. Soc.* **101**, 1–17.

Burt, W. H. (1976). "A Field Guide to the Mammals." Houghton Mifflin, Boston.

Burton, P. S. (1993). Intertemporal preferences and intergenerational equity considerations in optimal resource harvesting. *J. Environ. Econ. Manage.* **24**, 119–132.

Burton, T. M., and Likens, G. E. (1975a). Energy flow and nutrient cycling in salamander populations in the Hubbard Brook Experimental Forest, New Hampshire. *Ecology* **56**, 1068–1080.

Burton, T. M., and Likens, G. E. (1975b). Salamander populations and biomass in the Hubbard Brook Experimental Forest, New Hampshire. *Copeia*, pp. 541–546.

Bury, B. (1983). Differences in amphibian populations in logged and old-growth redwood forest. *Northwest Sci.* **57**, 167–178.

Bury, R. B., and Corn, P. S. (1988a). Responses of aquatic and streamside amphibians to timber harvest: A review. *In* "Streamside Management: Riparian Wildlife and Forestry Interactions," pp. 165–181. Institute of Forest Resources, University of Washington, Seattle.

Bury, R. B., and Corn, P. S. (1988b). Douglas-fir forests in the Oregon and Washington Cascades: Abundance of terrestrial herpetofauna related to stand age and moisture. *In* "Management of Amphibians, Reptiles, and Small Mammals in North America," USDA For. Serv. Gen. Tech. Res. Rep. RM-166, pp. 11–22. Rocky Mountain Forest and Range Experiment Station, Fort Collins, CO.

Busby, D. G., Pearce, P. A., and Garrity, N. R. (1981). Brain cholinesterase response in songbirds exposed to experimental fenitrothion spraying in New Brunswick, Canada. *Bull. Environ. Contam. Toxicol.* **26**, 401–406.

Busby, D. G., Pearce, P. A., and Garrity, N. R. (1982). Brain cholinesterase inhibition in forest passerines exposed to experimental aminocarb spraying. *Bull. Environ. Contam. Toxicol.* **28**, 225–229.

Busby, D. G., Pearce, P. A., Garrity, N. R., and Reynolds, L. M. (1983a). Effect of an organophosphorus insecticide on brain cholinesterase activity in white-throated sparrows exposed to aerial forest spraying. *J. Appl. Ecol.* **20**, 255–264.

Busby, D. G., Pearce, P. A., and Garrity, N. R. (1983b). Brain ChE response in forest songbirds exposed to aerial spraying of aminocarb and possible influence of application methodology and insecticide formulation. *Bull. Environ. Contam. Toxicol.* **31**, 125–131.

Busby, D. G., White, L. M., Pearce, P. A., and Mineau, P. (1989). Fenitrothion effects on forest songbirds: A critical new look. *In* "Environmental Effects of Fenitrothion Use in Forestry," pp. 43–108. Conservation and Protection, Environment Canada, Dartmouth, Nova Scotia.

Busby, D. G., White, L. M., and Pearce, P. A. (1990). Effects of aerial spraying of fenitrothion on breeding white-throated sparrows. *J. Appl. Ecol.* **27**, 743–755.

Busch, B. C. (1985). "The War Against the Seals: A History of the North American Seal Fishery." McGill-Queen's Univ. Press, Kingston, Ont.

Buskirk, S. W. (1992). Conserving circumboreal forests for martens and fisher. *Conserv. Biol.* **6**, 318–320.

Butler, J. N., Morris, B. F., and Sass, J. (1973). "Pelagic tar from Bermuda and the Sargasso Sea," Spec. Publ. No. 10. Bermuda Biological Station for Research, St. George's West.

Buttrick, P. L. (1917). A forester at the fighting front. *Am. For.* **23**, 710–716.

Cabioch, L. (1980). Pollution of subtidal sediments and disturbance of benthic animal communities. *Ambio* **9**, 294–296.

Cade, T. J. (1988). Using science and technology to re-establish species lost in nature. *In* "Biodiversity" (E. O. Wilson, ed.), pp. 279–288. National Academy Press, Washington, DC.

Cade, T. J., and Bird, D. M. (1990). Peregrine falcons, *Falco peregrinus*, nesting in an urban environment: A review. *Can. Field Nat.* **104**, 209–218.

Cade, T. J., and Fyfe, R. (1970). The North American peregrine survey, 1970. *Can. Field Nat.* **84**, 231–240.

Cadle, S. H., Dasch, J. M., and Grossnickle, N. E. (1984). Retention and release of chemical species by a northern Michigan snowpack. *Water, Air, Soil Pollut.* **22**, 303–319.

Cain, M. D., and Yaussy, D. A. (1984). Can hardwoods be eradicated from pine sites? *South. J. Appl. For.* **8**, 7–13.

Cairns, D. K., and Elliot, R. D. (1987). Oil spill impact assessment for seabirds: The role of refugia and growth centers. *Biol. Conserv.* **40**, 1–9.

Cairns, J., McCormick, P., and Niederlehner, B. (1991). "A Proposed Framework for Developing Indicators of Ecosystem Health for the Great Lakes Region." International Joint Commission, Ann Arbor, MI.

Campbell, B., and Lack, E. (1985). "A Dictionary of Birds." British Ornithologists' Union, London.

Campbell, P. G. C., Bisson, M., Bengie, R., Tessier, A., and Villeneuve, J. P. (1983). Speciation of aluminum in acidic waters. *Anal. Chem.* **55**, 2246–2252.

Campbell, R. A. (1990). Herbicide use for forest management in Canada: Where we are and where we are going? *For. Chron.* **66**, 355–360.

Campbell, W. B., Harris, R. W., and Benoit, R. E. (1973). Response of Alaskan tundra microflora to crude oil spill. *In* "Proceedings of the Symposium on the Impact of Oil Resource Development on Northern Plant Communities," pp. 53–62. Institute of Arctic Biology, Fairbanks, AK.

Canadian Environmental Assessment Research Council/National Research Council CEARC/NRC (1986). "Cumulative Environmental Effects." CEARC/NRC, Board on Basic Ecology, Ottawa, Ont.

Canfield, D. E., and Bachman, R. W. (1981). Prediction of total phosphorus concentrations, chlorophyll a, and secchi depths in natural and artificial lakes. *Can. J. Fish. Aquat. Sci.* **38**, 414–423.

Cann, D. B., MacDougall, J. I., and Hilchey, J. D. (1965). "Soil Survey of Kings County, Nova Scotia," Rep. No. 15. Nova Scotia Soil Survey, Nova Scotia Department of Agriculture and Marketing, Truro.

Cannon, H. L. (1960). Botanical prospecting for ore deposits. *Science* **132**, 591–598.

CANSAP (1983). "CANSAP Data Summary 1982." Environment Canada, Ottawa, Ont.

CANSAP (1984). "CANSAP Data Summary 1983." Environment Canada, Ottawa, Ont.

Canter, L. (1977). "Environmental Impact Assessment." McGraw-Hill, New York.

Cantwell, E. N., Jacobs, E. S., Gunz, W. G., and Liberi, V. E. (1972). Control of particulate lead emissions from automobiles. In "Cycling and Control of Metals" (M. G. Curry and G. M. Gigliotti, eds.), pp. 95–107. U.S. Environmental Protection Agency, Cincinnati, OH.

Capuzzo, J. M. (1987). Biological effects of petroleum hydrocarbons: Assessments from experimental results. In "Long-term Environmental Effects of Offshore Oil and Gas Development" (D. F. Boesch and N. N. Robelais, eds.), pp. 343–410. Elsevier, London.

Carbon Dioxide Information Center (CDIC) (1990). "Glossary. Carbon Dioxide and Climate." Oak Ridge National Laboratory, CDIC, Oak Ridge, TN.

Carey, A. B., Reid, J. A., and Horton, S. P. (1990). Spotted owl home range and habitat use in southern Oregon Coast Ranges. J. Wildl. Manage. 54, 11–17.

Carey, A. B., Horton, S. P., and Biswell, B. L. (1992). Northern spotted owls: Influence of prey base and landscape pattern. Ecol. Monogr. 62, 223–250.

Carey, A. C., Miller, E. A., Geballe, G. T., Wargo, P. M., Smith, W. H., and Siccama, T. G. (1984). Armillaria mellea and decline of red spruce. Plant Dis. 68, 794–795.

Carrier, L. (1986). "Decline in Quebec's forests. Assessment of the Situation." Service de la Recherche Appliqué, Ministère de l'Energie et des Ressources, Québec.

Carson, R. (1962). "Silent Spring." Houghton-Mifflin, Boston.

Carter, N. E. (1992). "Protection Spraying Against Spruce Budworm in New Brunswick, 1987 to 1992." Timber Management Branch, New Brunswick Department of Forests, Mines, and Energy, Fredericton.

Carter, N. E., and Lavigne, D. R. (1986). Protection Spraying Against Spruce Budworm in New Brunswick 1985." Timber Management Branch, New Brunswick Department of Forests, Mines, and Energy, Fredericton.

Case, A. B., and Donnelly, R. G. (1979). "Type and Extent of Ground Disturbance Following Skidder Logging in Newfoundland and Labrador," Inf. Rep. N-X-176. Newfoundland Forest Research Centre, St. John's.

Caseley, J., and Coupland, D. (1985). Environment and plant factors affecting glyphosate uptake, movement, and activity. In "The Herbicide Glyphosate" (E. Grossbard and D. Atkinson, eds.), pp. 92–123. Butterworth, Toronto, Ont.

Castell, C. P. (1944). The ponds and their vegetation. London Nat. 24, 15–22.

Castell, C. P. (1954). The bomb-crater ponds of Brookham Common. London Nat. 34, 16–21.

Catling, P. M., Freedman, B., Stewart, C., Kerekes, J. J., and Lefkovitch, L. P. (1986). Aquatic plants of acid lakes in Kejimkujik National Park; Floristic composition and relation to water chemistry. Can. J. Bot. 64, 724–729.

Cess, R. D. (1985). Nuclear war: Illustrative effects of atmospheric smoke and dust upon solar radiation. Clim. Change 7, 237–251.

Chambers, L. A. (1976). Classification and extent of air pollution problems. In "Air Pollution" (A. C. Stern, ed.), 3rd ed., Vol. 1, pp. 3–22. Academic Press, New York.

Chan, W. H., and Lusis, M. A. (1985). Post-superstack Sudbury smelter emissions and their fate in the atmosphere: An overview of the Sudbury environmental study. Water, Air, Soil Pollut. 20, 211–220.

Chan, W. H., Ro, C. U., Lusis, M. A., and Vet, R. J. (1982). Impact of the INCO nickel smelter emissions on precipitation quality in the Sudbury area. Atmos. Environ. 16, 801–814.

Chan, W. H., Vet, R. J., Lusis, M. A., and Skelton, G. B. (1983). Airborne particulate size distribution measurements in nickel smelter plumes. Atmos. Environ. 17, 1173–1181.

Chan, W. H., Vet, R. J., Ro, C. U., Tang, A. J. S., and Lusis, M. A. (1984a). Impact of INCO smelter emissions on wet and dry deposition in the Sudbury area. Atmos. Environ. 18, 1001–1008.

Chan, W. S., Vet, R. J., Ro, C. U., Tang, A. J. S., and Lusis, M. A. (1984b). Long-term precipitation quality and wet deposition fields in the Sudbury basin. Atmos. Environ. 18, 1175–1188.

Chapin, F. S., III, and Shaver, G.R. (1985). Individualistic growth response of tundra plant species to environmental manipulation in the field. Ecology 66, 564–576.

Chapin, G., and Wasserström, R. (1981). Agricultural production and malaria resurgence in Central America and India. Nature (London) 293, 181–185.

Chapman, A. R. O. (1981). Stability of sea urchin dominated barren grounds following destructive grazing of kelp in St. Margaret's Bay, eastern Canada. Mar. Biol. 62, 307–311.

Charlson, R. J., Lovelock, J. E., Andreae, M. O., and Warren, S. G. (1987). Oceanic phytoplankton, atmospheric sulfur, cloud albedo, and climate. Nature (London) 326, 655–661.

Charlton, M. N. (1979). "Hypolimnetic Oxygen Depletion in Central Lake Erie: Has There Been Any Change?" Sci. Ser. No. 110. Inland Waters Directorate, National Water Research Institute, Environment Canada, Burlington, Ont.

Charlton, M. N. (1980). Oxygen depletion in Lake Erie: Has there been any change? Can. J. Fish. Aquat. Sci. 37, 72–81.

Cheliak, W. M., and Rogers, D. L. (1990). Integrating biotechnology into tree improvement programs. Can. J. For. Res. 20, 452–463.

Cheliak, W. M., Wang, J., and Pitel, J. A. (1988). Population structure and genic diversity in tamarack, Larix laricina (Du Roi) K. Koch. Can. J. For. Res. 18, 1318–1324.

Chestnut, L. G., and Rowe, R. D. (1989). Economic measures of the impacts of air pollution on health and visibility. In "Air Pollution's Toll on Forests and Crops" (J.J. MacKenzie and M.T. Al-Ashry, eds.), pp. 316–342. Yale Univ. Press, New Haven, CT.

Chianelli, R. R., Aczel, T., Bare, R. E., George, G. N., Genowitz, M. W., Grossman, M. L., Haith, C. E., Kaiser, F. J., Cessard, R. R., Liotta, R., Masrtaccio, R. L., Minak-Bernero, V., Prince, R. C., Robbins, W. K., Stiefel, E. I., Wilkinson, J. B., Hinton, S. M., Bragg, J. R., McMillen,

S. J., and Atlas, R. M. (1991). "Bioremediation Technology Development and Application to the Alaskan Spill." Exxon Co., Houston, TX.

Christensen, N. L., Agee, J. K., Brussard, P. F., Hughes, J., Knight, D. H., Minshall, G. W., Peek, J. M., Pyne, S. J., Swanson, F. J., Thomas, J. W., Wells, S., Williams, S. E., and Wright, H. A. (1989). Interpreting the Yellowstone fires of 1988. *BioScience* **39,** 678–685.

Chung, Y. S., and Dann, T. (1985). Observations of stratospheric ozone at the ground level in Regina, Canada. *Atmos. Environ.* **19,** 157–162.

Cicerone, R. J. (1987). Changes in stratospheric ozone. *Science* **237,** 35–42.

Clair, T. A., and Komadina, V. (1984). "Aluminum Speciation in Waters of Nova Scotia and Their Impact on WQB Analytical and Field Method," IWD-AR-WQB-84–69. Inland Waters Directorate, Atlantic Region, Environment Canada, Moncton, New Brunswick.

Clark, D. R. (1979). Lead concentrations: Bats vs. terrestrial small mammals collected near a major highway. *Environ. Sci. Technol.* **13,** 338–340.

Clark, J. S. (1991). Ecosystem sensitivity to climate change and complex responses. *In* "Global Climate Change and Life on Earth" (R. L. Wyman, ed.), pp. 65–98. Routledge, Chapman, and Hall, New York.

Clark, R. B. (1984). Impact of oil pollution on seabirds. *Environ. Pollut., Ser. A* **33,** 1–22.

Clark, R. C., and Brown, D. W. (1977). Petroleum: Properties and analyses in biotic and abiotic systems. *In* "Effects of Petroleum on Arctic and Subarctic Marine Environments and Organisms" (D. C. Malins, ed.), Vol. 1, pp. 1–90. Academic Press, New York.

Clark, R. C., and MacLeod, W. D. (1977). Inputs, transport mechanisms, and observed concentrations of petroleum in the marine environment. *In* "Effects of Petroleum on Arctic and Subarctic Marine Environments and Organisms" (D. C. Malins, ed.), Vol. 1, pp. 91–224. Academic Press, New York.

Clark, R. S., Chair. (1947). "Scientific Meeting on the Effect of the War on the Stocks of Commercial Food Fishes." Copenhagen.

Clark, T. H., and Stern, C. W. (1968). "Geological Evolution of North America." Ronald Press, New York.

Clark, C. W. (1981). Bioeconomics. *In* "Theoretical Ecology. Principles and Applications" (R.M. May, ed.), pp. 387–423. Sinauer Assoc., Sunderland, MA.

Clark, W. C. (1986). Sustainable development of the biosphere: Themes for a research program. *In* "Sustainable Development of the Biosphere" (W. C. Clark and R. E. Munn, eds.), pp. 5–48. Cambridge Univ. Press, New York.

Clark, W. C. (1989). Clear-cut economies. Should we harvest everything now? *Sciences* (*N.Y.*), Jan./Feb., pp. 16–19.

Clough, G. C. (1987). Relations of small mammals to forest management in northern Maine. *Can. Field Nat.* **101,** 40–48.

Clowater, W. G., and Andrews, P. W. (1981). "An Assessment of Damage Caused by the Spruce Budworm on Spruce and Balsam Fir Trees in New Brunswick." New Brunswick Department of Natural Resources, Fredericton.

Clymo, R. S. (1963). Ion exchange in *Sphagnum* and its relation to bog ecology. *Ann. Bot. (London)* [N.S.] **27,** 309–324.

Clymo, R. S. (1964). The origin of acidity in *Sphagnum* bogs. *Bryologist* **67,** 427–431.

Clymo, R. S. (1973). The growth of *Sphagnum*: Some effects of environment. *J. Ecol.* **61,** 849–869.

Clymo, R. S. (1984). *Sphagnum*-dominated peat bogs: A naturally acidic ecosystem. *Philos. Trans. R. Soc. London* **305,** 487–499.

Coffin, D. L., and Stokinger, H. E. (1977). Biological effects of air pollutants. *In* "Air Pollution" (A. C. Stern, ed.), 3rd ed., Vol. 2, pp. 232–360. Academic Press, New York.

Cogbill, C. V. (1976). The history and character of acid precipitation in eastern North America. *Water, Air, Soil Pollut.* **6,** 407–413.

Cogbill, C. V. (1977). The effect of acid precipitation on tree growth in eastern North America. *Water, Air, Soil Pollut.* **8,** 89–93.

Cogbill, C. V., and Likens, G. E. (1974). Acid precipitation in the northeastern United States. *Water Resour. Res.* **10,** 1133–1137.

Cohen, C. J., Grothaus, L. C., and Perrigan, S. C. (1981). "Effects of Simulated Acid Rain on Crop Plants," Spec. Rep. 619. Agricultural Experiment Station, Oregon State University, Corvallis.

Cohen, D. B., Webber, M. D., and Bryant, D. N. (1978). Land application of chemical sewage sludge—lysimeter studies. *Conf. Proc.—Res. Program Abatement Munic. Pollut. Prov. Can.–Ont. Agreement Great Lakes Water Qual.* **6,** 108–137.

Colbeck, I., and Harrison, R. M. (1985). The photochemical pollution episode of 5–16 July 1983 in north-west England. *Atmos. Environ.* **19,** 1921–1929.

Cole, D. W., and Gessel, S. P. (1965). Movement of elements through forest soils as influenced by tree removal and fertilizer additions. *In* "Forest Soil Relationships in North America" (C. T. Youngberg, ed.), pp. 95–104. Oregon State Univ. Press, Corvallis.

Cole, D. W., Crane, W. J. B., and Grier, C. C. (1975). The effect of forest management practices on water chemistry in a second-growth Douglas-fir ecosystem. *In* "Forest Soils and Forest Land Management" (B. Bernier and C. H. Winget, eds.), pp. 195–207. Laval Univ. Press, Quebec.

Coleman, R. (1966). The importance of sulfur as a plant nutrient in world crop production. *Soil Sci.* **101,** 230–239.

Collar, N. J., and Andrew, P. (1988). "Birds to Watch: The ICBP World Checklist of Threatened Species." International Council for Bird Protection, Cambridge, UK.

Collins, S. L., James, F. C., and Risser, P. G. (1982). Habitat relationships of wood warblers (Parulidae) in northern central Wisconsin. *Oikos* **39,** 50–58.

Conner, R. N. (1978). Snag management for cavity-nesting birds. *In* "Proceedings of the Workshop on Management of Southern Forests for Non-game Birds," USDA For. Serv. Gen. Tech. Rep. SE-14, pp. 120–128. Southeastern Forest Experiment Station, Asheville, NC.

Conner, R. N., and Rudolph, D. C. (1989). "Red-cockaded Woodpecker Colony Status and Trends on the Angelina, Davy Crockett, and Sabine National Forests," USDA For. Serv. Res. Pap. SO-250. Southern Forest Experiment Station, New Orleans, LA.

Conner, R. N., and Rudolph, D. C. (1991). Forest habitat loss, fragmentation, and red-cockaded woodpecker populations. *Wilson Bull.* **103,** 446–457.

Conover, S. A. M., Strong, K. W., Hickey, T. W., and Sander, F. (1985). An evolving framework for environmental impact analysis. I. Methods. *J. Environ. Manage.* **21,** 343–358.

Conroy, N., Hawley, K., Keller, W., and LaFrance, C. (1975). Influence of the atmosphere on lakes in the Sudbury area. *Proc. Spec. Symp. Atmos. Contrib. Chem. Lake Waters, 1st,* Toronto, Ont., *1975,* pp. 146–165.

Conroy, N., Hawley, K., Keller, W., and LaFrance, C. (1976). Influences of the atmosphere on lakes in the Sudbury area. *J. Great Lakes Res.* **2,** Suppl. 1, 146–165.

Conway Morris, S., and Whittington, H. B. (1979). The animals of the Burgess Shale. *Sci. Am.* **240**(1), 122–133.

Cook, R. B., and Schindler, D. W. (1983). The biogeochemistry of sulfur in an experimentally acidified lake. *Ecol. Bull.* **35,** 115–127.

Cook, R. B., Kelly, C. A., Schindler, D. W., and Turner, M. A. (1986). Mechanisms of hydrogen ion neutralization in an experimentally acidified lake. *Limnol. Oceanogr.* **31,** 134–148.

Cooke, A. S. (1979). Changes in egg shell characteristics of the sparrowhawk (*Accipiter nisus*) and peregrine (*Falco peregrinus*) associated with exposure to environmental pollutants during recent decades. *J. Zool.* **187,** 245–263.

Cooke, A. S., and Frazer, J. F. D. (1976). Characteristics of newt breeding sites. *J. Zool.* **178,** 223–236.

Cooke, G. D., Welch, E. B., Peterson, S. A., and Newroth, P. R. (1986). "Lake and Reservoir Restoration." Butterworth, London.

Cooper, C. F. (1983). Carbon storage in managed forests. *Can. J. For. Res.* **13,** 155–166.

Cooper, K. (1991). Effects of pesticides on wildlife. *In* "Handbook of Pesticide Toxicology. Vol. 1. General Principles" (W. C. Hayes and E. R. Laws, eds.), pp. 463–496. Academic Press, San Diego.

Corbett, E. S., Lynch, J. A., and Sopper, W. E. (1978). Timber harvesting practices and water quality in the eastern United States. *J. For.* **76,** 484–488.

Corbett, J. R. (1974). "The Biological Mode of Action of Pesticides." Academic Press, New York.

Cordone, A. J., and Kelley, D. W. (1961). The influence of inorganic sediment on the aquatic life of streams. *Calif. Fish Game* **47,** 189–228.

Cormack, D. (1983). "Response to Oil and Chemical Marine Pollution." Appl. Sci. Publ., London.

Corn, P. S., and Bury, R. B. (1989). Logging in western Oregon: Responses of headwater habitats and stream amphibians. *For. Ecol. Manage.* 29, 39–57.

Corn, P. S., and Bury, R. B. (1991a). Small mammal communities in the Oregon Coast Range. *In* "Wildlife and Vegetation of Unmanaged Douglas-fir Forests," USDA For. Serv. Gen. Tech. Rep. PNW-285, pp. 241–254. Pacific Northwest Forest and Range Experiment Station, Portland, OR.

Corn, P. S., and Bury, R. B. (1991b). Terrestrial amphibian communities in the Oregon Coast Range. *In* "Wildlife and Vegetation of Unmanaged Douglas-fir Forests" USDA For. Serv. Gen. Tech. Rep. PNW-285, pp. 315–317. Pacific Northwest Forest and Range Experiment Station, Portland, OR.

Corner, E. D. S. (1978). Pollution studies with marine plankton. Part 1. Petroleum hydrocarbons and related compounds. *Adv. Mar. Biol.* **15,** 289–380.

Costanza, R., ed. (1991). "Ecological Economics. The Science and Management of Sustainability." Columbia Univ. Press, New York.

Costanza, R., and Daly, H. E. (1992). Natural capital and sustainable development. *Conserv. Biol.* **6,** 37–46.

Costanza, R., Daly, H. E., and Bartholemew, J. A. (1991). Goals, agenda, and policy reccomendations for ecological economics. *In* "Ecological Economics. The Science and Management of Sustainability" (R. Costanza, ed.), pp. 1–20. Columbia Univ. Press, New York.

Costescu, L. W. (1974). The ecological consequences of airborne metallic contaminants from the Sudbury smelters. Ph.D. Thesis, Department of Botany, University of Toronto, Toronto, Ont.

Council on Environmental Quality (CEQ) (1975). "Environmental Quality." U.S. Govt. Printing Office, Washington, DC.

Council on Environmental Quality (CEQ) (1979). "Environmental Quality." Executive Office of the President, Washington, DC.

Council on Environmental Quality (CEQ) (1992a). "Environmental Quality: 22nd Annual Report." CEQ, Washington, DC.

Council on Environmental Quality (CEQ) (1992b). Biodiversity. *In* "Environmental Quality: 22nd Annual Report," pp. 19–26. CEQ, Washington, DC.

Covey, C. (1985). Climatic effects of nuclear war. *BioScience* **35,** 563–569.

Covington, W. W., and Aber, J. D. (1980). Leaf production during secondary succession in northern hardwoods. *Ecology* **61,** 200–204.

Cowling, E. B. (1985). Comparison of regional declines of forests in Europe and North America: The possible role of airborne chemicals. *In* "Air Pollutant Effects on Forest Ecosystems," pp. 217–234. Acid Rain Foundation, St. Paul, MN.

Cowling, E. B. (1992). The performance and legacy of NAPAP. *Ecol. Appl.* **2,** 111–116.

Cox, R. M., and Hutchinson, T. C. (1979). Metal co-tolerances in the grass *Deschampsia caespitosa*. *Nature (London)* **279**, 231–233.

Cox, R. M., and Hutchinson, T. C. (1980). Multiple metal tolerances in the grass *Deschampsia caespitosa* (L.) Beauv. from the Sudbury smelting area. *New Phytol.* **84**, 631–647.

Cox, R. M., and Hutchinson, T. C. (1981). Environmental factors influencing the rate of spread of the grass *Deschampsia caespitosa* invading areas agound the Sudbury nickel-copper smelters. *Water, Air, Soil Pollut.* **16**, 83–106.

Crawford, H. S., and Jennings, D. T. (1989). Predation by birds on spruce budworm *Choriustoneura fumiferana*: Functional, numerical, and total responses. *Ecology* **70**, 152–163.

Crawford, H. S., and Titterington, R. W. (1979). Effects of silvicultural practices on bird communities in upland spruce-fir stands. *In* "Management of North Central and North-eastern Forests for Non-game Birds," USDA For. Serv. Gen. Tech. Rep. NC-51, pp. 110–119. North Central Forest Experiment Station, St. Paul, MN.

Crawford, H. S., Titterington, R. W., and Jennings, D. T. (1983). Bird predation and spruce budworm populations. *J. For.* **81**, 433–435.

Crete, M., and Bédard, J. (1975). Daily browse consumption by moose in the Gaspe Peninsula, Quebec. *J. Wildl. Manage.* **39**, 368–373.

Crocker, R. L., and Major, J. (1955). Soil development in relation to vegetation and surface age at Glacier Bay, Alaska. *J. Ecol.* **43**, 427–448.

Cromartie, E., Reichel, W. L., Locke, N. L., Belisle, A. A., Kaiser, T. E., Lamont, T. G., Mulhern, B. M., Prouty, R. M., and Swineford, D. M. (1975). Residues of organochlorine pesticide and polychlorinated biphenyls and autopsy data for bald eagles, 1971–1972. *Pestic. Monit. J.* **9**, 11–14.

Cronan, C. S., and Reiners, W. A. (1983). Canopy processing of acid precipitation by coniferous and hardwood forests in New England. *Oecologia* **59**, 216–223.

Cronan, C. S., and Schofield, C. L. (1979). Aluminum leaching response to acid precipitation: Effects on high-elevation watersheds in the northeast. *Science* **204**, 304–305.

Crouch, G. L. (1985). "Effects of Clearcutting a Subalpine Forest in Central Colorado on Wildlife Habitat," USDA For. Serv. Gen Tech. Res. Rep. RM-258. Rocky Mountain Forest and Range Experiment Station, Fort Collins, CO.

Crow, T. R., Mroz, G. D., and Gale, M. R. (1991). Regrowth and nutrient accumulations following whole-tree harvesting of a maple-oak forest. *Can. J. For. Res.* **21**, 1305–1315.

Crowell, M., and Freedman, B. (1993). Vegetation development across a post-disturbance chronosequence of hardwood forest in Nova Scotia. *Can. J. For. Res.* **24**, 260–271.

Crutzen, P. J., and Arnold, F. (1986). Nitric acid cloud formation in the cold Antarctic stratosphere: A major cause for the springtime "ozone hole." *Nature (London)* **324**, 651–655.

Crutzen, P. J., and Birks, J. W. (1982). The atmosphere after a nuclear war: Twilight at noon. *Ambio* **11**, 114–125.

Cullis, C. F., and Hirschler, M. M. (1980). Atmospheric sulfur: Natural and man-made sources. *Atmos. Environ.* **14**, 1263–1278.

Cumming, B. F., Smol, J. P., Kingston, J. C., Charles, D. F., Birks, H. J. B., Camburn, K. E., Dixit, S. S., Uutala, A. J., and Selle, A. R. (1992). How much acidification has occurred in Adirondack region lakes (New York, USA) since pre-industrial times? *Can. J. Fish. Aquat. Sci.* **49**, 128–141.

Cumming, H. G., and Beange, D. B. (1987). Dispersion and movements of woodland caribou near Lake Nipigon, Ontario. *J. Wildl. Manage.* **51**, 69–79.

Cunningham, J. C. (1985). Biorationals for control of spruce budworms. *In* "Recent Advances in Spruce Budworms Research," pp. 320–349. Canadian Forestry Service, Ottawa, Ont.

Cunningham, J. C. and van Frankenhuyzen, K. (1991). Microbial insecticides in forestry. *For. Chron.* **67**, 473–480.

Cunningham, J. D., Keeney, D. R., and Ryan, J. A. (1975a). Yield and metal composition of corn and rye grown on sewage sludge-amended soil. *J. Environ. Qual.* **4**, 448–454.

Cunningham, J. D., Ryan, J. A., and Keeney, D. R. (1975b). Phytotoxicity in and metal uptake from soil treated with metal-amended sewage sludge. *J. Environ. Qual.* **4**, 455–460.

Cure, J. D. (1985). Carbon dioxide doubling responses: A crop survey. *In* "Direct Effects of Increasing Carbon Dioxide on Vegetation," DOE/ER-0238, pp. 99–116. U.S. Department of Energy, Washington, DC.

Currie, D. J., and Kalff, J. (1984). The relative importance of bacterioplankton and phytoplankton in phosphorus uptake in freshwater. *Limnol. Oceanogr.* **29**, 311–321.

Currie, D. J., Bentzen, E., and Kalff, J. (1986). Does algal-bacterial phosphorus partitioning vary among lakes? A comparative study of orthophosphate uptake and alkaline phosphatase activity in freshwater. *Can. J. Fish. Aquat. Sci.* **43**, 311–318.

Currier, H. B., and Peoples, S. A. (1954). Phytotoxicity of hydrocarbons. *Hilgardia* **23**, 155–173.

Curry-Lindahl, K. (1971). War and the white rhinos. *Oryx* **11**, 263–267.

Curry-Lindahl, K. (1972). "Let Them Live. A Worldwatch Survey of Animals Threatened With Extinction." Morrow, New York.

Curtis, P. D., Drake, B. G., and Whigham, D. F. (1989). Nitrogen and carbon dynamics in C3 and C4 estuarine marsh plants grown under elkevated CO_2 *in situ*. *Oecologia* **78**, 297–301.

Cushing, D. H. (1988). "The Provident Sea." Cambridge Univ. Press, Cambridge, UK.

Cwynar, L. C., and MacDonald, G. M. (1987). Geographic variation of lodgepole pine in relation to population history. *Am. Nat.* **129**, 463–469.

Czapowskyj, M. M. (1977). Status of fertilization and nutrition research in northern forest types. *In* "Proceedings of the Symposium on Intensive Culture of Northern Forest Types,"

USDA For. Serv. Gen. Tech. Rep. NE-29, pp. 185–203. Northeastern Forest Experiment Station, Broomall, PA.

Daan, N. (1980). A review of replacement of depleted stocks by other species and the mechanisms underlying such replacement. *Rapp. P.-V. Reun., Cons. Int. Explor. Mer* **177,** 405–421.

Dale, J. M., and Freedman, B. (1982). Lead and zinc contamination of roadside soil and vegetation in Halifax, Nova Scotia. *Proc. N.S. Inst. Sci.* **32,** 327–336.

Dale, J. M., Freedman, B., and Kerekes, J. (1985). Acidity and associated water chemistry of amphibian habitats in Nova Scotia. *Can. J. Zool.* **63,** 97–105.

Dale, J. M., Freedman, B., and Kerekes, J. (1986). Experimental studies of the effects of acidity and associated water chemistry on amphibians. *Proc. N.S. Inst. Sci.* **35,** 35–54.

Dale, V. H., Houghton, R. A., and Hall, C. A. S. (1991). Estimating the effects of land-use change on global atmospheric CO_2 concentrations. *Can. J. For. Res.* **21,** 87–90.

Daly, H. E., and Townsend, K. N. (1993). "Valuing the Earth. Economics, Ecology, Ethics." MIT Press, Cambridge, MA.

Dana, S. T. (1914). French forests in the war zone. *Am. For.* **20,** 769–785.

Daniel, T. W., Helms, J. A., and Baker, F. S. (1979). "Principles of Silviculture," 2nd ed. McGraw-Hill, Toronto, Ont.

D'Anieri, P., Leslie, D. M., and McCormack, M. L. (1987). Small mammals in glyphosate-treated clearcuts in northern Maine. *Can. Field Nat.* **101,** 547–550.

Darby, W. R., and Pruitt, W. O. (1984). Habitat use, movements, and grouping behaviour of woodland caribou (*Rangifer tarandus caribou*) in southwestern Manitoba. *Can. Field Nat.* **98,** 184–190.

Davidson, A. (1990). "In the Wake of the Exxon Valdez." Douglas & McIntyre, Toronto, Ont.

Davidson, C. I. (1979). Air pollution in Pittsburgh: A historical perspective. *J. Air Pollut. Control Assoc.* **29,** 1035–1042.

Davis, A. M. (1972). Selenium accumulation in *Astragalus* species. *Agron. J.* **64,** 751–754.

Davis, J. W., Gregory, G. A., and Ockenfels, R. A., eds. (1983). "Snag Habitat Management," USDA For. Serv. Tech. Rep. RM-99. Rocky Mountain Forest and Range Experiment Station, Fort Collins, CO.

Davis, M. B. (1978). Climatic interpretation of pollen in quaternary sediments. *In* "Biology and Quaternary Environments" (D. Walker and J. C. Guppy, eds.), pp. 35–51. Australian Academy of Science, Canberra.

Davis, M. B. (1981). Outbreaks of forest pathogens in quaternary history. *Proc. Int. Conf. Palynol., 4th,* pp. 216–228.

Davis, M. B. (1983). Quaternary history of the deciduous forests of eastern North America. *Ann. Mo. Bot. Gard.* **70,** 550–563.

Dawson, W. R., Ligon, T. D., Murphy, J. R., Myers, J. P., Simberloff, D., and Vernon, J. (1987). Report of the scientific advisory panel on the spotted owl. *Condor* **89,** 205–229.

Day, R. T. (1983). A survey and census of the endangered Furbish lousewort, *Pedicularis furbishae,* in New Brunswick. *Can. Field Nat.* **93,** 325–327.

Daye, P. G., and Garside, E. T. (1979). Development and survival of embryos and alevins of the Atlantic salmon, *Salmo salar* L., continuously exposed to acidic levels of pH, from fertilization. *Can. J. Zool.* **57,** 1713–1718.

Dean, L. C., Havens, R., Harper, K. T., and Rosenbaum, J. B. (1973). Vegetative stabilization of mill mineral wastes. *In* "Ecology and Reclamation of Devastated Land" (R. J. Hutnik and G. Davis, eds.), Vol. 2, pp. 119–136. Gordon & Breach, New York.

deBauer, M., Tejeda, T. H., and Manning, W. J. (1985). Ozone causes needle injury and tree decline in *Pinus hartwegii* at high altitudes in the mountains around Mexico City. *J. Air Pollut. Control Assoc.* **35,** 838.

DeByle, N. V. (1990). Aspen ecology and management in the western United States. *In* "Aspen Symposium '89," USDA For. Serv. Gen. Tech. Rep. NC-140, pp. 11–20. North Central Forest Experiment Station, St. Paul, MN.

DeByle, N. V., and Packer, P. E. (1972). Plant nutrient and soil losses in overland flow from burned forest clear-cuts. *In* "Watersheds in Transition," pp. 296–307. American Water Resources Association, Urbana, IL.

DeCosta, J., Janicki, A., Shellito, G., and Wilcox, G. (1983). The effect of phosphorus additions in enclosures on the phytoplankton and zooplankton of an acid lake. *Oikos* **40,** 283–294.

Dederen, L. H. T., Wendelaar, S. E., Bonga, A., and Leuven, R. S. E. W. (1987). Ecological and physiological adaptations of the acid-tolerant mudminnow, *Umbra pygmaea* (DeKay). *Ann. Soc. R. Zool. Belgium* **116,** 277–284.

DeGraaf, R. M., and Payne, B. R. (1975). Economic values of non-game birds and some urban wildlife research needs. *Trans. North Am. Wildl. Nat. Resour. Conf.* **40,** 281–287.

DeGraaf, R. M., and Shigo, A. L. (1985). "Management of Cavity Trees for Wildlife in the Northeast," USDA For. Serv. Gen. Tech. Rep. NE-101. Northeastern Forest Experiment Station, Broomall, PA.

DeGraaf, R. M., Yamasaki, M., Leak, W. B., and Lanier, J. W. (1992). "New England Wildlife: Management of Forested Habitats," USDA For. Serv. Gen. Tech. Rep. NE-144. Northeastern Forest Experiment Station, Broomall, PA.

de Gusmao-Camara, T. (1983). Tropical moist forest conservation in Brazil. *In* "Tropical Rain Forest: Ecology and Management" (S. L. Sutton, T. C. Whitmore, and A. C. Chadwick, eds.), pp. 413–421. Blackwell, Boston.

Dekker, W. L., Jones, V., and Achutuni, R. (1985). The impact of CO_2-induced climate change on U.S. agriculture. *In* "Characterization of Information Requirements for Studies of CO_2 Effects: Water Resources, Agriculture, Fisheries, Forests, and Human Health," DOE/ER-0236, pp. 69–94. U.S. Department of Energy, Washington, DC.

DeLaune, R. D., Patrick, W. H., and Buresh, R. J. (1979). Effect of crude oil on a Lousiana *Spartina alterniflora* salt marsh. *Environ. Pollut., Ser. A* **20,** 21–31.

Del Moral, R. (1983). Initial recovery of vegetation on Mount St. Helena, Washington. *Am. Midl. Nat.* **109**, 72–80.

Deloitte and Touche Management Consultants (1992). "An Assessment of the Economic Benefits of Pest Control in Forestry." Deloitte and Touche Management Consultants, Guelph, Ont.

Delorme, L. D. (1982). Lake Erie oxygen: The prehistoric record. *Can. J. Fish. Aquat. Sci.* **39**, 1021–1029.

Deneke, F. J., McCown, B. H., Coyne, P. I., Rickard, W., and Brown, J. (1975). "Biological Aspects of Terrestrial Oil Spills," Res. Rep. 346. Cold Regions Research and Engineering Laboratory, U.S. Army Corps of Engineers, Hanover, NH.

Dennis, B., Munholland, P. L., and Scott, J. M. (1991). Estimates of growth and extinction parameters for endangered species. *Ecol. Monogr.* **61**, 115–143.

DesGranges, J. L., and Houde, B. (1989). "Effects of Acidity and Other Environmental Parameters on the Distribution of Lacustrine Birds in Quebec," Occas. Pap. 67. Canadian Wildlife Service, Sainte-Foy, Quebec.

DesGranges, J. L., and Hunter, M. L., Jr. (1987). Duckling response to lake acidification. *Trans. North Am. Wildl. Nat. Resour. Conf.*, **52**, 636–644.

Detwiler, R. P., and Hall, C. A. S. (1988). Tropical forests and the global carbon cycle. *Science* **239**, 42–47.

de Zafra, R. L., Jaramillo, M., Parrish, A., Solomon, P., Connor, P., and Barrett, J. (1987). High concentrations of chlorine monoxide at low altitudes in the Antarctic spring atmosphere: Diurnal variation. *Nature (London)* **328**, 408–411.

Diamond, J. M. (1972). Biogeographic kinetics: Estimation of relaxation times for avifaunas of southwest Pacific islands. *Proc. Natl. Acad. Sci. U.S.A.* **69**, 3199–3203.

Diamond, J. M. (1975). The island dilemma: Lessons of modern biogeographic studies for the design of nature reserves. *Biol. Conserv.* **7**, 129–146.

Diamond, J. M. (1982). Man the exterminator. *Nature (London)* **298**, 787–789.

Diamond, J. M. (1984). "Natural" extinctions of isolated populations. *In* "Extinctions" (M. H. Nitecki, ed.), pp. 191–246. Univ. of Chicago Press, Chicago.

Diamond, J. M., and May, R. M. (1976). Island biogeography and the design of nature reserves. *In* "Theoretical Ecology: Principles and Applications" (R. M. May, ed.), pp. 163–186. Saunders, Philadelphia.

Dicks, B. (1977). Changes in the vegetation of an oiled Southampton Water salt marsh. *In* "Recovery and Restoration of Damaged Ecosystems" (J. Cairns, K. L. Dickson, and E. E. Herricks, eds.), pp. 208–240. Univ. Press of Virginia, Charlottesville.

Dickson, J. G., Conner, R. N., and Williamson, J. H. (1983). Snag retention increases bird use of a clear-cut. *J. Wildl. Manage.* **47**, 799–804.

Dillon, P. J., and Rigler, F. H. (1974). The phosphorus-chlorophyll relationship in lakes. *Limnol. Oceanogr.* **19**, 767–773.

Dillon, P. J., and Smith, P. J. (1984). Trace metal and nutrient accumulation in the sediments of lakes near Sudbury, Ontario. *In* "Environmental Impacts of Smelters" (J. O. Nriagu, ed.), pp. 375–416. Wiley, New York.

Dillon, P. J., Jefferies, D. S., Snyder, W., Reid, R., Yan, N. D., Evans, D., Moss, J., and Scheider, W. A. (1977). "Acid Rain in South-central Ontario: Present Conditions and Future Consequences." Ontario Ministry of the Environment, Rexdale.

Dillon, P. J., Yan, N. D., Schieder, W. A., and Conroy, N. (1979). Acidic lakes in Ontario, Canada: Characterization, extent, and responses to base and nutrient additions. *Arch. Hydrobiol., Suppl.* **13**, 317–336.

Dilworth, T. G., Pearce, P. A., and Dobell, J. V. (1974). DDT in New Brunswick woodcocks. *J. Wildl. Manage.* **38**, 331–337.

Dimitriades, B. (1981). The role of natural organics in photochemical air pollution. *J. Air Pollut. Control Assoc.* **31**, 229–235.

Dixit, S. S., Dixit, A. S., and Smol, J. P. (1991). Multivariate environmental inferences based on diatom assemblages from Sudbury (Canada) lakes. *Freshwater Biol.* **26**, 251–266.

Doane, C. C., and McManus, M. L., eds. (1981). "The Gypsy Moth: Research Toward Integrated Pest Management," Tech. Bull. 1584. U.S. Department of Agriculture, Forest Service, Washington, DC.

Dobson, A. P., and Hudson, P. J. (1986). Parasites, disease, and the structure of ecological communities. *TREE* **1**, 11–15.

Dobzhansky, T., Ayala, F. J., Stebbins, G. L., and Valentine, J. W. (1977). "Evolution." Freeman, San Francisco.

Dodds, D. G. (1974). Distribution, habitat, and status of moose in the Atlantic Provinces of Canada and northeastern United States. *Nat. Can.* **101**, 51–65.

Dolan, D. M. (1993). Point source loadings of phosphorus to Lake Erie: 1986–1990. *J. Great Lakes Res.* **19**, 212–223.

Donaghho, W. R. (1950). Observations of some birds of Guadalcanal and Tulagi. *Condor* **52**, 127–132.

Donaldson, D., ed. (1978). "Forestry. Sector Policy Paper." World Bank, Washington, DC.

dos Santos, J. M. (1987). Climate, natural vegetation, and soils in Amazonia: An overview. *In* "The Geophysiology of Amazonia" (R. E. Dickinson, ed.), pp. 25–36. Wiley, New York.

Dost, H., ed. (1973). "Acid Sulphate Soils." Pudoc, Wageningen.

Douglas, R. J. W. (1970). "Geology and Economic Minerals of Canada." Department of Energy, Mines and Resources, Ottawa, Ont.

Douglass, J. E., and Swank, W. T. (1972). "Streamflow Modification Through Management of Eastern Forests," USDA For. Serv. Res. Pap. SE-94. Southeast Forest Experiment Station, Asheville, NC.

Douglass, J. E., and Swank, W. T. (1975). Effects of management practices on water quality and quantity: Coweeta Laboratory, North Carolina. *In* "Municipal Watershed Management," USDA For. Serv. Gen. Tech. Rep. NE-13, pp. 1–13. Northeastern Forest Experiment Station, Broomall, PA.

Dreisinger, B. R. (1967). "The Impact of Sulfur Pollution of Crops and Forests," Conf. Background Pap. A4–2-1. Canadian Council of Resource Ministers, Montreal.

Dreisinger, B. R., and McGovern, P. C. (1971). "Sulfur Dioxide Levels and Vegetation Injury in the Sudbury Area During the 1970 Season, Annual Report." Air Management Branch, Ontario Department of Energy and Resources Management, Sudbury.

Driscoll, C. T., Baker, J. P., Bisogni, J. J., and Schofield, C. L. (1980). Effect of aluminum speciation on fish in dilute acidified waters. *Nature (London)* **284,** 161–164.

Drolet, C. A. (1978). Use of forest clear-cuts by white-tailed deer in southern New Brunswick and central Nova Scotia. *Can. Field Nat.* **92,** 275–282.

Duce, R. A. (1978). Speculations on the budget of particulate and vapour phase non-methane organic carbon in the global troposphere. *Pure Appl. Geophys.* **116,** 244–273.

Dugle, J. R., and Mayoh, K. R. (1984). Responses of 56 naturally-growing shrub taxa to chronic gamma irradiation. *Environ. Exp. Bot.* **24,** 267–276.

Durning, A. T., and Brough, H. B. (1992). Reforming the livestock economy. *In* "State of the World 1992," pp. 66–82. Norton, New York.

Duthie, J. R. (1972a). Detergents: Nutrient considerations and total assessment. *In* "Nutrients and Eutrophication" (G. E. Likens, ed.), pp. 205–216. Am. Soc. Limnol. Oceanogr., Lawrence, KS.

Duthie, J. R. (1972b). Detergent developments and their impact on water quality. *In* "Nutrients in Natural Waters" (H. E. Allen and J. R. Kramer, eds.), pp. 333–352. Wiley, New York.

Dvorak, A. J., and Lewis, B. G. (1978). "Impacts of Coal-fired Power Plants on Fish, Wildlife, and Their Habitats," FWS/OB7–78. U.S. Department of the Interior, Fish and Wildlife Service, Washington, DC.

Dyck, W. J., Messina, M. G., and Hunter, I. R. (1986). Predicting the consequences of forest harvesting on site productivity—A review by the managing agent of IEA/FE project CPC-10. *In* "Predicting the Consequences of Intensive Forest Harvesting on Long-term Producticity" (G. I. Agren, ed.), Rep. No. 26, pp. 9–30. Swedish Univ. Agric. Sci., Uppsala.

Dyer, G. (1991). Environmental warfare in the Gulf. *Ecodecision* **1,** 21–31.

Dyrness, C. T. (1967). "Mass Soil Movements in the H.J. Andrews Experimental Forest," USDA For. Serv. Res. Pap. PNW-42. Pacific Northwest Forest and Range Experiment Station, Portland, OR.

Earle, S. (1992). Assessing the damage one year later. *Nat. Geogr.* **179**(2), 122–134.

Eaton, J. S., Likens, G. E., and Bormann, F. H. (1973).

Throughfall and stemflow chemistry in a northern hardwoods forest. *J. Ecol.* **61,** 495–508.

Eaton, J. S., Likens, G. E., and Bormann, F. H. (1980). Wet and dry deposition of sulfur at Hubbard Brook. *In* "Effects of Acid Precipitation on Terrestrial Ecosystems" (T.C. Hutchinson and M. Havas, eds.), pp. 69–75. Plenum, New York.

Ebregt, A., and Boldewijn, J. M. A. M. (1977). Influence of heavy metals in spruce forest soil on amylase activity, CO_2 evolution from starch, and soil respiration. *Plant Soil* **47,** 137–148.

Edminster, F. C. (1947). "The Ruffed Grouse: Its Life History, Ecology, and Management." Macmillan, New York.

Edmonds, E. J. (1988). Population status, distribution, and movements of woodland caribou in west central Alberta. *Can. J. Zool.* **66,** 817–826.

Edmonds, J. A., and Reilly, J. M. (1985). Future global energy and carbon dioxide emissions. *In* "Atmospheric Carbon Dioxide and the Global Carbon Cycle," DOE/ER-0239, pp. 215–245. U.S. Department of Energy, Washington, DC.

Edmondson, W. T. (1972). Nutrients and phytoplankton in Lake Washington. *In* "Nutrients and Eutrophication" (G. E. Likens, ed.), pp. 172–188. Am. Soc. Limnol. Oceanogr., Lawrence, KS.

Edmondson, W. T. (1977). Recovery of Lake Washington from eutrophication. *In* "Recovery and Restoration of Damaged Ecosystems" (J. Cairns, K. L. Dickson, and E. E. Herricks, eds.), pp. 102–110. Univ. Press of Virginia, Charlottesville.

Edmondson, W. T., and Lehman, J. T. (1981). The effect of changes in the nutrient income on the condition of Lake Washington. *Limnol. Oceanogr.* **26,** 1–29.

Edmondson, W. T., and Litt, A. H. (1982). *Daphnia* in Lake Washington. *Limnol. Oceanogr.* **27,** 272–293.

Edwards, C. A. (1975). "Persistent Pesticides in the Environment." CRC Press, Cleveland, OH.

Ehrlich, A. H. (1985). The human population: Size and dynamics. *Am. Zool.* **25,** 395–406.

Ehrlich, P. R. (1983). Genetics and the extinction of butterfly populations. *In* "Genetics and Conservation" (C. M. Schonewald-Cox, S. M. Chambers, B. MacBryde, and W. L. Thomas, eds.), pp. 152–163. Benjamin/Cummings, Menlo Park, CA.

Ehrlich, P. R. (1988). The loss of diversity. Causes and consequences. *In* "Biodiversity" (E. O. Wilson, ed.), pp. 21–27. National Academy Press, Washington, DC.

Ehrlich, P. R. (1989). Facing the habitat crisis. *BioScience* **39,** 480–482.

Ehrlich, P. R. (1992). One ecologist's opinion on the so-called Stanford scandals and social responsibility. *BioScience* **42,** 702–705.

Ehrlich, P. R., and Ehrlich, A. (1981). "Extinction. The Causes and Consequences of the Disappearance of Species." Ballantine, New York.

Ehrlich, P. R., and Wilson, E. O. (1991). Biodiversity studies: Science and policy. *Science* **253**, 758–762.

Ehrlich, P. R., Breedlove, D. E., Brussard, P. F., and Sharpe, M. A. (1972). Weather and the "regulation" of subalpine populations. *Ecology* **53**, 243–247.

Ehrlich, P. R., Ehrlich, A. H., and Holdren, J. P. (1977). "Ecoscience. Population, Resources, Environment." Freeman, San Francisco.

Eide, S. H., Miller, S. D., and Chihuly, M. A. (1986). Oil pipeline crossing sites utilized in winter by moose, *Alces alces,* and caribou, *Rangifer tarandus,* in southcentral Alaska. *Can. Field Nat.* **100**, 197–207.

Eidt, D. C. (1975). The effect of fenitrothion from large-scale forest spraying on benthos in New Brunswick headwater streams. *Can. Entomol.* **107**, 743–760.

Eidt, D. C. (1977). "Effects of Fenitrothion on Benthos in the Nashwaak Project Study Streams in 1976," Inf. Rep. M-X-70. Maritimes Forest Research Centre, Fredericton, New Brunswick.

Eidt, D. C. (1985). Toxicity of *Bacillus thuringiensis* var. *kurstaki* to aquatic insects. *Can. Entomol.* **117**, 829–837.

Eidt, D. C., and Mallet, V. N. (1986). Accumulation and persistence of fenitrothion in needles of balsam fir and possible effects on abundance of *Neodiprion abietis* (Harris) (Hymenoptera: Diprionidae). *Can. Entomol.* **118**, 69–77.

Eidt, D. C., and Pearce, P. A. (1986). The biological consequences of lingering fenitrothion residues in conifer foliage—a synthesis. *For. Chron.* **62**, 246–249.

Eidt, D. C., and Weaver, C. A. A. (1986). "Frequency of Forest Respraying and Use of *B.t.* in New Brunswick, 1975–1986," Inf. Rep. M-X-158. Maritimes Forest Research Centre, Fredericton, New Brunswick.

Eisenreich, S. J., Thornton, J. D., Munger, J. W., and Gorham, E. (1980). Impact of land-use on the chemical composition of rain and snow in northern Minnesota. *In* "Ecological Impact of Acid Precipitation" (C. D. Drablos and A. Tollan, eds.), pp. 110–111. SNSF Project, Oslo, Norway.

Eisenreich, S. J., Metzer, N. A., Urban, N. R., and Robbins, J. A. (1986). Responses of atmospheric lead to decreased use of lead in gasoline. *Environ. Sci. Technol.* **20**, 171–174.

El-Badry, M. A. (1992). World population change: A long-range perspective. *Ambio* **21**, 18–23.

Eldredge, N., and Gould, S. J. (1972). Punctuated equilibria: An alternative to phyletic gradualism. *In* "Models in Paleobiology" (T. J. M. Schopf, ed.), pp. 82–115. Freeman, Cooper, San Francisco.

Ellenberg, H. (1986). The effects of environmental factors and use alternatives upon the species diversity and regeneration of tropical rain forests. *Appl. Geogr. Dev.* **28**, 20–36.

Ellis, R. (1991). "Men and Whales." Knopf, New York.

Elsom, D. (1987). "Atmospheric Pollution." Blackwell, Oxford.

Endler, J. A. (1982). Pleistocene forest refuges: Fact or fancy. *In* "Biological Diversification in the Tropics" (G.T. Prance, ed.), pp. 641–657. Columbia Univ. Press, New York.

Enge, K. M., and Marion, W. R. (1986). Effects of clearcutting and site preparation on herpetofauna of a north Florida flatwoods. *For. Ecol. Manage.* **14**, 177–192.

Engelhardt, F. R., ed. (1985). "Petroleum Effects in the Arctic Environment." Am. Elsevier, New York.

Enger, L., and Hogström, U. (1979). Dispersion and wet deposition of sulfur from a power plant. *Atmos. Environ.* **13**, 797–810.

Ennis, T., and Caldwell, E. T. N. (1991). Bpruce budworm, chemical and biological control. *In* "Tortricid Pests, Their Biology, Natural Enemies, and Control" (L. P. S. van der Geest and H.H. Evenhuis, eds.), pp. 621–641. Elsevier, Amsterdam.

Enoch, H. Z., and Kimball, B. A. (1986). "Carbon Dioxide Enrichment of Greenhouse Crops." CRC Press, Boca Raton, FL.

Entomological Society of America (ESA) (1980). "Pesticide Handbook." ESA, College Park, MD.

Environmental and Social Systems Analysis Ltd. (ESSA) (1982). "Review and Evaluation of Adaptive Environmental Assessment and Management." Environment Canada, Ottawa, Ont.

Erickson, J. D., Mulinare, J., McClain, P. W., Fitch, T. G., James, L. M., McClearn, A. B., and Adams, M. J. (1984). "Vietman Veterans' Risks for Fathering Babies with Birth Defects." U.S. Department of Health and Human Services, Centers for Disease Control, Atlanta, GA.

Eriksson, F., Harnström, E., Mossberg, P., and Nyberg, P. (1983). Ecological effects of lime treatment of acidified lakes and rivers in Sweden. *Hydrobiologia* **101**, 145–164.

Eriksson, M. O. G. (1984). Acidification of lakes: Effects on waterbirds in Sweden. *Ambio* **13**, 260–262.

Eriksson, O. (1976). Silvicultural practices and reindeer grazing in northern Sweden. *Ecol. Bull.* **21**, 107–120.

Erwin, D. A. (1990). The end-Permian mass extinction. *Annu. Rev. Ecol. Syst.* **21**, 69–91.

Erwin, T. L. (1982). Tropical forests: Their richness in Coleoptera and other arthropod species. *Coleopt. Bull.* **36**, 74–75.

Erwin, T. L. (1983a). Beetles and other insects of tropical forest canopies at Manaus, Brazil, sampled by insecticidal fogging. *In* "Tropical Rain Forest: Ecology and Management" (S. L. Sutton, T. C. Whitmore, and A. C. Chadwick, eds.), pp. 59–75. Blackwell, Boston.

Erwin, T. L. (1983b). Tropical forest canopies: The last biotic frontier. *Bull. Entomol. Soc. Amer.* **29**(1), 14–19.

Erwin, T. L. (1988). The tropical forest canopy. The heart of biotic diversity. *In* "Biodiversity" (E. O. Wilson, ed.), pp. 123–129. National Academy Press, Washington, DC.

Erwin, T. L. (1991). How many species are there? Revisited. *Conserv. Biol.* **5**, 330–333.

Eto, M. (1974). "Organophosphorus Pesticides: Organic and Biological Chemistry." CRC Press, Cleveland, OH.

Euler, D. (1978). "Vegetation Management for Wildlife in Ontario." Ministry of Natural Resources, Toronto, Ont.

Evans, C. S., Asher, C. J., and Johnson, C. M. (1968). Isolation of dimethyl diselenide and other volatile selenium compounds from *Astragalus racemosus* (Pursh.). *Aust. J. Biol. Sci.* **21**, 13–20.

Evans, K. E. (1978). Forest management opportunities for songbirds. *Trans. North Am. Wildl. Nat. Resour. Conf.* **43**, 69–77.

Evans, K. E., and Conner, R. N. (1979). Snag management. *In* "Management of North Central and Northeastern Forests for Non-game Birds," USDA For. Serv. Gen. Tech. Rep. NC-51, pp. 214–225. North Central Forest Experiment Station, St. Paul, MN.

Evans, L. S. (1982). Effects of acidity in precipitation on terrestrial vegetation. *Water, Air, Soil Pollut.* **18**, 395–403.

Evans, L. S. (1986). Proposed mechanisms of initial injury-causing apical dieback in red spruce at high elevation in eastern North America. *Can. J. For. Res.* **16**, 1113–1116.

Evans, M. S. (1992). Historic changes in Lake Michigan zooplankton community structure: The 1960s revisited with implications for top-down control. *Can. J. Fish. Aquat. Sci.* **49**, 1734–1749.

Evans, P. G. H., and Nettleship, D. M. (1985). Conservation of the Atlantic Alcidae. *In* "The Atlantic Alcidae" (D. N. Nettleship and T. R. Birkhead, eds.), pp. 427–488. Academic Press, New York.

Exxon (1990). "How Tiny Organisms Helped Clean Prince William Sound." Exxon Co., Houston, TX.

Fabacher, D. L., Schmitt, C. J., Besser, J. M., Baumann, P. C., and Mac, M. J. (1986). Great Lakes fish—neoplasia investigations. *Toxicol. Chem. Aquat. Life: Res. Manage., 1986*, p. 87.

Fairchild, W. L., Ernst, W. R., and Mallet, V. N. (1989). Fenitrothion effects on aquatic organisms. *In* "Environmental Effects of Fenitrothion Use in Forestry," pp. 109–166. Conservation and Protection, Environment Canada, Dartmouth, Nova Scotia.

Fairfax, J. A. W., and Lepp, N. W. (1975). Effect of simulated "acid rain" on cation loss from leaves. *Nature (London)* **255**, 324–325.

Farman, J. C., Gradiner, B. G., and Shanklin, J. D. (1985). Large losses of total ozone in Antarctica reveals seasonal ClO_x/NO_x interaction. *Nature (London)* **315**, 207–210.

Farmer, C. B., Toon, G. C., Schaper, P. W., Blavien, J. F., and Lowes, L. L. (1987). Stratospheric trace gases in the spring 1986 Antarctic atmosphere. *Nature (London)* **329**, 126–130.

Farnsworth, N. R. (1988). Screening plants for new medicines. *In* "Biodiversity" (E. O. Wilson, ed.), pp. 83–97. National Academy Press, Washington, DC.

Farrell, E. P., Nilsson, I., Tamm, C. O., and Wiklander, G. (1980). Effects of artificial acidification with sulfuric acid on soil chemistry in a Scots pine forest. *In* "Ecological Impact of Acid Precipitation" (D. Drablos and A. Tollan, eds.), pp. 186–187. SNSF Project, Oslo, Norway.

Fay, J. (1978). Hawaii: Extinction unmerciful. *Garden (London)* **103**(7), 22–27.

Feare, C. J. (1991). Control of pest bird populations. *In* "Bird Population Studies" (C.M. Perrins, J.D. Lebreton, and G.

J. M. Hirons, eds.), pp. 463–478. Oxford University Press, Oxford, UK.

Federal Environmental Assessment and Review Office (FEARO) (1990). "Northern Diseased Bison. Report of the Environmental Assessment Panel." FEARO, Ottawa, Ont.

Feduccia, A. (1985). "Catesby's Birds of Colonial America." Univ. of North Carolina Press, Chapel Hill.

Fee, E. J. (1979). A relation between lake morphometry and primary productivity and its use in interpreting whole-lake eutrophication experiments. *Limnol. Oceanogr.* **24**, 401–416.

Feller, M. C. (1988). Relationships between fuel properties and slashburning-induced fire losses. *For. Sci.* **34**, 998–1015.

Feller, M. C. (1989). Estimation of nutrient loss to the atmosphere from slashburns in British Columbia. *Conf. Fire For. Meteorol., 10th*, pp. 126–135.

Feng, J. C., and Thompson, D. G. (1990). Fate of glyphosate in a Canadian forest watershed. 2. Persistence in foliage and soils. *J. Agric. Food Chem.* **38**, 1118–1125.

Feng, J. C., Thompson, D. G., and Reynolds, P. E. (1989). Fate of glyphosate in a Canadian forest stream. *In* "Proceedings of the Carnation Creek Workshop," pp. 45–64. FRDA Rep. 063, Forest Pest Management Institute, Sault Ste Marie, Ont.

Feng, J. C., Thompson, D. G., and Reynolds, P. E. (1990). Fate of glyphosate in a Canadian forest watershed. 1. Aquatic residues and off-target deposit assessment. *J. Agric. Food Chem.* **38**, 1110-1118.

Ferman, M. A., Wolff, G. T., and Kelly, N. A. (1981). The nature and sources of haze in the Shenandoah Valley/Blue Ridge Mountains area. *J. Air Pollut. Control Assoc.* **31**, 1074–1082.

Fimreite, N. (1970). Mercury uses in Canada and their possible hazards as sources of mercury contamination. *Environ. Pollut.* **1**, 119–131.

Fimreite, N., Feif, R. W., and Keith, J. A. (1970). Mercury contamination of Canadian prairie seed eaters and their avian predators. *Can. Field Nat.* **84**, 269–276.

Finch, D. M. (1991). "Population Ecology, Habitat Requirements, and Conservation Status of Neotropical Migrating Birds," USDA For. Serv. Gen. Tech. Rep. RM-205. Rocky Mountain Forest and Range Experiment Station, Fort Collins, CO.

Findlay, D. L., and Kasian, S. E. M. (1986). Phytoplanklton community responses to acidification of Lake 223, Experimental Lakes Area, northwestern Ontario. *Water, Air, Soil Pollut.* **30**, 719–726.

Findlay, D. L., and Kasian, S. E. M. (1987). Phytoplanklton community responses to nutrient addition in Lake 226, Experimental Lakes Area, northwestern Ontario. *Can. J. Fish. Aquat. Sci.* **44**, Suppl., 35–46.

Findlay, D. L., and Kling, H. J. (1975). "Seasonal successions of phytoplankton in seven lake-basins in the Experimental Lakes Area, northwestern Ontario, following artificial eutrophication," Fish. Mar. Serv., Tech. Rep. No. 513. Environment Canada, Freshwater Institute, Winnipeg, Manitoba.

Findlay, D. L., and Saesura, G. (1980). "Effects on phytoplankton biomass, succession, and composition in Lake 223 as a result of lowering pH levels from 7.0 to 5.6. Data from 1974 to 1979," Fish. Mar. Serv., Manuscr. Rep. No. 1585. Department of Fisheries and Oceans, Winnipeg, Manitoba.

Findlay, D. L., Hecky, R. E., Hendzel, L. L., Stainton, M. P., and Regehr, G. W. (1993). The relationship between nitrogen fixation and heterocyst abundance and its relevance to the nitrogen budget of Lake 227. *Can. J. Fish. Aquat. Sci.* (submitted for publication).

Finkelstein, K., and Gundlach, E. R. (1981). Method of estimating spilled oil quantity on the shoreline. *Environ. Sci. Technol.* **15**, 545–549.

Fisher, H. I., and Baldwin, P. H. (1946). War and the birds of Midway Atoll. *Condor* **48**, 3–15.

Fisher, J., Simon, N., and Vincent, J. (1969). "Wildlife in Danger." Viking Press, New York.

Fisher, K. C., and Elson, P. F. (1950). The selected temperature of the Atlantic salmon and speckled trout and the effect of temperature on the response to electrical stimulus. *Physiol. Zool.* **1**, 27–34.

Fishhoff, D. A., Bowdish, K. S., Perlak, F. J., Morrone, P. G., McCormick, S. M., Niedermeyer, J. G., Dean, D. A., Kusmo-Kretzmen, K., Mayer, E. J., Rochester, D. E., Rogers, S. G., and Fraley, R. J. (1987). Insect tolerant transgenic tomato plants. *Bio Technology* **5**, 807–814.

Fitter, R. (1968). "Vanishing Wild Animals of the World." Kaye & Ward, London.

Fitter, R. (1974). Most endangered mammals: An action program. *Oryx* **12**, 436–449.

Flannigan, M. D., and Van Wagner, C. E. (1991). Climate change and wildfire in Canada. *Can. J. For. Res.* **21**, 66–72.

Fletcher, J. L., and Busnel, R. G., eds. (1978). "Effects of Noise on Wildlife." Academic Press, New York.

Flick, W. A., Schofield, C. L., and Webster, D. A. (1982). Remedial actions for interim maintenance of fish stocks in acidified water. *Acid Rain/Fish., Proc. Int. Symp., 1981,* pp. 287–306.

Flint, M. L., and van den Bosch, R. (1983). "Introduction to Integrated Pest Management." Plenum, New York.

Floc'h, J.-Y., and Diouris, M. (1980). Initial effects of Amoco Cadiz oil on intertidal algae. *Ambio* **9**, 284–286.

Flower, R. J., and Battarbee, R. W. (1983). Diatom evidence for recent acidification of two Scottish lochs. *Nature (London)* **305**, 130–133.

Flower, R. J., Battarbee, R. W., and Appleby, P. G. (1987). The recent palaeolimnology of acid lakes in Galloway, southwest Scotland: Diatom analysis, pH trends and the role of afforestation. *J. Ecol.* **75**, 797–824.

Flower, R. J., Battarbee, R. W., Natkanski, J., Rippey, B., and Appleby, P. G. (1988). The recent acidification of a large Scottish loch located partly within a national nature reserve and site of special scientific interest. *J. Appl. Ecol.* **25**, 715–724.

Folkeson, L. (1984). Deterioration of the moss and lichen vegetation in a forest polluted by heavy metals. *Ambio* **13**, 37–39.

Food and Agriculture Organization (FAO) (1982). "World Forest Products Demand and Supply 1990 and 2000." FAO, Rome, Italy.

Food and Agriculture Organization (FAO) (1986). "Yearbook of Forest Products. 1973–1984," For. Ser. 19. FAO, Rome, Italy.

Ford, E. D., and Newbould, P. J. (1977). The biomass and production of ground vegetation and its relation to tree cover through a deciduous woodland cycle. *J. Ecol.* **65**, 201–212.

Forestry Canada (1992a). "Selected Forestry Statistics, Canada, 1991," Inf. Rep. E-X-46. Policy and Economics Directorate, Forestry Canada, Ottawa, Ont..

Forestry Canada (1992b). "Compendium of Canadian Forestry Statistics." National Forestry Database Program, Forestry Canada, Ottawa, Ont.

Forgeron, F. D. (1971). Soil geochemistry in the Canadian Shield. *CIM Bull.* **64**(715), 37–42.

Forman, R. T. T., and Godron, M. (1986). "Landscape Ecology." Wiley, New York.

Forman, R. T. T., Galli, A. E., and Leck, C. F. (1976). Forest size and avian diversity in New Jersey woodlots with some land use implications. *Oecologia* **26**, 1–8.

Forsyth, D. J., and Martin, P. A. (1993). Effects of fenitrothion on survival, behaviour, and brain cholinesterase activity of white-throated sparrows (*Zonotrichia albicollus*). *Environ. Contam. Toxicol.* **12**, 91–103.

Fortescue, J. A. C. (1980). "Environmental Geochemistry: A Holistic Approach." Springer-Verlag, New York.

Fosberg, F. R. (1959). Plants and fall-out. *Nature (London)* **183**, 1448.

Foster, M. S., and Holmes, R. W. (1977). The Santa Barbara oil spill: An ecological disaster? *In* "Recovery and Restoration of Damaged Ecosystems" (J. Cairns, K. L. Dickson, and E. E. Herricks, eds.), pp. 166–190. Univ. Press of Virginia, Charlottesville.

Foster, N. W. (1983). Acid precipitation and soil solution chemistry within a maple-birch forest in Canada. *For. Ecol. Manage.* **12**, 215–231.

Foster, N. W., and Morrison, I. K. (1983). "Soil Fertility, Fertilization, and Growth of Canadian Forests," Inf. Rep. O-X-353. Great Lakes Forest Research Centre, Sault Ste. Marie, Ont.

Fowells, H. A., comp. (1965). "Silvics of Forest Trees in the United States," Agric. Handb. No. 271. U.S. Department of Agriculture, Forest Service, Washington, DC.

Fowler, D. (1980a). Wet and dry deposition of sulfur and nitrogen compounds from the atmosphere. *In* "Effects of Acid Precipitation on Terrestrial Ecosystems" (T. C. Hutchinson and M. Havas, eds.), pp. 9–28. Plenum, New York.

Fowler, D. (1980b). Removal of sulfur and nitrogen compounds from the atmosphere in rain and by dry deposition. *In* "Ecological Impact of Acid Precipitation" (D. Drablos and A. Tollan, eds.), pp. 22–32. SNSF Project, Oslo, Norway.

Fowler, D. P., and Morris, R. W. (1977). Genetic diversity in red pine: Evidence for low genetic heterozygosity. *Can. J. For. Res.* **7**, 343–347.

Fox, D. L. (1986). The transformation of pollutants. *In* "Air Pollution" (A. C. Stern, ed.), 3rd ed., Vol. 6, pp. 61–94. Academic Press, New York.

Fox, G. A., Collins, B., Hayakawa, E., Weseloh, D. V., Ludwig, J. P., Kubiak, T. B., and Erdman, T. C. (1991). Reproductive outcomes in colonial fish-eating birds: A biomarker for developmental toxicants in Great Lakes food chanins. *J. Great Lakes Res.* **17,** 158–167.

Foy, C. D., Chaney, R. L., and White, M. C. (1978). The physiology of metal toxicity in plants. *Annu. Rev. Plant Physiol.* **29,** 511–566.

France, R. L. (1987). Reproductive impairment of the crayfish *Orconectes virilis* in response to acidification of Lake 223. *Can. J. Fish. Aquat. Sci.* **44,** Suppl. 1, 97–106.

France, R. L., and Collins, N. C. (1993). Extirpation of crayfish in a lake affected by long-range anthropogenic acidification. *Conserv. Biol.* **7,** 184–188.

France, R. L., and Graham, L. (1985). Increased microsporidian parasitism of the crayfish *Oronectes virilis* in an experimentally acidified lake. *Water, Air, Soil Pollut.* **26,** 129–136.

Frank, R., Ishida, K., and Suda, P. (1976a). Metals in agricultural soils in Ontario. *Can. J. Soil Sci.* **56,** 181–196.

Frank, R., Braun, H. E., Ishida, K., and Suda, P. (1976b). Persistent organic and inorganic pesticide residues in orchard soils and vineyards of southern Ontario. *Can. J. Soil Sci.* **56,** 463–484.

Frank, R., Braun, H. E., Van Hove Holdrinet, M., Sirans, S. J., and Ripley, B. D. (1982). Agriculture and water quality in the Canadian Great Lakes basin. V. Pesticide use in 11 agricultural watersheds and presence in streamwater, 1975–1977. *J. Environ. Qual.* **11,** 497–505.

Franklin, I. R. (1980). Evolutionary change in small populations. *In* "Conservation Biology" (M. E. Soule and B. A. Wilcox, eds.), pp. 135–149. Sinauer Assoc., Sunderland, MA.

Franklin, J. F. (1989). Importance and justification of long-term studies in ecology. *In* "Long-term Studies in Ecology. Approaches and Alternatives" (G. E. Likens, ed.), pp. 3–19. Springer-Verlag, New York.

Franklin, J. F., and Spies, T. A. (1991a). Ecological definitions of old-growth Douglas-fir forests. *In* "Wildlife and Vegetation of Unmanaged Douglas-fir Forests," USDA For. Serv. Gen. Tech. Rep. PNW-285, pp. 61–69. Pacific Northwest Forest and Range Experiment Station, Portland, OR.

Franklin, J. F., and Spies, T. A. (1991b). Composition, function, and structure of old-growth Douglas-fir forests. *In* "Wildlife and Vegetation of Unmanaged Douglas-fir Forests," USDA For. Serv. Gen. Tech. Rep. PNW-285, pp. 71–80. Pacific Northwest Forest and Range Experiment Station, Portland, OR.

Franklin, J. F., Bledsoe, C. S., and Callahan, J. T. (1990). Contributions of the Long-term Ecological Research Program. *BioScience* **40,** 509–524.

Franzreb, K. E., and Ohmart, R. D. (1978). The effects of timber harvesting on breeding birds in a mixed-coniferous forest. *Condor* **80,** 431–444.

Freda, J. (1986). The influence of acidic pond water on amphibians: A review. *Water, Air, Soil Pollut.* **30,** 439–450.

Freda, J., Sadinski, W. J., and Dunson, W. A. (1991). Long term monitoring of amphibians populations with respect to the effects of acidic deposition. *Water, Air, Soil Pollut.* **55,** 445–462.

Fredriksen, R. L., Moore, D. G., and Norris, L. A. (1975). The impact of timber harvest, fertilization, and herbicide treatment on streamwater quality in western Oregon and Washington. *In* "Forest Soils and Forest Land Management" (B. Bernier and C. H. Winget, eds.), pp. 283–313. Laval Univ. Press, Quebec.

Freed, L. A., Conant, S., and Fleischer, R. C. (1987). Evolutionary ecology and radiation of Hawaiian passerine birds. *TREE* **2,** 196–203.

Freedman, B. (1981a). "Intensive Forest Harvest—A Review of Nutrient Budget Considerations," Inf. Rep. M-X-121. Maritimes Forest Research Centre, Fredericton, New Brunswick.

Freedman, B. (1981b). Trace elements and organics associated with coal combustion in power plants: Emissions and environmental impacts. *Coal: Phoenix '80s [Eighties], Proc. CIC Coal Symp., 64th, 1981,* pp. 616–631.

Freedman, B. (1982). "An Overview of the Environmental Impacts of Forestry, with Particular Reference to the Atlantic Provinces." Institute for Resource and Environmental Studies, Dalhousie University, Halifax, Nova Scotia.

Freedman, B. (1991). "Vanadium and Nickel in the Vicinity of Oil-fired Power Plants: An Interpretive Literature Review." P. Lane & Associates, Halifax, Nova Scotia.

Freedman, B. (1992). Environmental stress and the management of ecological reserves. *In* "International Conference on Science and the Management of Protected Areas" (J. H. M. Willison, S. Bondrup-Nielsen, C. Drysdale, T. B. Herman, N. W. P. Munro, and T. L. Pollock, eds.), pp. 383–388. Elsevier, Amsterdam.

Freedman, B., and Clair, T. A. (1987). Ion mass balances and seasonal fluxes from four acidic brownwater streams in Nova Scotia. *Can. J. Fish. Aquat. Sci.* **44,** 538–548.

Freedman, B., and Hutchinson, T. C. (1976). Physical and biological effects of experimental crude oil spills on low arctic tundra in the vicinity of Tuktoyaktuk, N.W.T., Canada. *Can. J. Bot.* **54,** 2219–2230.

Freedman, B., and Hutchinson, T. C. (1980a). Pollutant inputs from the atmosphere and accumulations in soils and vegetation near a nickel-copper smelter at Sudbury, Ontario, Canada. *Can. J. Bot.* **58,** 108–132.

Freedman, B., and Hutchinson, T. C. (1980b). Effects of smelter pollutants on forest leaf litter decomposition near a nickel-copper smelter at Sudbury, Ontario. *Can. J. Bot.* **58,** 1722–1736.

Freedman, B., and Hutchinson, T. C. (1980c). Long-term effects of smelter pollution at Sudbury, Ontario, on forest community composition. *Can. J. Bot.* **58,** 2123–2140.

Freedman, B., and Hutchinson, T. C. (1981). Sources of metal and elemental contamination of terrestrial environments. *In*

"Effects of Heavy Metal Pollution on Plants," Vol. 2, pp. 35–94. Appl. Sci. Publ., London.

Freedman, B., and Hutchinson, T. C. (1986). Aluminum in terrestrial ecosystems. *In* "Aluminum in the Canadian Environment," pp. 129–152. National Research Council of Canada, Ottawa, Ont.

Freedman, B., and Prager, U. (1986). Ambient bulk deposition, throughfall, and stemflow in a variety of forest stands in Nova Scotia. *Can. J. For. Res.* **16,** 854–860.

Freedman, B., and Svoboda, J. (1982). Populations of breeding birds at Alexandra Fiord, Ellesmere Island, Northwest Territories, compared with other arctic localities. *Can. Field Nat.* **96,** 56–60.

Freedman, B., Beauchamp, C., McLaren, I. A., and Tingley, S. I. (1981a). Forestry management practices and populations of breeding birds in Nova Scotia. *Can. Field Nat.* **95,** 307–311.

Freedman, B., Morash, R., and Hanson, A. J. (1981b). Biomass and nutrient removals by conventional and whole-tree clearcutting of a red–balsam fir stand in central Nova Scotia. *Can. J. For. Res.* **11,** 249–257.

Freedman B., Duinker, P. N., Morash, R., and Prager, U. (1982). "Standing Crops of Biomass and Nutrients in a Variety of Forest Stands in Central Nova Scotia," Inf. Rep. M-X-134. Maritimes Forest Research Center, Fredericton, New Brunswick.

Freedman, B., Svoboda, J., Labine, C., Muc, M., Henry, G., Nams, M., Stewart, J., and Woodley, E. (1983). Physical and ecological characteristics of Alexandra Fiord, a high arctic oasis on Ellesmere Island. *Int. Conf. Permafrost, 4th,* pp. 301–306.

Freedman, B., Stewart, C., and Prager, U. (1985). "Patterns of Water Chemistry of Four Drainage Basins in Central Nova Scotia," Tech. Rep. IWD-AR-WQB-85-93. Water Quality Branch, Inland Waters Directorate, Environment Canada, Moncton, New Brunswick.

Freedman, B., Duinker, P. N., and Morash, R. (1986). Biomass and nutrients in Nova Scotia forests, and implications of intensive harvesting for future site productivity. *For. Ecol. Manage.* **15,** 103–127.

Freedman, B., Poirier, A. M., and Scott, F. (1988). Effects of 2,4,5-T on breeding birds, small mammals, and their habitat in conifer clearcuts in Nova Scotia. *Can. Field Nat.* **102,** 6–11.

Freedman, B., Kerekes, J., and Howell, G. (1989). Patterns of water chemistry among twenty-seven oligotrophic lakes in Kejimkujik National Park, Nova Scotia. *Water, Air, Soil Pollut.* **46,** 119–130.

Freedman, B., Zobens, V., and Hutchinson, T. C. (1990). Intense, natural pollution affects arctic tundra vegetation at the Smoking Hills, Canada. *Ecology* **71,** 492–503.

Freedman, B., Meth, F., and Hickman, C. (1992). Temperate forest ad a carvon-storage reservoir for carbon dioxide emitted by coal-fires generating stations. A case study for New Brunswick, Canada. *For. Ecol. Manage.* **15,** 103–127.

Freedman, B., Maass, W., and Parfenov, P. (1993a). Assessment of risks to *Drosera filiformis* of a proposal to mine fuel-peat from its habitat in Shelburne County, Nova Scotia. *Can. Field Nat.* **106,** 534–542.

Freedman, B., Morash, R., and MacKinnon, D. (1993b). Short-term changes in vegetation after the silvicultural spraying of glyphosate herbicide onto regenerating clearcuts in Nova Scotia, Canada. *Can. J. For. Res.* (in press).

Freedman, B., Staicer, C., and Shackell, N. (1993c). "Recommendations for a National Ecological-Monitoring Program," Occas. Pap. Ser. No. 2. State of the Environment Reporting Organization, Environment Canada, Ottawa, Ont.

Freedman, B., Svoboda, J., and Henry, G. (1993d). Alexandra Fiord—An ecological oasis in the polar desert. *In* "Ecology of a Polar Oasis: Alexandra Fiord, Ellesmere Island, Canada." Captus Univ. Press, Toronto, Ont.

Freedman, B., Woodley, S., and Loo, J. (1994). Forestry practices and biodiversity, with particular reference to the Maritime Provinces of eastern Canada. *Environ. Rev.* **2,** 33–77.

Freeman, M. M. R. (1985). Effects of petroleum activities on the ecology of Arctic man. *In* "Petroleum Effects in the Arctic Environment" (F. R. Engelhardt, ed.), pp. 245–273. Am. Elsevier, New York.

Freemark, K. (1988). Landscape ecology of forest birds in the northeast. *In* "Is Forest Fragmentation a Management Issue in the Northeast?" pp. 7–12. USDA For. Serv., Gen. Tech. Rep. NE-140. Northeastern Forest Experiment Station, Broomall, PA.

Freemark, K., and Boutin, C. (1993). Impacts of agricultural herbicide use on terrestrial wildlife: A review. *Agric. Ecosyst. Environ.* (submitted for publication).

Freemark, K., and Merriam, H. G. (1986). Importance of area and habitat heterogeneity to bird assemblages in temperate forest fragments. *Biol. Conserv.* **36,** 115–141.

Freeth, S. J., and Kay, R. L. F. (1987). The Lake Nyos gas disaster. *Nature (London)* **325,** 104–105.

Freidel, F. (1964). "Over There. The Story of America's First Great Overseas Crusade." Little, Brown, Boston.

Friedland, A. J., and Battles, J. J. (1987). Red spruce (*Picea rubens*) decline in the northeastern United States: Review and recent data from Whiteface Mountain. *In* "Proceedings of the Workshop on Forest Decline and Reproduction: Regional and Global Consequences" (L. Kairiukstis, S. Nilsson, and A. Stroszak, eds.), pp. 287–296. IIASA, Laxenburg, Austria.

Friedland, A. J., and Johnson, A. H. (1985). Lead distribution and fluxes in a high-elevation forest in northern Vermont. *J. Environ. Qual.* **14,** 332–336.

Friedland, A. J., Johnson, A. H., Siccama, T. G., and Mader, D. L. (1984). Trace metal profiles in the forest floor of New England. *Soil Sci. Soc. Am. J.* **48,** 422–425.

Friedland, A. J., Hawley, G. J., and Gregory, R. A. (1985). Investigations of nitrogen as a possible contributor to red spruce (*Picea rubens* Sarg.) decline. *In* "Air Pollutant Effects on Forest Ecosystems," pp. 95–106. Acid Rain Foundation, St. Paul, MN.

Friedland, A. J., Hawley, G. J., and Gregory, R. H. (1988). Red spruce (*Picea rubens* Sarg.) foliar chemistry in northern Vermont and New York, USA. *Plant Soil* **105,** 189–195.

Friedland, A. J., Craig, B. W., Miller, E. K., Herrick, G. T., Siccama, T. G., and Johnson, A. H. (1992). Decreasing lead levels in the forest floor of the northeastern USA. *Ambio* **21,** 400–403.

Friedman, S. T. (1991). Genetic implications of New Forestry practices: Concerns and evidence. Presented at the Vegetation Management Workshop, Redding, CA.

Frink, C. R., and Voigt, G. K. (1977). Potential effects of acid precipitation on soils of the humid temperate zone. *Water, Air, Soil Pollut.* **7,** 371–388.

Fritts, T. H. (1988). "The Brown Tree Snake, *Boiga irregularis,* a Threat to Pacific Islands," Biol. Rep. 88(31). U.S. Fish and Wildlife Service, Washington, DC.

Fritz, L., Quilliam, M. A., Wright, J. L. C., Beale, A. M., and Work, T. M. (1992). An outbreak of domoic acid poisoning attributed to the pennate diatom *Pseudonitzschia australis. J. Phycol.* **28,** 439–442.

Frost, D. V., and Lish, P. M. (1975). Selenium in biology. *Annu. Rev. Pharmacol.* **15,** 259–284.

Fry, D. M. (1992). Point-source and non-point-source problems affecting seabird populations. *In* "Wildlife 2001: Populations," pp. 547–562. Elsevier, New York.

Fry, F. E., Hart, J. S., and Walder, K. F. (1946). Lethal temperature for a sample of young speckled trout. *Ont. Fish. Res. Lab.* **66,** 5–35.

Fyfe, R. W. (1976). Rationale and success of the Canadian Wildlife Service peregrine breeding project. *Can. Field Nat.* **90,** 308–319.

Fyfe, R. W., Temple, S. A., and Cade, T. J. (1976). The 1975 North American peregrine falcon survey. *Can. Field Nat.* **90,** 228–273.

Fyfe, R. W., Banasch, U., Benarides, V., Hilgert de Benarides, N., Luscombe, A., and Sanchez, J. (1990). Organochlorine residues in potential prey of peregrine falcons, *Falco peregrinus,* in Latin America. *Can. Field Nat.* **104,** 285–292.

Gage, S. H., and Miller, C. A. (1978). "A Long-term Bird Census in Spruce Budworm–Prone Balsam Fir Habitats in Northwestern New Brunswick," Inf. Rep. M-X-84. Maritimes Forest Research Centre, Fredericton, New Brunswick.

Galli, A. E., Leck, C. F., and Formann, R. T. T. (1976). Avian distribution patterns in forest islands of different sizes in central New Jersey. **Auk 93,** 356–364.

Gallo, M. A., and Lawryk, N. J. (1991). Organic phosphorus pesticides. *In* "Handbook of Pesticide Toxicology. Vol. 3. Classes of Pesticides" (W. C. Hayes and E. R. Laws, eds.), pp. 917–1124. Academic Press, San Diego.

Galloway, J. N., Likens, G. E., Keene, W. C., and Miller, J. M. (1982a). The composition of precipitation in remote areas of the world. *J. Geophys. Res.* **87,** 8771–8786.

Galloway, J. N., Thornton, J. D., Norton, S. A., Volchok, H. L., and McLean, R. A. N. (1982b). Trace metals in atmospheric deposition: A review and assessment. *Atmos. Environ.* **16,** 1677–1700.

Galloway, J. N., Hendry, G. R., Schofield, C. L., Peters, N. E., and Johannes, A. H. (1987a). Processes and causes of lake acidification during spring snowmelt in the west-central Adirondack Mountains, New York. *Can. J. Fish. Aquat. Sci.* **44,** 1595–1602.

Galloway, J. N., Dianwu, Z., Jiling, X., and Likens, G. E. (1987b). Acid rain: China, United States, and a remote area. *Science* **236,** 1559–1562.

Galston, A. W., and Richards, P. W. (1984). Terrestrial plant ecology and forestry: An overview. *In* "Herbicides in War. The Long-term Ecological and Human Consequences" (A. H. Westing, ed.), pp. 39–42. Taylor & Francis, London.

Galt, J. A., Lehr, W. J., and Payton, D. L. (1991). Fate and transport of the *Exxon Valdez* oil spill. *Environ. Sci. Technol.* **25,** 202–209.

Gammon, R. H., Sundquist, E. T., and Fraser, P. J. (1985). History of carbon dioxide in the atmosphere. *In* "Atmospheric Carbon Dioxide and the Global Carbon Cycle," DOE/ER-0239, pp. 25–62. U.S. Department of Energy, Washington, DC.

Ganning, B., Reish, D. J., and Straughan, D. (1984). Recovery and restoration of rocky shores, sandy beaches, tidal flats, and shallow subtidal bottoms impacted by oil spills. *In* "Restoration of Habitats Impacted by Oil Spills" (J. L. Cairns and A. L. Buikema Jr., eds.), pp. 7–36. Butterworth, Boston.

Ganther, H. E., and Sundi, M. L. (1974). Effect of tuna fish and selenium on the toxicity of methylmercury: A progress report. *J. Food Sci.* **39,** 1–5.

Ganther, H. E., Goudie, C., Sundi, M. L., Kopecky, M. J., Wagner, P., Oh, S., and Hoekstra, W. G. (1972). Selenium: Relation to decreased toxicity of methylmercury added to diets containing tuna. *Science* **175,** 1122–1124.

Gaskin, D. E. (1991). An update of the status of the right whale, *Eubalaena glacialis,* in Canada. *Can. Field Nat.* **105,** 198–205.

Gaskin, D. E., Ishida, K., and Frank, R. (1972). Mercury in harbour porpoises (*Phoecena phoecena*) in the Bay of Fundy region. *J. Fish. Res. Board Can.* **29,** 1644–1646.

Gaston, K. J. (1991). The magnitude of global insect species richness. *Conserv. Biol.* **5,** 283–296.

Gates, D. M. (1962). "Energy Exchange in the Biosphere." Harper & Row, New York.

Gates, D. M. (1985). "Energy and Ecology." Sinauer Assoc., New York.

Gatz, D. F., Barnard, W. R., and Stensland, G. J. (1986). The role of alkaline materials in precipitation chemistry: A brief review of the issues. *Water, Air, Soil Pollut.* **30,** 245–251.

Gaynor, J. D., and Halstead, R. L. (1976). Chemical and plant extractability of metals and plant growth on soils amended with sludge. *Can. J. Soil Sci.* **11**, 1194–1201.

General Accounting Office (GAO) (1981). "Better Data Needed to Determine the Extent to Which Herbicides Should be Used on Forest Lands." GAO, Washington, DC.

Gentry, A. H. (1986). Endemism in tropical vs. temperate plant communities. *In* "Conservation Biology" (M. E. Soule, ed.), pp. 153–181. Sinauer Assoc., Sunderland, MA.

Gentry, A. H. (1988). Tree species of upper Amazonian forests. *Proc. Natl. Acad. Sci. U.S.A.* **85**, 156–159.

George, J. L., Snyder, A. P., and Hanley, G. (1981). The value of the wild-bird products industry. *Trans. North Am. Wildl. Nat. Resour. Conf.* **46**, 463–471.

Geraci, J. R., Anderson, D. M., Timperi, R. J., St. Aubin, D. J., Early, G. A., Prescott, J. H., and Mayo, C. A. (1989). Humpback whales (*Megaptera novaeangliae*) fatally poisoned by dinoflagellate poison. *Can. J. Fish. Aquat. Sci.* **46**, 1895–1898.

Gerardi, M. H., and Grimm, J. K. (1979). "The History, Biology, Damage, and Control of the Gypsy Moth." Associated University Press, London.

Gerrish, G., Mueller-Dombois, D., and Bridges, K. W. (1988). Nutrient limitation and *Metrosideros* forest dieback in Hawaii. *Ecology* **69**, 723–727.

Getter, C. D., Cintron, G., Dicks, B., Lewis, R. R., and Seneca, E. D. (1984). The recovery and restoration of salt marshes and mangroves following an oil spill. *In* "Restoration of Habitats Impacted by Oil Spills" (J. L. Cairns and A. L. Buikema, Jr., eds.), pp. 65–113. Butterworth, Boston.

Ghassemi, M., Quinlivan, S., and Dellarco, M. (1982). Environmental effects of new herbicides for vegetation control in forestry. *Environ. Int.* **7**, 389–401.

Gillis, A. M. (1990). The new forestry. An ecosystem approach to land management. *BioScience* **40**, 558–562.

Gillis, G. F. (1980). Assessment of the effects of insecticide contamination in streams on the behaviour and growth of fish. *In* "Environmental Surveillance in New Brunswick" (I. W. Varty, ed.), pp. 49–50. Department of Forest Resources, University of New Brunswick, Fredericton.

Gilmour, C. C., and Henry, E. A. (1991). Mercury methylation in aquatic systems affected by acid deposition. *Environ. Pollut.* **71**, 131–169.

Gilroy, N. T. (1983). More than just pipes in the ground. *Ambio* **12**, 245–251.

Gizyn, W. I. (1980). The chemistry and environmental impact of the bituminous shale fires at the Smoking Hills, N.W.T. M.Sc. Thesis, Department of Botany, University of Toronto, Toronto, Ont.

Gladstone, H. S. (1919). "Birds and the War." Skeffington & Son, London.

Glass, G. E., Brydges, T. G., and Loucks, O. L., eds. (1981). "Impact Assessment of Airborne Acidic Deposition on the Aquatic Environment of the United States and Canada," EPA-600/10–81–000. U.S. Environmental Protection Agency, Duluth, MN.

Glemarec, M., and Hussenot, E. (1982). Réponses des peuplements subtidaux à la perturbation créé par L'Amoco Cadiz dans les Abers Benoit et Wrac'h. *In* "Ecological Study of the Amoco Cadiz Oil Spill," pp. 191–203. U.S. Department of Commerce, National Oceanic and Atmospheric Administration, Washington, DC.

Glenn, S. M. (1990). Regional analysis of mammal distributions among Canadian parks: Implications for parks planning. *Can. J. Zool.* **68**, 2457–2464.

Gochfeld, M. (1980). Mercury levels in some seabirds of the Humboldt Current, Peru. *Environ. Pollut.* **22**, 197–205.

Goldemberg, J. (1992). Energy, technology, development. *Ambio* **21**, 14–17.

Golding, D. L., and Swanson, R. H. (1986). Snow distribution patterns in clearings and adjacent forest. *Water Resour. Res.* **22**, 1931–1940.

Goldman, C. R. (1961). The contribution of alder trees (*Alnus tenuifolia*) to the primary productivity of Castle Lake, California. *Ecology* **42**, 282–288.

Goldman, C. R. (1965). Micronutrient limiting factors and their detection in natural phytoplankton populations. *In* "Primary Productivity in Aquatic Environments," pp. 123–135. Univ. of California Press, Berkeley.

Goldman, C. R. (1972). The role of minor nutrients in limiting the productivity of aquatic ecosystems. *In* "Nutrients and Eutrophication" (G. E. Likens, ed.), pp. 21–33. Am. Soc. Limnol. Oceanogr., Lawrence, KS.

Goldsmith, J. R. (1986). Effects on human health. *In* "Air Pollution" (A.C. Stern, ed.), 3rd ed., Vol. 6, pp. 391–463. Academic Press, New York.

Goldstein, N. (1993). EPA releases final sludge management rule. *BioCycle* **1**, 59–63.

Goldwater, L. J., and Clarkson, T. W. (1972). Mercury. *In* "Metallic Contaminants and Human Health" (D. H. K. Lee, ed.), pp. 17–56. Academic Press, New York.

Gollop, J. B., Barry, T. W., and Iversen, E. H. (1986). "Eskimo Curlew. A Vanishing Species?" Spec. Publ. No. 17. Saskatchewan Natural History Society, Regina.

Gooders, J. (1983). "Birds That Came Back." Tanager Books, Dover, NH.

Goodison, B. E., Louie, P. Y. T., and Metcalf, J. R. (1986). Snowmelt acidic shock study in south central Ontario. *Water, Air, Soil Pollut.* **31**, 131–138.

Goodland, R., Daly, H., and El Serafy, S., eds. (1991). "Environmentally Sustainable Economic Development. Building on Brundtland," Environ. Work. Pap. No. 46. The World Bank, Environment Department, Washington, DC.

Goodman, G. T., Pitcairn, C. E. R., and Gemmell, R. P. (1973). Ecological factors affecting growth on sites contaminated with heavy metals. *In* "Ecology and Reclamation of Devastated Land" (R. J. Hutnik and G. Davis, eds.), Vol. 2, pp. 149–171. Gordon & Breach, New York.

Goodwin, J. G., and Hungerford, C. R. (1979). "Rodent Population Densities and Food Habits in Arizona Ponderosa Pine Forests," USDA For. Serv. Res. Pap. RM-214. Rocky

Mountain Forest and Range Experiment Station, Fort Collins, CO.

Gordon, A. G. (1981). Impacts of harvesting on nutrient cycling in the boreal mixedwood forest. *In* "Boreal Mixedwood Symposium," COJFRC Symp. Proc. O-P-9, pp. 121–140. Great Lakes Forest Research Centre, Sault Ste. Marie, Ont.

Gordon, A. G., and Gorham, E. (1963). Ecological aspects of air pollution from an iron-sintering plant at Wawa, Ontario. *Can. J. Bot.* **41**, 1063–1078.

Gorham, E. (1976). "Acid Precipitation and Its Influence upon Aquatic Ecosystems: An Overview, USDA For. Serv. Gen. Tech. Rep. NE-23, pp. 425–458. Northeastern Forest Experiment Station, Broomall, PA.

Gorham, E., and Gordon, A. G. (1960a). Some effects of smelter pollution northeast of Falconbridge, Ontario. *Can. J. Bot.* **38**, 307–312.

Gorham, E., and Gordon, A. G. (1960b). The influence of smelter fumes upon the chemical composition of lake waters near Sudbury, Ontario and upon the surrounding vegetation. *Can. J. Bot.* **38**, 477–487.

Gorham, E., and Gordon, A. G. (1963). Some effects of smelter pollution upon aquatic vegetation near Sudbury, Ontario. *Can. J. Bot.* **41**, 371–378.

Gorham, E., Bayley, S. E., and Schindler, D. W. (1984). Ecological effects of acid deposition upon peatlands—a neglected field in "acid rain" research. *Can. J. Fish. Aquat. Sci.* **41**, 1256–1268.

Gosner, K. L., and Black, I. H. (1957). The effects of acidity on the development and hatching of New Jersey frogs. *Ecology* **38**, 256–262.

Gould, S. J. (1989). "Wonderful Life: The Burgess Shale and the Nature of History." Norton, New York.

Gould, S. J. (1993). A special fondness for beetles. *Nat. Hist.* **102**(1), 4–12.

Gould, S. J., and Eldredge, N. (1980). Punctuated equilibria: The tempo and mode of evolution reconsidered. *Paleobiology* **3**, 115–151.

Graber, R. E., and Thompson, D. F. (1978). "Seeds in the Organic Layers and Soil of Four Beech-birch-maple Stands," USDA For. Serv. Res. Pap. NE-401. Northeastern Forest Experiment Station, Broomall, PA.

Graham, J. A. (principal author) (1990). Direct health effects of air pollutants associated with acidic precursor emissions. *In* "Acidic Deposition: State of Science and Technology. Vol. III. Terrestrial, Materials, Health, and Visibility Effects," pp. 22–3 to 22–133. Superintendent of Documents, U.S. Govt. Printing Office, Washington, DC.

Grahn, O. (1977). Macrophyte succession in Swedish lakes caused by deposition of airborne acid substances. *Water, Air, Soil Pollut.* **7**, 295–305.

Grahn, O. (1986). Vegetation structure and primary production in acidified lakes in southwestern Sweden. *Experientia* **42**, 465–470.

Grahn, O., and Sangfors, O. (1988). A comparative study of macrophytes in Lake Gardsjon, during acid and limed conditions. *In* Liming of Lake Gardsjon: An Acidified Lake in Sweden" (W. Dickson, ed.), Rep. 3426, pp. 281–308. National Swedish Environmental Protection Board, Solna.

Grahn, O., Hultberg, H., and Landner, L. (1974). Oligotrophication—a self-accelerating process in lakes subjected to excessive supply of acid substances. *Ambio* **3**, 93–94.

Grainger, A. (1983). Improving the monitoring of deforestation in the humid tropics. *In* "Tropical Rain Forest: Ecology and Management" (S. L. Sutton, T. C. Whitmore, and A. C. Chadwick, eds.), pp. 387–395. Blackwell, Boston.

Granat, L., and Rodhe, H. (1973). A study of fallout by precipitation around an oil-fired power plant. *Atmos. Environ.* **7**, 781–792.

Graves, H. S. (1918). Effect of the war on forests of France. *Am. For.* **24**, 707–717.

Green, R. E. (1984). The feeding ecology and survival of partridge chicks (*Alectoris rufa* and *Perdix perdix*) on arable farmland in East Anglia. *J. Appl. Ecol.* **21**, 817–830.

Greenwood, P. J. (1980). Mating systems, philopatry, and dispersal in birds and mammals. *Anim. Behav.* **28**, 1140–1162.

Gregor, D. J., and Johnson, M. G. (1980). Nonpoint source phosphorus inputs to the Great Lakes. *In* "Phosphorus Management Strategies for Lakes" (R. C. Loehr, C. S. Martin, and W. Rast, eds.), pp. 37–59. Ann Arbor Sci. Publ., Ann Arbor, MI.

Gregory, R. P. E., and Bradshaw, A. O. (1965). Heavy metal tolerance in populations of *Agrostis tenuis* and other grasses. *New Phytol.* **69**, 131–143.

Grennfelt, P., and Hultberg, H. (1986). Effects of nitrogen deposition on the acidification of terrestrial and aquatic ecosystems. *Water, Air, Soil Pollut.* **30**, 945–965.

Grennfelt, P., and Schjoldager, J. (1984). Photochemical oxidants in the troposphere: A mounting menace. *Ambio* **13**, 61–67.

Grennfelt, P., Bengston, C., and Skarby, L. (1980). An estimation of the atmospheric input of acidifying substances to a forest ecosystem. *In* "Effects of Acid Precipitation on Terrestrial Ecosystems" (T. C. Hutchinson and M. Havas, eds.), pp. 29–40. Plenum, New York.

Grieb, T. M., Driscoll, C. T., Gloss, S. P., Schofield, C. L., Bowie, G. L., and Porcella, D. B. (1990). Factors affecting mercury accumulation in fish in the upper Michigan peninsula. *Environ. Toxicol. Chem.* **9**, 919–930.

Grier, C. C. (1975). Wildfire effects on nutrient distribution and leaching in a coniferous ecosystem. *Can. J. For. Res.* **5**, 599–607.

Grier, J. W. (1982). Ban of DDT and subsequent recovery of reproduction in bald eagles. *Science* **218**, 1232–1235.

Griffiths, R. W., Schloesser, D. W., Leach, J. H., and Kovalak, W. P. (1988). Distribution and dispersal of the zebra mussel (*Dreissena polymorpha*) in the Great Lakes region. *Can. J. Fish. Aquat. Sci.* **48**, 1381–1388.

Grigg, G. (1989). Kangaroo harvesting and the conservation of arid and semi-arid rangelands. *Conserv. Biol.* **3**, 194–197.

Grignon, T. (1992). The dynamics of *Rubus strigosus* (Michx.) in post-clearcut mixedwood and softwood forests of Nova Scotia. Unpublished M.Sc. Thesis, Department of Biology, Dalhousie University, Halifax, Nova Scotia.

Grime, J. P. (1979). "Plant Strategies and Vegetation Processes." Wiley, Toronto, Ont.

Grime, J. P. (1986). Predictions of terrestrial vegetation responses to nuclear winter conditions. *Int. J. Environ. Stud.* **28,** 11–19.

Grodzinski, B. (1992). Plant nutrition and growth regulation by CO_2 enrichment. *BioScience* **42,** 517–525.

Groombridge, B. (1982). "The IUCN Amphibia-Reptilia Red Data Book." International Union for Conservation of Nature and Natural Resources, Gland, Switzerland.

Groombridge, B. (1992). "Global Biodiversity." World Conservation Monitoring Center, Chapman & Hall, London.

Grose, P. L., and Matton, J. S. (1977). "The Argo Merchant oil spill. A preliminary scientific report." U.S. Department of Commerce, National Oceanic and Atmospheric Administration, Washington, DC.

Grossbard, E. and Atkinson, D., eds. (1985). "The Herbicide Glyphosate." Butterworth, Toronto, Ont.

Group of Experts on the Scientific Aspects of Marine Pollution, Joint (GESAMP) (1991). "Carcinogens: Their Significance as Marine Pollutants," Rep. No. 46. International Marine Organization, London.

Group of Experts on the Scientific Aspects of Marine Pollution, Joint (GESAMP) (1993). "Impact of Oil and Related Chemicals and Wastes in the Marine Environment," Rep. No. 50. International Marine Organization, London.

Grover, H. D., and Harwell, M. A. (1985). Biological effects of nuclear war. II. Impact on the biosphere. *BioScience* **35,** 576–583.

Grover, H. D., and White, G. F. (1985). Toward understanding the effects of nuclear war. *BioScience* **35,** 552–556.

Grue, C. E., Fleming, W. J., Busby, D. G., and Hill, E. F. (1983). Assessing hazards of organophosphate pesticides to wildlife. *Trans. North Am. Wildl. Nat. Resour. Conf.* **48,** 200–220.

Grzimek, B. (1972). "Grzimek's Animal Life Encyclopedia." Van Nostrand-Reinhold, Toronto, Ont.

Gschwandtner, G., and Wagner, J. K. (1988). "Historic Emissions of Volatile Organic Compounds in the United States from 1900 to 1985," EPA-600/7-88-008a. U.S. Environmental Protection Agency, Research Triangle Park, NC.

Gschwandtner, G., Gschwandtner, K. C., and Eldridge, K. (1985). "Historic Emissions of Sulfur and Nitrogen Oxides in the United States from 1900 to 1980," EPA-600/7-85-009a. U.S. Environmental Protection Agency, Research Triangle Park, NC.

Gschwandtner, G., Wagner, J. K., and Husar, R. B. (1988). "Comparison of Historic SO_2 and NO_x Emission Data Sets," EPA-600/7-88-009a. U.S. Environmental Protection Agency, Research Triangle Park, NC.

Guicherit, R., and van den Hout, D. (1982). The global NO_x cycle. *In* "Air Pollution by Nitrogen Oxides" (T. Schneider and L. Grant, eds.), pp. 15–29. Elsevier, Amsterdam.

Gulland, J. A., and Garcia, S. (1984). Observed patterns in multispecies fisheries. *In* "Exploitation of Marine Communities" (R. M. May, ed.), pp. 155–190. Springer-Verlag, New York.

Gullion, G. W. (1967). Selection and use of drumming sites by male ruffed grouse. *Auk* **84,** 87–112.

Gullion, G. W. (1969). Aspen—ruffed grouse relationships. *Midwest Wildl. Conf., 31st,* St. Paul, Minnesota, *1969,* (cited in Johnsgard, 1983).

Gullion, G. W. (1977). Forest manipulation for ruffed grouse. *Trans. North Am. Wildl. Nat. Resour. Conf.* **42,** 449–458.

Gullion, G. W. (1986). "Northern Forest Management for Wildlife," For. Ind. Lect. No. 17. Faculty of Forestry, University of Alberta, Edmonton.

Gullion, G. W. (1988). Aspen management for ruffed grouse. *In* "Integrating Forest Management for Wildlife and Fish." USDA For. Serv. Gen. Tech. Rep. NC-122, pp. 9–12. North Central Forest Experiment Station, St. Paul, MN.

Guo-hui, K., De-qiang, Z., Chun, L., Qing-fa, Y., Feng, D., Chong-hui, L., and Guo-liang, S. (1991). The impacts of human activities on the forest and environment in Dinghushan Biosphere Reserve and our countermeasures. *In* "Natural Resource Management and Conservation in Chinese Tropical and Subtropical Regions." Shaoyuan, China.

Guries, R. P. (1989). Genetic structure of forest tree populations: Measurement and interpretation. *In* "Proceedings of the 31st Northeastern Forest Tree Improvement Conference," pp. 5–13. Pennsylvania State University, University Park.

Gutierrez, R. J., and Carey, A. B., eds. (1985). "Ecology and Management of the Spotted Owl in the Pacific Northwest," USDA For. Serv. Gen. Tech. Rep. PNW-185. Pacific Northwest Forest and Range Experiment Station, Portland, OR.

Haagen-Smit, A. J., and Wayne, L. G. (1976). Atmospheric reactions and scavenging processes. *In* "Air Pollution" (A.C. Stern, ed.), 3rd ed., Vol. 1, pp. 235–288. Academic Press, New York.

Haapala, H., Seepuren, E., and Meskus, E. (1975). Effects of spring floods on water acidity in the Kiiminkijoki area, Finland. *Oikos* **26,** 26–31.

Hadley, J. L., Friedland, A. J., Herrick, G. T., and Amundson, R. G. (1991). Winter desiccation and solar radiation in relation to red spruce decline in the northern Appalachians. *Can. J. For. Res.* **21,** 269–272.

Haefner, J. D., and Wallace, J. B. (1981). Shifts in aquatic insect populations in a first-order southern Appalachian stream following a decade of old field succession. *Can. J. Fish. Aquat. Sci.* **38,** 353–359.

Haertl, L. (1976). Nutrient limitation of algal standing crops in shallow prairie lakes. *Ecology* **57,** 664–678.

Haffer, J. (1969). Speciation in Amazonian forest birds. *Science* **165,** 131–137.

Haffer, J. (1982). General aspects of the refuge theory. *In* "Biological Diversification in the Tropics" (G. T. Prance, ed.), pp. 6–24. Columbia Univ. Press, New York.

Haig, S. M., Ballou, J. D., and Derrickson, S. R. (1990). Management options for preserving genetic diversity: Reintroduction of Guam rails to the wild. *Conserv. Biol.* **4,** 290–300.

Haines, T. A., and Baker, J. P. (1986). Evidence of fish population responses to acidification in the eastern United States. *Water, Air, Soil Pollut.* **31,** 605–629.

Hairston, N. G. (1987). "Community Ecology and Salamander Guilds." Cambridge Univ. Press, Cambridge, UK.

Hakman, I., and van Arnold, S. (1985). Plantlet regeneration through somatic embryogenesis in *Picea abies* (Norway spruce). *Can. J. Bot.* **121,** 149–158.

Hall, G.A. (1984). Population decline of neotropical migrants in an Appalachian forest. *Am. Birds* **38,** 14–18.

Hall, H. A., Eidt, D. C., Symons, P. E. K., and Banks, D. (1975). Biological consequences in streams of aerial spraying with fenitrothion against spruce budworm in New Brunswick. *Water Pollut. Res. Can.,* **10,** 84–88.

Hall, I. V. (1955). Floristic changes following the cutting and burning of a woodlot for blueberry production. *Can. J. Agric. Sci.* **35,** 143–152.

Hall, J. D., and Lantz, R. L. (1969). Effects of logging on the habitat of coho salmon and cutthroat trout in coastal streams. *In* "Symposium on Salmon and Trout in Streams" (T. G. Northcotte, ed.), pp. 355–375. University of British Columbia, Vancouver.

Hall, R. J., and Ide, F. P. (1987). Evidence of acidification on stream insect communities in central Ontario between 1937 and 1985. *Can. J. Fish. Aquat. Sci.* **44,** 1652–1657.

Hall, R. J., and Likens, G. E. (1980). Ecological effects of experimental acidification on a stream ecosystem. *In* "Ecological Impact of Acid Precipitation" (D. Drablos and A. Tollan, eds.), pp. 375–376. SNSF Project, Oslo, Norway.

Hall, R. J., and Likens, G. E. (1984). Effect of discharge rate on biotic and abiotic chemical flux in an acidifed stream. *Can. J. Fish. Aquat. Sci.* **41,** 1132–1138.

Hall, R. J., Likens, G. E., Fiance, S. B., and Hendry, G. R. (1990). Experimental acidification of a stream in Hubbard Brook Experimental Forest. *Ecology* **61,** 976–989.

Hall, S. (1989). Natural toxins. *In* "Microbiology of Marine Food Products" (D. R. Ward and C. Hackney, eds.), pp. 301–330. Van Nostrand Reinhold, New York.

Hallbacken, L., and Tamm, C. O. (1985). Changes in soil acidity from 1927 to 1982–84 in a forest area of south-west Sweden. *Soil Sci. Soc. Am. J.* **49,** 1280–1282.

Halls, L. K. (1978). Effect of timber harvesting on wildlife, wildlife habitat, and recreation values. *In* "Complete Tree Utilization of Southern Pine," Proc. Symp., pp. 108–114. Forest Products Research Laboratory, Madison, WI.

Halls, L. K., and Epps, E. A. (1969). Browse quality influenced by tree overstory in the south. *J. Wildl. Manage.* **32,** 1028–1031.

Hamburg, S. P., and Cogbill, C. V. (1988). Historical decline of red spruce populations and climatic change. *Nature (London)* **331,** 428–431.

Hamel, P. B. (1986). "Bachman's Warbler: A Species in Peril." Smithsonian Institution Press, Washington, DC.

Hamilton, G. A., Hunter, K., and Ruthven, A. D. (1981). Inhibition of brain acetylcholinesterase activity in songbirds exposed to fenitrothion during aerial spraying of forests. *Bull. Environ. Contam. Toxicol.* **27,** 856–863.

Hammer, U. T. (1969). Blue-green algal blooms in Saskatchewan lakes. *Verh.—Int. Ver. Theor. Angew. Limnol.* **17,** 116–125.

Hanley, P. T., Hemming, J. E., Morsell, J. W., Morehouse, T. A., Leask, L. E., and Harrison, G. S. (1981). "Natural Resource Protection and Petroleum Development in Alaska," Publ. FWS/OBS-80/22. U.S. Department of the Interior, Fish and Wildlife Service, Washington, DC.

Hansen, A. J., Spies, T. A., Swanson, F. J., and Ohmann, J. L. (1991). Conserving biodiversity in managed forests. *BioScience* **41,** 382–392.

Hansen, J., and Lebedev, S. (1987). Global trends of measured surface air temperature. *J. Geophys. Res.* **92,** 13,345–13,372.

Hansen, J., and Lebedev, S. (1988). Global surface air temperatures: Update through 1987. *Geophys. Res. Lett.* **15,** 323–326.

Hardin, G. (1968). The tragedy of the commons. *Science* **162,** 1243–1248.

Hargreaves, J. W., and Whitten, B. A. (1976). Effect of pH on growth of acid stream algae. *Br. Phycol. J.* **11,** 215–223.

Hargreaves, J. W., Lloyd, E. J. H., and Whitten, B. A. (1975). Chemistry and vegetation of highly acidic streams. *Freshwater Biol.* **5,** 563–576.

Harmon, M. E., Ferrell, W. K., and Franklin, J. F. (1990). Effects on carbon storage of conversion of old-growth forests to young forests. *Science* **247,** 699–702.

Harper, D. (1992). "Eutrophication of Freshwaters. Principles, Problems, and Restoration." Chapman & Hall, New York.

Harrington, J. B. (1987). Climatic change: A review of causes. *Can. J. For. Res.* **17,** 1313–1339.

Harris, G. P., and Vollenweider, R. A. (1982). Palaeolimnological evidence of early eutrophication in Lake Erie. *Can. J. Fish. Aquat. Sci.* **39,** 618–626.

Harris, L. D. (1984). "The Fragmented Forest." Univ. of Chicago Press, Chicago.

Harris, M. M., and Jurgensen, M. F. (1977). Development of *Salix* and *Populus* mycorrhizae in metallic mine tailings. *Plant Soil* **47,** 509–517.

Harris, M. P., and Birkhead, T. R. (1985). Breeding ecology of the Atlantic Alcidae. *In* "The Atlantic Alcidae" (D. N. Nettleship and T. R. Birkhead, eds.), pp. 156–204. Academic Press, New York.

Harrison, A. D. (1958). The effects of sulfuric acid pollution on the biology of streams in the Transvaal, South Africa. *Verh.—Int. Ver. Theor. Limnol. Agnew.* **13**, 603–610.

Hart, D. (1991). "The Hidden Seal Hunts." International Wildlife Coalition, Mississauga, Ont.

Hartman, W. L. (1972). Lake Erie: Effects of exploitation, environmental changes and new species on the fishery resources. *J. Fish. Res. Board Can.* **29**, 899–912.

Hartman, W. L. (1973). "Effects of Exploitation, Environmental Changes, and New Species on the Fish Habitats and Resources of Lake Erie," Tech. Rep. No. 22. Great Lakes Fisheries Commission, Ann Arbor, MI.

Hartman, W. L. (1988). Historical changes in the major fish resources of the Great Lakes. *In* "Toxic Contaminants and Ecosystem Health: A Great Lakes Perspective" (M.S. Evans, ed.). Wiley, New York.

Hartshorn, G. S. (1983). Wildlands conservation in Central America. *In* "Tropical Rain Forest: Ecology and Management" (S. L. Sutton, T. C. Whitmore, and A. C. Chadwick, eds.), pp. 423–444. Blackwell, Boston.

Harvey, A. E., Jurgensen, N. F., and Larsen, M. J. (1976). "Intensive Fibre Utilization and Prescribed Fire: Effects on the Microbial Ecology of Forests," USDA For. Serv. Gen. Tech. Rep. INT-28. Intermountain Forest and Range Experiment Station, Ogden, UT.

Harvey, A. E., Jurgensen, M. F., and Larsen, M. J. (1980). Biological implications of increasing harvest intensity on the maintenance of productivity of forest soils. *In* "Environmental Consequences of Timber Harvesting in Rocky Mountain Coniferous Forests," USDA For. Serv. Gen. Tech. Rep. INT-90, pp. 211–220. Intermountain Forest and Range Experiment Station, Ogden, UT.

Harvey, H. H., and Lee, C. (1982). Historical fisheries changes related to surface water pH changes in Canada. *Acid Rain/Fish., Proc. Int. Symp., 1981*, pp. 45–55.

Harvey, H. H., Pierce, R. C., Dillon, P. J., Kramer, J. P., and Whelpdale, D. M. (1981). "Acidification of the Canadian Environment: Scientific Criteria for Assessment of the Effects of Acidic Deposition on Aquatic Ecosystems," Rep. No. 18475. National Research Council of Canada, Ottawa, Ont.

Harwell, M. A., and Grover, H. D. (1985). Biological effects of nuclear war. I. Impact on humans. *BioScience* **35**, 570–575.

Harwell, M. A., and Hutchinson, T. C. (1985). "Environmental Consequences of Nuclear War," Vol. 2, SCOPE Rep. 28. Wiley, Toronto, Ont.

Hasler, A. D., Brynildson, O. M., and Helm, W. T. (1951). Improving conditions for fish in brown-water bog lakes by alkalization. *J. Wildl. Manage.* **15**, 347–352.

Hatcher, J. D., and White, F. M. M. (1984). "Task Force on Chemicals in the Environment and Human Reproductive Problems in New Brunswick," Report to Department of Health. Province of New Brunswick, Faculty of Medicine, Dalhousie University, Halifax, Nova Scotia.

Hauhs, M., and Wright, R. F. (1986). Regional pattern of acid deposition and forest decline along a cross section through Europe. *Water, Air, Soil Pollut.* **31**, 463–474.

Haupt, H. F., and Kidd, W. J. (1965). Good logging practices reduce sedimentation in central Ohio. *J. For.* **63**, 664–670.

Hausenbuiller, R. L. (1978). "Soil Science." Wm. C. Brown, Dubuque, IA.

Havas, M. (1986). Aluminum in the aquatic environment. *In* "Aluminum in the Canadian Environment," pp. 79–126. National Research Council of Canada, Associate Committee on Scientific Criteria for Environmental Quality, Ottawa, Ont.

Havas, M., and Hutchinson, T. C. (1982). Aquatic invertebrates from the Smoking Hills, N.W.T.: Effect of pH and metals on mortality. *Can. J. Fish. Aquat. Sci.* **39**, 890–893.

Havas, M., and Hutchinson, T. C. (1983a). The Smoking Hills: Natural acidification of an aquatic ecosystem. *Nature (London)* **301**, 23–27.

Havas, M., and Hutchinson, T. C. (1983b). Effect of low pH on the chemical composition of aquatic invertebrates from tundra ponds at the Smoking Hills, N.W.T., Canada. *Can. J. Zool.* **61**, 241–249.

Havas, M., Hutchinson, T. C., and Likens, G. E. (1984). Red herrings in acid rain research. *Environ. Sci. Technol.* **18**, 176A–186A.

Hawksworth, D. L., and Rose, F. (1976). "Lichens as Air Pollution Monitors," Stud. Biol. No. 66. Institute of Biology, Arnold, London.

Hay, A. (1982). "The Chemical Scythe: Lessons of 2,4,5-T and Dioxin." Plenum, New York.

Hayes, W. J. (1991). Introduction. *In* "Handbook of Pesticide Toxicology. Vol. 1. General Principles" (W. C. Hayes and E. R. Laws, eds.), pp. 1–37. Academic Press, San Diego.

Heath, R. T. (1986). Dissolved organic phosphorus compounds: Do they satisfy planktonic phosphate demands in summer? *Can. J. Fish. Aquat. Sci.* **43**, 343–350.

Hebert, P. D. N., Muncaster, B. W., and Mackie, G. L. (1989). Ecological and genetic studies on *Dreissena polymorpha* (Pallas): A new mollusc in the Great Lakes. *Can. J. Fish. Aquat. Sci.* **46**, 1587–1591.

Hebert, P. D. N., Wilson, C. C., Murdoch, M. H., and Lazar, R. (1991). Demography and ecological impact of the invading mollusc *Dreissena polymorpha*. *Can. J. Zool.* **69**, 405–409.

Hecht, S. B. (1989). The sacred cow in the green hell: Livestock and forest conversion in the Brazilian Amazon. *Ecologist* **19**, 229–234.

Heck, W. W. (1989). Assessment of crop losses from air pollutants in the United States. *In* "Air Pollution's Toll on Forests and Crops" (J. J. MacKenzie and M. T. El-Ashry, eds.), pp. 235–315. Yale Univ. Press, New Haven, CT.

Heck, W. W., and Brandt, C. S. (1978). Effects on vegetation: Native, crops, forest. *In* "Air Pollution" (A. C. Stern, ed.), 3rd ed., Vol. 2, pp. 158–231. Academic Press, New York.

Heck, W. W., Taylor, O. C., Adams, R., Bingham, G., Preston, E., and Weinstein, L. (1982). Assessment of crop loss from ozone. *J. Air Pollut. Control Assoc.* **32**, 353–361.

Heck, W. W., Adams, R. M., Cure, W. W., Heagle, A. S.,

Heggestad, H. E., Kohut, R. J., Kress, L. W., Rawlings, J. O., and Taylor, O. C. (1983). A reassessment of crop loss from ozone. *Environ. Sci. Technol.* **17**, 573A-581A.

Heck, W. W., Heagle, A. S., and Shriner, D. S. (1986). Effects on vegetation: Native, crops, and forests. *In* "Air Pollution" (A. C. Stern, ed.), 3rd ed., Vol. 6, pp. 247–350. Academic Press, New York.

Hecky, R. E., and Kilham, P. (1988). Nutrient limitation of phytoplankton in freshwater and marine environments: A review of recent evidence on the effects of enrichment. *Limnol. Oceanogr.* **33**(4, Part 2), 796–782.

Heggestad, H. E. (1980). Field assessment of air pollution impacts on the growth and productivity of crop species. Presented at the annual meeting of the Air Pollution Control Association, Montreal (cited in Roberts, 1984).

Heinrichs, H., and Mayer, R. (1977). Distribution and cycling of major and trace elements in two central European forest ecosystems. *J. Environ. Qual.* **6**, 402–406.

Hellquist, C. B., and Crow, G. E. (1980). "Aquatic Vascular Plants of New England. Part 1. Zosteraceae, Potamogetonaceae, Zannichelliaceae, Najadaceae," State Bull. 515. New Hampshire Agricultural Experiment Station, Durham.

Henderson, J. W. (1978). "The Effects of Forest Operations on the Water Resources of the Shubenacadie-Stewiacke River Basin," Tech. Rep. No. 11. Shubenacadie-Stewiacke River Basin Board, Truro, Nova Scotia.

Henderson-Sellers, B. (1984). "Pollution of Our Atmosphere." Adam Hilger, Techno House, Bristol, UK.

Hendry, G. R., and Vertucci, F. (1980). Benthic plant communities in acidic Lake Colden, New York: *Sphagnum* and the algal mat. *In* "Ecological Impact of Acid Precipitation" (D. Drablos and A. Tollan, eds.), pp. 314–315. SNSF Project, Oslo, Norway.

Hendry, G. R., and Wright, R. F. (1976). Acid precipitation in Norway: Effects on aquatic fauna. *J. Great Lakes Res.* **2**, Suppl., 192–207.

Hendry, G. R., Yan, N. D., and Baumgartner, B. J. (1980a). Responses of freshwater plants and invertebrates to acidification. *In* "Restoration of Lakes and Inland Rivers," EPA 440/5–81/010, pp. 457–466. U.S. Environmental Protection Agency, Washington, DC.

Hendry, G. R., Galloway, J. N., Norton, S. A., Schofield, C. L., Schoffer, P. W., and Burns, D. A. (1980b). "Geological and Hydrochemical Sensitivity of the Eastern United States to Acid Precipitation," EPA-600/3–80–024. U.S. Environmental Protection Agency, Environmental Research Laboratory, Corvallis, OR.

Hendzel, L. L., Hecky, R. E., and Findlay, D. L. (1993). Recent changes of nitrogen fixation in Lake 227 in response to reduction of N:P loading ratio. *Can. J. Fish. Aquat. Sci.* (submitted for publication).

Hengeveld, H. (1992). Global warming potentials. *CO₂ Clim. Rep.* **92-1**, 5–6.

Henricksen, L., Oscarson, H. G., and Stenson, J. A. E. (1984). "Development of the Crustacean Zooplankton Community after Lime Treatment of the Fishless Lake Gardsjon, Sweden," Rep. No. 61. Institute of Freshwater Research, National Swedish Board of Fisheries, Lund (cited in Olem, 1990).

Henriksen, A. (1980). Acidification of freshwaters—a large scale titration. *In* "Ecological Impacts of Acid Precipitation" (D. Drablos and A. Tollan, eds.), pp. 68–74. SNSF Project, Oslo, Norway.

Henriksen, A. (1982). Susceptibility of surface waters to acidification. *Acid Rain/Fish., Proc. Int. Symp., 1981*, pp. 103–121.

Henriksen, A., Lien, L., Rosseland, B. O., Traaen, T. S., and Sevaldrud, I. S. (1989). Lake acidification in Norway: Present and predicted fish status. *Ambio* **18**, 314–321.

Henry, G. H. R., Freedman, B., and Svoboda, J. (1986a). Effects of fertilization on three tundra plant communities of a polar desert oasis. *Can. J. Bot.* **64**, 2502–2507.

Henry, G. H. R., Freedman, B., and Svoboda, J. (1986b). Survey of vegetated areas and muskox populations in east-central Ellesmere Island. *Arctic* **39**, 78–81.

Hepting, G. H. (1971). "Diseases of Forest and Shade Trees of the United States," Agric. Handb. No. 386. U.S. Department of Agriculture, Forest Service, Washington, DC.

Hermens, J., Konemann, H., Leeuwangh, P., and Mursch, A. (1985). Quantitative structure-activity relationships in aquatic toxicity studies and complex mixtures of chemicals. *Environ. Toxicol. Chem.* **4**, 273–279.

Heron, J. (1961). The seasonal variation of phosphate, silicate, and nitrate in waters of the English Lake District. *Limnol. Oceanogr.* **6**, 338–346.

Hesser, R., Hooper, R., Weirich, C. B., Selcher, J., Hollender, B., and Snyder, R. (1975). The aquatic biota. *In* "Clearcutting in Pennsylvania," pp. 9–20. Pennsylvania State School of Forest Resources, University Park.

Hetherington, E. D. (1976). "Dennis Creek. A look at water quality following logging in the Okanagan Basin," Rep. BC-X-147. Pacific Forest Research Centre, Victoria, British Columbia.

Hewlett, J. F., and Helvey, J. D. (1970). Effects of forest clearfelling on the storm hydrograph. *Water Resour. Res.* **6**, 768–782.

Heybroek, H. M. (1982). Monoculture versus mixture: Interactions between susceptible and resistant trees in a mixed stand. *In* "Resistance to Disease and Pests in Forest Trees," pp. 226–241. Pudoc, Wageningen.

Heyerdahl, T. (1971). Atlantic Ocean pollution and biota observed by the "Ra" expeditions. *Biol. Conserv.* **3**, 164–167.

Hibber, C. R. (1964). Identity and significance of certain organisms associated with sugar maple decline in New York woodlands. *Phytopathology* **74**, 1389–1392.

Hibbert, A. R. (1967). Forest treatment effects on water yield. *In* "International Symposium on Forest Hydrology Proceedings" (W. E. Sopper and H. W. Hull, eds.), pp. 527–543. Pergamon, New York.

Hibbs, D. E. (1983). Forty years of forest succession in central New England. *Ecology* **64**, 1394–1401.

Hickey, J. J., ed. (1969). "Peregrine Falcon Populations: Their Biology and Decline." Univ. of Wisconsin Press, Madison.

Hickey, J. J., and Anderson, D. W. (1968). Chlorinated hydrocarbons and eggshell changes in raptorial and fish-eating birds. *Science* **162**, 271–273.

Hicks, B. B. (principal author) (1990). Atmospheric processes research and process model development. *In* "Acidic Deposition: State of Science and Technology. Vol. I. Emissions, Atmospheric Processes, and Deposition," pp. 2–1 to 2–298. Superintendent of Documents, U.S. Govt. Printing Office, Washington, DC.

Hiep, D. (1984). Long-term changes in the mangrove habitat following herbicidal attack. *In* "Herbicides in War. The Long-term Ecological and Human Consequences" (A. H. Westing, ed.), pp. 89–90. Taylor & Francis, London.

Hileman, B. (1983). Acid fog. *Environ. Sci. Technol.* **17**, 117A–120A.

Hill, D. A. (1985). The feeding ecology and survival of pheasant chicks on arable farmland. *J. Appl. Ecol.* **22**, 645–654.

Hinrichsen, D. (1986). Multiple pollutants and forest decline. *Ambio* **15**, 258–265.

Hinrichsen, D. (1987). The forest decline enigma. *BioScience* **37**, 542–546.

Hobaek, A., and Raddum, G. G. (1980). "Zooplankton Communities in Acidified Lakes in South Norway," IR 75/80. SNSF Project, Oslo, Norway.

Hobbs, R. J. (1992). The role of corridors in conservation: Solution or bandwagon? *TREE* **7**, 389–392.

Hodges, C. S., Adee, K. T., Stein, J. D., Wood, H. B., and Doty, R. D. (1986). "Decline of Ohia (*Metrosideros polymorpha*) in Hawaii: A Review," USDA For. Serv. Gen. Tech. Rep. PSW-86. Pacific Southwest Forest and Range Experiment Station, Berkeley, CA.

Hodgeman, T. P., and Bowyer, R. T. (1985). Winter use of arboreal lichens, ascomycetes, by white-tailed deer, *Odocoileus virginianus,* in Maine. *Can. Field Nat.* **99**, 313–316.

Hodgson, B. (1990). Alaska's big spill. *Nat. Geogr.* **177**(1), 5–43.

Hoffman, D. J., Harder, J. W., Rolf, R. S., and Rosen, J. M. (1987). Balloon-borne observations of the development and vertical structure of the Antarctic ozone hole in 1986. *Nature (London)* **326**, 59–62.

Hoffman, D. J., Ohlendorf, H. M., and Aldrich, T. W. (1988). Selenium teratogenesis in natural populations of aquatic birds in central California. *Arch. Environ. Contam. Toxicol.* **17**, 519–525.

Hoffman, M. (1991). Taking stock of Saddam's fiery legacy in Kuwait. *Science* **253**, 971.

Hogan, G. D., Courtin, G. M., and Rauser, W. E. (1977a). The effects of soil factors on the distribution of *Agrostis gigantea* on a mine waste site. *Can. J. Bot.* **55**, 1038–1042.

Hogan, G. D., Courtin, G. M., and Rauser, W. E. (1977b). Copper tolerance in clones of *Agrostis gigantea* from a mine waste site. *Can. J. Bot.* **55**, 1043–1050.

Hoggan, M. C., Davidson, A., Brunelle, M. F., Neuitt, J. S., and Gins, J. D. (1978). Motor vehicle emissions and atmospheric lead concentrations in the Los Angeles area. *J. Air Pollut. Control Assoc.* **28**, 1200–1206.

Hokenson, K. E. F., McCormick, J. H., Jones, B. R., and Tucker, J.H. (1973). Thermal requirements for maturation, spawning, and embryo survival of the brook trout (*Salvelinus fontinalis*). *J. Fish. Res. Board Can.* **30**, 975–984.

Holdaway, R. N. (1989). New Zealand's pre-human avifauna and its vulnerability. *N.Z. J. Ecol.* **12** Suppl., 11–25

Holling, C. S., and Walters C. J. (1977). Fenitrothion or not fenitrothion: That is not the question. *In* "Proceedings of a Symposium on Fenitrothion: The Long-term Effects of Its Use in Forest Ecosystems," pp. 279–297. National Research Council of Canada, Ottawa, Ont.

Holloway, G. T. (1917). "Report of the Ontario Nickel Commission." Legislative Assembly of Ontario, Toronto.

Holloway, M., and Horgan, J. (1991). Soiled shores. *Sci. Am.* **265**(4), 103–116.

Holmes, R. T., Sherry, T. W., and Sturges, F. W. (1986). Bird community dynamics in a temperate deciduous forest: Long-term trends at Hubbard Brook. *Ecol. Monogr.* **56**, 201–220.

Holmes, S. B. (1979). "Aquatic Impact Studies of a Spruce Budworm Control Program in the Lower St. Lawrence Region of Quebec in 1978," Rep. FPM-X-26. Forest Pest Management Institute, Canadian Forestry Service, Sault Ste. Marie, Ont.

Holmes, S. B., and Boag, P. T. (1990). Effects of the organophosphorus pesticide fenitrothion on behaviour and reproduction in zebra finches. *Environ. Res.* **53**, 62–75.

Holmes, S. B., and Sundaram, K. M. S. (1992). Insecticide residues and cholinesterase inhibition in zebra finches orally dosed with fenitrothion. *J. Environ. Sci. Health, Part A* **A27**, 889–902.

Holmes, W. N. (1984). Petroleum pollutants in the marine environment and their possible effects on seabirds. *Rev. Environ. Toxicol.* **1**, 251–317.

Holmes, W. N., and Cronshaw, J. (1977). Biological effects of petroleum on marine birds. *In* "Effects of Petroleum on Arctic and Subarctic Marine Environments" (D. C. Malins, ed.), Vol. 2, pp. 359–398. Academic Press, New York.

Holroyd, G. L. (1993). "Status of Peregrine Falcon Recovery in Canada During 1993." Canadian Wildlife Service, Western and Northern Region, Edmonton, Alberta.

Holroyd, G. L., and Banasch, U. (1990). The reintroduction of the peregrine falcon, *Falco peregrinus anatum,* into southern Canada. *Can. Field Nat.* **104**, 203–208.

Holtby, L. B. (1988). Effects of logging on stream temperatures in Carnation Creek, British Columbia, and associated impacts on the coho salmon (*Onchorhynchus kisutch*). *Can. J. Fish. Aquat. Sci.* **45**, 502–515.

Holtby, L. B., and Baillie, S. D. (1989a). Litter fall and detrital decomposition rates in a tributary of Carnation Creek, British Columbia, over-sprayed with the herbicide Roundup (glyphosate). *In* "Proceedings of the Carnation Creek Workshop," pp. 232–249. FRDA Rep. 063, Forest Pest Management Institute, Sault Ste Marie, Ont.

Holtby, L. B., and Baillie, S. D. (1989b). Effects of the herbicide Roundup on coho salmon fingerings in an over-sprayed tributary of Carnation Creek, British Columbia. *In* "Proceedings of the Carnation Creek Workshop," pp. 273–285. FRDA Rep. 063, Forest Pest Management Institute, Sault Ste Marie, Ont.

Honer, T. G., and Bickerstaff, A. (1985). "Canada's Forest Area and Wood Volume Balance 1977–1981," BC-X-272. Canadian Forestry Service, Pacific Forestry Centre, Victoria, British Columbia.

Hong, P. N. (1984a). Characteristics of mangroves in the region of Mekong River mouths. *Vietnam Natl. Symp. Mangrove Ecosyst., 1st 1984*, pp. 55–69.

Hong, P. N. (1984b). Effects of chemical warfare on mangrove forests on tip of the Canau Peninsula, Minhhai Province. *Vietnam Natl. Symp. Mangrove Ecosyst., 1st, 1984*, pp. 163–174.

Hong, P. N. (1987). Mangrove ecology in Viet Nam. Presented at the International Conference on Ecology in Viet Nam, 1987, New Paltz, NY.

Hongfa, C. (1989). Air pollution and its effects on plants in China. *J. Appl. Ecol.* **26**, 763–773.

Hooper, M. J., Brewer, L. W., Cobb, G. P., and Hendall, R. J. (1990). An integrated laboratory and field approach for assessing hazards of pesticide exposure to wildlife. *In* "Pesticide Effects on Terrestrial Wildlife" (L. Somerville and C. H. Walker, eds.), pp. 271–283. Taylor & Francis, New York.

Hooper, R. G., Robinson, A. F., and Jackson, J. A. (1980). "The Red-cockaded Woodpecker: Notes on Life History and Management," USDA For. Serv. Gen. Rep. SA-GR-9. Southeastern Forest Experiment Station, Asheville, NC.

Hoover, E. F. (1973). A wildlife brief for the clearcut logging of Douglas-fir. *J. For.* **71**, 210–214.

Hopner, T., Felzmann, H., and Struck, H. (1992). The Gulf oil pollution. *In* "Remediation of Oil Spills," pp. 183–192. German Soc. Petroleum & Coal Sci. & Technol., Hamburg.

Hopwood, D. (1991). "Principles and Practices of New Forestry," Land Manage. Rep. No. 71. Research Branch, B.C., Ministry of Forests, Victoria, British Columbia.

Horgan, J. (1991). The muddled cleanup in the Persian Gulf. *Sci. Am.* **265**(9), 107–110.

Horn, M. H., Teal, J. M., and Backus, R. H. (1970). Petroleum lumps on the surface of the sea. *Science* **168**, 245–246.

Hornbeck, J. W. (1973). Storm flow from hardwood-forested and cleared watersheds in New Hampshire. *Water Resour. Res.* **9**, 346–354.

Hornbeck, J. W. (1977). "Nutrients: A Major Consideration for Intensive Forest Management," USDA For. Serv. Gen. Tech. Rep. NE-29, pp. 241–250. Northeastern Forest Experiment Station, Broomall, PA.

Hornbeck, J. W., and Smith, R. B. (1985). Documentaion of red spruce growth decline. *Can. J. For. Res.* **15**, 1199–1201.

Hornbeck, J. W., and Ursic, S. J. (1979). Intensive harvest and forest sresams: Are they compatible? *In* "Impacts of Intensive Harvesting on Forest Nutrient Cycling," pp. 249–262. College of Environmental Science and Forestry, State University of New York, Syracuse.

Hornbeck, J. W., Likens, G. E., Pierce, R. S., and Bormann, F. H. (1975). Strip-cutting as a means of protecting site and streamflow quality when clear-cutting northern hardwoods. *In* "Forest Soils and Forest Land Management" (B. Bernier and C.W. Winget, eds.), pp. 209–225. Laval Univ. Press, Quebec.

Hornbeck, J. W., Smith, R. B., and Federer, C. A. (1986). Growth decline in red spruce and balsam fir relative to natural processes. *Water, Air, Soil Pollut.* **31**, 425–430.

Hornbeck, J. W., Martin, C. W., Pierce, R. S., Bormann, F. H., Likens, G. E., and Eaton, J. S. (1987a). "The Northern Hardwood Forest Ecosystem: Ten Years of Recovery from Clearcutting," USDA For. Serv. Publ. NE-RP-596. Northeastern Forest Experiment Station, Broomall, PA.

Hornbeck, J. W., Smith, R. B., and Federer, C. A. (1987b). Extended growth decreases in New England are limited to red spruce and balsam fir. *In* "Proceedings of the International Symposium on Ecological Aspects of Tree-ring Analysis," CONF-8608144, pp. 38–44. U.S. Department of Commerce, Springfield, VA.

Horsley, S. B. (1981). Control of herbaceous weeds in Allegheny hardwood forests with herbicides. *Weed Sci.* **29**, 655–662.

Horwood, J. (1990). "Biology and Exploitation of the Minke Whale." CRC Press, Boca Raton, FL.

Hosker, R. P., and Lindberg, S. E. (1982). Review: Atmospheric deposition and plant assimilation of gases and particles. *Atmos. Environ.* **16**, 889–910.

Houghton, H. G. (1955). On the chemical composition of fog and cloud water. *J. Meteorol.* **12**, 355–357.

Houghton, R. A. (1990). The future role of tropical forests in affecting the carbon dioxide concentration of the atmosphere. *Ambio* **19**, 204–209.

Houghton, R. A. (1991). The role of forests in affecting the greenhouse gas composition of the atmosphere. *In* "Global Climate Change and Life on Earth" (R. C. Wyman, ed.), pp. 43–56. Routledge, Chapman, and Hall, New York.

Houghton, R. A., Hobbie, J. E., Melillo, J. M., Moore, B., Peterson, B. J., Shaver, G. R., and Woodwell, G. M. (1983). Changes in the carbon content of terrestrial biota and soils between 1860 and 1980: A net release of CO_2 to the atmosphere. *Ecol. Monogr.* **53**, 235–262.

Houghton, R. A., Schlesinger, W. H., Brown, S., and Richards, J. F. (1985). Carbon dioxide exchange between the atmosphere and terrestrial ecosystems. *In* "Atmospheric Carbon Dioxide and the Global Carbon Cycle," DOE/ER-0239, pp. 113–140. U.S. Department of Energy, Washington, DC.

Houghton, R. A., Lefkowitz, D. S., and Skole, D. L. (1991). Changes in the landscape of Latin America between 1850 and 1985. I. Progressive loss of forest. *For. Ecol. Manage.* **38,** 143–172.

Houston, D. B. (1974). Response of selected *Pinus strobus* L. clones to fumigations with sulfur dioxide and ozone. *Can. J. For. Res.* **4,** 65–68.

Houston, D. B., and Stairs, G. R. (1973). Genetic control of sulfur dioxide and ozone tolerance in eastern white pine. *For. Sci.* **19,** 267–271.

Houston, D. R., Allen, D. C., and LaChance, D. (1990). "Sugarbush Management: A Guide to Maintaining Tree Health," USDA For. Serv. Gen. Tech. Rep. NE-129. Northeastern Forest Experiment Station, Radnor, PA.

Hov, O. (1984). Ozone in the troposphere: High level pollution. *Ambio* **13,** 73–79.

Howard, C. (1992). Field worker injury in vegetation management programs. *For. Pest Manage. Inst.* **10**(1), 2.

Howell, R. K., Koch, E. J., and Rose, L. (1979). Field assessment of air pollution induced soybean yield losses. *Agron. J.* **71,** 285–288.

Howitt, R. E., Gossard, T. W., and Adams, R. M. (1984). Effects of alternative ozone concentrations and response data on economic assessments: The case of California crops. *J. Air Pollut. Control Assoc.* **34,** 1122–1127.

Hubbell, S. P., and Foster, R. B. (1983). Diversity of canopy trees in a neotropical forest and implications for conservation. *In* "Tropical Rain Forest: Ecology and Management" (S. L. Sutton, T. C. Whitmore, and A. C. Chadwick, eds.), pp. 25–41. Blackwell, Boston.

Hubendick, B. (1987). Tropical diseases and human ecology. *Ambio* **16,** 218–219.

Huckabee, J. W., Mattice, J. S., Pitelka, L. F., Porcella, D. B., and Goldstein, R. A. (1989). An assessment of the ecological effects of acidic deposition. *Arch. Environ. Contamin. Toxicol.* **18,** 3–27.

Hudak, J. (1991). Integrated pest management and the eastern spruce budworm. *For. Ecol. Manage.* **39,** 313–337.

Hudson, R. H., Tucker, R. K., and Haegele, M. A. (1984). "Handbook of Toxicity of Pesticides to Wildlife," Resour. Publ. 153. U.S. Department of Interior, Fish and Wildlife Service, Washington, DC.

Huettl, R. F. (1986a). "New type" of forest decline and diagnostic fertilization. *Int. Conf. Environ. Contam., 2nd, 1986.*

Huettl, R. F. (1986b). "Forest Decline and Nutritional Disturbances." IUFRO World Congress, Ljubjana, Yugoslavia.

Huettl, R. F. (1989a). "New types" of forest damages in central Europe. *In* "Air Pollution's Toll on Forests and Crops" (J. J. MacKenzie and M. T. El-Ashry, eds.), pp. 22–74. Yale Univ. Press, New Haven, CT.

Huettl, R. F. (1989b). Liming and fertilization as mitigation tools in declining forest ecosystems. *Water, Air, Soil Pollut.* **44,** 93–118.

Huettl, R. F., and Wisniewski, J. (1987). Fertilization as a tool to mitigate forest decline. (manuscript).

Huey, N. A. (1968). The lead dioxide estimation of sulfur dioxide pollution. *J. Air Pollut. Control Assoc.* **18,** 610–611.

Huffaker, C. B., and Kennett, C. E. (1959). A ten-year study of vegetational changes associated with biological control of klamath weed. *J. Range Manage.* **12,** 69–82.

Huhn, F. J. (1974). Lake sediment records of industrialization in the Sudbury area of Ontario. M.Sc. Thesis, Department of Botany, University of Toronto, Toronto, Ont.

Hulme, M. A., Ennis, T. J., and Lavallee, A. (1983). Current status of *Bacillus thuringiensis* for spruce budworm control. *For. Chron.* **59,** 58–61.

Hultberg, H., and Grahn, O. (1975a). Effects of acidic precipitation on macrophytes in oligotrophic Swedish lakes. *In* "Atmospheric Contribution to the Chemistry of Lake Waters," pp. 208–221. Int. Assoc. Great Lakes Res.

Hultberg, H., and Grahn, O. (1975b). "Some Effects of Adding Lime to Lakes in Western Sweden," Transl. Ser. No. 3607. Department of the Environment, Fisheries and Marine Services, Ottawa, Ont.

Hultberg, H., and Grennfelt, P. (1986). Gardsjon project: Lake acidification, chemistry in catchment runoff, lake liming and microcatchment manipulations. *Water, Air, Soil Pollut.* **30,** 31–46.

Humane Society of the United States (1989). "Purse-seining on Dolphins." Humane Society of the United States, Washington, DC.

Hunsaker, C. T., and Carpenter, D. E., eds. (1990). "Ecological Indicators for the Environmental Monitoring and Assessment Program." EPA 600/3–90/060. Office of Research and Development, U.S. Environmental Protection Agency, Research Triangle Park, NC.

Hunsaker, C. T., Carpenter, D., and Messer, J. (1990). Ecological indicators for regional monitoring. *Bull. Ecol. Soc. Am.* **71,** 165–172.

Hunt, E. G., and Bischoff, A. I. (1960). Inimical effects on wildlife of periodic DDD applications to Clear Lake. *Calif. Fish Game* **46,** 91–106.

Hunter, B. A., Johnson, M. S., and Thompson, D. J. (1987a). Ecotoxicology of copper and cadmium in a contaminated grassland ecosystem. I. Soil and vegetation contamination. *J. Appl. Ecol.* **24,** 573–586.

Hunter, B. A., Johnson, M. S., and Thompson, D. J. (1987b). Ecotoxicology of copper and cadmium in a contaminated grassland ecosystem. II. Invertebrates. *J. Appl. Ecol.* **24,** 587–599.

Hunter, B. A., Johnson, M. S., and Thompson, D. J. (1987c). Ecotoxicology of copper and cadmium in a contaminated grassland ecosystem. III. Small mammals. *J. Appl. Ecol.* **24,** 601–614.

Hunter, B. A., Johnson, M. S., and Thompson, D. J. (1989). Ecotoxicology of copper and cadmium in a contaminated grassland ecosystem. IV. Tissue distribution and age accumulation in small mammals. *J. Appl. Ecol.* **26,** 89–99.

Hunter, J. G., and Vergnano, O. (1952). Nickel toxicity in plants. *Ann. Appl. Biol.* **39,** 279–284.

Hunter, M. L. (1990). "Wildlife, Forests, and Forestry: Principles of Managing Forests for Biological Diversity." Prentice-Hall, Englewood Heights, NJ.

Hunter, M. L., Jones, J. J., Gibbs, K. E., Moring, J. R., and Brett, M. (1985). "Interactions among Waterfowl, Fishes, Invertebrates, and Macrophytes in Four Maine Lakes of Different Acidity," Biol. Rep. No. 80. Eastern Energy and Land Use Team, U.S. Fish and Wildlife Service, Washington, DC.

Huntsman, A. G. (1942). Death of salmon and trout with high temperature. *J. Fish. Res. Board Can.* **5,** 485–501.

Hurlbert, S. H. (1971). The nonconcept of species diversity: A critique and alternative parameters. *Ecology* **52,** 577–586.

Hussell, D. J. J., Mather, M. H., and Sinclair, P. H. (1992). Trends in numbers of tropical- and temperate-wintering landbirds in migration at Long Point, Ontario, 1961–1988. *In* "Ecology and Conservation of Neotropical Migrant Landbirds" (J. M. Hagan and D. W. Johnson, eds.,) pp. 101–114. Smithsonian Institution Press, Washington, DC.

Hutcheson, M. R., and Hall, F. P. (1974). Sulphate washout from a coal-fired power plant plume. *Atmos. Environ.* **8,** 23–28.

Hutchings, J. A., and Myers, R. A. (1993). "What can be Learned from the Collapse of a "Renewable" Resource? Atlantic Cod, *Gadus morhua,* of Newfoundland and Labrador." Department of Fisheries and Oceans, Science Branch, St. John's, Newfoundland.

Hutchinson, G. E. (1957). "A Treatise on Limnology," Vol. 1. Wiley, New York.

Hutchinson, G. E. (1959). Homage to Santa Rosalia, or why are there so many kinds of animals? *Am. Nat.* **93,** 145–159.

Hutchinson, G. E. (1969). Eutrophication, past and present. *In* "Eutrophication: Causes, Consequences, Correctives," pp. 17–26. National Academy of Sciences, Washington, DC.

Hutchinson, G. E. (1975). "A Treatise on Limnology," Vol. 3. Wiley, New York.

Hutchinson, T. C. (1967). Comparative studies of the ability of species to withstand prolonged periods of darkness. *J. Ecol.* **55,** 291–299.

Hutchinson, T. C. (1972). "The Occurrence of Lead, Cadmium, Nickel, Vanadium, and Chloride in Soils and Vegetation of Toronto in Relation to Traffic Density," Publ. EH-2. Institute for Environmental Studies, University of Toronto, Toronto, Ont.

Hutchinson, T. C., and Freedman, B. (1978). Effects of experimental crude oil spills on subarctic boreal forest vegetation near Norman Wells, N.W.T., Canada. *Can. J. Bot.* **56,** 2424–2433.

Hutchinson, T. C., and Havas, M. (1985). Recovery of previously acidified lakes near Coniston, Canada following reduc-

tions in atmospheric sulfur and metal emissions. *Water, Air, Soil Pollut.* **20,** 20–32.

Hutchinson, T. C., and Whitby, L. M. (1974). Heavy metal pollution in the Sudbury mining and smelting region of Canada. 1. Soil and vegetation contamination by nickel, copper, and other metals. *Environ. Conserv.* **1,** 123–132.

Hutchinson, T. C., Gizyn, W., Havas, M., and Zobens, V. (1978). Effect of long-term lignite burns on arctic ecosystems at the Smoking Hills, N.W.T. *Trace Subst. Environ. Health* **12,** 317–332.

Hutchinson, T. C., Hellebust, J. A., Mackay, D., Tam, D., and Kauss, P. (1979). Relationships of hydrocarbon solubility to toxicity in algae and cellular membrane effects. *API Publ.* **4308,** 541–547.

Hutchinson, T. C., Nakatsu, C., and Tam, D. (1981). Multiple metal tolerances and co-tolerances in algae. *Heavy Met. Environ., Int. Conf., 3rd, 1981,* pp. 300–304.

Hutchinson, T. C., Bozic, L., and Muñoz-Vega, G. (1986). Responses of five species of conifer seedlings to aluminum stress. *Water, Air, Soil Pollut.* **31,** 283–294.

Hutton, M. (1980). Metal contamination of feral pigeons *Columba livia* from the London area. Biological effects of lead exposure. *Environ Pollut., Ser. A* **22,** 281–293.

Hutton, M. (1984). Impact of airborne metal contamination on a deciduous woodland system. *In* "Effects of Pollutants at the Ecosystem Level" (P. J. Sheehan, D. R. Miller, G. C. Butler, and P. Bourdeau, eds.), Scope Rep. 22, pp. 365–375. Wiley, New York.

Hutton, M., and Goodman, G. T. (1980). Metal contamination of feral pigeons *Columba livia* from the London area. Part 1. Tissue accumulation of lead, cadmium, and zinc. *Environ. Pollut.* **22,** 207–217.

Huynh, D. H., Can, D. N., Anh, Q., and Thang, N. V. (1984). Long-term changes in the mammalian fauna following herbicidal attack. *In* "Herbicides in War: The Long-term Ecological Consequences" (A. H. Westing, ed.), pp. 49–52. Taylor & Francis, London.

Hynes, H. B. M. (1971). "The Biology of Polluted Waters." Univ. of Toronto Press, Toronto, Ont.

Ilmavirta, V. (1980). Phytoplankton in 35 Finnish brown-water lakes of different trophic status. *Dev. Hydrobiol.* **3,** 121–130.

Imai, M., Yoshida, K., Kotchmar, D. J., and Lee, S. D. (1985). A survey of health effects studies of photochemical air pollution in Japan. *J. Air Pollut. Control Assoc.* **35,** 103–108.

Ingelog, T. (1978). Effects of the silvicultural use of phenoxy acid herbicides on forest vegetation in Sweden. *Ecol. Bull.* **27,** 240–254.

International Electric Research Exchange (IERE) (1981). "Effects of SO$_2$ and Its Derivatives on Health and Ecology," Vol. 1. IERE, Canadian Electrical Association, Montreal.

International Joint Commission (1987). "Report on Great Lakes Water Quality." Great Lakes Water Board, IJC, Windsor, Ont.

International Marine Organiztion (IMO) (1990). "Petroleum in the Marine Environment," Doc. MEPC 30/INF-13. Submitted by the United States to the International Marine Organization (IMO), London (cited in GESAMP, 1993).

Inving, P. M. (1983). Acidic precipitation effects on crops: A review and analysis of research. *J. Environ. Qual.* **12,** 442–453.

Irwin, L. L. (1985). Foods of moose, *Alces alces,* and white-tailed deer, *Odocoileus virginianus,* on a burn in boreal forest. *Can. Field Nat.* **99,** 240–245.

Irwin, L. L., and Peek, J. M. (1983). Elk, *Cervus elaphus* foraging related to forest management and succession in Idaho. *Can. Field Nat.* **97,** 443–447.

Isidorov, V. A., Zenkevich, I. G., and Ioffe, B. V. (1985). Volatile organic compounds in the atmosphere of forests. *Atmos. Environ.* **19,** 1–8.

Jablonski, D. (1991). Extinctions: A paleontological perspective. *Science* **253,** 754–757

Jacobson, J. (1980). The influence of rainfall composition on the yield and quality of agricultural crops. *In* "Ecological Impacts of Acid Precipitation" (D. Drablos and A. Tollan, eds.), pp. 41–46. SNSF Project, Oslo, Norway.

Jacobson, J. (1982a). Ozone and the growth and productivity of agricultural crops. *In* "Effects of Gaseous Air Pollution in Agriculture and Horticulture" (M. H. Unsworth and D. P. Ormrod, eds.), pp. 293–304. Butterworth, London.

Jacobson, J. (1982b). Economics of biological assessment. *J. Air Pollut. Control Assoc.* **32,** 145–146.

Jacobson, J., and Hill, A. C., eds. (1970). "Recognition of Air Pollution Injury to Plants: A Pictorial Atlas." Air Pollution Control Association, Pittsburgh.

Jacobson, J., and Showman, R. E. (1984). Field surveys of vegetation during a period of rising electric power generation in the Ohio Valley. *J. Air Pollut. Control Assoc.* **34,** 48–51.

Jacoby, J. M., Lynch, D. D., Welch, E. B., and Perkins, M. A. (1982). Internal phosphorus loading in a shallow eutrophic lake. *Water Res.* **16,** 911–919.

Jaffre, T., Brooks, R. R., Lee, J., and Reeves, R. D. (1976). *Sebertia acuminata*: A hyperaccumulator of nickel from New Caledonia. *Science* **193,** 579–580.

Jain, S. K., and Bradshaw, A. D. (1966). Evolutionary divergence among adjacent plant populations. 1. The evidence and its theoretical analysis. *Heredity* **21,** 407–441.

James, F. C. (1971). Ordinations of habitat relationships among breeding birds. *Wilson Bull.* **83,** 215–236.

James, G. I., and Courtin, G. M. (1985). Stand structure and growth form of the birch transition community in an industrially damaged ecosystem, Sudbury, Ontario. *Can. J. For. Res.* **15,** 809–817.

Janzen, D. H. (1987). Insect diversity in a Costa Rican dry forest: Why keep it, and how. *Biol. J. Linn. Soc.* **30,** 343–356.

Jeffries, D. S., Cox, C. M., and Dillon, P. J. (1976). Depression of pH in lakes and streams in central Ontario during snowmelt. *J. Fish. Res. Board Can.* **36,** 640–646.

Jennings, D. T., and Crawford, H. S. (1985). "Predators of Spruce Budworm," USDA For. Serv., Agric. Handb. No. 644. Northeastern Forest Experiment Station, Orono, ME.

Jennings, J. D. (1979). "The Prehistory of Polynesia." Harvard Univ. Press, Cambridge, MA.

Jenny, H. (1980). "The Soil Resource." Springer-Verlag, New York.

Jensen, K. W., and Snekvik, E. (1972). Low pH levels wipe out salmon and trout populations in southernmost Norway. *Ambio* **1,** 223–225.

Johannessen, M., and Henriksen, A. (1978). Chemistry of snow meltwater: Changes in concentration during melting. *Water Resour. Res.* **14,** 615–619.

Johnels, A., Tyler, G., and Westermark, T. (1979). A history of mercury levels in Swedish fauna. *Ambio* **8,** 160–168.

Johns, B. W. (1993). Whooping crane conservation. *In* "Proceedings of the Third Prairie Conservation and Endangered Species Workshop" (G. L. Holroyd, H. L. Dickson, M. Regnier, and H. C. Smith, eds.), Nat. Hist. Occas. Pap. No. 19, pp. 344–347. Provincial Museum of Alberta, Edmonton.

Johnsen, I., and Sochting, V. (1980). Distribution of cryptogamic epiphytes in a Danish city in relation to air pollution and bark properties. *Bryologist* **79,** 86–92.

Johnsgard, P. A. (1983). "The Grouse of the World". Univ. of Nebraska Press, Lincoln.

Johnson, A. H. (1979). Evidence of acidification of headwater streams in the New Jersey pinelands. *Science* **206,** 834–836.

Johnson, A. H. (1983). Red spruce decline in the northeastern U.S.: Hypotheses regarding the role of acid rain. *J. Air Pollut. Control Assoc.* **33,** 1049–1054.

Johnson, A. H., and Siccama, T. G. (1984). Decline of red spruce in the northern Appalachians: Assessing the possible role of acid deposition. *Tappi J.* **67,** 68–72.

Johnson, A. H., and Siccama, T. G. (1989). Decline of red spruce in the high-elevation forests of the northeastern United States. *In* "Air Pollution's Toll on Forests and Crops" (J. J. MacKenzie and M. T. El-Ashry, eds.), pp. 191–234. Yale Univ. Press, New Haven, CT.

Johnson, A. H., Siccama, T. G., Wong, D., Turner, R. S., and Barringer, T. H. (1981). Recent changes in pattern of tree growth rate in the New Jersey pinelands: A possible effect of acid rain. *J. Environ. Qual.* **10,** 427–430.

Johnson, A. H., Siccama, T. G., and Friedland, A. J. (1982). Spatial and temporal patterns of lead accumulation in the forest floor in the northeastern United States. *J. Environ. Qual.* **11,** 577–580.

Johnson, A. H., Friedland, A. J., and Dushoff, J. G. (1986). Recent and historical red spruce mortality: Evidence of climatic influence. *Water, Air, Soil Pollut.* **30,** 319–330.

Johnson, A. H., Cook, E. R., and Siccama, T. G. (1988). Climate and red spruce growth and decline in the northern Appalachians. *Proc. Natl. Acad. Sci. U.S.A.* **85,** 5369–5373.

Johnson, D. W., Kilsby, C. G., McKenna, D. S., Saunders, R. W., Jenkins, G. J., Smith, F. B., and Foot, J. S. (1991). Airborne observations of the physical and chemical characteristics of Kuwaiti oil smoke plume. *Nature (London)* **353,** 617–621.

Johnson, N. E., and Lawrence, W. H. (1977). Role of pesticides in the management of American forests. *In* "Pesticides in the Environment" (R. White-Stevens, ed.), Vol. 3, pp. 135–255. Dekker, New York.

Johnson, N. M. (1979). Acid rain: Neutralization within the Hubbard Brook ecosystem and regional implications. *Science* **204**, 497–499.

Johnson, W. W., and Finley, M. T. (1980). "Handbook of Acute Toxicity of Chemicals to Fish and Aquatic Invertebrates," Resour. Bull. 137. U.S. Department of the Interior, Fish and Wildlife Service, Washington, DC.

Johnsson, B., and Sundberg, R. (1972). "Has the Acidification by Atmospheric Pollution Caused a Growth Reduction in Swedish Forests?" Res. Note 20. Department of Forest Yield Research, Royal College of Forestry, Stockholm, Sweden.

Johnston, D. W., and Odum, E. P. (1956). Breeding bird populations in relation to plant succession in the Piedmont of Georgia. *Ecology* **37**, 50–62.

Jones, A. W. (1957). The flora of the city of London bombed sites. *London Nat.* **37**, 189–210.

Jones, D., Ronald, K., Levigne, D. M., Frank, R., Holdrinet, M., and Uthe, J. F. (1975). Organochlorine and mercury residues in the harp seal (*Pagophilus groenlandicus*). *Sci. Total Environ.* **5**, 181–195.

Jones, J. B. (1992). Environmental impact of trawling the seabed: A review. *N.Z. J. Mar. Freshwater Res.* **26**, 59–67.

Jones, P. (1991). Gulf oil spill. *Mar. Pollut. Bull.* **22**(4), 164.

Jones, P., Allen, L. H., Jones, J. W., and Valle, R. V. (1985). Photosynthesis and transpiration responses of soybean canopies to short- and long-term CO_2 treatments. *Agron. J.* **77**, 119–126.

Jones, P. D., Wigley, T. M. L., and Wright, P. B. (1986). Global temperature variations between 1861 and 1984. *Nature (London)* **322**, 430–434.

Jordan, M. J. (1975). Effects of zinc smelter emissions and fire on a chestnut-oak woodland. *Ecology* **56**, 78–91.

Jordan, M. J., and Lechevalier, M. P. (1975). Effects of zinc-smelter emissions on forest soil microflora. *Can. J. Microbiol.* **21**, 1855–1865.

Jowett, D. (1964). Population studies on lead tolerant *Agrostis tenuis*. *Evolution (Lawrence, Kans.)* **18**, 70–80.

Joyal, R. (1976). Winter foods of moose in La Verendre Park, Quebec: An evaluation of two browse survey methods. *Can. J. Zool.* **54**, 1765–1770.

Jurgensen, M. F., Larsen, M. J., and Harvey, A. E. (1979). "Forest Soil Biology—Timber Harvesting Relationships," USDA For. Serv. Gen. Tech. Rep. INT-69. Intermountain Forest and Range Experiment Station, Ogden, UT.

Kadono, Y. (1982). Occurrence of aquatic macrophytes in replation to pH, alkalinity, Ca^{++}, Cl^-, and conductivity. *Jpn. J. Ecol.* **32**, 39–44.

Kalff, J., and Welch, H. E. (1974). Phytoplankton production in Char Lake, a natural polar lake, and in Meretta Lake, a polluted polar lake, Cornwallis Island, Northwest Territories. *J. Fish. Res. Board Can.* **31**, 621–636.

Kalff, J., Kling, H. J., Holmgren, S. H., and Welch, H. E. (1975). Phytoplankton, phytoplankton growth, and biomass cycles in an unpolluted and in a polluted polar lake. *Verh.—Int. Ver. Theor. Angew. Limnol.* **19**, 487–495.

Kandler, O. (1992). Historical declines and diebacks of central European forests and present conditions. *Environ. Toxicol. Chem.* **11**, 1077–1093.

Karr, J. R. (1968). Habitat and avian diversity on strip-mined land in east-central Illinois. *Condor* **70**, 348–367.

Karr, J. R. (1981). Assessment of biotic integrity using fish communities. *Fisheries* **6**, 21–27.

Karr, J. R. (1982). Avian extinction on Barro Colorado Island, Panama: A reassessment. *Am. Nat.* **119**, 220–239.

Karr, J. R. (1991). Biological integrity: A long-neglected aspect of water resource management. *Ecol. Appl.* **1**, 66–74.

Karr, J. R. (1993). Measuring biological integrity: Lessons from streams. *In* "Ecological Integrity and the Management of Ecosystems" (S. Woodley, J. Kay, and G. Francis, eds.), pp. 83–104. St. Lucie Press, Boca Raton, FL.

Karr, J. R., and Roth, R. R. (1971). Vegetation structure and avian diversity in several New World areas. *Am. Nat.* **105**, 423–435.

Katz, M., ed. (1939). "Effects of sulfur dioxide on vegetation." National Research Council of Canada, Ottawa, Ont.

Kaufman, L. (1992). Catastrophic change in species-rich freshwater ecosystems. *BioScience* **42**, 846–858.

Kaufmann, M. R., Moir, W. H., and Covington, W. W. (1992). Old-growth forests: What do we know about their ecology and management in the Southwest and Rocky Mountain Regions? *In* "Old-growth Forests in the Southwest and Rocky Mountain Regions," USDA For. Serv. Gen. Tech. Rep. RM-213, pp. 1–11. Rocky Mountain Forest and Range Experiment Station, Fort Collins, CO.

Kawasaki, T. (1989). Long-term variability in the pelagic fish populations. *In* "Long-term Variability of Pelagic Fish Populations and Their Environment" (T. Kawasaki, S. Tanaka, Y. Toba, and A. Taniguchi, eds.), pp. 47–60. Pergamon, New York.

Keeling, C. D., and Whorf, T. P. (1991). Mauna Loa. *In* "Trends '91. A Compendium of Data on Global Change," pp. 12–15. Carbon Dioxide Information Center, Oak Ridge, TN.

Keeney, D. R. (1975). Toxic elements in agriculture. Unpublished manuscript, Department of Soil Science, University of Wisconsin, Madison.

Keeves, A. (1966). Some evidence of loss of productivity with successive rotations of *Pinus radiata* in the south-east of south Australia. *Aust. For.* **30**, 51–63.

Kellner, O., and Marshagen, M. (1991). Effects of irrigation and fertilization on the ground vegetation in a 130-year-old stand of Scotch pine. *Can. J. For. Res.* **21**, 733–738.

Kellogg, W. W., Cadle, R. D., Allen, E. R., Lazrus, A. L., and Martell, E. A. (1972). The sulfur cycle. *Science* **175**, 587–596.

Kelly, C. A., Rudd, J. W. M., Furutani, A., and Schindler, D. W. (1984). Effects of lake acidification on rates of organic matter decomposition in sediments. *Limnol. Oceanogr.* **29,** 687–694.

Kelly, J. R., and Harwell, M. A. (1989). Indicators of ecosystem response and recovery. *In* "Ecotoxicology: Problems and Approaches" (S. A Levin, M. A. Harwell, J. R. Kelly, and K. D. Kimball, eds.), pp. 9–35. Springer-Verlag, New York.

Kelly, J. M. (1984). Power plant influences on bulk precipitation, throughfall, and stemflow nutrient inputs. *J. Environ. Qual.* **13,** 405–409.

Kelso, D. D., and Kendziorek, M. (1991). Alaska's response to the *Exxon Valdez* oil spill. *Environ. Sci. Technol.* **25,** 16–23.

Kelso, J. R. M., Shaw, M. A., Minns, C. K., and Mills, K. H. (1990). An evaluation of the effects of atmospheric deposition on fish and the fishery resource of Canada. *Can. J. Fish. Aquat. Sci.* **47,** 644–655.

Kemperman, J. A., and Barnes, B. V. (1976). Clone size in American aspens. *Can. J. Bot.* **54,** 2603–2607.

Kenaga, E. E. (1975). The evaluation of the safety of 2,4,5-T to birds in areas treated for vegetation control. *Residue Rev.* **59,** 1–19.

Kendall, R. J. (1982). Wildlife toxicology. *Environ. Sci. Technol.* **16,** 448a-453a.

Kendall, R. J., and Akerman, J. (1992). Terrestrial wildlife exposed to agrochemicals: An ecological risk assessment perspective. *Environ. Toxicol. Chem.* **11,** 1727–1749.

Kendall, R. J., and Scanlon, P. F. (1982). Tissue lead concentrations and blood characteristics of mourning doves from southwestern Virginia. *Arch. Environ. Contam. Toxicol.* **11,** 269–272.

Kendeigh, S. C. (1948). Bird populations and biotic communities in northern lower Michigan. *Ecology* **29,** 101–114.

Kerekes, J., and Freedman, B. (1988). Physical, chemical, and biological characteristics of three watersheds in Kejimkujik National Park, Nova Scotia. *Arch. Environ. Contam. Toxicol.* **18,** 183–200.

Kerekes, J., and Freedman, B. (1989). Seasonal variations of water chemistry in oligotrophic streams and rivers in Kejimkujik National Park, Nova Scotia. *Water, Air, Soil Pollut.* **46,** 131–144.

Kerekes, J., Howell, G., Beauchamp, S., and Pollock, T. (1982). Characterization of three lake basins sensitive to acid precipitation in central Nova Scotia. *Int. Rev. Gesamten Hydrobiol.* **67,** 679–694.

Kerekes, J., Freedman, B., Howell, G., and Clifford, P. (1984). Comparison of the characteristics of an acidic eutrophic and an acidic oligotrophic lake near Halifax, Nova Scotia. *Water Pollut. Res. J. Can.* **19,** 1–10.

Kerekes, J., Freedman, B., Beauchamp, S., and Tordon, R. (1989). Physical and chemical characteristics of three acidic, oligotrophic lakes and their watersheds in Kejimkujik National Park, Nova Scotia. *Water, Air, Soil Pollut.* **46,** 99–117.

Kernan, H. S. (1945). War's toll of French forests. *Am. For.* **51,** 442.

Kerr, R. A. (1988a). Is the greenhouse here? *Science* **239,** 559–561.

Kerr, R. A. (1988b). Stratospheric ozone is decreasing. *Science* **239,** 1489–1491.

Kettela, E. (1983). "A Cartographic History of Spruce Budworm Defoliation from 1967 to 1981 in Eastern North America," Inf. Rep. DPC-X-14. Maritimes Forest Research Centre, Canadian Forestry Service, Fredericton, New Brunswick.

Kevan, P. G. (1975). Forest application of the insecticide fenitrothion and its effect on wild bee pollinators (Hymenoptera: Apoidea) of lowbush blueberries (*Vaccinium* spp.) in southern New Brunswick, Canada. *Biol. Conserv.* **7,** 301–309.

Kevan, P. G., and Plowright, R. C. (1989). Fenitrothion and insect pollinators. *In* "Environmental Effects of Fenitrothion Use in Forestry," pp. 13–42. Conservation and Protection, Environment Canada, Dartmouth, Nova Scotia.

Kilpatrick, R., Chair. (1979). "Review of the Safety for Use of the Herbicide 2,4,5-T." Advisory Committee on Pesticides, U.K. Ministry of Agriculture, Fisheries, and Food, London.

Kilpatrick, R., Chair. (1980). "Further Review of the Safety for Use in the U.K. of the Herbicide 2,4,5-T." Advisory Committee on Pesticides, U.K. Ministry of Agriculture, Fisheries, and Food, London.

Kimmins, J. P. (1977). Evaluation of the consequences for future tree productivity of the loss of nutrients in whole-tree harvesting. *For. Ecol. Manage.* **1,** 169–183.

Kimmins, J. P. (1986). Predicting the consequences of intensive forest harvesting on long-term productivity: The need for a hybrid model such as FORCYTE-11. *In* "Predicting the Consequences of Intensive Forest Harvesting on Long-term Producticity" (G. I. Agren, ed.), Rep. No. 26, pp. 31–84. Swedish Univ. Agric. Sci., Uppsala.

Kimmins, J. P. (1987). "Forest Ecology." Macmillan, New York.

Kimmins, J. P., and Feller, M. C. (1976). Effect of clear-cutting and broadcast slash burning on nutrient budgets, streamwater chemistry, and productivity in western Canmada. *Proc. IUFRO World Congr., 16th,* pp. 186–197.

Kimmins, J. P., Scoular, K. A., and Feller, M. C. (1981). "FORCYTE—a Computer Approach to Evaluating the Effects of Whole Tree Harvesting on Nutrient Budgets and Future Tree Productivity." Report to Canadian Forestry Service, ENFOR Programme, Faculty of Forestry, University of British Columbia, Vancouver.

King, D. L., Simmler, J. J., Decker, D. S., and Ogg, C. W. (1974). Acid strip mine lake recovery. *J. Water Pollut. Control Fed.* **10,** 2301–2316.

King, W. B. (1981). "Endangered Birds of the World: The ICBP Bird Red Data Book." Smithsonian Institution Press, Washington, DC.

Kingsbury, P. D. (1975). Effects of aerial forest spraying on aquatic fauna. *In* "Aerial Control of Forest Insects in Canada" (M. L. Prebble, ed.), pp. 280–283. Department of the Environment, Ottawa, Ont.

Kingsbury, P. D. (1977). "Fenitrothion in a Lake Ecosystem," Rep. CC-X-146. Chemical Control Research Institute, Environment Canada, Ottawa, Ont.

Kingsbury, P. D., and McLeod, B. B. (1980). The Impact of Spruce Budworm Control Operations Involving Sequential Applications Upon Forest Avifauna in the Lower St. Lawrence Region of Quebec," Rep. FPM-X-34. Forest Pest Management Institute, Canadian Forestry Service, Sault Ste. Marie, Ont.

Kingsbury, P. D., and McLeod, B. B. (1981). "Fenitrothion and Forest Avifauna Studies on the Effects of High Dosage Applications," Rep. FPM-X-43. Forest Pest Management Institute, Canadian Forestry Service, Sault Ste. Marie, Ont.

Kingsbury, P. D., McLeod, B. B., and Millikin, R. L. (1981). "The Environmental Impact of Nonyl Phenol and the Matacil Formulation. Part 2. Terrestrial Ecosystems," Rep. FPM-X-36. Forest Pest Management Institute, Canadian Forestry Service, Sault Ste. Marie, Ont.

Kirkby, E. A. (1969). Ion uptake and ionic balance in plants in relation to the form of nitrogen nutrition. In "Ecological Aspects of the Mineral Nutrition of Plants" (I. H. Rorison, ed.), pp. 215–235. Blackwell, London.

Kirkham, I. R., and Montevecchi, W. A. (1982). The breeding birds of Funk Island: An historical perspective. *Am. Birds* **36,** 111–118.

Kirkland, G. L. (1977). Responses of small mammals to the clearcutting of northern Appalachian forests. *J. Mammal.* **58,** 600–609.

Kitzmiller, J. H. (1990). Managing genetic diversity in a tree improvement program. *For. Ecol. Manage.* **35,** 131–149.

Klein, D. R. (1974). Reaction of reindeer to obstructions and disturbances. *Science* **173,** 393–398.

Klein, R. M., and Perkins, T. D. (1987). Cascades of causes and effects of forest decline. *Ambio* **16,** 86–93.

Kling, G. W., Clark, M. A., Compton, H. R., Devine, J. D., Evans, W. C., Humphry, A. M., Koenigsberg, E. J., Lockewood, J. P., Tuttle, M. L., and Wagner, G. N. (1987). The 1986 Lake Nyos gas disaster in Cameroon, West Africa. *Science* **236,** 169–175.

Knoll, A. H. (1984). Patterns of extinction in the fossil record of vascular plants. In "Extinctions" (M. H. Nitecki, ed.), pp. 21–68. Univ. of Chicago Press, Chicago.

Kochenderfer, J. N. (1970). "Erosion Control on Logging Roads in the Appalachians," USDA For. Serv. Res. Pap. NE-158. Northeastern Forest Experiment Station, Broomall, PA.

Kochenderfer, J. N., and Aubertin, G. M. (1975). Effects of management practices on water quality and quantity: Fernow Experimental Forest, West Virginia. In "Municipal Watershed Management," USDA For. Serv. Gen. Tech. Rep. NE-13, pp. 14–24. Northeastern Forest Experiment Station, Broomall, PA.

Kochenderfer, J. N., and Wendel, G. W. (1983). Plant succession and hydrological recovery on a deforested and herbicided watershed. *For. Sci.* **29,** 545–558.

Koeman, J. H., Peeters, W. H. M., Koudstaal-Hol, C. H. M., Tjioe, P. S., and Degoeij, J. J. M. (1973). Mercury-selenium correlations in marine mammals. *Nature (London)* **245,** 385–386.

Koivusaari, J., Nuuja, I., Polokangas, R., and Finnlund, M. (1980). Relationships between productivity, eggshell thickness, and pollutant contents of addled eggs in the population of white-tailed eagle *Haliaeetus albicilla* L. in Finland during 1969–78. *Environ. Pollut., Ser. A* **23,** 41–52.

Koons, C. B. (1984). Input of petroleum to the marine environment. *Mar. Technol. Soc. J.* **18,** 97–112.

Koons, C. B., and Jahns, H. O. (1992). The fat of oil from the Exxon Valdez—a perspective. *Mar. Technol. Soc. J.* **26,** 61–69.

Kornberg, H., Chair. (1981). "Royal Commission on Environmental Pollution," 8th Rep. H.M. Stationery Office, London.

Kramer, J., and Tessier, A. (1982). Acidification of aquatic ecosystems: A critique of chemical approaches. *Environ. Sci. Technol.* **16,** 606A-615A.

Krause, G. H. M., Arndt, U., Brandt, C. J., Bucher, J., Kenk, G., and Matzner, E. (1986). Forest decline in Europe: Development and possible causes. *Water, Air, Soil Pollut.* **31,** 647–668.

Krause, H. H. (1982). Nitrate formation and movement before and after clear-cutting of a monitored watershed in central New Brunswick, Canada. *Can. J. For. Res.* **12,** 922–930.

Krause, S. D. (1990). Rates and potential causes of mortality in North Atlantic right whales (*Eubalaena glacialis*). *Mar. Mammal Sci.* **6,** 278–291.

Krebs, C. J. (1985). "Ecology: The Experimental Analysis of Distribution and Abundance.," 3rd ed. Harper & Row, New York.

Krefting, L. W. (1962). Use of silvicultural techniques for improving deer habitat in the United States. *J. For.* **16,** 40–42.

Krefting, L. W. (1974). Moose distribution and habitat selection in north central North America. *Nat. Can.* **101,** 81–100.

Krefting, L. W., and Hansen, H. L. (1969). Increasing browse for deer by aerial applications of 2,4-D. *J. Wildl. Manage.* **33,** 784–790.

Kreutzweiser, D. P., and Kingsbury, P. D. (1989). Drift of aquatic invertebrates in a glyphosate contaminated watershed. In "Proceedings of the Carnation Creek Workshop," pp. 250–256. FRDA Rep. 063, Forest Pest Management Institute, Sault Ste Marie, Ont.

Kruckeberg, A. R. (1954). The ecology of serpentine soils. III. Plant species in relation to serpentine soils. *Ecology* **35,** 267–274.

Kruckeberg, A. R. (1984). "California Serpentine: Flora, Vegetation, Geology, Soils, and Management Problems." Univ. of California Press, Los Angeles.

Krug, E. C., Isaacson, P. J., and Frink, C. R. (1985). Appraisal of some current hypotheses describing acidification of watersheds. *J. Air Pollut. Control Assoc.* **35,** 109–114.

Kuja, A. L. (1980). Revegetation of mine tailings using native species from disturbed sites in northern Canada. M.Sc. Thesis, Department of Botany, University of Toronto, Toronto, Ont.

Kulhavy, D. L., Ross, W. G., Conner, R. N., Mitchell, J. H., and Chrismen, G. M. (1990). "Silviculture and the Red-cockaded Woodpecker: Where Do We Go From Here? USDA For. Serv. Gen. Tech. Rep. 70, pp. 786–794. Southeastern Forest Experiment Station, Asheville, NC.

Kwain, W. H. (1975). Effects of temperature on development and survival of rainbow trout, *Salmo gairdneri*, in acid waters. *J. Fish. Res. Board Can.* **32**, 493–497.

LaBudde, S. (1989). "Stripmining the Seas: A Global Perspective on Driftnet Fisheries." Earthtrust, Honolulu, HI.

Lacroix, G. L. (1985). Survival of eggs and alevins of Atlantic salmon (*Salmo salar*) in relation to the chemistry of interstitial water in redds in some acidic streams of Atlantic Canada. *Can. J. Fish. Aquat. Sci.* **42**, 292–299.

Lacroix, G. L., and Townsend, D. R. (1987). Responses of juvenile Atlantic salmon (*Salmo salar*) to episodic increases in acidity of Nova Scotia rivers. *Can. J. Fish. Aquat. Sci.* **44**, 1475–1484.

Lagerwerff, J. V., and Specht, A. W. (1970). Contamination of roadside soils and vegetation with cadmium, nickel, lead, and zinc. *Environ. Sci. Technol.* **4**, 583–586.

Lagerwerff, J. V., Biersdorf, G. T., and Brower, D. L. (1976). Retention of metals in sewage sludge. I. Constituent heavy metals. *J. Environ. Qual.* **5**, 19–22.

Lamberson, R. H., McKelvey, R., Noon, B. R., and Voss, C. (1992). A dynamic analysis of northern spotted owl viability in a fragmented forest landscape. *Conserv. Biol.* **6**, 505–512.

Land Use Regulatory Commission (LURC) (1979). "A Survey of Erosion and Sedimentation Problems Associated with Logging in Maine." LURC, Maine Department of Conservation, Augusta.

Lappe, M. (1991). "Chemical Deception. The Toxic Threat to Health and the Environment." Sierra Club Books, San Francisco.

Larcher, W., and Bauer, H. (1981). Ecological significance of resistance to low temperature. In "Physiological Plant Ecology" (O. L. Lange, C. B. Osmond, and H. Ziegler, eds.), pp. 403–437. Springer-Verlag, New York.

Last, F. J. (1987). The nature, and elucidation of causes of forest decline. *In* "Proceedings of the Workshop on Forest Decline and Reproduction: Regional and Global Consequences" (L. Kairiukstis, S. Nilsson, and A. Stroszak, eds.), pp. 63–78. IIASA, Laxenburg, Austria.

Lathrop, G. D., Wolfe, W. H., Albanese, R. A., and Moynahan, P. M. (1984). "An Epidemiologic Investigation of Health Effects in Air Force Personnel Following Exposure to Herbicides. Baseline Morbidity Study Results." The Surgeon General, U.S. Air Force, Washington, DC.

Lavallee, F. C., and Fedoruk, L. P. (1989). "The Elimination of Leaded Motor Gasoline," Rep. EPS 3/TS/1. Environmental Protection Service, Environment Canada, Ottawa, Ont.

Lavelle, P. (1987). Biological processes and productivity of soils in the humid tropics. *In* "The Geophysiology of Amazonia" (R. E. Dickensen, ed.), pp. 175–223. Wiley, New York.

Laws, R. M. (1977). The significance of vertebrates in the Antarctic marine ecosystems. *In* "Adaptations Within Antarctic Systems" (G. A. Llano, ed.), pp. 411–438. Smithsonian Institution, Washington, DC.

Lazerte, B. D. (1984). Forms of aqueous aluminum in acidified catchments of central Ontario: A methodological analysis. *Can. J. Fish. Aquat. Sci.* **41**, 766–776.

Leader-Williams, N., Albou, S. D., and Berry, P. S. M. (1990). Illegal exploitation of black rhinoceros and elephant populations: Patterns of decline, law enforcement, and patrol effort in Luangwa Valley, Zambia. *J. Appl. Ecol.* **27**, 1055–1087.

Leaf, C. F., and Brink, G. E. (1972). Simulating effects of harvest cutting on snowmelt in Colorado subalpine forests. *In* "Watersheds in Transition," pp. 191–196. American Water Resources Association, Urbana, IL.

LeBlanc, F., and De Sloover, J. (1970). Relation between industrialization and the distribution and growth of epiphytic lichens and mosses in Montreal. *Can. J. Bot.* **48**, 1485–1496.

LeBlanc, F., and Rao, D. N. (1966). Reaction of several lichens and epiphytic mosses to sulfur dioxide in Sudbury, Ontario. *Bryologist* **69**, 338–346.

LeBlanc, F., Rao, D. M., and Comeau, G. (1972). The epiphytic vegetation of *Populus tremuloides* and its significance as an air pollution indicator in Sudbury, Ontario. *Can. J. Bot.* **50**, 519–528.

Lee, P. L. (1985). History and current status of spotted owl (*Strix occidentalis*) habitat management in the Pacific Northwest region, U.S.D.A., Forest Service. *In* "Ecology and Management of the Spotted Owl in the Pacific Northwest," USDA For. Serv. Gen. Tech. Rep. PNW-185, pp. 5–10. Pacific Northwest Forest and Range Experiment Station, Portland, OR.

Lee, R. F. (1977). Accumulation and turnover of petroleum hydrocarbons in marine organisms. *In* "Fate and Effects of Petroleum Hydrocarbons in Marine Ecosystems and Organisms" (D. A. Wolfe, ed.), pp. 60–70. Pergamon, New York.

Lees, J. C. (1981). "Three Generations of Red Maple Stump Sprouts," M-X-119. Canadian Forestry Service—Maritimes, Fredericton, New Brunswick.

Lefohn, A. S. (principal author) (1990). Air quality measurements and characterizations for terrestrial effects research. *In* "Acidic Deposition: State of Science and Technology. Vol. I. Emissions, Atmospheric Processes, and Deposition," pp. 7-1 to 7–192. Superintendent of Documents, U.S. Govt. Printing Office, Washington, DC.

Lehman, J. T. (1988). Algal biomass unaltered by food-web changes in Lake Michigan. *Nature (London)* **332**, 537–538.

Leigh, C. E. G. (1982). Why are there so many kinds of tropical trees? *In* "The Ecology of a Tropical Forest" (C. E. G. Leigh, A. S. Rand, and D. M. Windsor, eds.), pp. 63–66. Smithsonian Institution Press, Washington, DC.

Leighton, F. A., Butler, R. G., and Peakall, D. B. (1985). Oil and Arctic marine birds: An assessment of risk. *In* "Petroleum Effects in the Arctic Environment" (F. R. Engelhardt, ed.), pp. 183–215. Am. Elsevier, New York.

Leivestad, H., and Muniz, I. P. (1976). Fish kills at low pH in a Norwegian river. *Nature (London)* **259**, 391–392.

Leivestad, H., Hendry, G., Muniz, I. P., and Snekvik, E. (1976). Effects of acid precipitation on freshwater organisms. *In* "Impact of Acid Precipitation on Forest and Freshwater Ecosystems in Norway" (F. H. Brakke, ed.), Res. Rep. FR6/76, pp. 87–111. SNSF Project, Oslo, Norway.

Lepp, N. W., ed. (1981). "Effect of Heavy Metal Pollution on Plants," Vol. 1. Appl. Sci. Publ., London.

Lepp, N. W., and Fairfax, J. A. W. (1976). The role of acid rain as a regulator of foliar nutrient uptake and loss. *In* "Microbiology of Aerial Plant Surfaces" (C. H. Dickinson and T. F. Preece, eds.), pp. 107–118. Academic Press, London.

LeRoy, J. C., and Keller, H. (1972). How to reclaim mined areas, tailings ponds, and dumps into valuable land. *World Min.* **25**(1), 34–41.

Lessmark, O., and Thornelof, E. (1986). Liming in Sweden. *Water, Air, Soil Pollut.* **31**, 809–815.

Levasseur, J. E., and Jory, M. L. (1982). Rétablissement naturel d'une végétation de marais maritimes altérée par les hydrocarbures de l'Amoco Cadiz: Modalités et tendances. *In* "Ecological Study of the Amoco Cadiz Oil Spill," pp. 329–362. U.S. Department of Commerce, National Oceanic and Atmospheric Administration, Washington, DC.

Levin, S. A. (1992). Orchestrating environmental research and assessment. *Ecol. Appl.* **2**, 103–106.

Levine, S. N., and Schindler, D. W. (1989). Phosphorus, nitrogen, and carbon dynamics of Experimental Lake 303, during recovery from eutrophication. *Can. J. Fish. Aquat. Sci.* **46**, 2–10.

Levine, S. N., Stainton, M. P., and Schindler, D. W. (1986). A radiotracer study of phosphorus cycling in a eutrophic Canadian Shield lake, Lake 227, northwestern Ontario. *Can. J. Fish. Aquat. Sci.* **43**, 366–378.

Levinson, A. A. (1974). "Introduction to Exploration Geochemistry." Applied Publ. Ltd., Maywood, IL.

Lewis, W. M., and Grant, M. C. (1979). Relationships between stream discharges and yield of dissolved substances from a Colorado mountain watershed. *Soil Sci.* **128**, 353–363.

Li, P., and Adams, W. T. (1989). Range-wide patterns of allozyme variation in Douglas-fir (*Pseudotsuga menziesii*). *Can. J. For. Res.* **19**, 149–161.

Li, T. A., and Landsberg, H. E. (1975). Rainwater pH close to a major power plant. *Atmos. Environ.* **9**, 81–88.

Liebsch, E. J., and de Pena, R. G. (1982). Sulfate aerosol production in coal-fired power plant plumes. *Atmos. Environ.* **16**, 1323–1331.

Likens, G. E. (1985). An experimental approach for the study of ecosystems. *J. Ecol.* **73**, 381–396.

Likens, G. E., ed. (1989). "Long-term Studies in Ecology: Approaches and Alternatives." Springer-Verlag, New York.

Likens, G. E., and Butler, T. J. (1981). Recent acidification of precipitation in North America. *Atmos. Environ.* **15**, 1103–1110.

Likens, G. E., Bormann, F. H., Johnson, N. M., Fisher, D. W., and Pierce, R. S. (1970). Effects of forest cutting and herbicide treatment on nutrient budgets in the Hubbard Brook Watershed ecosystem. *Ecol. Monogr.* **40**, 23–47.

Likens, G. E., Bormann, F. H., Pierce, R. J., Eaton, J. S., and Johnson, N. M. (1977). "Biogeochemistry of a Forested Ecosystem." Springer-Verlag, New York.

Likens, G. E., Bormann, F. H., Pierce, R. S., and Reiners, W. A. (1978). Recovery of a deforested ecosystem. *Science* **199**, 492–496.

Likens, G. E., Bormann, F. H., Pierce, R. S., Eaton, J. S., and Munn, R. E. (1984). Long-term trends in precipitation chemistry at Hubbard Brook, New Hampshire. *Atmos. Environ.* **18**, 2641–2647.

Liljestrand, H. M. (1985). Average rainwater pH, concepts of atmospheric acidity, and buffering in open systems. *Atmos. Environ.* **19**, 487–499.

Lincer, J. L., Cade, T. J., and Devine, J. M. (1970). Organochlorine residues in Alaskan peregrine falcons, rough-legged hawks, and their prey. *Can. Field Nat.* **84**, 255–263.

Lincoln, R. J., Boxshall, G. A., and Clark, P. F. (1982). "A Dictionary of Ecology, Evolution, and Systematics." Cambridge Univ. Press, Cambridge, UK.

Lind, C. T., and Cottam, G. (1969). The submerged aquatics of University Bay: A study of eutrophication. *Am. Midl. Nat.* **81**, 353–369.

Lindberg, S. E., Lovett, G. M., Richter, D. D., and Johnson, D. W. (1986). Atmospheric deposition and canopy interactions of major ions in a forest. *Science* **231**, 141–145.

Linkins, A. E., Johnson, L. A., Everett, K. R., and Atlas, R. M. (1984). Oil spills: Damage and recovery in tundra and taiga. *In* "Restoration of Habitats Impacted by Oil Spills" (J. L. Cairns and A. L. Buikema, Jr., eds.), pp. 135–155. Butterworth, Boston.

Linzon, S. N. (1971). Economic effects of SO_2 on forest growth. *J. Air Pollut. Control Assoc.* **21**, 81–86.

Linzon, S. N., and Temple, P. J. (1980). Soil resampling and pH measurements after an 18-year period in Ontario. *In* "Ecological Impact of Acid Precipitation" (D. Drablos and A. Tollan, eds.) pp. 176–177. SNSF Project, Oslo, Norway.

Lioy, P. J., and Samson, P. J. (1979). Ozone concentration patterns observed during the 1976–1977 long range transport study. *Environ. Int.* **2**, 77–83.

Little, P., and Martin, H. (1972). A survey of zinc, lead, and cadmium in soil and natural vegetation around a smelting complex. *Environ. Pollut.* **3**, 241–254.

Lloyd-Smith, J., and Piene, H. (1981). "Snowshoe Hare Girdling of Balsam Fir on the Cape Breton Highlands," Inf. Rep. M-X-124. Maritimes Forest Research Centre, Fredericton, New Brunswick.

Lockie, J. D., Ratcliffe, D. A., and Balharry, R. (1969). Breeding success and dieldrin contamination of golden eagles in west Scotland. *J. Appl. Ecol.* **6,** 381–389.

Loehr, R. C., Martin, C. S., and Rast, W., eds. (1980). "Phosphorus Management Strategies for Lakes." Ann Arbor Sci. Publ., Ann Arbor, MI.

Logan, J. A. (1983). Nitrogen oxides in the atmosphere: Global and regional budgets. *J. Geophys. Res.* **88,** 785–807.

Logie, R. R. (1975). Effects of aerial spraying of DDT on salmon populations of the Miramichi River. *In* "Aerial Control of Forest Insects in Canada" (M. L. Prebble, ed.), pp. 293–300. Department of the Environment, Ottawa, Ont.

Longstreth, J. (1991). Global climate change: Potential impacts on human health. *In* "Global Climate Change and Life on Earth" (R. L. Wyman, ed.), pp. 201–215. Routledge, Chapman, and Hall, New York.

Lorius, C., Barkov, N. I., Jouzel, J., Korotkevich, Y. S., Kotylakov, V. M., and Raymond, D. (1988). Antarctic ice core: CO_2 and climate change over the last climatic cycle. *Eos* **69,** 681–684.

Loucks, O. L. (1985). Looking for surprise in managed stressed systems. *BioScience* **35,** 428–432.

Lovejoy, T. E. (1985). Amazonia, people and today. *In* "Amazonia" (G. T. Prance and T. E. Lovejoy, eds.), pp. 328–338. Pergamon, New York.

Lovejoy, T. E., Bierregaard, R. O., Jr., Rylands, A. B., Malcolm, J. R., Quintela, C. E., Harper, L. H., Brown K. S., Jr., Powell, A. H., Schubart, H. O. R., and Hays, M. B. (1986). Edge and other effects of isolation on Amazon forest fragments. *In* "Conservation Biology" (M. E. Soule, ed.), pp. 257–285. Sinauer Assoc., Sunderland, MA.

Lovejoy, T. E., Ramkin, J. M., Bierregaard, R. O., Brown, K. S., Emmons, L. H., and Van der Voort, M. E. (1984). Ecosystem decay of Amazon forest remnants. *In* "Extinctions" (M. H. Nitecki, ed.), pp. 297–325. Univ. of Chicago Press, Chicago.

Lovett, G. M., Reiners, W. A., and Olson, R. K. (1982). Cloud droplet deposition in subalpine balsam fir forests: Hydrological and chemical budgets. *Science* **218,** 1303–1304.

Lozano, F. C., and Morrison, I. K. (1981). Disruption of hardwood nutrition by sulfur dioxide, nickel, and copper air pollution near Sudbury, Ontario. *J. Environ. Qual.* **10,** 198–204.

Lubchenko, J., Olson, A. M., Brubaker, L. B., Carpenter, S. R., Holland, M. J., Hubbell, S. P., Levin, S. A., MacMahon, J. A., Matson, P. A., Melillo, J. M., Mooney, H. A., Peterson, C. H., Pulliam, H. R., Real, L. A., Regal, P. J., and Risser, P. G. (1991). The sustainable biosphere initiative: An ecological research agenda. *Ecology* **72,** 371–412.

Lucas, G., and Synge, H. (1978). "The IUCN Plant Red Data Book." International Union for Conservation of Nature and Natural Resources, Morges, Switzerland.

Ludke, J. L., Hill, E. F., and Dieter, M. P. (1975). Cholinesterase (ChE) response and related mortality among birds fed ChE inhibitors. *Arch. Environ. Contamin. Toxicol.* **3,** 1–21.

Ludyanskiy, M. L., McDonald, D., and MacNeil, D. (1993). Impact of the zebra mussel, a bivalve indicator. *BioScience* **43,** 533–544.

Lugo, A. E. (1988). Estimating reductions in the diversity of tropical forest species. *In* "Biodiversity" (E. O. Wilson, ed.), pp. 58–70. National Academy Press, Washington, DC.

Lugo, A. E. (1992). Tree plantations for rehabilitating damaged forest lands in the tropics. *In* "Ecosystem Rehabilitation. Vol. 2. Ecosystem Analysis and Synthesis" (W. L. Wells, ed.), pp. 247–255. SPB Academic Publishers, The Hague, The Netherlands.

Lugo, A. E., and Brown, S. (1986). Steady state terrestrial ecosystems and the global carbon cycle. *Vegetatio* **68,** 83–90.

Luinenberg, O., and Osborne, S. (1990). "The Little Green Book: Quotations on the Environment." Pulp Press Book Publ., Vancouver, British Columbia.

Lund, J. W. G. (1950). Studies on *Asterionella formosa* Hass. II. Nutrient depletion and the spring maximum. *J. Ecol.* **38,** 1–35.

Lund, J. W. G., Mackereth, F. J. H., and Mortimer, C. H. (1963). Changes in depth and time of certain chemical and physical conditions and of the standing crop of *Asterionella formosa* Hass. in the North Basin of Windermere in 1947. *Philos. Trans. R. Soc. London, Ser. B* **246,** 255–290.

Lund-Hoie, K. (1984). Growth responses of Norway spruce (*Picea abies*) to different vegetation management programmes. *Aspects Appl. Biol.* **5,** 127–133.

Lund-Hoie, K. (1985). Efficacy of glyphosate in forest plantations. *In* "The Herbicide Glyphosate" (E. Grossbard and D. Atkinson, eds.), pp. 328–338. Butterworth, Toronto, Ont.

Lundquist, R. W., and Mariani, J. M. (1991). Nesting habitat and abundance of snag-dependent birds in the southern Washington Cascade Range. *In* "Wildlife and Vegetation of Unmanaged Douglas-fir Forests," USDA For. Serv. Gen. Tech. Rep. PNW-285, pp. 220–240. Pacific Northwest Forest and Range Experiment Station, Portland, OR.

Lusis, M. A., Chan, W. H., Tang, A. J. S., and Johnson, N. D. (1983). Scavenging rates of sulfur and trace metals from a smelter plume. *In* "Precipitation Scavenging, Dry Deposition, and Resuspension," pp. 369–382. Elsevier, London.

Luther, F. M., and Ellingson, R. G. (1985). Carbon dioxide and the radiation budget. *In* "Projecting the Climatic Effects of Increasing Carbon Dioxide," DOE/ER-0237, pp. 25–56. U.S. Department of Energy, Washington, DC.

Lynch, M., and Shapiro, J. (1981). Predation, enrichment, and phytoplankton community structure. *Limnol. Oceanogr.* **26,** 86–102.

Lyon, J. J. (1976). Elk use as related to characteristics of clearcuts in western Montana. *In* "Proceedings of the Elk—Logging—Roads Symposium," pp. 69–72. University of Idaho, Moscow.

Lyon, L. J., and Basile, J. V. (1980). Influences of timber harvesting and residue management on big game. *In* "Environmental Consequences of Timber Harvesting in Rocky Mountain Coniferous Forests," USDA For. Serv. Gen. Tech. Rep. INT-90, pp. 441–453. Intermountain Forest and Range Experiment Station, Ogden, UT.

Lyon, L. J., and Jensen, C. E. (1980). Management implications of elk and deer use of clear-cuts in Montana. *J. Wildl. Manage.* **44,** 352–362.

Lyon, L. J., and Mueggler, W. F. (1966). Herbicide treatment of north Idaho browse evaluated six years later. *J. Wildl. Manage.* **31,** 538–541.

Mabbutt, J. A. (1984). A new global assessment of the status and trends of desertification. *Environ. Conserv.* **11,** 103–113.

MacArthur, R. H. (1964). Environmental factors affecting bird species diversity. *Am. Nat.* **98,** 387–397.

MacArthur, R. H., and MacArthur, J. W. (1961). On bird species diversity. *Ecology* **42,** 594–598.

MacArthur, R. H., and Wilson, E. O. (1967). "The Theory of Island Biogeography." Princeton Univ. Press, Princeton, NJ.

MacArthur, R. H., Recher, H., and Cody, M. (1966). On the relation between habitat selection and species diversity. *Am. Nat.* **100,** 319–322.

MacCracken, M. C., and Luther, F. M., eds. (1985). "Projecting the Climatic Effects of Increasing Carbon Dioxide," DOE/ER-0237. U.S. Department of Energy, Washington, DC.

MacCrimmon, H. R., Wren, C. D., and Gots, B. L. (1983). Mercury uptake by lake trout, *Salvelinus namaycush,* relative to age, growth, and diet in Tadenac Lake with comparative data from other Canadian Shield lakes. *Can. J. Fish. Aquat. Sci.* **40,** 114–120.

Mackay, D., and Wells, P. G. (1981). Factors influencing the aquatic toxicity of chemically dispersed oils. *Proc. Arct. Mar. Oilspill Program Tech. Semin., 4th,* pp. 445–467.

MacKenzie, J. J., and El-Ashry, M. T. (1989). Tree and crop injury: A summary of the evidence. *In* "Air Pollution's Toll on Forests and Crops" (J. J. MacKenzie and M. T. El-Ashry, eds.), pp. 1–21. Yale Univ. Press, New Haven, CT.

Mackie, G. L. (1991). Biology of expotic zebra mussel, *Dreissena polymorpha,* in relation to native bivalves and its potential impact in Lake St. Clair. *Hydrobiologia* **219,** 251–268.

Mackinnon, D., and Freedman, B. (1992). "Effects of Silvicultural Use of the Herbicide Glyphosate on Breeding Birds of Regenerating Clearcuts in Nova Scotia," Research Report to World Wildlife Fund. Department of Biology, Dalhousie University, Halifax, Nova Scotia.

Mackinnon, D., and Freedman, B. (1993). Effects of silvicultural use of the herbicide glyphosate on breeding birds of regenerating clearcuts in Nova Scotia, Canada. *J. Appl. Ecol.* **30,** 395–406.

MacLean, A. J., and Dekker, A. J. (1978). Availability of zinc, copper, and nickel to plants grown in sewage-treated soils. *Can. J. Soil Sci.* **58,** 381–389.

MacLean, A. J., Store, B., and Cordukes, W. B. (1973).

Amounts of mercury in soil of some golf course sites. *Can. J. Soil Sci.* **53,** 130–132.

MacLean, D. A. (1980). Vulnerability of fir-spruce stands during uncontrolled spruce budworm outbreaks: A review and discussion. *For. Chron.* **56,** 213–221.

MacLean, D. A. (1984). Effects of spruce budworm outbreaks on the productivity and stability of balsam fir forests. *For. Chron.* **60,** 273–279.

MacLean, D. A. (1985). Effects of spruce budworm outbreaks on forest growth and yield. *In* "Recent Advances in Spruce Budworms Research," pp. 148–175. Canadian Forestry Service, Ottawa, Ont.

MacLean, D. A. (1988). Effects of spruce budworm outbreaks on vegetation, structure, and succession of balsam fir forests on Cape Breton Island, Canada. *In* "Plant Form and Vegetation Structure" (M. J. A. Werger, P. J. M. van der Aart, H. J. During, and J. J. A. Verhoeven, eds.), pp. 253–261. SPB Academic Publishers, The Hague, The Netherlands.

MacLean, D. A. (1990). Impact of forest pests and fire on stand growth and timber yield: Implications for forest management planning. *Can. J. For. Res.* **20,** 391–404.

MacLean, D. A., and Erdle, T. A. (1984). A method to determine effects of spruce budworm on stand yield and wood supply projections for New Brunswick. *For. Chron.* **60,** 167–173.

MacLean, D. A., and Morgan, M. G. (1983). Long-term growth and yield response of young fir to manual and chemical release from shrub competition. *For. Chron.* **59,** 177–183.

MacLean, D. A., and Ostaff, D. P. (1989). Patterns of balsam fir mortality caused by an uncontrolled spruce budworm outbreak. *Can. J. For. Res.* **19,** 1087–1095.

MacLean, D. A., and Wein, R. W. (1977). Changes in understory vegetation with increasing stand age in New Brunswick forests: Species composition, cover, biomass, and nutrients. *Can. J. Bot.* **7,** 2818–2831.

MacLean, D. A., Kline, A. W., and Levine, D. R. (1984). Effectiveness of spruce budworm spraying in New Brunswick in protecting the spruce component of spruce-fir stands. *Can. J. For. Res.* **14,** 163–176.

MacLean, D. C., and Schneider, R. E. (1976). Photochemical oxidants in Yonkers, New York. Effects on yield of bean and tomato. *J. Environ. Qual.* **5,** 75–78.

Maessen, O., Freedman, B., and Nams, M. L. N. (1983). Resource allocation in high-arctic vascular plants of differing growth form. *Can. J. Bot.* **61,** 1680–1691.

Magasi, L. P. (1983). "Forest Pest Conditions in the Maritimes in 1982," Inf. Rep. M-X-141. Forestry Canada—Maritimes Region, Fredericton, New Brunswick.

Magasi, L. P. (1990). "Forest Pest Conditions in the Maritimes in 1989," Inf. Rep. M-X-177. Forestry Canada—Maritimes Region, Fredericton, New Brunswick.

Magurran, A. E. (1988). "Ecological Diversity and Its Measurement." Princeton Univ. Press, Princeton, NJ.

Mahendrappa, M. K. (1983). Chemical characteristics of precipitation and hydrogen input in throughfall and stemflow under eastern Canadian forest stands. *Can. J. For. Res.* **13**, 948–955.

Makarewicz, J. C. (1993a). Phytoplankton biomass and species composition in Lake Erie, 1970 to 1987. *J. Great Lakes Res.* **19**, 258–274.

Makarewicz, J. C. (1993b). A lakewide comparison of zooplankton biomass and its species composition in Lake Erie, 1983 to 1987. *J. Great Lakes Res.* **19**, 275–290.

Makarewicz, J. C., and Bertram, P. (1991). Evidence for the restoration of the Lake Erie ecosystem. *BioScience* **41**, 216–223.

Maki, A. W. (1991). The *Exxon Valdez* oil spill: Initial environmental impact assessment. *Environ. Sci. Technol.* **25**, 24–29.

Malik, N., and Vanden Born, W. H. (1986). "Use of Herbicides in Forest Management," Inf. Rep. NOR-X-282. Northern Forest Research Centre, Edmonton, Alberta.

Malingreau, J. P., Stephens, G., and Fellows, L. (1985). Remote sensing of forest fires: Kalimantan and North Borneo in 1982–83. *Ambio* **14**, 314–321.

Malins, D. C., ed. (1977). "Effects of Petroleum on Arctic and Subarctic Marine Environments," Vol. 2. Academic Press, New York.

Malins, D. C., McCain, B. B., Brown, D. W., Chan, S.-L., Myers, M. S., Landahl, J. J., Prohaska, P. G., Friedman, A. J., Rhodes, L. D., Burrows, D. G., Gronlund, W. D., and Hodgins, H. O. (1983). Chemical pollutants in sediments and diseases of bottom-dwelling fish in Puget Sound, Washington. *Environ. Sci. Technol.* **18**, 705–713.

Malley, D. F., Findlay, D. L., and Chang, P. S. S. (1982). Ecological effects of acid precipitation on zooplankton. *In* "Acid Precipitation Effects on Ecological Systems" (F. M. D'Itri, ed.), pp. 297–327. Ann Arbor Sci. Publ., Ann Arbor, MI.

Malyuga, D. P. (1964). "Biogeochemical Methods of Prospecting." Academic Science Press, Moscow (Transl.: Consultants Bureau, New York).

Manion, P. D. (1985). Factors contributing to the decline of forests, a conceptual overview. *In* "Air Pollutant Effects on Forest Ecosystems," pp. 63–73. Acid Rain Foundation, St. Paul, MN.

Mann, K. H. (1977). Destruction of kelp-beds by sea urchins: A cyclical phenomenon or irreversible degradation. *Helgol. Wiss. Meeresunters.* **30**, 455–467.

Mann, L. K., Johnson, D. W., West, D. C., Cole, D. W., Hornbeck, J. W., Martin, C. W., Riekerk, H., Smith, C. T., Swank, W. T., Tritton, L. M., and Van Lear, D. H. (1988). Effects of whole-tree and stem-only clearcutting on postharvest hydrologic losses, nutrient capital, and regrowth. *For. Sci.* **34**, 412–428.

Marchand, P. T. (1984). Dendrochronology of a fir wave. *Can. J. For. Res.* **14**, 51–56.

Marks, P. L. (1974). The role of pin cherry (*Prunus pensylvanica*) in the maintenance of stability in orthern hardwood ecosystems. *Ecol. Monogr.* **44**, 73–88.

Marks, P. L., and Bormann, F. H. (1972). Revegetation following forest cutting: Mechanisms for return to steady-state nutrient cycling. *Science* **176**, 914–915.

Marland, G. (1988). "The Prospect of Solving the CO_2 Problem Through Global Reforestation," DOE/NBB-0082. U.S. Department of Energy, Oak Ridge, TN.

Marland, G., and Rotty, R. M. (1985). Greenhouse gases in the atmosphere: What do we know. *J. Air Pollut. Control Assoc.* **35**, 1033–1038.

Marquis, D. A. (1975). Seed storage and germination under northern hardwood forests. *Can. J. For. Res.* **5**, 478–484.

Marston, R. A., and Anderson, J. E. (1991). Watersheds and vegetation of the Greater Yellowstone Ecosystem. *Conserv. Biol.* **5**, 338–346.

Martell, A. E., and Motekaitis, R. J. (1989). Coordination chemistry and speciation of Al(III) in aqueous solution. *In* "Environmental Chemistry and Toxicology of Aluminum" (T. E. Lewis, ed.), pp. 3–17. Lewis Pubs., Boca Raton, FL.

Martell, A. M. (1984). Changes in small mammal communities after fire in northcentral Ontario. *Can. Field Nat.* **98**, 223–226.

Martin, A. C., Zim, H. S., and Nelson, A. L. (1951). "American Wildlife and Plants: A Guide to Wildlife Food Habits." Dover, New York.

Martin, C. W., and Hornbeck, J. W. (1989). "Revegetation After Strip Cutting and Block Clearcutting in Northern Hardwoods: A 10-year History," USDA For. Serv. Res. Pap. NE-625. Northeastern Forest Experiment Station, Broomall, PA.

Martin, C. W., Pierce, R. S., Likens, G. E., and Bormann, F. H. (1986). "Clearcutting Affects Stream Chemistry in the White Mountains of New Hampshire," USDA For. Serv. Res. Pap. NE-579. Notheastern Forest Experiment Station, Broomall, PA.

Martin, E. S., and Hiebert, M. (1985). Explosive remnants of the Second Indochina War in Viet Nam. *In* "Explosive Remnants of War: Mitigating the Environmental Impacts" (A. H. Westing, ed.), pp. 39–50. Taylor & Francis, Philadelphia.

Martin, N. D. (1960). An analysis of bird populations in relation to forest succession in Algonquin Provincial Park, Ontario. *Ecology* **41**, 126–140.

Martin, P. S. (1967). Prehistoric overkill. *In* "Pleistocene Extinctions; The Search for a Cause" (P. S. Martin and H. E. Wright, ed.), pp. 75–120. Yale Univ. Press, New Haven, CT.

Martin, P. S. (1984). Catastrophic extinctions and late Pleistocene blitzkrieg: Two radiocarbon tests. *In* "Extinctions" (M. H. Nitecki, ed.), pp. 153–189. Univ. of Chicago Press, Chicago.

Maser, C. (1990). "The Redesigned Forest." Stoddart Publ., Toronto, Ont.

Maser, C., Anderson, R. G., Cromack, K., Williams, J. T., and Martin, R. E. (1979). "Dead and Down Woody Material," Agric Handb. No. 553, pp. 78–95. U.S. Department of Agriculture, Forest Service, Washington, DC.

Maser, C., Tarant, R. F., Trappe, J. M., and Franklin, J. F. (1988). "From the Forest to the Sea: A Story of Fallen Trees,"

USDA For. Serv. Gen. Tech. Rep. PNW-GTR-229. Pacific Northwest Research Station, Portland, OR.

Mathews, W. H., and Bustin, R. M. (1984). Why do the Smoking Hills burn? *Can. J. Earth Sci.* **21,** 737–742.

Matida, Y., Furuta, Y., Kumada, H., Tanaka, H., Yokote, M., and Kimura, S. (1975). Effects of some herbicides applied in the forest to freshwater fishes and other aquatic organisms. I. Survey of the effects of aerially applied sodium chlorate and a mixture of 2,4-D and 2,4,5-T on the stream community. *Bull. Freshwater Fish. Res. Lab.* **25,** 41–52.

Matson, P. A., and Vitousek, P. M. (1990). Ecosystem approaches to a global nitrous oxide budget. *BioScience* **40,** 667–672.

Mattson, D. J., and Reid, M. M. (1991). Conservation of the Yellowstone grizzly bear. *Conserv. Biol.* **5,** 364–372.

Matzner, E., Murach, D., and Fortmann, H. (1986). Soil acidity and its relationship to root growth in declining forest stands in Germany. *Water, Air, Soil Pollut.* **31,** 273–282.

May, A. W. (1967). Fecundity of Atlantic cod. *J. Fish. Res. Board Can.* **24,** 1531–1541.

Mayer, F. L., and Ellersieck, M. R. (1986). "Manual of Acute Toxicity: Interpretation and Data Base for 410 Chemicals and 66 Species of Freshwater Animals," Publ. No. 160. U.S. Fish and Wildlife Service, Washington, DC.

Mayer, R., and Ulrich, B. (1974). Conclusions on the filtering action of forests from ecosystem analysis. *Oecol. Plant.* **9,** 157–168.

Mayer, R., and Ulrich, B. (1977). Acidity of precipitation as influenced by the filtering of atmospheric sulfur and nitrogen compounds—its role in the element balance and effect on soil. *Water, Air, Soil Pollut.* **7,** 409–416.

Mayfield, H. F. (1960). "The Kirtland's Warbler," Bull. No. 40. Cranbrook Institute of Science, Bloomfield Hills, MI.

Mayfield, H. F. (1961). Cowbird parasitism and the population of the Kirtland's warbler. *Evolution (Lawrence, Kans.)* **15,** 174–179.

Mayfield, H. F. (1977). Brood parasitism: Reducing interactions between Kirtland's warblers and brown-headed cowbirds. *In* "Endangered Birds, Management Techniques for Preserving Threatened Species" (S. A. Temple, ed.), pp. 85–91. Univ. of Wisconsin Press, Madison.

McBride, J. R., Miller, P. R., and Laver, R. D. (1985). Effects of oxidant air pollutants on forest succession in the mixed conifer forest type of southern California. *In* "Air Pollutant Effects on Forest Ecosystems," pp. 157–167. Acid Rain Foundation, St. Paul, MN.

McCabe, R. E., and McCabe, R. T. (1984). Of slings and arrows: An historical retrospection. *In* "White-tailed Deer: Ecology and Management" (L. K. Halls, ed.), pp. 19–72. Stackpole Books, Harrisburg, PA.

McClenahan, J. R. (1978). Community changes in a deciduous forest exposed to air pollution. *Can. J. For. Res.* **8,** 432–438.

McClung, R. M. (1969). "Lost Wild America. The Story of Our Extinct and Vanishing Wildlife." Morrow, New York.

McColl, J. G. (1978). Ionic composition of forest soil solutions and effects of clear-cutting. *Soil Sci. Soc. Am. J.* **42,** 358–363.

McColl, J. G., and Grigall, D. F. (1979). Nutrient losses by leaching and erosion by intensive forest harvesting. *In* "Impact of Intensive Harvesting on Forest Nutrient Cycling," pp. 231–248. College of Environmental Science and Forestry, State University of New York, Syracuse.

McCormick, J. F., and Platt, R. B. (1962). Effects of ionizing radiation on a natural plant community. *Radiat. Bot.* **3,** 161–188.

McCormick, L. H., and Steiner, K. C. (1978). Variation in aluminum tolerance among six genera of trees. *For. Sci.* **24,** 565–568.

McCown, B. H., Deneke, F. J., Rickard, W., and Tieszen, L. L. (1973a). The response of Alaskan terrestrial plant communities to the presence of petroleum. *In* "Proceedings of the Symposium on the Impact of Oil Resource Development on Northern Plant Communities," pp. 34–43. Institute of Arctic Biology, Fairbanks, AK.

McCown, B. H., Brown, J., and Barsdate, R. J. (1973b). Natural oil seeps at Cape Simpson, Alaska: Localized influences on terrestrial habitat. *In* "Proceedings of the Symposium on the Impact of Oil Resource Development on Northern Plant Communities," pp. 86–90. Institute of Arctic Biology, Fairbanks, AK.

McCoy, C. (1989). Broken promises—Alyeska record shows how Big Oil neglected Alaska environment. *Wall Street J.,* July 6.

McCracken, F. I. (1985a). Observations on the decline of black willow. *J. Miss. Acad. Sci.* **30,** 1–5.

McCracken, F. I. (1985b). Oak decline and mortality in the south. *Proc. Symp. Southeast. Hardwoods, 3rd, 1985,* pp. 77–81.

McCulloch, B. (1982). "No Moa." Canterbury Museum, Christchurch, N.Z.

McEvoy, A. F. (1986). "The Fisherman's Problem. Ecology and Law in the California Fisheries." Cambridge Univ. Press, Cambridge, UK.

McEvoy, P., Cox, C., and Coombs, E. (1991). Successful biological control of ragwort, *Senecio jacobea,* by introduced insects in Oregon. *Ecol. Appl.* **1,** 430–442.

McEwen, F. L., and Stephenson, G. R. (1979). "The Use and Significance of Pesticides in the Environment." Wiley, New York.

McFee, W. W., Kelley, J. M., and Beck, R. H. (1977). Acid precipitation effects on soil pH and base saturation of exchange sites. *Water, Air, Soil Pollut.* **7,** 401–408.

McGlone, M. S. (1989). The Polynesian settlement of New Zealand in relation to environmental and biotic changes. *N. Z. J. Ecol.* **12,** Suppl., 115–164.

McGovern, P. C., and Balsillie, D. (1972). "SO$_2$ Levels and Environmental Studies in the Sudbury Area during 1971." Ontario Ministry of the Environment, Air Quality Branch, Sudbury.

McIlveen, W. D., Rutherford, S. T., and Linzon, S. N. (1986).

"A Historical Perspective of Sugar Maple Decline Within Ontario and Outside of Ontario," ARB-141–86-Phyto. Ontario Ministry of the Environment, Toronto.

McIver, J. D., Parsons, G. L., and Moldenke, A. R. (1992). Litter spider succession after clearcutting in a western conifer forest. *Can. J. For. Res.* **22**, 984–992.

McKay, C. (1985). "Freshwater Fish Contamination in Canadian Waters," Unpublished report. Chemical Hazards Division, Fish Habitat Management Branch, Department of Fisheries and Oceans, Ottawa, Ont.

McKinnon, M., and Vine, P. (1991). "Tides of War." Boxtree Ltd., London.

McLaughlin, S. B. (1985). Effects of air pollution on forests. A critical review. *J. Air Pollut. Control Assoc.* **35**, 512–534.

McLaughlin, S. B., Anderson, C. P., Edwards, N. Y., Roy, W. K., and Layton, P. A. (1990). Seasonal patterns of photosynthesis and respiration of red spruce saplings from two elevations in declining Southern Appalachian stands. *Can. J. For. Res.* **20**, 485–495.

McLaughlin, S. B., Anderson, C. P., Hanson, P. J., Tjoelken, M. G., and Roy, W. K. (1991). Increased dark respiration and calcium deficiency of red spruce in relation to acidic deposition at high-elevation Southern Appalachian sites. *Can. J. For. Res.* **21**, 1234–1244.

McLeod, J. M. (1967). The effect of phosphamidon on bird populations in jack pine stands in Quebec. *Can. Field Nat.* **81**, 102–106.

McLeod, B., and Millikin, R. (1982). "Environmental Impact Assessment of Experimental Spruce Budworm Adulticide Trials. Part 1. Effects on Forest Avifauna," Rep. FPM-X-54. Forest Pest Management Institute, Canadian Forestry Service, Sault Ste. Marie, Ont.

McNeil, D. K., Bendell, B. E., and Ross, R. K. (1987). "Studies of the Effects of Acidification on Aquatic Wildlife in Canada: Waterfowl and Trophic Relationships in Small Lakes in Northern Ontario," Occas. Pap. No. 62. Canadian Wildlife Service, Ottawa, Ont.

McNicol, J. G., and Timmermann, H. R. (1981). Effects of forestry practices on ungulate populations in the boreal mixedwood forest. *In* "Boreal Mixedwood Symposium," CO-JFRC Symp. Proc. O-P-9, pp. 141–154. Great Lakes Forest Research Centre, Sault Ste. Marie, Ont.

McQueen, D. J. (1990). Manipulating lake community structure: Where do we go from here? *Freshwater Biol.* **23**, 613–620.

McQueen, D. J., Johannes, M. R. S., Post, J. R., Stewart, T. J., and Lean, D. R. S. (1989). Bottom-up and top-down impacts on freshwater pelagic community structure. *Ecol. Monogr.* **59**, 289–309.

McQueen, D. J., Johannes, M. R. S., Lafontaine, N. R., Young, A. S., Longbotham, E., and Lean, D. R. S. (1990). Effects of planktivore abundance on chlorophyll-*a* and secchi depth. *Hydrobiologia* **200/201**, 337–341.

McQueen, D. J., Mills, E. L., Farney, J. L., Johannes, M. R. S., and Post, J. R. (1992). Trophic level relationships in pelagic food webs: Comparisons derived from long-term data sets for Oneida Lake, New York (USA) and Lake St. George, Ontario (Canada). *Can. J. Fish. Aquat. Sci.* **49**, 1588–1596.

McQueen, E. G., Veale, A. M. O., Alexander, W. S., and Bates, M. N. (1977). "2,4,5-T and Human Birth Defects." Department of Health, New Zealand, Auckland.

Means, J. E., and Winjum, J. K. (1983). Road to recovery after eruption of Mt. St. Helens. *In* "Using Our Natural Resources: 1983 Yearbook of Agriculture," pp. 204–215. U.S. Department of Agriculture, Washington, DC.

Mearns, A. J., and Sherwood, M. J. (1977). Distributions of neoplasms and other diseases in marine fishes relative to the discharge of waste water. *Ann. N.Y. Acad. Sci.* **298**, 210–223.

Medin, D. E., and Booth, G. D. (1989). "Responses of Birds and Small Mammals to Single-tree Selection Logging in Idaho," Res. Pap. INT-408. Intermountain Forest and Range Experiment Station, Ogden, UT.

Megahan, W. F., and Kidd, W. J. (1972). Effects of logging and logging roads on erosion and sediment deposition from steep terrain. *J. For.* **70**, 136–141.

Melillo, J. M., Palm, C. A., Houghton, R. A., Woodwell, G. M., and Myers, N. (1985). A comparison of two recent estimates of disturbance in tropical forests. *Environ. Conserv.* **12**, 37–40.

Melillo, J. M., Callaghan, T. V., Woodward, F. I., Salati, E., and Sinha, S. K. (1990). Effects on ecosystems. *In* "Climate Change: The IPCC Scientific Assessment" (J. T. Houghton, G. J. Jenkins, and J. J. Ephraums, eds.), pp. 287–310. Cambridge Univ. Press, Cambridge, UK.

Messieh, S. N., Ravell, T. W., Peer, D. L., and Cranford, P. J. (1991). The effects of trawling, dredging, and ocean dumping on the eastern Canadian continental shelf seabed. *Cont. Shelf Res.* **20**, 1–27.

Meszaros, E. (1981). "Atmospheric Chemistry: Fundamental Aspects." Elsevier, New York.

Metcalf, R. L. (1971). The chemistry and biology of pesticides. *In* "Pesticides in the Environment," Vol. 1, Part 1, pp. 1–144. Dekker, New York.

Michael, A. D. (1977). The effects of petroleum hydrocarbons on marine populations and communities. *In* "Fate and Effects of Petroleum Hydrocarbons in Marine Ecosystems and Organisms" (D. A. Wolfe, ed.), pp. 129–137. Pergamon, New York.

Miegroet, H. V., Johnson, D. W., and Todd, D. E. (1993). Foliar response of red spruce saplings to fertilization with Ca and Mg in the Great Smoky Mountains National Park. *Can. J. For. Res.* **23**, 89–95.

Mierle, G. (1990). Aqueous inputs of mercury to Precambrian Shield lakes in Ontario. *Environ. Contam. Chem.* **9**, 843–851.

Millan, M., Barton, S. C., Johnson, N. D., Weisman, B., Lusis, M., Chan, W., and Vet, R. (1982). Rain scavenging from tall stack plumes: A new experimental approach. *Atmos. Environ.* **16**, 2709–2714.

Miller, C. A. (1975). The spruce budworm: How it lives and what it does. *For. Chron.* **51**, 136–138.

Miller, D. R. (1984). Chemicals in the environment. *In* "Effects of Pollutants at the Ecosystem Level" (P. J. Sheehan, D. R. Miller, G. C. Butler, and P. Bourdeau, eds.), pp. 7–14. Wiley, New York.

Miller, E., and Miller, D. R. (1980). Snag use by birds. *In* "Management of Western Forests and Grasslands for Nongame Birds," USDA For. Serv. Gen. Tech. Rep. INT-86, pp. 337–356. Intermountain Forest and Range Experiment Station, Ogden, UT.

Miller, E. L., Beasley, R. S., and Lawson, E. R. (1985). Stormflow, sedimentation, and water quality responses following silvicultural treatments in the Ouachita Mountains. *In* "Proceedings, Forestry and Water Quality: A Mid-South Symposium" (B. G. Blackman, ed.), pp. 117–129. University of Arkansas, Little Rock.

Miller, G. E., Grant, P. M., Kishore, R., Steinbruger, F. J., Rowland, F. S., and Guinn, V. P. (1972). Mercury concentrations in museum specimens of tuna and swordfish. *Science* **175**, 1121–1122.

Miller, G. T. (1985). "Living in the Environment." Wadsworth, Belmont, CA.

Miller, J. H., and Sirois, D. L. (1986). Soil disturbance by skyline yarding vs. skidding in a loamy hill forest. *Soil Sci. Soc. Am. J.* **50**, 1579–1583.

Miller, K., and Tangley, L. (1991). "Trees of Life." Beacon Press, Boston.

Miller, L. K., Lingg, A. J., and Bulla, L. A. (1983). Bacterial, viral, and fungal insecticides. *Science* **219**, 715–721.

Miller, M., Wasik, S. P., Huang, G. L., Shiu, W. Y., and McKay, D. (1985). Relationships between octanol-water partition coefficients and aqueous solubility. *Environ. Sci. Technol.* **19**, 522–529.

Miller, P. R. (1973). Oxidant-induced community change in a mixed conifer forest. *Adv. Chem. Ser.* **122**, 101–117.

Miller, P. R. (1989). Concept of forest decline in relation to western U.S. forests. *In* "Air Pollution's Toll on Forests and Crops" (J. J. MacKenzie and M. T. El-Ashry, eds.), pp. 75–112. Yale Univ. Press, New Haven, CT.

Miller, P. R., McCutchan, M. H., and Milligan, H. P. (1972). Oxidant air pollution in the Central Valley, Sierra Nevada foothills, and Mineral King Valley of California. *Atmos. Environ.* **6**, 623–633.

Millers, I., Allen, D. C., and Lachance, D. (1992). "Sugar Maple Crown Conditions Improve Between 1988 and 1990," USDA For. Serv. NA-TP-03-92. Northeastern Area, Durham, NH.

Millikin, R. L. (1990). Effects of fenitrothion on the arthropod food of tree-foraging forest songbirds. *Can. J. Zool.* **68**, 2235–2242.

Millikin, R. L., and Smith, J. N. M. (1990). Sublethal effects of fenitrothion on forest passerines. *J. Appl. Ecol.* **27**, 983–1000.

Mills, E. L., Leach, J. H., Carlton, J. T., and Secar, C. L.

(1993). Exotic species in the Great Lakes: A history of biotic crises amd anthropogenic introductions. *J. Great Lakes Res.* **19**, 1–54.

Mills, K. H. (1984). Fish population responses to experimental acidification of a small Ontario lake. *In* "Early Biotic Responses to Advancing Lake Acidification" (G. R. Hendry, ed.), pp. 117–131. Butterworth, Toronto, Ont.

Mills, K. H. (1985). Responses of lake whitefish (*Coregonus clupeaformis*) to fertilization of Lake 223, the Experimental Lakes Area. *Can. J. Fish. Aquat. Sci.* **42**, 129–138.

Mills, K. H., and Chalanchuk, S. M. (1987). Population dynamics of lake whitefish (*Coregonus clupeaformis*) during and after the fertilization of Lake 226, the Experimental Lakes Area. *Can. J. Fish. Aquat. Sci.* **44**, Suppl. 1, 55–63.

Mills, K. H., Chalanchuk, S. M., Mohr, L. C., and Davies, I. J. (1987). Responses of fish populations in Lake 223 to 8 years of experimental acidification. *Can. J. Fish. Aquat. Sci.* **44**, Suppl. 1, 114–125.

Milne, A. R., and Herlinveaux, R. H. (1977). "Crude Oil in Cold Water." Department of Fisheries and Department of the Environment, Sidney, British Columbia.

Mineau, P., ed. (1991). "Cholinesterase-Inhibiting Insecticides. Their Impact on Wildlife and the Environment." Elsevier, Amsterdam.

Mineau, P. (1993). "The Hazard of Carbofuran to Birds and Other Vertebrate Wildlife," Tech. Rep. No. 177. Environment Canada, Canadian Wildlife Service, Wildlife Toxicology Section, Ottawa, Ont.

Mineau, P., and Collins, B. T. (1988). "Avian Mortality in Agroecosystems. 2. Methods of Detection," BCPC Monogr. No. 40, Environmental Effects of Pesticides, British Crop Protection Conference, London.

Mineau, P., and Peakall, D. B. (1987). An evaluation of avian impact assessment techniques following broad-scale forest insecticide sprays. *Environ. Sci. Technol.* **6**, 781–791.

Ministry of Agriculture and Environment (MAEC) (1983). "Acidification Today and Tomorrow." '82 Committee, MAEC, Royal Ministry of Agriculture, Stockholm, Sweden.

Mishra, D., and Kar, M. (1974). Nickel in plant growth and metabolism. *Bot. Rev.* **40**, 395–452.

Mitchell, G. C. (1986). Vampire bat control in Latin America. *In* "Ecological Knowledge and Environmental Problem-Solving," pp. 151- 163. National Academy Press, Washington, DC.

Mitchell, M. F., and Roberts, J. R. (1984). A case study of the use of fenitrothion in New Brunswick: The evolution of an advanced approach to ecological monitoring. *In* "Effects of Pollutants at the Ecosystem Level" (P. J. Sheehan, D. R. Miller, and P. H. Boudreau, eds.), pp. 377–402. Wiley, Toronto, Ont.

Mittermeier, R. A. (1988). Primate diversity and the tropical forest. Case studies from Brazil and Madagascar and the importance of megadiversity countries. *In* "Biodiversity" (E. O. Wilson, ed.), pp. 145–154. National Academy Press, Washington, DC.

Mix, M. C. (1986). Cancerous diseases in aquatic animals and their association with environmental pollutants: A critical literature review. *Mar. Environ. Res.* **20,** 1–141.

Moir, W. H. (1992). Ecological concepts in old-growth forests definition. *In* "Old-growth Forests in the Southwest and Rocky Mountain Regions," USDA For. Serv. Gen. Tech. Rep. RM-213, pp. 18–23. Rocky Mountain Forest and Range Experiment Station, Fort Collins, CO.

Molden, B., and Schnoor, J. L. (1992). Czechoslovakia. Examining a critically ill economy. *Environ. Sci. Technol.* **26,** 14–21.

Moller, D. (1984). Estimation of the global man-made sulfur emission. *Atmos. Environ.* **18,** 19–27.

Mollitor, A. V., and Raynal, D. J. (1983). Atmospheric deposition and ionic input in Adirondack forests. *J. Air Pollut. Control Assoc.* **33,** 1032–1036.

Monk, C. D. (1966). Effects of short-term gamma irradiation on an old field. *Radiat. Bot.* **6,** 329–335.

Monthey, R. W. (1984). Effects of timber harvesting on ungulates in northern Maine. *J. Wildl. Manage.* **48,** 279–285.

Monthey, R. W. (1986). Responses of snowshoe hares, *Lepus americanus,* to timber harvesting in northern Maine. *Can. Field Nat.* **100,** 568–670.

Monthey, R. W., and Soutiere, E. C. (1985). Responses of small mammals to forest harvesting in northern Maine. *Can. Field Nat.* **99,** 13–18.

Mooney, H. A. (1991). Biological response to climate change: An agenda for research. *Ecol. Appl.* **1,** 112–117.

Mooney, H. A., Drake, B. G., Luxmoore, R. J., Oechel, W. C., and Pitelka, L. F. (1991). Predicting ecosystem responses to elevated CO_2 concentrations. *BioScience* **41,** 96–104.

Moore, B., and Bolin, B. (1987). The oceans, carbon dioxide, and global climate change. *Oceanus* **29,** 9–15.

Moore, G. C., and Boer, A. H. (1979). "New Brunswick Deer Yard Management Projects: A Summary." New Brunswick Department of Natural Resources, Fredericton.

Moore, N. W., and Hooper, M. D. (1975). On the number of bird species in British woods. *Biol. Conserv.* **8,** 239–250.

Moore, T. R., and Zimmermann, R. C. (1972). Establishment of vegetation on serpentine asbestos mine wastes, southeastern Quebec, Canada. *J. Appl. Ecol.* **14,** 589–599.

Morash, R., and Freedman, B. (1983). Seedbanks in several recently clearcut and mature forests in Nova Scotia. *Proc. N.S. Inst. Sci.* **33,** 85–94.

Morash, R., and Freedman, B. (1987). "Initial Impacts of Silvicultural Herbicide Spraying on the Vegetation of Regenerating Clearcuts in Central Nova Scotia," Research Report to Maritimes Forest Research Centre. School for Resource and Environmental Studies, Dalhousie University, Halifax, Nova Scotia.

Morash, R., and Freedman, B. (1989). Effects of several herbicides on the germination of seeds in the forest floor. *Can. J. For. Res.* **19,** 347–350.

Morash, R., Freedman, B., and McCurdy, R. (1987). Persistence of glyphosate in foliage and litter. *In* "Measurement of the Environmental Effects Associated with Forestry Use of Roundup," pp. 15–19. Environment Canada, Environmental Protection Service, Dartmouth, Nova Scotia.

Morgan, K., and Freedman, B. (1986). Breeding bird communities in a hardwood forest succession in Nova Scotia. *Can. Field Nat.* **100,** 506–519.

Moriarty, F. (1988). "Ecotoxicology: The Study of Pollutants in Ecosystems." Academic Press, London.

Morrin, D. J., Holland, M. M., and Lawrence, D. M. (1993). Profiles of ecologists: Results of a survey of the membership of the Ecological Society of America. Part III. Environmental science capabilities and funding. *Bull. Ecol. Soc. Am.* **74,** 227–249.

Morris, O. N. (1982). Bacteria as pesticides: Forest applications. *In* "Microbial and Viral Pesticides" (E. Kurstak, ed.), pp. 239–287. Dekker, New York.

Morris, R. F., Cheshire, W. F., Miller, C. A., and Mott, D. G. (1958). The numerical response of avian and mammalian predators during a gradation of the spruce budworm. *Ecology* **39,** 487–494.

Morrison, I. K. (1981). Effect of simulated acid precipitation on composition of percolate from reconstructed profiles of two northern Ontario forest soils. *Can. For. Serv. Res. Notes* **1,** 6–8.

Morrison, I. K. (1983). Composition of percolate from reconstructed profiles of jack pine forest soils as influenced by acid input. *In* "Effects of Accumulation of Air Pollutants in Forest Ecosystems" (B. Ulrich and J. Pankrath, eds.), pp. 195–206. Reidel, Berlin.

Morrison, I. K. (1984). A review of literature on acid deposition effects on forest ecosystems. *Commonw. For. Bur. For. Abstr.* **45,** 483–505.

Morrison, M. L., and Meslow, E. C. (1983). Impacts of forest herbicides on wildlife: Toxicity and habitat alteration. *Trans. North Am. Wildl. Nat. Resour. Conf.* **48,** 175–185.

Morrison, M. L., and Meslow, E. C. (1984a). Response of avian communities to herbicide-induced habitat changes. *J. Wildl. Manage.* **48,** 14–22.

Morrison, M. L., and Meslow, E. C. (1984b). Effects of the herbicide glyphosate on bird community structure, western Oregon. *For. Sci.* **30,** 95–106.

Morrison, R. G., and Yarranton, G. A. (1973). Diversity, richness, and evenness during a primary sand dune succession at Grand Bend, Ontario. *Can. J. Bot.* **51,** 2401–2411.

Morse, D. H. (1989). "American Warblers. An Ecological and Behavioural Perspective." Harvard Univ. Press, Cambridge, MA.

Mowat, F. (1984). "Sea of Slaughter." McLelland-Stewart, Toronto, Ont.

Mueller-Dombois, D. (1980). The Ohia'a dieback phenomenon in the Hawaiian rain forest. *In* "The Recovery Process in Damaged Ecosystems" (J. Cairns, ed.), pp. 153–161. Ann Arbor Sci. Publ., Ann Arbor, MI.

Mueller-Dombois, D. (1986). Perspectives for an etiology of stand-level dieback. *Annu. Rev. Ecol. Syst.* **17,** 221–243.

Mueller-Dombois, D. (1987a). Natural dieback in forests. *Bio-Science* **37**, 575–583.

Mueller-Dombois, D. (1987b). Forest dynamics in Hawaii. *TREE* **2**, 216–220.

Mueller-Dombois, D. (1992). A global perspective on forest decline. *Environ. Contam. Chem.* **11**, 1069–1076.

Mueller-Dombois, D., Canfield, J. E., Holt, R. A., and Buelow, G. P. (1983). Tree-group death in North America and Hawaiian forests: A pathological problem or a new problem for vegetation ecology? *Phytocoenologia* **11**, 117–137.

Muir, P. S., Kimberley, K. A., Carter, B. H., Armentano, T. V., and Pribush, R. A. (1986). Fog chemistry at an urban midwestern site. *J. Air Pollut. Control Assoc.* **36**, 1359–1361.

Muller, E. F., and Kramer, J. R. (1977). Precipitation scavenging in central and northern Ontario. *ERDA Symp. Ser.* **41**, 590–601.

Muller, P. (1980). Effects of artificial acidification on the growth of periphyton. *Can. J. Fish. Aquat. Sci.* **37**, 355–363.

Mullinson, W. R. (1981). Public Concerns About the Herbicide 2,4-D." Dow Chemical, Midland, MI.

Munger, J. W. (1982). Chemistry of atmospheric precipitation in the north-central United States: Influence of sulfate, nitrate, ammonium, and calcareous soil particulates. *Atmos. Environ.* **16**, 1633–1645.

Munger, J. W., and Eisenreich, S. J. (1983). Continental-scale variations in precipitation chemistry. *Environ. Sci. Technol.* **17**, 32A–42A.

Munn, R. E., Likens, G. E., Weisman, B., Hornbeck, J. W., Martin, C. W., and Bormann, F. H. (1984). A meteorological analysis of precipitation chemistry event samples at Hubbard Brook, N.H. *Atmos. Environ.* **18**, 2775–2779.

Muona, O. (1990). Population genetics in forest tree improvement. *In* "Plant Population Genetics, Breeding, and Genetic Resources" (A. D. H. Brown, M. T. Clegg, A. L. Kahlen, and B. S. Weir, eds.), pp. 282–298. Sinauer Assoc., Sunderland, MA.

Murphy, D. D., and Noon, B. R. (1992). Integrating scientific methods with habitat conservation planning: Reserve design for northern spotted owls. *Ecol. Appl.* **2**, 3–17.

Murphy, J. E. (1990). The 1985–86 Canadian peregrine falcon, *Falco peregrinus,* survey. *Can. Field Nat.* **104**, 182–192.

Murphy, J. E. (1987). "The 1985–86 Canadian Peregrine Falcon Survey." Canadian Wildlife Service, Ottawa, Ont.

Murphy, M. L., and Hall, J. D. (1983). Varied effects of clear-cut logging on predators and their habitat in small streams in the Cascade Mountains, Oregon. *Can. J. Fish. Aquat. Sci.* **38**, 137–145.

Murphy, M. L., Hawkins, C. P., and Anderson, N. H. (1981). Effects of canopy modification and accumulated sediment on stream communities. *Trans. Am. Fish. Soc.* **110**, 469–478.

Murphy, M. L., Heifetz, J., Johnson, S. W., Koski, K. V., and Thedinga, J. F. (1986). Effects of clear-cut logging with and without buffer strips on juvenile salmonids in Alaskan streams. *Can. J. Fish. Aquat. Sci.* **43**, 1521–1533.

Murphy, S. D. (1980). Pesticides. *In* "Toxicology: The Basic Science of Poisons" (J. Doull, C. D. Klaassen, and M. O. Amdur, eds.), pp. 357–408. Macmillan, New York.

Murray, F. (1981). Effects of fluorides on plant communities around an aluminum smelter. *Environ. Pollut., Ser. A* **24**, 45–56.

Myers, N. (1979). "The Sinking Ark: A New Look at the Problem of Disappearing Species." Pergamon, Oxford.

Myers, N. (1983). "A Wealth of Wild Species." Westview Press, Boulder, CO.

Myers, N. (1988). Tropical forests and their species. Going . . . going . . . *In* "Biodiversity" (E. O. Wilson, ed.), pp. 28–35. National Academy Press, Washington, DC.

Myers, R. A., Drinkwater, K. F., and Baird, J. W. (1993). Salinity and recruitment of Atlantic cod (*Gadus morhua*) in the Newfoundland region. *Can. J. Fish. Aquat. Sci.* **50** (in press).

Nakatsu, C. H. (1983). The algal flora and ecology of three brown water systems. M.Sc. Thesis, Department of Botany, University of Toronto, Toronto, Ont.

Nams, M. L. N., and Freedman, B. (1987a). Ecology of heath communities dominated by *Cassiope tetragona* at Alexandra Fiord, Ellesmere Island, Canada. *Holarctic Ecol.* **10**, 22–32.

Nams, M. L. N., and Freedman, B. (1987b). Phenology and resource allocation in a high arctic evergreen dwarf shrub, *Cassiope tetragona. Holarctic Ecol.* **10**, 128–136.

Narver, D. W. (1970). Effects of logging debris on fish production. *In* "Forest Land Uses and Stream Environment," pp. 100–111. Oregon State University, Corvallis.

National Academy of Sciences (NAS) (1971). "Fluorides." NAS, Washington, DC.

National Academy of Sciences (NAS) (1972). "Lead—Airborne Lead in Perspective." Committee on Biological Effects of Atmospheric Pollutants, NAS, Washington, DC.

National Academy of Sciences (NAS) (1974). "The Effects of Herbicides in South Vietnam, Part A." Committee on the Effects of Herbicides in Vietnam, NAS, Washington, DC.

National Academy of Sciences (NAS) (1975a). "Forest Pest Control." NAS, Washington, DC.

National Academy of Sciences (NAS) (1975b). "Underexploited Tropical Plants." NAS, Washington, DC.

National Academy of Sciences (NAS) (1975c). "Contemporary Pest Control Practices and Prospects." NAS, Washington, DC.

National Academy of Sciences (NAS) (1975d). "Petroleum in the Marine Environment." NAS, Washington, DC.

National Academy of Sciences (NAS) (1976). "Halocarbons: Effects on Stratospheric Ozone." NAS, Washington, DC.

National Academy of Sciences (NAS) (1979a). "Stratospheric Ozone Depletion by Halocarbons: Chemistry and Transport." NAS, Washington, DC.

National Academy of Sciences (NAS) (1979b). "Tropical Legumes: Resources for the Future." NAS, Washington, DC.

National Academy of Sciences (NAS) (1985). "Oil in the Sea." NAS, Washington, DC.

National Marine Fisheries Service (NMFS) (1991a). "Endangered Whales: Status Update." NMFS, National Oceanic and Atmospheric Administration, U.S. Department of Commerce, Silver Spring, MD.

National Marine Fisheries Service (NMFS) (1991b). "Proposed Regime to Govern Interactions Between Marine Mammals and Commercial Fishing Operations." NMFS, National Oceanic and Atmospheric Administration, U.S. Department of Commerce, Silver Spring, MD.

National Oceanic and Atmospheric Administration (NOAA-CNEXO) (1982). "Ecological Study of the Amoco Cadiz Oil Spill," Report of the NOAA-CNEXO Joint Scientific Commission. U.S. Department of Commerce, Washington, DC.

National Research Council (NRC) (1976). "Vapour-Phase Organic Pollutants." Committee on Medical and Biologic Effects of Environmental Pollutants, NRC, Washington, DC.

National Research Council (NRC) (1982). "Carbon Dioxide and Climate: A Second Assessment." NRC, National Academy Press, Washington, DC.

National Research Council (NRC) (1983). "Changing Climate." NRC, National Academy Press, Washington, DC.

National Research Council (NRC) (1985a). "The Effects on the Atmosphere of a Major Nuclear Exchange." NRC, National Academy Press, Washington, DC.

National Research Council (NRC) (1985b). "Oil in the Sea. Inputs, Fates, and Effects." NRC, National Academy Press, Washington, DC.

National Research Council (NRC) (1986). "Pesticide Resistance." NRC, National Academy Press, Washington, DC.

National Research Council (NRC) (1989). "Using Oil Spill Dispersants in the Sea." NRC, National Academy Press, Washington, DC.

Nauwerck, A. (1963). Die neziehungen zwischen zooplankton und phytoplankton in See Erken. *Symb. Bot. Ups.* **17**, 1–163 (cited in Ostrovsky and Duthie, 1975).

Nearing, B. P., and Leach, J. H. (1992). Mapping the potential spread of the zebra mussel (*Dreissena polymorpha*) in Ontario. *Can. J. Fish. Aquat. Sci.* **49**, 406–415.

Neary, D. G., and Currier, J. B. (1982). Impact of wildfire and watershed management on water quality in South Carolina's Blue Ridge Mountains. *South. J. Appl. For.* **6**, 81–90.

Neff, J. M., and Anderson, J. W. (1981). "Response of Marine Mammals to Petroleum and Specific Petroleum Hydrocarbons." Appl. Sci. Publ., London.

Neff, J. M., and Haensley, W. E. (1982). Long-term impact of the Amoco Cadiz crude oil spill on oysters *Crassostrea gigas* and plaice *Pleuronectes platessa* from Aber Benoit and Aber Wrac'h, Brittany, France. *In* "Ecological Study of the Amoco Cadiz Oil Spill." U.S. Department of Commerce, National Oceanic and Atmospheric Administration, Washington, DC.

Neilson, M. M. (1985). Spruce budmoth—A case history. Issues and constraints. *For. Chron.* **61**, 252–255.

Nelson, P. F., and Quigley, S. M. (1984). The hydrocarbon composition of exhaust emitted from gasoline fueled vehicles. *Atmos. Environ.* **18**, 79–87.

Nelson, R. W. (1976). Behavioural aspects of egg breakage in peregrine falcons. *Can. Field Nat.* **90**, 320–329.

Nelson-Smith, A. (1977). Recovery of some British rocky seashores from oil spills and cleanup operations. *In* "Recovery and Restoration of Damaged Ecosystems" (J. Cairns, K. L. Dickson, and E. E. Herricks, eds.), pp. 191–207. Univ. Press of Virginia, Charlottesville.

Nero, R. W., and Schindler, D. W. (1983). Decline of *Mysis relicta* during the acidification of Lake 223. *Can. J. Fish. Aquat. Sci.* **40**, 1905–1911.

Nettleship, D. N., and Evans, P. G. H. (1985). Distribution and status of the Atlantic Alcidae. *In* "The Atlantic Alcidae" (D. N. Nettleship and T. R. Birkhead, ed.), pp. 54–154. Academic Press, New York.

Newhook, F. J. (1989). Indigenous forest health in the South Pacific—A plant pathologist's view. *N. Z. J. For. Sci.* **19**, 231–242.

Newman, A. (1990). "Tropical Rainforests." Facts on File, New York.

Newman, H. C., and Schmidt, W. C. (1980). Silviculture and residue treatment affect water used by a larch/fir forest. *In* "Environmental Consequences of Timber Harvesting in Rocky Mountain Coniferous Forests," USDA For. Serv. Gen. Tech. Rep. INT-90, pp. 75–100. Intermountain Forest and Range Experiment Station, Ogden, UT.

Newman, L. (1981). Atmospheric oxidation of sulfur dioxide: A review as viewed from power plant and smelter plume studies. *Atmos. Environ.* **15**, 2231–2239.

Newton, I., and Wyllie, I. (1992). Recovery of a sparrowhawk population in relation to declining pesticide contamination. *J. Appl. Ecol.* **29**, 476–484.

Newton, M. (1975). Constructive use of herbicides in forest resource management. *J. For.* **73**, 329–336.

Newton, M., and Knight, F. B. (1981). "Handbook of Weed and Insect Control Chemicals for Forest Resource Managers." Timber Press, Beaverton, OR.

Newton, M., Howard, L., Kelpas, B., Danhaus, R., Lottman, C., and Dubelman, S. (1984). Fate of glyphosate in an Oregon forest ecosystem. *J. Agric. Food Chem.* **32**, 1144–1151.

Newton, M., Cole, E. C., Lautenschlager, R. A., White, D. E., and McCormack, M. L. (1989). Browse availability after conifer release in Maine's spruce-for forests. *J. Wildl. Manage.* **53**, 643–649.

Nicholas, N. S., and Zedaker, S. M. (1989). Ice damage in spruce-fir forests of the Black Mountains, North Carolina. *Can. J. For. Res.* **19**, 1487–1491.

Niemi, G. J., and Hanowski, J. M. (1984). Relationships of breeding birds to habitat characteristics in logged areas. *J. Wildl. Manage.* **48**, 438–443.

Nigam, P. C. (1975). Chemical insecticides. *In* "Aerial Control of Forest Insects in Canada" (M. L. Prebble, ed.), pp. 8–24. Department of the Environment, Ottawa, Ont.

Nihlgard, B. (1985). The ammonium hypothesis—an additional explanation to the forest dieback in Europe. *Ambio* **14**, 2–8.

Nilssen, J. P. (1980). Acidification of a small watershed in southern Norway and some characteristics of acidic aquatic environments. *Int. Rev. Gesamten Hydrobiol.* **65**, 177–207.

Nilsson, S. I., and Duinker, P. (1987). The extent of forest decline in Europe. *Environment* **29**(9), 4–31.

Nilsson, S. I., Miller, H. G., and Miller, J. D. (1982). Forest growth as a possible cause of soil and water acidification: An examination of the concepts. *Oikos* **39**, 40–49.

Nisbet, I. C. T. (1993). Effects of pollution on marine birds. *In* "Seabirds on Islands: Threats, Case Studies, and Action Plans" (D. N. Nettleship, J. Burger, and M. Gochfeld, eds.), pp. 11–43. International Council for Bird Preservation, Cambridge, UK.

Noble, D. G., and Elliott, J. E. (1990). Levels of contaminants in Canadian raptors, 1966 to 1988; Effects and temporal trends. *Can. Field Nat.* **104**, 222–243.

Nolan, V. (1978). "The Ecology an Behaviour of the Prairie Warbler." Ornithol. Monogr. No. 26. American Ornithologists' Union, Washington, D. C.

Noon, B. R., Bingham, V. P., and Noon, J. P. (1979). The effects of changes in habitat on northern hardwood forest bird communities. *In* "Management of North Central and Northeastern Forests for Non-game Birds," USDA For. Serv. Gen. Tech. Rep. NC-51, pp. 33–48. North Central Forest Experiment Station, St. Paul, MN.

Norheim, G., Somme, L., and Holt, G. (1982). Mercury and persistent chlorinated hydrocarbons in Antarctic birds from Bouvetoya and Dronning Maud Land. *Environ. Pollut., Ser. A* **28**, 233–240.

Norris, L. A. (1981). The movement, persistence, and fate of the phenoxy herbicides and TCDD in the forest floor. *Residue Rev.* **80**, 65–135.

Norris, L. A., Montgomery, M. L., and Johnson, E. R. (1977). The persistence of 2,4,5-T in a Pacific Northwest forest. *Weed Sci.* **25**, 417–422.

Norris, L. A., Montgomery, M. L., Warren, L. E., and Mosher, W. D. (1982). Brush control with herbicides on hill pasture sites in southern Oregon. *J. Range Manage.* **35**, 75–80.

Norris, L. A., Lorz, H. W., and Gregory, S. V. (1983). "Influence of Forest and Rangeland Management on Anadromous Fish Habitat in North America," USDA For. Serv. Gen. Tech. Rep. PNW-149. Pacific Northwest Forest and Range Experiment Station, Portland, OR.

Northridge, S. P. (1984). "World Review of Interactions Between Marine Mammals and Fisheries," FAO Fish. Tech. Pap. 251. Food and Agricultural Organization of the United Nations, Rome, Italy.

Norton, D. A. (1991). *Trilepidea adamsii*: An obituary for a species. *Conserv. Biol.* **5**, 52–57.

Norton, D. A. (1992). Disruption of natural ecosystems and biological invasion. *In* "Science and the Management of Protected Areas." pp. 309–320. Elsevier, Amsterdam.

Norton, S. A., and Young, H. E. (1976). Forest biomass utilization and nutrient budgets. *In* "Oslo Biomass Studies," pp. 56–73. University of Maine at Orono, Orono.

Noss, R. F. (1983). A regional landscape approach to maintain diversity. *BioScience* **33**, 700–706.

Noss, R. F. (1987). Corridors in real landscapes: A reply to Simberloff and Cox. *Conserv. Biol.* **1**, 159–164.

Nounou, P. (1980). The oil spill age. *Ambio* **9**, 297–302.

Nunn, M. D. (1983). "Decision in the Supreme Court of Nova Scotia, Trial Division, between Palmer, Googoo, Mullendore, Francis, MacGillivary, Sampson, Shaw, MacIntyre, MacInnes, Mustard, Dauphney, Haldeman, Grose, Epifano, and Calvert; and Nova Scotia Forest Industries." Supreme Court of Nova Scotia, Trial Division, Halifax.

O'Brien, P. Y., and Dixon, P. S. (1976). The effects of oils and oil components on algae: A review. *Br. Phycol. J.* **11**, 115–142.

O'Brien, R. D. (1967). "Insecticides: Action and Metabolism." Academic Press, New York.

O'Connor, R. J., and Shrubb, M. (1986). "Farming and Birds." Cambridge Univ. Press, Cambridge, UK.

Oden, S. (1976). "The Acidity Problem—An Outline of Concepts," USDA For. Serv. Gen. Tech. Rep. NE-23, pp. 1–36. Northeastern Forest Experiment Station, Broomall, PA.

Oden, S., and Andersson, R. (1971). "The longterm changes in the chemistry of soils in Scandinavia due to acid precipitation," Sect. 5.1. Sweden's Case Study for United Nations Conference on the Human Environment, Stockholm.

Odum, E. P. (1981). The effects of stress on the trajectory of ecological succession. *In* "Stress Effects on Natural Ecosystems" (G. W. Barrett and R. Rosenberg, eds.), pp. 43–47. Wiley, New York.

Odum, E. P. (1983). "Basic Ecology." Saunders College Publishing, New York.

Odum, E. P. (1985). Trends expected in stressed ecosystems. *BioScience* **35**, 419–422.

Oechel, W. C., and Strain, B. R. (1985). Native species responses to increased atmospheric carbon dioxide concentration. *In* "Direct Effects of Increasing Carbon Dioxide on Vegetation," DOE/ER-0238, pp. 117–155. U.S. Department of Energy, Washington, DC.

Office of Technology Assessment (OTA) (1987). "Technologies to Maintain Biological Diversity," OTA-F-330. Congress of the United States, Washington, DC.

Ogner, G., and Tiegen, O. (1980). Effects of acid irrigation and liming on two clones of Norway spruce. *Plant Soil* **57**, 305–321.

Ohlendorf, H. M. (1986). Embryonic mortality and abnormalities of aquatic birds: Apparent impacts of selenium from irrigation water. *Sci. Total—Environ.* **52**, 49–63.

Ohlendorf, H. M. (1989). Bioaccumulation and effects of selenium in wildlife. *In* "Selenium in Agriculture and the Environment," pp. 133–177. Soil Science Society of America, Madison, WI.

Ohlendorf, H. M., Risebrough, R. W., and Vermeer, K. (1978). "Exposure of Marine Birds to Environmental Pollutants," Wildl. Res. Rep. 9. U.S. Department of the Interior, Fish and Wildlife Service, Washington, DC.

Okaichi, T., Anderson, D. M., and Nemoto, T., eds. (1989). "Red Tides: Biology, Environmental Science, and Toxicology." Elsevier, New York.

Okland, J., and Okland, K. A. (1986). The effects of acid deposition on benthic animals in lakes and streams. *Experientia* **42**, 471–486.

Olem, H. (principal author) (1990). Liming acidic surface waters. *In* "Acidic Deposition: State of Science and Technology. Vol. II. Aquatic Processes and Effects," pp. 15–1 to 15–149. Superintendent of Documents, U.S. Govt. Printing Office, Washington, DC.

Olem, H., and Flock, G., eds. (1990). "The Lake and Reservoir Restoration Guidance Manual," EPA-440/4–90–006. U.S. Environmental Protection Agency, Office of Water, Washington, DC.

Oliver, B. G., Thurman, E. M., and Malcolm, R. L. (1983). The contribution of humic substances to the acidity of coloured natural waters. *Geochim. Cosmochim. Acta* **47**, 2031–2035.

Olmstead, N. W., and Curtis, J. D. (1947). Seeds of the forest floor. *Ecology* **28**, 48–59.

Olson, J. S. (1958). Rates of succession and soil changes on southern Lake Michigan sand dunes. *Bot. Gaz. (Chicago)* **119**, 125–170.

Olson, J. S., Garrels, R. M., Berner, R. A., Armentano, T. V., Dyer, M. I., and Yaalen, D. H. (1985). The natural carbon cycle. *In* "Atmospheric Carbon Dioxide and the Global Carbon Cycle," Tech. Rep. DOE/ER-0239, pp. 175–214. Office of Energy Research, U.S. Department of Energy, Washington, DC.

Olson, S. L., Swift, C. C., and Mokhiber, C. (1979). An attempt to determine the prey of the great auk, *Pinguinis impennis. Auk* **96**, 790–792.

Olsson, H., and Rapp, A. (1991). Dryland degradation in central Sudan and conservation for survival. *Ambio* **20**, 192–195.

Organization for Economic Cooperation and Development (OECD) (1985). "The State of the Environment 1985." OECD, Paris.

Organization of Economic Cooperation and Development (OECD) (1991a). "The State of the Environment." OECD, Paris.

Organization of Economic Cooperation and Development (OECD) (1991b). "Environmental Indicators." OECD, Paris.

Orians, G. H., and Pfeiffer, E. W. (1970). Ecological effects of the war in Vietnam. *Science* **168**, 544–554.

Ormerod, S. J., Donald, A. P., and Brown, S. J. (1989). The influence of plantation forestry on the pH and aluminum concentration of upland Welsh streams: A re-examination. *Environ. Pollut.* **62**, 47–62.

Ostaff, D. P. (1985). Quantifying effects of spruce budworm damage in eastern Canada. *In* "Recent Advances in Spruce Budworms Research," pp. 247–248. Canadian Forestry Service, Ottawa, Ont.

Ostaff, D. P., and MacLean, D. A. (1989). Spruce budworm populations, defoliation, and changes in stand condition during an uncontrolled spruce budworm outbreak on Cape Breton Island, Nova Scotia. *Can. J. For. Res.* **19**, 1077–1086.

Ostrovsky, M. L., and Duthie, H. C. (1975). Primary productivity and phytoplankton of lakes on the eastern Canadian Shield. *Verh.—Int. Ver. Theor. Angew. Limnol.* **19**, 732–738.

Otvos, I. S., and Raske, A. G. (1980). "Effects of Aerial Application of Matacil on Larval and Pupal Parasites of the Eastern Spruce Budworm," Inf. Rep. N-X-189. Newfoundland Forest Research Centre, St. John's.

Ouimet, R., and Fortin, J. M. (1992). Growth and foliar nutrient status of sugar maple: Incidence of forest decline and reaction to fertilization. *Can. J. For. Res.* **22**, 699–706.

Overrein, L. N. (1972). Sulphur pollution patterns observed; Leaching of calcium in forest soil determined. *Ambio* **1**, 145–147.

Ovington, J. D. (1959). Mineral content of plantations of *Pinus sylvestris. Ann. Bot. (London)* [n.s.] **23**, 75–88.

Ovington, J. D. (1962). Quantitative ecology and the woodland ecosystem concept. *Adv. Ecol. Res.* **1**, 103–192.

Owadally, A. W. (1979). The dodo and the tambalaque tree. *Science* **203**, 1363–1364.

Owens, E. H. (1991). "Shoreline Conditions Following the Exxon Valdez Spill as of Fall 1990." Woodward-Clyde Consultants, Seattle, WA (presented at 14th annual Arctic and Marine Oilspill Program Technical Seminar, June, 1991).

Packer, P. E. (1967a). Criteria for designing and locating logging roads to control sediment. *For. Sci.* **13**, 2–18.

Packer, P. E. (1967b). Forest treatment effects on water quality. *Int. Symp. For. Hydrol.*, pp. 687–699.

Packer, P. E., and Williams, B. D. (1976). Logging and prescribed burning effects on the hydrologic and soil stability behaviour of larch/Douglas-fir forests in the northern Rocky Mountains. *Proc. Tall Timbers Fire Conf.* **14**, 465–479.

Page, A. L. (1974). "Fate and Effects of Trace Elements in Sewage Sludge Applied to Agricultural Lands," EPA-670/2–74–005. Office of Research and Development, U.S. Environmental Protection Agency, Cincinnati, OH.

Page, A. L., Ganje, T. J., and Joshi, M. S. (1971). Lead quantities in plants, soil, and air near some major highways in southern California. *Hilgardia* **41**, 1–31.

Paijmans, J. (1970). An analysis of four tropical rain forest sites in New Guinea. *J. Ecol.* **58**, 77–101.

Paine, R. T. (1969). A note on trophic complexity and community stability. *Am. Nat.* **103**, 91–93.

Palumbo, R. F. (1962). Recovery of the land plants at Eniwetok Atoll following a nuclear explosion. *Radiat. Bot.* **1**, 182–189.

Parker, C. A., Freegarde, M., and Hatchard, C. G. (1971). The effect of some chemical and biological factors on the degradation of crude oil at sea. *In* "Water Pollution by Oil," pp. 237–244. Institute of Petroleum, London.

Parker, G. R. (1984). Use of spruce plantations by snowshoe hare in New Brunswick. *For. Chron.* **60**, 162–166.

Parker, G. R. (1986). The importance of cover on use of conifer plantations by snowshoe hares in northern New Brunswick. *For. Chron.* **62**, 159–163.

Parker, G. R. (1989). Effects of reforestation upon small mammal communities in New Brunswick. *Can. Field Nat.* **103**, 509–519.

Parker, G. R., and Morton, L. D. (1978). The estimation of winter forage and its use by moose on clearcuts in north-central Newfoundland. *J. Range Manage.* **31**, 300–304.

Parker, G. R., Petrie, M. J., and Sears, D. T. (1992). Waterfowl distribution relative to wetland acidity. *J. Wildl. Manage.* **56**, 268–274.

Parker, G. R., Kimball, D. G., and Dalzell, B. (1993). "Richness and Diversity of Bird Communities Breeding in Selected Spruce and Pine Plantations in New Brunswick," Manuscript report. Canadian Wildlife Service, Sackville, New Brunswick.

Pascual, J. A., and Peris, S. J. (1992). Effects of forest spraying with two applications of cypermethrin on food supply and on breeding success of the blue tit (*Parus caerulus*). *Environ. Toxicol. Chem.* **11**, 1271–1280.

Pastor, J., Gardner, R. H., Dale, V. H., and Post, W. M. (1987). Seasonal changes in nitrogen availability as a potential factor contributing to spruce declines in boreal North America. *Can. J. For. Res.* **17**, 1394–1400.

Patil, G. P., and Taillie, C. (1982). Diversity as a concept, and its measures. *J. Am. Stat. Assoc.* **77**, 548–561.

Patton, D. R. (1974). Patch cutting increased deer and elk use of a pine forest in Arizona. *J. For.* **72**, 764–766.

Paul, G. S. (1989). Giant meteor impacts and great eruptions: Dinosaur killers? *BioScience* **39**, 162–172.

Pauli, B. D., Holmes, S. B., Sebastien, R. J., and Rawn, G. P. (1993). "Fenitrothion Risk Assessment," Tech. Rep. Ser. No. 165. Canadian Wildlife Service, Ottawa, Ont.

Paveglio, F. L., Bunck, C. M., and Heinz, G. M. (1992). Selenium and boron in aquatic birds from central California. *J. Wildl. Manage.* **56**, 31–42.

Payne, B. R., and DeGraaf, R. M. (1975). Economic values and recreational trends associated with human enjoyment of nongame birds. *In* "Management of Forest and Range Habitats for Non-game Birds," USDA For. Serv. Gen. Tech. Rep. WO-1, pp. 6–10. U.S. Department of Agriculture, Washington, DC.

Payne, J. R., and McNabb, G. D. (1984). Weathering of petroleum in the marine environment. *Mar. Technol. Soc. J.* **18**(3), 24–42.

Peakall, D. B. (1976). The pergrine falcon (*Falco peregrinus*) and pesticides. *Can. Field Nat.* **90**, 301–307.

Peakall, D. B. (1990). Prospects for the pergrine falcon, *Falco peregrinus,* in the nineties. *Can. Field Nat.* **104**, 168–173.

Peakall, D. B., Noble, D. G., Elliott, J. E., Somers, J. D., and

Erickson, G. (1990). Environmental contaminants in Canadian pergrine falcons, *Falco peregrinus*: A toxicological assessment. *Can. Field Nat.* **104**, 244–254.

Pearce, D., and Turner, R. K. (1990). "Economics of Natural Resources and the Environment." Harvester Wheatsheaf, New York.

Pearce, D., Markandya, A., and Barbier, E. B. (1989). "Blueprint for a Green Economy." Earthscan Publ. Ltd., London.

Pearce, P. A. (1975). Effects on birds. *In* "Aerial Control of Forest Insects in Canada" (M. L. Prebble, ed.), pp. 306–313. Department of the Environment, Ottawa, Ont.

Pearce, P. A., and Busby D. G. (1980). Research on the effects of fenitrothion on the white-throated sparrow. *In* "Environmental Surveillance in New Brunswick," pp. 24–28. EMO-FICO, University of New Brunswick, Fredericton.

Pearce, P. A., and Peakall, D. B. (1977). The impact of fenitrothion on bird populations in New Brunswick. *Natl. Res. Counc. Can., NRC Assoc. Comm. Sci. Criter. Environ. Qual., [Rep.] NRCC* **NRCC-16073**, 299–305.

Pearce, P. A., Peakall, D. B., and Erskine, A. J. (1979). "Impact on Forest Birds of the 1976 Spruce Budworm Spray Operation in New Brunswick," Prog. Note No. 97. Canadian Wildlife Service, Environment Canada, Ottawa, Ont.

Peek, J. M. (1974). A review of moose food habits in North America. *Nat. Can.* **101**, 195–215.

Pellew, R. (1991). Disaster in the Gulf. *I.U.C.N. Bull.* **22**(3), 17–18.

Pengelly, W. L. (1972). Clearcutting: Detrimental aspects for wildlife resources. *J. Soil Water Conserv.* **27**, 255–258.

Penner, J. E., Haselman, L. C., and Edwards, L. L. (1986). Smoke-plume distribution above large-scale fires: Implications for simulations of "nuclear winter." *J. Clim. Appl. Meteorol.* **25**, 1434–1444.

Perala, D. A. (1977). "Manager's Handbook for Aspen in the North-central States," USDA For. Serv. Gen. Tech Rep. NC-36. North Central Forest Experiment Station, Saint Paul, MN.

Percy, J. A., and Wells, P. G. (1984). Effects of petroleum in polar marine environments. *Mar. Technol. Soc. J.* **18**(3), 51–61.

Percy, K. E. (1983). Sensitivity of eastern Canadian forest tree species to simulated acid precipitation. *Aquilo, Ser. Bot.* **19**, 41–49.

Percy, K. E. (1986). The effects of simulated acid rain on germinative capacity, growth, and morphology of forest tree seedlings. *New Phytol.* **104**, 473–484.

Perins, P. J., and Mautz, W. W. (1989). Forage-nutritional advantages of small fuelwood cuts for deer. *North. J. Appl. For.* **6**, 72–74.

Perkins, H. C. (1974). "Air Pollution." McGraw-Hill, New York.

Perkins, T. D., Klein, R. M., Badger, G. T., and Easter, M. J. (1992). Spruce-fir decline and gap dynamics on Camels Hump, Vermont. *Can. J. For. Res.* **22**, 413–422.

Perry, I., and Moore, P. D. (1987). Dutch elm disease as an

analogue of Neolithic elm decline. *Nature (London)* **326**, 72–73.

Perry, R., Kirk, P. W. W., Stephenson, T., and Lester, J. N. (1984). Environmental aspects of the use of NTA as a detergent builder. *Water Res.* **18**, 255–276.

Persson, G., chair. (1982). "Acidification Today and Tomorrow." Ministry of Agriculture, Environment '82 Committee, Stockholm, Sweden.

Persson, H. (1948). On the discovery of *Merceya ligulata* in the Azores with a discussion of the so-called copper mosses. *Rev. Bryol. Lichenol.* **17**, 75–78.

Persson, R. (1974). "Review of the World's Forest Resources in the Early 1970s." Skogshogsholm, Stockholm, Sweden (cited in Donaldson, 1978).

Peters, R. L. (1991). Consequences of global warming for biological diversity. *In* "Global Climate Change and Life on Earth" (R. L. Wyman, ed.), pp. 99–118. Routledge, Chapman, and Hall, New York.

Peters, R. L., and Darling, J. D. S. (1985). The greenhouse effect and nature reserves. *BioScience* **35**, 707–717.

Peters, T. H. (1984). Rehabilitation of mine tailings: A case of complete ecosystem reconstruction and revegetation of industrially stressed lands in the Sudbury area, Ontario, Canada. *In* "Effects of Pollutants at the Ecosystem Level" (P. J. Sheehan, D. R. Miller, and P. Bourdeau, eds.), pp. 403–421. Wiley, New York.

Petersen, J., Hargeby, A., and Kullberg, A. (1987). "The Biological Importance of Humic Material in Acidified Waters," Rep. No. 3388, National Swedish Environmental Protection Boards, Solna.

Petersen, K. L., and Best, L. B. (1987). Effects of prescribed burning on non-game birds in a sagebrush community. *Wildlife Soc. Bull.* **15**, 317–329.

Peterson, H. B., and Nielson, R. F. (1973). Toxicities and deficiencies in mine tailings. *In* "Ecology and Reclamation of Devastated Land" (R. J. Hutnik and G. Davis, eds.), Vol. 1, pp. 15–24. Gordon & Breach, New York.

Peterson, J. (1984). Global population projections through the 21st century. *Ambio* **13**, 134–141.

Peterson, K. M., Billings, W. D., and Reynolds, D. N. (1984). Influence of water table and atmospheric CO_2 concentration on the carbon balance of arctic tundra. *Arct. Alp. Res.* **16**, 331–355.

Peterson, P. J. (1977). Element accumulation by plants and their tolerance of toxic mineral soils. *Int. Conf. Heavy Met. Environ. [Symp. Proc.], 1st 1975*, Vol. II, pp. 39–59.

Peterson, P. J. (1978). Land and vegetation. *In* "The Biogeochemistry of Lead in the Environment" (J. O. Nriagu, ed.), Part B, pp. 355–384. Elsevier/North-Holland Biomedical Press, Amsterdam.

Peterson, P. J., and Butler, G. W. (1962). The uptake and assimilation of selenite by higher plants. *Aust. J. Bot.* **15**, 126–146.

Peterson, P. J., and Butler, G. W. (1967). Significance of selenocystathionine in an Australian selenium-accumulating

plant, *Neptunia amplexicaulis*. *Nature (London)* **213**, 599–600.

Peterson, R. H., Daye, P. G., and Metcalfe, J. L. (1980). Inhibition of Atlantic salmon (*Salmo salar*) hatching at low pH. *Can. J. Fish. Aquat. Sci.* **37**, 770–774.

Peterson, R. J. (1969). Population trends of ospreys in the northeastern U.S. *In* "Peregrine Falcon Populations: Their Biology and Decline" (J. J. Hickey, ed.), pp. 341–350. Univ. of Wisconsin Press, Madison.

Petranka, J. A., Eldridge, M. E., and Haley, K. E. (1993). Effects of timber harvesting on southern Appalachian salamanders. *Conserv. Biol.* **7**, 363–370.

Phillips, R. W. (1970). Effects of sediment on the gravel environment and fish production. *In* "Proceedings of the Conference on Forest Land Uses and Stream Environments," pp. 64–74. Oregon State University, Corvallis.

Piatt, J. F., and Lensink, C. J. (1989). *Exxon Valdez* oil spill. *Nature* (London) **342**, 865–866.

Piatt, J. F., Lensink, C. J., Butler, W., Kendziarek, M., and Nysewander, D. R. (1990). Immediate effect of the Exxon Valdez oil spill on marine birds. *Auk* **107**, 387–397.

Piatt, J. F., Carter, H. R., and Nettleship, D. N. (1991). Effects of oil pollution on marine bird populations. *In* "The Effects of Oil on Wildlife: Research, Rehabilitation, and General Concerns" (J. White, ed.), pp. 125–141. Sheridan Press, Hanover, PA.

Pielou, E. C. (1975). "Ecological Diversity." Wiley, New York.

Pielou, E. C. (1977). "Mathematical Ecology." Wiley, New York.

Pielou, E. C. (1991). "After the Ice Age. The Return of Life to Glaciated North America." Univ. of Chicago Press, Chicago.

Piene, H. (1974). "Factors Influencing Organic Matter Decomposition and Nutrient Turnover in Cleared and Spaced, Young Conifer Stands on the Cape Breton Highlands, Nova Scotia," Inf. Rep. M-X-41. Maritimes Forest Research Centre, Fredericton, New Brunswick.

Pierce, R. C., Martin, C. W., Reeves, C. C., Likens, G. E., and Bormann, F. H. (1972). Nutrient loss from clearcuttings in New Hampshire. *In* "Watersheds in Transition," pp. 285–295. American Water Resources Association, Urbana, IL.

Pimentel, D., McLaughlin, L., Zepp, A., Lakitan, B., Kraus, T., Kleinman, P., Vancini, F., Roach, W. J., Graap., E., Keeton, W. S., and Selig, G. (1991). Environmental end economic effects of reducing pesticide use. *BioScience* **41**, 402–409.

Pimentel, D., Acquay, H., Biltonen, M., Rice, P., Silva, M., Nelson, J., Lipner, V., Giordano, S., Horowitz, A., and D'Amare, M. (1992). Environmental end economic costs of pesticide use. *BioScience* **41**, 402–409.

Pimlott, D., Brown, D., and Sam, K. (1976). "Oil Under the Ice." Canadian Arctic Resources Committee, Ottawa, Ont.

Pimm, S. L. (1987). The snake that ate Guam. *TREE* **2**, 293–294.

Pitcher, T. J., and Hart, P. J. B. (1982.) "Fisheries Ecology." Croom Helm, London.

Pitelka, F. A. (1941). Distribution of birds in relation to major biotic communities. *Am. Midl. Nat.* **25**, 113–137.

Pitelka, L. F., and Raynal, D. J. (1990). Forest decline and acidic deposition. *Ecology* **70**, 2–10.

Pittock, A. B., Ackerman, T. B., Crutzen, P. J., MacCracken, M. C., Shapiro, C. S., and Turco, R. P. (1985). "Environmental Consequences of Nuclear War," Vol. 1, SCOPE 28. Wiley, Toronto, Ont.

Placet, M. (principal author) (1990). Emissions involved in acidic deposition processes. *In* "Acidic Deposition: State of Science and Technology. Vol. I. Emissions, Atmospheric Processes, and Deposition," pp. 1-3 to 1-183. Superintendent of Documents, U.S. Govt. Printing Office, Washington, DC.

Planas, D., and Hecky, R. E. (1984). Comparison of phosphorus turnover times in northern Manitoba reservoirs with lakes of the Experimental Lakes Area. *Can. J. Fish. Aquat. Sci.* **41**, 605–612.

Platt, R. B. (1965). Ionizing radiation and homeostasis of ecosystems. *In* "Ecological Effects of Nuclear War" (G. M. Woodwell, ed.), Brookhaven Natl. Lab., AEC Rep. BNC-917 (C-43), pp. 39–60. U.S. Atomic Energy Commission, Washington, DC.

Plotkin, M. J. (1988). The outlook for new agricultural and industrial products from the tropics. *In* "Biodiversity" (E. O. Wilson, ed.), pp. 106–116. National Academy Press, Washington, DC.

Plowright, R. C., Pendrel, B. A., and McLaren, I. A. (1978). The impact of aerial fenitrothion spraying upon the population biology of bumble bees (*Bombus* LATR: Hym.) in southwestern New Brunswick. *Can. Entomol.* **110**, 1145–1156.

Pollard, D. (1985). A forestry perspective on the carbon dioxide issue. *For. Chron.* **61**, 312–318.

Pomerleau, R. (1991). "Experiments on the Causal Mechanisms of Dieback of Deciduous Forests in Quebec," LAU-X-96. Forestry Canada, Quebec Region, St. Foy, Que.

Ponting, C. (1991). "A Green History of the World." Penguin Books, Middlesex, England.

Popkin, R. (1991). Responding to eco-terrorism. *EPA J.*, July/Aug., pp. 23–26.

Porcella, D. B. (1978). Eutrophication. *J. Water Pollut. Control Fed.* **50**, 1313–1319.

Porter, M. A., and Grodzinski, B. (1985). CO_2 enrichment of protected plants. *Hortic. Rev.* **7**, 345–398.

Postel, S. (1987). Stabilizing chemical cycles. *In* "State of the World 1987," pp. 157–176. Worldwatch Institute, Norton, New York.

Potts, G. R. (1977). Population dynamics of the grey partridge: Overall effects of herbicides and insecticides on chick survival rates. *Proc. Int. Congr. Game Biol.* **13**, 203–211.

Potts, G. R. (1980). The effects of modern agriculture, nest predation, and game management on the population ecology of partridges (*Perdix perdix* and *Alectoris rufa*). *Adv. Ecol. Res.* **8**, 2–79.

Potts, G. R. (1985). Herbicides and the decline of partridge: An international perspective. *Br. Crop Prot. Conf., 1985*, pp. 983–990.

Pough, F. H., Smith, E. M., Rhodes, D. H., and Collazo, A. (1987). The abundance of salamanders in forest stands with different histories of disturbance. *For. Ecol. Manage.* **20**, 1–9.

Prager, U., and Goldsmith, F. B. (1977). Stump sprout formation by red maple (*Acer rubrum* L.) in Nova Scotia. *Proc. N.S. Inst. Sci.* **28**, 93–99.

Prakash, A. (1976). "NTA (Nitrilotriacetic Acid)—an Ecological Appraisal," Rep. EPA 3-WP-76-8. Environmental Protection Service, Environment Canada, Ottawa, Ont.

Prance, G. T., ed. (1982). "Biological Diversification in the Tropics." Columbia Univ. Press, New York.

Primack, R. B., and Hall, P. (1992). Biodiversity and forest change in Malaysian Borneo. *BioScience* **42**, 829–837.

Prinz, B. (1984). "Recent Forest Decline in the Federal Republic of Germany and Contribution of the LIS for its Explanation." German/American Information Exchange on Forest Dieback, Excursion Guide, Essen, F.R.G.

Probst, J. R. (1988). Kirtland's warbler breeding biology and habitat management. *In* "Integrating Forest Management for Wildlife and Fish" USDA For. Serv. Gen. Tech. Rep. NC-122, pp. 28–35. North Central Forest Experiment Station, St. Paul, MN.

Probst, J. R., and Rakstad, D. S. (1987). Small mammal communities in three aspen stand-age classes. *Can. Field Nat.* **101**, 362–368.

Proctor, J. (1971a). The plant ecology of serpentine. II. Plant response to serpentine soils. *J. Ecol.* **59**, 397–410.

Proctor, J. (1971b). The plant ecology of serpentine. III. The influence of a high magnesium/calcium ratio and high nickel and chromium levels in some British and Swedish serpentine soils. *J. Ecol.* **59**, 827–842.

Proctor, J., and Woodell, S. R. J. (1975). The ecology of serpentine soils. *Adv. Ecol. Res.* **9**, 255–265.

Quiring, D., Turgeon, J., Simpson, D., and Smith, A. (1991). Genetically based differences in susceptibility of white spruce to the spruce bud moth. *Can. J. For. Res.* **21**, 42–47.

Quy, V. (1984). Birds in the Mekong Delta. *Vietnam Natl. Symp. Mangrove Ecosyst., 1st, 1984*, pp. 175–181.

Quy, V., Huynh, D. H., Landa, V., Leighton, M., Medvedev, L. W., Mototani, I., Pfeiffer, E. W., Sokolov, V. E., and Thang, T. T. (1984). Terrestrial animal ecology: Symposium summary. *In* "Herbicides in War: The Long-term Ecological and Human Consequences" (A. H. Westing, ed.), pp. 45–47. Taylor & Francis, London.

Raddum, G. G. (1978). Invertebrates: Quality and quantity as fish food. *In* "Limnological Aspects of Acid Precipitation" (G. R. Hendry, ed.), Rep. No. 51074. Brookhaven National Laboratory, Brookhaven, NY.

Raddum, G. G., and Fjellheim, A. (1984). Acidification and early warning organisms in freshwater in western Norway. *Int. Assoc. Theor. Appl. Limnol., 2nd, Proc.*, 1973–1980.

Rahel, F. J., and Magnuson, J. J. (1983). Low pH and the absence of fish species in naturally acidic Wisconsin lakes: Inferences for cultural acidification. *Can. J. Fish. Aquat. Sci.* **40**, 3–9.

Ramanathan, V. (1988). The greenhouse theory of climate changes: A test by an inadvertent global experiment. *Science* **240**, 293–299.

Rampino, M. R., and Stothers, R. B. (1984). Geological rhythms and cometary impacts. *Science* **226**, 1427–1431.

Rands, M. R. W. (1985). Pesticide use on cereals and the survival of young grey partridge chicks: A field experiment. *J. Appl. Ecol.* **22**, 49–54.

Rands, M. R. W. (1986a). Effects of hedgerow characteristics on partridge breeding densities. *J. Appl. Ecol.* **23**, 479–487.

Rands, M. R. W. (1986b). The survival of gamebird (Galliformes) chicks in relation to pesticide use on cereals. *Ibis* **128**, 57–64.

Rands, M. R. W. (1987). Hedgerow management for the conservation of partridges, *Perdix perdix* and *Alectoris rufa*. *Biol. Conserv.* **40**, 127–139.

Raphael, M. G., and White, M. (1984). Use of snags by cavity nesting birds in the Sierra Nevada. *Wildl. Monogr.* **86**, 1–66.

Rapport, D. J. (1983). The stress-response environmental statistical system and its applicability to the Laurentian Lower Great Lakes. *Stat. J. U. N. ECE1*, 377–405.

Rapport, D. J. (1990). Challenges in the detection and diagnosis of pathological change in aquatic ecosystems. *J. Great Lakes Res.* **16**, 609–618.

Rapport, D. J. (1992). Evaluating ecosystem health. *J. Aquat. Ecosyst. Health* **1**, 15–24.

Rapport, D. J., Regier, H. A., and Hutchinson, T. C. (1985). Ecosystem behaviour under stress. *Am. Nat.* **125**, 617–640.

Ratcliffe, D. A. (1967). Decrease in eggshell weight in certain birds of prey. *Nature (London)* **215**, 208–210.

Ratcliffe, D. A. (1970). Changes attributable to pesticides in egg breakage frequency and eggshell thickness in some British birds. *J. Appl. Ecol.* **7**, 67–115.

Raup, D. M. (1984). Death of species. *In* "Extinctions" (M. H. Nitecki, ed.), pp. 1–19. Univ. of Chicago Press, Chicago.

Raup, D. M. (1986a). Biological extinctions in Earth history. *Science* **231**, 1528–1533.

Raup, D. M. (1986b). "The Nemesis Affair." Norton, New York.

Raup, D. M. (1988). Diversity crises in the geological past. *In* "Biodiversity" (E. O. Wilson, ed.), pp. 51–57. National Academy Press, Washington, DC.

Raven, P. H. (1990). The politics of preserving biodiversity. *BioScience* **40**, 769–774.

Ream, C. H., and Gruel, G. E. (1980). Influences of harvesting and residue treatments on small mammals and, implications for forest management. *In* "Environmental Consequences of Timber Harvesting in Rocky Mountain Coniferous Forests." USDA For. Serv. Gen. Tech. Rep. INT-90, pp. 455–467.

Intermountain Forest and Range Experiment Station, Ogden, UT.

Reams, G. A., and Huso, M. M. P. (1990). Stand history: An alternative explanation of red spruce radial growth reduction. *Can. J. For. Res.* **20**, 250–253.

Reay, R. F. (1972). The accumulation of arsenic from arsenic-rich natural waters by aquatic plants. *J. Appl. Ecol.* **9**, 557–565.

Redford, K. H. (1992). The empty forest. *BioScience* **42**, 412–422.

Redford, K. H., and Stearman, A. M. (1993). Forest-dwelling native Amazonians and the conservation of biodiversity: Interests in common or in collision? *Conserv. Biol.* **7**, 248–255.

Reeders, H. H., and Bij der Vaate, A. (1990). Zebra mussels (*Dreissena polymorpha*): A new prospective for water quality management. *Hydrobiologia* **200**, 437–450.

Reeders, H. H., Bij der Vaate, A., and Slim, F. J. (1989). The filtration rate of *Dreissena polymorpha* in three Dutch lakes with reference to biological water quality. *Freshwater Biol.* **22**, 133–141.

Rees, W. E. (1990). The ecology of sustainable development. *Ecologist* **20(1)**, 18–23.

Reeves, R. D., Brooks, R. R., and MacFarlane, R. M. (1981). Nickel uptake by *Streptanthus* and *Caulanthus* with particular reference to the hyperaccumulator *S. polygaloides* Gray (Brassicaceae). *Am. J. Bot.* **68**, 708–712.

Regier, H. A. (1993). The notion of natural and cultural integrity. *In* "Ecological Integrity and the Management, of Ecosystems," (S. Woodley, J. Kay, and G. Francis, eds.), pp. 3–18. St. Lucie Press, Boca Raton, FL.

Regier, H. A., and Hartman, W. L. (1973). Lake Erie's fish community: 150 years of cultural stresses. *Science* **180**, 1248–1255.

Regier, H. A., and Loftus, K. H. (1972). Effects of fisheries exploitation on salmonid communities in oligotrophic lakes. *J. Fish. Res. Board Can.* **29**, 959–968.

Reich, P. B., and Amundson, R. G. (1985). Ambient levels of ozone reduce net photosynthesis in tree and crop species. *Science* **230**, 566–570.

Reid, A. (1964). Light intensity and herb growth in white oak forests. *Ecology* **45**, 396–398.

Reid, W., Barker, C., and Miller, K. (principal authors) (1992). "Global Biodiversity Strategy." World Resources Institute, Washington, DC.

Reilly, A., and Reilly, C. (1973). Copper-induced chlorosis in *Becium homblei*. *Plant Soil* **38**, 671–674.

Reilly, C. (1967). Accumulation of copper by some Zambian plants. *Nature (London)* **215**, 666–669.

Reiners, W. A. (1992). Twenty years of ecosystem reorganization following experimental deforestation and regrowth suppression. *Ecol. Monogr.* **62**, 503–523.

Reiners, W. A., and Lang, G. E. (1979). Vegetational patterns and processes in the balsam fir zone, White Mountains, New Hampshire. *Ecology* **60**, 403–417.

Reinhart, K. G. (1973). "Timber-harvest, Clearcutting, and Nutrients in the Northeastern United States," USDA For. Serv. Res. Rep. NE-170. Northeastern Forest Experiment Station, Broomall, PA.

Reinhart, K. G., Eschner, A. R., and Trimble, G. R. (1963). "Effect on Streamflow of Four Forest Practices in the Mountains of West Virginia," USDA For. Serv. Res. Pap. NE-1. Northeastern Forest Experiment Station, Broomall, PA.

Reuss, J. R. (1975). Chemical/biological Relationships Relevant to Ecological Effects of Acid Rainfall," EPA-660/3-75-032. U.S. Environmental Protection Agency, Corvallis, OR.

Reuss, J. R. (1976). "Chemical and Biological Relationships Relevant to the Effect of Acid Rainfall on the Soil-plant System," USDA For. Serv. Gen. Tech. Rep. NE-23, pp. 791–813. Northeastern Forest Experiment Station, Broomall, PA.

Reuss, J. R., and Johnson, D. W. (1985). Effect of soil processes on the acidification of water by acid precipitation. *J. Environ. Qual.* **14**, 26–31.

Reuss, J. R., Cosby, B. J., and Wright, R. F. (1987). Chemical processes governing soil and water acidification. *Nature (London)* **329**, 27–32.

Reynolds, H. G. (1962). "Use of Natural Openings in a Ponderosa Pine Forest of Arizona by Deer, Elk, and Cattle," USDA For. Serv. Res. Note RM-78. Rocky Mountain Forest and Range Experiment Station, Fort Collins, CO.

Reynolds, H. G. (1966). "Use of Openings in Spruce-fir Forests of Arizona by Elk, Deer, and Cattle," USDA For. Serv. Res. Note RM-66. Rocky Mountain Forest and Range Experiment Station, Fort Collins, CO.

Reynolds, H. G. (1969). Improvement of deer habitat on southwestern forest lands. *J. For.* **67**, 803–805.

Reynolds, H. W., and Hawley, A. W. L. (1987). "Bison ecology in relation to agricultural development in the Slave River lowlands, N.W.T.," Occas. Pap. No. 63. Canadian Wildlife Service, Ottawa, Ont.

Reynoldson, T. B., and Hamilton, A. L. (1993). Historic changes in populations of burrowing mayflies (*Hexagenia limbata*) from Lake Erie based on sediment tusk profiles. *J. Great Lakes Res.* **19**, 250–257.

Rice, R. M., Rothacher, J. S., and Megahan, W. F. (1972). Erosional consequences of timber harvesting: An appraisal. *In* "Watersheds in Transition," pp. 321–329. American Water Resources Association, Urbana, IL.

Richards, J. F. (1986). World environmental history and economic development. *In* "Sustainable Development of the Biosphere" (W. C. Clark and R. E. Munn, eds.), pp. 53–74. Cambridge Univ. Press, New York.

Richards, P. W. (1952). "The Tropical Rain Forest." Cambridge Univ. Press, Cambridge, London and New York.

Richardson, C. J., and Lund, J. A. (1975). Effects of clearcutting on nutrient losses in aspen forests on three soil types in Michigan. *In* "Mineral Cycling in Southeastern Ecosystems" (F. G. Howell, J. B. Gentry, and M. H. Smith, eds.),

ERDA Symp. Ser. CONF-740513, pp. 673–686. Energy Research and Development Administration, Washington, DC.

Richman, A. D., Case, T. J., and Schwaner, T. D. (1988). Natural and unnatural extinction rates of reptiles on islands. *Am. Nat.* **131**, 611–630.

Richmond, M. L., Henny, C. J., Floyd, R. L., Mannan, R. W., Finch, D. M., and DeWeese, L. R. (1979). "Effects of Sevin-4-oil, Dimilin, and Orthene on Forest Birds in Northeastern Oregon," USDA For. Serv. Res. Pap. PSW-148. Pacific Southwest Forest and Range Experiment Station, Berkeley, CA.

Riley, C. E. (1960). The ecology of water areas associated with coal strip-mined lands in Ohio. *Ohio J. Sci.* **60**, 106–121.

Risdale, P. S. (1916). Shot, shell and soldiers devastate forests. *Am. For.* **22**, 333–340.

Risdale, P. S. (1919a). Forest casualties of our allies. *Am. For.* **25**, 899–906.

Risdale, P. S. (1919b). French forests for our army. *Am. For.* **25**, 963–972.

Risdale, P. S. (1919c). War's destruction of British forests. *Am. For.* **25**, 1027–1040.

Risdale, P. S. (1919d). Belgium's forests blighted by the Hun. *Am. For.* **25**, 1251–1258.

Risebrough, R. W., Florant, G. L., and Berger, D. D. (1970). Organochlorine pollutants in peregrines and merlins migrating through Wisconsin. *Can. Field Nat.* **84**, 247–253.

Risser, P. G. (1991). "Long-term Ecological Research. An International Perspective," SCOPE 47. Wiley, West Sussex, England.

Risser, P. G., Karr, J. R., and Forman, R. T. T. (1984). Landscape ecology: Direction and approaches. *Spec. Publ.—Ill. Nat. Hist. Surv.* **2**.

Rivers, J. B., Pearson, J. E., and Schultz, C. D. (1972). Total and organic mercury in marine fish. *Bull. Environ. Contam. Toxicol.* **8**, 257–266.

Rizzo, D. G., and Harrington, T. C. (1988). Root movement and root damage of red spruce and balsam fir on subalpine sites in the White Mountains, New Hampshire. *Can. J. For. Res.* **18**, 991–1001.

Robarge, W. P., Pye, J. M., and Bruck, R. I. (1989). Foliar elemental composition of spruce-fir in the southern Blue Ridge province. *Plant Soil* **113**, 39–43.

Robbins, C. S. (1979). Effects of forest fragmentation on bird populations. *In* "Management of North Central and Northeastern Forests for Non-game Birds," USDA For. Serv. Gen. Tech. Rep. NC-51, pp. 198–212. North Central Forest Experiment Station, St. Paul, MN.

Robbins, C. S., Sauer, J. R., Greenberg, R. S., and Droege, S. (1989). Population declines in North American birds that migrate to the tropics. *Proc. Natl. Acad. Sci. U.S.A.* **86**, 7658–7662.

Roberds, J. H., Namkoong, G., and Skroppa, T. (1990). Genetic analysis of risk in clonal populations of forest trees. *Theor. Appl. Genet.* **79**, 841–848.

Roberts, B. A., and Proctor, J., eds. (1992). "Ecology of Areas

With Serpentinized Rocks: A World View." Kluwer Academic Press, Amsterdam.

Roberts, T. M. (1984). Effects of air pollutants in agriculture and forestry. *Atmos. Environ.* **18**, 629–652.

Robinson, E., and Robbins, R. C. (1972). Emissions, concentrations, and fate of gaseous atmospheric pollutants. *In* "Air Pollution Control" (W. Strauss, ed.). Part II, pp. 1–94. Wiley (Interscience), New York.

Robinson, S. K. (1992). Population dynamics of breeding neotropical migrants in a fragmented Illinois landscape. *In* "Ecology and Conservation of Neotropical Migrant Landbirds," pp. 408–418. Smithsonian Institution Press, Washington, DC.

Robinson, S. M., Kendall, D. J., Robinson, R., Driver, C. J., and Lachen, T. E., Jr. (1988). Effects of agricultural spraying of methyl parathion on cholinesterase activity and reproductive success of wild starlings (*Sturnus vulgaris*). *Environ. Toxicol. Chem.* **7**, 343–349.

Robinson, W. L., and Bolen, E. G. (1989). "Wildlife Ecology and Management." Macmillan, New York.

Rodhe, H. (1990). A comparison of the contributions of various gases to the greenhouse effect. *Science* **248**, 1217–1219.

Rodhe, H., and Granat, L. (1984). An evaluation of sulfate in European precipitation 1955–1982. *Atmos. Environ.* **18**, 2627–2639.

Rodricks, J. V. (1992). "Calculated Risks. The Toxicity and Human Health Risks of Chemicals in Our Environment." Cambridge Univ. Press, Cambridge, UK.

Roff, D. A., and Bowen, W. D. (1983). Population dynamics and management of the northwest Atlantic harp seal (*Phoca groenlandica*). *Can. J. Fish. Aquat. Sci.* **40**, 919–932.

Roff, J. C., and Kwiatkowski, R. E. (1977). Zooplankton and zoobenthos communities of selected northern Ontario lakes of differing acidities. *Can. J. Zool.* **55**, 899–911.

Rogers, L. L., Mooty, J. J., and Dawson, D. (1981). "Foods of the White-tailed Deer in the Upper Great Lakes Region—a Review," USDA For. Serv. Gen. Tech. Rep. NC-65. North Central Forest Experiment Station, St. Paul, MN.

Rohlich, G. A., and Uttormark, P. D. (1972). Wastewater treatment and eutrophication. *In* "Nutrients and Eutrophication" (G. E. Likens, ed.), pp. 231–243. Am. Soc. Limnol. Oceanogr., Lawrence, KS.

Romme, W. H., and Knight, D. H. (1982). Landscape diversity: The concept applied to Yellowstone National Park. *BioScience* **32**, 664–670.

Roose, M., Roberts, T. M., and Bradshaw, A. D. (1982). Evolution of resistance to gaseous air pollutants. *In* "Effects of Gaseous Air Pollution in Agriculture and Horticulture" (M. H. Unsworth and D. P. Ormrod, eds.), pp. 379–410. Butterworth, London.

Rorison, I. H. (1980). The effects of soil acidity on nutrient availability and plant response. *In* "Effects of Acid Precipitation on Terrestrial Ecosystems" (T. C. Hutchinson and M. Havas, eds.), pp. 283–304. Plenum, New York.

Rosen, K., Gunderson, P., Tegnhammar, L., Johansson, M.,

and Frogner, T. (1992). Nitrogen enrichment of Nordic forest ecosystems. *Ambio* **21**, 364–368.

Rosenberg, C. R., Hutnik, R. J., and Davis, D. D. (1979). Forest composition at varying distances from a coal-fired power plant. *Environ. Pollut.* **19**, 307–317.

Rosenberg, R., Lindahl, O., and Blank, H. (1988). Silent spring in the sea. *Ambio* **17**, 287–290.

Rosenfeld, C. L. (1980). Observations on the Mount St. Helena's eruption. *Am. Sci.* **68**, 494–509.

Rosenfeld, I., and Beath, O. A. (1964). "Selenium: Geobotany, Biochemistry, Toxicity, and Nutrition." Academic Press, New York.

Rosenqvist, I. T. (1978a). Alternative sources for acidification of river water in Norway. *Sci. Total Environ.* **10**, 39–49.

Rosenqvist, I. T. (1978b). Acid precipitation and other possible sources for acidification of rivers and lakes. *Sci. Total Environ.* **10**, 271–272.

Rosenqvist, I. T. (1980). Influence of forest vegetation and agriculture on the acidity of freshwater. *In* "Advances in Environmental Science and Engineering" (J. R. Pfallin and E. W. Ziegler, eds.), pp. 56–79. Gordon & Breach, New York.

Rosenqvist, I. T., Jorgensen, P., and Rueslatter, H. (1980). The importance of natural H^+ production for acidity in soil and water. *In* "Ecological Impact of Acid Precipitation" (C. D. Drablos and A. Tollan, eds.), pp. 240–242. SNSF Project, Oslo, Norway.

Ross, S. L., Logan, W. J., and Rowland, W. (1977). "Oil Spill Countermeasures." Department of Fisheries and the Environment, Sidney, British Columbia.

Rosseland, B. O., Skogheim, O. K., and Sevaldrud, I. H. (1986). Acid deposition and effects in northern Europe. Damage to fish populations in Scandinavia continue to apace. *Water, Air, Soil Pollut.* **30**, 65–74.

Roth, R. R. (1976). Spatial heterogeneity and bird species diversity. *Ecology* **57**, 773–782.

Rothacher, J. (1970). Increases in water yield following clearcut logging in the Pacific northwest. *Water Resour. Res.* **6**, 653–657.

Rothewell, R. L. (1971). "Watershed Management Guidelines for Logging and Road Construction," Inf. Rep. A-X-42. Northern Forest Research Centre, Edmonton, Alberta.

Rothschild, B. J. (1986). "Dynamics of Marine Fish Populations." Harvard Univ. Press, Cambridge, UK.

Rotty, R. M., and Masters, C. D. (1985). Carbon dioxide from fossil fuel combustion: Trends, resources, and technological implications. *In* "Atmospheric Carbon Dioxide and the Global Carbon Dioxide Cycle," pp. 63–80. U.S. Department of Energy, Washington, DC.

Row, C., and Phelps, R. B. (1990). Carbon cycle impacts of improving forest products utilization and recycling. Presented at the North American Conference on Forestry Responses to Climate Change, Washington, DC, 1990.

Rowe, R. D., and Chestnut, L. G. (1985). Economic assessment of the effects of air pollution on agricultural crops in the San Joaquin Valley. *J. Air Polut. Control Assoc.* **35**, 728–734.

Rowland, F. S. (1988). Chlorofluorocarbons, stratospheric ozone, and the Antarctic "ozone hole." *Environ. Conserv.* **15,** 101–116.

Roy, D. N., Konar, S. K., Banerjee, S., Charles, D. A., Thompson, D. G., and Prasad, R. (1989). Uptake and persistence of the herbicide glyphosate in fruit of wild blueberry and red raspberry. *Can. J. For. Res.* **19,** 842–847.

Royal Ministry of Foreign Affairs, Royal Ministry of Agriculture (RMFA/RMA) (1971). "Air Pollution Across National Boundaries: The Impact on the Environment of Sulfur in Air and Precipitation." RMFA/RMA, Stockholm, Sweden.

Royal Society of New Zealand (RSNZ) (1980). "Assessment of Toxic Hazards of the Herbicide 2,4,5-T in New Zealand," Misc. Ser. No. 4. RSNZ, Auckland, New Zealand.

Royama, T. (1984). Population dynamics of the spruce budworm, *Choristoneura fumiferana. Ecol. Monogr.* **54,** 429–462.

Royce, W. F., Schroeder, T. R., Olsen, A. A., and Allenden, W. J. (1991). "Alaskan Fisheries—Two Years After the Oil Spill." Cook Inlet Fisheries Consultants, Homer, AL.

Rozencranz, A. (1988). Bhopal, transnational corporations, and hazardous technologies. *Ambio* **17,** 336–341.

Rubner, H. (1985). Greek thought and forest science. *Environ. Rev.* **9,** 277–295.

Rudd, J. W. M., Kelly, C. A., Schindler, D. W., and Turner, M. A. (1988). Disruption of the nitrogen cycle in acidified lakes. *Science* **240,** 1515–1517.

Ruhling, A., Baath, E., Nordgren, A., and Soderström, B. (1984). Fungi in metal-contaminated soil near the Gusum brass mill, Sweden. *Ambio* **13,** 34–36.

Russell, E. W. (1973). "Soil Conditions and Plant Growth," 10th ed. Longman, London.

Ryan, D. F., and Bormann, F. H. (1982). Nutrient resorption in northern hardwood forests. *BioScience* **32,** 29–32.

Ryder, R. A. (1990). Ecosystem health, a human perspective: Definition, detection, and the dichotomous key. *J. Great Lakes Res.* **16,** 619–624.

Ryding, S.-O. (1985). Chemical and microbiological processes as regulators of the exchange of substances between sediments and water in shallow eutrophic lakes. *Int. Rev. Gesamten Hydrobiol.* **70,** 657–702.

Ryding, S.-O., and Rast, W. (1989). "The Control of Eutrophication of Lakes and Rivers." Butler and Tanner, London.

Saber, P. A., and Dunson, W. A. (1978). Toxicity of bog water to embryonic and larval anuran amphibians. *J. Exp. Zool.* **204,** 33–42.

Safina, C. (1993). Bluefin tuna in the West Atlantic: Negligent management and the making of an endangered species. *Conserv. Biol.* **7,** 229–234.

Sakamoto, M. (1966). Primary production by phytoplankton community in some Japanese lakes and its dependence on lake depth. *Arch. Hydrobiol.* **63,** 1–28.

Salisbury, E. J. (1943). The flora of bombed areas. *Nature (London)* **151,** 462–466.

Salop, J., Levy, G. F., Wakelyn, N. T., Middleton, E. M., and Gervin, J. C. (1983). The application of forest classification from Landsat data as a basis for natural hydrocarbon emission estimation and photochemical oxidant model simulations in southern Virginia. *J. Air Pollut. Control Assoc.* **33,** 17–22.

Sanders, C. J. (1991). Biology of North American spruce budworms. *In* "Tortricid Pests, Their Biology, Natural Enemies, and Control" (L. P. S. van der Geest and H. H. Evenhuis, eds.), pp. 579–620. Elsevier, Amsterdam.

Sanders, C. J., Stark, R. W., Mullins, E. J., and Murphy, J., eds. (1985). "Recent Advances in Spruce Budworms Research." Canadian Forestry Service, Ottawa, Ont.

Sanderson, G. C., and Bellrose, F. C. (1986). "Lead Poisoning in Waterfowl," Spec. Publ. No. 4. Illinois Natural History Society, Urbana.

Santillo, D. J., Brown, P. W., and Leslie, D. M. (1989a). Response of songbirds to glyphosate-induced habitat changes on clearcuts. *J. Wildl. Manage.* **53,** 64–71.

Santillo, D. J., Brown, P. W., and Leslie, D. M. (1989b). Response of small mammals and habitat to glyphosate application on clearcuts. *J. Wildl. Manage.* **53,** 164–172.

Sassman, J., Pienta, R., Jacobs, M., and Cioffi, J. (1984). "Pesticide Background Statements. Vol. 1. Herbicides." U.S. Department of Agriculture Forest Service, Washington, DC.

Saunders, D. A., Hobbs, R. J., and Margules, C. R. (1991). Biological consequences of ecosystem fragmentation: A review. *Biol. Conserv.* **5,** 18–32.

Savidge, J. A. (1978). Wildlife in a herbicide-treated Jeffrey pine plantation in eastern California. *J. For.* **76,** 476–478.

Savidge, J. A. (1984). Guam: Paradise lost for wildlife. *Biol. Conserv.* **30,** 305–317.

Savidge, J. A. (1987). Extinction of an island forest avifauna by an introduced snake. *Ecology* **68,** 660–668.

Savile, D. B. O. (1972). "Arctic Adaptations in Plants," Monogr. No. 6. Research Branch, Canada Department of Agriculture, Ottawa, Ont.

Scale, P. R. (1982). The effects of emissions from an iron-sintering plant in Wawa, Canada on forest communities. MES Thesis, Institute for Environmental Studies, University of Toronto, Toronto, Ont.

Scarborough, A. M., and Flanagan, P. W. (1973). Observations on the effects of mechanical disturbance and oil on soil microbial populations. *In* "Proceedings of the Symposium on the Impact of Oil Resource Development on Northern Plant Communities," pp. 63–71. Institute of Arctic Biology, Fairbanks, AK.

Scavia, D., Fahnenstiel, G. L., Evans, M. S., Jude, D. J., and Lehman, J. T. (1986). Influence of salmonine predation and weather on long-term water quality in Lake Michigan. *Can. J. Fish. Aquat. Sci.* **43,** 435–443.

Schaeffer, D. J., Herricks, D. E., Adwin, E., and Kerster, H. W. (1988). Ecosystem health: I. Measuring ecosystem health. *Environ. Manage.* **12,** 445–455.

Scheafer, J. A., and Pruitt, W. O. (1991). "Fire and Woodland

Caribou in Southeastern Manitoba," Wildl. Monogr. No. 116. Wildlife Society, Bethesda, MD.

Scheibling, R. E. (1984). Echinoids, epizootics, and ecological stability in the rocky subtidal off Nova Scotia, Canada. *Helgol. Meeresunters.* **37**, 233–242.

Scheibling, R. E., and Stephenson, R. L. (1984). Mass mortality of *Strongylocentrotus droebachiensis* (Echinodermata: Echinoidea) off Nova Scotia, Canada. *Mar. Biol.* **78**, 153–164.

Scheider, W. A., Adamski, J., and Paylor, M. (1975). "Reclamation of Acidified Lakes Near Sudbury, Ontario." Ontario Ministry of the Environment, Rexdale.

Scheider, W. A., Cave, B., and Jones, J. (1976). "Reclamation of Acidified Lakes Near Sudbury, Ontario by Neutralization and Fertilization." Ontario Ministry of the Environment, Rexdale.

Scheider, W. A., Jeffries, D. S., and Dillon, P. J. (1980). "Bulk Deposition in the Sudbury and Muskoka-Haliburton Areas of Ontario During the Shutdown of INCO Ltd. in Sudbury." Ontario Ministry of the Environment, Rexdale.

Schelske, C. L., and Stoermer, E. F. (1972). Phosphorus, silica, and eutrophication of Lake Michigan. *In* "Nutrients and Eutrophication" (G. E. Likens, ed.), pp. 157–170. Am. Soc. Limnol. Oceanogr., Lawrence, KS.

Schelske, C. L., Stoermer, E. F., Fahnenstiel, G. L., and Haibach, M. (1986). Phosphorus enrichment, silica utilization, and biogeochemical silica depletion in the Great Lakes. *Can. J. Fish. Aquat. Sci.* **43**, 407–415.

Scheuhammer, A. M. (1987). The chronic toxicity of aluminum, cadmium, mercury, and lead to birds: A review. *Environ. Pollut.* **46**, 263–295.

Scheuhammer, A. M. (1991). Effects of acidification on the availability of toxic metals and calcium to wild birds and mammals. *Environ. Pollut.* **71**, 329–375.

Schindler, D. W. (1977). Evolution of phosphorus limitation in lakes: Natural mechanisms compensate for deficiencies of nitrogen and carbon in eutrophied lakes. *Science* **195**, 260–262.

Schindler, D. W. (1978). Factors regulating phytoplankton production and standing crop in the world's freshwaters. *Limnol. Oceanogr.* **23**, 478–486.

Schindler, D. W. (1985). The coupling of elemental cycles by organisms: evidence from whole-lake chemical perturbations. *In* "Chemical Processes in Lakes" (W. Stumm, ed.), pp. 225–250. Wiley, New York.

Schindler, D. W. (1987). Detecting ecosystem responses to anthropogenic stress. *Can. J. Fish. Aquat. Sci.* **44**, 6–25.

Schindler, D. W. (1988). Effects of acid rain on freshwater ecosystems. *Science* **239**, 149–157.

Schindler, D. W. (1990). Experimental perturbations of whole lakes as tests of hypotheses concerning ecosystem structure and function. *Oikos* **57**, 25–41.

Schindler, D. W. (1991). Comments on the *Sustainable Biological Initiative*. *Conserv. Biol.* **5**, 550–551.

Schindler, D. W., and Fee, E. J. (1974). Experimental Lakes

Area: whole-lake experiments in eutrophication. *J. Fish. Res. Board Can.* **31**, 937–953.

Schindler, D. W., and Holmgren, S. K. (1971). Primary production and phytoplankton of the Experimental Lakes Area, northwestern Ontario, and other low-carbonate waters, and a liquid scintillation method for determining ^{14}C activity in photosynthesis. *J. Fish. Res. Board Can.* **28**, 189–201.

Schindler, D. W., and Noven, B. (1971). Vertical distribution and seasonal abundance of zooplankton in two shallow lakes of the Experimental Lakes Area. *J. Fish. Res. Board Can.* **28**, 245–256.

Schindler, D. W., and Turner, M. A. (1982). Biological, chemical, and physical responses of lakes to experimental acidification. *Water, Air, Soil Pollut.* **18**, 259–271.

Schindler, D. W., Brunskill, G. J., Emerson, S., Broecker, W. S., and Perg, T. H. (1972). Atmospheric carbon dioxide: Its role in maintaining phytoplankton standing crop. *Science* **177**, 1192–1194.

Schindler, D. W., Kling, H., Schmidt, R. V., Prokopowich, J., Frost, V. E., Reid, R. A., and Capel, M. (1973). Eutrophication of Lake 227 by addition of phosphate and nitrate: The second, third, and fourth years of enrichment, 1970, 1971, 1972. *J. Fish. Res. Board Can.* **30**, 1415–1440.

Schindler, D. W., Kalff, J., Welch, H. E., Brunskill, G. J., Kling, H., and Kritsch, N. (1974). Eutrophication in the high arctic—Meretta Lake, Cornwallis Island (75°N Lat.). *J. Fish. Res. Board Can.* **31**, 647–662.

Schindler, D. W., Newbury, R. W., Beaty, K. G., and Campbell, P. (1976). Natural water and chemical budgets for a small Precambrian lake basin in eastern Canada. *J. Fish. Res. Board Can.* **33**, 2526–2543.

Schindler, D. W., Newbury, R. W., Beaty, K. G., Prokopowich, J., Ruszczynski, T., and Dulton, J. A. (1980). Effects of a windstorm and forest fire on chemical losses from forested watersheds and on the quality of receiving streams. *Can. J. Fish. Aquat. Sci.* **37**, 328–334.

Schindler, D. W., Mills, K. H., Malley, D. F., Findlay, D. L., Shearer, J. A., Davies, I. J., Turner, M. A., Linsey, G. A., and Cruikshank, D. A. (1985). Long-term ecosystem stress: The effects of years of experimental acidification on a small lake. *Science* **228**, 1395–1401.

Schindler, D. W., Turner, M. A., Stainton, M. P., and Linsey, G. A. (1986). Natural sources of acid neutralizing capacity in low alkalinity lakes of the Precambrian Shield. *Science* **232**, 843–847.

Schlegel, H., Amundson, R. G., and Huttermann, A. (1992). Element distribution in red spruce (*Picea rubens*) fine roots: Evidence for aluminum toxicity at Whiteface Mountain. *Can. J. For. Res.* **22**, 1132–1138.

Schlesinger, W. H. (1977). Carbon balance in terrestrial detritus. *Annu. Rev. Ecol. Syst.* **8**, 51–81.

Schlesinger, W. H. (1989). The role of ecologists in the face of global change. *Ecology* **70**, 1.

Schmidt, W. C., Felton, D. G., and Carlson, C. E. (1983).

Alternatives to chemical insecticides in budworm-susceptible forests. *West. Wildl.* **9**, 13–19.

Schneider, S. H. (1989). The changing climate. *Sci. Am.* **261**(3), 70–79.

Schneider, S. H. and Mesirow, L. E. (1976). "The Genesis Strategy." Plenum, New York.

Schneider, S. H., and Thompson, S. L. (1988). Simulating the climatic effects of nuclear war. *Nature (London)* **333**, 221–227.

Schnitzer, M. (1978). Humic substances: Chemistry and reactions. *In* "Soil Organic Matter" (M. Schnitzer and S. U. Khan, eds.), pp. 1–64. Am. Elsevier, New York.

Schoenwald-Cox, C. M., Chambers, S. M., MacBryde, B., and Thomas, W. L., eds., (1983). "Genetics and Conservation." Benjamin/Cummings, Menlo Park, CA.

Schofield, C. L. (1976a). "Lake Acidification in the Adirondack Mountains of New York: Causes and Consequences," USDA For. Serv. Gen. Tech. Rep. NE-23, p. 477. Northeastern Forest Experiment Station, Broomall, PA.

Schofield, C. L. (1976b). Acid precipitation: Effects on fish. *Ambio* **5**, 228–230.

Schofield, C. L. (1982). Historical fisheries changes in the United States related to decreases in surface water pH. *Acid Rain/Fish., Proc. Int. Symp., 1981*, pp. 57–67.

Schoonmaker, P., and McKee, A. (1988). Species composition and diversity during secondary succession of coniferous forests in the western Cascade Mountains of Oregon. *For. Sci.* **34**, 960–979.

Schorger, A. W. (1955). "The Passenger Pigeon." Univ. of Wisconsin Press, Madison, WI.

Schroeder, T. R. (1991). "Alaskan Fisheries Two Years After the Spill." Exxon Co., Houston, TX.

Schultze, E. D., De Vries, W., Hauhs, M., Rosen, S., Rasmussen, L., Tamm, C. O., and Nilsson, J. (1989). Critical loads for nitrogen deposition on forest ecosystems. *Water, Air, Soil Pollut.* **48**, 451–456.

Scoggan, H. J. (1950). The flora of Bic and the Gaspé Peninsula, Quebe. *Bull.—Natl. Mus. Can.* **115**, Biol. Ser. No. 39.

Scott, J. T., Siccama, T. G., Johnson, A. H., and Breish, A. R. (1984). Decline of red spruce in the Adirondacks. *Bull. Torrey Bot. Club* **111**, 438–444.

Scott, V. E., Whelan, J. A., and Svoboda, P. L. (1980). Cavity-nesting birds and forest management. *In* "Management of Western Forests and Grasslands for Non-game Birds," USDA For. Serv. Gen. Tech. Rep. INT-86, pp. 311–324. Intermountain Forest and Range Experiment Station, Ogden, UT.

Scotter, G. W. (1967). The winter diet of barren-ground caribou in northern Canada. *Can. Field Nat.* **81**, 33–39.

Scrivener, J. C., and Carruthers, S. (1989). Changes in the invertebrate populations of the main stream and back channels of Carnation Creek, British Columbia, following spraying with the herbicide Roundup (glyphosate). *In* "Proceedings of the Carnation Creek Workshop," pp. 263–272. FRDA Rep. 063, Forest Pest Management Institute, Sault Ste Marie, Ont.

Seddon, B. (1972). Aquatic macrophytes as limnological indicators. *Freshwater Biol.* **2**, 107–130.

Sedjo, R. A. (1989a). Forests to offset the greenhouse effect. *J. For.*, pp. 12–16.

Sedjo, R. A. (1989b). Forests: A tool to moderate global warming? *Environment* **31**, 14–20.

Seip, H. M., and Freedman, B. (1980). Effects of acid precipitation on soils. *In* "Effects of Acid Precipitation on Terrestrial Ecosystems" (T. C. Hutchinson and M. Havas, eds.), pp. 591–595. Plenum, New York.

Seip, H. M., and Tollan, A. (1978). Acid precipitation and other possible sources for acidification of rivers and lakes. *Sci. Total Environ.* **10**, 253–270.

Seneca, E. D., and Broome, S. W. (1982). Restoration of marsh vegetation impacted by the Amoco Cadiz oil spill and subsequent cleanup operations at Ile Grande, France. *In* "Ecological Study of the Amoco Cadiz Oil Spill," pp. 363–419. U.S. Department of Commerce, National Oceanic and Atmospheric Administration, Washington, DC.

Sergeant, D. E. (1991). "Harp Seals, Man, and Ice," Spec. Publ. Fish & Aquat. Sci., p. 114. Fisheries and Oceans Canada, Ottawa, Ont.

Serra, R. (1989). Long-term variability of the Chilean sardine. *In* "Long-term Variability of Pelagic Fish Populations and Their Environment" (T. Kawasaki, S. Tanaka, Y. Toba, and A. Taniguchi, eds.), pp. 165–172. Pergamon, New York.

Shacklette, H. T. (1965). Bryophytes associated with mineral deposits and solutions in Alaska. *Geol. Surv. Bull. (U.S.)* **1198c**.

Shafer, C. L. (1990). "Nature Reserves. Island Theory and Conservation Practice." Smithsonian Institution Press, Washington, DC.

Shaffer, M. L. (1985). The metapopulation and species conservation: The special case of the northern spotted owl. *In* "Ecology and Management of the Spotted Owl in the Pacific Northwest," USDA For. Serv. Gen. Tech. Rep. PNW-185, pp. 86–99. Pacific Northwest Forest and Range Experiment Station, Portland, OR.

Shafi, M. I., and Yarranton, G. A. (1973). Diversity, floristic richness, and species evenness during a secondary (postfire) succession. *Ecology* **54**, 897–902.

Shannon, C. E., and Weaver, W. (1949). "The Mathematical Theory of Communication." Univ. of Illinois Press, Urbana.

Shannon, J. D., and Lecht, B. M. (1986). Estimation of source-receptor matrices for deposition of NO_x-N. *Water, Air, Soil Pollut.* **30**, 815–824.

Shapiro, J. (1980). The importance of trophic-level interactions to the abundance and species composition of algae in lakes. *In* "Hypertrophic Ecosystems" (J. Barica and L. R. Mur, eds.), pp. 105–116. Junk, The Hague, The Netherlands.

Shapiro, J., and Wright, D. I. (1984). Lake restoration by biomanipulation: Round Lake, Minnesota, the first two years. *Freshwater Biol.* **14**, 371–383.

Shearer, J. A., Fee, E. J., DeBruyn, E. R., and DeClercq, D. R.

(1987a). Phytoplankton productivity changes in a small, double-basin lake in response to termination of experimental fertilization. *Can. J. Fish. Aquat. Sci.* **44,** Suppl. 1, 47–54.

Shearer, J. A., Fee, E. J., DeBruyn, E. R., and DeClercq, D. R. (1987b). Phytoplankton primary production and light attenuation responses to the experimental acidification of a small Canadian Shield lake. *Can. J. Fish. Aquat. Sci.* **44,** Suppl. 1, 83–90.

Sheath, R. G., Havas, M., Hellebust, J. A., and Hutchinson, T. C. (1982). Effects of long-term natural acidification on the algal communities of tundra ponds at the Smoking Hills, N.W.T., Canada. *Can. J. Bot.* **60,** 58–72.

Sheridan, D. (1981). "Desertification of the United States." Council on Environmental Quality, Washington, DC.

Sherwood, M. J., and Mearns, A. J. (1977). Environmental significance of fin erosion in southern California demersal fishes. *Ann. N.Y. Acad. Sci.* **298,** 177–189.

Shields, L. M., and Wells, P. V. (1962). Effects of nuclear testing on desert vegetation. *Science* **135,** 38–40.

Shields, L. M., Wells, P. V., and Rickard, W. H. (1963). Vegetational recovery on atomic target areas in Nevada. *Ecology* **44,** 697–705.

Shipp, A. M., Hogg, M. L., Crump, K. S., and Kodell, R. L. (1986). "Worst Case Analysis Study on Forest Plantation Herbicide Use." Forest Land Management Division, State of Washington, Seattle.

Short, H. L., Blair, R. M., and Epps, E. A. (1975). "Composition and Digestibility of Deer Browse in Southern Forests," USDA For. Serv. Res. Pap. SO-111.

Shortle, W. C., and Smith, K. T. (1988). Aluminum-induced calcium deficiency syndrome in declining red spruce. *Science* **240,** 1017–1018.

Showman, R. E. (1975). Lichens as indicators of air quality around a coal-fired power generating station. *Bryologist* **78,** 1–6.

Showman, R. E. (1981). Lichen recolonization following air quality improvement. *Bryologist* **84,** 492–497.

Shrift, A. (1969). Aspects of selenium metabolism in higher plants. *Annu. Rev. Plant Physiol.* **20,** 475–494.

Shriner, D. S. (principal author) (1990). Responses of vegetation to atmospheric deposition and air pollution. *In* "Acidic Deposition: State of Science and Technology. Vol. III. Terrestrial, Materials, Health, and Visibility Effects," pp. 18-1 to 18-206. Superintendent of Documents, U.S. Govt. Printing Office, Washington, DC.

Shugart, H. H. (1984). "A Theory of Forest Dynamics: The Ecological Implications of Forest Succession Models." Springer-Verlag, New York.

Shugart, H. H., Smith, T. M., Kitchings, J. T., and Kroodsma, R. L. (1978). The relationship of nongame birds to southern forest types and successional stages. *In* "Proceedings of the Workshop on Management of Southern Forests for Nongame Birds," USDA For. Serv. Gen. Tech. Rep. SE-14, pp. 5–16. Southeastern Forest Experiment Station, Asheville, NC.

Shuman, T. (1984). The aging of the world's population. *Ambio* **13,** 175–181.

Siccama, T. G. (1974). Vegetation, soil, and climate on the Green Mountains of Vermont. *Ecol. Monogr.* **44,** 325–345.

Siccama, T. G., and Smith, W. H. (1978). Lead accumulation in a northern hardwood forest. *Environ. Sci. Technol.* **12,** 593–594.

Siccama, T. G., Bliss, M., and Vogelmann, H. W. (1982). Decline of red spruce in the Green Mountains of Vermont. *Bull. Torrey Bot. Club* **109,** 162–168.

Sidle, R. C., Hook, J. E., and Kardos, L. T. (1976). Heavy metals application and plant uptake in a land disposal system for waste water. *J. Environ. Qual.* **5,** 97–102.

Sigal, L. L., and Nash, T. H., III (1983). Lichen communities on conifers in southern California mountains: An ecological survey relative to oxidant air pollution. *Ecology* **64,** 1343–1354.

Sigman, J. T., Gilliam, F. S., and Partin, M. E. (1989). Precipitation and throughfall chemistry for a montane hardwood forest ecosystem: Potential contributions from cloud water. *Can. J. For. Res.* **19,** 1240–1247.

Silver, J. (1924). Rodent enemies of fruit and shade trees. *J. Mammal.* **5,** 165–173.

Simberloff, D. (1988). The contribution of population and community biology to conservation science. *Annu. Rev. Ecol. Syst.* **19,** 473–511.

Simberloff, D., and Cox, J. (1987). Consequences and costs of conservation corridors. *Conserv. Biol.* **1,** 63–71.

Simberloff, D., Farr, J. A., Cox, J., and Mehlman, D. W. (1992). Movement corridors: Conservation bargains or poor investment. *Conserv. Biol.* **6,** 493–504.

Simmons, F. C. (1979). "Handbook for Eastern Timber Harvesting." Superintendent of Documents, U.S. Govt. Printing Office, Washington, DC.

Simon, C. (1987). Hawaiian evolutionary biology: An introduction. *TREE* **2,** 175–178.

Simpson, V. R., Hunt, A. E., and French, M. C. (1979). Chronic lead poisoning in a herd of mute swans. *Environ. Pollut.* **18,** 187–202.

Singer, R., Roberts, D. A., and Boylen, C. W. (1983). The macrophytic community of an acidic lake in Adirondack (New York: USA): A new depth record for aquatic angiosperms. *Aquat. Bot.* **16,** 49–57.

Singh, H. B., Viezee, W., Johnson, W. B., and Ludwig, F. L. (1980). The impact of stratospheric ozone on tropospheric air quality. *J. Air Pollut. Control Assoc.* **30,** 1009–1017.

Singh, T., and Wheaton, E. E. (1991). Boreal forest sensitivity to global warming: Implications for forest management in western interior Canada. *For. Chron.* **67,** 342–348.

Sisterton, D. L. (principal author) (1990). Deposition monitoring: Methods and results. *In* "Acidic Deposition: State of Science and Technology. Vol. I. Emissions, Atmospheric Processes, and Deposition," pp. 6-1 to 6-338. Superintendent of Documents, U.S. Govt. Printing Office, Washington, DC.

Sivard, R. L. (1982). "World Military and Social Expenditures 1982." World Priorities, Leesburg, VA.

Sivard, R. L. (1986). "World Military and Social Expenditures 1986," 11th ed. World Priorities, Washington, DC.

Sivard, R. L. (1987). "World Military and Social Expenditures 1987–1988." World Priorities, Washington, DC.

Sivard, R. L. (1989). "World Military and Social Expenditures 1989." World Priorities, Washington, DC.

Sivard, R. L. (1993). "World Military and Social Expenditures 1983." World Priorities, Washington, DC.

Skarby, L., and Sellden, G. (1984). The effects of ozone on crops and forests. *Ambio* **13**, 68–72.

Skene, M. (1915). The acidity of *Sphagnum* and its relation to chalk and mineral salts. *Ann. Bot. (London)* [n.s.] **29**, 65–90.

Slagsvold, T. (1977). Bird population changes after clearance of deciduous scrub. *Biol. Conserv.* **12**, 229–244.

Slaney, P. A., Smith, T. A., and Halsey, T. G. (1977a). Physical Alterations to Small Stream Channels Associated with Streamside Logging Practices in the Central Interior of British Columbia," Fish. Manage. Rep. No. 31. Fish and Wildlife Branch, Fisheries Research and Technical Services Division, Environment Canada, Vancouver, British Columbia.

Slaney, P. A., Smith, T. A., and Halsey, T. G. (1977b). "Some Effects of Forest Harvesting on Salmonid Rearing Habitat in Two Streams in the Central Interior of British Columbia," Fish. Manage. Rep. No. 71. Fish and Wildlife Branch, Fisheries Research and Technical Services Division, Environment Canada, Vancouver, British Columbia.

Small, R. D. (1991). Environmental impacts of the fires in Kuwait. *Nature (London)* **350**, 11–12.

Small, R. D., and Bush, B. W. (1985). Smoke production from multiple nuclear explosions in nonurban areas. *Science* **229**, 465–469.

Smith, A. E. (1982). Herbicides and the soil environment in Canada. *Can. J. Soil Sci.* **62**, 433–460.

Smith, A. G., and Pilcher, J. R. (1973). Radiocarbon dates and vegetational history of the British Isles. *New Phytol.* **72**, 903–914.

Smith, C. T. (1985). "Literature Review and Approaches to Studying the Impacts of Forest Harvesting and Residue Management on Forest Nutrient Cycles," Inf. Rep. No.13. College of Forest Resources, University of Maine at Orono, Orono.

Smith, D. W. (1970). Concentration of soil nutrients before and after fire. *Can. J. Soil Sci.* **50**, 17–29.

Smith, H. J., Archibald, R. M., and Corner, A. H. (1964). Elaphostrongylosis in Maritime Moose and deer. *Can. Vet. J.* **5**, 287–296.

Smith, J. E. (1968). "Torrey Canyon Pollution and Marine Life." Cambridge Univ. Press, London and New York.

Smith, R. A., Alexander, R. B., and Wolman, M. G. (1987). Water-quality trends in the Nation's rivers. *Science* **235**, 1607–1615.

Smith, R. C., Prezelin, B. B., Baker, K. S., Bidigare, R. R., Boucher, N. P., Coley, T., Karentz, D., MacIntyre, S., Matlick, H. A., Menzies, D., Ondrusek, M., Wan, Z., and Waters, K. J. (1992). Ozone depletion: Ultraviolet radiation and phytoplankton biology in Antarctic waters. *Science* **255**, 952–958.

Smith, S. H. (1972a). Factors of ecologic succession in oligotrophic fish communities of the Laurentian Great Lakes. *J. Fish. Res. Board Can.* **29**, 717–730.

Smith, S. H. (1972b). The future of Salmonid communities in the Laurentian Great Lakes. *J. Fish. Res. Board Can.* **29**, 951–957.

Smith, S. M., Carrow, J. R., and Laing, J. E. (1990). Inundative release of the egg parasitoid, *Trichogramma minuta* (Hymenoptera: Trichogrammatidae), against forest pests such as the Spruce Budworm, *Choristoneura fumiferana* (Lepidoptera: Tortricidae): The Ontario Project 1982–1986. *Mem. Entomol. Soc. Can.*, 153.

Smith, S. V., and Buddemeier, R. W. (1992). Global change and coral reef ecosystems. *Annu. Rev. Ecol. Syst.* **23**, 89–118.

Smith, T. G., and Armstrong, F. A. G. (1978). Mercury and selenium in ringed and bearded seal tissues. *Arctic* **31**, 75–84.

Smith, V. H. (1980). Nutrient dependence of primary productivity in lakes. *Limnol. Oceanogr.* **24**, 1051–1064.

Smith, V. H. (1983). Low nitrogen to phosphorus ratios favour dominance by bluegreen algae in lake phytoplankton. *Science* **221**, 669–671.

Smith, V. H., and Shapiro, J. (1981). Chlorophyll-phosphorus relations in individual lakes. Their importance to lake restoration strategies. *Environ. Sci. Technol.* **15**, 444–451.

Smith, W. H. (1981). "Air Pollution and Forests." Springer-Verlag, New York.

Smith, W. H. (1984). Ecosystem pathology: A new perspective for phytopathology. *For. Ecol. Manage.* **9**, 193–219.

Smith, W. H., and Siccama, T. G. (1981). The Hubbantrd Brook ecosystem study: Biogeochemistry of lead in the northern hardwood forest. *J. Environ. Qual.* **10**, 323–333.

Smith, W. P. (1991). *Odocoileus virginianus. Mamm. Species* **388**, 1–13.

Smol, J. P., and Glew, J. R. (1992). Paleolimnology. *In* "Encyclopedia of Earth System Science" (W. Nierenberg, ed.), Vol. 3, pp. 551–564. Academic Press, San Diego.

Snedaker, S. C. (1984). Coastal, marine, and aquatic ecology: An overview. *In* "Herbicides in War: The Long-term Ecological and Human Consequences" (A. H. Westing, ed.), pp. 95–107. Taylor & Francis, London.

Soane, B. O., and Saunder, D. H. (1959). Nickel and chromium toxicity of serpentine soils in southern Rhodesia. *Soil Sci.* **88**, 322–330.

Soikkeli, S., and Karenlampi, L. (1984). The effects of nitrogen fertilization on the ultrastructure of mesophyll cells of conifer needles in northern Finland. *Eur. J. For. Pathol.* **14**, 129–136.

Solbrig, O. T. (1992). The IUBC-SCOPE-UNESCO Program of research on biodiversity. *Ecol. Appl.* **2**, 131–138.

Solomon, A. M., and West, D. C. (1985). Potential responses of forests to CO_2-induced climate change. *In* "Characterization of Information Requirements for Studies of CO_2 Effects: Water Resouces, Agriculture, Fisheries, Forests, and Human Health," DOE/ER-0236, pp. 145–169. U.S. Department of Energy, Washington, DC.

Solomon, A. M., Trabolka, J. R., Reichle, D. E., and Voorhees, L. D. (1985). The global cycle of carbon. *In* "Atmospheric Carbon Dioxide and the Global Carbon Cycle," DOE/ER-0239, pp. 1–13. U.S. Department of Energy, Washington, DC.

Solomon, P. M., Connor, B., R. L., de Zafra, Parrish, A., Barrett, J., and Jaramillo, M. (1987). High concentrations of chlorine monoxide at low altitudes in the Antarctic spring stratosphere: Secular variation. *Nature (London)* **328**, 411–413.

Solomon, S. (1987). More news from Antarctica. *Nature (London)* **326**, 20.

Solomon, S. (1990). Progress towards a quantitative understanding of Antarctic ozone depletion. *Nature (London)* **347**, 347–354.

Sonstegard, R. A. (1977). Environmental carcinogenesis studies in fishes of the Great Lakes of North America. *Ann. N.Y. Acad. Sci.* **298**, 261–269.

Sonzogni, W. C., Robertson, A., and Beeton, A. M. (1983). Great Lakes management: Ecological factors. *Environ. Manage.* **7**, 531–542.

Sopper, W. E. (1975). Effects of timber harvesting and related management practices on water quality in forested watersheds. *J. Environ. Qual.* **4**, 24–29.

Sorkhoh, N., Al-Hasain, R., Radwan, S., and Hopner, T. (1992). Self-cleaning of the Gulf. *Nature (London)* **359**, 109.

Sotherton, N. W., and Rands, M. R. W. (1986). Predicting, measuring, and minimizing the effects of pesticides on farmland wildlife on intensively managed arable land in Britain. *In* "Pesticide Science and Technology" (R. Greenhalgh and T. M. Roberts, eds.), pp. 433–436. Blackwell, London.

Sotherton, N. W., Rands, M. R. W., and Moreby, S. J. (1985). Comparison of herbicide treated and untreated headlands for the survival of game and wildlife. *Br. Crop Prot. Conf.,* **1985**, pp. 991–998.

Soule, M. E. (1983). What do we really know about extinction? *In* "Genetics and Conservation" (C. M. Schonewald-Cox, S. M. Chambers, B. MacBryde, and W. L. Thomas, eds.), pp. 111–124. Benjamin/Cummings, Menlo Park, CA.

Soule, M. E., ed. (1986). "Conservation Biology: The Science of Scarcity and Diversity." Sinauer Assoc., Sunderland, MA.

Soule, M. E. (1991). Conservation: Tactics for a constant crisis. *Science* **253**, 744–750.

Southward, A. J., and Southward, E. C. (1978). Recolonization of rocky shores in Cornwall after use of toxic dispersants to clean up the Torrey Canyon spill. *J. Fish. Res. Board Can.* **35**, 682–706.

Southwood, T. R. E., and Cross, D. J. (1969). The ecology of the partridge. III. Breeding success and the abundance of insects in natural habitats. *J. Anim. Ecol.* **38**, 497–509.

Soutiere, E. C. (1979). Effects of timber harvesting on marten in Maine. *J. Wildl. Manage.* **43**, 850–860.

Sparling, J. H. (1967). Assimilation rates of some woodland herbs in Ontario. *Bot. Gaz. (Chicago)* **128**, 160–168.

Sparrow, A. H., Schwemmer, S. S., Klug, E. E., and Pugliellio, L. (1970). Woody plants: Changes in survival in response to long-term (8 years) chronic gamma irradiation. *Science* **169**, 1082–1084.

Speake, D. W., Hill, E. P., and Carter, V. E. (1975). Aspects of land management with regard to production of wood and wildlife in the southwestern United States. *In* "Forest Soils and Forest Land Management" (B. Bernier and C. H. Winget, eds.), pp. 333–349. Laval Univ. Press, Quebec, Que.

Spencer, C. N., and King, D. L. (1984). Role of fish in regulation of plant and animal communities in eutrophic ponds. *Can. J. Fish. Aquat. Sci.* **41**, 1851–1855.

Spies, T. A. (1991). Plant species diversity and occurrence in young, mature, and old-growth Douglas-fir stands in Western Orgeon and Washington. *In* "Wildlife and Vegetation of Unmanaged Douglas-fir Forests," USDA For. Serv. Gen. Tech. Rep. PNW-285, pp. 111–121. Pacific Northwest Forest and Range Experiment Station, Portland, OR.

Spies, T. A., and Cline, S. P. (1988). Coarse woody debris in forests and plantations of coastal Oregon. *In* "From the Forest to the Sea: A Story of Fallen Trees," USDA For. Serv. Gen. Tech. Rep. PNW-GTR-229, pp. 5–12. Pacific Northwest Forest and Range Experiment Station, Portland, OR.

Spies, T. A., Franklin, J. F., and Thomas, T. B. (1988). Coarse woody debris in Douglas-fir forests of western Oregon and Washington. *Ecology* **69**, 1689–1702.

Spitzer, P. R., and Poole, A. (1980). Coastal ospreys between New York City and Boston: A decade of reproductive recovery 1969–1979. *Am. Birds* **34**, 234–241.

Spitzer, P. R., Risebrough, R. W., Walker, W., Hernandez, R., Poole, A., Puleston, D., and Nisbet, I. C. T. (1978). Productivity of ospreys in Connecticut—Long Island increases as DDE residues decline. *Science* **202**, 333–335.

Sprankle, P., Meggitt, W., and Penner, D. (1975). Absorption, mobility, and microbial degradation of glyphosate in the soil. *Weed Sci.* **3**, 229–234.

Spray, C. J., Crick, H. Q. P., and Harty, A. D. M. (1987). Effects of aerial applications of fenitrothion on bird populations of a Scottish pine plantation. *J. Appl. Ecol.* **24**, 29–47.

Sprugel, D. G. (1976). Dynamic structure of wave-regenerated *Abies balsamea* forests in the northeastern United States. *J. Ecol.* **64**, 889–911.

Sprugel, D. G., and Bormann, F. H. (1981). Natural disturbance and the steady state in high-altitude balsam fir forests. *Science* **211**, 390–393.

Sprules, W. G. (1975a). Midsummer crustacean zooplankton communities in acid-stressed lakes. *J. Fish. Res. Board Can.* **32**, 389–395.

Sprules, W. G. (1975b). Factors affecting the structure of limnetic crustacean zooplankton communities in central Ontario lakes. *Verh.—Int. Ver. Theor. Angew. Limnol.* **19,** 635–643.

Sprules, W. G. (1977). Crustacean zooplankton communities as indicators of limnological conditions: An approach using principal components analysis. *J. Fish. Res. Board Can.* **34,** 962–975.

Spurr, S. H., and Barnes, B. V. (1980). "Forest Ecology," 3rd ed. Wiley, Toronto, Ont.

Spry, D. J., and Wiener, J. G. (1991). Metal bioavailability and toxicity to fish in low-alkalinity lakes: A critical review. *Environ. Pollut.* **71,** 243–304.

Squire, R. O., Flinn, D. W., and Campbell, R. G. (1991). Silvicultural research for sustained wood production and biosphere conservation in the pine plantations and native eucalypt forests of south-eastern Australia. *In* "Long-term Field Trials to Assess Environmental Impacts of Harvesting," IEA/BE,T6/A6 Rep. No. 5, pp. 3–28. Forest Research Institute, Rotorua, New Zealand.

Stadtman, T. C. (1974). Selenium biochemistry. *Science* **183,** 915–922.

Staicer, C. A., Freedman, B., and Shackell, N. (1993). "Results of Regional Workshops on Ecological Monitoring and Indicators." Environment Canada, State of the Environment Reporting Organization, Ottawa, Ont.

Staicer, C. A., Freedman, B., Srivastava, D., Dowd, N., Kilgar, J., Hayden, J., Payne, F., and Pollock, T. (1993). Characteristics of Nova Sctian lakes in relation to their use by black duck broods. *Hydrobiologica* **279**/**280,** 185–199.

Staley, J. M. (1965). Decline and mortality of red and scarlet oaks. *For. Sci.* **11,** 2–17.

Stanley, S. M. (1984). Marine mass extinctions. *In* "Extinctions" (M. H. Nitecki, ed.), pp. 69–118. Univ. of Chicago Press, Chicago.

Stark, N. (1980). Changes in soil water quality resulting from three timber cutting methods and three levels of fibre utilization. *J. Soil Water Conserv.* **35,** 183–187.

Steadman, D. W. (1991). Extinction of species: Past, present, and future. *In* "Global Climate Change and Life on Earth" (R. C. Wyman, ed.), pp. 156–169. Routledge, Chapman, and Hall, New York.

Steedman, R. J. (1988). Modification and assessment of an index of biotic integrity to quantify stream water quality in southern Ontario. *Can. J. Fish. Aquat. Sci.* **45,** 492–501.

Steedman, R. J., and Regier, H. A. (1990). Ecological bases for an understanding of ecosystem integrity in the Great Lakes basin. *In* "An Ecosystem Approach to the Integrity of the Great Lakes in Turbulent Times," Spec. Publ. 90-4, pp. 257–270. Great Lakes Fishery Commission, Ann Arbor, MI.

Steele, D. H., Anderson, R., and Green, J. M. (1992). The managed commercial annihilation of northern cod. *Newfoundland Stud.* **8,** 34–68.

Steinbeck, J. (1954). "Sweet Thursday." Viking Press, New York.

Steiner, K. C., McCormick, L. H., and Canavera, D. J. (1980). Differential response of paper birch provenances to aluminum in solution culture. *Can. J. For. Res.* **10,** 25–29.

Steinhart, C. E., and Steinhart, J. S. (1972). "Blowout: A Case Study of the Santa Barbara Oil Spill." Duxbury Press, Belmont, CA.

Stephens, S. L., and Birks, J. W. (1985). After nuclear war: Perturbations in atmospheric chemistry. *BioScience* **35,** 557–562.

Stephenson, G. R. (1983). "Expert Report in the Case of Victoria Palmer *et al.* v. Nova Scotia Forest Industries." Supreme Court of Nova Scotia, Trial Division, Halifax.

Stephenson, M., and Mackie, G. L. (1986). Lake acidification as a limiting factor in the distribution of the freshwater amphipod *Hyalella azteca. Can. J. Fish. Aquat. Sci.* **43,** 288–292.

Stevens, J. T., and Sumner, D. D. (1991). Herbicides. *In* "Handbook of Pesticide Toxicology. Vol. 3. Classes of Pesticides" (W. L. Hayes and E. R. Laws, eds.), pp. 1317–1408. Academic Press, San Diego.

Stewart, C. C., and Freedman, B. (1989). Comparison of the macrophyte communities of a clearwater and a brownwater oligotrophic lake in Kejimkujik National Park, Nova Scotia. *Water, Air, Soil Pollut.* **46,** 335–341.

Stewart, G. H. (1989). Ecological considerations of dieback in New Zealand's indigenous forests. *N. Z. J. For. Sci.* **19,** 243–249.

Stewart, G. H., and Veblen, T. T. (1983). Forest instability and canopy tree mortality in Westland, New Zealand. *Pac. Sci.* **37,** 427–431.

Stewart, R. E., and Aldrich, J. W. (1951). Removal and repopulation of breeding birds in a spruce-fir forest community. *Auk* **68,** 471–482.

Stewart, R. E., Gross, L. L., and Hankala, B. H. (1984). "Effects of Competing Vegetation on Forest Trees: A Bibliography with Abstracts," USDA For. Serv. Gen. Tech. Rep. WO-43. U.S. Department of Agriculture, Washington, DC.

Stigliani, W. M., and Shaw, R. W. (1990). Energy use and acid deposition: The view from Europe. *Annu. Rev. Energy* **15,** 201–216.

Stocks, B. J. (1985). Forest fire behaviour in spruce budworm-killed balsam fir. *In* "Recent Advances in Spruce Budworms Research," pp. 188–199. Canadian Forestry Service, Ottawa, Ont.

Stocks, B. J. (1987). Fire potential in the spruce budworm-killed forests of Ontario. *For. Chron.* **63,** 8–14.

Stokes, P. M. (1980). Benthic algal communities in acidic lakes. *In* "Effects of Acidic Precipitation on Benthos" (R. Singer, ed.), pp. 119–133. North American Benthological Society, Springfield, IL.

Stone, E. L., and Timmer, V. R. (1975). On the copper content of some northern conifers. *Can. J. Bot.* **53,** 1453–1456.

Stothers, R. B. (1984). The great Tambora eruption in 1815 and its aftermath. *Science* **224,** 1191–1198.

Stover, E., and Charles, D. (1991). The killing minefields of Cambodia. *New Sci.*, Oct., pp. 26–30.

Strain, B., and Cure, J. D., eds. (1985). "Direct Effects of Increasing Carbon Dioxide on Vegetation," DOE/ER-0238. U.S. Department of Energy, Washington, DC.

Strand, L. (1983). Acid precipitation and forest growth in Norway. *Aquilo, Ser. Bot.* **19**, 32–39.

Strange, N. E., Bodaly, R. A., and Fudge, R. J. P. (1991). "Mercury Concentrations of Fish in Southern Indian Lake and Issett Lake, Manitoba, 1975–1988: The Effect of Lake Impoundment and Churchill River Diversion," Can. Tech. Rep. Fish Aquat. Sci. 1824. Department of Fisheries and Oceans, Winnipeg, Manitoba.

Straughan, D., and Abbott, B. C. (1971). The Santa Barbara oil spill: Ecological changes and natural oil leaks. *In* "Water Pollution by Oil," pp. 257–262. Institute of Petroleum, London.

Strayer, D. L. (1991). Projected distribution of the zebra mussel, *Dreissena polymorpha,* in North America. *Can. J. Fish. Aquat. Sci.* **48**, 1389–1395.

Strijbosch, H. (1979). Habitat selection of amphibians during their aquatic phase. *Oikos* **33**, 363–372.

Strojan, C. L. (1978a). The impact of zinc smelter emissions on forest litter arthropods. *Oikos* **31**, 41–46.

Strojan, C. L. (1978b). Forest leaf litter decomposition in the vicinity of a zinc smelter. *Oecologia* **32**, 203–212.

Stross, R. G., and Hasler, A. D. (1960). Some lime-induced changes in lake metabolism. *Limnol. Oceanogr.* **5**, 265–272.

Stross, R. G., Neess, J. C., and Hasler, A. D. (1961). Turnover time and production of planktonic crustacea in limed and reference portion of a bog lake. *Ecology* **42**, 237–245.

Stroud, R. H. (1967). "Water Quality Criteria to Protect Aquatic Life—A Summary," Spec. Publ. No. 4. American Fisheries Society, Washington, DC.

Stuanes, A. O. (1980). Effects of acid precipitation on soil and forest. 5. Release and loss of nutrients from a Norwegian forest soil due to artificial rain of varying acidity. *In* "Ecological Impact of Acid Precipitation" (D. Drablos and A. Tollan, eds.), pp. 198–199. SNSF Project, Oslo, Norway.

Sullivan, T. J. (principal author) (1990). Historical changes in surface water acid-base chemistry in response to acidic deposition. *In* "Acidic Deposition: State of Science and Technology. Vol. II. Aquatic Processes and Effects," pp. 11-1 to 11-181. Superintendent of Documents, U.S. Govt. Printing Office, Washington, DC.

Sullivan, T. P. (1990). Influence of forest herbicide on deer mouse and Orgeon vole populations. *J. Wildl. Manage.* **54**, 566–576.

Sullivan, T. P., and Sullivan, D. S. (1981). Barking damage by snowshoe hare and red squirrels in lodgepole pine stands in central British Columbia. *Can. J. For. Res.* **12**, 443–448.

Sullivan, T. P., and Sullivan, D. S. (1982). Responses of small-mammal populations to a forest herbicide application in a 20-year- old conifer plantation. *J. Appl. Ecol.* **19**, 95–106.

Sullivan, T. P., and Sullivan, D. S. (1988). Influence of stand thinning on snowshoe hare population dynamics and feeding damage in Lodgepole Pine forest. *J. Appl. Ecol.* **25**, 791–805.

Sullivan, T. P., and Sullivan, D. S. (1991). Responses of a deer mouse population to a forest herbicide application: Reproduction, growth, and survival. *Can. J. Zool.* **59**, 1148–1154.

Sundaram, A., Sundaram, K. M. S., and Cadogan, B. L. (1985). Influence of formulation properties on droplet spectra and soil residues of aminocarb aerial sprays in conifer forests. *J. Environ. Sci. Health* **2**, 167–186.

Sundaram, K. M. S., and Nott, R. (1984). "Fenitrothion Residues in Selected Components of a Conifer Forest Following Aerial Application of Tank Mixes Containing Triton X-100," Inf. Rep. FPM-X-65. Forest Pest Management Institute, Canadian Forestry Service, Ottawa, Ont.

Sundaram, K. M. S., and Nott, R. (1985). "Distribution and Persistence of Aminocarb in Terrestrial Components of the Forest Environment after Semi-operational Application of Two Mixtures of Matacil 180F," Inf. Rep. FPM-X-67. Forest Pest Management Institute, Canadian Forestry Service, Sault Ste. Marie, Ont.

Sundaram, K. M. S., and Nott, R. (1986). "Persistence of Fenitrothion Residues in a Conifer Forest Environment," Inf. Rep. FPM-X-75. Forest Pest Management Institute, Sault Ste. Marie, Ont.

Sussman, V. H., and Mulhern, J. J. (1964). Air pollution from coal refuse disposal areas. *J. Air Pollut. Control Assoc.* **14**, 279–284.

Sutcliffe, D. W., and Carrick, T. R. (1973). Studies on mountain streams in the English Lake District. I. pH, calcium, and the distribution of invertebrates in the River Duddon. *Freshwater Biol.* **3**, 437–462.

Sutcliffe, W. H., Loucks, R. H., Drinkwater, K. F., and Coote, A. R. (1983). Nutrient flux onto the Labrador Shelf from Hudson Strait and its biological significance. *Can. J. Fish. Aquat. Sci.* **40**, 1962–1971

Sutton, R. (1978). Glyphosate herbicide: An assessment of forestry potential. *For. Chron.* **541**, 24–28.

Sutton, R. (1985). "Vegetation Management in Canadian Forestry," Inf. Rep. O-X-369, Great Lakes Forest Research Station, Sault Ste. Marie, Ont.

Svoboda, J., and Freedman, B. (1994). "Ecology of a Polar Oasis. Alexandra Fiord, Ellesmere Island, Canada." Captus University Publications, Toronto, Ont.

Swan, D., Freedman, B., and Dilworth, T. (1984). Effects of various hardwood forest management practices on small mammals in central Nova Scotia. *Can. Field-Nat.* **98**, 362–364.

Swan, L. A., and Papp, C. S. (1972). "The Common Insects of North America." Harper & Row, New York.

Swanson, F. J., and Franklin, J. F. (1992). New forestry principles from ecosystem analysis of Pacific Northwest forests. *Ecol. Appl.* **2**, 262–274.

Swanson, R. H., Golding, D. L., Rothwell, R. L., and Bernier, P. Y. (1986). "Hydrologic Effects of Clear-cutting at Marmot Creek and Streeter Watersheds, Alberta," Inf. Rep. NOR-

X-278. Northern Forestry Research Centre, Edmonton, Alberta.

Swift, L. W., and Messer, J. B. (1971). Forest cutting raises water temperatures of small streams in the southern Appalachians. *J. Soil Water Conserv.* **26,** 111–115.

Swift, M. J., ed. (1984). "Soil Biological Processes and Tropical Soil Fertility," News Mag., Spec. Issue. International Union of Biological Sciences.

Sykes, J. M., Lowe, V. P. W., and Briggs, D. R. (1989). Some effects of afforestation on the flora and fauna of an upland sheepwalk during 12 years after planting. *J. Appl. Ecol.* **26,** 299–320.

Szaro, R. C., and Balda, R. P. (1979). "Effects of Harvesting Ponderosa Pine on Non-game Bird Populations," Res. Pap. RM-212. Rocky Mountain Forest and Range Experiment Station, Fort Collins, CO.

Tabatabai, M. A. (1987). Physicochemical fate of sulfate in soils. *J. Air Pollut. Control Assoc.* **37,** 34–38.

Takano, B. (1987). Correlation of volcanic activity with sulfur oxyanion speciation in a crater lake. *Science* **235,** 1633–1635.

Tamm, C. O., and Cowling, E. B. (1976). "Acidic Precipitation and Forest Vegetation," USDA For. Serv. Gen. Tech. Rep. NE-23, pp. 845–855. Northeastern Forest Experiment Station, Upper Darby, PA.

Tamm, C. O., and Hallbacken, L. (1986). Changes in soil pH over a 50-year period under different canopies in SW Sweden. *Water, Air, Soil Pollut.* **31,** 337–341.

Tamm, C. O., and Hallbacken, L. (1988). Changes in soil acidity in two forest areas with different acid deposition: 1920s to 1980s. *Ambio* **17,** 56–61.

Tamm, C. O., and Wiklander, G. (1980). Effects of artificial acidification with sulphuric acid on tree growth in Scotch pine forest. *In* "Ecological Impacts of Acid Precipitation" (D. Drablos and A. Tollan, ed.), pp. 188–189. SNSF Project, Oslo, Norway.

Tamm, C. O., Holmen, H., Popovic, B., and Wiklander, G. (1974). Leaching of plant nutrients from soils as a consequence of forestry operations. *Ambio* **3,** 211–221.

Tang, A. J. S., Yap, D., Kurtz, J., Chan, W. H., and Lusis, M. A. (1987). An analysis of the impact of the Sudbury smelters on wet and dry deposition in Ontario. *Atmos. Environ.* **21,** 813–824.

Tannock, J., Howells, W. W., and Phelps, R. J. (1983). Chlorinated hydrocarbon pesticide residues in eggs of some birds in Zimbabwe. *Environ. Pollut., Ser. B* **5,** 147–155.

Taylor, A. W. (1978). Post-application volatilization of pesticides under field conditions. *J. Air Pollut. Control Assoc.* **28,** 922–927.

Taylor, O. C., Thompson, C. R., Tingey, D. T., and Reinert, R. A. (1975). Oxides of nitrogen. *In* "Responses of Plants to Air Pollution" (T. T. Kozlowski, ed.). Academic Press, San Diego.

Taylor, R. J., and Basake, F. A. (1984). Patterns of fluoride accumulation and growth reduction exhibited by Douglas-fir

in the vicinity of an aluminum-reduction plant. *Environ. Pollut., Ser. A* **33,** 221–235.

Tejning, S. (1967). Mercury in pheasant (*Phasianus colchicus* L.) deriving from seed grain dressed with methyl and ethyl mercury compounds. *Oikos* **18,** 334–344.

Telfer, E. S. (1967a). Comparison of a deer yard and a moose yard in Nova Scotia. *Can. J. Zool.* **45,** 485–490.

Telfer, E. S. (1967b). Comparison of a moose and deer winter range in Nova Scotia. *J. Wildl. Manage.* **31,** 418–425.

Telfer, E. S. (1970). Winter habitat selection by moose and deer. *J. Wildl. Manage.* **34,** 553–559.

Telfer, E. S. (1972). Browse selection by deer and hares. *J. Wildl. Manage.* **36,** 1344–1349.

Telfer, E. S. (1978a). Cervid distribution, browse, and snow cover in Alberta. *J. Wildl. Manage.* **42,** 352–361.

Telfer, E. S. (1978b). Silviculture in eastern deer yards. *For. Chron.* **54,** 203–208.

Telfer, E. S. (1978c). Habitat requirements of moose—the principal taiga range animal. *Proc. Int. Rangeland Congr.,* 1st, pp. 462–465.

Temple, S. A. (1977). Plant-animal mutualism: Coevolution with dodo leads to near extinction of plant. *Science* **197,** 885–886.

Temple, S. A. (1979). Reply to: The dodo and the tambaloque tree, by Owadally (1979). *Science* **203,** 1363–1364.

Temple, S. A., Mossman, M. J., and Ambuel, B. (1979). The ecology and management of avian communities in mixed hardwood-coniferous forest. *In* "Management of North Central and Northeastern Forests for Non-game Birds," USDA For. Serv. Gen. Tech. Rep. NC-51, pp. 132–151. North Central Forest Experiment Station, St. Paul, MN.

Terborgh, J. (1974). Preservation of natural diversity: The problem of extinction prone species. *BioScience* **24,** 715–722.

Terborgh, J. (1979). "Where Have All the Birds Gone?" Princeton Univ. Press, Princeton, NJ.

Terborgh, J. (1992). Perspectives on the conservation of neotropical migrant landbirds. *In* "Ecology and Conservation of Neotropical Migrant Landbirds," pp. 7–12. Smithsonian Institution Press, Washington, DC.

Terborgh, J., and Winter, B. (1980). Some causes of extinction. *In* "Conservation Biology" (M. E. Soule and B. A. Wilcox, eds.), pp. 119–133. Sinauer, Sunderland, MA.

Terborgh, J., Robinson, S. K., Parker, T. A., Muna, C. A., and Pierpont, N. (1990). Structure and organization of an Amazonian forest bird community. *Ecol. Monogr.* **60,** 312–238.

Terman, G. L. (1978). "Atmospheric Sulfur: The Agronomic Aspects," Tech. Bull. No. 23. Sulfur Institute, Washington, DC.

Terraglio, F. P., and Manganelli, R. M. (1967). The absorption of atmospheric SO_2 by water solutions. *J. Air Pollut. Control Assoc.* **17,** 403–406.

Tevis, L. (1956). Responses of small mammal populations to logging of Douglas-fir. *J. Mammal.* **37,** 189–196.

Thaler, G. R., and Plowright, R. C. (1980). The effect of aerial insecticide spraying for spruce budworm control on the fecundity of entomophilous plants in New Brunswick. *Can. J. Bot.* **58**, 2022–2027.

Thalken, C. E., and Young, A. L. (1983). Long-term field studies of a rodent population continuously exposed to TCDD. *In* "Human and Environmental Risks of Chlorinated Dioxins and Related Compounds" (R. E. Tucker, A. L. Young, and A. P. Gray, eds.), pp. 357–372. Plenum, New York.

The Institute of Ecology (TIE) (1973). "An Ecological Glossary for Engineers and Resource Managers." TIE, Madison, WI.

The Times (1918). "The Times History of the War," Vol. XVI. The Times, London.

The Times (1919). "The Times History of the War," Vol. XIX. The Times, London.

Thill, R. E., Mossis, H. F., and Harrel, A. T. (1990). Nutritional quality of deer diets from southern pine-hardwood forests. *Am. Midl. Nat.* **124**, 413–417.

Thiollay, J.-M. (1992). Influence of selective logging on bird species diversity in a Guaianan rain forest. *Conserv. Biol.* **6**, 47–63.

Thirgood, J. V. (1981). "Man and the Mediterranean Forest: A History of Resource Depletion." Academic Press, New York.

Thomas, J. W., Anderson, R. G., Maser, C., and Bull, E. L. (1979). "Snags," Agric. Handb. No. 553, pp. 60–77. U.S. Department of Agriculture, Washington, DC.

Thomas, M. L. H. (1977). Long term biological effects of bunker C oil in the intertidal zone. *In* "Fate and Effect of Petroleum Hydrocarbons in Marine Ecosystems and Organisms" (D. A. Wolfe, ed.), pp. 238–245. Pergamon, New York.

Thomas, N. A., Robertson, A., and Sonzogni, W. C. (1980). Review of control objectives: New target loads and input controls. *In* "Phosphorus Management Strategies for Lakes" (R. C. Loehr, C. S. Martin, and W. Rast, eds.), pp. 61–90. Ann Arbor Sci. Publ., Ann Arbor, MI.

Thompson, C. R., and Taylor, O. C. (1969). Effects of air pollutants on growth, leaf drop, fruit drop, and yield of citrus trees. *Environ. Sci. Technol.* **3**, 934–940.

Thompson, D. R., Hamer, K. C., and Furness, R. W. (1991). Mercury accumulation in great skuas (*Catharacta skua*) of known age and sex, and its effect on breeding and survival. *J. Appl. Ecol.* **28**, 672–684.

Thompson, D. R., Furness, R. W., and Walsh, P. M. (1992). Historical changes in mercury concentrations in the marine ecosystem of the north and north-east Atlantic Ocean as indicated by seabird feathers. *J. Appl. Ecol.* **29**, 79–84.

Thompson, F. R., and Caper, D. E. (1988). Avian assemblages in seral stages of a Vermont forest. *J. Wildl. Manage.* **52**, 771–777.

Thompson, I. D. (1988). Habitat needs of furbearers in relation to logging in boreal Ontario. *For. Chron.* **64**, 251–261.

Thompson, L. K., Sidhu, S. S., and Roberts, B. A. (1979). Fluoride accumulations in soil and vegetation in the vicinity of a phosphorus plant. *Environ. Pollut., Ser. A* **18**, 221–234.

Thompson, M. E., Elder, F. C., Davis, A. R., and Whitlow, S. (1980). Evidence of acidification of rivers in eastern Canada.

In "Ecological Impact of Acid Precipitation" (D. Drablos and A. Tollan, eds.), pp. 244–245. SNSF Project, Oslo, Norway.

Thompson, S. L., Aleksandrov, V. V., Stenchikov, G. L., Schneider, S. H., Covey, C., and Chervin, R. M. (1984). Global climatic consequences of nuclear war: Simulations with three dimensional models. *Ambio* **13**, 236–243.

Thorington, R. W., Tannenbaum, B., Tarak, A., and Rudran, R. (1982). Distribution of trees on Barro Colorado Island: A five hectare sample. *In* "The Ecology of a Tropical Forest" (C. E. G. Leigh, A. S. Rand, and D. M. Windsor, eds.), pp. 83–94. Smithsonian Institution Press, Washington, DC.

Thornback, J., and Jenkins, M. (1982). "The IUCN Mammal Red Data Book." International Union for Conservation of Nature and Natural Resources, Gland, Switzerland.

Thornton, F. C., Schaedle, M., and Raynal, D. J. (1986). Effect of aluminum on the growth of sugar maple in solution culture. *Can. J. For. Res.* **16**, 892–896.

Thornton, I. W. B. (1984). Krakatau—the development and repair of a tropical ecosystem. *Ambio* **13**, 216–225.

Tiedemann, A. R., Helvey, J. D., and Anderson, T. D. (1978). Stream chemistry and watershed nutrient economy following wildfire and fertilization in eastern Washington. *J. Environ. Qual.* **7**, 580–588.

Titterington, R. W., Crawford, H. S., and Burgason, B. N. (1979). Songbird responses to commercial clear-cutting in Maine spruce–fir forests. *J. Wildlife Manag.* **43**, 602–609.

Tobalske, B. W., Shearer, R. C., and Hutto, R. L. (1991). "Bird Populations in Logged and Unlogged Western Larch/Douglas-fir Forest in Northwestern Montana," USDA For. Serv. Res. Pap. INT-442. Intermountain Forest and Range Experiment Station, Ogden, UT.

Tome, M. A., and Pough, F. H. (1982). Responses of amphibians to acid precipitation. *Acid Rain/Fish., Proc. Int. Symp., 1981*, pp. 245–253.

Torstensson, L. and Stark, J. (1979). Persistence of glyphosate in forest soils. *Swed. Weed Conf. [Proc.]* **20**, 145–149.

Trail, P. W., and Baptista, L. F. (1993). The impact of brown-headed cowbird parasitism on populations of the Nuttall's white-crowned sparrow. *Conserv. Biol.* **7**, 309–315.

Tredici, P. D., Ling, H., and Yang, G. (1992). The *Gingko*s of Tian Mu Shan. *Conserv. Biol.* **6**, 202–209.

Trelease, S. F., and Martin, A. L. (1936). Plants made poisonous by selenium absorbed from the soil. *Bot. Rev.* **2**, 373–396.

Trelease, S. F., and Trelease, H. M. (1938). Selenium as a stimulating and possible essential element for indicator plants. *Am. J. Bot.* **25**, 373–380.

Trelease, S. F., and Trelease, H. M. (1939). Physiological differentiation in *Astragalus* with reference to selenium. *Am. J. Bot.* **26**, 530–535.

Trelease, S. F., di Somma, A. A., and Jacobs, A. L. (1960). Seleno-amino acid found in *Astragalus bisulcatus. Science* **132**, 168.

Trimble, G. R., and Sartz, J. R. (1957). How far from a stream should a logging road be located. *J. For.* **55**, 339–341.

Tristram, H. B. (1873). "The Natural History of the Bible." Society for Promoting Christian Knowledge, London.

Troedsson, T. (1980). Ten year acidification of Swedish forest soils. *In* "Ecological Impact of Acid Precipitation" (D. Drablos and A. Tollan, eds.), p. 184. SNSF Project, Oslo, Norway.

Troendle, C. A. (1987). "The Potential Effect of Partial Cutting and Thinning on Streamflow from the Subalpine Forest," USDA For. Serv. Res. Pap. RM-274. Rocky Mountain Forest and Range Experiment Station, Fort Collins, CO.

Tu, A. T., ed. (1988). "Handbook of Natural Toxins," Vol. 3. Dekker, New York.

Tukey, H. B. (1980). Some effects of rain and mist on plants, with implications for acid precipitation. *In* "Effects of Acid Precipitation on Terrestrial Ecosystems" (T. C. Hutchinson and M. Havas, eds.), pp. 140–150. Plenum, New York.

Turco, R. P., Toon, O. B., Ackerman, T. P., Pollack, J. B., and Sagan, C. (1990). Climate and smoke: An appraisal of nuclear winter. *Science* **247**, 166–176.

Turgeon, J. J. (1986). The phenological relationship between the larval development of the spruce budmoth, *Zeiraphera canadensis* (Lepidoptera: Olethreutidae), and white spruce in northern New Brunswick. *Can. Entomol.* **118**, 345–350.

Turner, D. J. (1977). "The Safety of the Herbicides 2,4-D and 2,4,5-T," For. Comm. Bull. 57. H. M. Stationery Office, London.

Turner, M. A., Jackson, M. B., Findlay, D. L., Graham, R. W., DeBruyn, E. R., and Vandermeer, E. M. (1987). Early responses of periphyton to experimental lake acidification. *Can. J. Fish. Aquat. Sci.* **44**, Suppl. 1, 135–149.

Turner, M. G. (1989). Landscape ecology: The effect of pattern on process. *Annu. Rev. Ecol. Syst.* **20**, 171–197.

Turner, R. S. (principal author) (1990). Watershed and lake processes affecting surface water acid-base chemistry. *In* "Acidic Deposition: State of Science and Technology. Vol. II. Aquatic Processes and Effects," pp. 10–1 to 10–167. Superintendent of Documents, U.S. Govt. Printing Office, Washington, DC.

Tveite, B. (1980a). Effects of acid precipitation on soil and forest. 8. Foliar nutrient concentrations in field experiments. *In* "Ecological Impact of Acid Precipitation" (D. Drablos and A. Tollan, eds.), pp. 204–205. SNSF Project, Oslo, Norway.

Tveite, B. (1980b). Effects of acid precipitation on soil and forest. 9. Tree growth in field experiments. *In* "Ecological Impact of Acid Precipitation" (D. Drablos and A. Tollan, eds.) pp. 206–207. SNSF Project, Oslo, Norway.

Tver, D. F. (1981). "Dictionary of Dangerous Pollutants, Ecology, and Environment." Industrial Press, New York.

Twery, M. J. (1990). Effects of defoliation by gypsy moth. *In* "Interagency Gypsy Moth Research Review," USDA For. Serv. Gen. Tech. Rep. NE-146, pp. 27–39. Northeastern Forest Experiment Station, Radnot, PA.

Tyler, G. (1974). Heavy metal pollution and soil enzymatic activity. *Plant Soil* **41**, 303–311.

Tyler, G. (1975). Heavy metal pollution and mineralization of nitrogen in forest soils. *Nature (London)* **255**, 701–702.

Tyler, G. (1976). Heavy metal pollution, phosphatase activity, and mineralization of organic phosphorus in forest soils. *Soil Biol. Biochem.* **8**, 327–332.

Tyler, G. (1981). Leaching of metals from the A-horizon of a spruce forest soil. *Water, Air, Soil Pollut.* **15**, 353–369.

Tyler, G. (1984). The impact of heavy metal pollution on forests: a case study of Gusum, Sweden. *Ambio* **13**, 18–24

Ueno, M. (1958). The disharmonious lakes of Japan. *Verh.— Int. Ver. Theor. Angew. Limnol.* **13**, 217–226.

Uhl, C. (1983). You can keep a good forest down. *Nat. Hist.* **4**, 71–78.

Uhl, C. (1987). Factors controlling succession following slash-and-burn agriculture in Amazonia. *J. Ecol.* **75**, 377–407.

Uhlmann, D. (1980). Stability and multiple steady states of hypereutrophic ecosystems. *In* "Hypertrophic Ecosystems" (J. Barica and L. R. Mur, eds.), pp. 235–247. Junk, The Hague, The Netherlands.

Ulrich, B. (1983). Soil acidity and its relation to acid deposition. *In* "Effect of Accumulation of Air Pollutants in Forest Ecosystems" (B. Ulrich and J. Pankrath, eds.), pp. 127–146. Reidel, Boston, MA.

Ulrich, B., Mayer, R. and Khanna, P. K. (1980). Chemical changes due to acid precipitation in a loess-derived soil in central Europe. *Soil Sci.* **130**, 193–199.

United Nations Environment Program (UNEP) (1982). "English-Russian Glossary of Selected Terms in Preventive Toxicology." UNEP, Moscow.

Unsworth, M. H. (1979). Dry deposition of gases and particulate inputs onto vegetation: A review. *In* "Methods Involved in Studies of Acid Precipitation to Forest Ecosystems," pp. 9–15. Edinburgh.

Urone, P. (1976). The primary air pollutants—gaseous. Their occurrence, sources, and effects. *In* "Air Pollutants, Their Transformation and Transport" (A. C. Stern, ed.), 3rd ed., pp. 23–75. Academic Press, New York.

Urone, P. (1986). The pollutants. *In* "Air Pollution" (A. C. Stern, ed.), 3rd ed., Vol. 4, pp. 1–60. Academic Press, Orlando, FL.

U.S. Department of Agriculture (USDA) (1984). "Pesticide Background Statements. Herbicides," Agric. Handb. No. 633. U.S.D.A., Washington, DC.

Usher, M. B. (1986). Wildlife conservation evaluation: Attributes, criteria, and value. *In* "Wildlife Conservation Evaluation" (M. B. Usher, ed.), pp. 1–17. Chapman & Hall, London.

Vallentyne, J. R. (1974). "The Algal Bowl. Lakes and Man," Misc. Spec. Publ. No. 22. Department of the Environment, Fisheries and Marine Service, Ottawa, Ont.

Van Cleve, K., and Martin, S. (1991). "Long-term Ecological Research in the United States." Long-term Ecological Research Network Office, University of Washington, Seattle.

Van Dach, H. (1943). The effect of pH on pure cultures of *Euglena mutabilis. Ohio J. Sci.* **43**, 47–48.

Vandermeulen, J. A. (1987). Toxicity and sublethal effects of petroleum hydrocarbons in freshwater biota. *In* "Oil in Freshwater: Chemistry, Biology, Countermeasure Technology," pp. 267–303. Pergamon, New York.

van Frankenhuyzen, K. (1990). Development and current status of *Bacillus thuringiensis* for control of defoliating forest insects. *For. Chron.* **66,** 498–507.

Van Groenwoud, H. (1977). "Interim Recommendations for the Use of Buffer Strips for the Protection of Small Streams in the Maritimes," Inf. Rep. M-X-74. Canadian Forestry Service, Maritimes Forest Research Centre, Fredericton, New Brunswick.

Van Loon, J. C. (1974). Analysis of heavy metals in sewage sludge and liquids associated with sludges. Presented at the Canada/Ontario Sludge Handling and Disposal Seminar, Toronto, Ontario, 1974.

Varty, I. W. (1975). Side effects of pest control projects on terrestrial arthropods other than the target species. *In* "Aerial Control of Forest Insects in Canada" (M. L. Prebble, ed.), pp. 266–275. Department of the Environment, Ottawa, Ont.

Varty, I. W. (1977). Long-term effects of fenitrothion spray programs on non-target terrestrial arthropods. *Natl. Res. Counc. Can., NRC Assoc. Comm. Sci. Criter. Environ. Qual. [Rep.] NRCC* **NRCC-16073,** 343–375.

Veith, G. D., DeFoe, D. L., and Bergstedt, B. V. (1979). Measuring and estimating the bioconcentration factor of chemicals in fish. *J. Fish. Res. Board Can.* **36,** 1040–1048.

Veneman, P. L. M., Murray, J. M., and Baker, J. H. (1983). Spatial distribution of pesticide residues in a former apple orchard. *J. Environ. Qual.* **12,** 101–104.

Venkatram, A. (principal author) (1990). Relationships between atmospheric emissions and deposition/air quality. *In* "Acidic Deposition: State of Science and Technology. Vol. I. Emissions, Atmospheric Processes, and Deposition," pp. 8-1 to 8-110. Superintendent of Documents, U.S. Govt. Printing Office, Washington, DC.

Vergnano, O., and Hunter, J. G. (1952). Nickel and cobalt toxicities in oat plants. *Ann. Bot. (London)* [n.s.] **17,** 317–328.

Vermeij, G. J. (1986). The biology of human-caused extinctions. *In* "The Preservation of Species" (B.G. Norton, ed.), pp. 29–49. Princeton Univ. Press, Princeton, NJ.

Verner, J. (1975). Avian behaviour and habitat management. *In* "Symposium on Management of Forest and Range Habitats for Nongame Birds," USDA For. Serv. Gen. Tech. Rep. WO-1, pp. 39–58. Southwestern Forest and Range Experiment Station, Tuscon, AZ.

Verry, E. S. (1972). Effect of aspen clearcutting on water yield and quality in northern Minnesota. *In* "Watersheds in Transition" pp. 276–284. American Water Resources Association, Urbana, IL.

Vet, R. J., Ro, C. U., Sukoff, W. B., and Ord, D. (1993). "NAt Chem 1990. Annual Report—Acid Precipitation in Eastern North America." Environment Canada, Atmospheric Environment Service, Downsview, Ont.

Virupaksha, T. K., and Shrift, A. (1965). Biochemical differ-

ences between selenium accumulator and non-accumulator *Astragalus* species. *Biochim. Biophys. Acta* **107,** 69–80.

Vitousek, P. M. (1988). Diversity and biological invasions of oceanic islands. *In* "Biodiversity" (E. O. Wilson, ed.), pp. 181–189. National Academy Press, Washington, DC.

Vitousek, P. M. (1991). Can planted forests counteract increasing atmospheric carbon dioxide? *J. Environ. Qual.* **20,** 348–354.

Vitousek, P. M., Gosz, J. R., Grier, C. C., Melillo, J. M., Reiners, W. A., and Todd, R. C. (1979). Nitrate losses from disturbed ecosystems. *Science* **204,** 469–474.

Vitousek, P. M., Ehrlich, P. R., Ehrlich, A. H., and Matson, P. A. (1986). Human appropriation of the products of photosynthesis. *BioScience* **36,** 368–373.

Vogelmann, H. W., Siccama, T., Leedy, D., and Ovitt, D. C. (1968). Precipitation from fog moisture in the Green Mountains of Vermont. *Ecology* **49,** 1205–1207.

Vogelmann, H. W., Badger, G. T., Bliss, M., and Klein, R. M. (1985). Forest decline on Camels Hump, Vermont. *Bull. Torrey Bot. Club* **112,** 274–287.

Vollenweider, R. A., and Kerekes, J. J. (1981). Background and summary results of the OECD cooperative program on eutrophication. *In* "Restoration of Inland Lakes and Waters," pp. 25–36. U.S. Environmental Protection Agency, Washington, DC.

Vollenweider, R. A., and Kerekes, J. J. (1982). "Eutrophication of Waters. Monitoring, Assessment and Control." Organization for Economic Co-operation and Development, Paris.

Wagg, J. W. B. (1963). Notes on food habits of small mammals of the white spruce forest. *For. Chron.* **39,** 436–445.

Wagle, R. F., and Kitchen, J. H. (1972). Influence of fire on soil nutrients in a ponderosa pine type. *Ecology* **53,** 118–125.

Walker, A. (1984). Extinction in hominid evolution. *In* "Extinctions" (M. H. Nitecki, ed.), pp. 119–152. Univ. of Chicago Press, Chicago.

Walker, E. P., Warnick, F., Hamlet, S. E., Lange, K. I., Davis, M. A., Vible, H. E., and Wright, P. F. (1968). "Mammals of the World," 2nd ed. Johns Hopkins Press, Baltimore, MD.

Walker, H. M. (1985). Ten-year ozone trends in California and Texas. *J. Air Pollut. Control Assoc.* **35,** 903–912.

Walkinshaw, L. H. (1983). "The Kirtland's Warbler." Cranbrook Institute of Science, Brookfield Hills, MI.

Wallace, E. S., and Freedman, B. (1986). Forest floor dynamics in a chronosequence of hardwood stands in Nova Scotia. *Can. J. For. Res.* **16,** 293–302.

Wallace, H. L., Good, J. E. G., and Williams, T. G. (1992). The effects of afforestation on upland plant communities: An application of the British National Vegetation Classification. *J. Appl. Ecol.* **29,** 180–194.

Wallin, K. (1984). Decrease and recovery patterns of some raptors in relation to the introduction and ban of alkyl-mercury and DDT in Sweden. *Ambio* **13,** 263–265.

Walsh, R. J., Donovan, J. W., Adena, M. A., Rose, G., and Battistutta, D. (1983). "Case-control Studies of Congenital Anomalies and Vietnam Service (Birth Defects Study)," Report to Minister for Veteran's Affairs. Australian Government Publishing Service, Canberra.

Walstad, J. D., and Dost, F. N. (1984). "The Health Risks of Herbicides in Forestry: A Review of the Scientific Literature," Spec. Publ. No. 10. Forest Research Laboratory, University of Oregon, Corvallis.

Walters, J. R. (1991). Application of ecological principles to the management of endangered species: The case of the red-cockaded woodpecker. *Annu. Rev. Ecol. Syst.* **22,** 505–523.

Walters, J. R., Doerr, P. D., and Carter, J. H. (1988). The cooperative breeding system of the red-cockaded woodpecker. *Ethology* **78,** 275–305.

Ward, D. M., Atlas, R. M., Boehm, P. D., and Calder, J. A. (1980). Microbial degradation and chemical evolution of oil from the Amoco spill. *Ambio* **9,** 277–283.

Ward, N. I., Brooks, R. R., Roberts, E., and Boswell, C. R. (1977). Heavy metal pollution from automotive emissions and its effect on roadside soils and pasture species in New Zealand. *Environ. Sci. Technol.* **11,** 917–923.

Wardle, J. A., and Allen, R. B. (1983). Dieback in New Zealand *Nothofagus* forests. *Pac. Sci.* **37,** 397–404.

Wargo, P. M. (1985). Interactions of stress and secondary organisms in decline of forest trees. *In* "Air Pollutant Effects on Forest Ecosystems," pp. 75–86. Acid Rain Foundation, St. Paul, MN.

Wargo, P. M., and Houston, D. R. (1974). Infection of defoliated sugar maple trees by *Armillaria mellea. Phytopathology* **64,** 817–882.

Warner, F. (1991). The environmental consequences of the Gulf War. *Environment* **33**(5), 7–26.

Warnock, J. W., and Lewis, J. (1978). "The Other Face of 2,4-D. A Citizens Report." South Okanagan Environmental Coalition, Penticton, British Columbia.

Warren, H. V., Delevault, R. E., and Barakso, J. (1966). Some observations on the geochemistry of mercury as applied to prospecting. *Econ. Geol. Ser. Can.* **61,** 1010–1028.

Waters, T. F. (1956). The effects of lime application to acid bog lakes in northern Michigan. *Trans. Am. Fish. Soc.* **86,** 329–344.

Waters, T. F., and Ball, R. C. (1957). Lime application to a soft-water, unproductive lake in northern Michigan. *J. Wildl. Manage.* **21,** 385–391.

Watson, W. Y., and Richardson, D. H. (1972). Appreciating the potential of a devastated land. *For. Chron.* **48,** 312–315.

Watt, K. E. F. (1968). "Ecology and Resource Management." McGraw-Hill, New York.

Watt, W. D., Scott, D., and Ray, S. (1979). Acidification and other chemical changes in Halifax County lakes after 21 years. *Limnol. Oceanogr.* **24,** 1154–1161.

Watt, W. D., Scott, C. D., and White, W. J. (1983). Evidence of acidification of some Nova Scotia rivers and its impact on Atlantic salmon, *Salmo salar. Can. J. Fish. Aquat. Sci.* **40,** 462–473.

Webb, J. M., Alexander, S. K., and Winters, J. K. (1985). Effects of autumn application of oil on *Spartina alterniflora* in a Texas salt marsh. *Environ. Pollut., Ser. A* **24,** 321–337.

Webb, W. L., Behrend, D. F., and Saisorn, B. (1977). Effect of logging on songbird populations in a northern hardwood forest. *Wildlife Monogr.* **55,** 1–35.

Webber, L. R., and Beauchamp, E. G. (1977). Heavy metals in corn grown on waste amended soils. *Int. Conf. Heavy Met. Environ. [Symp. Proc.], 1st, 1975,* Vol. 2, pp. 443–452.

Webster, F. B., Roberts, D. R., McInnes, S. N., and Sutton, B. C. S. (1990). Propagation of interior spruce by somatic embryogenesis. *Can. J. For. Res.* **20,** 1759–1765.

Weetman, G. F., and Algar, D. (1983). Low site-class black spruce and jack pine nutrient removals after full-tree and tree-length logging. *Can. J. For. Res.* **13,** 1030–1036.

Weetman, G. F., and Webber, B. (1972). The influence of wood harvesting on the nutrient status of two spruce stands. *Can. J. For. Res.* **2,** 351–369.

Weetman, G. F., Krause, H. H., Timmer, V. R., and Hoyt, J. S. (1974). "Some Fertilization Response Data for the Maritimes and Quebec," For. Tech. Rep. 5. Canadian Forestry Service, Great Lakes Forest Research Centre, Sault Ste. Marie, Ont.

Wein, R. W., and MacLean, D. A. (1983). An overview of fire in northern ecosystems. *In* "The Role of Fire in Northern Circumpolar Ecosystems" (R. W. Wein and D. A. MacLean, eds.), pp. 1–18. Wiley, New York.

Weiss, M. J., and Rizzo, D. M. (1987). Forest declines in major forest types of the eastern United States. *In* "Proceedings of the Workshop on Forest Decline and Reproduction: Regional and Global Consequences" (L. Kairiukstis, S. Nilsson, and A. Stroszak, eds.), pp. 297–305. IIASA, Laxenburg, Austria.

Welch, H. E., Symons, P. E. K., and Narver, D. W. (1977). "Some Effects of Potato Farming and Forest Clear-cutting on Small New Brunswick Streams," Tech. Rep. 745. Environment Canada, Fisheries and Marine Service, St. Andrews, New Brunswick.

Wells, P. G. (1984). Marine ecotoxicological tests with zooplankton. *Ecotoxicol. Test. Mar. Environ.* **1,** 215–256.

Wells, P. G., and Percy, J. A. (1985). Effects of oil on Arctic invertebrates. *In* "Petroleum Effects in the Arctic Environment" (F. R. Engelhardt, ed.), pp. 101–156. Am. Elsevier, New York.

Wells, S. M., Pyle, R. M., and Collins, N. M. (1983). "The IUCN Invertebrate Red Data Book." International Union for Conservation of Nature and Natural Resources, Gland, Switzerland.

Welsh, D., and Fillman, D. R. (1980). The impact of forest cutting on boreal bird populations. *Am. Birds* **34,** 84–94.

Wendel, G. W., and Kochenderfer, J. (1982). "Glyphosate Controls Hardwoods in West Virginia," USDA For. Serv. Res. Pap. NE-497. Northeastern Forest Experiment Station, Broomall, PA.

Wenger, K. F., ed. (1984). "Forestry Handbook," 2nd ed. Wiley, New York.

West, D. C., Shugart, H. H., and Botkin, D. B. (1981). "Forest Succession. Concepts and Application." Springer-Verlag, New York.

Westing, A. H. (1966). Sugar maple decline: An evaluation. *Econ. Bot.* **20**, 196–212.

Westing, A. H. (1971). Ecological effects of military defoliation on the forests of South Vietnam. *BioScience* **21**, 893–898.

Westing, A. H. (1976). "Ecological Consequences of the Second Indochina War." Almqvist & Wiksell, Stockholm, Sweden.

Westing, A. H. (1977). "Weapons of Mass Destruction and the Environment." Crane, Russak, New York.

Westing, A. H. (1980). "Warfare in a Fragile World. Military Impact on the Human Environment." Taylor & Francis, London.

Westing, A. H. (1982). The environmental aftermath of warfare in Viet Nam. *In* "SIPRI Yearbook 1982," pp. 363–389. Stockholm International Peace Research Institute, Stockholm, Sweden.

Westing, A. H. (1984a). The remnants of war. *Ambio* **13**, 14–17.

Westing, A. H., ed. (1984b). "Herbicides in War: The Long-term Ecological and Human Consequences." Taylor & Francis, London.

Westing, A. H. (1984c). Herbicides in war: Past and present. *In* "Herbicides in War: The Long-term Ecological and Human Consequences" (A. H. Westing, ed.), pp. 3–24. Taylor & Francis, London.

Westing, A. H. (1985a). Explosive remnants of war: An overview. *In* "Explosive Remnants of War: Mitigating the Environmental Impacts" (A. H. Westing, ed.), pp. 1–16. Taylor & Francis, Philadelphia.

Westing, A. H. (1985b). Misspent energy: Munitions expenditures past and future. The world annual arsenal of nuclear weapons. *Bull. Peace Proposals* **16**, 9–10.

Westing, A. H. (1987). "The Ecological Dimension of Nuclear War." Stockholm International Peace Research Institute, Stockholm, Sweden.

Wetherald, R. T. (1991). Changes of temperature and hydrology caused by an increase of atmospheric carbon dioxide as predicted by general circulation models. *In* "Global Climate Change and Life on Earth" (R. L. Wyman, ed.), pp. 1–17. Routledge, Chapman, and Hall, New York.

Wetmore, S. P., Keller, R. A., and Smith, G. E. G. (1985). Effects of logging on bird populations in British Columbia as determined by a modified point-count method. *Can. Field Nat.* **99**, 224–233.

Wetzel, R. G. (1975). "Limnology." Saunders, Toronto, Ont.

Whelpdale, D. M., and Munn, R. E. (1976). Global sources, sinks, and transport of air pollution. *In* "Air Pollution" (A. C. Stern, ed.), 3rd ed., Vol. 1, pp. 289–324. Academic Press, New York.

Whicker, F. W., and Schultz, V. (1982). "Radioecology: Nuclear Energy and the Environment." CRC Press, Boca Raton, FL.

Whitby, L. M., and Hutchinson, T. C. (1974). Heavy metal pollution in the Sudbury mining and smelting region of Canada. II. Soil toxicity tests. *Environ. Conserv.* **1**, 191–200.

Whitby, L. M., Stokes, P. M., Hutchinson, T. C., and Myslik, G. (1976). Ecological consequence of acidic and heavy-metal discharges from the Sudbury smelters. *Can. Mineral.* **14**, 47–57.

White, A. S. (1991). The importance of different forms of regeneration to secondary succession in a Maine hardwood forest. *Bull. Torrey Bot. Club* **118**, 303–311.

White, C. M., Fyfe, R. W., and Leman, D. B. (1990). The 1980 North American peregrine falcon, *Falco peregrinus,* survey. *Can. Field Nat.* **104**, 174–181.

White, F. M. M., Cohen, F. G., McCurdy, R. F., and Sherman, G. (1984). The chlorophenoxy herbicides and human reproductive problems: A review of the epidemiological evidence. *In* "Chlorophenates in the Wood Industry," pp. 41–50. University of British Columbia, Vancouver.

White, L. (1967). The historical roots of our ecological crisis. *Science* **155**, 1203–1207.

Whiteside, T. (1979). "The Pendulum and the Toxic Cloud: The Course of Dioxin Contamination." Yale Univ. Press, New Haven, CT.

Whittaker, R. H. (1954). The ecology of serpentine soils. IV. The vegetational response to serpentine soils. *Ecology* **35**, 275–288.

Whittaker, R. H., Likens, G. E., Bormann, F. H., Eaton, J. J., and Siccama, T. G. (1979). The Hubbard Brook ecosystem study. Nutrient cycling and element behaviour. *Ecology* **60**, 203–220.

Whitten, A. J., Damanik, S. J., Anwar, J., and Hisyam, N. (1987). "The Ecology of Sumatra." Gadjah Mada Univ. Press, Yogyokarta, Indonesia.

Whyte, A. G. D. (1973). Productivity of first and second crops of *Pinus radiata* on the Moutere gravel soils of Nelson. *N. Z. J. For.* **18**, 87–103.

Wiemeyer, S. N., Scott, J. M., Anderson, M. P., Bloom, P. H., and Stafford, C. J. (1988). Environmental contaminants in California condors. *J. Wildl. Manage.* **52**, 238–247.

Wiener, J. G., Fitzgerald, W. F., Watras, C. J., and Rada, R. G. (1990). Partitioning and bioavailability of mercury in an experimentally acidified Wisconsin lake. *Environ. Toxicol. Chem.* **9**, 909–918.

Wiens, J. A., and Rotenberry, J. T. (1985). Response of breeding birds to rangeland alteration in a North American shrub-steppe locality. *J Appl. Ecol.* **22**, 655–668.

Wiens, J. A., Rotenberry, J. T., and Van Horne, B. (1986). A lesson in the limitations of field experiments: Shrubsteppe birds and habitat alteration. *Ecology* **67**, 365–376.

Wigington, P. J. (principal author) (1990). Episodic acidification of surface waters due to acidic deposition. *In* "Acidic Deposition: State of Science and Technology. Vol. II. Aquatic Processes and Effects," pp. 12-1 to 12-200. Superintendent of Documents, U.S. Govt. Printing Office, Washington, DC.

Wigley, T. M. L., and Raper, S. C. B. (1992). Implications for climate and sea level of revised IPCC emissions scenarios. *Nature (London)* **357**, 293–300.

Wiklander, L. (1975). The role of neutral salts in the ion exchange between acid precipitation and soil. *Geoderma* **14**, 93–105.

Wiklander, L. (1980). Interaction between cations and anions influencing adsorption and leaching. *In* "Effects of Acid Precipitation on Terrestrial Ecosystems" (T. C. Hutchinson and M. Havas, eds.), pp. 239–254. Plenum, New York.

Wilander, A., and Ahl, T. (1972). The effects of lime treatment to a small lake in Bergslagen, Sweden. *Vatten* **5**, 431–445.

Wilcove, D. S. (1985). Nest predation in forest tracts and the decline in migratory songbirds. *Ecology* **66**, 1211–1214.

Wilcove, D. S. (1988). Forest fragmentation as a wildlife management issue in the eastern United States. *In* "Is Forest Fragmentation a Management Issue in the Northeast?" USDA For. Serv. Gen. Tech. Rep. NE-140, pp. 1–6. Northeastern Forest Experiment Station, Broomall, PA.

Wile, I., and Miller, G. (1983). "The Macrophyte Flora of 46 Acidified and Acid-sensitive Soft Water Lakes in Ontario." Ontario Ministry of the Environment, Rexdale.

Wile, I., Miller, G. E., Hitchin, G. G., and Yan, N. D. (1985). Species composition and biomass of the macrophyte vegetation of one acidified and two acid-sensitive lakes in Ontario. *Can. Field Nat.* **99**, 308–312.

Wilkie, D. S., Sidle, J. G., and Boundzanga, G. C. (1992). Mechanized logging, market hunting, and a bank loan in Congo. *Conserv. Biol.* **6**, 570–580.

Will, G. (1985). Productivity forcasting—the need to consider factors other than nitrogen and direct nutrient removal in produce. *In* "Proceedings, Workshop on Nutritional Consequences of Intensive Harvesting on Site Quality," pp. 27–30. Forest Research Institute, Rotorua, New Zealand.

Williams, B. L., Cooper, J. M., and Pyatt, D. F. (1979). Effects of afforestation with *Pinus contorta* on nutrient content, acidity, and exchangeable cations in peat. *Forestry* **51**, 29–35.

Williams, W. T. (1980). Air pollution disease in the Californian forests. A base line for smog disease on ponderosa and jeffrey pines in the Sequoia and Los Padres National Forests, California. *Environ. Sci. Technol.* **14** 179–182.

Williams, W. T. (1983). Tree growth and smog disease in the forests of California: Case history, ponderosa pine in the southern California Nevada. *Environ. Pollut., Ser. A* **30**, 59–75.

Williams, W. T., and Williams, J. A. (1986). Effects of oxidant air pollution on needle health and annual-ring width in a ponderosa pine forest. *Environ. Conserv.* **13**, 229–234.

Williams, W. T., Brady, M., and Willison, S. C. (1977). Air pollution damage to the forests of the Sierra Nevada Mountains of California. *J. Air Pollut. Control Assoc.* **27**, 230–234.

Will-Wolf, S. (1980a). Structure of corticolous lichen communities before and after exposure to emissions from a "clean" coal-fired generating station. *Bryologist* **83**, 281–295.

Will-Wolf, S. (1980b). Effects of a "clean" coal-fired power generating station on four common Wisconsin lichen species. *Bryologist* **83**, 296–300.

Wilson, E. O., ed. (1988a). "Biodiversity." National Academy Press, Washington, DC.

Wilson, E. O. (1988b). The current state of biological diversity. *In* "Biodiversity" (E. O. Wilson, ed.), pp. 3–18. National Academy Press, Washington, DC.

Wilson, W. E. (1981). Sulfate formation in point source plumes: A review of recent field studies. *Atmos. Environ.* **15**, 2573–2581.

Windholz, M., ed. (1983). "The Merck Index," 10th ed. Merck, Rahway, NJ.

Winfrey, M. R., and Rudd, J. W. M. (1990). Environmental factors affecting the formation of methylmercury in low pH lakes. *Environ. Contam. Chem.* **9**, 853–869.

Winterhalder, E. K. (1978). A historical perspective of mining and reclamation in Sudbury. *Proc. 3rd Annu. Meet., Can. Land Reclam. Assoc.*, pp. 1–13.

Wise, W. (1968). "Killer Smog." Ballantyne Books, New York.

Wisniewski, R. (1990). Shoals of *Dreissena polymorpha* as bioprocessor of seston. *Hydrobiology* **200**, 451–458.

Witter, J. A., and Ragenovich, I. R. (1986). Regeneration of Fraser fir at Mount Mitchell, North Carolina, after depredations of the balsam wooly adelgid. *For. Sci.* **32**, 585–594.

Wolfe, W. J. (1971). Biogeochemical prospecting in glaciated terrain of the Canadian Precambrian Shield. *CIM Bull.* **64**(715), 72- 80.

Wolff, L. (1958). "In Flanders Fields." Viking, New York.

Wood, B. W., Wittwer, R. F., and Carpenter, S. B. (1977). Nutrient element accumulation and distribution in an intensively cultured American sycamore plantation. *Plant Soil* **48**, 417–433.

Wood, C. W., and Nash, T. N., III (1976). Copper smelter effluent effects on Sonoran Desert vegetation. *Ecology* **57**, 1311–1316.

Wood, G. L. (1972). "Animal Facts and Figures." Guinness Superlatives Limited, Enfield, UK.

Wood, G. W. (1979). Recuperation of native bee populations in blueberry fields exposed to drift of fenitrothion from forest spray operations in New Brunswick. *J. Econ. Entomol.* **72**, 36–39.

Wood, T., and Bormann, F. H. (1975). Increases in foliar leaching caused by acidification of an artificial mist. *Ambio* **4**, 169–171.

Wood, T., and Bormann, F. H. (1976). "Short-term Effects of A Simulated Acid Rain Upon the Growth and Nutrient Relations of *Pinus strobus* L." USDA For. Serv. Gen. Tech. Rep. NE-23, pp. 815–824. Northeastern Forest Experiment Station, Upper Darby, PA.

Woodley, S. (1991). "A Data Base for Ecological Monitoring in Canadian National Parks." Heritage Resources Center, University of Waterloo, Waterloo, Ont.

Woodley, S. (1993). Monitoring and measuring ecosystem integrity in Canadian national parks. *In* "Ecological Integrity and the Management of Ecosystems" (S. Woodley, J. Kay, and G. Francis, eds.), pp. 155–176. St. Lucie Press, Boca Raton, FL.

Woods, D. B., and Turner, N. C. (1971). Stomatal response to changing light by four tree species of varying shade tolerance. *New Phytol.* **70**, 77–84.

Woodwell, G. M. (1970). Effects of pollution on the structure and physiology of ecosystems. *Science* **168**, 429–433.

Woodwell, G. M. (1982). The biotic effects of ionizing radiation. *Ambio* **11**, 143–148.

Woodwell, G. M., and Whittaker, R. H. (1968). Effects of chronic gamma irradiation on plant communities. *Q. Rev. Biol.* **43,** 42–55.

Woodwell, G. M., Wurster, C. F., and Isaacson, P. A. (1967). DDT residues in an East Coast estuary: A case of biological concentration of a persistent insecticide. *Science* **156,** 821–824.

Woodwell, G. M., Whittaker, R. H., Reiners, W. A., Likens, G. E., Delwiche, C. C., and Botkin, D. B. (1978). The biota and the world carbon budget. *Science* **199,** 141–146.

World Commission on Environment and Development (WCED) (1987). "Our Common Future." WCED, Oxford Univ. Press, Oxford, UK.

World Meteorological Organization (WMO) (1986). Tropospheric trace gases. *In* "Atmospheric Ozone: Global Research and Monitoring." WMO, Geneva, Switzerland.

World Resources Institute (WRI) (1986). "World Resources 1986. An Assessment of the Resource Base That Supports the Global Economy." WRI, New York.

World Resources Institute (WRI) (1987). "World Resources 1987. An Assessment of the Resource Base That Supports the Global Economy." WRI, New York.

World Resources Institute (WRI) (1988). "Power Company to Fund Reforestation to Offset Carbon Dioxide Emissions, Slow Greenhouse Effect." WRI, New York.

World Resources Institute (WRI) (1990). "World Resources 1990–91. An Assessment of the Resource Base That Supports the Global Economy." WRI, New York.

World Resources Institute (WRI) (1992). "World Resources 1992–93. An Assessment of the Resource Base That Supports the Global Economy." WRI, New York.

World Wildlife Fund (WWF) (1984). "Plantpac Number 4. Threatened Plants." WWF, London.

Wormington, A., and Leach, J. H. (1992). Concentrations of migrant diving ducks at Point Pelee National Park, Ontario, in response to invasion of zebra mussels, *Dreissena polymorpha. Can. Field Nat.* **106,** 376–380.

Wright, R. F. (1976). The impact of forest fire on the nutrient fluxes to small watersheds in northeastern Minnesota. *Ecology* **57,** 649–663.

Wright, R. F. (1985). Chemistry of Lake Hovvatn, Norway, following liming and re-acidification. *Can. J. Fish. Aquat. Sci.* **42,** 1103–1113.

Wright, R. F., and Gjessing, E. T. (1976). Acid precipitation: Changes in the chemical composition of lakes. *Ambio* **5,** 219–223.

Wright, R. F., and Snekvik, E. (1978). Acid precipitation: Chemistry and fish populations in 700 lakes in southernmost Norway. *Verh.—Int. Ver. Theor. Angew. Limnol.* **20,** 765–

Wrighton, F. E. (1947). Plant ecology at Crippengate, 1947. *London Nat.* **27,** 44–48.

Wrighton, F. E. (1948). Plant ecology at Crippengate, 1948. *London Nat.* **28,** 29–44.

Wrighton, F. E. (1949). Plant ecology at Crippengate, 1949. *London Nat.* **29,** 85–88.

Wrighton, F. E. (1950). Plant ecology at Crippengate, 1950. *London Nat.* **30,** 73–79.

Wrighton, F. E. (1952). Plant ecology at Crippengate, 1951–2. *London Nat.* **32,** 90–93.

Wu, L., and Bradshaw, A. D. (1972). Aerial pollution and the rapid evolution of copper tolerance. *Nature (London)* **238,** 167–169.

Wu, L., Bradshaw, A. D., and Thompson, D. A. (1975). The potential for evolution of heavy metal tolerance in plants. III. The rapid evolution of copper tolerance in Agrostis stolonifera. *Heredity* **34,** 165–187.

Wurster, D. H., Wurster, C. F., and Strickland, W. N. (1965). Bird mortality following DDT spray for Dutch elm disease. *Ecology* **46,** 488–499.

Yahner, R. H., and Scott, D. P. (1988). Effects of forest fragmentation on depredation of artificial nests. *J. Wildl. Manage.* **52,** 158–161.

Yan, N. D. (1979). Phytoplankton community of an acidified, heavy metal-contaminated lake near Sudbury, Ontario: 1973–1977. *Water, Air, Soil Pollut.* **11,** 43–55.

Yan, N. D., and Lafrance, C. (1984). Response of acidic and neutralized lakes near Sudbury, Ontario, to nutrient enrichment. *In* "Environmental Impacts of Smelters" (J. O. Nriagu, ed.), 457–521. Wiley (Interscience), Toronto, Ont.

Yan, N. D., and Miller, G. E. (1984). Effects of deposition of acids and metals on chemistry and biology of lakes near Sudbury, Ontario. *In* "Environmental Impacts of Smelters" (J. O. Nriagu, ed.), pp. 243–282. Wiley, New York.

Yan, N. D., and Stokes, P. M. (1978). Phytoplankton of an acidic lake, and its response to experimental alterations of pH. *Environ. Conserv.* **5,** 93–100.

Yan, N. D., and Strus, R. (1980). Crustacean zooplankton communities of acidic metal-contaminated lakes near Sudbury, Ontario. *Can. J. Fish. Aquat. Sci.* **37,** 2282–2293.

Yan, N. D., Girard, R. E., and Lafrance, C. J. (1979). "Survival of Rainbow Trout, *Salmo gairdneri,* in Submerged Enclosures in Lakes Treated with Neutralizing Agents near Sudbury, Ontario." Ontario Ministry of the Environment, Rexdale.

Yevich, P. P., and Barszcz, C. A. (1977). Neoplasia in soft-shell clams (*Mya arenaria*) collected from oil-impacted sites. *Ann. N.Y. Acad. Sci.* **298,** 409–426.

Yokouchi, Y., Okaniwa, M., Ambe, Y., and Fuwa, K. (1983). Seasonal variations of monoterpenes in the atmosphere of a pine forest. *Atmos. Environ.* **17,** 743–750.

Yoshimura, S. (1933). Kata-numa, a very strong acid-water lake on Volcano Katanuma, Miyagi Prefecture, Japan. *Arch. Hydrobiol.* **26,** 197–202.

Yoshimura, S. (1935). Lake Akanuma, a siderotrophic lake at the foot of Volcano Baande, Hikisuma Prefecture, Japan. *Proc. Imp. Acad. (Tokyo)* **11,** 426–428.

Yule, W. N. (1975). Persistence and dispersal of insecticide residues. *In* "Aerial Control of Forest Insects in Canada" (M. L. Prebble, ed.) pp. 263–266. Department of the Environment, Ottawa, Ont.

Zabinski, C. (1992). Isozyme variation in eastern hemlock. *Can. J. For. Res.* **22,** 1838–1842.

Zahner, R., Saucier, J. R., and Myers, R. K. (1989). Tree-ring model interprets growth decline in natural stands of loblolly pine in the southeastern United States. *Can. J. For. Res.* **19,** 612–621.

Zar, H. (1980). Point source loads of phosphorus to the Great Lakes. *In* "Phosphorus Management Strategies for Lakes" (R. C. Loehr, C. S. Martin, and W. Rast, eds.), pp. 27–36. Ann Arbor Sci. Publ., Ann Arbor, MI.

Zavitkovski, J. (1976). Ground vegetation biomass, production, and efficiency of energy utilization in some northern Wisconsin forest ecosystems. *Ecology* **57,** 694–706.

Zhu, J. (1989). Nature conservation in China. *J. Appl. Ecol.* **26,** 825–833.

Zieman, J. C., Orth, R., Phillips, R. C., Thayler, G., and Thorhaug, A. (1984). The effects of oil on seagrass ecosystems. *In* "Restoration of Habitats Impacted by Oil Spills" (J. L. Cairns and A. L. Buikema Jr., eds.), pp. 37–64. Butterworth, Boston.

Ziff, S. (1985). "The Toxic Time Bomb." Aurora Press, New York.

Zimmer, C. (1992). Ecowar. *Discover* **1,** 37–40.

Zimmerman, P. R. (1979). "Testing of Hydrocarbon Emissions from Vegetation, Leaf Litter, and Aquatic Surfaces, and Development of a Methodology for Compiling Biogenic Emission Inventories," EPA-450/4–79–004. U.S. Environmental Protection Agency, Washington, DC. (cited by Yokouchi *et al.*, 1983).

Ziswiler, V. (1967). "Extinct and Vanishing Animals." Springer-Verlag, New York.

Zoettl, H. W., and Huettl, R. H. (1986). Nutrient supply and forest decline in southwest Germany. *Water, Air, Soil Pollut.* **31,** 449–462.

GLOSSARY

This glossary defines a number of terms and phrases that are specific to ecology, forestry, wildlife biology, environmental studies, and other disciplines relevant to the ecological effects of pollution, disturbance, and other stressors. However, the names of specific chemicals and organisms are not defined here. The primary intent of this section is to make the contents of this book relatively easily accessible to readers who may not have had an introductory class or other exposures to ecology. References consulted during the preparation of parts of this glossary were The Institute of Ecology (TIE) (1973), Abercrombie *et al*. (1978), Bonnor (1978), Hausenbuiller (1978), Newton and Knight (1981), Tver (1981), Lincoln *et al*. (1982), United Nations Environment Program (UNEP) (1982), Miller (1985), and Carbon Dioxide Information Center (CDIC) (1990).

aboriginal   The original peoples inhabiting a region prior to European contact and/or colonization.

above ground   All biomass or nutrient capital that occurs above the surface of the ground.

acaricide   A pesticide used against spiders and mites.

acclimation   A phenotypically plastic response to exposure to some environmental stress, characterized by an enhanced degree of physiological tolerance of that stress. See also **phenotypic plasticity**.

accretion   An increase in the quantity of biomass or nutrients in an ecosystem.

accumulator   An organism that sequesters nutrients or toxic substances from the environment to an unusually large concentration. See also **hyperaccumulator**.

accuracy   The degree to which a measurement of the value of an environmental or biological variable reflects the true value of that variable. Compare with **precision**.

acidic deposition   A shorter but less accurate version of the phrase, "the deposition of acidifying substances from the atmosphere."

acid-neutralization capacity (ANC)   The sum of all titratable bases in solution, which are available to neutralize inputs of acids. In most freshwaters, the most important sources of ANC are $HCO_3^- + CO_3^{2-} + OH^-$.

acid rain   See **acidic precipitation**.

acidic precipitation   (1) Rain, snow, or fog water having a pH less than 5.65. (2) The deposition of acidifying substances from the atmosphere during a precipitation event.

acidification   An increase over time in the content of acidity in a system, accompanied by a decrease in the acid-neutralizing capacity of that system.

acidifying substance   Any substance that causes acidification. The substance may have an acidic character and therefore act directly, or it may be nonacidic but generate acidity as a result of its chemical transformation, as happens when ammonium is nitrified to nitrate and when sulfides are oxidized to sulfate.

acidity   The ability of a solution to neutralize an input of hydroxide ion. Acidity is usually measured as the concentration of hydrogen ion, in logarithmic pH units (see also **pH**). Strictly speaking, an acidic solution has a pH < 7.0.

acid-mine drainage   Surface water or ground water that has been acidified by the oxidation of pyrite and other reduced-sulfur minerals that occur in coal mines and coal-mine wastes.

acidophilous   Refers to organisms that occur only in acidic habitats.

acid shock   A short-term event of great acidity. This phenomenon regularly occurs in freshwater systems that receive a large pulse of relatively acidic water when an accumulated snowpack melts rapidly in the spring.

acid–sulfate soil   An edaphic condition that occurs when a shallow-water marine, estuarine, or salt-marsh system is drained for agriculture. When the previously anaerobic soil, rich in pyrites and other reduced-sulfur species is exposed to the atmosphere, oxidation of the sulfides by chemoautotrophic bacteria generates a great deal of acidity.

active ingredient (a.i.)   Refers to the particular chemical within a pesticidal formulation that is actually intended to cause toxicity to the pest. Nonpesticidal chemicals in the formulation, the so-called "inert" ingredients, are used to dilute the active ingredient, to enhance its spread or adherence on foliar surfaces, and for other purposes associated with increased efficacy.

acute   Refers to a relatively large or severe effect caused by a relatively shorter-term exposure to a toxic environmental agent. Compare with **chronic**.

acute injury   See **acute toxicity**.

acute toxicity   A poisonous effect produced by a single, shorter-term exposure to some toxic agent and causing measurable biochemical or anatomical damages, or even death of the organism.

advance regeneration   Young and/or small individuals of tree species that are established naturally in a stand of mature trees and comprise or contribute to the development of the next forest that regenerates after the overstory trees are harvested.

aerial spray   A pesticidal application that is delivered to the spray site by a fixed-wing airplane or helicopter.

aerobic   Refers to an environment in which oxygen is present.

aerosol   A suspension of fine solid or liquid particulates within an atmospheric or aquatic medium. Examples include haze, smoke, and fog.

afforestation   The conversion of a nonforested ecosystem to a forest by the planting of trees.

aggrading   See **accretion**.

agroecosystem   An ecosystem that is intensively managed to optimize the productivity of agricultural plants and/or animals.

agronomic   Pertaining to an agricultural context.

air burst   A nuclear explosion occurring in the atmosphere close to the ground surface, generally at about 500–1000 m.

air mass   A portion of the atmosphere with relatively homogeneous conditions of temperature, pressure, and chemistry that moves as a cohesive unit.

airshed   A more or less definable portion of the atmosphere that overlies a particular landscape.

albedo   The ratio of reflected to incident electromagnetic radiation for some surface.

alevin   A young salmonid fish, recently hatched and still having a yolk sac.

algal bloom   See **bloom**.

alkaline   Refers to a nonacidic solution with pH >7.0.

alkalinity   The amount of alkali in a solution. In freshwater, alkalinity is mainly composed of bicarbonates, carbonates, and hydroxides, and it is generally measured by titration with acid to a fixed end point.

allochthonous   Refers to an influx of biomass, nutrients, or some other material to an ecosystem, but coming from some other place. Compare with **autochthonous**.

ambient   Refers to the environmental conditions that affect a body or system but are not affected by the body or system.

ammonia volatilization   The flux of ammonia to the atmosphere from sewage or after treatment of land or water with sewage, ammonium nitrate, urea, or some other nitrogen-containing chemical.

ammonification   The conversion of organic nitrogen to ammonium by soil or aquatic microbes.

amphi-Atlantic   A biogeographic distribution that encompasses both western Europe and eastern North America.

anaerobic   Refers to an environment in which oxygen is not present or occurs only in a small concentration.

angiosperm   A flowering plant in the subdivision Spermatophyta, which is distinguishable from the Gymnospermae by having ovules borne within a specialized closed structure, the ovary, which develops into a fruit.

anion   An ion with a negative charge.

annual   An organism that completes its life cycle in 1 year.

annual-allowable cut (AAC)   The amount of wood volume or biomass that can be removed each year from a forested area that is being managed so as to allow for a continuously sustained cropping of tree biomass. Strictly speaking, the AAC should not exceed the annual volume increment of an area that is being managed for sustainable production.

anoxic   See **anaerobic**.

anthropogenic   Occurring because of or influenced by the activities of humans.

anticoagulant   A substance that interferes with the normal clotting of blood that has been exposed to the atmosphere.

arboreal lichen   An epiphytic lichen that occurs on an aerial tree substrate.

Arctic (adj. **arctic**)   Any area north of the Arctic circle, or $66° 32'$ N. Sometimes defined as all terrain north of the boreal forest.

atmospheric deposition The influx of a substance from the atmosphere, occurring both with precipitation and in its absence. See also **acidic deposition, dry deposition**, and **wet deposition**.

atomic density The ratio of atomic mass to atomic volume.

autecology The ecology of individuals or of populations of a particular species. Compare with **synecology**.

autochthonous Refers to an influx of biomass, nutrients, or some other material to an ecosystem, resulting from processes internal to that system. Compare with **allochthonous**.

autotroph An organism that can synthesize its biochemical constituents using inorganic precursors and an external source of energy. Photoautotrophs make use of sunlight through the process of photosynthesis, while chemoautotrophs harness some of the energy content of inorganic chemicals through the process of chemosynthesis.

available concentration The portion of the total concentration of a nutrient or toxic substance, present in soil or water, that can be assimilated by organisms. Availability of many chemicals is determined by the degree of solubility in an aqueous solution of some sort, and it is measured by various indices. The simplest index used in studies of soil is the water-soluble concentration. See also **total concentration**.

avifauna All of the species of birds in a specified area.

background concentration The concentration of a substance in a particular environment that has not been measurably influenced by anthropogenic sources.

baleen whale A whale that filter feeds on planktonic crustacea or small fish.

barrens A generic term used to describe a situation where there is a degraded ecosystem of much simpler structure and function than could potentially occur, considering the climatic and edaphic characteristics of the site. Usually refers to a nonforest vegetation occurring in a landscape that could potentially support forest.

basal area The cross-sectional area of the stem of a tree or of a stand of trees.

base saturation The absolute or relative degree of occupation of the total soil cation exchange capacity (CEC) by calcium, magnesium, potassium, and sodium ions. The remaining saturation of CEC is usually dominated by hydrogen and aluminum ions. See also **cation exchange capacity**.

basic cation (base cation, base) Refers to Ca, Mg, K, and Na ions in soil or freshwater.

bedrock The solid rock that underlies the soil or deeper surficial deposits covering a site.

below-ground All biomass occurring below the surface of the soil.

benthos The biota living on or in the surface sediment of a freshwater or marine ecosystem.

bilge washings Hydrocarbon-contaminated water that has been used to rinse the petroleum or fuel-holding compartments of a tanker and is then discharged to the environment. See also **load-on-top**.

bimodal A distribution of observations that is characterized by two distinct peaks of frequency.

binomial The proper, scientific name of a species, composed of two latinized words. The first word identifies the genus and the second the species. Sometimes a third latinized word identifies a distinct subspecies.

bioaccumulation. The biological sequestering of a substance at a larger concentration than that at which it occurs in the inorganic environment, such as soil or water.

bioassay. A quantitative estimation of the intensity or concentration of a biologically active environmental factor, measured using some biological response under standardized conditions or varied exposure.

bioavailable See **available concentration**.

biochemical oxygen demand (BOD). A test that measures the amount of oxygen consumed by microorganisms in an aqueous solution in a standardized, closed incubation. If all oxygen consumption can be attributed to the oxidation of organic matter, then BOD is effectively biological oxygen demand. However, if oxygen-consuming inorganic chemicals are also occurring, then biochemical oxygen demand is measured.

bioconcentration See **bioaccumulation**.

biodegradation The process by which biological metabolism, usually microbial, transforms certain toxic chemicals or sewage into other, relatively innocuous chemicals.

biodiversity Often considered to be related to the number of species occurring in some area. Biodiversity is better defined, however, as being composed of the totality of the richness of biological variation, ranging from population-based genetic variation, through subspecies and species, to communities and the pattern and dynamics of these on the landscape.

biogenic The emission of toxic or other substances by organisms.

biogeochemical prospecting   The chemical analysis of biota, soil, sediment, or water to discover surface or near-surface occurrences of commercially valuable minerals. May also be supplemented by information about the distribution of indicator species or varieties whose occurrence is specific to environmental conditions associated with the mineral.

biogeochemistry   The study of the quantity and cycling of chemicals in communities, landscapes, or the Earth as a whole.

biogeography   The study of the larger-scale distributions of taxa of plants and animals and the factors that influence those distributions.

biological control   The control of a pest or pathogen by a biological method, such as the use of a tolerant crop genotype or the introduction of a predator, parasite, or disease, etc.

biomagnification   See **bioaccumulation**.

biomass   The quantity of living and/or dead organic matter, usually standardized per-unit area in a terrestrial ecosystem and per-unit volume in an aquatic ecosystem.

biome   A geographically extensive ecosystem, which is usually characterized by its dominant vegetation, for example, coniferous boreal forest, tropical rain forest, or arctic graminoid tundra.

biosphere   The thin veneer of the Earth in which organisms occur and where biological processes take place.

biota   The assemblage of organisms in some defined site or geographical area.

bird watching   The recreational observation of birds.

birth defect   The occurrence of a biochemical disorder or anatomical defect in a newly born animal.

black rain   (1) A sooty, radioactive rain that is induced by the convective uplift of an air mass, induced by the thermal energy of an above-ground nuclear explosion. (2) Other sooty rains, associated perhaps with a burning blowout of an oil well.

blast   The kinetic energy of a moving shock wave of air, caused by an explosion.

bloom   An event of productivity of phytoplankton, resulting in a large standing crop of algal biomass and of chlorophyll and a reduced transparency of the surface water.

blowout   An uncontrolled emission of natural gas, petroleum, or water from an oil well.

bog   An oligotrophic, acidic wetland that receives all of its input of nutrients by atmospheric deposition and that has a plant community dominated by acidophilous species of *Sphagnum* moss.

bole   The stem or trunk of a tree.

boreal forest   The conifer forest that occurs in the low Arctic in regions with a long and cold winter and a short growing season. Boreal forest gives way to tundra at more northern latitudes.

boundary layer   A thin layer of fluid or air that is adjacent to a solid surface, in which there is no turbulent transport of the medium and through which dissolved substances must move by diffusion.

box model   A nondynamic assessment of the quantity of material or energy in various compartments of an ecosystem and of the fluxes between compartments.

broadcast spraying   The treatment of an area so as to achieve an even coverage of a pesticide. This method of application is not pest specific, and it exposes many nontarget organisms to the pesticide.

broad spectrum   Refers to a pesticide or other toxic agent that affects a wide variety of biota.

browse   The young twigs, foliage, and reproductive tissues of woody plants that are eaten by mammalian herbivores such as rabbits and deer.

brown water   Water that is colored with a brownish hue by the presence of dissolved organic matter. Sometimes termed a dystrophic system. Compare with **clear water**.

bryophyte   A moss or liverwort.

budget   An analysis of the quantity of material or energy in the compartments of a system, of the exchanges among compartments, and of the total inputs to and outputs from the system.

buffer   A chemical compound that has the capacity to absorb or exchange hydrogen or hydroxide ions and allows a system to assimilate a limited amount of these ions without changing substantially in pH.

buffering capacity   The quantitative ability of a solution to absorb hydrogen or hydroxide ions without undergoing a substantial change in pH.

buffer strip   An uncut border of forest that is left beside water bodies or roads during a forest harvest.

bulk collector   A sampling device that is continuously open to the atmosphere and samples both wet and dry depositions of substances.

bulk deposition   The deposition of substances from the atmosphere with precipitation and with dry depositions, as measured with an open-top bulk collector.

bulk precipitation (bulk-collected precipitation)   See **bulk deposition**.

bunker-C fuel oil   A relatively heavy, liquid hydrocarbon fraction that is used as a fuel for ships and oil-fired power plants.

cable logging   A system used to remove logs from a harvested forest. Usually, cables are attached to a tall tree or spar and are used to winch logs to that central place, where they are later loaded onto a truck.

calcareous   A situation rich in calcium carbonate, with a relatively high pH and often having a distinctive flora.

cancer   A malignant growth or tumor caused by an uncontrolled and undifferentiated division of cells.

canopy   The predominant layer of foliage of an ecological community.

canopy closure   A point in succession where almost all of the sky is obscured by foliage, when viewed upward from the ground surface.

capital   (1) The quantity of biomass or nutrient on a site. (2) In economic terms, capital is actual or potential wealth that is capable of being applied toward the production of further wealth. Capital can be divided into three types: manufactured capital is the industrial means of production, human capital is the cultural means of production, and natural capital is the quantity of natural resources that can be harvested and processed to yield a flow of goods or services. Natural capital can be of two types: nonrenewable and renewable.

carnivore   An animal that eats other animals.

carrying capacity   The maximum abundance of a species that can be sustained in an area and beyond which degradation of the habitat will be caused.

case study   Consideration of the pertinent aspects of a particularly exemplary situation.

catalyst   A substance that decreases the activation energy of a chemical reaction and thereby increases its rate without being consumed in itself. See also **enzyme**.

cation   An ion with a positive charge.

cation-exchange capacity (CEC)   The total amount of exchangeable cations that a given quantity of soil can absorb. See also **base saturation**.

cavity nester   A bird that builds its nest in a hollow in a tree.

census plot   A designated area or volume of habitat in which the abundance of animals is determined.

chapparal   Vegetation dominated by evergreen angiosperm shrubs, and found in an area with a mediterranean climate, i.e., with a hot and dry growing season and a mild and wet winter.

chelation   The sequestering of a metal by a chemical that contains a closed ring of atoms, one of which must be a metal.

chemoautotroph   See **autotroph**.

chemosynthesis   See **autotroph**.

chlorofluorocarbon   A carbon-based compound containing chlorine and/or fluorine atoms.

chlorophyll   A green pigment found in almost all algae

and higher plants that is responsible for the light capture that drives photosynthesis.

chlorosis   A symptom of stress in a vascular plant, characterized by a general or mottled yellow or light-green coloration of the foliage, when the healthy condition would be a dark-green color.

chronic injury   See **chronic toxicity**.

chronic toxicity   A poisonous effect produced by a long period of exposure to a small or moderate level of some toxic agent and causing measurable biochemical or anatomical damage. Compare with **acute toxicity**.

chronosequence   A time series of stands or soils of different age that originated after a similar type of disturbance.

circumneutral   Having a pH of about 7.

clear-cut   A forest harvest in which all merchantable trees have been removed. Usually the trees are delimbed and branches and foliage are left on the site. See also **whole-tree** and **complete-tree clear-cut**.

clearwater   Water that is not colored with a brownish hue by dissolved organic matter. Compare with **brown water**.

climate   The typical, longer-term, prevailing conditions of temperature, precipitation, insolation, and cloud cover of an area. Compare with **weather**.

climax   The more or less stable plant and animal community that culminates succession under a given set of conditions of climate, site, and biota.

cluster analysis   A multivariate, mathematical analysis of data that groups cases on the basis of their correlated attributes.

clutch   The eggs laid by a particular bird or by a number of birds in a particular nest.

codistillation.   A process in which dissolved substances evaporate with water at a small concentration. The codistillant occurs in a trace concentration in the condensed vapor of the evaporation.

coevolution   The intrinsically linked evolution of two or more species, caused by some close ecological relationship such as predation, herbivory, pollination, symbiosis.

cohort   An even-aged group of individuals of the same generation.

community   An assemblage of interacting plants, animals, and microbes on a shared site. Sometimes only a part of the larger community is considered, as in bird or plant communities.

compaction   See **soil compaction**.

compartment   A designated unit of an ecosystem that contains a quantity of material and energy.

competition   An interaction between organisms of the

same or different taxa, associated with the need for a shared resource that occurs in an insufficient supply relative to the biological demand.

competitive release A situation in which an organism or taxon is relieved of stresses associated with competition, allowing it to grow more freely.

complete-tree clear-cut Removal of the above-ground and below-ground tree biomass from a harvested site.

complex A chemical compound in which atoms are attached to a central metal atom by coordinate bonds. See also **chelation**.

concentration The amount of a substance per-unit volume or per-unit weight of the matrix.

conceptual model A nonquantitative assessment of the relationships among important compartments and fluxes of material or energy in a system or of biological interactions.

congeneric (congeners) Two or more species of the same genus.

conifer A gymnosperm plant. See also **gymnosperm**.

conservation The careful management of a natural resource, to ensure that its use is sustainable.

conservation of electrochemical neutrality Refers to an aqueous solution, in which the number of cation equivalents equals the number of anion equivalents, so that the solution does not have a net electrical charge.

contact herbicide An herbicide that causes injury at the point of entry to the plant, which is usually the foliage.

contamination The occurrence of a relatively large concentration of some toxic substance, compared with the normal ambient condition. Contamination does not necessarily imply that ecological damage can be demonstrated. Compare with **pollution**.

content The total amount of a substance in an organism, compartment, or ecosystem. Sometimes standardized per-unit area or per-unit volume in aquatic studies.

continental Refers to a climate that is not moderated by proximity to a large body of water and is characterized by relatively cold winters and hot summers.

control (1) An experimental treatment in which a variable is held constant, for the purpose of statistical comparison with a parallel treatment in which that variable is altered. (2) A standard of comparison used in the statistical analysis of a scientific experiment. Compare with **reference**.

conventional munitions Weapons in which the explosive mechanism does not involve a nuclear reaction.

conversion A longer-term change in character of the ecosystem at some place, as when a natural forest is harvested and the land developed into a forest plantation or an agroecosystem.

correlation A statistical relationship that shows the degree of quantitative association between two or more variables.

correlation coefficient A statistic that quantitatively expresses the degree of correlation between two or more variables.

cover A measure or estimate of the relative or absolute amount of ground surface that is obscured by plant foliage.

crop Harvested biomass of plants or animals.

crown closure See **canopy closure**.

crude oil See **petroleum**.

cull trees Trees that are removed during a silvicultural thin or shelterwood cut. See also **thin** and **shelterwood cut**.

cultural eutrophication Eutrophication caused by an anthropogenic input of nutrients. See also **eutrophication**.

culvert An enclosed water-drainage channel that crosses beneath a road, railway, etc.

cuticle A waxy, superficial layer that covers the foliage of vascular plants. The cuticle is continuous except for microscopic stomata, through which gases and water vapor pass.

cutover A tract of land from which trees have been harvested.

cycle A closed and continuous transfer of material or energy among the compartments of a system. Except at the level of the biosphere, few ecological cycles are truly closed.

cyclic succession A succession that occurs repeatedly on the landscape, as a result of a disturbance that occurs at regular intervals.

deciduous A plant in which all leaves dehisce before a dormant season associated with extreme cold or drought. More properly termed "seasonally deciduous," since the foliage of most so-called "evergreen" plants is also deciduous, but not all at once.

decline See **forest decline**.

decomposition The heterotrophic, mostly microbial oxidation of dead organic matter.

decrement A decrease in biomass or productivity.

deer yard A wintertime, forested habitat of deer, which is often dominated by coniferous trees and has a relatively shallow accumulation of snow.

defoliation The removal of plant foliage by mechanical means, acute or chronic injury, or herbivory.

deforestation   The longer-term removal of forest from a site, usually to gain new agricultural land, or for some other anthropogenic conversion.

deglaciation   The retreat of glaciers from the landscape by the mass wasting of ice.

degradation   (1) The process by which a toxic chemical is converted metabolically or abiotically to a less toxic or nontoxic metabolite. See also **detoxification**. (2) A reduction in the quality or quantity of a natural resource.

demography   The science of population statistics.

dendrochronology   The study of patterns of tree-ring width toward the determination of historical changes in climate, defoliation, pollution, or some other environmental factor.

denitrification   The microbial reduction of nitrate, usually to dinitrogen or nitrous oxide.

density   The number of organisms per-unit area or per-unit volume in aquatic studies.

density-independent   A population change that is not influenced by the density of organisms. A density-independent effect is often associated with a catastrophic advent of stress or disturbance.

depauperate   An ecosystem or community with a small richness of species.

deposition   The rate of influx of material per-unit area and time. Frequently pertains to influx from the atmosphere.

deposition velocity   A constant that allows calculation of the rate of deposition of an atmospheric chemical constituent of known concentration to a vegetated or other surface with particular physical and chemical characteristics.

desertification   A climatic change that involves a decreased precipitation input, causing a progressive diminution or destruction of the biological productivity of the landscape and leading ultimately to desert-like conditions.

desulfurization   The removal of some or all of the sulfur-containing fractions of coal or oil to reduce the emissions of sulfur dioxide during combustion.

detection limit   See **undetectable concentration**.

detergent   A surfactant chemical used as a cleaning agent because it facilitates the formation of an oil-in-water emulsion.

detoxification   Reduction of the toxic quality of a chemical by its microbial or inorganic transformation to another, less toxic chemical. See also **degradation**.

detritivore   A heterotrophic microbe or animal that feeds on dead biomass.

devegetated   A site that has been severely degraded by the removal of all or most of its biomass by disturbance or toxic stress.

development   In ecological economics, development is achieved through efficiency of use, renewal, and fostering of renewable resources, and occurring without degradation of those resources. If the natural-resource base does not change over time, then the economy is in a steady-state condition. Compare with **growth.**

diameter-limit cut.   A forest harvest in which all trees larger than a commercially determined diameter are removed.

dicotyledonous (dicot)   Refers to an angiosperm plant of the Dicotyledonae, characterized by two seed leaves or cotyledons, among other characteristics. The other class of angiosperms is the Monocotyledonae, with a single cotyledon.

dieback   A progressive dying from the extremity of any part of a plant. See also **forest decline** and **forest dieback.**

dinitrogen fixation.   See **nitrogen fixation.**

dioecious   A botanical term, referring to species in which individuals have female or male reproductive organs, but not both. Compare with **monoecious.**

dioxin   A class of chlorinated hydrocarbon, including the toxic isomer TCDD (2,3,7,8-tetrachlorodibenzo-p-dioxin).

direct seeding   A method of regeneration of a clear-cut that involves the natural input of seed from nearby uncut forest, or the artificial application of tree seed or seed-bearing cones.

disclimax   An unusual, climax community that occurs because of atypical, anthropogenic or natural stressors in some place or region.

disease   Any impairment of the normal physiological function of a plant or animal, caused by inorganic or pathogenic stressors.

disjunct   A biogeographic term that refers to a spatially discontinuous distribution.

dispersal   The act of scattering of individuals or propagules.

dispersant   A chemical agent that encourages the formation of an oil-in-water emulsion of spilled hydrocarbons so as to reduce the amount of pollution on a shore or on the water surface. See also **detergent.**

disturbance   An episodic but intense environmental influence, usually physical, that causes a substantial ecological change. Can act at the larger, stand or landscape level, or more locally as microdisturbance.

diurnal   Happening daily or during the day.

diversity   An ecological concept that incorporates both the number of species in a particular sampling area, and the evenness with which individuals are distributed among the various species (the latter is sometimes expressed as the probability of randomly encountering an individual of a particular species). See also **species richness**.

dominance   The degree of ecological influence of a particular species within a community. In plant ecology, dominance is usually indexed as the relative biomass or foliage area of particular species, or as the relative basal area of particular tree species.

dose   The quantity of toxic chemical or radiation impinging on or absorbed by an organism, often standardized per unit weight or surface area.

drift   (1) The off-site occurrence or deposition of a pesticide-spray aerosol. (2) The rate of flux of invertebrate biomass or individuals on the surface of a flowing water body, such as a stream.

drought   A prolonged period of water availability that is insufficient to sustain the vegetation normally characteristic of an area.

dry deposition   The influx of gaseous or particulate nutrients or toxic substances from the atmosphere, occurring in the intervals between precipitation events.

dry weight (d.w.)   The weight of a substance after water has been removed.

duff   A layer of the forest floor that is largely composed of partially decomposed leaf litter, which is no longer recognizable to species, but still contains obvious leaf fragments. See also **forest floor**.

dwarf shrub   A low-growing woody plant whose height is genetically constrained to less than about 0.5 m. The usual habitat is tundra or some other highly stressed environment.

dynamic model   A model that is capable of responding to temporal and spatial changes in key variables.

dynamic response   A rapid change in abundance or some ecosystem function, caused by the advent or a substantial change in intensity of stress. See also **numerical response**.

ecocide   An intentional, anti-environmental action carried out over a large area, for example, as a military strategy.

ecological economics   Unlike conventional economics, ecological economics attempts to find a non-anthropocentric system of valuation, so that accountings can be made of important social and environmental costs that are associated with resource depletion and ecological degradation. Compare with **economics**.

ecological impact   A measurable effect on some ecosystem characteristic that is caused by a change in an environmental factor.

ecological integrity   An undefined notion, but ecological integrity is greater in systems that: are relatively resilient and resistant to changes in the intensity of environmental stressors, are biodiverse, are structurally and functionally complex, have large species, have higher-order predators, have controlled nutrient cycling, and are components of a "natural" sere that is not strongly influenced by human activities.

ecological monitoring   Within the larger context of environmental monitoring, ecological monitoring focuses on temporal changes in the structure and function of ecosystems. See also **environmental monitoring**.

ecological-monitoring program   As considered in this book, refers to two, integrated activities—monitoring and research. The monitoring investigates questions that are conceptually simple, involving changes over time. The hypotheses investigated in ecological research are more diverse and complex, but often they deal with important changes detected during monitoring.

ecological release   See **competitive release**.

ecological reserve   A tract of land that has been set aside for the purposes of conservation or preservation, often because it is the habitat of rare or endangered species or because it is representative of a rare or threatened type of ecological community.

ecological rotation   A harvest cycle that is sustainable on the longer term without causing degradations of the economic forest resource or of site quality.

ecologist   A scientist who studies ecology.

ecology   The study of the relationships between and among organisms and their environment. The environment consists of both non-living factors and other organisms.

economics   A social science that examines the allocation of scarce resources among competing uses. Economics is concerned with understanding and predicting the consumption of goods and services, as well as the broader, commercial activities of societies. Compare with **ecological economics**.

ecosystem   A generic term for a system that includes one or more communities of organisms and their interactions among themselves and their environment.

ecosystem function   Refers to ecological processes such as production, predation, nutrient transformation, and the influx and efflux of energy and material.

ecosystem structure   Refers to ecosystem characteristics such as the vertical and horizontal distributions of plant biomass, the quantities of biomass and nutrients in various trophic levels, the size distribution of organisms, and species richness and diversity.

ecotone   A zone of abrupt transition between distinct communities or habitats.

ecotoxicology   The study of the effects of toxic stressors on the structure and function of ecosystems. Includes the direct effects of toxic substances, but also the indirect effects occurring, for example, through changes in habitat structure or in the abundance of food. See also **toxicology**.

ecotype   A local population of a wide-ranging species that is genetically adapted to coping with particular environmental stresses, such as the edaphic toxicity of a site with a large concentration of toxic elements in soil.

edaphic   Pertaining to the soil environment.

edge   See **ecotone**.

edge effect   A change in the abundance and species richness of wildlife at a habitat or community discontinuity.

efficacy   The effectiveness of a pesticide in achieving its intended effect, i.e., in terms of killing insects, weeds, etc.

electrochemical   A chemical effect that produces a heterogenous distribution of electric charge.

electromagnetic energy   The energy of photons, having properties of both particles and waves. Sometimes abbreviated as "light" energy.

electrostatic precipitator   A pollution-control device that removes particulates from industrial waste gases by conferring on them an electric charge and then collecting them at an electrode. The typical collection efficiency is about 99% of the particulate mass.

eluviation   The process by which soluble materials are transported downward through soil in conjunction with percolating rainwater.

embryotoxic   Pertaining to a substance that causes injury or death of a developing embryo.

emission   The release of a substance to the environment.

empirical   Derived from experiment or observation, rather than from theory.

emulsion   A colloidal suspension in which both phases are liquids, as in an oil-in-water emulsion.

endangered   A designation of conservation status, in which a species of wildlife is at risk of imminent extirpation or extinction throughout all or a significant portion of its range.

endemic   A distinct race or species that originated locally and has a geographically restricted distribution.

endemic phase   A stage of low population density of an irruptive species.

energy budget   An analysis of the inputs and outputs of energy for a designated system and of the internal transformations and storage within the system.

enrichment   Enhancement of the rate of supply of nutrients to a system, causing an increase in productivity.

entrainment   The process by which small concentrations of material are transported within a larger, moving matrix, as when soil particles are picked up by and carried within an energetic air mass.

environment   The complex of all external, biotic, and abiotic influences on an organism or group of organisms.

environmental degradation   A decrease in the quality of the environment, as judged from an anthropic perspective.

environmental factor   Any biotic or abiotic influence on an organism or group of organisms.

environmental impact assessment (EIA)   An interdisciplinary process by which the environmental consequences of proposed actions and various alternatives are presented and considered. May involve assessments of ecological, sociological, anthropological, geological, and other environmental effects.

environmental monitoring   A multi- and interdisciplinary activity, involving repeated measurements of inorganic, ecological, social, and economic factors, with a view to documenting or predicting important changes. See also **ecological monitoring**.

enzyme   A proteinaceous, biological catalyst, whose activity is determined by its three-dimensional structure.

epicenter   (1) The site immediately proximate to a nuclear explosion. (2) The site of origin of earthquake tremors.

epidemic   The widespread occurrence of a pathogenic disease.

epidemiology   The study of the occurrence, transmission, and control of epidemic diseases and other health-related factors and of their influences on the distribution and productivity of populations of organisms.

epilimnion   The relatively warm surface water of a thermally stratified lake. See also **stratified**.

epiphyte   A nonparasitic plant that grows on another plant.

episode   An incident of some ecological phenomenon.

epizootic   A disease epidemic affecting a large number of animals.

equilibrium   A condition in which the magnitudes or rates of all influencing forces, chemical processes, or demographic variables cancel each other out, creating a stable and unchanging state.

equivalent.   Abbreviation for mole equivalent and calculated as the molecular or atomic weight multiplied times the number of charges on the ion. Equivalent units are necessary for a charge-balance calculation. See also **conservation of electrochemical neutrality**.

erosion   The mass wasting or wearing away of rock or surficial deposits by the actions of water and/or wind.

estuary   The widening channel of a river where it meets the sea, or a semi-enclosed coastal embayment, characterized by tidal-influenced spatial and temporal variations in salinity of the water.

etiology   The cause or causes of a disease.

euphotic zone   The upper portion of a water column, where light intensity is sufficient to allow photosynthesis to occur. The depth of this zone varies depending on the angle of incidence of solar irradiation and on factors that influence water clarity, particularly turbidity, dissolved organic matter, and the abundance of phytoplankton.

eutrophic   The characteristic of being productive, as a result of a large rate of nutrient input. Usually refers to an aquatic ecosystem. See also **mesotrophic** and **oligotrophic**.

eutrophication   The process by which an aquatic ecosystem increases in productivity as a result of an increase in the rate of nutrient input.

evapotranspiration   The evaporation of water from a landscape, including evaporation from inorganic surfaces (including bodies of water) and transpirational water losses from foliage.

even aged   Refers to a population of organisms in which all individuals are approximately the same age.

evolution   Gradual changes over time in the genetically based characteristics of a population of organisms.

exchangeable   In soil science, refers to an ion that can take the place of another ion at an ion-exchange site on organic-matter or clay surfaces. Most important in determining whether exchange will occur are ionic density, the relative number of charges on the ions (related to the ion exchange series), and relative concentrations of the various ions.

exothermic   A chemical reaction that has a positive net production of thermal kinetic energy.

experiment   A scientific test or manipulation that is designed to provide evidence in support of a null hypothesis. See also **hypothesis**.

exploitation   The harvest of a natural resource for subsistence or for a commercial purpose.

explosive yield   The quantity of energy that is produced by the explosion of a bomb.

exponential   Refers to a logarithmic relationship, or a mathematical relationship involving a number raised to the power of an exponent. See also **logarithmic**.

exposure   In ecological monitoring, exposure indicators measure changes in the intensity of stressors or in the accumulated dose.

extant   A species or other taxon that survives at the present time.

externalities   In economics, externalities are actions undertaken by individuals or corporations that result in costs or benefits to others. In resource economics, externalities include nonvalued costs that do not have to be directly borne by those who are using the resource.

extinct   A designation of conservation status, in which a species or other taxon of wildlife is no longer known to exist anywhere.

extinction   An event whereby there are no surviving individuals of a species or other taxon.

extinction coefficient   An index of the transparency of water.

extirpated   A designation of conservation status, in which a species of wildlife was formerly indigenous to some area, but now only survives elsewhere.

extractable   In soil science, refers to the quantity of a chemical that is solubilized by a particular type of extracting solution, such as a mildly acidic solution.

fallout   Particulates that settle gravitationally from the atmosphere or that are removed by precipitation. Sometimes specifically refers to radioactive particulates.

fauna   The assemblage of animal taxa in some defined site or geographical area.

fecal-pellet group   A cluster of animal excrement. In some studies of animals that are difficult to census directly, the density of fecal-pellet groups may be surveyed as an index of abundance.

fecundity   The capacity to produce offspring. Measured as the total number of propagules potentially produced during the lifetime of an individual or population. Compare with **fertility**.

fen   A circumneutral to slightly acidic, oligotrophic to mesotrophic wetland that receives most of its input of nutrients by slowly percolating surface waters. Fens have a plant community dominated by short-statured sedges and rushes and by nonacidophilous *Sphagnum* mosses.

fermentation   An anaerobic biochemical process by which organic molecules are split into simpler substances, as when yeasts metabolize sucrose and produce ethanol as a by-product.

fertility   (1) The inherent capacity of a site to sustain the production of biomass, usually of plants. (2) The actual number of viable propagules produced during the lifetime of an individual or population. Compare with **fecundity**.

fertilization   A management practice that involves the enhancement of nutrient supply to stimulate a larger rate of productivity.

fertilization trial   A field experiment that is used to evaluate the response of a community to various rates of fertilization with particular types or combinations of nutrients.

fire protection   The active suppression of fire in a stand, by preventing the buildup of a large biomass of fuel through the use of prescribed burns and by the quenching of fires that accidentally ignite.

fixed   Refers to nutrients or toxic substances that are incorporated into organic matter, as in organic nitrogen. See also **organically bound**.

flora   The assemblage of plant taxa in some defined site or geographical area.

flue gases   Waste gases from an industrial process. The flue gases may be vented directly to the atmosphere or they may have some or virtually all of the harmful pollutants removed by pollution-control devices, as in flue-gas desulfurization to remove sulfur dioxide or the use of an electrostatic precipitator or physical filter to remove particulates.

flux   The rate of movement of a quantity of material or energy, usually standardized per unit of area and time.

fog   (1) A meteorological condition characterized by a ground-level aerosol of water droplets and limited visibility. Fogwater aerosols are generally smaller than 40 μm in diameter. (2) A type of pesticide application in which the above-ground atmosphere of the sprayed community is saturated with a fine, pesticidal aerosol.

food chain   A linear sequence of organisms that are linked by trophic interactions, as in grass–cattle–people.

food web   A complex assemblage of organisms that are interlinked by trophic interactions.

food-web-accumulation   The tendency of certain chemicals to occur in their largest concentration in predators at the top of the ecological food web. As such, chemicals such as DDT, PCBs, and mercury in the aquatic environment have their largest concentrations in top predators, in comparison with the nonliving environment, and with plants and herbivores that are the food of the predators.

forb   An herbaceous plant that is not a grass (Poaceae), sedge (Cyperaceae), or rush (Juncaceae).

forest   A stand that is dominated structurally by tree-sized plants.

forest decline   A syndrome of group- or stand-level loss of vigor of one or all tree species, caused by an unknown environmental agent or combination of agents and often leading to dieback and synchronous mass mortality. Often, the etiology is unknown. See also **dieback, etiology**.

forester   A person trained in the inventory, harvesting, and management of forested land.

forest floor   The organic layer that overlies the mineral soil in a forest. The forest floor is composed of three layers: the litter (top), duff (middle), and humus (bottom). See also **litter, duff**, and **humus**.

forest-management area   A tract of land that is being managed so as to provide a supply of tree biomass for some commercial purpose.

forestry   The science of the harvesting and management of forests for a sustained yield of products.

formulation   A pesticidal spray, composed of the active pesticidal ingredient, a solvent, and miscellaneous other chemicals that enhance the effectiveness of the active ingredient.

fossil fuel   A mined fuel that is primarily composed of hydrocarbons, such as coal, petroleum, and natural gas.

frequency   The number of occurrences of a particular observation within a larger population of observations, usually standardized as a percentage.

freshwater   Nonsaline water.

frond   The leaf of a fern or palm or the thallus of a seaweed.

frugivore   An animal that eats fruit.

frustule   The siliceous, outer cell wall of a diatom.

fry   The young of various species of fish.

fumigation   An air-pollution incident characterized by a large concentration of toxic gas or particulates.

fungicide   A pesticide used against fungi.

game   Hunted wildlife.

gamma radiation   High-energy electromagnetic radiation with a frequency greater than about 3×10^{19} Hz. See also **radioactivity**.

general circulation model (GCM) Atmospheric GCMs are dynamic models that predict the surface, horizontal, and/or temporal distributions of wind velocity, air temperature, atmospheric density, and water vapor. GCMs are largely developed from an understanding of the physical laws governing the conservation of mass, hydrostatics and water-vapor conservation, thermodynamics, motion, and the equation of state. Oceanic GCMs predict spatial and temporal variations of water temperature, density, and circulation.

generation (1) The average span of time of the life cycle of a species. (2) All of the individuals produced within one contemporaneous life cycle.

genetic Refers to information that is contained within the base series of DNA, usually in chromosomes.

genetic drift A random change in the collective genetic information, occurring in small, isolated populations.

genetic intergradation A mixing of the genetic information of two populations or distinct taxa as a result of interbreeding.

genotype The genetic constitution of an organism. Compare with **phenotype**

geochemistry The chemistry of the crust of the Earth.

geothermal Relates to heat present in the interior of the Earth.

girdle To cause a lethal injury to a tree, usually by cutting several centimeters into the stem around the circumference of the trunk and thereby completely severing the vascular tissues.

glacial drift See **glacial till**.

glacial till A variable and unconsolidated mineral debris that remains after the meltback of a glacier.

glaciation The process by which the landscape becomes covered with glacial ice.

gradient Continuous changes with distance in the magnitude of a variable or variables.

graminoid A generic term for plants with a grass-like growth form, such as grasses (Poaceae), sedges (Cyperaceae), and rushes (Juncaceae).

greenhouse effect A climatic effect that occurs in a greenhouse, in which solar electromagnetic radiation passes freely through the encasing glass. Much of the transmitted energy is then absorbed within the greenhouse and transformed into thermal kinetic energy. This causes a reradiation of much of the absorbed energy in the form of longer-wave, infrared electromagnetic radiation, but much of this is trapped within the greenhouse because it is absorbed by atmospheric moisture and carbon dioxide and because it cannot pass through the glass, and there is a resulting warming effect. It is believed that carbon dioxide and other radiatively active gases in the Earth's atmosphere may act somewhat like the glass and atmosphere of a greenhouse, so that their increasing concentrations could cause an enhancement of the already-existing greenhouse effect, resulting in a global warming. See also **radiatively active gas**.

greenhouse gases Atmospheric gases that are transparent to most incoming solar radiation, but that absorb longer-wave radiations of the type that are emitted by Earth's surface. The most important greenhouse gases are water vapor, carbon dioxide, ozone, nitrous oxide, methane, and chlorofluorocarbons. See also **radiatively active gas**.

gross primary production (GPP) The amount of carbon fixed by photosynthesis, including that which is used up in respiration. See also **production**.

ground spray A pesticide spray that is applied by a hand-held or vehicle-mounted apparatus.

ground vegetation Vegetation with a height less than about 1 m, occurring within a stand of much taller stature, such as a forest.

groundwater Water occurring below the soil surface that is held in the soil itself or in a deeper aquifer composed of fractures and pores in the bedrock.

groundwater storage The quantity or volume of water that occurs as groundwater.

growth In ecological economics, growth occurs through the net consumption of natural capital, including the mining of NNC and RNC. Because there are physical and thermodynamic limits to growth, it cannot be sustained over the longer term. Compare with **development**.

guard cell One of a pair of cells that controls the aperture width of a stoma. See also **stomata**.

gymnosperm A flowering plan, in the subdivision Spermatophyta, distinguishable from the Angiospermae by having ovules borne naked on the surface of the megasporophylls and often aggregated into cones. See also **conifer**.

gyre A circular or spiral oceanic current.

habitat edge See **ecotone**.

habitat island A spatially isolated habitat that occurs within a much larger matrix of a different kind of habitat, as where a forest remnant is surrounded by agricultural fields.

habitat structure The horizontal and vertical distribution of biomass and species within a habitat. See also **ecosystem** structure.

half-life The time required for the disappearance of one-half of an initial quantity of material, energy, or radioactivity.

halophyte A plant with a tolerance of, and requirement for, a saline environment.

hardening A physiological process that occurs in plants in the autumn and early winter and that results in the progressive development of a tolerance to cold temperatures.

hardwood An angiosperm, woody plant. Not an accurate term, because some angiosperm species of tree have softer wood than some conifer or softwood trees. Compare with **softwood**.

harvest The removal of biomass for some anthropic purpose.

harvest rotation The length of time between successive harvests of biomass.

haze A condition of reduced atmospheric visibility caused by large concentrations of fine inorganic particulates.

headwater The uppermost surface water in a watershed, which does not receive a hydrologic input from lands higher in altitude.

heat island The mass of air over a large city, which averages slightly warmer than the surrounding, ambient condition.

herbaceous Refers to a nonwoody perennial plant in which the above-ground biomass dies back each year.

herbarium A collection of pressed, dried, and identified plant specimens, arranged in a systematic order.

herbivore An animal that feeds on plants.

heritable A characteristic of an organism that has a genetic basis and can be passed along to offspring.

heterotroph An organism that requires a source of organic matter as food. Compare with **autotroph**.

hibernaculum The place where a hibernating animal or group of animals spends the winter.

hidden injury Plant damage, such as decreased productivity, that occurs after exposure to an intensity of toxic stress that is not sufficient to cause an obvious, acute injury.

histopathology Tissue damage that is observable at a microscopic level.

host An organism that is susceptible to infection by a particular pathogen or to infestation by a pest.

humus (1) Amorphous, partly decomposed organic matter. (2) The deepest layer of the forest floor, which lies just above the mineral soil, and in which the humified organic matter no longer contains easily identified leaf fragments. See also **forest floor**.

hybrid An offspring that arises from the interbreeding of two dissimilar taxa. See also **genetic intergradation**.

hydric Refers to wet site conditions, which are usually manifest by at least a brief period during the growing season in which the water table occurs above the soil surface.

hydrocarbon An organic molecule that contains only atoms of hydrogen and carbon.

hydroelectric Refers to electricity that is generated by harnessing the kinetic energy of falling water.

hydrology The study of the distribution, movement, and properties of water in Earth's atmosphere, on its surface, and in its near-surface crust.

hydrolysis A chemical reaction in which a substrate reacts with water to produce a product.

hydrophobicity The characteristic of insolubility in an aqueous medium.

hygroscopic Refers to the tendency of a substance to absorb water from the atmosphere.

hyperaccumulator A species that accumulates a nutrient or toxic chemical to a very large concentration. See also **accumulator**.

hypertrophic Refers to an aquatic system with a large nutrient supply and a large rate of productivity.

hypolimnion The deeper, relatively cool water of a thermally stratified waterbody. See also **stratified**.

hypothesis A suggested explanation for a phenomenon or assemblage of observations, which can be experimentally tested in various ways. Usually, experiments are designed to disprove hypotheses, rather than to prove them.

impaction A particulate-removal mechanism that involves the physical interaction with a solid surface by particulates suspended in water or air.

imperfectly drained Refers to a site in which the water table occurs in a near-surface soil horizon for at least part of the growing season.

incident precipitation The rate of influx of water, and the concentrations and influx of nutrients via precipitation, prior to hydrologic and chemical alterations by interaction with the biota and inorganic surfaces of an ecosystem.

indicator Surrogate measurements considered to be related to important aspects of environmental quality. For example, a self-maintaining population of spotted owl might be considered to be an indicator of the health of its habitat of old-growth forest and its other species.

indicator plant An identifiable plant taxon that is specific to particular site characteristics and indicates those characteristics by its presence or abundance. See also **biogeochemical prospecting**.

infestation   The occurrence of a large population of an irruptive pest.

inflorescence   The part of an angiosperm that consists of the flowering structures.

infrared   Refers to electromagnetic radiation with a wavelength longer than that of visible red or about 0.7 μm and extending to about 1000 μm. Most of the energy emitted by Earth's surface and vegetation is in this wavelength range and is effectively absorbed by radiatively active gases, leading to the so-called greenhouse effect, which keeps Earth warmer than it would otherwise be. See also **greenhouse effect** and **radiatively active gases**.

insecticide   A pesticide that is used against insects.

insectivorous   An organism that feeds on insects.

insolation   The amount of incoming, solar radiation impinging upon a horizontal surface.

integrated pest management (IPM)   A relatively complex pest-management scheme that may incorporate the use of resistant host genotypes, cultural practices that reduce vulnerability, biological control, and the use of pesticides when necessary.

intensive harvesting   The removal of a large fraction of the biomass of a stand, as in clear-cutting.

interbreeding   See **genetic intergradation** and **hybrid**.

intertidal   Refers to a marine shore zone occurring between the high-water mark and the low-water mark.

intolerant   (1) Refers to a species or genotype that is vulnerable to the effects of stressors. (2) Refers to a species that cannot occur in the understory of a forest because of the stresses associated with the restricted availabilities of light, water, and nutrients.

intoxication   Any case of acute injury caused by a toxic agent.

inversion   An atmospheric condition in which air temperature increases with increasing altitude, instead of the usual decrease of temperature with increasing altitude. The presence of a temperature inversion above a lower air mass leads to stable atmospheric conditions, which can be accompanied by the accumulation of large concentrations of air pollutants if emissions continue during the inversion event.

invertebrate   Any animal that does not have an internal skeleton of bone or cartilage.

ion   An electrically charged atom or molecule.

ion exchange   The process by which ionic species are exchanged between an aqueous solution and an insoluble exchange surface, such as a resin, organic matter, or certain clay minerals.

ionizing radiation   Electromagnetic energy or corpuscular radiation (i.e., beta and gamma radiation) that is sufficiently energetic to cause ionization. See also **radioactivity**.

irrigation   The practice of supplying semiarid land with water to promote the growth of agricultural crops.

irruption   A sporadic or rare occurrence of a great abundance of a species.

island biogeography   An ecological theory that predicts a relatively large rate of extinction and a small species richness on small and isolated oceanic or habitat islands.

juvenile   A sexually immature individual.

lamination   Refers to the occurrence of distinctly layered sediment, as in a meromictic lake. See also **meromictic**.

landscape   An extensive area of terrain, encompassing many discrete ecological communities.

large animal   An animal that weighs more than 44 kg (100 lb).

lateral stem bud   A bud that occurs below the shoot apex. Lateral stem buds generally remain dormant unless the terminal bud is killed or injured, after which dormancy is broken and new shoots emerge from the lateral buds. See also **stump sprouting**.

latex   A whitish fluid found in diverse plants that variously contains proteins, starch, alkaloids, and other biochemicals.

LC_{50}   The concentration of a toxic agent that is required to kill one-half of the organisms in a bioassay.

LD_{50}   The dose or amount of a toxic agent that is required to kill one-half of the organisms in a bioassay.

leaching   The process by which dissolved substances are removed from soil by a percolating water solution.

leaf-area index   An estimation of the collective surface area of all foliage in a stand, standardized per square meter of the ground surface.

life form   A classification of plants that is based on the characteristic structural features, including the position with respect to the ground surface of the vegetatively perennating tissue.

limiting factor   The metabolically essential environmental factor that is present in least supply relative to biological demand and thereby restricts the rate of productivity.

limnology   The study of freshwater bodies.

line-of-vision   An unobstructed view.

litter (1) Recently dead, largely undecomposed, plant biomass. (2) The surface layer of the forest floor, in which leaf detritus is identifiable to species. See also **forest floor**.

litterfall The input of dead biomass to the surface of an ecosystem.

littoral Refers to the relatively shallow, nearshore zone of an aquatic system.

load-on-top (LOT) A simple technique for reducing the operational discharges of petroleum to the environment by oceanic tankers. In LOT, the bilge washings are retained for some time in holding tanks, allowing the oil and water to separate. The relatively clean water is discharged to the sea, and the residual oil is combined with the next load. See also **bilge washings**.

logarithmic Refers to a mathematical relationship in which the distance or difference between points is proportional to the logarithms of the numbers, as in $Y = \log X$.

long-distance transport See **long-range transport**.

long-range transport The atmospheric transport of pollutants within a moving air mass for a distance greater than 100 km. Can also refer to long-distance transport in aquatic systems.

longer term (long term) In ecology, this refers to a period of time that is longer than several life spans of an organism or of the dominant species in an ecological community. Can also be considered relative to the time frame of ecological processes.

lysimeter An instrument that is used to sample soil water to determine the concentration or rate of leaching of dissolved substances.

macroalga An alga with a thallus that is visible without magnification, as in seaweeds.

macronutrient A nutrient that is required by biota in a relatively large quantity, including C, H, O, N, P, K, Ca, Mg, and S.

macrophyte A large plant, visible without magnification. Usually refers to aquatic plants.

macroscopic Visible without the aid of magnification.

macrozooplankton A large species of zooplankton, such as krill (Euphasiacea) or shrimp (Decapoda).

management The care and tending of a renewable natural resource.

mangrove A marine-coastal forest at low latitude, usually dominated by tree species in the families Rhizophoraceae and/or Avicenniaceae.

manual spraying The application of a pesticide using a hand-held spray apparatus.

manual treatment The nonherbicidal suppression of weeds using a hand-held cutting tool or apparatus.

maritime Refers to the moderating climate influence that occurs near large bodies of water, especially an ocean.

marsh A relatively productive wetland through which water flows, and in which the vegetation is dominated by tall, emergent graminoids such as rushes, cattails, and certain grasses called reeds.

mass extinction The extinction of an unusually large number of taxa in a geologically short period of time.

mass mortality An event in which a large fraction of the total number of individuals of a population or community dies.

mature (1) An organism that is old enough to be capable of sexual reproduction. (2) A stand in which the dominant individuals are sexually mature.

maximum sustainable yield The largest amount of harvest that can be taken from a population without causing a diminution of its overall productivity.

membrane A thin, pliable, and sometimes fibrous tissue that covers, lines, or connects biological tissue or cells. See also **plasma membrane**.

merchantable stem The part of the bole of a tree that is of commercial value. The diameter limit for merchantable varies depending on tree species, geographic area, and intended use.

merchantable timber The quantity of tree-bole biomass that is valuable for the manufacture of sawn timber.

meristem A plant tissue that is responsible for growth, by rapid cell division and differentiation into organ tissue.

meromictic Refers to a lake in which the water column is stratified permanently or for at least several years, causing anoxia in the hypolimnion, an absence of benthic invertebrates, and sediment occurring in laminations because of the absence of bioturbation.

mesic Refers to moderate site conditions, particularly with respect to water availability.

mesotrophic Refers to an aquatic condition of moderate nutrient supply and productivity.

metalliferous Refers to minerals, soils, or organisms with large concentrations of a metal or metals.

microbe A microscopic organism.

microclimate The climatic conditions that immediately affect an organism or a small group of organisms.

microcosm A miniaturized representation of an ecosystem, usually used for some experimental purpose.

microfauna The assemblage of microscopic animal taxa in some defined site or geographical area.

microflora   The assemblage of microscopic plant taxa in some defined site or geographical area.

micronutrient   A nutrient that is required by biota in a relatively small quantity, including metals such as Cu, Fe, and Zn.

microscopic   Refers to an object that is visible only with the aid of magnification.

migration   A periodic, long-distance movement undertaken by animals.

mill   (1) An industrial facility where raw ore is crushed, followed by the separation of a valuable metal-containing concentrate from the waste tailings, which are disposed in a tailings dump. (2) An industrial facility where harvested logs are manufactured into sawn timber, or are pulped for processing into paper.

mineral soil   The inorganic matrix that serves as the substrate of a terrestrial ecosystem. If the ecosystem has developed a distinct, organic surface layer such as a forest floor, then the mineral soil is considered to begin immediately below it.

mineralization   (1) An occurrence at the soil surface of minerals containing large concentrations of toxic elements. See also **metalliferous**. (2) The conversion of a nutrient or toxic chemical from an organically bound form to a water-soluble inorganic form, as a result of either inorganic or biological chemical reactions.

minerotrophic   Refers to a wetland or surface water body that receives much of its nutrient supply as substances dissolved in water draining from a part of the watershed that is higher in altitude. Compare with **ombrotrophic**.

mining   Refers to the exploitation of a nonrenewable natural resource, or the overexploitation of a potentially renewable natural resource.

miscarriage   The spontaneous expulsion of a fetus from the womb.

mitigation   An environmental-management activity that is intended to reduce the intensity of environmental damage associated with a stressor. During environmental impact assessments, mitigations are the most common mechanisms of resolution of potential conflicts between stressors and valued ecosystem components. See also **environmental impact assessment** and **valued ecosystem component**.

mixed-hardwood forest   A forest composed of a mixture of two or more angiosperm tree species.

model   A conceptual, quantitative, or mathematical representation of an ecological or environmental process, system, or theory. See also **box model, con-**ceptual model, dynamic model,** and **simulation model**.

monitoring   The repeated measurement of indicators to detect changes over time or to predict future changes. See also **environmental monitoring** and **ecological monitoring**.

monoculture   The cultivation of an agricultural or tree species in the absence or virtual absence of strongly competing species.

monoecious   A botanical term, referring to species in which individuals have both female and male reproductive organs. Compare with **dioecious**.

monospecific   Refers to a community that is composed of or strongly dominated by a single species.

morbidity   The rate of incidence of a particular disease in a particular place or area.

mortality   The rate of death within a population of a species or group of species in a particular place or area.

mosaic   A landscape that is characterized by a complex spatial arrangement of distinct habitat or community types.

mousse   A water-in-oil emulsion that is formed by surface-water turbulence after a marine petroleum spill.

multiple use   The use of a natural resource for more than one purpose, as where a forested watershed is managed for the harvest of wood for commercial purposes, to provide and regulate the supply of water, for hunting and fishing, and for nonconsumptive recreational purposes such as camping and bird watching.

multivariate   Refers to a system in which there is simultaneous variation in more than one independent variables.

municipal watershed   A watershed that is used as a source of water for a centralized system of drinking water treatment and distribution, usually for an urban area.

mutation   A sudden change in the genetic information encoded in the DNA of an individual.

mutualism   A mutually beneficial relationship between two or more organisms.

mycorrhizal fungus (mycorrhiza; pl. mycorrhizae)   A fungus that lives in a symbiotic association with the roots of a vascular plant. The heterotrophic fungus benefits from access to fixed carbon exuded by or contained in the plant roots, while the host plant benefits from a markedly enhanced supply of inorganic nutrients, especially phosphorus.

natural   A situation that is not measurably influenced by humans, or is relatively uninfluenced.

natural resource   Any naturally occurring resource that can be used by people. Renewable natural resources can potentially be exploited indefinitely if they are not degraded by use that is too intensive. Nonrenewable natural resources are of finite supply and can only be mined. See also **capital**.

natural source   A nonanthropogenic emission of toxic or nontoxic substances.

necrosis   The death of plant or animal tissue caused by disease or exposure to a toxic stress.

nematicide   A pesticide used against nematodes.

neoplasia   The occurrence of a new, undifferentiated tissue growth or tumor.

neotropics   A biogeographical zone consisting of tropical regions within Central and South America and the West Indies.

nest parasitism   An avian breeding system in which the parasitic species lays an egg or eggs in the nest of an unsuspecting host, which then incubates the egg and raises the parasitic young.

net flux   The difference between the total incoming quantity of a material or energy and the total outgoing quantity. Net flux can be positive, negative, or zero.

net primary production (NPP)   The difference between gross primary production and autotrophic respiration, which if positive leads to an accumulation of organic matter by an individual or population. See also **production**.

neurological   Pertaining to the nervous system.

neutralization   The process by which an acid interacts with a base to form a neutral salt.

niche   The role that an organism or species plays in its ecological community, including its activities, resource use, and interactions with other organisms.

nitrification   The chemoautotrophic process by which certain bacteria oxidize ammonium to nitrite and then to nitrate.

nitrogen fixation (dinitrogen fixation)   The conversion of atmospheric dinitrogen (i.e., N_2) to ammonia or an oxide of nitrogen, occurring inorganically at high temperature and/or pressure, and biologically via the action of the microbial enzyme, nitrogenase.

nitrogen saturation   A condition occurring when the biological capacity for nitrate uptake within the watershed is exceeded, usually because of atmospheric inputs or fertilization, so that nitrate leaches readily from the system.

no-effect level (NOEL)   The chronic or acute dose of some chemical that is tolerated in a bioassay without a detectable biological effect.

nongame wildlife   Species of wildlife that are not hunted, although they may be of economic importance for other reasons.

nongovernmental organization (NGO)   A public-sector organization without direct links to government.

nonrenewable natural resource   See **natural resource**.

nonselective   Refers to a nonspecific pesticide that is toxic to and kills many organisms in addition to the intended, pest target.

nontarget   Organisms other than the intended pest target of a control action, such as pesticide spraying.

nuclear aftermath   The sociological and environmental conditions that would occur after a large-scale exchange of nuclear weapons.

nuclear holocaust   A large-scale exchange of nuclear weapons, causing tremendous losses of human life and of wildlife, destruction of the built environment, climatic change, and other great ecological and sociological effects.

nuclear winter   A widespread climatic cooling caused by the probable effects of nuclear warfare on atmospheric conditions, which would reduce the amount of sunlight penetrating to the Earth's surface.

numerical response   A large increase or decrease in the abundance of a population of organisms, occurring in response to a change in environmental conditions. Often refers to a change in the abundance of predators in response to a change in the abundance of their prey.

nutrient   Any chemical that is required for life. See also **macronutrient** and **micronutrient**.

nutrient capital   The total quantity of a nutrient that is present in all site compartments. In a terrestrial ecosystem, the site nutrient capital occurs in the living plus dead biomass and the mineral soil.

nutrient cycling   A generic term to describe the influxes and effluxes of nutrients from some designated system, as well as all of the internal fluxes, compartments, and transformations.

nutrient loading   The influx of nutrients to a system.

nutrient-use efficiency   The amount of nutrient taken up or required by a plant or crop, per unit of biomass that is produced. See also **water-use efficiency**.

occupational   Pertaining to the working environment of people.

oil   See **petroleum**.

oil seep A natural emission of liquid petroleum.

old growth An old-growth forest is late successional and is characterized by great age, an uneven-aged population structure, domination by long-lived species, and with a complex physical structure, including multiple layers in the canopy, large trees, and many large-dimension snags and dead logs. In some ecological contexts, the term old-growth could also refer to senescent stands of shorter-lived species.

oligotrophic Refers to a condition of a restricted supply of nutrients and small productivity. Usually refers to an aquatic system. See also **eutrophic** and **mesotrophic**.

ombrotrophic Refers to a situation where there is no input of nutrients from ground water or surface water, so that all nutrient supply arrives from the atmosphere with wet and dry deposition.

open-top chamber An experimental apparatus used to simulate the effects on plants of a fumigation by an air pollutant. The chamber is open to the atmosphere on the top, but incursions of ambient air are prevented by maintaining a positive air pressure within the enclosure.

operational Refers to conditions occurring during an actual commercial procedure, as opposed to a simulation experiment or some other model system.

opportunistic Refers to an organism that takes advantage of an ephemeral condition of great resource availability, especially soon after disturbance. See also **weed**.

opportunity cost In economics, this refers to the next best, foregone alternative to some particular investment. Whenever money is invested in some enterprise, there are costs of foregone opportunities, because that money is not available to invest in other ventures.

order of magnitude A factor of 10.

organic anion An organic molecule with an overall negative ionic charge.

organic matter Living and dead biomass.

organically bound Refers to a nutrient or toxic chemical that is structurally incorporated into, or electrochemically bound to, organic matter. See also **fixed**.

organometallic complex See **organically bound**. Refers specifically to metals.

orographic effect The cooling of a moving air mass as it is forced to rise in altitude when it encounters a mountain range, often causing its load of water vapor to condense and leading to a relatively large rate of precipitation on the windward side, and a rain shadow on the leeward side.

outbreak The occurrence of a great abundance of an irruptive pest or pathogen.

outfall A site where there is a large point loading of domestic or industrial waste material or heat to an aquatic system.

overexploitation The unsustainable exploitation of a potentially renewable natural resource. See also **natural resource**.

overharvesting The overexploitation of plants or animals. See also **overexploitation**.

overhunting The overexploitation of animals that are hunted. See also **overharvesting, overexploitation**.

overland water flow The horizontal flow of water above the soil, forest floor, or bedrock. Can be encouraged by compaction of soil.

overstory The highest level of foliage cover in a vertically complex stand or community.

oxidation A chemical reaction in which an oxidant causes another substance to undergo a decrease in its number of electrons. In ecological systems, the most frequent oxidant is molecular oxygen.

oxidizing smog See **photochemical smog**.

oxygen tension The concentration of molecular oxygen in water.

palaeotropics A biogeographical zone consisting of tropical regions within Europe, Africa, and Asia.

paludification The process by which upland bogs form, including a gradual raising of height of the water table.

palynology The study of the historical occurrence of local plant communities, as reconstructed by inferences made from examination of the record of fossil pollen preserved in lake and bog sediments.

pandemic A widespread disease epidemic.

parameter One or more constants that determine the form of a mathematical equation. In the linear equation $Y = aX + b$, a and b are parameters, and Y and X are variables.

parameterize To adjust the parameters of a mathematical equation or computer program to simulate a particular situation or condition.

parasite An organism that derives benefit from its relationship with a host organism, which is either unaffected or suffers detriment.

parasitoid A parasitic wasp (Hymenoptera).

parent material The original mineral substrate from which a soil has been developed. See also **soil**.

parr   A salmon up to 2 years of age.

particulate   A small particle. If sufficiently small, it can be suspended in an atmospheric or aquatic medium. If larger, it will settle gravitationally.

pathogen   A disease-causing organism.

pelagic   Refers to an open-water system.

perennating tissue   The meristematic tissue from which vegetative plant regeneration occurs.

perennial   A long-lived organism. Usually refers to plants.

periphyton   Algae that occur on the inorganic and macrophytic surfaces of an aquatic ecosystem.

permafrost   Ground that remains at a temperature less than 0°C for more than several years.

permanent plot   A sampling area that can be exactly relocated and that can be resurveyed over time.

persistence   The length of time that a toxic agent occurs in a compartment or environment. Depending on the toxic agent, persistence can be affected by chemical breakdown, radioactive decay, or mass transport processes such as volatilization and erosion of toxin-containing particles.

perturbation   The stressing or disturbance of a system.

pest   An organism that is considered to be undesirable, from the anthropic perspective.

pesticide   A chemical or a microbial pathogen that is toxic to pests.

petroleum   A naturally occurring liquid composed primarily of a mixture of hydrocarbon species, which is mined and refined for many uses in the energy and chemical manufacturing sectors. Also known as crude oil.

pH   The negative logarithm to the base 10 of the aqueous concentration of hydrogen ion in units of moles per liter. An acidic solution has pH less than 7, while an alkaline solution has pH greater than 7. Note that a one-unit difference in pH implies a 10-fold difference in the concentration of hydrogen ion.

phenology   Seasonal changes in biological processes, usually studied in relation to changes in climate-related variables.

phenotype   The actual, biological expression in an individual organism of genetically based information, as influenced by environmental conditions.

phenotypic plasticity   The degree of variable expression of genetically based biological potential, in terms of growth form, biochemistry, etc. Some genotypic traits can have a broad latitude of expression; e.g., individual size of many plants can vary widely depending on soil fertility, moisture availability, and other constraints on productivity. Other traits can vary little, e.g., many physiological processes, and the size and shape of certain plant and animal organs.

phosphatase enzyme   An enzyme that produces phosphate by the hydrolysis of organically bound phosphorus. Phosphatase enzymes can be extracellular or intracellular.

photoautotroph   See **autotroph**.

photochemical   Refers to a chemical reaction in which the rate is enhanced by particular wavelengths of electromagnetic radiation.

photochemical smog   Air pollution caused by complex reactions involving emitted chemicals, chemicals formed secondarily in the atmosphere, and sunlight. The most important pollutants in photochemical smog are oxidants such as ozone and peroxyacetyl nitrate. Sometimes called oxidizing or Los Angeles-type smog. See also **smog**.

photodissociation   A chemical reaction in which the rate of breakdown of a substrate into simpler products is enhanced by particular wavelengths of electromagnetic radiation.

photosynthesis   A set of coordinated biochemical reactions that occur in plants, in which visible electromagnetic radiation absorbed by chlorophyll and other pigments is used to synthesize organic compounds from carbon dioxide and water, with the release of molecular oxygen as a waste product.

phylogenetic   Refers to the putative sequence of historical, evolutionary relationships within and among species or higher taxa.

phytoplankton   Microscopic photoautotrophs that occur suspended in the water column.

phytotoxicity   Acute injury to a plant caused by some toxic agent.

piscivorous   An animal that feeds on fish.

planktivorous   An animal that feeds on plankton. Usually refers to feeding on zooplankton or small fish.

plankton   Plants and animals that are suspended in the water column.

plantation   A tract of land on which commercially desirable trees have been planted and tended, often in a monoculture.

plant available   See **available concentration**.

plasma membrane   A thin membrane formed of a protein–lipid bilayer that surrounds all cells and that is responsible for the integrity of the cellular contents and the restricted access of extracellular materials.

Physical disruption of the plasma membrane quickly causes death of the cell.

plutonic Refers to igneous rocks that have derived from intrusive magma that has cooled and solidified beneath the Earth's surface.

podsolization A soil-forming process characterized by acidification of the upper A horizon, the downward leaching of basic cations, metals, and humic substances from the A horizon, and their deposition in the B horizon. Podsolization is most prominent in cool and wet climates and under coniferous trees, oaks, and heaths.

point source A situation where a large quantity of pollutants is emitted from a single source, such as a smokestack, a volcano, or a sewage outfall.

pollution The occurrence of toxic substances or energy in a larger quantity than the ecological communities or particular species can tolerate without suffering measurable detriment. Compare with **contamination**.

polynya Open water surrounded by sea ice in the Arctic.

population (1) The abundance of a species in a specified area. (2) An interbreeding group of individuals of the same species.

population dynamics The study of the temporal changes in the size and other characteristics of populations.

prairie A treeless community dominated by grasses and forbs that occurs in a moderately dry, temperate climate.

precipitation (1) The deposition of rain, snow, fog droplets, or any other type of water from the atmosphere. (2) A chemical process in which an insoluble, crystalline product is formed by the reaction of soluble ions, or from a saturated solution of nonionic molecules.

precision The degree of repeatability of a measurement or observation. Compare with **accuracy**.

predation A biological interaction in which a predator kills and eats its prey. Usually refers to a carnivore eating another animal, but some insect-trapping and -digesting plants can also be considered to be carnivorous.

premature abscission The shedding of foliage at an earlier time than normal. Premature abscission is often induced by disease, exposure to toxic stress, or partial defoliation.

premature senescence The senescence of tissues or an organism before the passing of the usual life span. Premature senescence is usually caused by disease or exposure to toxic stress. See also **senescence**.

prescribed burning The controlled burning of a community as a management practice to encourage a desirable type of regeneration, to decrease the abundance of a pathogen, or to prevent a large buildup of fuel that could lead to a more catastrophic wildfire.

primary Refers to a factor that is first in importance in initiating an ecological phenomenon or first in a sequence of events.

primary production Production by autotrophs. See also **production**.

principle components analysis (PCA) A mathematical technique of ordering samples along multivariate axes.

production The quantity of organic matter that is produced by biological activity per unit area or volume. Gross production (GP) refers to all production, without accounting for respiratory losses (R). Net production (NP) is GP − R. Primary production is production by autotrophs. Secondary production refers to herbivores, and tertiary production refers to carnivores.

productivity Production standardized per unit of time and area.

prominence The relative importance of an individual or species in an ecosystem, often indexed by relative biomass, relative cover, or relative basal area of trees.

protection (1) Refers to a management practice in which a forest stand is sprayed with insecticide to reduce the abundance of an injurious insect, or where there is an active quenching of fire. (2) Refers to a context in which elements of biodiversity are defended from human exploitation, and not conserved for sustainable use as a natural resource. For example, nonexploited ecological reserves are protected, rather than sustainably utilized as a source of natural commodities. Compare with **conservation**.

pyrite A metal sulfide mineral, most frequently an iron sulfide.

qualitative Distinctions that are not based on measurement. Compare with **quantitative**.

quantitative Distinctions that are based on measurement and that can be represented by numerical values. See also **qualitative**.

radial growth The growth in diameter of a woody plant.

radiation sickness An illness caused by overexposure of an organism to X rays or to radioactive materials.

radiatively active gas A gas that absorbs electromagnetic radiation, converts it to kinetic thermal energy and thereby assumes a higher temperature, and then reradiates some of the absorbed energy as longer-wave electromagnetic radiation. Usually refers to gases that absorb infrared, electromagnetic radiation and that can contribute to the greenhouse effect, especially carbon dioxide. See also **greenhouse effect**.

radioactivity The property of a radioactive material that is characterized by the emission of alpha, beta, or gamma radiation as a result of nuclear transformation. An alpha particle consists of two neutrons and two protons, i.e., a helium nucleus. A beta particle is a high-speed electron. Gamma radiation is high-energy electromagnetic radiation, with a frequency greater than about 3×10^{19} Hz.

radiotracer A radioactive isotope of an ecologically important element or molecule that is used in a small but easily measured concentration to study processes that are important in physiology and biogeochemical cycling.

rain forest A mature forest with large biomass occurring in the temperate or tropical zone, in a site with a large rate of precipitation.

rainout The removal of particulate air pollutants from the atmosphere by their serving as condensation nuclei for the formation of raindrops, or of gases by their dissolution in raindrops within clouds, followed by the removal of the raindrops from the atmosphere as precipitation. See also **washout**.

rangeland An extensive track of open land on which livestock graze.

raptor A bird of prey.

rare A species or other taxon that can be of localized or wide distribution, but that is encountered infrequently.

reclamation A practice in which toxic or otherwise degraded land or water is managed in order to restore a more acceptable level of ecological development. May involve liming to reduce acidity, fertilizing to treat nutrient deficiency, and other ameliorative practices.

recovery The natural or anthropogenic restoration of a degraded site to a previous, more acceptable ecological condition.

recruitment The rate at which new individuals enter a population of organisms.

reducing smog An air pollution episode characterized by the occurrence of large concentrations of sulfur dioxide and smoke (i.e., particulate aerosol). Sometimes called London-type smog.

reduction A chemical reaction in which there is a net gain of electrons.

reference A monitored ecological situation that does not undergo experimental manipulation and is used for the purpose of comparison with an experimental treatment. Reference comparisons are used where the experimental design precludes the statistical use of a control treatment. Compare with **control**.

refinery An industrial facility in which a crude raw material is processed into purer products, as in a metal refinery or a petroleum refinery.

refoliation The process by which a defoliated plant produces new leaves.

refugium A geographic area or region that has remained unaltered by climatic change or glaciation and thereby serves as a haven for biota.

regeneration Refers to the regrowth of vegetation on a disturbed site, either by the incursion of sexual propagules or by vegetative growth of surviving individuals.

relaxation The progressive loss of species richness in a newly isolated habitat island.

release A silvicultural term that refers to the response of conifer vegetation to the suppression of weeds by herbiciding or manual cutting.

renewable natural resource A natural resource that potentially can be harvested indefinitely, unless the rate of exploitation is sufficiently intense to reduce the capability for regeneration.

replicate The repetition of an experimental procedure or measurement in order to measure the experimental error and variation. Replication is required to test for statistically significant differences between treatments using commonly used procedures such as t tests and analysis of variance.

reproduction A sexual or asexual process by which organisms produce new and discrete individuals that are similar to the parent.

reproductive failure A toxic or otherwise stressful situation in which reproduction is not possible, so that there is no recruitment of young individuals, even though the intensity of stress may be tolerated by adult organisms.

reproductive potential The number of offspring that a mature adult could potentially produce if there was no mortality of the young. See also **fecundity**.

reproductive success The number of offspring that a mature individual successfully produces.

reradiation The process by which a physical body absorbs electromagnetic radiation, converts it to thermal kinetic energy and thereby assumes a higher temperature, and then reradiates some of the absorbed energy as longer-wave, electromagnetic radiation. See also **radiatively active gas**.

reserve   See **ecological reserve**.

residence time   The length of time that a quantity of substance remains in an environmental compartment. Residence time is influenced by chemical degradation and transformation, by the mass or volume of the reservoir, and by the rates of mass transport processes such as erosion and leaching that can remove the substance from the compartment.

resident   A nonmigratory individual or species.

residue   The quantity of a pesticide or other pollutant that remains in a particular environmental compartment at various times after application.

resilience   The ability of a system to recover from disturbance.

resistance   (1) The ability of an organism or population to tolerate a toxic stress without undergoing significant change. If there is genetic variance for resistance to a toxic stress, then evolution will occur when the stress is encountered, and the trait will rapidly become more frequent in the population. (2) Refers to the ability of a population or community to avoid displacement from some state of ecoogical development as a result of a disturbance or intensification of stress. There are thresholds of resistance to stressor intensity which can be exeeded, after which change occurs.

resorption   The mobilization of a nutrient from a plant tissue that is about to senesce or become litter, and its translocation to a perennial tissue for reuse or storage.

resource development   The socioeconomic process in which a natural resource is identified, quantified, exploited, and, if the resource is potentially renewable, managed to ensure regeneration.

respiration   A biochemical process that occurs in all organisms in which complex organic molecules are broken down by various enzymatic reactions in order to derive energy to support metabolism. The ultimate products of respiration are carbon dioxide, water, and other simple inorganic molecules.

response   In ecological monitoring, response indicators measure effects on organisms, communities, processes, or landscapes that are caused by exposure to stressors.

revegetation   The regrowth of vegetation on a previously devegetated site.

richness   See **species richness**.

right-of-way   The corridor of terrain through which a road, power line, or railroad passes.

riparian habitat   Habitat occurring beside flowing water, such as a river or stream.

roast bed   A primitive and intensely polluting smelting technique in which a ground-level pile of wood and sulfide-metal ore is ignited and allowed to oxidize for several months, after which the metal concentrate is collected and taken away for refining.

rocky intertidal   A hard-rock habitat zone that occurs between the high-water and the low-water levels of the seashore.

rodenticide   A pesticide that is used against rodents.

Roengten (R)   A unit of exposure to ionizing radiation, measured as the ability of the radiation to induce an electrical charge.

rooting depth   The soil depth within which the roots of plants are encountered.

rotation   Refers to the time period between harvests of a biological resource. The rotation is usually 1 year in agriculture, but 50–100 years in forestry.

r-strategist   An organism that produces a large number of offspring during its life span.

ruderal   Refers to a plant that occurs on recently disturbed sites, until the intensification of competition-related stresses associated with succession eliminates them from the community. See also **weed**.

salinization   The process by which irrigation degrades semiarid land by causing a buildup of salts in surface soils, as a result of the accumulation of soluble, inorganic chemicals that remain behind when water evaporates.

salt marsh   A brackish wetland that is periodically inundated by oceanic water and is dominated by graminoid vegetation.

saltwater intrusion   A phenomenon that occurs near the ocean, when the withdrawal or fresh groundwater causes a landward movement of saline water, resulting in a degradation of the aquifer.

saprophyte   An organism that feeds on dead organic matter.

savannah   A tropical or subtropical grassland with scattered trees or shrubs.

sawlog   A log that is sufficiently large to be sawn into timber.

scarification   (1) A mechanical disruption of the organic surface of a site to facilitate the establishment of tree seedlings, accomplished by the dragging of heavy chains or barrels, ploughing, etc. (2) The mechanical or chemical abrasion of a hard seedcoat in order to stimulate or allow germination.

scenario   A proposed sequence of events.

science assessment   A process of review and evaluation

of scientific and technological information and understanding in order to predict the outcomes of alternative courses of action. A science assessment can, for example, contribute to decisions to require reductions of emissions of air pollutants or to restrict the use of certain pesticides.

sclerophyll   A growth form of woody plant that is frequent in the vegetation of certain dry habitats and is characterized by small, leathery, evergreen leaves.

secchi depth   An index of the transparency of a water column, measured as the depth at which a round, black-and-white disc is no longer visible from the surface.

secondary   An indirect effect or relationship.

sector   A subdivision of an economy.

seed bank   The density of viable seeds that is present in the surface organic layer and soil of an ecosystem.

seed rain   The input of viable seeds from beyond the stand, often expressed on an annual basis.

seedling   A recently germinated plant.

seismic   Relating to earth tremors caused by earthquakes, volcanoes, or underground explosions.

selection cut   A forest harvest in which only trees of a desired species and size class are removed.

selection-tree cutting   See **selection cut**.

seleniferous   Refers to a soil rich in selenium.

semidesert   A region where precipitation is barely sufficient to support the growth of prairie or savannah.

senescence   The process of growing old or by which there is a general loss of vigor.

sequester   The removal of a substance from the general environment and into a relatively immobile compartment or chemical complex. See also **complex** and **fixed**.

sere   A successional series of communities that occurs as a result of a particular combination of types of disturbance, vegetation, and site conditions.

serpentine   Refers to a group of hydrated magnesium silicate-based minerals, which when they occur at the surface cause the development of a depauperate vegetation because of metal toxicity and deficiency and imbalance of nutrients.

sewage   Wastewater containing human and animal fecal wastes and industrial effluent.

sewage sludge   A dry or semisolid, organic-rich material produced by the anaerobic digestion of sewage and often applied to agricultural or forested land as a soil conditioner.

shade tolerant   Refers to a plant that can survive in the stressful conditions of the understory of a closed forest, where there is a limited availability of light, water, and nutrients. See also **tolerant**.

Shannon–Weiner diversity   An index of species diversity that incorporates both species richness and the relative abundance of species. See also **species diversity** and **species richness**.

shelterwood cut   A partial harvest of a forest, in which selected, large trees are left on the site to favor particular species in the regeneration and to stimulate the growth of the uncut trees so as to produce high-quality sawlogs at the time of the next cut.

shifting cultivation   A tropical agricultural system in which one to several hectares of forest are cleared and the biomass removed or burned, followed by use of the site for the production of a mixed agricultural crop for several years, until declining fertility and a vigorous development of weeds require abandonment for a 15 to 30-year fallow period and the clearing of a new tract of forest. Compare with **slash and burn**.

shorter term (shorter term)   In ecology this is a relative term that depends on the life span of the organisms that are being considered, but it generally refers to a time period of less than one generation. Can also be considered relative to the time frame of ecological processes.

shortgrass prairie   Prairie in a semidesert region, which is dominated by graminoids less than about 50 cm in height.

significant   See **statistically significant**.

siltation   The infilling of a shallow water body or wetland as a result of the deposition of the sediment load of a river or stream.

silviculture   The branch of forestry that is concerned with the cultivation of trees.

simulation experiment   The use of a dynamic computer model of an ecological or environmental phenomenon or process to simulate the effects of a change in one or more parameters.

simulation model   A dynamic computer model of an ecological or environmental phenomenon, process, or theory.

sink   An environmental or biological repository for a substance, characterized by an output that is nil or small relative to the rate of input.

site   The particular piece of terrain where a community occurs.

site capability   In forestry, a term that is related to the inherent capacity of a site to support the growth of trees. Sometimes called site productivity or site quality, and also relevant to agriculture. See also **fertility**.

site class   An index of site capability in forestry, measured as the number of years that it takes an unsuppressed tree to reach a given height, often about 15 m.

site impoverishment   A reduction in site capability.

site preparation   The treatment of a harvested site prior to the planting of tree seedlings. Site preparation can involve prescribed burning, the crushing or raking of slash, herbiciding, or scarification.

site productivity   See **site capability**.

site quality   See **site capability**.

skid trail   The heavily disturbed path along which harvested logs are dragged (or skidded) from the forest.

skidder   A machine used to drag logs during a forest harvest.

slash   The debris of branches, foliage, and tree tops that is left behind during a forest harvest in which the trees were delimbed on the site and dragged away as logs.

slash and burn   A subsistence or local-market agricultural system used mostly in tropical countries in which forest is cut and burned and the land used to grow crops for many years. The agricultural conversion is relatively stable and longer term in nature. Compare with **shifting cultivation**.

slick   A patch of oil floating on water.

slurry   A dense but flowable suspension of solids in water.

smelter   An industrial facility where a metal-rich concentrate is produced by heating and oxidizing an ore or some other metal-containing material. A primary smelter treats raw, mined ore. A secondary smelter treats discarded, metal-containing wastes that are being recycled.

smog   An amalgam of the words "smoke" and "fog" that refers to an atmospheric condition of poor visibility and a large concentration of air pollutants. See also **photochemical smog** and **reducing smog**.

snag   An erect but dead tree.

snowpack   The depth or volume of snow that has accumulated on the landscape.

softwood   A gymnosperm tree. Not an accurate term, because some coniferous species of tree have harder wood than some angiosperm or hardwood trees. Compare with **hardwood**.

soil   A terrestrial, surface substrate that has been chemically, structurally, and biologically modified by organisms and inorganic processes and is ultimately derived from the parent, inorganic materials that lie on Earth's surface.

soil compaction   A process in which the pore space of soil is reduced because of compression caused by the heavy weight of machinery or livestock, or because of a loss of tilth caused by the oxidation of soil organic matter.

soil horizon   A visually and chemically distinct, horizontal stratum of soil that develops over time as a result of biological and climatic processes. The A horizon lies beneath the surface organic layer and is characterized by acidification and the out-leaching of metal cations and dissolved organic matter. The B horizon is a zone of deposition of materials leached from the A horizon. The C horizon is undifferentiated parent material.

soil profile   Refers to the arrangement and characteristics of soil horizons. See also **soil horizon**.

soil sterilant   A pesticide or steam that is used to fumigate soil to kill all organisms.

solar constant   The input of electromagnetic radiation measured at the outer surface of the atmosphere at the Earth's average distance from the sun, and having a value of 2.00 cal/cm^2-min, equivalent to 0.140 W/cm^2.

solar radiation   Electromagnetic radiation that has been emitted by the sun.

solubilization   The process by which materials are made water soluble, as in the use of detergent to solubilize an oil into water.

somatic   Relating to the body of an organism.

source category   A distinct emissions sector that is used during the quantitative budgeting of the emissions of pollutants for a large geographic area.

spatial   Referring to space.

species   Populations of organisms that actually or potentially interbreed and produce fertile hybrids. Species are named with a latinized binomial.

species composition   The species present in a defined area. See also **species richness**.

species diversity   See **diversity**.

species richness   The number of species present in a defined area, irrespective of their relative abundance.

spongy mesophyll   An internal tissue of a leaf that occurs between the cuticle layers.

sporeling   A recently germinated spore of a fungus, fern, or other simple plant.

spray swath   The width of the area directly affected by the spray deposited by a ground or aerial application of pesticide.

spring meltwater flush   The hydrologic peak of water flow that occurs when an accumulated snowpack melts during a brief period of time in the spring.

stability   The tendency of a system to persist relatively unchanged over time.

stand   A forest or other plant community in a defined area.

standard   A required or recommended intensity of pollution that should not be exceeded.

standing crop   The quantity of biomass per unit area or volume of an ecosystem.

statistically significant   Refers to a level of probability that two or more sampled observations come from different populations, so that in a statistical sense they can be considered to be different. In most ecological studies, the cutoff level is a likelihood of 95% that such a conclusion is correct (i.e., $P = 0.05$). Often abbreviated as "significant."

steady state   A flow-through situation in which the influx of material or energy equals the efflux, so that the net quantity occurring in the compartment does not change over time.

stemflow   Precipitation water that runs down the stem of trees to the ground surface.

stocking   The density of trees in a stand.

stomata (sing. **stoma**)   Microscopic pores in the leaf cuticle, bordered by guard cells. Stomata occur in a large density, and carbon dioxide, oxygen, and water vapor diffuse through these minute pores. See also **guard cell**.

stormflow   The peak of water flow that occurs after a large precipitation event.

strategic nuclear weapon   A nuclear weapon that is delivered over a distance of thousands of kilometers. Compare with **tactical weapon**.

strategy   The syndrome of adaptive physiological, anatomical, and behavioral traits that characterizes an organism or species.

stratified   Refers to a water body that develops a layered water column during the growing season, with relatively dense and cool water in the deeper hypolimnion and warmer, less dense water in the upper epilimnion. These layers are separated by a zone of rapid change in temperature called the thermocline.

stratosphere   The upper atmospheric layer that caps the troposphere, above an altitude of 8–17 km and extending to about 50 km, depending on season and latitude. Within the stratosphere, air temperature varies little with altitude, and there are few convective air currents.

streamwater yield   See **water yield**.

stress   Physical, chemical, and biological constraints that limit the potential productivity of the biota. A **stressor** is any environmental influence that causes measurable ecological detriment or change.

stress tolerator   A plant that is well suited anatomically and physiologically to coping with climatic, toxic, or other stresses.

strip-cut   Refers to a silvicultural system in which there is a series of long and narrow clear-cuts, with alternating uncut strips of forest left in between. Several years after the first strip-cuts were made, sufficient tree regeneration should have established as a result of seeding-in from the uncut strips. At that time, the uncut strips are harvested as well.

structure   See **ecosystem structure**.

stump sprouting   The production of a large number of vegetative sprouts from the base or roots of a cut stump of certain angiosperm tree species. These shoots progressively self-thin, so that eventually there are only one to three mature, tree-sized stems left.

subalpine   Refers to a coniferous forest zone that occurs beneath the higher-elevation alpine tundra of mountainous terrain.

sublittoral   Refers to an intermediate water depth that occurs between the nearshore shallow littoral zone and the deepwater zone.

subspecies   A genetically and anatomically distinct, subspecific taxon occurring in reproductive isolation but capable of successfully interbreeding with other subspecies. See also **binomial**.

subtropical   Refers to a region lying between the tropical and temperate latitudes.

succession   A process that occurs subsequent to disturbance and involves the progressive replacement of earlier biotic communities with others over time. In the absence of further disturbance, this process culminates in a stable climax community that is determined by climate, soil, and the nature of the participating biota. Primary succession occurs on a bare substrate that has not previously been modified by organisms. Secondary succession follows a less intensive disturbance, and it occurs on substrates that have been modified biologically and have a residual, regenerative capacity of organisms that survived the disturbance.

successional trajectory   The likely sequence of plant and animal communities that is predicted to occur on a site at various times after a particular type of disturbance.

suite   A group of species that tend to cooccur.

sulfuric-acid plant An industrial facility that reduces the emission of sulfur dioxide to the atmosphere by using the gas to manufacture sulfuric acid, which can be sold as a commodity.

surface water Freshwater occurring freely exposed to the atmosphere, as in lakes, ponds, rivers, streams, etc.

surfactant A substance that serves as a wetting agent, by reducing the surface tension of a liquid such as water and thereby allowing it to foam or penetrate the small pores of certain solids such as cloth. Detergent is an example of a surfactant.

susceptible Apt to be infected by a particular pathogen, infested by a particular pest, or affected by a particular intensity of a stressor.

suspended sediment Particulate matter occurring in the water column.

sustainable development In ecological economics, this notion refers to an economic system based on the longer-term exploitation of natural capital in ways that does not compromise the availability of those resources for use by future generations. Ecological degradation must also be managed within acceptable limits. See also **growth** and **development**.

sustainable resource use Exploitation of a renewable natural resource in such a way that does not degrade its capacity to regenerate. See also **renewable natural resource** and **capital**.

swamp A forested wetland.

symbiosis An obligate mutualism, in which two dependent organisms occur in a mutually beneficial relationship. See also **mutualism**.

syndrome A set of conditions that is indicative of a particular disease or disorder.

synecology The ecology of communities and ecosystems. Compare with **autecology**.

synergism A more-than-additive effect that can occur when two or more environmental influences are operating simultaneously.

systematic error A source of error that carries through an entire calculation or measurement.

systemic Refers to a physiological effect that occurs throughout an organism.

tableland A relatively flat expanse of terrain occurring at high altitude.

tactical weapon A weapon that is used locally in a battlefield, over a range of less than about 100 km.

tailings The waste material from an industrial process called milling. See also **mill**.

tainting The occurrence of a small residue of hydrocar-

bons in an edible product, which results in a discernible and unpleasant taste.

tallgrass prairie A relatively mesic prairie dominated by graminoids and forbs that reach a height of up to about 3 m. In the absence of periodic wildfire, a tallgrass prairie can be replaced by an open forest.

tar balls Dense, semisolid, asphaltic residuum of the weathering of crude oil spilled at sea.

taxon (pl. taxa) Any identifiable group of taxonomically related organisms.

taxonomy The discipline within biology that is concerned with the classification of organisms based on their phylogenetic relationships.

temperate Refers to a climate that is intermediate between that of the tropics and the polar regions, characterized by long, warm summers and moderately short, but cold winters.

temporal Refers to variation with time.

tephra Solid material ejected into the atmosphere during a volcanic eruption.

teratogenic An environmental influence that causes a deformity of a developing fetus.

terrestrial Refers to the land, as opposed to the aquatic or marine environment.

thermal energy Kinetic energy of molecular vibration, sometimes called heat energy.

thermocline A vertical zone of rapid temperature change in a thermally stratified water body. See also **stratified**.

thin A silvicultural treatment in which an overstocked stand has some trees removed to enhance the growth of the remaining trees.

threatened A designation of conservation status in which a species of wildlife is likely to become endangered if the factors affecting its vulnerability do not become reversed.

threshold An intensity of an environmental factor at which a measurable biological effect begins to occur.

throughfall Water of precipitation that penetrates the canopy and arrives at the forest floor without running down the trunks of trees. See also **stem flow**.

till See **glacial till**.

titration A chemical measurement in which a measured volume of one solution is added to a known volume of another, until the reaction between the two is complete or it has been taken to some desired end point.

tolerance (1) Refers to a genetically based physiological tolerance to the effects of an environmental stressor or combination of stressors. (2) Refers specifically to tolerance of the stressful environmental conditions of the understory of a closed forest. See also **shade tolerant**.

total concentration The amount of the nutrient- or toxic-element content of a sample that is solubilized by a heat-assisted, strong-acid digestion. This represents virtually all of the quantity present in the sample. See also **available concentration**.

toxic element Toxic metals such as Cu, Hg, and Ni and toxic nonmetals such As and Se.

toxic threshold. See **threshold**.

toxicity Refers to the quality of being poisonous. See also **acute toxicity** and **chronic toxicity**.

toxicology The study of the biological effects of poisons.

trajectory analysis The prediction of the future location of a moving object or air or water mass, based on measurements of the location in the recent past.

transect A straight sampling line that runs in a particular direction from a central place.

transformation (1) A change in physical state or in chemical structure. (2) In statistics, a calculated change of a variable, usually used to transform the actual frequency distribution to a normal distribution, so that a parametric statistical test can be applied.

transition See **ecotone**.

transparency Relates to the degree to which visible radiation can pass through a water column.

transpiration The evaporation of water from foliage.

tree ring An annual, concentric growth ring of a tree that is growing in a site with a strongly seasonal climate. Growth rings are easily visible in a cross section of the stem, or in a horizontal core taken from the stem.

trophic response Refers to the response of an ecosystem to the addition of a nutrient or combination of nutrients. Can also be induced by large changes in feeding relationships through herbivory or carnivory. Trophic response is usually measured in terms of productivity of plants, and sometimes of herbivores and carnivores.

trophic status Refers to the rate of nutrient supply and productivity of a system. See also **hypertrophic, eutrophic, mesotrophic, oligotrophic,** and **ultraoligotrophic**.

tropical Refers to a low-latitude climate that is characterized by consistently warm and humid conditions.

troposphere The atmospheric layer that extends from the ground surface up to the lower limit of the stratosphere at an altitude of 8–17 km. Within the troposphere, air temperature typically decreases with increasing altitude, and there are many convective air currents. As a result, the troposphere is sometimes called the "weather atmosphere." Compare with **stratosphere**.

tumor An undifferentiated tissue mass formed by a recent growth of cells. See also **cancer** and **neoplasia**.

tundra Treeless vegetation that occurs at high latitude or high altitude.

turbidity Refers to a condition of poor transparency caused by a large concentration of suspended particulates.

turnover time The time required for the replacement of the quantity of material or energy in a compartment, with an equivalent quantity of material or energy.

ultraoligotrophic Refers to extremely unproductive aquatic systems with a restricted nutrient supply.

ultraviolet radiation Electromagnetic radiation having a wavelength less than about 0.4 μm, but longer than that of X rays.

unavailable concentration See **available concentration**.

understory Vegetation that occurs beneath the principal canopy of a vertically complex ecosystem, such as a forest.

undescribed Refers to a taxon that has not yet been identified and given a scientific binomial by a taxonomist.

undetectable concentration A concentration of a substance that is smaller than the detection limit of the available analytical technology. This does not necessarily imply a concentration of zero.

upland game Hunted wildlife that does not occur in wetlands. Upland game usually refers to relatively small species, such as rabbits, hares, pheasants, and grouse.

vagile Refers to an organism that is easily capable of undertaking a long-distance movement.

valuation The assignment of economic worth.

valued environmental component (VEC) During an impact assessment, VECs are selected for consideration of the risks associated with proposed environmental-management activities. VECs may be chosen because they are of economic value, endangered, of cultural or aesthetic importance, indicators of the potential effects on a complex of ecological variables, or otherwise identified as important during public consultations.

variable A measurement that has a range of possible values. See also **parameter**.

variance   A statistical measurement that is related to the amount of variation within a population of measurements.

variety   A subspecific taxonomic category.

vascular plant   A plant with specialized tissue for the internal transport of water and dissolved substances.

vector   A mobile species that transports a pathogen among hosts. For example, mosquitoes are the vector between the malaria-causing *Plasmodium* parasite and humans.

vegetation   A generic term to describe communities of plants on the landscape.

vegetative regeneration   Asexual propagation.

vigor   The capacity of an organism for healthy growth and survival.

visible radiation   Electromagnetic radiation within the wavelength band of about 0.4–0.7 μm, comprising "light" that is visually detected by the human eye.

visual predator   A predator that locates its prey by sight.

volatilization   The process of evaporation from a solid or liquid state to a vapor.

volcanic   Referring to material ejected from an erupting volcano or some other volcanic influence.

volume growth   The increase in volume or weight of biomass of a tree or forest.

volume weighted   Refers to a calculation that involves the summation or averaging of values of chemical concentration of a number of water samples, in which the contribution of each sample is assigned a weighting factor determined by its volume.

vulnerable   (1) A designation of conservation status, in which a species of wildlife is at risk because of small or declining numbers, occurrence at the fringe of its range or in restricted areas, habitat fragmentation, or for some other reason. (2) Apt to suffer damage from exposure to some stressful environmental agent.

washout   The removal of particulate or gaseous air pollutants by their impaction with falling droplets of precipitation. Compare with **rainout**.

water body   Any depression of the landscape that is filled with water, such as a lake, pond, river, or stream.

water column   A unit area of water that extends from the surface to the bottom of a water body.

watershed   The total expanse of terrain from which water flows into a water body or stream.

water table   The height of the water-saturated level of the soil.

water-use efficiency   The amount of water transpired by a plant or crop, per unit of biomass that is produced. See also **nutrient-use efficiency**.

water vapor   Water present in the atmosphere in gaseous form. Water vapor is the immediate source of all condensation and precipitation of water.

water yield   The amount or volume of water that flows in a given period of time from a watershed.

weather   The relatively short-term, instantaneous or day-to-day meteorological conditions at a place. Compare with **climate**.

weathering   (1) Refers to chemical and biological processes by which inorganic nutrients are transformed from a water-insoluble to a water-soluble form. (2) Refers to the progressive volatilization of the relatively light hydrocarbon fractions of spilled petroleum, so that the residues become progressively heavier and more viscous and eventually transform into a tarry substance.

weed   Any organism that occurs in a situation where people regard it as a pest. Usually refers to plants.

wet deposition   The flux of chemicals to the surface of the Earth during an event of precipitation. Compare with **dry deposition**.

wetland   Any habitat that is regularly saturated by water at or near to the surface. The vegetation of wetlands must be adapted to the conditions of saturated soils. The most common types of wetland are bogs, fens, marshes, salt marshes, swamps, and wet meadows.

wet meadow   A hydric, graminoid-dominated community.

wet scrubber   A device used to remove pollutants from the waste flue gases of a smelter or fossil-fueled power plant. A wet scrubber traps particulates by impacting them with small water droplets and removes sulfur dioxide by causing it to form calcium sulfate by reaction with calcium carbonate in solution.

whole lake   Refers to a situation where an entire lake is manipulated for some experimental purpose, as in a whole-lake fertilization experiment.

whole-tree clear-cut   A clear-cut in which all of the above-ground biomass is removed during the harvest. See also clear-cut and complete-tree clear-cut.

wildlife   Any microbes, plants, or animals that live free of human husbandry.

windbreak   A line of trees or a fence that gives protection from the prevailing wind.

wood supply   The rate at which forest biomass can be

provided for an industrial or other anthropic purpose.

xeric   Refers to dry site conditions.

yard   See **deer yard**.

yarding   The practice of dragging or skidding logs to a central place (the landing) during a forest harvest.

yield   (1) The amount of biomass that is harvested. (2) The amount of energy liberated during an explosion.

yield decrement   A decrease in yield caused by some environmental stressor.

zooplankton   Small animals that occur in the water column.

BIOLOGICAL INDEX

CHEMICAL INDEX

GEOGRAPHICAL INDEX

SUBJECT INDEX

A

Aboriginal hunters, 350–351, 378, 384
Accumulator species (see also
    bioaccumulation), 26, 67, 69–71
Acidification, 8, 9, 37, 39, 94–143,
    150, 327, 451, 455, 458, 460, 462
    acidic (or acidifying) deposition, 12,
        32, 95, 96, 108, 110, 114–116,
        120, 121, 124, 125, 128, 135,
        139, 142, 144, 151, 154, 155,
        157, 289, 451, 456, 462, 463,
        465
    acidic drainage, 31, 32
    acidic fumes, 29, 30, 142
    acidic mist and fog, 34, 39, 155, 157
    acidic precipitation, 13, 14, 18, 36,
        94–96, 101, 103, 108, 114, 116,
        127, 136, 143, 149, 152, 155,
        157, 458, 465
    acidic soil, 32, 108, 149, 150, 151,
        157, 294
    acidic water, 31, 32, 41, 73, 95, 107,
        130, 133, 138, 139, 208, 240
    acid mine drainage (AMD), 31–33,
        111, 121, 130
    acid neutralization (and neutralizing
        capacity), 100, 112, 114, 116,
        121–125, 130, 140, 141
    acidity, 64, 66, 111
        of soil, 104, 108, 111, 281
        of water, 117, 118, 131
    "acid-shock" (acid pulses), 118, 134,
        138
    acid sulfate soil, 111
    effects on amphibians, 139
    effects on benthic invertebrates, 133–
        135
    effects on crustacean zooplankton,
        131, 133
    effects on fish, 135–139, 465
    effects on vegetation
        aquatic, 128–130
        terrestrial, 126–128, 465
    effects on waterfowl, 140

exchange acidity, 113
management and mitigation, 114
    fertilization,
        soil, 151
        water, 142, 453
    liming treatment,
        soil, 87, 114, 151
        freshwater, 119, 140–142, 452,
        453
    of freshwater, 31, 33, 40, 111, 120–
        125, 136, 138, 143, 190, 451–
        453, 457, 465
    of soil, 31, 81, 82, 87, 110, 111,
        113–116, 151, 154, 157
Acne, 399
Aerosols, 24, 103, 117, 120
    marine derived, 100
    sulfate, 56, 107, 108
Aesthetics, 190, 280, 438, 440, 442,
    452
Agriculture, 3, 17, 25, 60, 61, 73–77,
    79, 95, 98, 100, 101, 111, 115,
    126, 127, 152, 189–191, 195–196,
    201–204, 206, 209, 213, 217–220,
    226, 229, 240, 241, 253, 275, 276,
    278–281, 285, 329, 334–335, 339,
    343, 345, 348, 352, 353, 355, 360,
    362, 363, 366, 369, 370, 379–381,
    385 390, 391, 395, 397, 410, 414,
    415, 431, 433, 434, 438, 443, 452,
    459, 463
    abandonment of lands, 305, 306, 310
    agricultural ecosystem, 218–220,
        229, 242, 275, 354–356, 442
    agricultural failure, 401
    crop rotation, 218
    fallow period, 369
    intensive management, 218, 229,
        242, 301, 432
    monocultural agroecosystems, 218,
        225
    no-tillage cultivation, 219
    plowing, 219
    site impoverishment, 281
    slash and burn, 318, 329, 369

subsistence, 329, 368
Airshed, 144
Albedo, 7, 52, 56, 401, 409
Algae, 32, 33, 57, 78, 128–130, 169,
    170–173, 182, 190, 192, 196, 228,
    231, 407
    algal toxins, 228
    epiphyton, 142
    filamentous algae, 254
    macroalgae, 8, 167, 170, 228
    nanoplankton, 198
    periphyton (benthic), 129, 130, 134,
        142, 323
    phytoplankton, 15, 119, 128, 129,
        133, 141, 142, 178, 189, 190,
        192, 194, 196–208, 210–212,
        228, 450, 457
Alkalinity, 110, 117, 118, 122–125,
    131, 134, 140
    bicarbonate, 124
    interference with alkalinity
        generation, 130
Alleles, 300
Allozyme heterozygosity, 300
Amelioration, 451
Ammonification, 90, 104, 110, 294
Amoco Cadiz, 163, 168, 171–173
Amphibians, 303, 321–322, 333–335,
    339, 347, 349, 407
Angiosperm, 107, 127, 150, 221, 222,
    225, 257, 260, 261, 299, 304, 305,
    313, 331, 342, 343, 348, 355, 366,
    398
    graminoid, 266, 269, 274, 304, 308–
        309, 313
    shrub, 263, 264, 269, 273, 274, 304,
        449
    tree, 144, 145, 222, 287, 288, 292,
        301, 304, 313, 439, 443, 449,
        456
Anions, 96, 100
Antibiotics, 431
Antimycin, 216
Aquaculture, 395, 423, 431, 432
Arachnids, 216, 252

KING ALFRED'S COLLEGE
LIBRARY